S0-BLI-606

COMPUTATIONAL MODELING OF POLYMERS

PLASTICS ENGINEERING

Series Editor

Donald E. Hudgin

Professor
Clemson University
Clemson, South Carolina

1. Plastics Waste: Recovery of Economic Value, *Jacob Leidner*
2. Polyester Molding Compounds, *Robert Burns*
3. Carbon Black-Polymer Composites: The Physics of Electrically Con- ducting Composites, *edited by Enid Keil Sichel*
4. The Strength and Stiffness of Polymers, *edited by Anagnostis E. Zachariades and Roger S. Porter*
5. Selecting Thermoplastics for Engineering Applications, *Charles P. MacDermott*
6. Engineering with Rigid PVC: Processability and Applications, *edited by I. Luis Gomez*
7. Computer-Aided Design of Polymers and Composites, *D. H. Kaelble*
8. Engineering Thermoplastics: Properties and Applications, *edited by James M. Margolis*
9. Structural Foam: A Purchasing and Design Guide, *Bruce C. Wendle*
10. Plastics in Architecture: A Guide to Acrylic and Polycarbonate, *Ralph Montella*
11. Metal-Filled Polymers: Properties and Applications, *edited by Swa- pan K. Bhattacharya*
12. Plastics Technology Handbook, *Manas Chanda and Salil K. Roy*
13. Reaction Injection Molding Machinery and Processes, *F. Melvin Sweeney*
14. Practical Thermoforming: Principles and Applications, *John Florian*
15. Injection and Compression Molding Fundamentals, *edited by Avraam I. Isayev*

16. Polymer Mixing and Extrusion Technology, *Nicholas P. Cheremisinoff*
17. High Modulus Polymers: Approaches to Design and Development, *edited by Anagnostis E. Zachariades and Roger S. Porter*
18. Corrosion-Resistant Plastic Composites in Chemical Plant Design, *John H. Mallinson*
19. Handbook of Elastomers: New Developments and Technology, *edited by Anil K. Bhowmick and Howard L. Stephens*
20. Rubber Compounding: Principles, Materials, and Techniques, *Fred W. Barlow*
21. Thermoplastic Polymer Additives: Theory and Practice, *edited by John T. Lutz, Jr.*
22. Emulsion Polymer Technology, *Robert D. Athey, Jr.*
23. Mixing in Polymer Processing, *edited by Chris Rauwendaal*
24. Handbook of Polymer Synthesis, Parts A and B, *edited by Hans R. Kricheldorf*
25. Computational Modeling of Polymers, *edited by Jozef Bicerano*

Additional Volumes in Preparation

Plastics Technology Handbook: Second Edition, Revised and Expanded, *Manas Chanda and S. K. Roy*

COMPUTATIONAL MODELING OF POLYMERS

EDITED BY
JOZEF BICERANO
The Dow Chemical Company
Midland, Michigan

Marcel Dekker, Inc. New York • Basel • Hong Kong

Library of Congress Cataloging-in-Publication Data

Computational modeling of polymers / edited by Jozef Bicerano.
p. cm. -- (Plastics engineering ; 25)
Includes bibliographical references and index.
ISBN 0-8247-8438-3 (acid-free paper)
1. Polymers--Mathematical models. I. Bicerano, Jozef.
II. Series: Plastics engineering (Marcel Dekker, Inc.) ; 25.
QD381.9.M3C65 1992
668.9--dc20 92-3395
CIP

This book is printed on acid-free paper.

MARCEL DEKKER, INC.
270 Madison Avenue, New York, New York 10016

Current printing (last digit):
10 9 8 7 6 5 4 3 2 1

PRINTED IN THE UNITED STATES OF AMERICA

Preface

The use of computational modeling to study the structures and properties of polymers is reviewed in this book, with particular emphasis on non-crystalline polymers. Both the computational techniques and their applications to specific types of polymers are covered in detail. The book is intended for those interested in understanding the structures and properties of polymers—including workers and students in the fields of polymer science, plastics engineering, materials science, organic chemistry, chemical engineering, theoretical chemistry, and theoretical solid-state physics.

The contributors have attempted to (a) present the similarities and differences among various types of techniques, (b) explore how these techniques can be combined with one another and with experimental results to provide more complete descriptions of the structures and properties of specific materials, (c) bridge the gap between molecular-level calculations and techniques for the prediction of the properties of bulk materials, (d) present critical state-of-the-art reviews of their topics, and (e) indicate what they perceive to be the most promising directions for future research.

Emphasis is placed on the contribution of molecular-level theory to understanding the structures and physical and mechanical properties of *synthetic amorphous polymers* used in or intended for plastics technology.

The following topics have, therefore, either been excluded or are not discussed in great detail: (a) detailed treatment of the effects of semicrystallinity in polymers, (b) the modeling of biological polymers, such as proteins and nucleic acids, (c) the properties of conducting polymers, such as polyacetylene and its derivatives, (d) the computational techniques used only for these three types of polymers or problems, and (e) computer-aided design and manufacture of plastics by the techniques of plastics engineering not involving molecular-level descriptions.

The introductory chapter is an overview that provides a perspective of certain aspects of the field, both as a subfield of polymer science and within its own historical context. Each successive part of the book discusses computational techniques more complex than those covered in the preceding parts. This increasing "complexity" can be due to the increasing sophistication of the formalisms used, or to more computer time being required to solve the equations of the formalisms, or to a combination of these two factors.

Several types of computational techniques are often combined in order to provide schemes that enable the study of complicated polymer structures and properties. The partitioning of the book into five parts should therefore be viewed mainly as a heuristic device, rather than as a reflection of a real division of the techniques into different classes.

In Part One, Chapter 1, the application of group contribution techniques for the correlation of the properties of polymers with their chemical structures is reviewed. The very simple but useful group contribution techniques are based on the empirical observation that the values of certain important properties can be approximated by additive contributions from structural subunits of the polymer.

The use of force-field techniques is covered in Part Two. These techniques are all based on a simple classic mechanical model of molecular structure, in which atoms are treated as point masses connected by springs. Chapter 2 begins with a general introduction to force-field techniques. It then goes on to describe the force-field hamiltonians most commonly used, and to give examples of their use in calculating the conformational stabilities of polymers. In Chapter 3, a detailed example of the use of rotational isomeric state theory for conformational analysis and for calculating configurational properties is presented. The placement of this chapter in Part Two is simply in recognition of the fact that rotational isomeric-state calculations most often start from the results of conformational-energy calculations performed by force-field techniques on model molecules representing polymer-chain segments. The "Strophon" model for deformation of glassy amorphous polymers is presented in Chapter 4. This model is based on a force-field description of intrachain and interchain interactions in polymers. It has been quite successful in spite of its apparent simplicity.

Thermodynamic and statistical mechanical techniques are covered in Part Three. Chapter 5 discusses Monte Carlo and molecular dynamics simulations of amorphous polymers, giving several examples of the applications of such simulations. Free volume, the relationship between free-volume fluctuations and localized molecular rearrangements, and the application of free-volume theory to polymer relaxation in the glassy state, are discussed in Chapter 6. In Chapter 7, an example is given of the utilization of a combination of force-field, free-volume, and statistical mechanical techniques to study the structures of a family of semicrystalline copolymers in relation to a technologically important property, namely the transport of penetrant molecules through plastics. In Chapter 8, an atomistic model is presented for the structure and mechanical properties of polymeric glasses.

In Part Four, Chapter 9 treats the modeling of polymer-surface interactions, which are extremely important in many technological applications. These interactions are most easily modeled using crystalline systems. Much can be learned from the results of calculations such as those reported in Chapter 9, justifying the inclusion of a chapter on crystalline polymers in a book whose main emphasis is on noncrystalline polymers.

The use of quantum mechanical techniques to study noncrystalline polymers is covered in Part Five. It is only in recent years that the increase in speed, available storage space, and reliability, and the decrease in size and price of computers, coupled with the development of powerful new integrated software packages, have enabled these techniques to begin making significant contributions to understanding the structures and properties of polymers. Although this field is still in its infancy, it holds great promise for the future. Chapter 10 provides a general introduction to quantum mechanical techniques, including brief discussions of some of the specific quantum mechanical techniques useful for studying noncrystalline polymers. Some of the specific applications of quantum mechanical techniques to the study of noncrystalline polymers are presented in Chapters 11 to 13.

I thank the authors whose contributions made this book possible, and the management of The Dow Chemical Company for allowing me to edit this book.

Jozef Bicerano

Contributors

Jean-Marie André Professor, Chemistry Department, Facultés Universitaires Notre-Dame de la Paix, Namur, Belgium

Jozef Bicerano Central Research, Materials Science and Development Laboratory, The Dow Chemical Company, Midland, Michigan

Kurt Binder Professor, Institute of Physics, Johannes-Gutenberg-Universität, Mainz, Germany

Raymond F. Boyer Distinguished Research Fellow, Michigan Molecular Institute, Midland, Michigan

Hayden A. Clark Materials Science and Development Laboratory, The Dow Chemical Company, Midland, Michigan

Stelian Grigoras Senior Research Specialist, Dow Corning Company, Midland, Michigan

Anton J. Hopfinger Department of Medicinal Chemistry, Chemistry and Chemical Engineering, The University of Illinois, Chicago, Illinois

Alan Letton Associate Professor, Polymer Technology Consortium, Polymer Science and Engineering Program, Texas A&M University, College Station, Texas

Peter J. Ludovice Polymer Project Manager, Polygen Corporation, Waltham, Massachusetts

Robert R. Luise Research Associate, Polymer Products Department, R&D Division, DuPont Experimental Station, E. I. DuPont de Nemours and Company, Wilmington, Delaware

James E. Mark Department of Chemistry, University of Cincinnati, Cincinnati, Ohio

Richard E. Robertson Professor, Department of Materials Science and Engineering, The University of Michigan, Ann Arbor, Michigan

Ulrich W. Suter Professor, Institut für Polymere, ETH, Zürich, Switzerland

Dirk W. Van Krevelen Professor, Delft University of Technology, Delft, The Netherlands

William J. Welsh Associate Professor, Department of Chemistry, University of Missouri-St. Louis, St. Louis, Missouri

Ioannis V. Yannas Professor of Polymer Science and Engineering, Department of Mechanical Engineering, Massachusetts Institute of Technology, Cambridge, Massachusetts

Contents

COMPUTATIONAL MODELING OF POLYMERS

Multiple Transitions and Relaxation in Synthetic Organic Amorphous Polymers and Copolymers: An Overview

Raymond F. Boyer
Michigan Molecular Institute
Midland, Michigan

The general subject matter of this review began appearing in the world literature—England, Europe, the Soviet Union, Japan, and the United States—with increasing frequency for at least 25 years prior to the beginning of our following account. Dielectric and dynamic mechanical loss, ultrasonic attenuation, and, more recently, NMR, became familiar topics. Suddenly there was an apparent uncoordinated implosion of widely scattered publications into a proliferation of highly coordinated reviews, starting in 1963. Their specific subject matter and sequence appears in the introduction.

Our single directive from the editor was to be concerned with amorphous polymers. This has been interpreted broadly as covering amorphous polymers and copolymers, the amorphous phase of semicrystalline polymers, and the amorphous material above T_m in crystalline polymers. Thus, crystalline melting points per se, the kinetics of crystallization and melting, and crystal morphology have been avoided, and yet the ratio of T_g/T_m, is included. The existence of local structure in amorphous polymers is reviewed in some detail. We differ with conventional opinion and cite diverse evidence supporting the presence of local structure in amorphous polymers.

We consider three amorphous phase events, T_g, T_{ll} at $1.2T_g$, and T_u at $1.2T_m$, as indicative of local structure in amorphous polymers.

Considerable attention focuses on the three common $T < T_g$ transitions, T_β, T_γ, T_δ; T_g itself; and the four liquid state transitions: T_{ll}, $T_{l\rho}$, and $T_{l\phi} > T_g$; and $T_u > T_m$: their positions along the temperature scale; their relative intensities; apparent activation energies; and, still largely speculative, their molecular origins. A possible general phenomenon, designated the *triumvirate* concerns T_β at $0.75T_g$; T_g as a dominant amorphous phase event; and T_{ll} at $1.2T_g$, with T_β and T_{ll} being weak satellites of T_g. Double glass temperatures in semicrystalline polymers, block copolymers, and incompatible polyblends are not covered.

The use of relaxation maps, log frequency in hertz versus T^{-1}, are considered for representing the frequency dependence of each of the several common multiple transitions as used by McCrumm et al. and by McCall. The use of apparent activation energies, ΔH_a, in kcal/mole or kJ/mole for the same purpose, is also used but questioned by some.

We finally list some typical group motions which are common to several polymers. Also given are some common empirical equations that have emerged in connection with polymer transitions. We are possibly derelict in not distinguishing between transitions such as melting, and relaxations such as T_g. Once T_g was dignified with the title of a transition when, in fact, it exhibited rate effects like a relaxation, we settled for transitions until nomenclature experts decided otherwise.

INTRODUCTION: HISTORY: THE 1963–1967 IMPLOSION

Two unrelated reviews on the named subject matter appeared in 1963. The one by Saito et al. [1], "Molecular Motion in Solid State Polymers," covered amorphous and semicrystalline polymers. The four authors were active in various aspects of the related subject matter and familiar with relevant world literature. Saito, for example, played a key role in developing a theory for local mode motion in the glassy state below T_g.

The author of the second paper [2], "The Relation of Transition Temperatures to Chemical Structure in High Polymers," was, by contrast, an amateur except with regard to glass-transition phenomena [3], which subject bridged a path for him into this new field. Still a third manuscript by Allen [4] was essentially ready for publication but was withdrawn after Ref. 2 appeared in print.

Our own interest arose as a result of a main lecture at the 1960 Gordon Conference on Polymers by the late polymer physicist, Dr. Karl Wolf, then from BASF in Ludwigshafen, Germany. He first discussed the torsion pendulum developed by Schmieder and himself [5a] and then reviewed extensive measurements on a wide variety of elastomers, amorphous polymers, copolymers, and crystalline polymers [5b]. As numerous "spectra" showing tan δ or $\lambda = \pi \tan \delta$, and shear modulus G, flashed on the screen in rapid succession, our familiarity with T_g for most of these polymers allowed us to concentrate on T_g as the dominant amorphous phase transition and thence to compare the location and intensity of secondary transitions in the glassy state with the T_g loss peak.

The following morning we awoke with a clear conception that if different "spectra" were superimposed, with T_g's overlapping, the spectra became remarkably similar, differing only in fine details. Dr. Wolf lectured the following week at the Dow Chemical Company in Midland, MI [6]. This gave us a second opportunity to review his extensive presentation of torsion pendulum data. This reinforced the earlier conclusion.

Subsequent to the July 1960 Gordon Conference, we began an in-depth review of all pertinent subject matter:

1. Dynamic mechanical loss, both torsion pendulum and other techniques
2. Dielectric and NMR data
3. Ultrasonic loss
4. Volume-temperature and specific heat–temperature

This was supplemented by discussions with scientists active in the field, primarily American and British polymer scientists. The practice of plotting loss data on a relative temperature scale was pursued. A provisional 100-page typed report was circulated to peer groups for comments and became the basis for an IUPAC lecture [7], which was the precursor to Ref. 2.

The T/T_g plotting technique quickly led to the recognition that there was evidence for a loss process lying above T_g and hence in the liquid state [2]. This new event was considered to represent a transition between two liquid states, liquid$_1$ $\leftrightarrow$ liquid$_2$, both of unknown nature. We designated this transition event as T_{ll}, with evidence for it from thermal expansion, torsion pendulum, penetrometry, x-ray diffraction, and NMR data [2]. The transition $T_\beta < T_g$ was widely believed to represent a local motion of 1 to 2 monomers, and T_g the micro-Brownian motion of 10–20 monomers. It was natural to postulate that T_{ll} involved the motion of the entire polymer chain in an unspecified manner.

The T_{ll} concept was to encounter resistance on several grounds:

1. T_{ll} was a new concept not hitherto noted by peer groups.
2. T_{ll} was very weak, generally 5–20% of the strength of T_g, as we were to discover much later. Its detection was therefore prone to artifacts inherent in the several test methods used to show T_{ll}.
3. We overlooked some of the best evidence for it, widely dispersed in the literature.

4. We cited as evidence for T_{ll} data later found to be inappropriate, i.e., relaxation data rather than transition data, and crystalline loss peaks.
5. We failed to recognize the pioneering work of Kurt Ueberreiter on structure in the liquid state of amorphous polymers, and a relaxation called T_f associated therewith [8]. Incidentally, his work was also ignored by his peer groups in Germany and elsewhere.
6. Only much more recently did we discover that many prominent polymer scientists had extended physical measurements into the $T > T_g$ region and found evidence for T_{ll} which was either ignored or dismissed as insignificant by the authors. For example, Ferry and his students published dynamic mechanical loss data on polyisobutylene as a function of temperature and frequency [9]. A plot of tan $\delta - T$ showed a double loss peak: T_g and $T > T_g$ [9a]. They considered the double peaks to be real and part of the process of going from the glassy to the rubbery state. We did not locate this reference until about 1975. A plot against log frequency [9b] showed T_{ll} lying below T_g and hence labeled by the authors as "the slow process" [9b].

Our 1963 paper [2] actually covered several different areas:

1. Liquid state transition (relaxation) above T_g, originally T_{ll}
2. The plotting of multiple loss peaks on a relative temperature scale
3. Free-volume considerations
4. Dependence of mechanical properties such as impact strength on glassy state loss peaks, a subject later treated in detail by Heijboer [10] and this writer [11]

We failed to address the problem as to how a polymer molecule could move as a unit in the molten state. The reptation model of deGennes was to provide this answer almost 10 years later.

It is not our intent to convert this review into a treatise on T_{ll}. However, we shall later propose that T_{ll} is more of an equilibrium transition than is T_g and may be considered the precursor of T_g on cooling. It deserves the attention of theoreticians from this point of view, and also because it implies structure in amorphous polymers. Detailed reviews on T_{ll} appear elsewhere [8,11–15]. Pertinent features of T_{ll} will be discussed when appropriate.

Returning to the historical introduction, our paper [2] was translated into Chinese by a professor from China. It was, for a period, duplicated in lots of 100 for classroom use in a key East Coast technological university. Years later it was granted a Citation Index Award as one of the most cited publications in its field [16]. Requests for reprints of our paper continued for many years. A printing of 500 copies was made by Dow Chemical with permission of *Rubber Chemistry and Technology*. A sequel, as it were, by

Eisenberg and Shen also appeared in *Rubber Chemistry and Technology* [17,18].

In 1965, the High Polymer Division of the American Chemical Society sponsored a three-day symposium on multiple transitions at its fall meeting. Professor Robert Simha [19], as chairman of the polymer division, conceived the idea and asked this writer to organize a program. It was published by Interscience as "Transitions and Relaxations in Polymers" [20]. This publication was authored largely by Americans, one Englishman, and Saito as a coauthor. This volume was translated into Russian by the Soviet Scientific Press [21].

In 1967 the classical book, *Anelastic and Dielectric Effects in Polymeric Solids* by McCrum et al., appeared [22]. This book contained introductory chapters on the nature of polymers, principles of dielectric and mechanical loss data, and key apparatus involved. There then followed seven chapters reviewing published data arranged by key polymer types: methacrylates and acrylates, polyvinylesters, hydrocarbons, halogen polymers, polyamides and urethane, polyesters and carbonates, and oxide polymers. World literature was reviewed, particularly British, Dutch, German, Italian, Japanese, Soviet, and American. Brief tutorials concerning the nature of each polymer type were helpful even for professionals.

Many original figures are reproduced showing loss and modulus data versus temperature and/or frequency [22]. One very important feature of each chapter is the inclusion of relaxation maps, i.e., log frequency–1000/T plots, showing loci for T_m, T_g, T_β, T_γ, T_δ, with data based primarily on dynamic mechanical, dielectric, and ultrasonic loss, bulk modulus, and occasionally NMR data. More rarely, a $T > T_g$ locus was shown and indicated as part of the T_g process. We consider this as evidence of T_{ll}. This book can be extended to cover newer data on old and new polymer types, but can hardly be improved in concept or execution. It must remain the classic in its field.

Incidentally, the latest references cited in [22] appear to be 1965 and these mostly to papers by the three authors. This amount of delay between the latest citation and the date of publication is reasonable in a work of this magnitude. It had to have been conceived and begun several years prior to 1965.

We next note a publication by McCall [23], with an introductory tutorial followed by numerous relaxation maps on all key polymer types, with emphasis on dielectric, mechanical, and NMR data, the latter being a specialty of the author. We consider this compilation to be complementary to the McCrum et al. volume because of its inclusion of —CH_3 group motion as detected by NMR measurements. The symposium at which McCall presented his paper was held in 1967. The book *Dielectric Spectroscopy of*

Polymers, by Hedvig [24], overlaps earlier books and reviews but also contains topics of special interest to its author. Plazek [25] has cautioned about usage of the term "spectroscopy," particularly in regard to dynamic mechanical loss, which is an energy dissipative process. Optical spectroscopy is a resonant process. A pertinent multiauthor book edited by Meier concerned a 1975 international symposium not published until 1978 [26] on the general area of this preface.

We end this introduction by emphasizing the "implosion" of reviews over the four year period 1963–1967 in this specialized field covering data which was accumulating over a period of at least 25 years. Experimental and/or theoretical results feeding this "implosion" had been accumulating for at least 20 years and most frequently in the 10 years prior to 1963–1967. Such work originated primarily in Belgium, Czechoslovakia, England, Germany, Holland, Italy, Japan, the United States, and the Soviet Union. One can get an excellent impression of this from the references section of McCrum et al., pp. 575–596 of [22], arranged alphabetically by authors. Prominent workers in the field are thereby clearly recognized, especially the pioneers.

NATURE OF THE POLYMERIC AMORPHOUS STATE

One of the key unresolved issues currently facing polymer scientists concerns the nature of the amorphous polymeric state: is it structureless or structured? If the latter, to what extent? Lack of x-ray diffraction lines eliminates the existence of long-range structure, while x-ray scattering halos suggest short range or local order, to be discussed later.

Flory postulated in his *Principles* [27] that polymer chains in the solid state behave as if in a theta solvent with unperturbed dimensions and hence random coil dimensions: i.e., there is no order. Conversely, Soviet polymer scientists, under the leadership of the late Valentine Kargin, believed both on general and on experimental grounds that short-range order does exist. The Soviet position has been summarized by three of Kargin's last students: Arzhakov, Bakeev, and Kabanov [29]. They start with several simple premises: (1) order has been proven to exist in simple liquids; (2) amorphous polymers are liquid; (3) therefore polymers should possess as much order, and possibly more, than do simple liquids. The Soviet Press has never published a translation of Flory's *Principles* because of his stand on this issue of order so contrary to their own extensive data.

Many polymer scientists, before and after Flory, have offered logical evidence for local order from various points of view. Kurt Ueberreiter, a German polymer physicist, proposed in the early 1940s that an amorphous polymer above T_g was not a true liquid but a liquid with fixed structure [30a]. Later he defined the nature of this structure as arising from an intermolecular attraction between low-energy "cold" spots along polymer chains [30b]. His original position seemed an intuitive one, although he later offered experimental evidence. (For details see pp. 145–147 of [8b]).

In 1953, Bunn [31] presented experimental evidence for an empirical relationship:

$$T_g\ (\mathrm{K}) = kT_m\ (\mathrm{K}) \tag{1}$$

where k is a constant less than unity. He explained this as arising from order in the glassy state much shorter in range than the known long-range order in the crystalline state. This empirical rule was discovered independently by this writer [32,33] and by Beaman [34]. The value of k was commonly found to be about 2/3 for substituted polymers and 1/2 for unsubstituted ones [32]. Following Van Krevelen [35], Boyer later showed, after correcting for errors in T_g, a refined version of Van Krevelen's S-shaped cumulative plot (see Fig. 18, p. 769 of [28]). A slightly modified form appears as Figure 1.

Figure 1 is divided into regions A, B, and C. Region A contains unsubstituted polymers, especially the group shown later in Figure 14. Region B contains most common polymers with an inflection at $T_g/T_m \sim 0.667 \pm 0.05$. Region C contains nontypical polymers: for poly(2,6-dimethylphenyleneoxide) the ratio is 0.92; for poly(4-methylpentene-1) the ratio is actually 1.01.

It has long been recognized that T_g increases with chain stiffness, σ, and with cohesive energy density (CED) [36]. (See also Figures 43–45 and pages 801–804 of [28], which lists primary sources). For polymers in A, σ and CED are low, and hence T_g should be low. Bunn presented evidence to show that T_m depends on σ and CED, but also on a shape factor which influences packing [36c]. We suggest that unsubstituted polymers pack very efficiently leading to abnormally high values of T_m.

Polymers in group B are characterized by higher values of σ and CED but inefficient packing because of side groups, and hence generally higher values of T_g/T_m.

Krimm and Tobolsky [37–39] presented x-ray scattering data for randomly and uniaxially oriented atactic PS showing, in their opinion, evidence for local order which disappeared near our T_{ll} temperature. Their experiments were repeated and extended many years later by Hatakeyama [40] on at-PS of several molecular weights.

As noted in the introduction, this author [2] reported the existence of a $T > T_g$ transition labeled T_{ll} which was considered to represent the disruption of some kind of local order. T_{ll} has only reluctantly gained partial but not complete acceptance, as evidenced in part by an invited review article in a prestigious encyclopedia [15].

In 1965 Robertson [41] published a paper in which he noted that the ratio of amorphous density, ρ_a, to crystalline density, ρ_c, approached unity

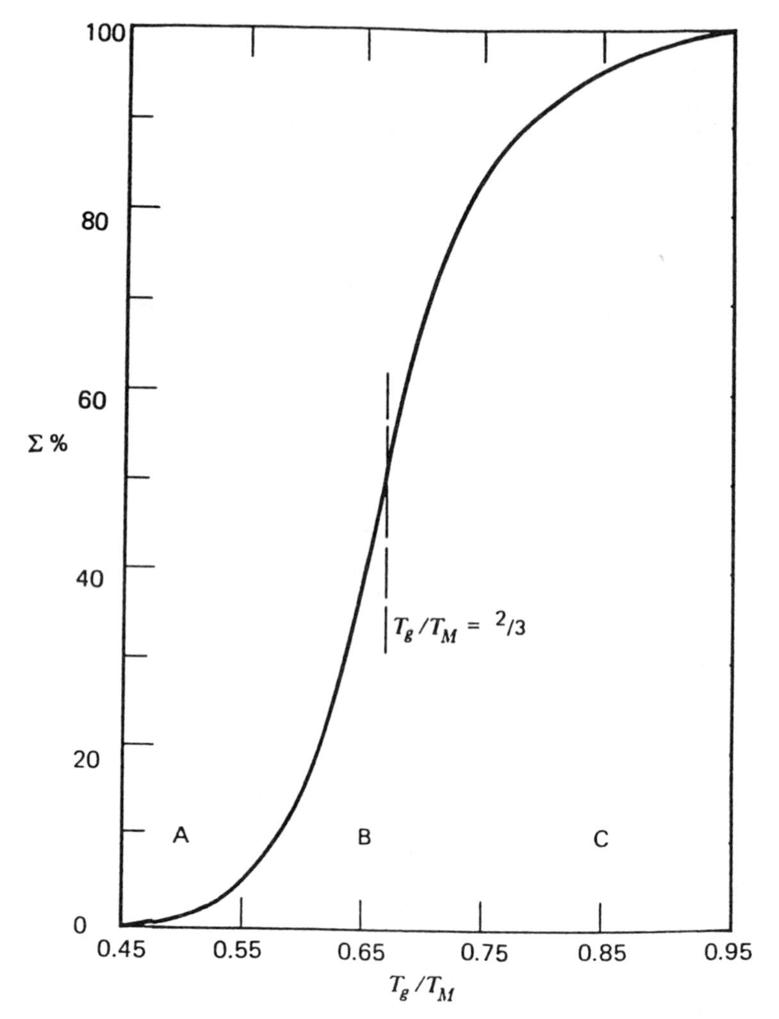

Figure 1 Van Krevelen's cumulative plot of T_g/T_m. Region *A* is for polymers without side groups. Regions *B* and *C* are for polymers with side groups. Cohesive energy density and chain stiffness tend to increase from *B* to *C*. Polymers in group *C* have unusual structures. The ratio, d_a/d_c, of amorphous to crystalline density increases with increasing T_g/T_m. (Modified from Ref. 28.)

for some polymers, especially those with stiff chains, compared to a theoretical value of 0.62 which he calculated for random coils. He concluded that local order existed in amorphous polymers, the more so the higher is ρ_a/ρ_c. This is analogous to Bunn's conclusion based on T_g/T_m.

Next we direct attention to the historical conference on polymers sponsored by The Welsh Foundation on Chemical Research, *X*, *Polymers*, in

1967 [42]. Among the speakers were Flory and Corradini, who discussed his x-ray studies with Natta on isotactic polyalphaolefins. He also mentioned x-ray scattering results on amorphous polyalphaolefins which were indicative of local order proportional in spacing to the length of the side group, as Corradini stated.

All comments, questions, and answers are published [42]. Flory expressed his opinion that there was no such thing as local order in amorphous polymers. Several distinguished polymer scientists in the audience proposed a compromise between the conclusions of Corradini and the views of Flory. Flory insisted that a compromise was impossible; there were two states, ordered and random, but nothing in between.

Under such pressure from Flory, Corradini seemed to retreat [43]. Corradini (p. 335 of [43]), after presenting evidence supporting the random coil concept for bulk amorphous polymers, concludes as follows:

> The only point which is left unresolved by the above discussion is the extent to which there is orientational correlation among the neighboring and strongly asymmetric polymer molecules in the condensed amorphous phase due to space filling requirements while still leaving equally probable the possibility of access to each type of the whole set of unperturbed conformations. When strong directional forces are present among atoms of different macromolecules, restrictions on the availability of conformations are expected. Such a case is provided, for instance, by the polyamides, where hydrogen bonding appears to be maintained almost completely above the melting point.

Concerning the polyamides, he writes (p. 337 of [43]):

> In the approximation of infinite length of the macromolecules, each hydrogen bond which is established between two amide groups leads to the closure of an independent ring of bonded atoms. Constraints on the internal coordinates of the macromolecules are introduced, even in the case of imperfect or incomplete hydrogen bonding. Correspondingly, we propose that the variability of internal rotation angles will be strongly restricted and the melting entropy decrease with respect to the value calculated according to the rotational isomeric model for an isolated macromolecule.

To be sure, the polyamides represent an extreme case of intermolecular forces compared to the dispersive and dipole forces also represented by cohesive energy density.

In 1973 at an IUPAC meeting in Helsinki, Flory presented four "proofs" of his views on random coil structure [44].

1. Thermoelastic retraction
2. Thermodynamics of concentrated solutions
3. Macrocyclization equilibria of polysiloxanes
4. Neutron scattering data

The first three sets of experimental data were carried out in solution and/or at high temperature where any structure, if ever present, could have been destroyed. The neutron results are, to our knowledge, still controversial. In Tables 5 and 6 of his Helsinki paper, Flory presented evidence for intermolecular correlations (order) for PE and PBD as greater but no more than in simple liquids.

A 1975 American Chemical Society Symposium on order in amorphous polymers [45] was dominated by the views of the Flory school, with only Geil as a dissenting voice (pp. 173–208 of Ref. 45). While not an invited speaker, we were permitted at our request to contribute a dissenter's views to the published proceedings [46]. We covered different types of evidence believed by the several quoted authors to indicate local order.

Miller and the writer have recently published the results of extensive x-ray scattering data on a large variety of amorphous polymers. One paper with Heijboer [47] contains new data on polyalkylmethacrylates and acrylates, measured at MMI; a second paper surveys literature data [46]; a third paper is a summary paper with some new data on at-PS [48].

Figure 2 summarizes the major findings of these x-ray studies in a plot of equivalent Bragg spacings as a function of cross-sectional area per polymer chain. The bottom line represents intramolecular van der Waals (VDW) distances which are almost independent of area; the middle line represents intermolecular distance or large van der Waals (LVDW) distances. The top line is also intermolecular for simple polymer chains, and corresponds to spacings of second-nearest neighbors.

Miller then concludes [49] inter alia that

1. The appearance of LVDW peaks on scattering patterns seems to be the norm for the amorphous state of polymers.
2. From the observed intensities one can infer a significant concentration of atoms preferring the LVDW spacing in many polymers, at least 15%.
3. It would seem that a marked LVDW peak is a "polymer" property.
4. Miller concluded that the above observations are explainable on the basis of the existence of a nonvanishing degree of local order in amorphous polymers as normally encountered (p. 46 of [49]).

We now believe that there can be a compromise between the two extremes of randomness and crystalline order best described by the phase dualism hypothesis of Baranov and Frenkel [50]. This view states that while a polymer chain overall behaves as a gas, with the density of a gas, at a local level chain segments behave like a liquid and associate with neighboring segments under the influence of intermolecular forces proportional to cohesive energy density. This clearly implies local order in contrast with random coil behavior.

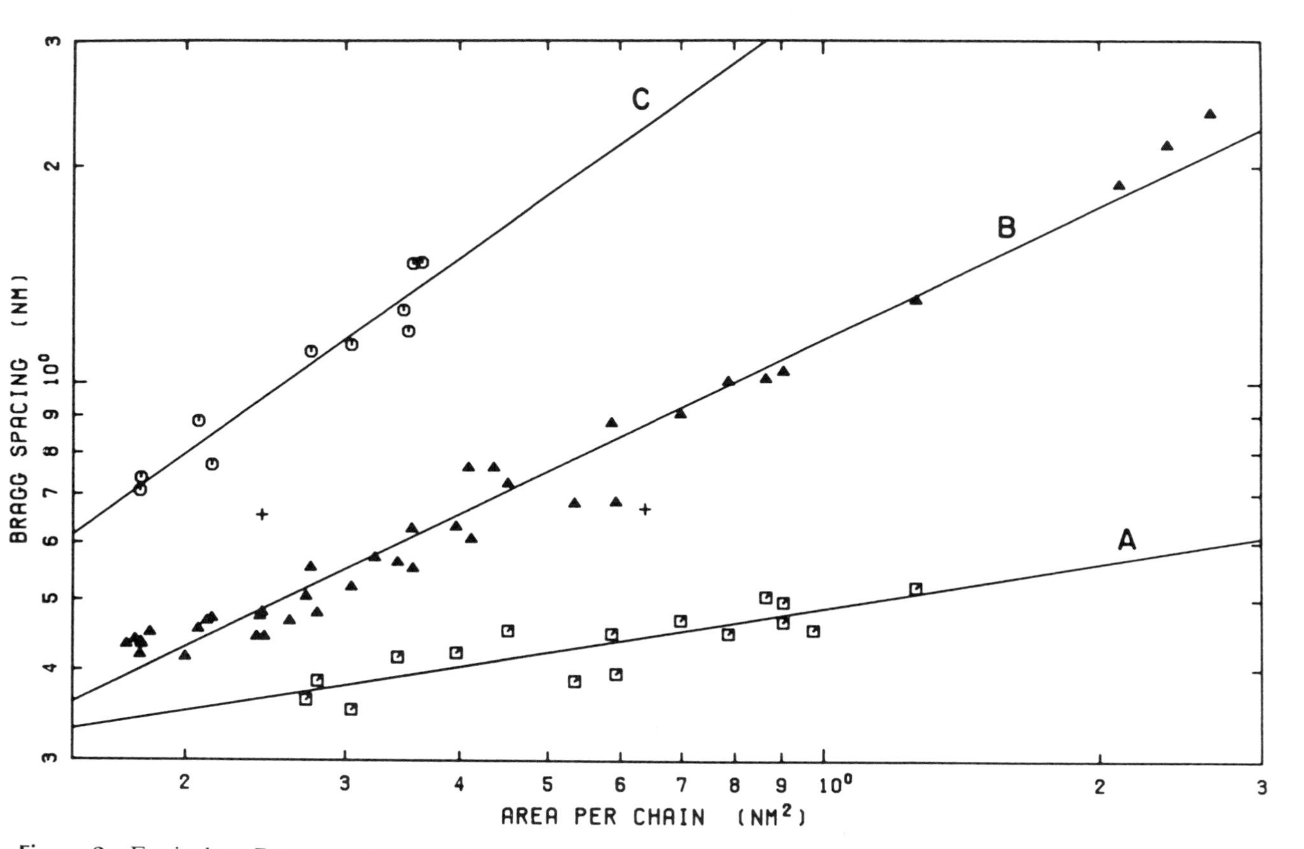

Figure 2 Equivalent Bragg spacings, d_b, in nm, as a function of cross-sectional area per polymer chain, A, in $(nm)^2$. Bottom line is normal intramolecular van der Waals distances; middle and top lines are intermolecular distances whose presence implies some local structure. (After Miller et al., based on amorphous x-ray scattering halos [47–49]. Numerical values on which this figure is based are from Ref. 48, Table I.)

This phase dualism concept is reminiscent in part of Corradini's analysis given earlier that strong intermolecular forces as in polyamides can perturb random coil behavior.

One must ask in the more general case, say at-PS, as to what extent intermolecular attractions, in this case between phenyl groups, can perturb random coil behavior.

We have recently discovered [51] an empirical pair of correlations involving T_g (K) and T_{ll} (K) as a function of chain stiffness times cohesive energy density which, we believe, are consistent with phase dualism. In this case, chain stiffness activates a persistence length which could accentuate a correlation of orientations between chains, with cohesive energy density much less than that of the polaymides but able to "bond" together these adjacent segments.

The foregoing discussion raises another issue: if structure is present in amorphous bulk polymers and copolymers, at what temperature will it disappear? We suggest that amorphous polymers and copolymers will be free of local structure only at $T > T_{ll}$. In this regard we cite high-precision volume-temperature data of Höcker, Blake, and Flory [52] on at-PS of narrow-molecular-weight distribution. Specimens were first heated to above 200°C, step cooled to $T > T_g$, $T = T_g + 10$ K and then step-heated in increments of 5 K. A third-order transition in the V-T data was noted by them near 170°C (our T_{ll}). On p. 2254 these authors state: "These observations suggest, but do not prove, that the apparent discontinuity in α (coefficient of cubical expansion) in Figure 2 is due to a "freezing-in" which precludes full attainment of equilibrium below about 170°C" (not a full quote). We take this finding to imply that structure or local order in at-PS ceases at T_{ll} with no further evidence for other ordered structures at least up to 220°C. We note in Figures 2 and 3 two other events in the liquid state, $T_{l\rho}$ and $T_{l\phi}$, but each of these appears to represent very local motion, (less than one monomer unit) and hence no long-range order effects.

Hence structureless atactic PS can be found at $T > T_{ll}$ and most certainly at $T_{ll} + 40$–60 K. This should be the ideal liquid state for atactic polymers and copolymers.

Temperatures T_{ll} and T_g have recently been found on DSC cooling traces (annealed specimen cooled at 30 K/min from ~200°C) followed at lower temperature by T_g), and hence T_{ll} can be considered a precursor of T_g on cooling.

Alternative Explanations

Assuming that Flory and others who postulate a structureless amorphous state are correct, it would be desirable to find alternative explanations for many empirical rules, such as those of Bunn [31] and Robertson [41], as well as the whole body of experimental findings associated with the x-ray

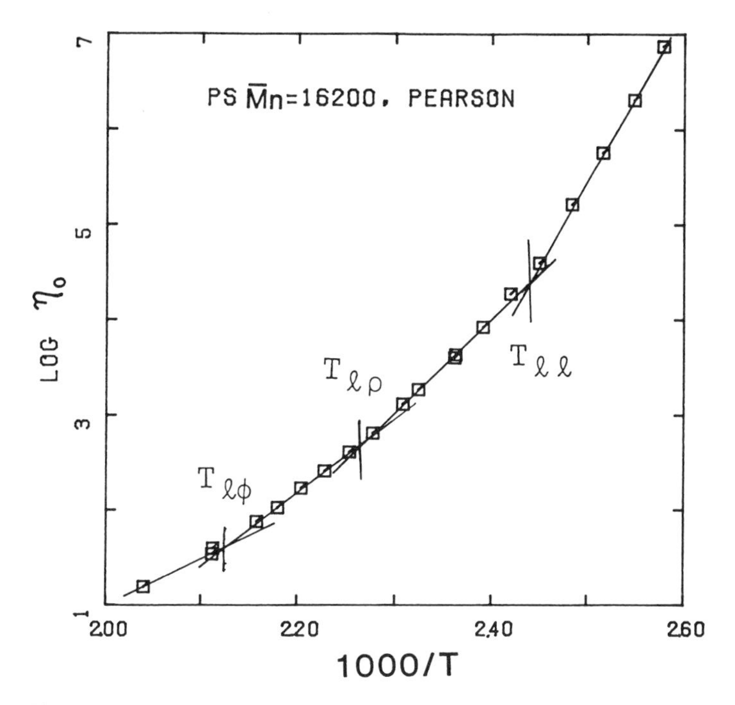

Figure 3 Log zero shear melt viscosity, η_o, plotted against T^{-1} with slope changes indicating the several liquid state transitions. (at-PS, $\overline{M}_n$ = 16,200). ΔH_a values in *k*cal/mol. calculated from slopes are $T < T_{ll}$, 80; $T > T_{ll}$, 47; $T > T_{l\rho}$, 36; $T > T_{l\phi}$, 26. (From Ref. 68.)

studies of Miller [47–49] and the T_{ll} phenomena reported by this author [11–15], as well as the third-order transition at circa 170°C in at-PS reported by Flory and his coworkers [52].

Order in the Melt of Crystalline Polymers

There is another amorphous state, namely the melt of crystalline polymers. By definition, long-range order disappears at T_m, but how about a residual local order? We earlier reviewed scattered evidence in the literature suggesting the persistence of local order above T_m (pp. 826–892 of [28]). It is certain that other examples can be found. We particularly cite here recent studies of Krüger et al. [53,54] reporting a transition above T_m in various crystalline polymers. Using Brillouin spectroscopy, they examined linear PE of $\overline{M}_w$ = 6600. Slope changes in plots of ultrasonic velocity, frequency, refractive index, and density as functions of temperature were noted at the $T > T_m$ transition, designated T_u. They ascribed T_u to disruption of a nematic phase which persists above T_m. A subsequent review article by Krüger updates his results (pp. 281–304 of [8b]). Denny and this writer conducted exploratory DSC studies on a series of n-alkanes and a few crystalline polymers [55]. Our data suggested a tendency expressed by

$$T_u\ (\mathrm{K}) \sim 1.2T_m\ (\mathrm{K}) \tag{2}$$

In the case of crystalline *trans*-polyisoprene, [56], T_u was sufficiently strong to show up in dynamic mechanical analysis (see also Fig. 40, p. 313 of [14], and related discussion).

We suggest that the region above T_m is ideal for causing local order. Polymer chains have been in regular crystalline arrays up to T_m. They should still be juxtaposed just above T_m even though long-range order has vanished. Cohesive energy density should attract adjacent chains to each other. Eventually such attraction must be disrupted as the chains tend to assume random coil conformations.

Conclusions for the Second Section

Finally we urge that investigators interested in computer modeling of motion in polymers not accept blindly the prevailing view that atactic polymers are free of local structure. A case in point is the computer Monte Carlo studies of Kolinski et al. [57]. Their original model showed no evidence of structure or transitions. Introduction of a chain stiffness parameter led to a prediction of local order but not long-range order. Further introduction of an intermolecular force led to a prediction of a transition, "probably of first order."

Returning to the concerns of Corradini on p. 9, values of the stiffness parameters, σ or C_∞, are generally obtained in solution, either in a Θ solvent or extrapolated to Θ conditions. Neither reflects the situation characteristic of bulk state packing. Neutron scattering does reflect the bulk state, but concern has been expressed about the ability of this technique to measure fine details. (See, for example, Mansfield [57a].)

CONVENTIONS, DEFINITIONS, NOMENCLATURE

General

This section summarizes the key terms employed in this overview, primarily by means of illustrative figures. It was decided to employ atactic polystyrene and derivatives as a model system for several reasons:

1. It is commercially available over a wide range of molecular weights with $\overline{M}_w/\overline{M}_n \sim 1.05$.

2. Many derivatives are available from chemical supply houses, both ring- and side-chain-substituted.
3. Isotactic PS has long been available, and syndiotactic PS has now been synthesized [58] and characterized [59].
4. The phenyl ring is an excellent probe, especially when halogenated for dielectric studies, deuterated for NMR purposes, or hydrogenated for steric effects.
5. PS has remarkable thermal stability in the absence of oxygen.
6. T_g's of styrene copolymers are well known. Only recently has T_{ll} of styrene and other copolymers been determined with interesting results; i.e. T_{ll} mimics T_g in shape, whether showing a maximum or minimum or linearity vs composition [60].

All of the above factors have resulted in an extensive literature on the glassy and liquid states of PS. Examples will be selected from other polymers to illustrate specific points as necessary.

Types of Amorphous Polymeric Materials

One can distinguish five types of amorphous regions in vinyl polymers:

1. Atactic polymers
2. Random copolymers
3. The amorphous regions in semicrystalline polymers which may give rise to two T_g's; T_g(L) for a lower T_g, uninfluenced by crystallinity; and T_g(U) > T_g(L) influenced by fractional crystallinity, χ_c, and by morphology [62]
4. The same two T_g's as in 3 but for polymers without side groups, to be shown later in Figure 14
5. Crystalline polymers at $T > T_m$

This last case was considered near the end of the second [53–56]. A possible T_u in Teflon was recently reported [56a]. The $T_u > T_m$ transition is characteristic of the final disappearance of local order. The T_u phenomena may well be connected with persistent nuclei thought to influence recrystallization after quench cooling.

In addition to these standard amorphous states, there are several induced states such as

6. Densified by application of hydrostatic pressure above T_g, followed by cooling under pressure. There is extensive literature, with studies by Roe et al. being the most recent [60a].
7. Rarefaction as a consequence of cross-linking which freezes volume at gelation, while polymerization continues, as with styrene-divinylbenzene, reviewed by us on an earlier occasion [60b].

Such special amorphous states are useful in studies of physical aging and the nature of the glassy state.

Nomenclature for the Glass Transition T_g

The glass transition, now generally designated T_g, after Fox and Flory [61], was originally called a second-order transition under the Ehrenfest scheme because of the discontinuity in dV/dT and dH/dT characteristic of T_g [3]. Flory [63a] and Rehage [63b] later stated that T_g is not a true second-order transition and should be called a glass transition. Some authors prefer the term "relaxation" or "dispersion," because of frequency dependence [64]. The events covered herein are all frequency-dependent, as will be shown. Several recent authors choose to discuss the glass transition T_g even when defining second-order transitions. Lobanov and Frenkel [65] suggest the term "behavioral transition" because there is frequently a change in physical behavior at a major transition such as T_g.

The three recent reviews on T_g may be summarized briefly as follows, in alphabetical order of authors [66]:

Bendler is mainly concerned with a detailed examination of available theories of T_g: free volume and entropy models. He concedes that all current theories possess a number of shortcomings but considers the entropy model of Gibbs and DiMarzio somewhat more correct [66a].

McKenna has two broad topics: thermodynamics of the glass transition and kinetics of glass-forming systems, including two pages on physical aging [66b].

Roe gives a condensed account (12 pages) with definitions, key issues, reference background and theory relating to T_g [66c].

There is of necessity overlap between the above three treatments, but the examples employed and approaches to the subject matter differ.

$T < T_g$ and $T > T_g$

Figure 4 is a schematic mechanical loss "spectrum," tan δ, against a relative temperature scale, T/T_g (K/K) with the designation α for T_g and Greek letters, β, γ, δ, etc., for $T < T_g$. This is now conventional usage, although exceptions, even reversals, are found in the older literature. Above T_g, nomenclature is new and generally that of this author. This schematic plot is patterned after actual values for at-PS, but is believed to be general.

Below the figure are numerical values indicative of the number of monomer units involved in each event. The symbol ∞ for T_{ll} simply implies that the entire polymer molecule is involved, regardless of molecular weight, but preferably above the oligomeric range. Activation energies appear later in Figure 7(a).

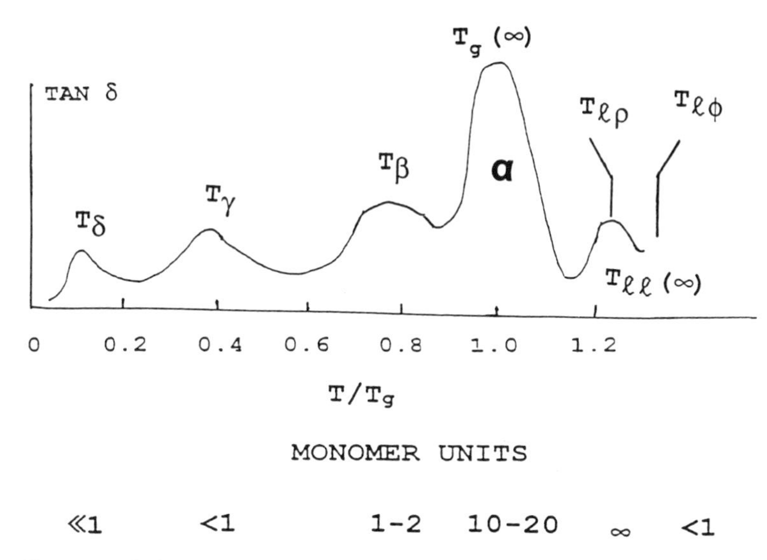

Figure 4 Schematic "spectrum" of mechanical loss, tan δ (or related quantity) plotted against a reduced temperature scale showing conventional designations below T_g and this author's designations above T_g. Plot is based on numerical data for at-PS but is considered to be general except for fine details. Also shown are number of monomer units involved in each process. $T_{l\rho}$ and $T_{l\phi}$ involve less than one monomer unit in the polymer chains, T_{ll} involves the entire chain, designated as ∞. See also Figure 8 for additional details.

Figure 5(a) shows thermal expansion data for at-PS based on recent measurements of Greiner and Schwarzl [67]. Several methods were employed: linear thermal expansion at low temperature, mercury dilatometry above the freezing point of *Hg*. All transitions indicated in Figure 1 are present: T_δ, T_γ, T_β, T_g, and T_{ll}. These data cover the widest span of temperature for thermal expansivity known to us. They verify a principle long recognized: any mechanical loss peak should be revealed by thermal expansion [13, Fig. 1]. Similar plots by the same authors [67] show similar data for PMMA, PVC, and polycarbonate. The general case for thermal expansion near T_g is shown in Figure 5(b). Reference 52 involves precision *V-T* data in at-PS above T_g.

Thermal expansion is an ideal absolute method for locating transitions both below and above T_g. Another absolute method, but only for the liquid state, is zero shear melt viscosity plotted as log $\eta_0 - T^{-1}$, or as the derivative, $d(\log \eta_0)/dT^{-1}$ [68]. Slope changes locate the several liquid state transitions as shown in Figure 3. We first noted the potential of this in 1966 (see p. 269, Fig. 1 of [20]), further tested it again for PS in Figure 8, p.

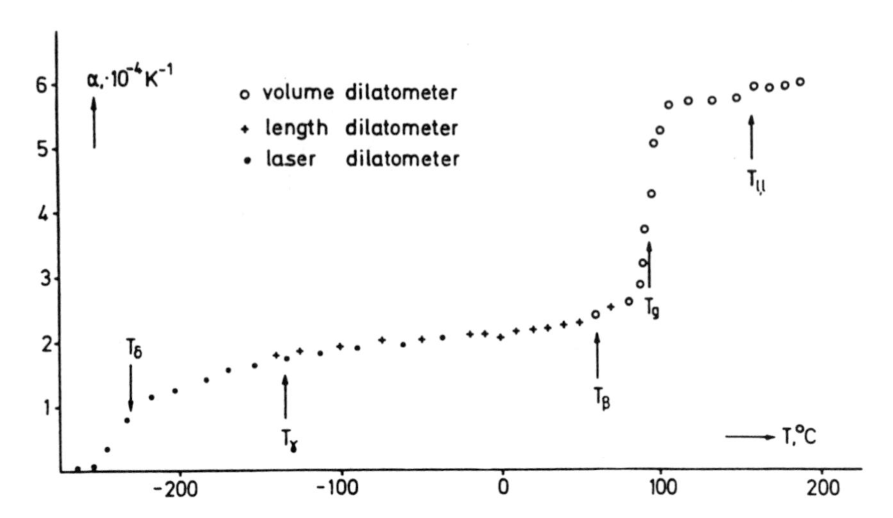

Figure 5(a) Thermal expansion data for at-PS with step increases corresponding to loss peaks in Figure 4 from T_γ to T_{ll}. α is the coefficient of cubical expansion, $(1/v)(dv/dT)$. (From Ref. 67.)

357 of [71], and proved its values including use of the derivative method in Ref. [68]. Its application to other polymers was presented in Table III, p. 150 of [8b] and in Figure 1 of [15]. Figure 2(b) with the three liquid state events, was not hitherto shown. On older plots, $T_{l\phi}$ was designated $T_{l\rho}$. This current system of designation appears first in Figure 1 of [15].

Relaxation Maps

Figure 6 is a relaxation map, log f (Hz) – $1000/T$ (K), again based on at-PS data (Table 1), for each of the loss peaks in Figure 1. In general, each locus except T_{ll} tends to be linear at $f \geq 1$ Hz. An apparent energy of activation, ΔH_a in kcal or kJ/mole, can be estimated from these Arrhenius-type plots. The quantity T_{ll} is nonlinear. In general, T_{ll} follows Vogel behavior, which means that it can be linearized using a log f – $(T - T_0)^{-1}$ plot, where T_0 is a reference temperature, as shown by Lacabanne et al. for T_{ll} [69] by thermally stimulated current (TSC) techniques.

Such nonlinear behavior was also found by this writer and Enns [70] on dynamic mechanical loss data of Fitzgerald et al. [9a,b] for PIB. Curvature for T_{ll} is probably general.

Figure 6 is patterned after Figure 10.35 for McCrumm et al. [22] with several modifications.

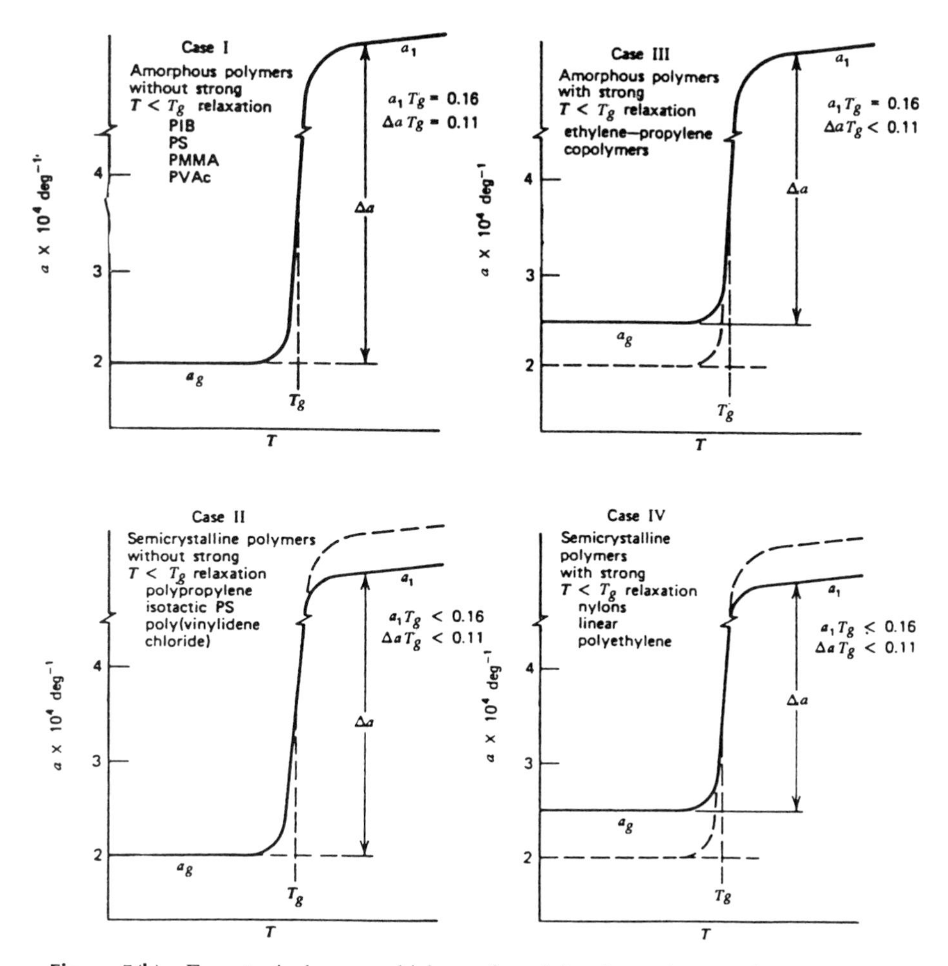

Figure 5(b) Four typical cases which are found for thermal expansion across T_g in amorphous and semicrystalline polymers. (From Ref. 28, Fig. 10.)

1. A locus for T_{ll} was added based on data of Lacabanne et al. [69], while loci for T_γ and T_δ are taken from Figure 13 of McCall [23].
2. Loci for T_β and T_g were extended by us until they intersect at a log f of ~6 at a reciprocal temperature designated T^*. Lobanov and Frenkel [65] have found empirically that $T^*(f \sim 10^6) = T_{ll}$ (static), although they did not give such a plot for at-PS. We showed this on a prior occasion but without T_{ll} (Fig. 18 of [8b]).

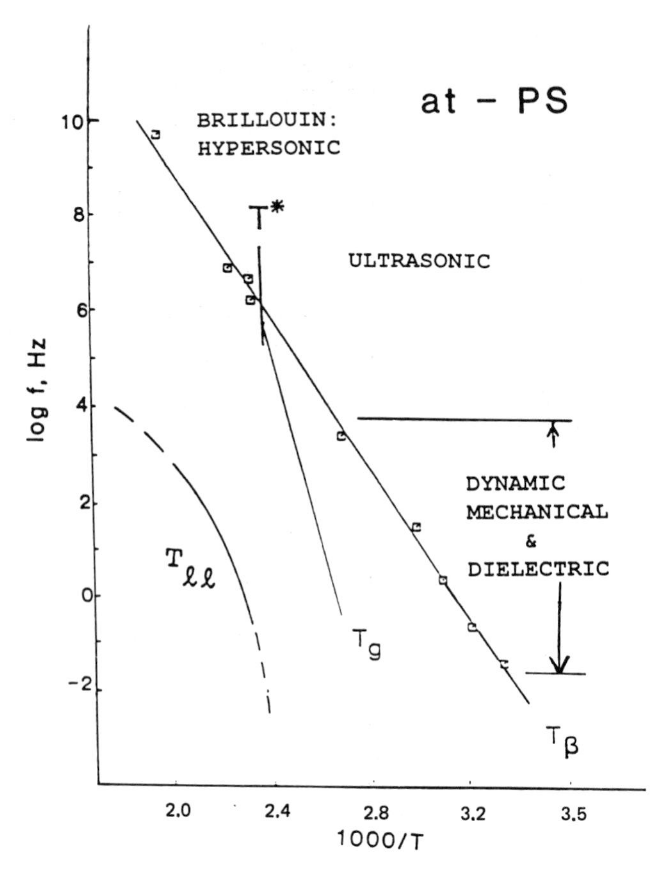

Figure 6 Relaxation map for at-PS by methods and sources given in Table 1. Points for T_g omitted for clarity. T_g and $T < T_g$ events follow Arrhenius behavior over the frequency range indicated. T_{ll} shows non-Arrhenius behavior. Linear least-squares line for T_β, using all nine data points, shows a correlation coefficient of $R^2 = 0.99686$ and a slope corresponding to $\Delta H_a = 35.68$ kcal/mol. T^* is the Lobanov-Frenkel temperature at which T_β and T_g merge. They show that T^* ($f \sim 10^6$) $= T_{ll}$ (static). (From Ref. 65.)

Figure 7(a) is a plot of log ΔH_a against transition temperatures for at-PS as first noted by this author for $T \leq T_g$ [71]. The linearity of this plot can hardly be fortuitous. It probably depends on two factors: the increase in size of the moving group from T_δ to T_g, as suggested in Figure 7(b), and the increase in amplitude of the respective motions. The size of the error bars allows the possibility for nonlinearity.

Table 1 Sources of Data for at-PS (Fig. 6) (mol. wts. ~ 10^5)

Authors	Process	Type of data	Frequency Hz	Refs.
Broens and Müller	T_g	Dielectric loss[a]	$<10^5$	c
Becker	T_β	Dynamic mechanical[b]	$<10^5$	c
Illers	T_β	Dynamic mechanical[b]	$<10^5$	c
Thurn	T_β & T_g	Ultrasonic attenuation	2×10^6	d
Boyd and Biliyar	T_β & T_g	Ultrasonic attenuation	9×10^6	e
Shen et al.	T_β & T_g	Ultrasonic attenuation	9×10^6	e
Adachi et al.	T_β & T_g	Ultrasonic attenuation	5.5×10^6	f
Patterson and Latham	T_β & T_g	Brillouin scattering	5.5×10^9	g
Lacabanne et al.	T_{ll}	Thermally stimulated current	calculated	h

[a]Data points for T_g not shown.
[b]Datum point for T_β not shown.
Sources:
[c]See Figure 10.35 of McCrum et al. [22] and related references.
[d]H. Thurn, *Z. Angew. Phys. 7*, 44 (1955); H. Thurn and K. Wolf, *Kol. Z.*, *148*, 16 (1956).
[e](a) R. H. Boyd and K. Biliyar, *Polym. Preprints, Am. Chem. Soc.*, *14*, 329 (1973); (b) M. Shen, V. A. Kaniskin, K. Biliyar, and R. H. Boyd, *J. Polym. Sci., Polym. Phys. Ed.*, *11*, 2261 (1973).
[f]K. Adachi, A. M. North, R. A. Pethrick, G. Harrison, and J. Lamb, *Polymer*, *23*, 1451, (1982).
[g]G. D. Patterson and J. P. Latham, *J. Polym. Sci.*, *Macromol. Rev.*, *15*, 1–27 (1980);
[h]See Ref. 69.
For general background, see R. A. Pethrick, in *Comprehensive Polymer Science*, Vol. 2, pp. 571–600 (C. Booth and C. Price, eds.), Pergamon Press (1989).

Multiple Transitions Versus Molecular Weight

Figure 8 is a transition temperature map versus molecular weight, again based on at-PS data obtained under quasi-static conditions, basically at $f \leq 1$ Hz (Fig. 1 of [28]).

The glass transition T_g was found to depend on $\overline{M}_n$ by Fox and Flory [61]; T_{ll} is a function of $\overline{M}_n$, as first demonstrated by Glandt et al. [72] for at-PS blends of two narrow distribution PS, each with $\overline{M}_n \leq 20{,}000$. The temperature T_f is from data based on low frequency but involving viscous flow such as with the torsion pendulum or torsion braid (Fig. 17 of Stadnicki et al. [73]). T_f is an isoviscous state, viscosity ~10^6 poise [73]. Hence it is plotted against $\overline{M}_w$. $T_{l\rho}$ and $T_{l\phi}$ are both independent of molecular weight, implying that they involve local motion of one monomer or less. Our current speculation is that $T_{l\rho}$ involves CH_2 rotation about the chain axis while $T_{l\phi}$ involves CHϕ in PS, or CHR in general. This unpub-

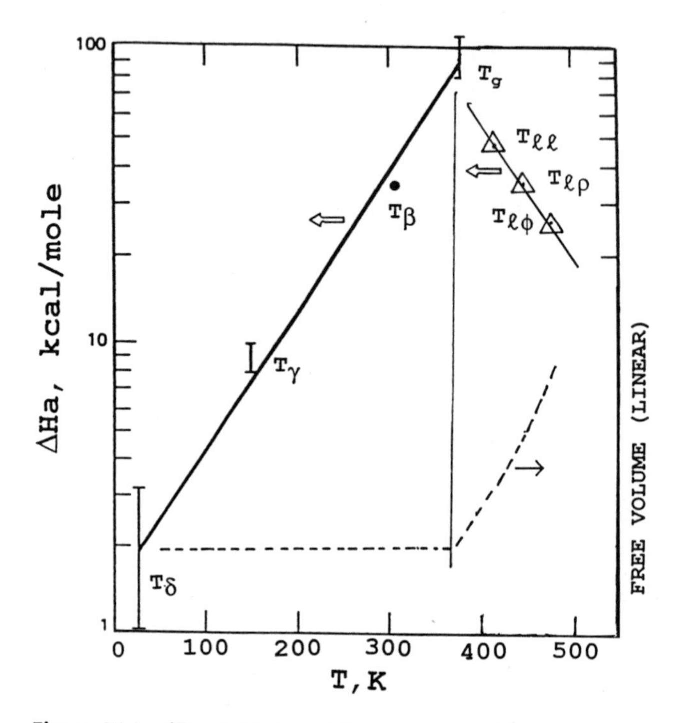

Figure 7(a) (Top left) Log ΔH_a versus transition temperature for the four glassy state transitions in at-PS of $\overline{M}_n \sim 10^5$; (top right) ΔH_a for the three liquid state transitions in at-PS of $\overline{M}_n = 16{,}500$, estimated by us from the log $\eta_o - T^{-1}$ data of Figure 3; (bottom) schematic free volume versus T assuming iso free volume in the glass, increasing free volume in the liquid.

lished speculation by RFB is based on carbon 13 NMR data in the literature [74a-c].

Loci for T_β, T_γ, and T_δ are not indicated. They will all lie below T_g and should be independent of $\overline{M}_n$ because they involve local motion. We are not aware of actual data as a function of molecular weight.

Evolution of the Transition Map

It has long been of interest to us to chart the historical development of the current multiple transition spectrum for key polymer types: PE, PMMA, polycarbonate, and at-PS. The latter appears as Figure 9. In general the tendency is for a linear growth. The question to be raised concerns whether this trend will continue or level off. In some cases, prevention of leveling off has involved showing that a given transition is complex, i.e., T_β for

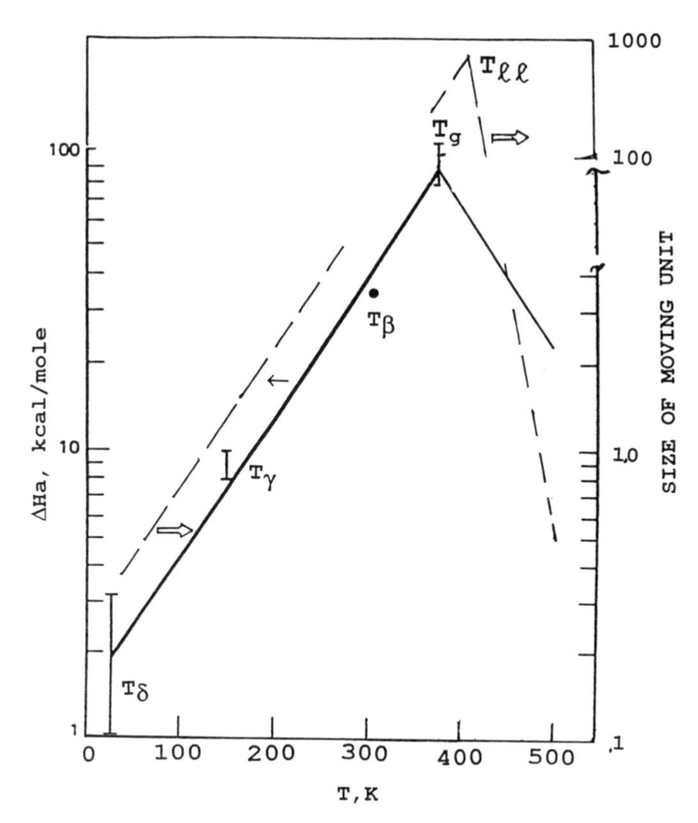

Figure 7(b) Suggested size in carbon atoms of the moving group at each transition peaking at T_{ll} where the assumed motion of the entire polymer molecule is that of reptation. Bottom lines repeated from Figure 7(a).

PMMA [75], T_γ for bisphenol polycarbonate [76]. Either may result from a new technique such as torsional creep for PMMA [75] or multiple techniques for polycarbonate [76].

It is interesting to note that the first liquid state event, $T_{l\rho}$, was found by Fox and Flory as early as 1950 [61]. We were aware of this $T > T_g$ event since it was first published. It was long ignored by us and others, and was later dismissed by Flory et al. as incorrect [52]. Only recently did we find evidence for $T_{l\rho}$ by several other techniques, namely zero shear melt viscosity and depolarization of fluorescence in PS [77]. $T_{l\phi}$ seems common to polymers with bulky side groups (see J. B. Enns and R. F. Boyer, Table IV, p. 242 of [8b]).

Continuing the historical theme in Figure 9, the very strong T_g [78] and

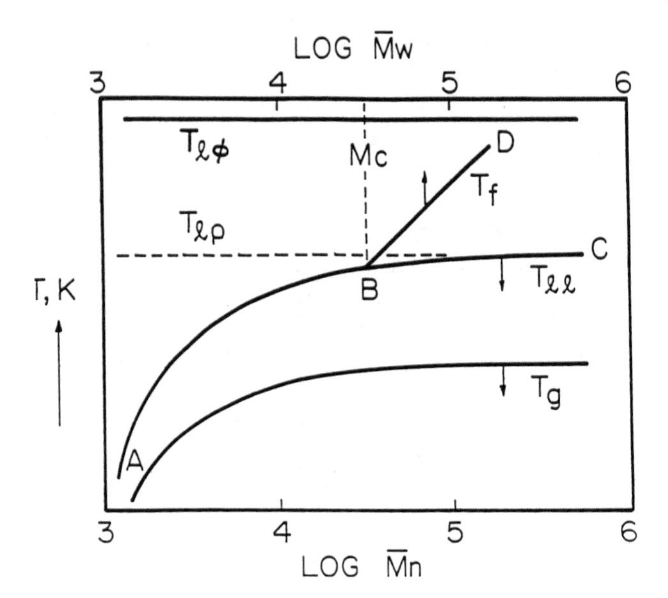

Figure 8 Transition temperature as a function of $\overline{M}_n$ for the several loss processes in at-PS. T_f, plotted against $\overline{M}_w$, represents an isoviscous state event found by torsion braid at log η_o (poise) ~ 6 (A-B-D). Below entanglement molecular weight, $\overline{M}_c$, $T_f \equiv T_{ll}$. $T_{l\rho}$, found only in at-PS by several methods at $M < M_c$ is asymptotic to T_{ll} (∞). $T_{l\phi}$ is found in many polymers at T_{ll} + ~40 K. (See Enns and Boyer, pp. 221–249, Table IV of Ref. 8b. Transition temperature from Fig. 1 of Ref. 15. Curves A, B, D from Fig. 17 of Ref. 73.)

$T_{l\rho}$ [61] depended on classical V-T. T_β and T_γ resulted from the advent of torsion pendulum with liquid N_2 [79]. T_δ used torsion pendulum with liquid helium [80]. T_{ll} was possible with the inverted torsion pendulum [81]. $T_{l\phi}$ was first found with torsional braid equipment by Gillham et al. [82] and very soon thereafter by Enns et al. [83], using DSC.

The important role played by the torsion pendulum, inverted torsion pendulum and torsional braid is evident. A history by Heijboer of the development of these three instruments gives details [84].

Mechanical Dipoles

Gisolf introduced the concept of mechanical dipoles for interpreting dynamic mechanical loss data. He and his coworkers studied $T < T_g$ events in several polymers, including effects of crystallinity, as follows: general [85,86]; PVC [87,88]; polycarbonate [89].

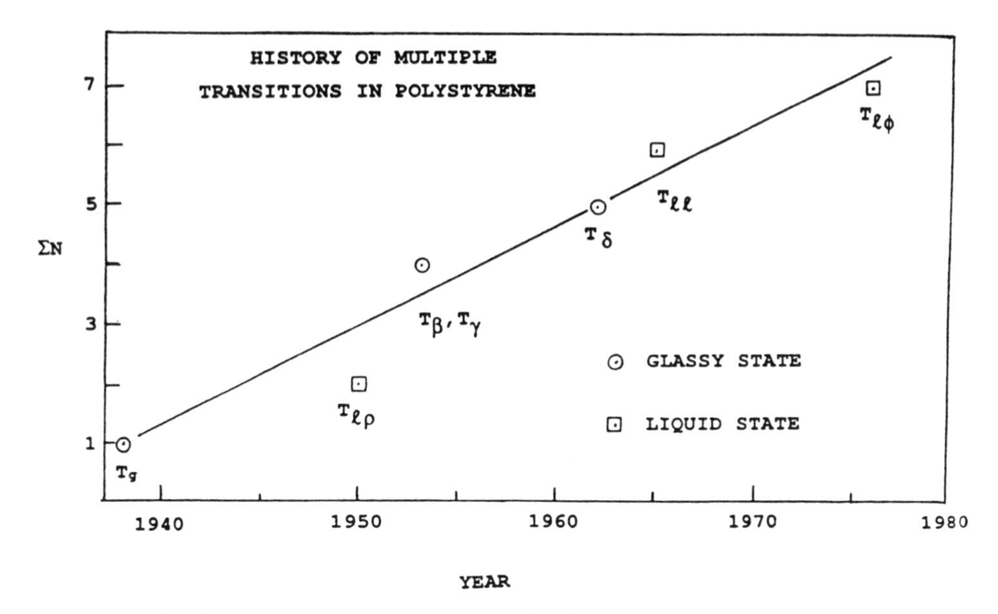

Figure 9 Historical sequence of discovery for the several transitions in at-PS as a function of time in years. Similar patterns hold for other polymers, PE and PMMA. Such patterns evolve from use of wider temperature spans and/or improved instruments and techniques. ΣN is cumulative number of transitions.

We consider that ultrasonic absorption might be interpreted on such a basis. Attenuation losses at $f = 2 \times 10^6$ Hz are shown as Table 2. Attenuation is least for PE and highest for PVC and PIB.

Multiple Transitions in Polystyrene—Diluent Systems

Early background concerning the effect of diluents in lowering T_g and T_{ll} is found on pp. 246–247 of [14]. Here we stress the torsion braid study of Gillham et al. [91] with narrow distribution PS, $\overline{M}_n \sim 37{,}000$ plus an aromatic diluent (see caption). All compositions are for $\overline{M}_w$ below M_c. Results are shown in Figure 10. It is clear from this that T_g, T_{ll}, and $T_{l\phi}$ are not restricted to being macromolecular events, as had been recognized before. A different version of Figure 10 is given as Figure 6 of [15] to bring out other implications of such data. $T_{l\phi}$ was designated as T_{ll} in the original publication [91]. While data points are not shown in Figure 10, tabulated data is given in Ref. 12. $T_{l\rho}$ is likely too close to T_{ll} at $\overline{M}_n = 37{,}000$ to be resolved.

Table 2 Ultrasonic Attenuation for Some Common Polymers

Polymer	Temp. of maximum attenuation (°C)	Maximum attenuation, D_L, in decibels per cm at 2 MHz	Source of data	Loss peak	Area per chain ($Å^2$)[f]
PVC	106	125	Thurn[c]	T_g	27.0
PIB	26	85	Ivey[d]	T_g	41.2
PS	160	83	Thurn[c]	T_g	69.8
PVAC	93	65	Thurn & Wolf[e]	T_g	59.3
PMA	68	64	Thurn & Wolf[e]	T_g	64.0
PVPR[a]	53	55	Thurn & Wolf[e]	T_g	92.0[b]
Hevea	−12	44	Ivey[d]	T_g	28.0
SBR	4	28	Ivey[d]	T_g	29.0[b]
PE	25	13	Thurn[c]	Chain motion?	18.3

[a]Polyvinylpropionate.
[b]Estimated.
Sources:
[c]H. Turn, *Z. Angew. Phys. 7*, 44 (1955).
[d]K. Ivey, Ph.D. Thesis, Notre Dame University (1949); D. G. Ivey, B. A. Mrowca, and E. Guth, *J. Appl. Phys.*, *20*, 486 (1949).
[e]H. Thurn and K. Wolf, *Kol. Z.*, *148*, 16 (1956).
[f]R. F. Boyer and R. L. Miller, *Macromolecules*, *10*, 1167 (1977).
For background on ultrasonic attenuation see footnote i to Table 1.

Effects of Hydrostatic Pressure on Transition Temperatures

It has long been recognized that hydrostatic pressure increases T_g and in some cases $T < T_g$. (See, for example, Table 5, p. 793 of [28].) Only recently have values of dT_{ll}/dP become available. The best documented of these is that of Solc et al. [91a] for iso-PMMA and of Boyer and Miller for at-PS [91b], whose results are shown in Figure 11 at a single isotherm, 196.6°C. In general, based on data for various polymers, $dT_{ll}/dP \sim 2 \times dT_g/dP$, which is probably a consequence of the greater amount of free volume at T_{ll} than at T_g.

TYPICAL GROUP MOTIONS AND EMPIRICAL CORRELATIONS

As the field of multiple transitions evolved, several typical group motions, i.e. the CH_3, cyclohexyl, etc., were noted in different polymers; and specific

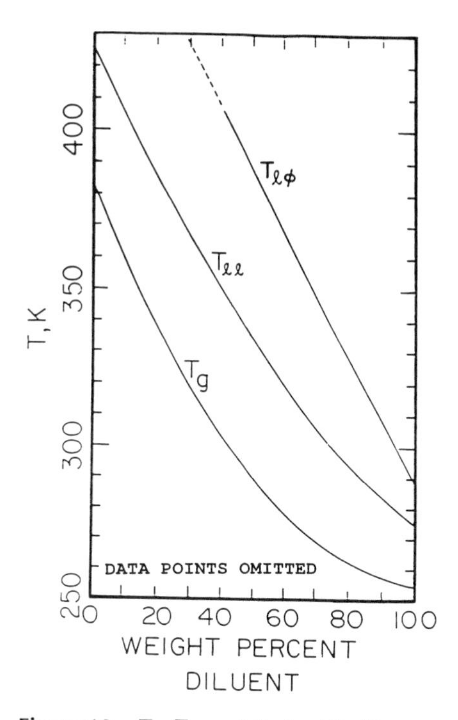

Figure 10 T_g, T_{ll}, and $T_{l\phi}$ for PS-diluent systems over the entire composition range by Gillham et al. using torsional braid. This work resulted in discovery of $T_{l\phi}$ (originally called $T_{ll'}$). Specific diluent used in this example is $(C_6H_5OC_6H_4O)_2 \cdot C_6H_4$. Such patterns for T_g and T_{ll} were first reported by Ueberreiter and by Colborne. (See pp. 246–247 and Refs. 59–60 of Ref. 14. T_g, T_{ll}, $T_{l\phi}$ from Ref. 91.)

empirical correlations were discovered, i.e., $(\Delta\alpha)\ T_g$ = constant. These two classes of phenomena will now be detailed in that order. Ultimately they should receive a molecular interpretation. For the moment we simply catalog them.

Group Motions

Willbourn [92] arrived at the generalization that the group $(CH_2)_n$ with $n \geq 4$ in the main chain would lead to a glassy state relaxation which is near the γ peak in PE, aliphatic polyesters, and aliphatic nylons.

A side group of $(CH_2)_nO$ would likewise generate a transition if $n \geq 3$.

CH_3 Group

Another specific group frequently encountered is the methyl group, CH_3—, readily detected by NMR, on some occasions by dynamic mechanical loss.

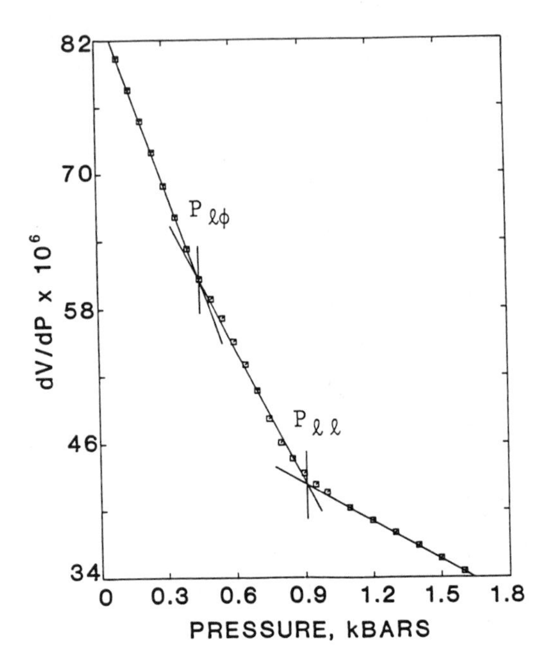

Figure 11 Effect of hydrostatic pressure on isothermal compressibility of at-PS, $\overline{M}_n$ = 20,200 T = 196.6°C. $P_{l\phi}$ is the pressure at which $T_{l\phi}$ occurs, and similarly for P_{ll}. (Results from Ref. 91b.)

Its frequency response differs greatly depending on its specific environment. Figure 12 collects many common examples. Shown therein are the T^{-1} values at $f = 10^6$ Hz read from relaxation maps published by McCall [23]. Presently missing from this compilation are values for polyalphamethyl styrene and any of the ring methyl polystyrenes. Presence of crowding is evident in PIB and polycarbonate, and its absence in PDMS or in chain end CH_3 groups.

Cyclohexyl Group

Still another group with a characteristic response is the cyclohexyl moiety studied in detail by Heijboer [93,94]. It exhibits a loss maximum at T = 190 K, $f \sim 1$ Hz for the cyclohexyl ester of polymethyacrylic acid. Heijboer has reported a log $f - T^{-1}$ plot linear over 10 decades of frequency with ΔH_a = 11.3 kcal/mol [94]. Heijboer has also studied other cyclic esters of PM acrylic acid varying in ring size from 5 to 18 [94].

Polyvinylcyclohexane is an interesting variant on the cyclohexyl group. Figure 13 adapted from a plot by Heijboer [95] shows a double γ loss peak, a β peak, and an implied T_g peak. Abe and Hama have also studied this polymer and show the actual loss peak at T_g = 423 K [96]. This high T_g compared to T_g of PS can be ascribed to the bulky cyclohexyl group.

The γ region double peak is explained as follows: the loss peak at 193 K in Figure 13 arises from the chair-chair transition of the cyclohexyl group. The peak at 153 K reflects an oscillation or rotation of the cyclohexyl group about its bond to the backbone carbon. It is thus analogous to the phenyl group γ motion in at-PS.

Polymers Without Side Groups

One characteristic attribute of side groups is that they tend to inhibit crystallinity in the absence of stereospecific catalysis, because they are, in fact, random copolymers of iso-, syndio-, and atactic placements. At-PS, for example, tends to be about 20% iso, 54% syndio, and 26% atactic when prepared by anionic catalysis [73].

There is a class of polymers free from side groups except hydrogen. They are semicrystalline to highly crystalline and hence outside our purview. However, they do exhibit multiple amorphous transitions and this presents the issue that side groups are not a necessary condition for mutiple in-chain transitions.

We had been intrigued and baffled by this group of polymers for years because of the complicated nature of the multiple transition spectrum. A recent publication brought some clarification [97]. For definition, we accepted fluorine as a non–side group because its volume is similar to that of hydrogen, although its polarity is not. Aliphatic polyesters with

$$—\overset{\overset{\displaystyle O}{\|}}{C}—O— \quad \text{and nylons with} \quad —\overset{\overset{\displaystyle O}{\|}}{C}—\overset{H}{N}— \quad \text{were rejected.}$$

The polymers under consideration exhibit a $T < T_g$ event similar to T_β, and two T_g's: T_g (L) and T_g (U). T_g (L) is the glass temperature of the amorphous polymer uninfluenced by the presence of crystallinity, for example the T_g observed on quenching the polymer to the amorphous state. T_g (U) for upper is the glass temperature influenced by the presence of crystallinity and likely by morphology. T_g (U) usually increases with crystallinity, while its intensity shows a maximum. Double glass transitions have been discussed in detail on several occasions [98].

Figure 14 summarizes our study to date for nine polymers as indicated, with multiple transition temperatures on the ordinate plotted against melting temperatures. $T < T_g$ (L) = T_β and T_g (L) are only mildly dependent on T_m.

$T_g < T_g$ (L)/T_g (L) decreases from 0.802 to 0.705 as T_m ranges from 200 to 600 K, with a value of 0.745 near the midrange, $T_m \sim$ 400 K. Hence we classify it as a T_β as shown later in Eq. (8). The ratio of T_g (L) to T_m

SURVEY OF METHYL GROUPS BY NMR

LOCATION OF CH_3 - GROUP	POLYMER		1000/T(K) AT 10^6 Hz
ON MAIN CHAIN			
CH_3 - CARBON	Polypropylene (PP) (isotactic)	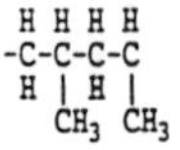 -C(H)(H)-C(H)(CH_3)-C(H)(H)-C(H)(CH_3)	9
	Polyisobutylene (PIB)	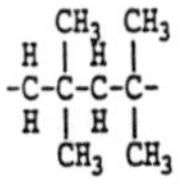 -C(H)(H)-C(CH_3)(CH_3)-C(H)(H)-C(CH_3)(CH_3)-	6.3
	cis Polyisoprene	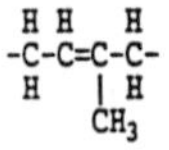 -C(H)(H)-C(H)=C(CH_3)-C(H)(H)-	11
	Polycarbonate (PC)	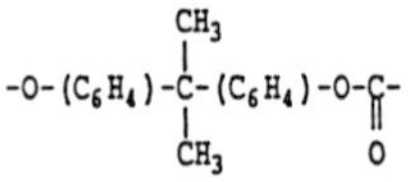 -O-(C_6H_4)-C(CH_3)(CH_3)-(C_6H_4)-O-C(=O)-	5.4
CH_3 - SILICON	Polydimethyl-Siloxane (PDMS)	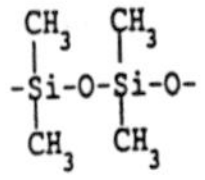 -Si(CH_3)(CH_3)-O-Si(CH_3)(CH_3)-O-	14.5
END OF SIDE CHAIN			
CH_3 - CARBON	Polybutene-1 (PB-1)	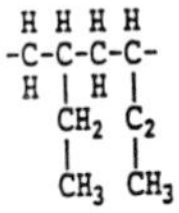 -C(H)(H)-C(H)(CH_2CH_3)-C(H)(H)-C(H)(C_2CH_3)-	11
CH_3 - OXYGEN	Polymethylacrylate (PMA)	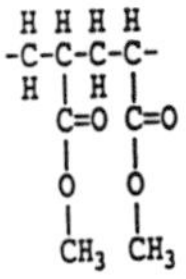 -C(H)(H)-C(H)(C(=O)-O-CH_3)-C(H)(H)-C(H)(C(=O)-O-CH_3)-	11.6

Figure 12 Ease of methyl group rotation expressed as $1000/T$ at $f = 10^6$ Hz. Estimated by us from relaxation maps of McCall except for PDMS from Litvinov et al. (From Ref. 23 and footnote.)

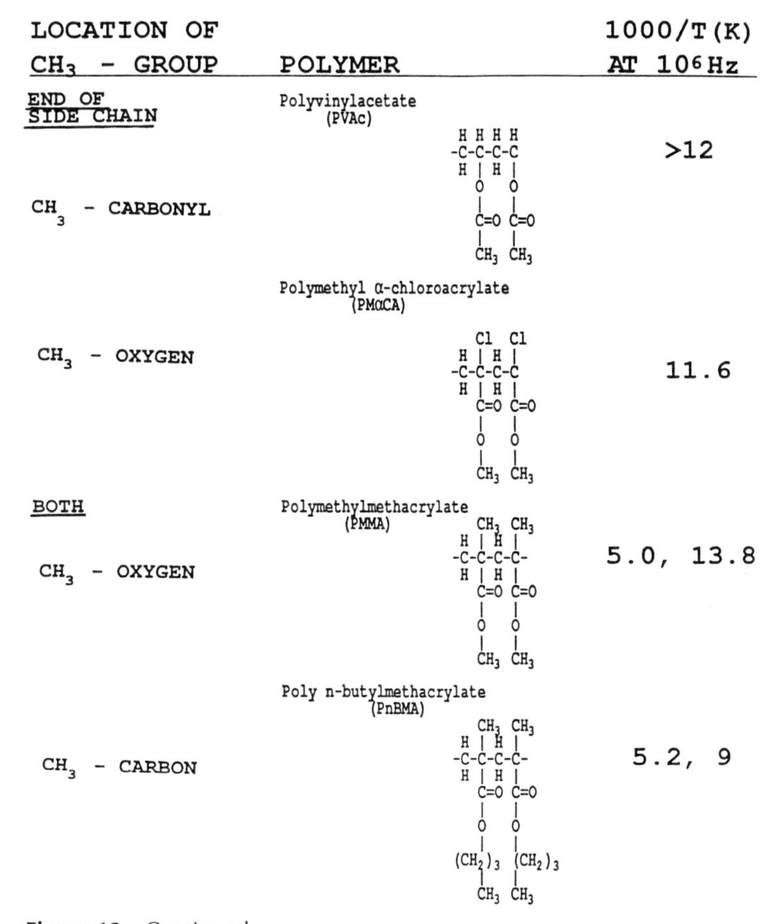

Figure 12 Continued

ranges from 0.700 for *cis*-PPE to 0.374 for PTFE, and hence generally below the 2/3 value for many sustituted polymers. (See Fig. 1.)

We ignore for present purposes T_g (U) since it is T_g (L) perturbed by adventitious factors, such as level of crystallinity and morphology. $T_{\alpha c}$ is likewise not pertinent because it occurs within crystalline regions and is commonly referred to as a premelting transition occurring at $\sim 0.85\ T_m$ (See also p. 760 of [28].)

Thus the key conclusion is that a side group is not necessary for T_g (L) and $T_\beta < T_g$ (L). We have not considered T_γ although it certainly occurs in PE and, by Willbourn's rule [92], in other polymers, i.e., $P[(CH_2)_nO]$ with $n \geq 3$.

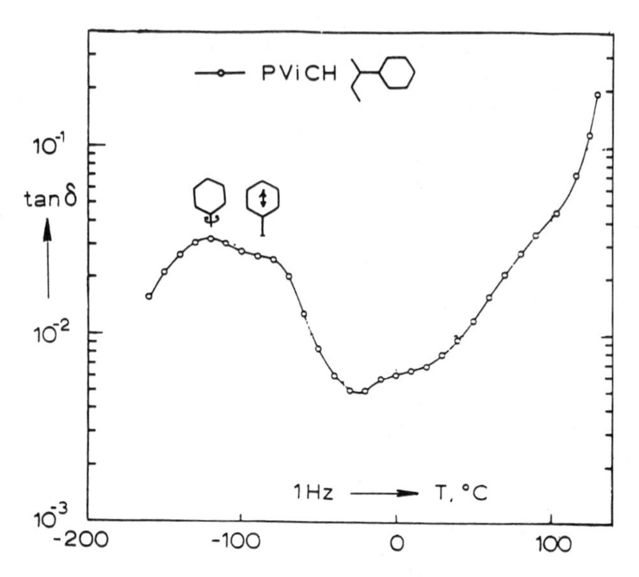

Figure 13 Dynamic mechanical loss for polyvinylcyclohexane as measured by Heijboer. T_g is off scale to right. Origins of the two gamma peaks are in a chair-chair cyclohexyl ring transition at $-80°C$ and oscillatory motion of the entire ring about the bond connecting it to the backbone at $-120°C$, similar to T_γ in at-PS. (From Ref. 95.)

Empirical Numerical Relationships

In addition to examples related to specific characteristic groups, there is another category of examples involving numerical relationships.

Thermal Expansion

For example, Simha and Boyer [99] found the empirical dimensionless equation

$$(\Delta\alpha)T_g = 0.113 \tag{3}$$

where $\Delta\alpha$ is the change in coefficient of cubical expansion across T_g, unit K^{-1}, and T_g is in kelvin. They also found

$$\alpha_l T_g = 0.18 \tag{4}$$

where α_l is the coefficient of cubical expansion above T_g. This expression is useful for polymers with abnormally high glassy state expansions, α_g.

Later studies indicated a slightly more accurate relationship [99]:

$$(\Delta\alpha)T_g = A + BT \tag{3a}$$

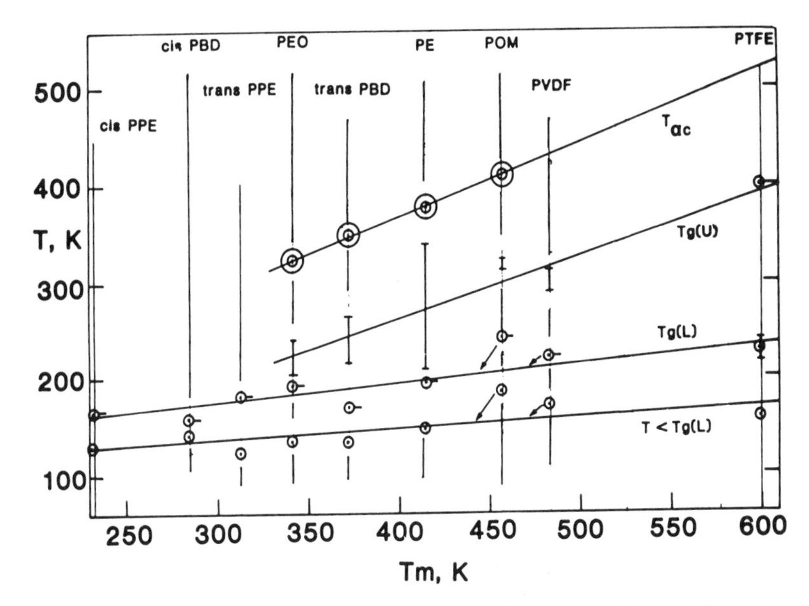

Figure 14 Several amorphous phase transitions, T_β, T_g(L), T_g(U) and intracrystalline $T_{\alpha c}$ process for unsubstituted polymers plotted against melting temperatures. Polymers from left to right: *cis*-polypentenamer, *cis*-polybutadiene, *trans*-polypentenamer, polyethyleneoxide, *trans*-polybutadiene, polyethylene, polyoxymethylene, polyvinylidenefluoride, and polytetrafluoroethylene.

where $A = 0.07$, $B = 1 \times 10^{-4}$. Figure 5(b) showed the most general situations where crystallinity and/or a strong $T < T_g$ interfere [14].

Specific Heat

This author found another empirical correlation,

$$(\Delta C_p)T_g = 27.5 \tag{5}$$

valid for many common polymers except polyalkyleneoxides, for which the constant is 40 [100].

Latent Heats of Vaporization

It follows from Eqs. (3) and (5) that

$$\frac{\Delta C_p}{\Delta \alpha} = 248 \tag{6}$$

The empirical relationship has not been published hitherto. It should be compared with a corresponding ratio examined by Kauzmann [101] which follows from the hole theory of liquids.

Hole Theory of Liquids

Kauzmann noted (p. 235 of [101]) that from the hole theory of liquids

$$\frac{\Delta C_p}{\Delta \alpha} \cong \Delta H_{\text{vap}} \tag{7}$$

where the rhs is the latent heat of vaporization. His Table 2 (p. 236 of [101]) shows that for many molecular glasses and polymers the *l*hs of Eq. (7) is several times greater than ΔH_{vap}. The hole theory of liquids postulates that the energy required to generate extra volume, hence holes, above T_g should be proportional to the latent heat of vaporization. The discrepancy posed above requires resolution.

T_β versus T_g

Matsuoka and Ishida [102] discovered the empirical correlation

$$T_\beta \text{ (K)} \sim 0.75 T_g \text{ (K)} \tag{8}$$

at $f = 100$ Hz. In their tabulation on p. 257 of [20], polymers which deviate most widely from the 0.75 value, such as PVAc and polycarbonate, presumably do so because the authors have used T_γ rather than T_β; or in the case of PE, both T_g and T_β are incorrect values. Figure 16 of [28] is a plot of T_β versus T_g from which we have eliminated such errors. Several loci on relaxation maps carry a T_β label when indeed they are T_γ (Fig. 8.40 of [22] for PMA; Fig. 9.4 of [22] for PVAc).

ΔH_a for $T < T_g$

Heijboer [103] found that the apparent activation energy related to secondary transitions below T_g is given by

$$\Delta H_a = 0.060 T \text{ (K)} \qquad (f = 1 \text{ Hz}) \tag{9}$$

We have verified this relationship for a variety of polymers (see Fig. 35 of [27]). ΔH_a is in kcal/mol.

The Liquid State

More recently, Enns et al. [104] showed empirically that

$$T_{ll} \text{ (K)} \sim 1.20 T_g \text{ (K)} \tag{10}$$

which is now found to be more accurately given by

$$T_{ll} \text{ (K)} = 49 + 1.041 T_g \tag{10a}$$

which leads at $T_g = 200$, 300, and 400 K to $T_{ll}/T_g = 1.23$, 1.19, and 1.16, respectively. (See Fig. 2 and related discussion of [14].) This is closer to reality. A rule of thumb then becomes

$$T_{ll} = (1.20 \pm 0.05)T_g \tag{11}$$

Dielectric Loss Data

Lobanov and Frenkel [65] have arrived at a significant empirical generalization based on dielectric loss data as follows: It is generally recognized that the T_g and T_β loci in a log f–$1/T$ plot intersect at high frequency and become a single locus, with the slope of T_β. Calling the intersection temperature T^* (K) (at frequency $f \sim 10^6$ Hz) they arrived at two relationships:

$$T^*(f \sim 10^6) = T_{ll} \quad \text{(static)} \tag{12}$$

and

$$T^*\ (\mathrm{K}) = 100 + 0.75T_g\ (\mathrm{K}) \tag{13}$$

which is of the form shown in Eq. (10a) for T_{ll} versus T_g.

They have presented at least 30 examples. In some cases which lack a T_β, such as plasticized PVC, the T_g locus experiences a sharp slope decrease at T_{ll}. We have extended their technique to other polymers not reported by them, relying in some cases on relaxation maps in McCrum et al. [22], using combined dielectric and dynamic data. We have also used dielectric, low-frequency dynamic loss, ultrasonic loss, various types of NMR data, and Brillouin loss data [15], as illustrated in Figure 6 for at-PS.

Equation (12) seems a contradiction in terms since T_{ll} is frequency-dependent. We have proposed a resolution for this dilemma. Collapse of the three-dimensional physical network at T_{ll} postulated by Lobanov and Frenkel [65] permits local motion such as the T_β motion to occur suddenly at T_{ll}, after the long-range micro-Brownian motion associated with T_g ceases. Incidentally, log f at T^* is in the range of 6 ± 1, being higher for flexible backbone polymers such as polybutadiene, lower for stiff chains such as PIB.

The fact noted earlier that T_β is not necessary to cause a slope change at T^*, as we have found for PDMS and Hevea rubber [15], suggests that the sudden loss of a three-dimensional physical network at T_{ll} "catalyzes" the sudden slope change.

Carbon 13 NMR

This writer and colleagues noted that carbon 13 spectral collapse temperatures, T_c, for the methylene carbon 13, as reported in the literature [74a,74b] follow the rule

$$T_c\ (f \sim 10^6\ \mathrm{Hz}) = 1.21T_g \qquad \text{(static)} \tag{14}$$

$$\therefore \equiv T_{ll} \qquad \text{(static)} \tag{14a}$$

This is analogous to Eq. (12) and again presents a paradox which we have not yet fully resolved. It follows that

$$T_c \equiv T^* \tag{14b}$$

Indeed we have prepared a linear plot of combined T_c or T^* on the ordinate against T_g (static) on the abscissa (Fig. 7 of [105b]).

One possible resolution of the above paradox is as follows: at temperatures well above $T^* = T_{ll}$, the ^{13}C Hz spectral line is moving at high frequency ($\gg 10^6$ Hz) and is quite sharp. On cooling, the frequency decreases exponentially and then suddenly collapses under restraint by the physical network which reforms at T_{ll}.

Zero Shear Melt Viscosity, η_0

As mentioned in the introduction, we found empirically that precision $\eta_0 - T$ data provides an absolute method for locating multiple liquid state transitions. The example chosen in Figure 5 with $\overline{M}_n = 16{,}200 < M_c$ (based on tabulated data of Pierson [68]) permitted detection of all three liquid state events for a reason apparent by inspection of Figure 5. At $\overline{M}_n \to M_c$, $T_{ll} \to T_{l\rho}$ and the two cannot be resolved.

Attention is directed to Figure 5 of [15] showing log ΔH_a below and above T_{ll} as a function of T_{ll}. We suggest that the difference in ΔH_a across T_{ll}, reflects the sharp disappearance of the three-dimension physical network at T_{ll}. It seems logical that this difference should increase with T_{ll} as both intermolecular forces and chain stiffness increase. (See the next section.)

Cohesive Energy Density/Chain Stiffness

The most recent empirical correlation of note (and still unpublished) involves linear relations between T_g or T_{ll} and the product of a cohesive energy density and a chain stiffness factor. For the former we use a dimensionless ratio, CED_X/CED (R), where (R) signifies a reference polymer such as PE and X is any polymer. Chain stiffness is taken as Flory's characteristic ratio, C_∞ [106d], also dimensionless. T_g and T_{ll} each require two lines. The difference(s) between polymer types in the upper and lower lines is not yet fully understood. Polymers in the upper line tend to show T_β but not T_γ. Conversely, polymers in the lower line lack T_β but possess a strong T_γ. Implications of this difference are being pursued.

Such linear relationships can be given an ad hoc rationalization as follows: there exists a persistence length, $p \sim C_\infty$, which can imply parallelization of chains, along which a force proportional to cohesive energy density is acting. See also Kolinski et al. [57]. Using computer modeling they find that chain stiffness leads to local order. Addition of an intermolecular force leads to a transition.

Other Characteristic Temperature Ratios

Okui [106e] has presented a series of correlations concerned with several relationships of interest to the present study: T_g/T_m as in Figure 1; T_β/T_g as in Eq. (8); T_{ll}/T_g as shown by Eqs. (10) and (11). He also reviews the relationship T_{max}/T_m, where T_{max} is the temperature at which rate of crystallization is a maximum. This ratio has a value of ~0.82 for many polymers. He notes that $T_{ll}/T_g \sim 1.2$ is the same as T_{max}/T_g. He suggests the possibility of a correlation between molecular motion at T_{ll} in the amorphous state and $T_{\alpha c}$ the crystalline premelting transition. We had also noted such a connection (Fig. 33b of [2]).

Most of the empirical correlations cited herein involve a sufficient number of polymer types as to suggest a fundamental molecular origin. It is assumed that these may become challenges to theoreticians or might fall out of theoretical studies.

THE TRIUMVIRATE: T_β, T_g, and T_{ll}

Figure 24 of [2] and Figure 1 of [20] are schematic sketches suggesting the possible generality of the above three events in addition polymers. T_β involves glassy state motion of 1 to 2 monomer units; T_g the coordinated motion of 10 to 20 consecutive monomer units; and T_{ll} the motion of all units in the chain, even if the chain were of infinite length. Figure 25 of [2] summarizes actual data on at-PP consistent with such a view.

The first impressive experimental evidence for the triumvirate is that of Wolf [107], based on torsion pendulum data for at-PMMA as illustrated in Figure 15. The ability of the hanging torsion pendulum to go so far above T_g and even above T_{ll} is considered by us to arise from the fact that melt viscosity of at-PMMA is about 100-fold that of at-PS of the same $\overline{M}_w$. (See Appendix I of Denny et al. [90].) Figure 16 shows torsion braid results on at-PP with T_β, T_g, and T_{ll} as indicated, found by Gillham and adapted from Figure 8 of [28]. Figure 17 is adapted from dielectric results on polypropyleneoxide as determined by Johari [108].

DSC of tactic PMMAs by Denny et al. [90] revealed T_β, T_g, and T_{ll} for at- and syndio-PMMAs but not for iso- (not low enough starting temperature). Figure 18 shows thermally stimulated current data on iso-PMMA by Gourari et al. [109].

Thus, there are numerous examples of T_β and T_g, of T_g and T_{ll}, but relatively few of all three in the same experiment. But the triumvirate concept appears sound, with exceptions arising from temperature limita-

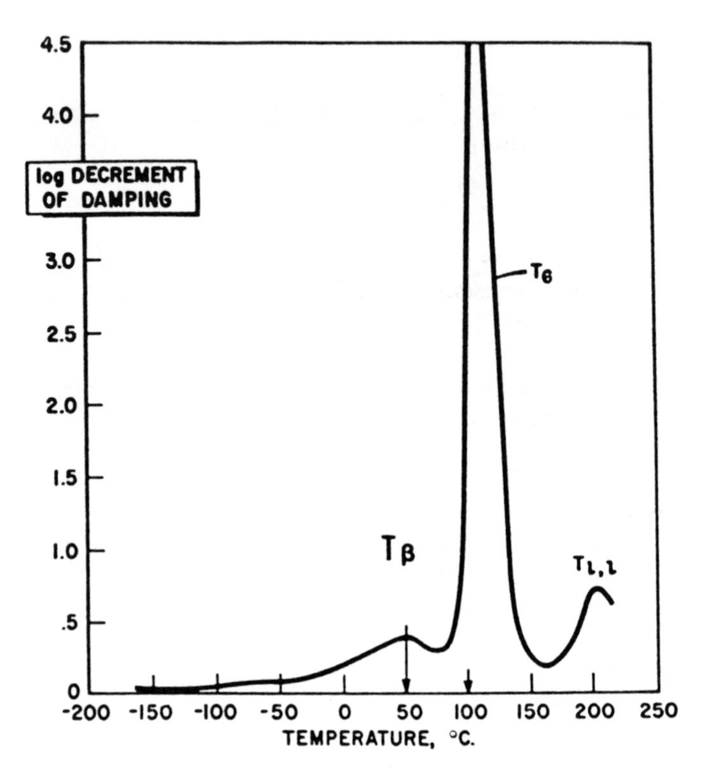

Figure 15 Torsion pendulum trace of log decrement – temperature (°C) for at-PMMA according to Wolf. T_β and T_{ll} designations added by us. (From Ref. 107.)

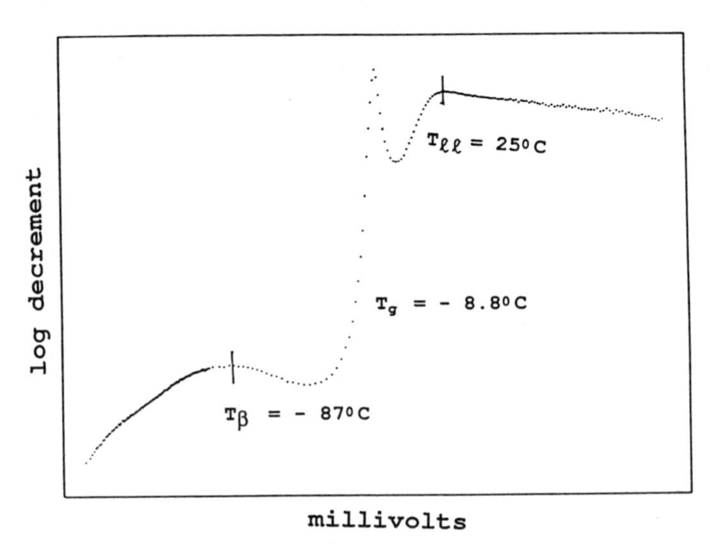

Figure 16 Dynamic mechanical loss data for $f \sim 1$ Hz in at-PP. (Adapted from Fig. 8 of Ref. 28 by Gillham, using his torsion braid (TBA) apparatus.)

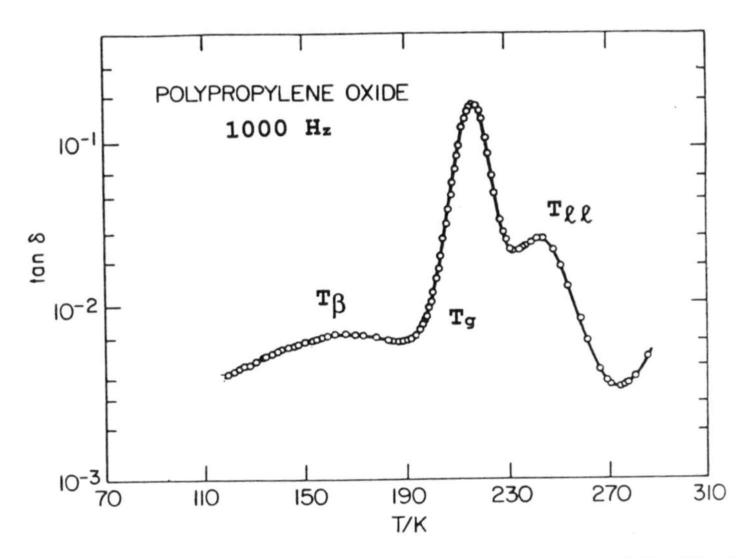

Figure 17 Dielectric loss data for polypropylene oxide at 1 kHz. The T_{ll} designation was added by us. (Adapted from Ref. 108.)

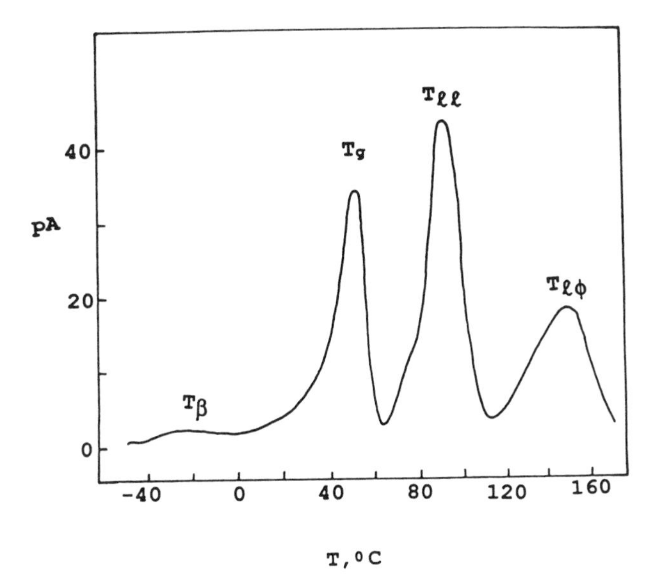

Figure 18 Thermally stimulated current (TSC) technique with current output in picoamperes plotted against temperature for amorphous isotactic PMMA. T_β is inherently weak in iso-PMMA. (Data from Ref. 109.)

tions, or equipment limitations and/or method limitations. Some polymers, such as *cis*-polyisoprene, lack a T_β.

PVC represents a case in which thermal decomposition occurs circa 125°C with loss of HCl, cross-linking, and carbonization, all tending to obscure T_{ll}. Enns measured PVC by DSC and located $T_g = 336$ K, $T_{ll} =$ 398 K, $T_{ll}/T_g = 1.18$, and $T_{l\phi} = 440$ K [104]. A heating rate of 40 K/min presumably permitted detection of the high-temperature events before decoposition became serious. Enns did not start at sufficiently low temperature to locate T_β.

T_{ll} from dielectric loss was obtained by the indirect T^* method of Lobanov and Frenkel [69], indicated earlier in Figure 6 with PS. Dielectric loss on pure atactic PVC by Stoll et al. [110] was analyzed, using $\epsilon'' - T$ plots. $T^* = T_{ll} = 400$ K, at an extrapolated log frequency of 6.5, was indicated. T_g was estimated at 352 K for $T_{ll}/T_g = 1.14$. Dielectric loss at T^* again is assumed to involve motion of the C—Cl dipole at $T > T_g$.

PVC with and without a diluent and measured electrically at 60 Hz by Fuoss [111] was discussed as Figure 11 of Ref. 28. Pure PVC shows T_g and a strong T_β, whereas the presence of 10% diphenyl as diluent shifts T_g and T_β to lower temperatures while supressing the intensity of T_β. The latter phenomenon is most likely caused by loss of free volume, as discussed for polycarbonate by Wyzgoski and Yeh [112]. It is accompanied in PVC by a decrease in impact strength [111]. The PVC loss data was obscured by dc conduction current below and into the T_{ll} regions.

An intimate connection between T_β and T_g was first established by Matsuoka and Ishida [102] and later amended and extended by this writer (p. 765 and Fig. 16 of [28]), as shown later in Figure 19.

More recently, Bershtein et al. [113,114] have examined other relationships between T_β and T_g; i.e., that ΔH_a for T_β tends to be (1/4) ΔH_a for T_g (Fig. 3 of [114]). Bach Van and Noel [115] conducted a DSC study of T_β in various at-PS derivatives, showing both T_β and T_g as well as thermal history effects on T_β. Their data appear to confirm the general 3/4 rule for T_β (K)/T_g (K). Concerning T_{ll} and T_g, the tendency for T_{ll} (K) $\sim 1.2T_g$ (K), as first noted by Enns, has been thoroughly reviewed. See Figure 2 of [14].

To us, the key issue is whether or not the existence of the triumvirate is fortuitous. Ibar (pp. 371–393 of [8b]) has expressed his conclusion that T_β and T_{ll} are both a consequence of T_g, a conclusion which follows from his energy-kinetic (EKNET) theory about T_{ll}. This implies a fundamental basis for the triumvirate.

Polymers lacking a T_β pose a problem (for example, PDMS and *cis*-polyisoprene). These are both low T_g polymers, and it may be difficult to quench rapidly enough to cause a T_β. Conversely the EKNET origin of T_β may not be correct.

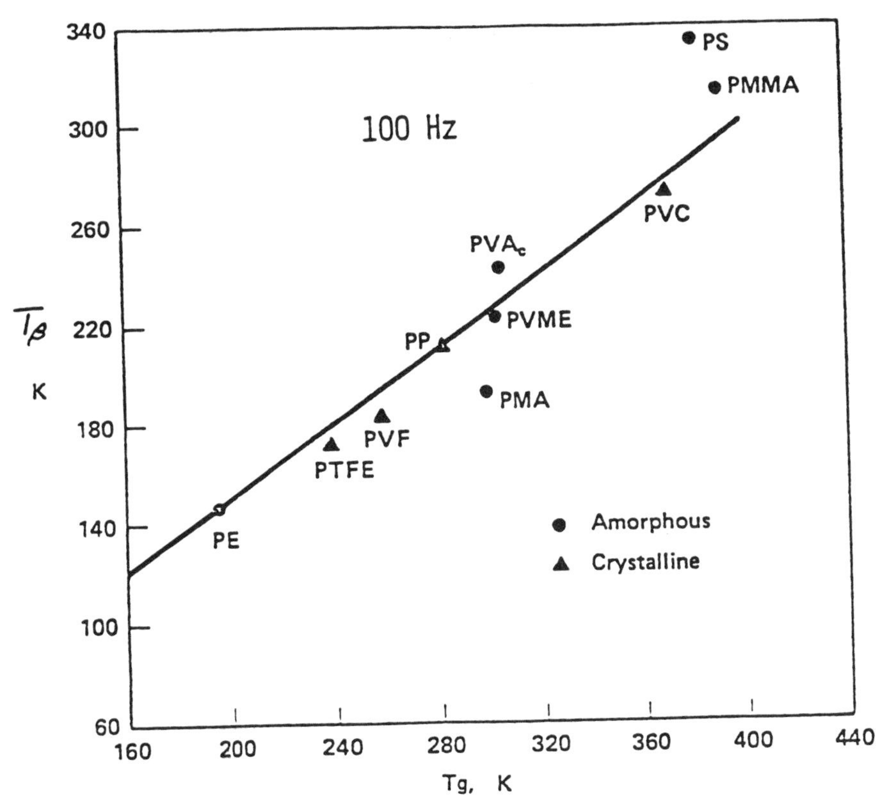

Figure 19 Linear correlation of T_β with T_g for some common semicrystalline and amorphous polymers from 1 to *r*: polyethylene, polytetrafluoroethylene, polyvinylfluoride, polypropylene, polymethylacrylate, polymethylvinylether, polyvinylacetate, polyvinylchloride, polystyrene, and polymethylmethacrylate. Slope of line ~0.75.

From a philosophical point of view, we visualize an experiment (not conducted as yet, to our knowledge). Styrene can be polymerized to completion at $T > T_g$ without cooling to T_g, or even $T > T_{ll}$. In the latter case, will T_{ll} be found on heating? More simply, any trace of T_g can be erased by heating to above T_{ll} and annealing. DSC experiments by Warner [51] have revealed T_{ll} on DSC cooling traces at a fixed rate of cooling from above T_{ll}.

We have not yet designed an experiment to find T_β without cooling from above T_g. There is no T_β in PS as a result of very slow cooling from above T_g. There is a plausible experimental basis for stating that T_g is a precursor of T_β. T_g is not unambiguously a precursor of T_{ll}.

SUMMARY AND CONCLUSIONS

This chapter has been concerned primarily with the glassy state $T < T_g$, the liquid state $T > T_g$, and the T_g transformation region from glass to liquid of synthetic amorphous polymers and copolymers. In addition, the amorphous state of semicrystalline polymers and transitions associated therewith have had to be considered only briefly. The particular body of knowledge treated herein appeared to erupt and then congeal into a highly organized form in the brief period 1963–1967 in England, Japan, and the United States, to be shortly followed with evidence from other countries. Knowledge about old and new polymers has proliferated. There are still many gaps readily apparent when one consults a reference source such as *Polymer Handbook*, third edition.

Phenomenology associated with T_g has proliferated, especially with regard to time effects and physical aging not covered herein. It is not clear to us, even after reviewing three authoritative articles on T_g [66a,b,c], that any explicit theory for T_g, similar to that for crystallization and melting, has yet emerged. The main contenders are free-volume theory and entropy with a slight preference for the latter.

Three distinct trends seem to have emerged in the past 25 years:

1. Enough diverse evidence for local order in amorphous polymers exists that present and future polymer scientists should not dismiss its existence.
2. Liquid state transitions, i.e., at $1.2T_g$ (K) for atactic polymers and $1.2T_m$ (K) for semicrystalline polymers, likewise should no longer be dismissed as artifacts.
3. Instrumental techniques have advanced to the point that guesswork and speculation can no longer be tolerated.

For many years the normal pattern consisted of three key steps:

1. The organic chemist made as many variants of a key polymer type as were feasible.
2. The physical chemist or physicist measured dynamic mechanical and/or dielectric loss, ultrasonic absorption, etc., as instrumentation of the period permitted.
3. Likely group and atomic motions were then proposed based on steric restraints and other plausible considerations.

Today modern instrumentation, especially spectroscopic, can determine the precise nature and temperature dependence of many specific motions. Recent literature is most instructive [118–121].

We were instructed to ignore crystalline polymers. This was done as regards the process of crystallization, the resulting morphology, and, to a large extent, the kinetics. The controversy over chain folding is ignored. But transitions in semicrystalline polymers cannot be ignored. The classical T_g/T_m relationship of Figure 1, ratio of amorphous to crystalline density, and/or a possible connection between $T_{\alpha c}$ and T_{ll} could not be forgotten.

The fact that $T_\beta \sim 0.75T_g$ holds not only for amorphous polymers but also for semicrystalline polymers, as illustrated in Figure 19 (adapted from Fig. 16 of [28]). PE is highly crystalline, but what is plotted here is the ratio of two amorphous phase events $T < T_g$ (L)/T_g (L). It is further known, as illustrated in Figure 20, that

1. The intensity of T_β decreases less rapidly with increasing crystallinity than does the intensity of T_g.
2. But even more significantly, the intensity of T_β is still finite at $\chi_c \rightarrow$ unity.

These facts led Takayanagi et al. [120] to postulate a three-"phase" model: amorphous material, crystalline material, and crystal defect regions.

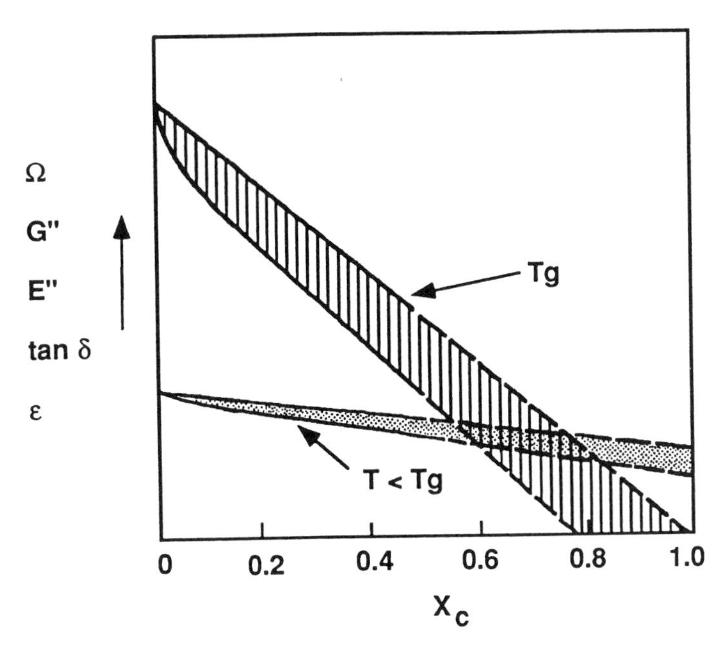

Figure 20 Schematic plot of dynamic mechanical loss and dielectric loss on ordinate as a function of fractional crystallinity, χ_c, for T_g and T_β based on actual data for several polymers. A finite intensity for T_β at $\chi_c = 1$ implies, according to Takayanagi [120], that T_β can be accommodated by crystal defect regions. (After Ref. 121.)

Alfrey and this writer [121] (see pp. 193–202 of Ref. 20) modeled a semicrystalline polymer with two assumptions:

1. The number of consecutive chain atoms needed for T_g is 10 times the number needed for T_β, i.e., 50/5.
2. Several different morphologies were assumed, but none with chain defects.

The conclusions reached by Alfrey and the writer based on experimental results on intensity of two amorphous phase transitions, T_β and T_g, are as follows.

1. The strengths of amorphous transitions in semicrystalline polymers should decrease monotonically with increasing χ_c.
2. With some types of morphology, the curves can be expected to fall somewhat below the straight lines representing simple proportionality with $1 - \chi_c$.
3. Such downward deviations from linearity would be expected to be more pronounced for the glass transition than for the $T < T_g$ transition, and more pronounced in fine-textured morphologies than in coarse-textured morphologies.
4. Predicted patterns are rather insensitive to details of the model. We would hesitate to draw any structural conclusions based on a deviation from simple proportionality.
5. At the same time, these calculations suggest that the type of behavior shown in Figure 20 should be general to all semicrystalline polymers, regardless of their morphologies, except for defects at high χ_c, which were not considered in our models.
6. The related problem of extensive confusion in the literature over interpretation of amorphous loss peaks based only on intensities (tan δ_{max}, E''_{max}, G''_{max}, etc.) has been discussed elsewhere [121].
7. The general question of the most significant measure of the "strength" of an amorphous transition is recognized, but is beyond the scope of this note.
8. The ad hoc assumption that relaxation strength is proportional to the number of chain atoms involved in a relaxation and also to the number of (amorphous) segments of that length yields predicted trends which are qualitatively similar to observed results.

Another characteristic feature of the period under study is as follows: polymer science under Staudinger survived one period of authoritarianism dominated by the colloid chemists. We suggest that polymer science itself is now dominated by a new age of authoritarianism, by schools of polymer science. One need only consider the controversy about chain folding with adjacent reentry to realize the gravity of the problem [122].

APPENDIX

After the above material was completed, an extensive study on thermal analysis of polymers by Wrasidlo [123] was examined by us. He reported a correlation of T_g (K) as a linear function of ε_h, the energy of hole formation for organic polymers and copolymers. Figure A-1 is a plot prepared by Miller [124] of T_g (K) versus ε_h in cal/mol, using numerical values from Tables 2a and 2b of [123].

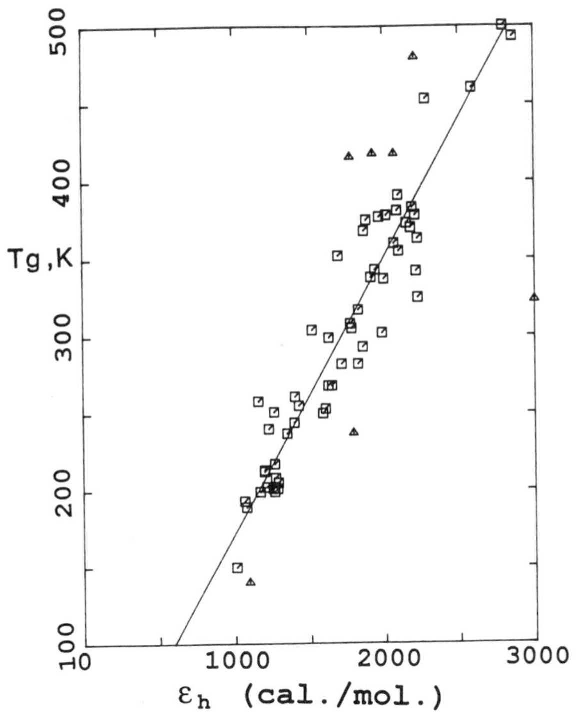

Figure A-1 Correlation of T_g with energy of hole formation ε_h in cal/mole (data from Tables 2a and 2b in Ref. 123. ΔC_p is increase in heat capacity at T_g, $\Delta\alpha$ increase in thermal expansivity at T_g; V_h is volume of molecules, v_o is volume of holes; ϕ_1^* specific free volume. $R^2 = 0.893$. Points designated Δ are considered deviant, and were not employed in linear least-squares correlation.

A linear least-squares line is given by Eq. (A-1):

$$T_g = -7.434 + 0.179\epsilon_h \qquad \text{(correlation coefficient } R^2 = 0.893) \tag{A-1}$$

with

$$\epsilon_h = \left(\frac{\Delta C_p}{\Delta\alpha^*}\right)\left(\frac{V_o}{V_h}\right)(1 - \phi_1^*) \tag{A-2}$$

where the first term is essentially the latent heat of vaporization defined earlier in connection with Eq. (7); V_0/V_h is the ratio of molecular volume to hole volume; and ϕ_1^* is the specific free volume as defined by Kanig [125]. ϕ^* averages 0.0235 but ranges from 0.0116 to 0.0455. Kanig concluded that T_g is not an iso-free volume state. ϕ_1^* is numerically equivalent to the WLF free volume available for transport.

There is considerable scatter with the standard deviation of T_g (calculated) = 28.2 K. Part of the scatter can be ascribed to use of polymer homologous series, i.e., seven alkylmethacrylates, five polyvinylethers, etc. Some deviant points involve incorrect values of T_g, i.e., for PE whose T_g we find to be 195 K, falling exactly on the line, rather than 140 and 247 K, used by Wrasidlo.

Returning to Figure A-1, the intercept is −7.4 K, but the standard error in the intercept is 15 K, a consequence of scatter in the data. Hence one cannot assume with confidence an intercept of 0, 0 K.

Equation (A-1) is derived from the hole theory of polymer liquids, by Hirai and Eyring [126] and its modifications by Kanig [125]. Thus far we have not completely reconciled Eq. A-1 with the earlier product, CED × C_∞ (see p. 36). Kanig [125] states that the free volume at T_g, ϕ_1^*, is a measure of the stiffness of the molecules i.e., small $\phi_1^* \sim$ flexible chains.

Note: Wrasidlo (mid-page 49) promised a forthcoming interpretation of his data, but a careful search of *Chemical Abstracts* has not located it.

ACKNOWLEDGMENTS

Mrs. Sara Macartney of MMI transcribed this manuscript on an IBM computer. Many illustrations have been copied, frequently with modifications, from earlier reviews and manuscripts by this author. Mrs. Macartney has assisted with various illustrations and entirely with Figure 12, which is presented for the first time. Dr. Robert L. Miller of MMI is responsible, at our request, for the computer printout and linear least-squares regression analysis for Figure A-1. Dr. Miller also provided us with a glossy print of Figure 2. We are indebted to Dr. J. Heijboer, TNO, Delft, Holland, for the torsion pendulum trace on which Figure 13 is based.

The author gratefully acknowledges critical proofreading of the page proofs by Professor Philip L. Kumler, SUNY College at Fredonia.

REFERENCES

1. N. Saito, K. Okano, S. Iwayangi, and T. Hideshima, in *Solid State Physics,* pp. 343–502 (F. Seitz and D. Turnbull, eds.), Academic Press, New York (1963).
2. R. F. Boyer, *Rubber Chem. Technol.*, *Rubber Rev. for 1963*, *36*, 1303–1421 (1963).
3. R. F. Boyer and R. S. Spencer, in *Advances in Colloid Science,* Vol. 2, pp. 1–55 (E. O. Kramer, ed.), Interscience, New York (1947).
4. G. Allen, Department of Chemistry, University of Manchester, Manchester, England. Private communication. We received a copy of the manuscript, which contained enough new data to justify publication, in our opinion.

5a. K. Schmieder and K. Wolf, *Kol. Z.*, *127*, 65 (1952).

5b. K. Schmieder and K. Wolf, *Kol. Z.*, *134*, 149 (1953).

6. Where R. F. Boyer was then employed.
7. IUPAC lecture, Montreal (1961). (Never published but utilized in (2)).

8a. Prof. Dr. Kurt Ueberreiter, Kaiser Wilhelm Institute, Berlin (Dahlem) retired. We have summarized his contributions to liquid state behavior on pp. 145–147 of Ref. 8b, and have corresponded with him during the past 10 years.

8b. R. F. Boyer, in *Order in the Amorphous 'State' of Polymers*, pp. 135–186 (S. E. Keinath, R. L. Miller, and J. K. Rieke, eds.), Plenum Press, New York (1987); J. B. Enns and R. F. Boyer, ibid., pp. 221–249.

9a. E. R. Fitzgerald, L. D. Grandine, and J. D. Ferry, *J. Appl. Phys.*, *24*, 650 (1953).

9b. J. D. Ferry, L. D. Grandine, and E. R. Fitzgerald, *J. Appl. Phys.*, *24*, 911 (1953).

10. J. Heijboer, *J. Polym. Sci.*, *C-16*, 3755 (1968).
11. R. F. Boyer, *Polym. Engng. Sci*, *8*, 161 (1968).
12. J. Gillham and R. F. Boyer, *J. Macrmol. Sci.*, *Phys.*, *B-13*, 497–535 (1977).
13. R. F. Boyer, *J. Macromol. Sci.*, *Phys.*, *B18*, 461–553 (1980).
14. R. F. Boyer, in *Polymer Yearbook*, Vol. 2 pp. 233–343 (R. A. Pethrick, ed.), Harwood Academic, New York (1985).
15. R. F. Boyer, in *Encyclopedia of Polymer Science and Engineering*, Vol. 17, pp. 23–47 (J. Kroschwitz, ed.), Wiley, New York (1989).
16. R. F. Boyer, *Current Contents*, *46*, 18 (1981).
17. M. C. Shen and A. Eisenberg, *Rubber Chem. Technol.*, *43*, 95–155 (1970). (This was reprinted from *Progress in Solid State Chem.*, *3*, 407 (1966).)
18. A. Eisenberg and M. S. Shen, *Rubber Chem. Technol.*, *43*, 156–170 (1970). (Note: These papers by Shen and Eisenberg concern T_g and not multiple transitions. They relate more to Ref. 3 than to the general topic of this chapter.)

19. Prof. R. Simha, current address: Case-Western Reserve University, Department of Macromolecular Science, Cleveland, OH 44106.
20. R. F. Boyer, ed., *J. Polym. Sci., Part C, Polym. Symp.*, *14*, 3–340 (1966).
21. Russian Translation. R. F. Boyer, *Perekhody* /\ *Relaksatsionye Lavlenya* \/ *Polimerakh*, Izd. "Mir," Moscow (1968).
22. N. G. McCrum, B. E. Read, and G. Williams, *Anelastic and Dielectric Effects in Polymeric Solids*, Wiley, New York (1967). Note: The three coauthors were cited in their references (pp. 575–596) nine, eight, and ten times, respectively.
23. D. McCall, in *Molecular Dynamics and Structure of Solids* (R. S. Carter and J. J. Rush, eds.), National Bureau of Standards, Special Publication, 301 (June 1969). For CH_3 motion in PDMS see Litvinov et al., *Polym. Sci.*, USSR, *27*, 2786 (1985).
24. P. Hedvig, *Dielectric Spectroscopy of Polymers*, Halsted, Wiley, New York (1977).
25. D. J. Plazek, *J. Polym. Sci., Polym. Phys. Ed.*, *20*, 1533 (1982).
26. D. J. Meier, ed., *Molecular Basis of Transitions and Relaxations*, Gordon and Breach, New York (1978).
27. P. J. Flory, *Principles of Polymer Chemistry*, p. 602, Cornell University Press, Ithaca, New York, (1953).
28. R. F. Boyer, in *Encyclopedia of Polymer Science and Technology*, Supplement Vol. II, pp. 745–839 (N. Bikales, ed.), Wiley, New York (1977). This publication, in paperback reprint form, has been used for graduate courses by Prof. R. Simha at Case, by Prof. John Gillham at Princeton, and by this writer at MMI, because equivalent material had not appeared in integrated condensed form.
29. S. A. Arzhakov, N. F. Bakeev, and V. A. Kabanov, *Polym. Sci. USSR*, *A-15*, 1296–1313 (1973); original in *Vysokomol. Soyed*, *A-15*, 1154–1167 (1973).
30a. K. Ueberreiter, *Kol. Z.*, *102*, 272 (1943). This refers to several earlier papers.
30b. K. Ueberreiter, *Kautschuk*, *19*, 17 (1943); *Chem. Abstract*, *38*, 4470 (1944).
31. C. W. Bunn, in *Fibers from Synthetic Polymers*, pp. 324 ff (R. Hill, ed.), Elsevier, Amsterdam (1953).
32. R. F. Boyer, Compes Rend. de la 2nd Reunion de Chimie Physique, June 2–7, Paris (1952).
33. R. F. Boyer, *J. Appl. Phys.*, *25*, 25 (1954).
34. R. G. Beaman, *J. Polym. Sci.*, *9*, 472 (1953).
35. D. W. Van Krevelen, *Properties of Polymers*, Elsevier, New York, First Ed., Chap. 7 (1972); Second Ed., Chap. 6 (1977).
36a. W. A. Lee and J. H. Sewell, *J. Appl. Polym. Sci.*, *12*, 1397 (1968).
36b. V. P. Privalko and Yu. S. Lipatov, *J. Macromol. Sci. Physics*, *B-9*, 551 (1974).
36c. C. W. Bunn, *J. Polym. Sci.*, *16*, 323 (1955).
37. S. Krimm and A. V. Tobolsky, *Textile Res. J.*, *21*, 805 (1951).
38. S. Krimm and A. V. Tobolsky, *J. Polym Sci.*, *6*, 667 (1951).
39. S. Krimm, *J. Phys. Chem.*, *57*, 22 (1953).

40. T. Hatakeyama, *J. Macrmol. Sci. Phys.*, *B-21*, 299 (1982).
41. R. E. Robertson, *J. Phys. Chem.*, *69*, 1575 (1965).
42. W. P. Milligan, ed., *Proc. Robert A. Welsh Foundation Conf. on Chem. Research X*, *Polymers*, pp. 163–167 Robert A. Welsh Foundation, Houston, TX (1967).
43. P. Corradini, *J. Poly. Sci.*, *C-50*, 327, 335, 337 (1975).
44. P. J. Flory, *Macromolecular Chemistry*, *8*, pp. 1–15, Butterworths, London (1973); reprinted *Rubber Chem. Technol. 48*, 513 (1975).
45. G. Allen and E. Petrie, eds., Symposium Issue, *J. Macrmol. Sci. Phys. B-12*, 1–301 (1976).
46. R. F. Boyer, *J. Macromol. Sci. Phys.* pp. 253 ff. See also R. F. Boyer, *Ann.*, *N.Y. Acad. Sci.*, *279*, 223 (1976) for additional views of this author on the same theme. The editors [45] conceded to this writer that for reasons beyond their wishes and control, the symposium had been unbalanced. Hence they accepted our review.
47. R. L. Miller, R. F. Boyer, and J. Heijboer, *J. Polym. Sci.*, *Polym. Phys. Ed.*, *22*, 20–21 (1984).
48. R. L. Miller and R. F. Boyer, *J. Polym. Sci.*, *Polym. Phys. Ed.*, 2043, (1984).
49. R. L. Miller, *Interrelationships between Molecular and Physical Structure in Amorphous Polymers*, pp. 33–51 of Ref. 8b.
50a. S. Ya. Frenkel and V. G. Baranov, *British Polym. J.*, *9*, 228 (1977).
50b. V. G. Baranov and S. Ya. Frenkel, *J. Polym. Sci.*, *Polym. Symp.*, *61*, 35 (1977).
51. R. F. Boyer, MMI, unpublished. See also page 36 for further details.
52. H. Höcker, G. J. Blake, and P. J. Flory, *Trans. Faraday Soc.*, *67*, Part 8, 2251 (1971).
53. J. K. Krüger, L. Peetz, W. Wildner, and M. Pietralla, *Polymer*, *21*, 620 (1980).
54. J. K. Krüger and M. Pietrella, Ref. 8b, pp. 281–304.
55. L. R. Denny and R. F. Boyer, *Polym. Bull.*, *4*, 527 (1981).
56. R. F. Boyer, K. M. Panichella, and L. R. Denny, *Polym. Bull.*, *9*, 344 (1983).
56a. Y. P. Khana, G. Chomyn, R. Kumar, N. Sanjeeva Murthy, K. P. O'Brien, and A. C. Reimschuessel, *Macromolecules*, *23*, 2488 (1990).
57. A. Kolinski, J. Skolnick, and R. Yares, *Macromolecules*, *19*, 2550 (1986); ibid., *19*, 2560 (1986).
57a. M. L. Mansfield, *Macromolecules*, *19*, 1421 (1986).
58. N. Ishiara, T. Seimiya, N. Kuramoto, and M. Uoi, *Macromolecules*, *19*, 2464 (1986).
59. North American Thermal Analysis Society (NATAS) Notes, Vol. *21(1)*; *43* (1989), Abstract "Thermal Properties of Syndiotactic Polystyrene," A. J. Parztor, Jr., B. J. Landes, and P. J. Karjala, Dow Chemical Co., Midland, MI 48667.
60. P. L. Kumler, M. H. Kailani, B. J. Schober, K. A. Kolasa, S. J. Dent, and R. F. Boyer, *Macromolecules*, *22*, 2994 (1989).
60a. R. J. Roe and H.-H. Song, *Macromolecules*, *18*, 1603 (1985); see also J.

M. O'Reilly, R. S. Stein, G. Hadziannou, and G. D. Wignall, *Polymer*, *24*, 1255 (1983).

60b. R. F. Boyer, *Polymer*, *17*, 996 (1976), especially p. 1000.
61a. T. G. Fox and P. J. Flory, *J. Appl. Phys.*, *21*, 581 (1950).
61b. T. G. Fox and P. J. Flory, *J. Polym. Sci. Lett. Ed.*, *14*, 315 (1954).
62. R. F. Boyer, *J. Macromol. Sci. Phys.*, *B-8*, 503 (1973).
63a. P. J. Flory, *Rec. Sci.*, 25A, 636 (1955), in Printed Proc. Milan, IUPAC Meeting, (1955).
63b. G. Rehage and W. Borchard, in *The Physics of Glassy Polymers*, pp. 54–107 (R. N. Haward, ed.), Applied Science Publishers, London (1973).
64. See, for example, Plazek [25] and also general comments on nomenclature in the editor's preface to Ref. 20.
65. A. M. Lobanov and S. Ya. Frenkel, *Vysokomol. Soedin*, *A22*, 1045 (1980); translated in *Polymer Science USSR*, *22*, 1150 (1980).
66a. J. T. Bendler, in *Encyclopedia of Polymer Science and Engineering*, Vol. 17, pp. 1–22 (J. M. Kroschwitz, ed.), Wiley, New York (1989).
66b. G. B. McKenna, in *Comprehensive Polymer Science*, Vol. 2, pp. 311–362, (C. Booth and C. Price, eds.), Pergamon, Oxford (1989).
66c. R. J. Roe, in *Encyclopedia of Polymer Science and Engineering*, Vol. 7, pp. 531–544 (J. I. Kroschwitz, ed), Wiley, New York (1987).
67. R. Greiner and F. R. Schwarzl, *Rheol. Acta*, *23*, 378 (1984).
68. R. F. Boyer, *European Polym. J.*, *17*, 661 (1981). Note: The η_o values came largely from the doctoral dissertation of J. F. Pierson, University of Strasbourg (1968). CNRS Centre de documentation, No. AO-2106, address 26 Rue Boyer, 20th Arrondissment, Paris, France, can provide a copy of thesis at nominal cost.
69. C. Lacabanne, P. Goyaud, and R. F. Boyer, *J. Polym. Sci. Polym. Phys. Ed.*, *17*, 277 (1980).
70. R. F. Boyer and J. B. Enns, *J. Appl. Polym. Sci.*, *32*, 4075 (1986).
71. R. F. Boyer, in *Encyclopedia of Polymer Science and Technology*, Vol. 13, p. 280, Fig. 3 (N. Bikales, ed.), Wiley, New York (1970).
72. C. A. Glandt, H. K. Toh, J. K. Gillham, and R. F. Boyer, *J. Appl. Polym. Sci.*, *20* 1277 (1976).
73. S. J. Stadnicki, J. K. Gillham, and R. F. Boyer, *J. Appl. Polym. Sci.*, *20*, 245ff (1976).
74a. D. E. Axelson and L. Mandelkern, *J. Polym. Sci. Phys. Ed.*, *16*, 1235 (1978).
74b. A. Dekmezian, D. E. Axelson, J. J. Dechter, B. Borah, and L. Mandelkern, *J. Polym. Sci. Phys. Ed.*, *23*, 367 (1985).
74c. R. F. Boyer, J. P. Heeschen, and J. K. Gillham, *J. Polym. Sci. Phys. Ed.*, *19*, 13 (1981). This paper first reports $T_c = 1.2T_g$.
75. L. LaMarre, H. P. Schreiber, M. R. Wertheimer, D. Chatain, C. Lacabanne, *J. Macromol. Sci. Phys.*, *B-18*, 195 (1980).
76. K. Varadarajan and R. F. Boyer, *J. Polym. Sci. Phys. Ed.*, *20*, 141 (1982). See Ref. 68.
77. R. F. Boyer, unpublished (1986).

78. E. Jenckel and K. Ueberreiter, circa 1940.
79a. K. Schmieder and K. Wolf, *Kol. Z. Z. Polym.*, *134*, 139 (1953).
79b. K. H. Illers and E. Jenckel, *Rheol. Acta*, *1*, 322 (1958).
79c. K. H. Illers and E. Jenckel, *J. Polym. Sci.*, *41*, 528 (1959).
79d. K. H. Illers, *Z. Elektrochem*, *65*, 679 (1961).
80. K. M. Sinnott, *Trans. Soc. Plastics Engrs. 2*, 65 (1962).
81. S. G. Turley, Dow Chemical Company, unpublished observations.
82. J. K. Gillham, J. Benci, and R. F. Boyer, *Polym. Eng. Sci.*, *16*, 357 (1976).
83. J. B. Enns and R. F. Boyer, pp. 221–249 of Ref. 8b. This paper references the original studies in 1977.
84. J. Heijboer, *Poly. Eng. Sci.*, *19*, 664 (1978).
85. J. H. Gisolf, Delft Progress Report Series A, *Chem. Phys. Eng.*, *1*, 85 (1974).
86. J. H. Gisolf, *Col. Polym. Sci.*, *253*, 185 (1975).
87. J. H. Gisolf, Delft Progress Report Series A, *Chem. Phys.*, *Chem. Phys. Eng.*, *1*, 125 (1976).
88. J. L. Kalwak, D. J. Van Dijk, and J. H. Gisolf, *Chem. Physics. Chem. Phys. Eng.*, *1*, 131–136 (1976).
89. J. H. Gisolf, *Col. Poly. Sci.*, *255*, 625 (1977).
90. L. R. Denny, R. F. Boyer, and H.-G. Elias, *J. Macromol. Sci. Phys.*, *B-25*, 227 (1988).
91. J. K. Gillham, J. A. Benci, and R. F. Boyer, *Polym. Eng. Sci.*, *16*, 357 (1976).
91a. K. Solc, S. E. Keinath, and R. F. Boyer, *Macromolecules*, *16*, 1645 (1983).
91b. R. F. Boyer and R. L. Miller, *Macromolecules*, *17*, 365 (1984).
92. A. H. Willbourn, *Trans. Faraday Soc.*, *54*, 717 (1958).
93. J. Heijboer, *Proc. Int. Conf. Non-Crystalline Solids*, pp. 231–254, North-Holland, Amsterdam (1965).
94. J. Heijboer, *Mechanical Properties of Glassy Polymers Containing Saturated Rings*, Doctoral Dissertation, Uitgeverij Wallman, Delft (1972).
95. J. Heijboer, private communication.
96. A. Abe and T. Hama, *Polym. Lett.*, *7*, 427 (1969).
97. R. F. Boyer, *British Poly. J.*, *4*, 163 (1982).
98. R. F. Boyer, *J. Macromol. Sci. Phys.*, *B-8*, 503 (1973). See also pp. 770–772 of [28].
99. R. F. Boyer and R. Simha, *J. Polym. Sci. Polym. Lett. Ed. 11*, 33 (1973).
100. R. F. Boyer, *J. Macromol. Sci. Phys.*, *B-7*, 487 (1973).
101. W. Kauzmann, *Chem Rev.*, *43*, 219–250 (1948).
102. S. Matsuoka and Y. Ishida, Ref. 20, pp. 247–260.
103. J. Heijboer, Atti. Del. 2° Convegno Della Societa Italiano Di Reologia Siena, 10–11 Maggio (1973).
104. J. B. Enns and R. F. Boyer, Ref. 8b, pp. 221–249.
105a. R. F. Boyer, J. P. Heeschen, and J. K. Gillham, *J. Polym. Sci.*, *Polym. Phys. Ed.*, *19*, 13 (1981).
105b. R. F. Boyer, *J. Polym. Sci.*, *Polym. Phys. Ed.*, *26*, 893 (1988).
106a. R. F. Boyer, *Europ. Polym. J.*, *17*, 661 (1981).

106b. K. S. Hyun and R. F. Boyer, in *Encyclopedia of Polymer Science and Technology*, Vol. 13, p. 357, Fig. 8 (N. Bikales, ed.), Wiley, New York (1970).
106c. Ref. 8b, pp. 148–150.
106d. P. J. Flory, *Statistical Mechanics of Chain Motion*, pp. 40–41, Interscience, New York (1969).
106e. N. Okui, *Polymer*, *31*, 92 (1990).
107. K. Wolf, *Physiker Tagung München Hauptvorträge*, 1957, p. 141, Fig. 9, Mosbach, Physikverlag (1957). See Fig. 9 of Ref. 13 for copy of original.
108. G. P. Johari, *Polymer* (*British*), *27*, 866 (1986). (See Fig. 1.)
109. A. Gourari, R. F. Boyer, C. Lacabanne, and G. M. Bendaoud, *J. Polym. Sci. Polym. Physics*, *Ed. 23*, 889 (1985).
110. B. Stoll, W. Pechhold, and S. Blasenbrey, *Kol. Z.*, *250*, 1111 (1972).
111. R. M. Fuoss, *J. Am. Chem. Soc.*, *63*, 378 (1941). (PVC has been measured numerous times (McCrum et al. [22], Chap. 11), but the Fuoss citation is a classic.
112. M. G. Wyzgoski and G. S. Y. Yeh, *Polym. J.*, (*Japan*), *4*, 29 (1973).
113. V. A. Bershtein, V. M. Egorov, Y. Emelyanov, and V. A. Stepanov, *Polym. Bull.*, *9*, 96 (1983).
114. V. A. Bershtein, V. M. Egorov, A. F. Podolsky, and V. A. Stepanov, *J. Polym. Sci. Polym. Lett. Ed.*, *23*, 371 (1985).
115. N. Bach Van and C. Noel, *J. Polym. Sci. Polym. Chem. Ed.*, *14*, 1627 (1976).
116. H. W. Spress, *Adv. Polym. Sci.*, *66* (1985).
117. M. A. Winnik, ed., *Photophysical and Photochemical Tools in Polymer Science—Conformation, Dynamics, Morphology*, Reidel, Dordrecht (1986).
118a. D. Richter and T. Springer, eds., *Polymer Motion in Dense Systems*, Springer-Verlag, Berlin (1988).
118b. M. Pietralla and W. Pechold, eds., *Progress in Colloid and Polymer Science*, *80*, 1–27 (1989).
119. V. D. Fedotov and H. Schneider, eds., *Structure and Dynamics of Bulk Polymers by NMR Methods*, Springer-Verlag, Berlin (1989).
120. M. Takayanagi, M. Yoshimo, and S. Minami, *J. Polym. Sci.*, *61*, S-7, (1962).
121. T. Alfrey and R. F. Boyer, pp. 193–202 of Ref. 26.
122. Discussions Royal Soc. Chem., No. 68, Royal Soc. Chem. London (1979).
123. W. Wrasidlo, *Adv. Polym. Sci.*, *13*, pp. 3–99 (1974). See pp. 39–49, Tables 2a and 2b, and Fig. 19 on p. 49.
124. R. L. Miller, Michigan Molecular Institute. Computer plot prepared at our request.
125. G. Kanig, *K. Z.*, *Z. Polym.*, *29*, 810, (1938).
126. N. Hirai and H. Eyring, *J. Appl. Phys.*, *29*, 810 (1938).

part one

GROUP CONTRIBUTION TECHNIQUES

1

Group Contribution Techniques for Correlating Polymer Properties and Chemical Structure

Dirk W. Van Krevelen
Delft University of Technology
Delft, The Netherlands

ADDITIVE FUNCTIONS IN POLYMER SCIENCE

The Approach

Science consists of *data* (experimental facts), *correlations* (rules, laws), and *theories* (concepts, models, hypotheses). *Basic science* is the search for insight and understanding; it tries to develop theories which merge data and correlations into a consistent whole in order to understand the properties of matter on the basis of a structural image. *Applied science*, on the other hand, uses the basic insights obtained as the steering instrument for technology and practice. It is useful as far as it enables us to estimate or predict the required numerical values of properties and correlations which are not yet measured or discovered. Technology and practice have a growing need for numerical data on all kinds of properties of materials expected to be forthcoming.

There are three approaches used to obtain the required useful information [1].

1. The purely empirical: on extrapolation this approach is often unreliable or even dangerous.

2. The purely theoretical: this is highly desired, but seldom adequately developed and suitable for use.
3. The semiempirical, i.e., empirical, but based on a theoretical concept: this is the most useful and reliable for practical purposes.

In the field of polymers the semiempirical approach is mostly necessary and sometimes the only possible way. What is needed in practice is a formulation which is designed to deal directly with the phenomena and make use of the language of observation. This approach is pragmatic and designed specifically for use; it is a completely nonspeculative procedure.

The value of estimation and prediction techniques largely depends on their simplicity. Complicated methods have to be rejected. This will be one of our guiding principles in this chapter.

The Additivity Principle

A powerful tool in the semiempirical approach to the study of physicochemical properties in general, and of polymer properties in particular, is the use of the *additivity principle*. This principle means that many properties, or combinations of properties, if expressed per mole of a substance, may be calculated by summation of either atomic, group, or bond contributions:

$$F = \sum_i N_i \cdot F_i$$

where F is an additive molar function, N_i is the number of components of type i, and F_i is the *numerical contribution* (also called *increment*) of the component (atom, atom group, bond), summed over all contributing components [2].

Due to their sequential structure, polymers are ideal materials for the application of the additivity principle. End groups play a negligible part compared with the many sequential units. Therefore the molar quantities can be expressed per mole of the structural unit [3,4].

Discrepancies between numerical values calculated by means of the additivity principle and experimental values are always caused by interactions.

Additivity and interaction—intrinsically polar concepts—are basic in physical sciences. Chemistry became a real science when the first additivity concept was introduced: the mass of a molecule as the sum of the additive masses A_i of the composing atoms (based on the law of conservation of mass). But from the beginning it was clear that interaction is the second pillar of chemistry and physics. Additivity is valid as long as interaction is weak or follows simple laws; discrepancies arise when interactions become strong or follow complicated laws.

In the field of polymers two types of interactions exist. In the first place we have *intramolecular interactions. Steric hindrance* of backbone groups in their torsional oscillation or of side groups in their rotation is a well-known example; through it the stiffness of the chain is increased. Another example is *conjugation of π-electrons* ("resonance") between double bonds and/or aromatic ring systems; this also increases the stiffness and thus decreases the flexibility of the chain backbone. The second type is *intermolecular interaction*; entanglements of long side chains, precise intermolecular fitting, and physical network formation by hydrogen bonding are typical examples.

The concept of additivity has already proved extremely fruitful for studying correlations between the chemical constitution of substances like mineral oils and coals and their physical and chemical properties [5a,5b]. Also, properties of homogeneous mixtures can often be calculated fairly accurately by means of additive molar quantities. Sometimes the discrepancies between numerical values estimated by means of additivities and the experimental values proved to be an important key to the disclosure of special constitutional effects.

Methods for Expressing the Additivity

Three methods of additivity calculations, according to the nature of the structural elements used, were mentioned already:

1. Use of *atomic* contributions. If the additivity is perfect, the relevant property of a molecule may be calculated from the contributions of the atoms from which it is composed. The *molar mass* (*molar weight*) is an example (the oldest additive molar quantity). This *simplest system of additivity*, however, has a restricted value. Accurate comparison of molar properties of related compounds reveals that contributions from the same atoms may have different values according to the nature of their neighbor atoms. The extent to which this effect is observed depends upon the importance of outer valence electrons upon the property concerned.
2. Use of *group* contributions. More sophisticated models start from the basic group contributions and hence have *built-in information* on the valence structure associated with a significant proportion of the atoms present. This is the most widely used method.

3. Use of *bond* contributions. A further refinement is associated with bond contributions in which specific differences between various types of carbon-carbon, carbon-oxygen, carbon-nitrogen bonds, etc., are directly included.

The use of atomic contributions is too simplistic in general; the use of bond contributions leads to an impractically large number of different bond types and, thus, to a very complicated notation.

For practical purposes the method of group contributions or group increments is preferred. This method is used in this chapter.

Discovery of Additive Functions and Derivation of Additive Group Contributions (Group Increments)

Additive functions are discovered sometimes by intuitive vision and sometimes along theoretical lines. A typical example is the molar refraction of organic compounds [6–10]. More than a century ago Gladstone and Dale (1858) found that the product $(n - 1)\ M/\rho$, when calculated for a series of aliphatic organic compounds with increasing chain lengths, grows with a constant increment for the CH_2-group; they also found that other nonpolar groups gave a characteristic increase. This resulted in the *purely empirical* additive function: molar refraction. Later Lorentz and Lorenz (independently of each other, 1880) derived from Maxwell's electromagnetic theory of light another form of the molar refraction: $(n^2 - 1)M/(n^2 + 2)\rho$. This more complex form of the refractive index function resulted from *theoretical* studies. In the hands of Eisenlohr and others, this additive function became the basis of a system of increments. Many years later, Vogel (1948) found that the simplest combination, namely nM, is also additive, though not temperature-independent. Finally, Looyenga (1965) showed that the combination $(n^{2/3} - 1)M/\rho$ has advantages over the former ones: it is nearly temperature-independent and can be used for pure substances and for homogeneous and heterogeneous mixtures.

So the additive functions must be *discovered*; the values of the atom group contributions or increments must be *derived*. This derivation of group contributions is relatively easy when the shape of the additive function is known and if sufficient experimental data for a fairly large number of substances are known. The derivation is mostly based on trial-and-error methods or linear programming; in the latter case the program contains the desired group increments as adjustable parameters. The objective function aims at minimum differences between calculated and experimental molar quantities.

The Concepts: "Polymer" and "Molar Structural Unit"

Polymers are macro*molecules*; a molecule is the smallest entity wherein all actual and potential properties of a substance or a material are present, "coded" in a structural or constitutional formula. In linear polymers the molecules consist of identical repeating units and end groups. In real "high polymers" the mass percentage of the end groups is so small that the repeating unit is determining for the properties. In our discussion on correlations between properties and constitution, based on the additivity principle, the repeating unit is the most basic concept. The repeating unit symbolizes the whole macromolecule. So, e.g., a molecule of poly(ethylene terephthalate), PETP, is represented as

$$\left\{ -CH_2-CH_2-O-\overset{O}{\overset{\|}{C}}-C_6H_4-\overset{O}{\overset{\|}{C}}-O- \right\}_p$$

where p = average degree of polymerization.

All additive molar functions will therefore be expressed on the basis of the repeating unit, the *molar structural* unit. All group contribution or group increment values refer to the constituent groups which together form the structural Unit.

Group increment values are always found in fixed combinations. This is a consequence of the fact that *polymers are generally linear combinations of bivalent groups*. Only for bivalent groups can increments be objectively calculated, i.e., without additional assumptions. Mono-, tri-, and tetravalent groups always occur in linear polymers in bivalent combinations. For instance, a trivalent —CH— group and a univalent $—CH_3$ group together form a bivalent $—CH(CH_3)—$ group. Separate values of the subgroups need a more or less arbitrary division.

In this chapter we shall consequently use bivalent group contributions. It must always be recommended that the accuracy of estimation (i.e., the mean difference between calculated and experimental data) is of the same order as the accuracy of the experimental data themselves.

Survey of the Additive Molar Functions

A survey of the additive molar functions (AMFs), discussed throughout the chapter is given in Table 1. There the names, symbols and definitions are given of the 21 AMFs from which the majority of the physical and physicochemical properties of polymers can be calculated or at least estimated. Table 1 is at the same time a condensed list of the nomenclature used.

Table 1 Additive Molar Functions (per structural unit)

Class	AMF	Name	Symbol	Formula	Main parameter	Symbol
I	1.	Molar mass	$\{M\}$	$\sum_i \{N_i M_i\}$	Mass of bivalent structural group	M_i
	2.	Molar number of backbone atoms	$\{Z\}$	$\sum_i \{N_i Z_i\}$	Number of backbone atoms per str. gr.	Z_i
	3.	Molar van der Waals volume	$\{V_W\}$	$\sum_i \{N_i V_{W,i}\}$	"Hard volume" per structural group	$V_{W,i}$
II	4.	Molar glass transition	$\{Y_g\}$	$\{M\}T_g$	Glass-transition temperature	T_g
	5.	Molar melt transition	$\{Y_m\}$	$\{M\}T_m$	Melt transition temperature	T_m
	6.	Molar unit volume	$\{V\}$	$\{M\}/\rho$	Density	ρ
III	7.	Molar heat capacity	$\{C_p\}$	$\{M\}c_p$	Specific heat capacity	c_p
	8.	Molar melt entropy	$\{\Delta S_m\}$	$\{\Delta H_m/T_m\}$	Heat of melting	ΔH_m
	9.	Molar cohesive energy	$\{E_{coh}\}$	$\{M\}e_{coh}$	Cohesive energy density	e_{coh}

IV	10.	Molar interaction	$\{F\}$	$\{V\}\delta$	Solubility parameter	δ
	11.	Molar Parachor	$\{P_S\}$	$\{V\}\gamma^{1/4}$	Surface tension	γ
	12.	Molar Permachor	$\{\Pi\}$	$\{N\}\pi$	Specific Permachor	Π
V	13.	Molar elastic wave velocity	$\{U\}$	$\{V\}u^{1/3}$	Sound wave velocity	u
	14.	Molar intrinsic viscosity	$\{\mathcal{J}\}$	$K_\theta^{1/2}\{M\} - 4.2\{Z\}$	Intrinsic viscosity coefficient	K_θ
	15.	Molar viscosity-temp gradient	$\{H_\eta\}$	$\{M\}E_\eta^{1/3}$	Activation energy of viscous flow	E_η
VI	16.	Molar polarization	$\{P\}$	$\{V\}\dfrac{\epsilon - 1}{\epsilon + 2}$	Dielectric constant	ϵ
	17.	Molar optical refraction	$\{R\}$	$\{V\}(n - 1)$	Index of light refraction	n
	18.	Molar magnetic susceptibility	$\{X\}$	$\{M\}\chi$	Magnetic susceptibility	χ
VII	19.	Molar free energy of formation	$\{\Delta G_f^\circ\}$	$A_i - B_iT$	Heat of formation entropy of format	$A_i B_i$
	20.	Molar thermal decomposition	$\{Y_{d,1/2}\}$	$\{M\}T_{d,1/2}$	Temperature of half-way decompos.	$T_{d,1/2}$
	21.	Molar char forming tendency	$\{C_{FT}\}$	$\{M\}$CR/1200	Char residue in wt %	CR

The Structural Unit contains by definition *N bivalent* Structural Groups.

Seven classes of additive molar functions can be distinguished, each containing three AMFs:

1. Those which are "exact" and "fundamental," since they are based on the mass and on the "hard" volume of the constituting atom groups; they are completely independent of temperature and time (age) (AMFs 1–3).
2. Those which are connected with phase transitions and phase states and thus are of paramount importance for nearly all properties (AMFs 4–6).
3. Those connected with the different forms of internal energy (AMFs 7–9).
4. Those connected with the interplay between polymers and liquids or gases: solubility, wetting and repulsion (AMFs 10–11) permeability, sorption, and diffusivity (AMF 12).
5. Those connected with elastic phenomena and molecular mobility: elastomechanical properties (AMF 13) viscometric and rheological properties (AMFs 14–15).
6. Those connected with electromagnetic phenomena (AMFs 16–18).
7. Those connected with thermal stability and decomposition (AMFs 19–21).

In this order the 21 additive molar functions will be discussed.

Catalog of Group Contributions or Group Increments

A catalog of the (bivalent) group contributions or group increments is given in the comprehensive Table 2.

ADDITIVE MOLAR FUNCTIONS OF A FUNDAMENTAL NATURE

The Mass of the Molar Unit $\{M\}$

$$AMF\ 1 \qquad \{M\} = \left\{\sum_i N_i M_i\right\} = \left\{\sum_i n_i A_i\right\}$$

where N = number of (bivalent) structural groups and
n = number of atoms

both per structural unit.

The Number of Backbone Atoms of the Molar Unit $\{Z\}$

$$AMF\ 2 \qquad \{Z\} = \left\{\sum_i N_i Z_i\right\}$$

Table 2 Catalog of Increments

AMF nr	Symbol	Dimensions used
1	M	$g \cdot mol^{-1}$
2	Z	mol^{-1}
3	V_W	$cm^3 \cdot mol^{-1}$
4	Y_g	$K \cdot kg \cdot mol^{-1}$
5	Y_m	$K \cdot kg \cdot mol^{-1}$
6	V_a	$cm^3 \cdot mol^{-1}$
7a	$C_{p,s}$	$J \cdot mol^{-1} \cdot K^{-1}$
7b	$C_{p,l}$	$J \cdot mol^{-1} \cdot K^{-1}$
8	ΔS_m	$J \cdot mol^{-1} \cdot K^{-1}$
9	E_{coh}	$kJ \cdot mol^{-1}$
10	F	$(J \cdot mol^{-1})^{1/2} \cdot (cm^3 mol^{-1})^{1/2}$
11	P_s	$(cm^3 \cdot mol^{-1})(mJ \cdot m^{-2})^{1/4}$
12	Π	mol^{-1}
13	U	$(cm^3 \cdot mol^{-1}) \cdot (cm \cdot s^{-1})^{1/3}$
14	ϑ	$K_0 \cdot (cm^3 \cdot g^{-1}) \cdot (g \cdot mol^{-1})^{1/2}$
15	H_η	$(g \cdot mol^{-1}) \cdot (J \cdot mol^{-1})^{1/3}$
16	P_{LL}	$cm^3 \cdot mol^{-1}$
17	R_{GD}	$cm^3 \cdot mol^{-1}$
18	X	$cm^3 \cdot mol^{-1} \cdot 10^{-6}$
19	ΔG_f° A	$kJ \cdot mol^{-1}$
	B	$kJ \cdot mol^{-1} \cdot K^{-1}$
20	$Y_{d,1/2}$	$K \cdot kg \cdot mol^{-1}$
21	G_{FT}	mol^{-1}

Arrangement of group increments
1. Carbon groups in main chain
1.1. Nonconjugating groups a. Unsubstituted $—CH_2—$ group b. Substituted $—CH_2—$ group c. Cyclic groups 1.2. Conjugating groups a. Double and triple bonded b. The $—C(=O)—$ group c. Aromatic groups
2. Heterogroups and carbon-heterocombi(nation) groups in main chain
2.1. Nonconjugating groups a. Acyclic groups b. Cyclic groups 2.2. Conjugating Groups a. Acyclic groups b. Cyclic groups
3. Frequently used bicyclic combigroups in main chain

Table 2 Continued

Additive molar function		1.1.a	1.1.b					
Nr.	Symbol	—CH_2—	—CH— | CH_3	—CH— | *i*-Prop	—CH— | *t*-But	—CH— [cyclopentyl]	—CH— [cyclohexyl]	—CH— [phenyl]
1	M	14.03	28.05	56.11	70.13	82.14	96.17	90.12
2	Z	1	1	1	1	1	1	1
3	V_W	10.23	20.45	40.90	51.12	53.28	63.58	52.62
4	Y_g	2.7[a] 4.3	8.0	19.9	25.6	30.7	41.3	36.1
5	Y_m	5.7[b]	13.0	35.3	(45)	—	—	48
6	V_a	16.15	33	65.5	82	85	101	82
7a	$C_{p,s}$	25.35	46.5	93	114.5	110.8	121.2	101.2
7b	$C_{p,l}$	30.4	57.85	(115)	139	147.5	173.9	144.15
8	ΔS_m	8.4	13.6	—	—	—	—	9.6
9	E_{coh}	4.19	10.06	(15.4)	(18)	(24)	(28)	31.4
10	F	272	495	(1000)	(1200)	1430	1680	1560
11	P_s	39.0	78.0	156	195	208.3	244.9	211.9
12	Π	15	15	—	—	—	—	39
13	U	880	1850	(3700)	—	(4600)	(5500)	5100
14	ϑ	2.35	4.70	9.4	11.8	—	11.2	19.4
15	H_η	420	1060	—	—	—	—	3600
16	P_{LL}	4.65	9.26	18.52	23.13	25.65	30.30	29.1
17	R_{GD}	7.83	15.62	31.24	38.98	43.77	51.75	50.97
18	X	11.35	23.5	47	59.5	63.5	74.5	62
19	ΔG_f° A	−22.0	−48.7	−97.4	−120.7	−73.4	−118.4	84.3
	B	−0.102	−0.215	−0.43	−0.545	−0.548	−0.680	−0.287
20	$Y_{d,1/2}$	9.5	18.5	—	—	—	60	56.5
21	C_{FT}	(−1)	(−1)	(−3)	(−4)	0	0	1

[a,b]See annotations to Table 2, pp. 73–74.

Table 2 Continued

Additive molar function		1.1.b						
Nr.	Symbol	—CH—(C$_6$H$_4$)—CH$_3$	—CH—OH	—CH—OCH$_3$	—CH—O—C(=O)—CH$_3$	—CH—C(=O)—OCH$_3$	—CH—C≡N	—CH—F
1	M	104.14	30.03	44.05	72.06	72.06	39.04	32.02
2	Z	1	1	1	1	1	1	1
3	V_W	63.77	14.82	25.45	36.95	36.95	22	13.0
4	Y_g	41.2	13	11.9	23.3	21.3	17.3	12.0
5	Y_m	54	18	18.7	38	38	26.9	15.5
6	V_a	102	21.5	40	59	59	32	20
7a	$C_{p,s}$	125.3	32.6	63.3	(92.5)	(92.5)	40.6	37.0
7b	$C_{p,l}$	171	65.8	93.5	(123)	(123)	(58)	42
8	ΔS_m	—	4.6	—	—	—	0	7.6
9	E_{coh}	(34)	(39)	(16)	(20.5)	(20.5)	(23.5)	4.89
10	F	1840	(900)	(800)	(1100)	(1100)	900	(310)
11	P_s	250.9	59	98	(143)	(143)	85.6	47.6
12	Π	—	255 wet 100	—	—	—	205	85
13	U	(6000)	(1050)	(2050)	(3100)	(3100)	(1750)	(950)
14	ϑ	21.0	9.15	4.8	11.1	13.7	16.2	5.4
15	H_η	—	—	—	—	—	—	—
16	P_{LL}	30.5	(10)	10.85	15.5	15.5	—	(5)
17	R_{GD}	59.27	10.75	18.58	26.38	26.38	15.9	7.68
18	X	73.5	16.5	28.5	37.5	37.5	20	15.6
19	ΔG_f° A	−51.3	−179	−169	−539	−539	120	−198
	B	−0.395	−0.17	−0.285	−0.40	−0.40	−0.07	−0.114
20	$Y_{d,1/2}$	—	14	—	37.5	42.5	28	18
21	C_{FT}	1	0.33	0	0	0	0	—

Table 2 Continued

Additive molar function		1.1.b						
Nr.	Symbol	$-CH(Cl)-$	$-C(CH_3)_2-$	$-C(CH_3)(C_6H_5)-$	$-C(CH_3)(C{=}O\,OCH_3)-$	$-CF_2-$	$-CClF-$	$-CCl_2-$
1	M	48.48	42.08	104.1	86.05	50.01	66.47	82.92
2	Z	1	1	1	1	1	1	1
3	V_W	19.0	30.67	62.84	46.7	16	21.57	27.8
4	Y_g	20	8.5^{15}_{26}	51	35.1	13.0	23	25
5	Y_m	27.5	12.1^{22}_{39}	—	41.5	30	32	41.5
6	V_a	29	50	101	73	25.5	35	44
7a	$C_{p,s}$	42.7	68.0	122.7	114.0	—	54.7	60.4
7b	$C_{p,l}$	(60.8)	81.2	167.5	146.2	—	(68)	87
8	ΔS_m	11.6	(10)	—	—	4.8	9.2	—
9	E_{coh}	13.4	13.7	35.1	(23.4)	3.36	11.88	20.4
10	F	610	686	1752	(1320)	(310)	(612)	914
11	P_s	76.2	117.0	250.9	181.8	56.2	84.8	113.4
12	Π	108	20	—	—	(100)	(130)	155
13	U	1600	2850	6100	4100	(1050)	(1700)	2350
14	ϑ	13.4	7.1	20	13.5	(8.4)	(16.5)	24.5
15	H_η	2330	1620	—	—	—	—	—
16	P_{LL}	13.7	13.86	33.7	(20)	6.25	(13.9)	17.7
17	R_{GD}	16.71	23.33	58.41	34.1	7.12	16.3	25.54
18	X	27.5	36	74.5	50.0	20.2	32.1	44
19	ΔG_f° A	−51.7	−72.0	61	−530	−370	−224	−78
	B	−0.11	−0.33	−0.402	−0.72	−0.13	−0.125	−0.122
20	$Y_{d,1/2}$	23.5	25.5	56	51	38.5	(39)	39
21	C_{FT}	—	(−3)	0	0	—	—	—

Table 2 Continued

Nr.	Symbol (Additive molar function)	1.1.c: $-CH_2-C_6H_4-CH_2-$	1.1.c: $-C_6H_{10}-$ (trans)	1.2.a: $-CH=CH-$	1.2.a: $-CH=C(CH_3)-$	1.2.a: $-CH=CCl-$	1.2.a: $-C\equiv C-$	1.2.b: $-C(=O)-$
1	M	104.14	82.14	26.04	40.06	60.49	24.02	28.01
2	Z	6	4	2	2	2	2	1
3	V_W	63.78	53.34	16.94	27.16	25.72	16.1	11.5
4	Y_g	35	27	cis 3.8 tr 7.4	cis 8.1 tr 9.1	15.2	11	9^{14}_{19}
5	Y_m	47[c]	45	cis 8.0 tr (11)	cis (10) tr (13)	22	(16.5)	$12^{(18)}_{(25)}$
6	V_a	97	86.5	27	43	41	25	16
7a	$C_{p,s}$	129.5	103.2	37.3	60.05	56.25	—	23.05
7b	$C_{p,l}$	173.9	147.5	42.8	74.32	(77)	—	52.8
8	ΔS_m	—	M		21.7	9.6	—	0?
9	E_{coh}	33.52	(23.5)	10.2	14.5	17.85	(7.6)	(17.5)
10	F	1890	1410	454	704	818	435	563
11	P_S	250.9	205.9	67.0	106	104.2	56.0	(48)
12	Π	90	−54	−12	−30	33	—	—
13	U	5860	(2900)	1400	2150	1900	1240	900
14	$\mathcal{J}$	21.0	8.0	0.5	2.9	11.6	—	(9)
15	H_η	—	—	—	—	—	—	—
16	P_{LL}	34.3	25.7	8.9	13.5	—	—	—
17	R_{GD}	59.32	44.00	15.50	23.24	24.33	—	8
18	X	73	63.5	13.2	25.6	29.6	14	6.5
19	ΔG°_f A	56.0	−100	73	39	39	230	−132
	B	−0.384	−0.58	−0.08	−0.186	−0.079	0.05	−0.04
20	$Y_{d,1/2}$	73	—	18	21.5	—	—	14^{20}_{26}
21	C_{FT}	(4)	0	0	(0)	—	(2)	(1)

Table 2 Continued

Nr.	Additive molar function Symbol	1.2.c 	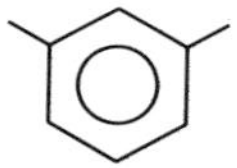		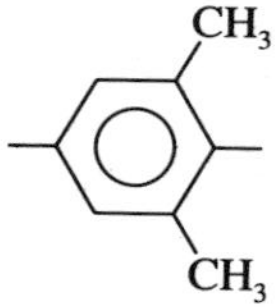	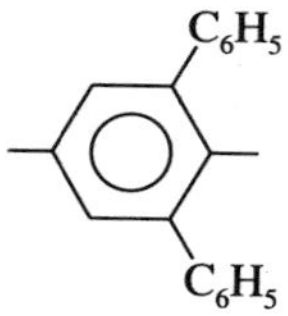	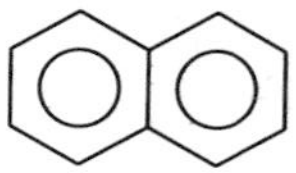	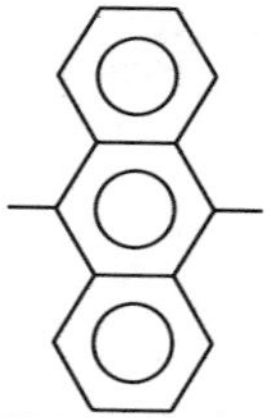
					CH_3 / CH_3	C_6H_5 / C_6H_5		
1	M	76.09	76.09	76.09	104.14	228.28	126.18	176.23
2	Z	2	3	4	4	4	(6)	(4)
3	V_W	43.32	43.32	43.32	65.62	132.48	71.45	94.5
4	Y_g	(9)	25_{34}^{29}	29.5_{41}^{35}	54	118	68	—
5	Y_m	(13)	$31_{(42)}^{36}$	38_{56}^{47}	(65)	173	(85)	—
6	V_a	68	68	65	104	(206)	(112.5)	(149)
7a	$C_{p.s}$	(78.8)	(78.8)	78.8	127	236	127	175
7b	$C_{p.l}$	(113.1)	(113.1)	113.1	167	340	(182)	(250)
8	ΔS_m	—	—	26	—	—	—	—
9	E_{coh}	(26)	(26)	25.14	(40)	(92.5)	(52.5)	(60)
10	F	(1346)	(1346)	1346	1900	(4000)	(2400)	(3000)
11	P_s	(173)	(173)	172.9	251	519	(282)	(391)
12	Π	—	—	60	−44	—	—	—
13	U	(3450)	(3500)	4100	(6150)	(12650)	(5200)	(6800)
14	ϑ	—	—	16.3	—	—	—	—
15	H_η	—	—	3200	—	—	—	—
16	P_{LL}	24.7	25.0	25.0	34.8	(75)	—	—
17	R_{GD}	44.2	44.7	44.8	61.0	(135)	—	—
18	X	(50)	(50)	50	75	(155)	88	(126)
19	ΔG_f° A	(100)	(100)	100	33	300	167	234
	B	(−0.18)	(−0.18)	−0.180	−0.39	−0.53	−0.66	−0.37
20	$Y_{d.1/2}$	—	—$_{65}$	54_{75}^{62}	82	—	85	110
21	C_{FT}	2	3	4	(2)	(8)	6	—

Table 2 Continued

Nr.	Additive molar function Symbol	2.1.a						2.1.b
		—O—	—S—	—Si(CH_3)$_2$—	—OC(=O)— (—OCO—)	—OC(=O)NH— (—OCNH—)	—NHC(=O)NH— (—NHCNH—)	—N(CO)$_2C_6H_2$(CO)$_2$N— (pyromellitimide)
1	M	16.00	32.06	58.15	60.01	59.03	58.04	214.13
2	Z	1	1	1	3	3	3	7
3	V_W	5	10.8	42.2	18.9	23	25	94.5
4	Y_g	4	8	7_{16}^{11}	20	20	20	175
5	Y_m	13.5	22.5	—	(30)	(43.5)	(60)	(225)
6	V_a	8	17.0	66.5	31	36	40	150
7a	$C_{p,s}$	16.8	24.05	—	(63)	(58)	(50)	191
7b	$C_{p,l}$	35.6	44.8	—	—	—	—	
8	ΔS_m	7	—	—	(7)	(7)	(0)	—
9	E_{coh}	6.3	8.8	—	(18)	(26.5)	(35)	(130)
10	F	145	460	—	775	(1200)	(1500)	(4300)
11	P_s	20.0	48.2	—	85	—	—	—
12	Π	70	—	−116	24	—	—	—
13	U	400	(550)	—	1600	(1800)	(2000)	(8500)
14	ϑ	0.1	—	12	(27.5)	(25)	—	—
15	H_η	480	—	1350	3150	—	—	—
16	P_{LL}	(5)	8	—	22	—	—	—
17	R_{GD}	(2.85)	14.44	(25)	13.2	(16.9)	(20.5)	(96)
18	X	5	16	40	19	20	(27)	(85)
19	ΔG_f° A	−120	40	—	−457	−279	−16	−209
	ΔG_f° B	−0.07	0.025	—	−0.19	−0.24	−0.28	−0.66
20	$Y_{d,1/2}$	8	(33)	(60)	(30)	32.5	40	200
21	C_{FT}	0.3	(1)	—	0	0	—	12

Table 2 Continued

Nr.	Additive molar function / Symbol		2.2.a				2.2.b		
			–C(=O)–O–	–C(=O)–NH–	–C(=O)–O–C(=O)–	–S(=O)(=O)–	1,3,4-oxadiazole ring (–C=N–N=C–, O)	1,3,4-thiadiazole ring (–C=N–N=C–, S)	thiazole ring (HC–S, –C=, N=C–)
1	M		44.01	43.03	72.02	64.06	68.04	84.11	83.12
2	Z		2	2	3	1	3	3	3
3	V_W		15.2	18.8	27	20.3	25.4	31.2	34.5
4	Y_g		$12.5^{13.5}_{15}$	$15^{21.5}_{30^d}$	22	32.5^{36}_{40}	(30)	(35)	(35)
5	Y_m		25^{29}_{25}	$45^{51}_{60^e}$	35	$56^{61}_{(66)}$	—	—	—
6	V_a		mc 24.0 sc 20.0	27.5	40	32	40	(50)	(55)
7a	$C_{p,s}$		(46)	(46)	(63)	(50)	72	79	81
7b	$C_{p,l}$		(65)	(85)	(114)	—	155	165	142
8	ΔS_m		0	0	(0)	—	—	—	—
9	E_{coh}		13.41	(45)	(40)	—	(56)	(60)	(37)
10	F		634	(1230)	1160	—	(1500)	(1700)	(1400)
11	P_s		64.8	(78)	(113)	—	—	—	—
12	Π		102	309 wet 210	—	—	—	—	—
13	U		1250	1700	(2150)	(1250)	(600)	(750)	(1400)
14	ϑ		9.0 acr 6.4	12.6	—	12	—	—	—
15	H_η		1450	1650	—	—	—	—	—
16	P_{LL}		15	30	(25)	—	—	—	—
17	R_{GD}		10.6	15.15	(18.4)	—	(34)	(45.5)	(45)
18	X		14	14	18	—	(38)	(49)	(44)
19	ΔG_f°	A	−337	−74	−470	−280	100	260	230
		B	−0.116	−0.16	−0.16	−0.15	−0.29	−0.19	−0.18
20	$Y_{d,1/2}$		20_{25}	30_{37}	(50)	(50)	(50)	60	—
21	C_{FT}		0	(0)	(0)	—	1	3	(4)

[d,e]See annotations. mc = main chain; sc = side chain.

Table 2 Continued

Additive molar function		2.2.b				
Nr.	Symbol	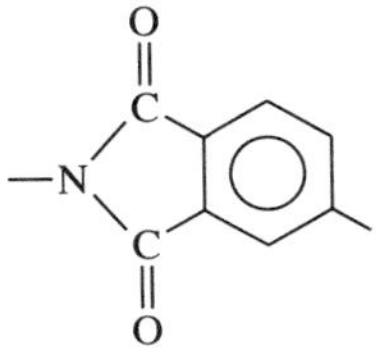	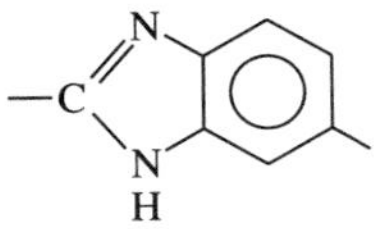	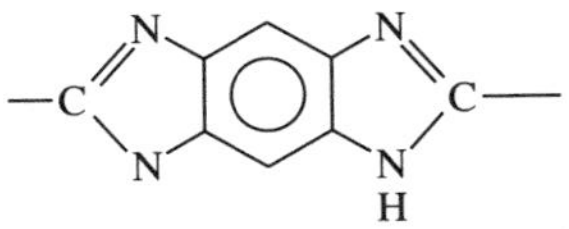	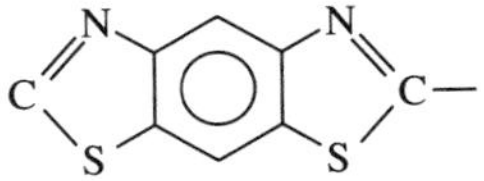	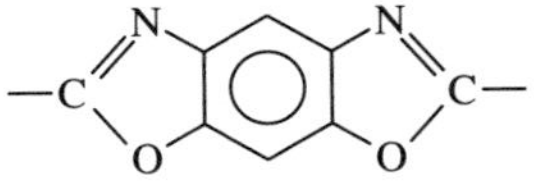
1	M	145.11	116.12	156.14	190.25	158.11
2	Z	5	5	7	7	7
3	V_W	61	59	75	80	69
4	Y_g	85_{95}	$(78)_{88}$	110	(130)	(110)
5	Y_m	125	—	—	—	—
6	V_a	97	94	120	130	110
7a	$C_{p,s}$	135	114	139	156	110
7b	$C_{p,l}$	253	195	276	300	215
8	ΔS_m	—	—	—	—	—
9	E_{coh}	(80)	(48)	(68)	(67)	(60)
10	F	(2825)	(2100)	(2850)	(2900)	(2500)
11	P_s	—	—	—	—	—
12	Π	—	—	—	—	—
13	U	(6300)	(4200)	(4200)	(4650)	(4350)
14	ϑ	—	—	—	—	—
15	H_η	—	—	—	—	—
16	P_{LL}	—	—	—	—	—
17	R_{GD}	(70)	(77)	(92.5)	(104)	(81)
18	X	(68)	(75)	(100)	(110)	(90)
19	ΔG_f° A	−55	290	443	407	87
	B	−0.42	−0.41	—	—	—
20	$Y_{d,1/2}$	—	105	130	—	—
21	C_{FT}	—	7	0	—	—

Table 2 Continued

Additive molar function			3						
Nr.	Symbol		$-C_6H_4-C_6H_4-$	$-C_6H_4-CH_2-C_6H_4-$	$-C_6H_4-C(CH_3)_2-C_6H_4-$	$-C_6H_4-O-C_6H_4-$	$-C_6H_4-S-C_6H_4-$	$-C_6H_4-CO-C_6H_4-$	$-C_6H_4-SO_2-C_6H_4-$
1	M		152.18	166.21	194.26	168.18	184.25	180.21	216.25
2	Z		8	9	9	9	9	9	9
3	V_W		86.64	96.87	117.31	91.6	97.4	98.1	106.9
4	Y_g		70	62	85	65	70	85	110
5	Y_m		94	80	115	100	105	115	140
6	V_a		140	155	185	145	153	155	170
7a	$C_{p,s}$		157.6	183	226	174	182	181	(230)
7b	$C_{p,l}$		226	257	307	262	271	279	(325)
8	ΔS_m		—	—	—	—	—	(55)	—
9	E_{coh}		50.3	54.5	(65)	(55)	59	68	—
10	F		2692	2964	(3400)	(2850)	3150	3250	—
11	P_s		346	385	463	366	(395)	394	—
12	Π		120	135	100	190	—	—	—
13	U		8200	9100	(11000)	8600	(5750)	(9100)	(9500)
14	ϑ		32.6	35	—	—	—	—	—
15	H_η		—	—	—	—	—	—	—
16	P_{LL}		50.0	54.6	63.9	(550	58	(55)	—
17	R_{GD}		89.6	96.9	113	(92.5)	104	98	—
18	X		100	111	136	105	116	107	—
19	ΔG_f°	A	200	178	128	80	240	68	−80
		B	−0.36	−0.46	−0.69	−0.43	−0.325	−0.40	−0.50
20	$Y_{d,1/2}$		122	114	—	—	(140)	(150)	(150)
21	C_{FT}		8	(7)	(5)	(10)	(9)	(9)	—

Annotations to Table 2

[a]*The Group Contribution $Y_g(CH_2)$*

1. The value 2.7 for $Y_g(CH_2)$ is valid only for main chains without strong intermolecular interaction.
2. If intermolecular hydrogen bonds are active (hydrogen bond networks, as is the case in polyamides, polyurethanes, and polyurea), a higher $Y_g(CH_2)$ value, namely 4.3, must be used.
3. For long side chains with sequences of CH_2 groups the situation is more complicated ("comb polymers"). Here the value of the CH_2 contribution depends on the distance from the main chain. The main representatives of these comb polymers are the vinyl polymers of the type

$$\{-CH_2-\underset{\underset{\underset{CH_3}{|}}{\overset{|}{(CH_2)_n}}}{T}-\}$$

where T stands for any trivalent or tetravalent substituted CH_2 group. The polymer with $n = 0$ will be called the base polymer, whose structural unit has an Y_g value equal to $\{Y_{g,0}\}$. It is a fortunate empirical fact that for all vinyl polymers T_g passes through a minimum of about 200 K when $n = 9$. So $Y_{g,9}$ equals $0.20\{M_g\}$.

For $n < 9$ the Y_g value may be interpolated between $\{Y_{g,0}\}$ and $\{Y_{g,9}\}$. If $n > 9$, the long side chains start to interact by forming side chain crystallization domains. This effect causes the increment $Y_g(CH_2)$ to be raised to the value of 7.5.

So the following rules may be used:

$$n < 9 \quad \{Y_g\} = \{Y_{g,0}\} + \frac{1}{9}n(0.20\,\{M_9\} - \{Y_{g,0}\})$$

$$n = 9 \quad \{Y_g\} = 0.20\{M_9\}$$

$$n > 9 \quad \{Y_g\} = 0.20\{M_9\} + 7.5(n - 9)$$

[b]*The group contribution $Y_m(CH_2)$*

1. Strictly, the value 5.7 for the CH_2 group increment is only valid for sequences of CH_2 groups larger than 5. For smaller sequences $Y_m(CH_2)$ depends on its location, namely its distance from the functional group which characterizes the polymer (see Fig. 2).

If the positional distance of the CH_2 group considered from the characteristic functional group is defined as α, β, γ, δ, etc., the respective $Y_m(CH_2)$ values are

$\alpha = 1.0 \quad \beta = 3.0$
$\gamma = 4.8 \quad \delta = 5.5$
$>\delta = 5.7$

2. In contradistinction with what we have seen in the case of Y_g of chain polymers, there is no extra effect due to network formation by hydrogen bonds.

However, there is another effect, caused by the symmetry or asymmetry of the structural unit in crystallization, namely the odd-even effect. If the CH_2 sequence between two functional groups is *odd*, the value of $\{Y_m\}$ is *lowered* (−); if it is *even*, the value of $\{Y_m\}$ is *raised* (+). The numerical value of the odd-even effect depends on the nature and polarity of the functional group:

—O—: ± 0.15
—S—: ± 0.5
—COO—: ± 0.75
—CONH—: ± 1.25

3. In long side chains of sequences of CH_2 groups one states a similar effect as we have met for the glass transition. If the structural unit is again characterized by the formula

$$\{-CH_2-\underset{\underset{\underset{CH_3}{|}}{(CH_2)_n}}{\overset{}{\underset{|}{T}}}-\}$$

the following empirical statement can be made: for all available vinyl polymer series T_m (as a function of n) passes through a minimum of about 235 K, so that for every family $\{Y_{m,5}\} \approx 0.235\{M_5\}$. For $n < 5$ interpolation between $\{Y_{m,0}\}$ and $\{Y_{m,5}\}$ may be applied, whereas for $n > 5$ each CH_2 group has the normal value of 5.7. So the following rules may be used:

$$n < 5 \quad \{Y_m\} = \{Y_{m,0}\} + \frac{1}{5}n(0.235\{M_5\} - \{Y_{m,0}\})$$

$$n = 5 \quad \{Y_m\} = 0.235\{M_5\}$$

$$n > 5 \quad \{Y_m\} = 0.235\{M_5\} + 5.7(n - 5)$$

[c]For *p*-xylylene this value is 70.
[d]The value 30 is valid for (rigid) arylates.
[e]The value 60 is valid for (rigid) aramids.

(end of annotations)

Remark. By "chain backbone" is understood the number of covalently bonded atoms that form the "naked" chain of the polymer, without side groups and side chains. Homochain polymers have a backbone consisting of C atoms only, heterochain polymers may also have O, N, S, Si, and P atoms in the chain backbone.

In the molar unit Z_i is counted "along the chain." So in aromatic rings $Z_i = 2$ in the *ortho*-phenylene group, $Z_i = 3$ in the *meta*-phenylene group, and $Z_i = 4$ in the *para*-phenylene group. For other ring systems (alicyclic, heterocyclic) a similar rule is applied. All vinyl polymers have $\{Z\}$ equal to 2. The $\{Z\}$ function was introduced by Weyland et al. [11] in a study on the correlation of the glass transition temperature with the chemical structure. Especially in additive functions connected with molecular mobility, $\{Z\}$ may play a part.

The van der Waals Volume of the Molar Unit $\{V_W\}$

AMF 3 $$\{V_W\} = \left\{\sum_i N_i V_{W,i}\right\}$$

The van der Waals volume of a molecule is its "hard" volume, the occupied space which is completely impenetrable to other molecules with normal thermal energies. The van der Waals volume is assumed to be bounded by the outer surface of the *interpenetrating atomic spheres* within the molecule. The radii of these spheres are the (constant) radii of the atoms involved; the distances of the centers of the spheres are the bond lengths.

Polymers are ideal substances for the application of the additive van der Waals volume per repeating unit and therefore also, if the average degree of polymerization is known, for the calculation of the "hard volume" of the whole polymer molecule in its free state, independent of its mode of packing. Group contributions for the additive van der Waals volume in polymers have been derived by Bondi [2] and by Slonimskii et al. [12]. The most reliable values are given in the Table 2.

The Actual Molecular Mass of the Polymer ($\overline{M}$)

The sum of the p repeating units is roughly equal to the actual molecule, so

$$\overline{M} = \overline{p}\{M\}$$

The actual molar mass plays a part in all phenomena where whole molecules or swarms of molecules are moving relative to each other, so in melts and solutions (rheological phenomena).

The degree of polymerization p is determined by the nature of the polymerization process and its reaction conditions. It cannot be predetermined but must be measured (predetermining is possible in very few cases only, e.g., for "living" polymers in anionic polymerization). p will always have an *average* value: polymers normally are *polydisperse*. Therefore also M has an average value. The ways in which $\overline{M}$ is determined leads to different averages: *number*, *weight*, *viscosity*, and *higher averages*. The following survey shows the methods, type of average and the upper limits [13], as shown in Table 3.

These different $\overline{M}$ values are special points on a *distribution curve*. During the last decades methods have been developed for determination of the whole distribution curve, giving immediately all characteristic $\overline{M}$ values (gel permeation chromatography, dynamic light scattering, sedimentation field flow fractionation).

The ratio Q of the weight average to number average is highly characteristic for all distribution curves published, as has been shown by Van Krevelen et al. [14]. The following expressions can be used:

$$\frac{\overline{M}_w}{\overline{M}_n} = Q, \quad \frac{\overline{M}_z}{\overline{M}_n} = Q^{1.75}, \quad \frac{\overline{M}_v}{\overline{M}_n} = \frac{1-a}{2} + \frac{1+a}{2} Q \approx Q^{0.75}$$

where a is the exponent in the well-known *Mark-Houwink equation* (see page 109). The value of Q depends on the nature of the polymerization process. The following survey gives some relevant information [13] shown in Table 4. Under conditions of high shear rates in rheological studies Q was found to be one of the most important parameters.

The Critical Molecular Mass ($\overline{M}_{cr}$)

In order to be a real high polymer the average molecular mass must have a value larger than $\overline{M}_{cr}$. For any polymer the value of $\overline{M}_{cr}$ is typical (it is

Table 3 Methods for the Determination of the Molecular Weight of Polymers, Types of Average Values, and Approximate Limits of Measurement

Method	Type of average	Upper limit of average
Ultracentrifuge	M_w	$\sim 10^7$
Light scattering	M_w	$\sim 10^7$
Solution viscometry	M_v	$\sim 10^7$
Melt viscosity	M_w	$\sim 10^6$
Osmometry	M_a	$\sim 10^6$
Vapor pressure osmometry	M_a	$\sim 10^5$
Cryoscopic measurements	M_a	$\sim 10^4$ Dep
Ebullioscopic measurements	M_a	$\sim 10^4$ solven

Table 4 Polydispersity for Various Types of Polymerization

Polymers	M_w/M_n
Living polymers	1.01–1.05
Polymer formed by the combination of two radicals	1.5
Polymer formed by the disproportionation, condensation or addition of two radicals	2.0
Vinyl polymers obtained from polymerization to a high degree of conversion	2–5
Polymers, synthesized by the 'Trommsdorf effect'	5–10
Coordination polymers	8–30
Branched polymers	20–50

a kind of "passport number"). $\overline{M}_{cr}$ is the value of $\overline{M}_w$ where the melt viscosity of the polymer (as a function of $\overline{M}_w$) changes rather drastically. Below $\overline{M}_{cr}$ the melt vicosity is proportional to $\overline{M}_w$; if $\overline{M}_w$ is higher than $\overline{M}_{cr}$ the melt viscosity becomes proportional to $\overline{M}_w^{3.4}$. A related effect is found in polymer solutions, though less spectacular. As a matter of fact a critical chain length $\overline{Z}_{cr}$ is directly related to $\overline{M}_{cr}$.

The value of $\overline{M}_{cr}$ varies for the different polymers from 10^3 to 10^5, the corresponding $\overline{Z}_{cr}$ varies from 50 to 1000. For PETP e.g., $\overline{M}_{cr} = 6000$ and $\overline{Z}_{cr} = 310$.

An interesting relationship was found between $\overline{M}_{cr}$ and the important viscometric quantity K_θ, which determines the viscosity of very diluted solutions (see page 108):

$$K_\theta \overline{M}_{cr}^{1/2} = 0.013 \text{ m}^3/\text{kg} \quad [3]$$

ADDITIVE MOLAR FUNCTIONS CONNECTED WITH PHASE TRANSITIONS AND PHASE STATES

The Molar Glass-Transition Function $\{Y_g\}$

$$AMF\ 4 \qquad \{Y_g\} = \left\{\sum_i N_i Y_{g,i}\right\} = \{M\}T_g$$

This function was introduced by Van Krevelen and Hoftyzer [3]. From this expression T_g can be readily calculated:

$$T_g = \frac{\{Y_g\}}{\{M\}}$$

Remarks.

1. *Other additive correlation methods*. Numerous methods have been proposed in the literature for the correlation of the glass-rubber transition temperature with the chemical structure. In general, they are all based on the formula

$$\sum_i s_i T_{g,i} = T_g \sum_i s_i$$

where $T_{g,i}$ and s_i are, respectively, the T_g contribution and the statistical "weight" of the groups i in the repeating unit. Different assumptions for s_i were proposed. Barton and Lee [15] suggested s_i to be equal to the weight fraction or mole fraction of the relevant group in relation to the structural unit. Weyland et al. [11] put s_i equal to Z_i, the number of backbone atoms of the contributing group. Kreibich and Batzer [16] identified s_i with the number of independently oscillating elements in the backbone of the structural unit. The same had been proposed by Becker [17]. In general, s_i is held as a kind of entropy of transition. With regard to the product $\Sigma_i\, s_i T_{g,i}$, most authors see it as proportional to the cohesion energy, so, e.g., Hayes [18], Wolstenholme [19], and Kreibich and Batzer [16]. It should be noted that the form of the afore mentioned equation is the same as the well-known thermodynamic expression for phase transitions: $\Delta H_{tr} = T_{tr}\, \Delta S_{tr}$.

A serious objection against this thermodynamic analogy for identifying $\Sigma_i\, s_i T_{g,i}$ with the cohesion energy is twofold. First, the glass-rubber transition is not a thermodynamic phase transition at all, neither a first- nor a second-order transition, as was proved by Staverman [20] and Breuer and Rehage [21]; the glassy state is not thermodynamically stable and thus not defined by the normal state variables; also its history and its age play a part. At the very best the T_g transition may be seen as a quasi-second-order transition but certainly not as a first-order one.

A second, even more serious, objection against the use of the cohesion energy is that the glass transition is a change from one condensed state (glass) to another condensed state (liquid) whereas the cohesion energy belongs to the change of the condensed state to the completely free state of the molecules (e.g., in ideally diluted solutions or in a hypothetical gaseous state). So it seems wiser not to load the AMF of the glass transition with pseudotheoretical assumptions, and to use the pragmatic definition given at the beginning of this section.

2. *Comments on the increments*. Van Krevelen's system is based on the T_g of the polymethylene chain, which, according to Boyer [22], is 195

K. This means that the contribution $(Y_g)_{CH_2}$ is $195 \cdot 14 \cdot 10^{-3} = 2.7$ K · kg/mol. This increment is found indeed if $\{Y_g\}$ values of condensation polymers are plotted as a function of the number of CH_2 groups (Fig. 1). In the usual way the contributions of all bivalent groups are then derived.

Two complications appeared. The *first* occurred in the side chains, where the CH_2 groups do not have the same value as in the main chain. Side chains may be considered as a monovalent end group (e.g., methyl, isobutyl, *tert*-butyl, etc.) and an inserted sequence of CH_2 groups between the main chain and end group. It is quite plausible that the value of the CH_2 increment will be influenced by its distance to the backbone of the chain. Hence, different values for $\{Y_g\}_{CH_2}$ as a function of this distance (see Table 2 Annotations).

The *second* complication is the influence of inter- and intramolecular interactions. The most important intermolecular interactions which influence the glass temperature are intermolecular hydrogen bonds. If they are present (e.g., in polyamides, polyurethanes, and polyureas) the CH_2 contribution is virtually enlarged to 4.3 (instead of 2.7). Intramolecular interactions influencing the glass-transition temperature are *steric hindrance*, caused by bulky groups, and *π-electron conjugation* between groups. These two effects stiffen the chain backbone, thus raise T_g and give an *exaltation* of the increments of the groups involved. For this reason several values for the increments of potentially conjugating groups and of groups responsible for steric hindrance are mentioned in Table 2. The first and lowest value is valid in case of absence of interaction, either steric or conjugational. Bulky groups thus may have three values; the second and third are for the cases that the bulky group has one or two bulky neighbors. Also for the potentially conjugating groups three values may be mentioned, namely the second and third for one- and two-sided conjugation.

Generally the exaltation of the group increment is of the order of 5 units per conjugation, or per steric hindrance (the —COO— group only gives a very small conjugation effect). So, e.g., the $-\overset{\displaystyle O}{\overset{\|}{C}}-$ group has the following Y_{gi} values:

in the combination $-CH_2-(CO)-CH_2-$	9
in the combination $-CH_2-(CO)-pC_6H_4-$	14
in the combination $-pC_6H_4-(CO)-pC_6H_4-$	19

3. *Verification*. The system of Y_g increments has been tested on about 600 polymers. About 80% of the T_g values calculated differed less than 20 K from the experimental ones. Keeping in mind that a certain percentage

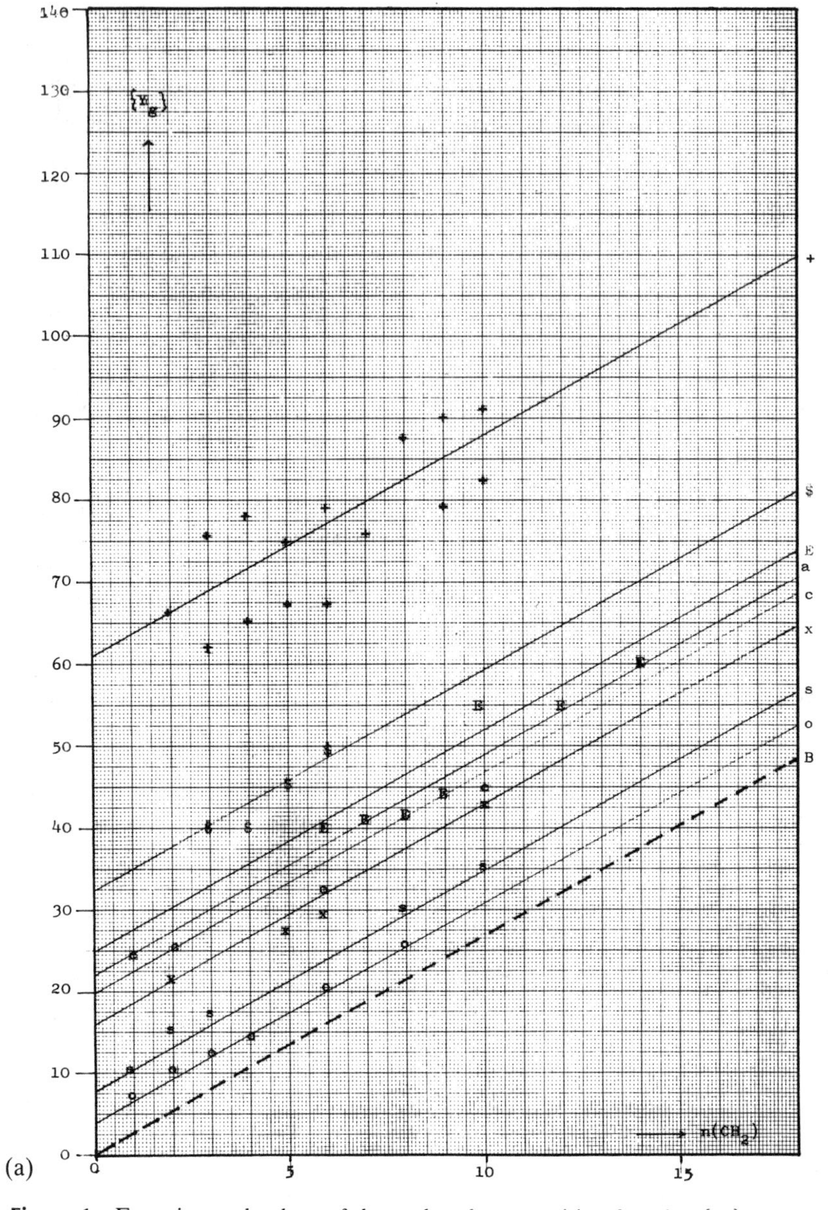

Figure 1 Experimental values of the molar glass-transition function $\{Y_g\}$ for some important polymer families, as a function of the number of main chain CH_2 groups in the structural unit. The empirical values are indicated by the different symbols; the drawn lines are calculated by means of group contributions. (a) Polymers without hydrogen bonding groups; (b) polymers *with* hydrogen bonding groups. *Note* that for *all* polymer families the relationship between Y_g and $n(CH_2)$ is a linear function (straight line); for polymers without hydrogen bonds the slope is 2.7 (standard value), whereas for polymers *with* hydrogen bonds the slope is in-

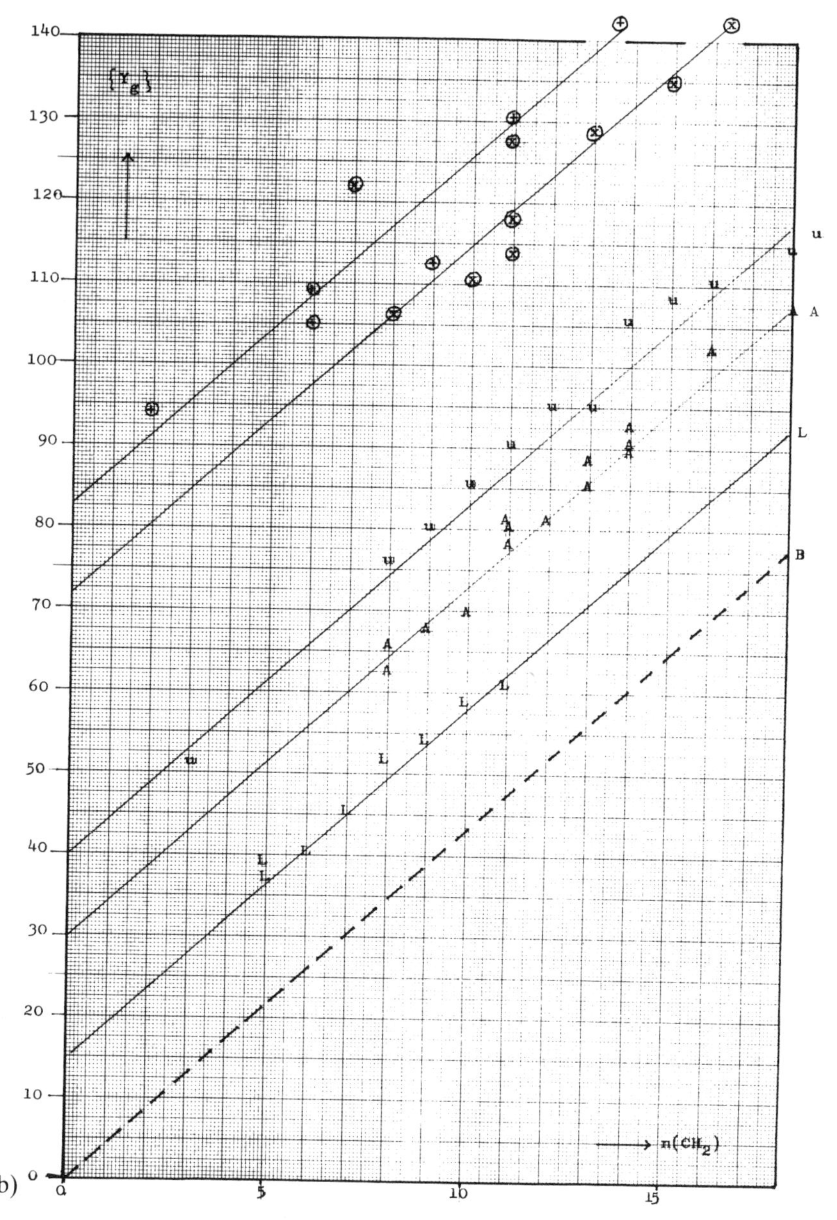

(b)

creased to 4.3, due to hydrogen bonded network formation. B = base line (indicating the slope), o = aliphatic polyoxides, s = aliphatic polysulfides, x = aliphatic polydisulfides, c = aliphatic polycarbonates, a = aliphatic polyanhydrides, E = alphatic polydiesters, \$ = aliphatic polysulfones, + = poly(terephthalates), L = aliphatic polylactams, A = aliphatic polydiamides, u = aliphatic polydiurethanes, U = aliphatic polydiurea, ⊕ = poly(terephthalamides), ⊗ = poly(xylylene diamides) and poly(phenylene diethylene diamides).

of the T_g data in the literature is unreliable (sometimes two or more values are published, which differ as much as 50 to 100 K!), this result may be considered satisfactory.

The Molar Melt Transition Function $\{Y_m\}$

$$AMF\ 5 \qquad \{Y_m\} = \left\{\sum_i N_i Y_{m,i}\right\} = \{M\}T_m$$

This function was also introduced by Van Krevelen and Hoftyzer [3].

Remarks.

1. *Other additive correlation methods.* It is remarkable that practically no T_m structure relationships have been published in the literature, although the number of experimental data available for T_m is at least equal to that of T_g. More than 30 years ago [23], a correspondence was observed between T_g and T_m, which suggests that a treatment analogous to that for T_g could also be used for the estimation of T_m. This leads to the formula

$$T_m = \frac{\sum_i n_i(s_i T_{\mathrm{mi}})}{\sum_i n_i s_i} \tag{a}$$

There is, however, a fundamental difference between T_g and T_m: the melting point is thermodynamically a *real first-order transition* temperature, at which the free energy of the two phases in equilibrium are equal, so that

$$T_m = \frac{\Delta H_m}{\Delta S_m} \tag{b}$$

where ΔH_m is the enthalpy and ΔS_m the entropy of fusion (molar). This equation suggests that an "exact" method for estimating or predicting T_m should be based on calculation of ΔH_m and ΔS_m, which both have an additive character. The lack of reliable data on these quantities (from which increments could be derived) makes this method as yet inapplicable.

In view of this, the correlation for T_m presented here will be based on equation (a). In analogy to the glass transition a *molar melt transition function* $\{Y_m\}$ is introduced, which shows an additive character: $\{Y_m\} = \Sigma_i N_i Y_{\mathrm{mi}}$, so that

$$T_m = \frac{\{Y_m\}}{\{M\}} = \frac{\left\{\sum_i N_i Y_{\mathrm{mi}}\right\}}{\left\{\sum_i N_i M\right\}}$$

This equation has been applied to about 800 crystalline melting points available in the literature. From this collection of Y_m values the rules of correlation and the increments have been derived.

2. *Comments on the increments.* The unbranched polymethylene chain was chosen again as the basis for all flexible polymers. Its melting point is 409 K, which gives for the —CH_2— increment to $\{Y_m\}$

$$Y_m(CH_2) = 409 \times 14.03 \times 10^{-3} = 5.74 \approx 5.7 \text{ K} \cdot \text{kg/mol}$$

Figure 2 shows that $\{Y_m\}$ values of different families of flexible condensation polymers as a function of the number of CH_2 groups in the repeating unit. It is obvious that for all polymer families with repeating units containing more than seven CH_2 groups the slope approaches the value 5.7. But in contradistinction to what was found for Y_g, the slope changes with decreasing numbers of CH_2 groups. This is a clear example of *interaction* effects: the *mobility* of the backbone atoms is changed (CH_2's in the neighborhood of functional groups (= intramolecular effect)), or by variation in *inter*molecular *fitting*. The latter effect is confirmed by the "odd-even effect"; odd numbers of CH_2 groups in the repeating unit show a lower, even numbers a higher, $\{Y_m\}$ value than expected. This odd-even effect is the cause of a *separate structural increment.*

Similar intramolecular interaction effects as found for the $Y_{g,i}$ increments, are also observed for the $Y_{m,i}$ group contributions. Conjugation of aromatic rings, mutual or with double bonds, raises the $\{Y_m\}$ value by about 9 units per conjugation. The influence of side chains on $\{Y_m\}$ is analogous to that on $\{Y_g\}$, though restricted to the first five CH_2 groups., whereas in long side chains the CH_2 increment approaches the normal value of 5.7. Group contributions and corrective structural increments are given again in Table 2 and the Annotations.

3. *Verification.* Of the nearly 800 polymers whose melting points are reported, about 75% gave calculated values which differed less than 20 K from the experimental ones. Part of the experimental values of the other 25% is not fully reliable. The result may be considered satisfactory.

Relationships Beween Transition Temperatures

In the pioneering years of polymer science a certain relationship was already found between T_g and T_m (see, e.g., Boyer [23], Beaman [24], and Bunn [25]). On the average the ratio T_g/T_m proved to be about 2/3.

Later, in an extensive study, Lee and Knight [26] found the mentioned relation to vary widely. The following general rules for the T_g/T_m ratio

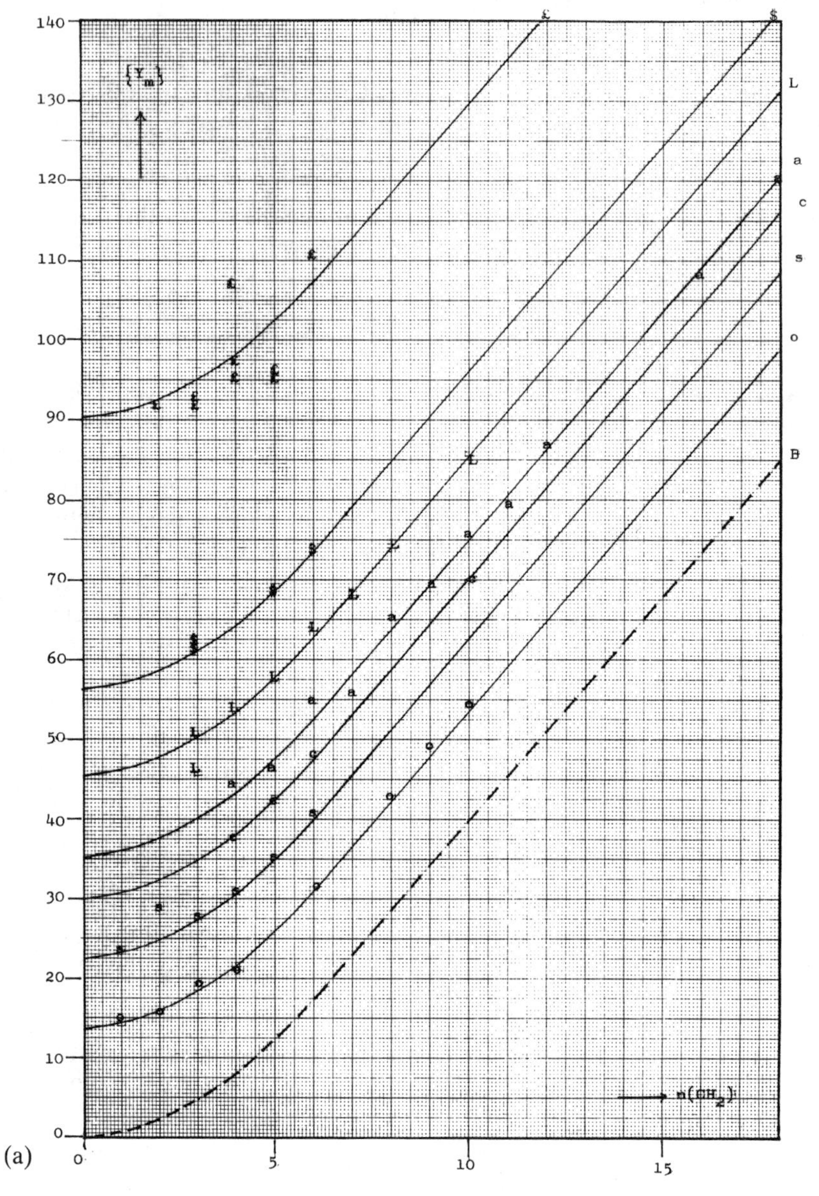

Figure 2 Experimental values of the molar melt transition function $\{Y_m\}$ for some important polymer families, as a function of the number of main chain CH_2 groups in the structural unit. The empirical values are indicated by the different symbols; the drawn lines are calculated by means of group contributions. (a) Polymers with one functional group per structural unit; (b) polymers with two or more functional groups per structural unit. *Note* that for all polymer families the relationship between $\{Y_m\}$ and $n(CH_2)$ is a nonlinear function (curved line) at low values of $n(CH_2)$, passing over into a linear function (straight line) at higher values of $n(CH_2)$. The

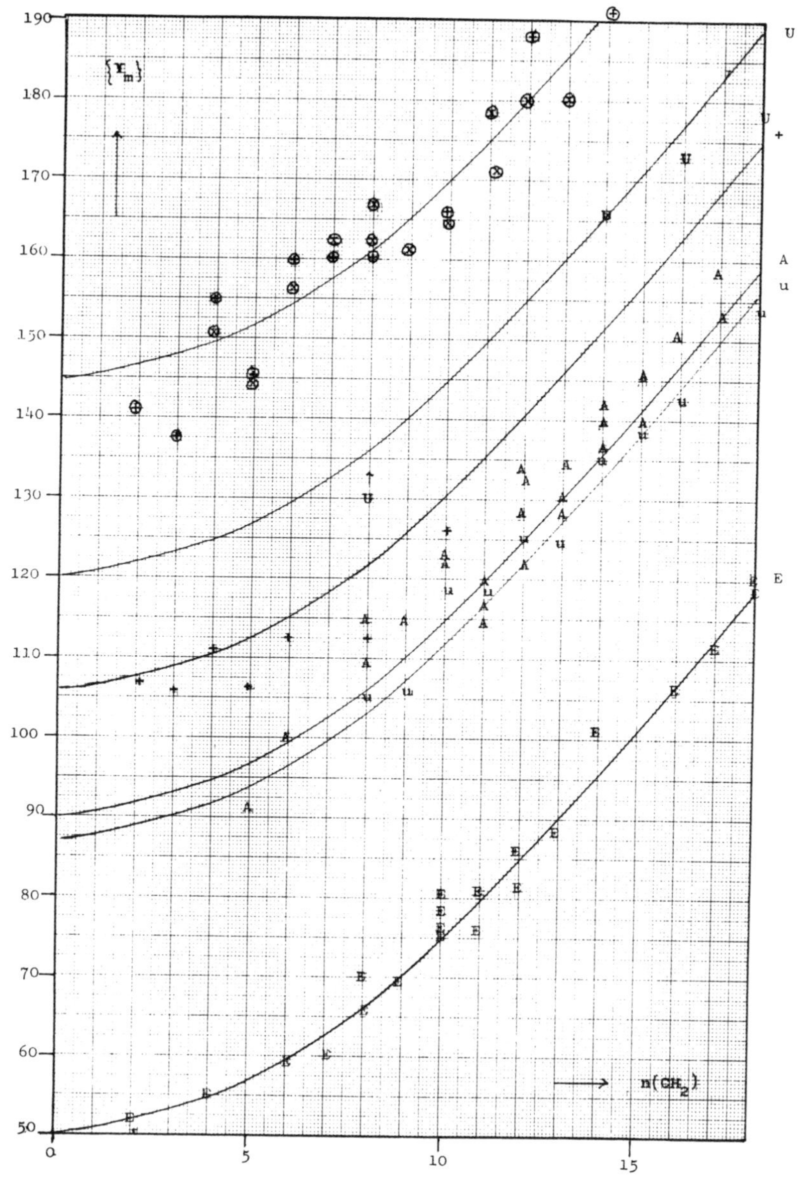

(b)

curvature is dependent on the number of functional groups per structural unit (see also Fig. 2(c)). B = base line, o = aliphatic polyoxides, s = aliphatic polysulfides, c = aliphatic polycarbonates, a = aliphatic polyanhydrides, L = aliphatic polylactams, $ = aliphatic polysulfones, £ = partly aromatic polylactams, E = aliphatic polydiesters, u = aliphatic polydiurethanes, A = aliphatic polydiamides, U = aliphatic polydiurea, + = poly(terephthalates), ⊗ = poly(xylylene diamides) and poly(phenylene diethylene diamides), ⊕ = poly(terephthalamides).

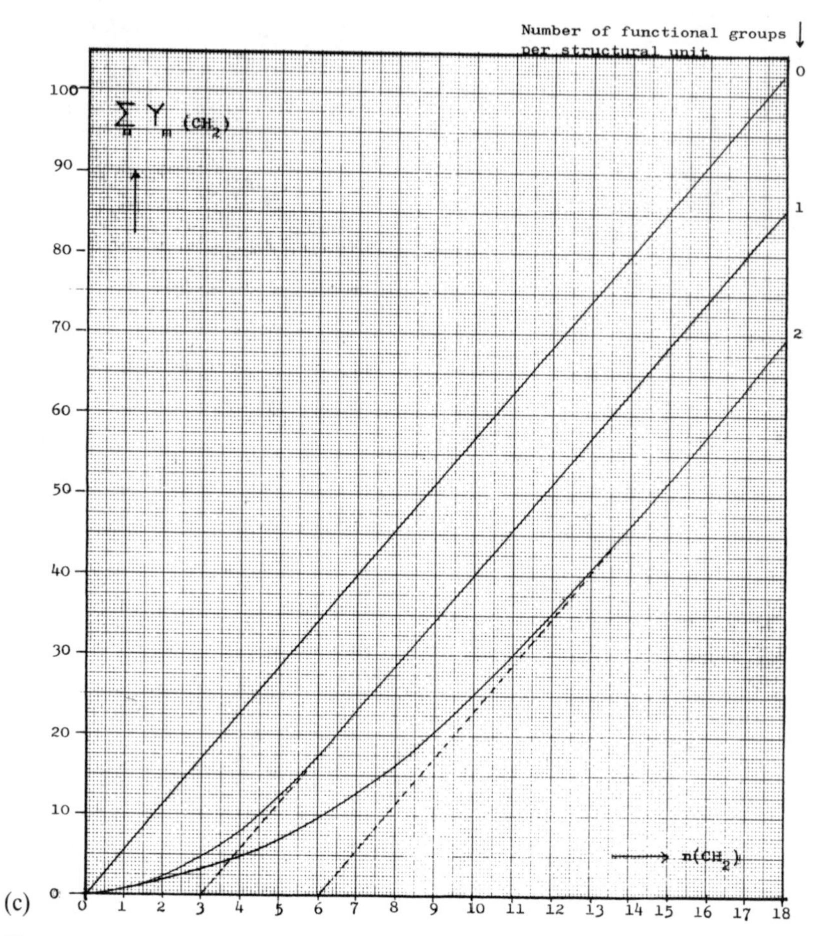

Figure 2(c) Cumulative contribution of the main-chain CH_2 groups in the structural unit to the molar melt transition function $\{Y_m\}$. The number of functional groups (such as —O—, —COO—, etc.) per structural unit of the polymer is the parameter in this graph. Note that for hydrocarbon chains the relationship between ΣY_{mi} and $n(CH_2)$ is a linear function (a straight line with a slope of 5.7). With functional groups in the main chain, the relationship starts as a nonlinear function (curve) passing over into a straight line with slope 5.7. The curvature varies with the number of functional groups per structural unit. It is obvious that the contribution per CH_2 group to $\{Y_m\}$ is "depressed" by the functional groups, and the more so as more functional groups per structural unit are present.

could be formulated:

1. Highly symmetrical polymers with short repeating units, consisting of one or two main-chain atoms each and carrying substituents of only a single atom, have T_g/T_m ratios of 0.5 or lower, and are markedly crystalline.
2. Very unsymmetrical polymers and those with a complex structure have T_g/T_m values of 0.77 or higher.
3. The majority of the polymers (about 80%) have a T_g/T_m ratio between 0.55 and 0.75 with a maximum around 2/3.

In some recent papers by Boyer [27] the transition temperatures of polymers were further discussed. Boyer found that some (semi)crystalline polymers have two glass transitions: T_{gL} and T_{gU} (lower and upper). He assumes that T_{gL} arises from purely amorphous material, while T_{gU} arises from amorphous material under restraint, due to the vicinity of crystallites; frequently T_{gU} is a function of the degree of crsytallinity. Some general rules are

$$\frac{T_{gU}}{T_{gL}} \approx 1.2 \pm 0.1; \qquad \frac{T_{gL}}{T_m} \approx 0.575 \pm 0.075$$

$$\frac{T_{gU}}{T_m} \approx 0.7 \pm 0.1$$

Most polymers show secondary phase transition below T_g, of which the "local mode relaxation" or β-transition ($T_\beta \approx 0.75T_g$) is the most important.

An interesting transition of amorphous polymers is a transition in the liquid phase: $T_{LL} \approx 1.2T_g$; it appears to be connected with the change from the viscoelastic to the normal viscous state.

Also semicrystalline polymers show transitions; at $T \approx 0.9T_m$ there is a mechanical loss peak, probably connected with a premelting transition. It might also be connected with the undercooling necessary for (spontaneous) crystallization.

In liquid crystalline polymers melt transitions of the liquid crystals may be observed.

Influence of T_g and T_m on Other Properties

The influence of T_g and of the dimensionless ratio T_g/T_m is paramount for almost every property of polymeric solids and melts.

Some rules of thumb will be given here; they are not very accurate, but are handy for a quick first orientation.

Relationships of the Ratio T_g/T_m

The ratio T_g/T_m is obviously of influence on all phenomena concerned with the transition polymer melt → (semi) crystalline polymer, where it is correlated with the density ratio (ρ_a/ρ_c), the attainable degree of crystallinity (x_c) and the maximum rate of linear growth of spherulitic crystallites (v_{max}).

$$\frac{\rho_a}{\rho_c} \approx 0.4\left(\frac{T_g}{T_m}\right) + 0.6 \tag{c}$$

$$x_c \approx 1.8 - 2\left(\frac{T_g}{T_m}\right) \tag{d}$$

$$\log v_{max}\ [\text{nm/s}] \approx 13.7 - 18\left(\frac{T_g}{T_m}\right) \quad [3] \tag{e}$$

Relationships with T_g

The direct influence of T_g will become manifest in later sections, namely on

1. The activation energy of diffusion of simple gases into the micropores of the polymer (E_D) (page 96)
2. The stiffness of the glassy state (G_g) (page 102)
3. The shift factor of mechanical damping peaks (a_T) (page 103)
4. The Newtonian viscosity of polymer melts (n_N) (page 113)

The Molar Unit Volume of the Phase States {*V*}

$$AMF\ 6 \qquad \{V\} = \left\{\sum_i N_i V_i\right\}$$

Remarks.

1. *Earlier propositions.* For small molecules the molar volume was one of the first discovered additive quantities. Traube [28] was its first investigator; he used atomic increments and a "residual volume." After him several scientists improved his method, namely Davis and Gottlieb [29], Harrison [30], and Rheineck and Linn [31]. For amorphous polymers Van Krevelen and Hoftyzer [32] gave the first list of increments. Some years later Fedors [33] provided a very extensive system. Table 2 contains the values of Van Krevelen and Hoftyzer [3,32] for amorphous polymers at room temperature (298 K).

2. *Correlation with the van der Waals volume.* A good correlation was found between the molar unit volume $\{V_a(298)\}$ of amorphous polymers and their van der Waals volume $\{V_W\}$:

$$\{V_a(298)\} \approx 1.6V_W \quad [32]$$

3. *Molar unit volume for the crystalline state* $\{V_c\}$. For a limited number of polymers the densities, and thus the molar unit volumes, of the crystalline state are known (mostly obtained by extrapolation of the density to 100% crystalline or from x-ray studies). Also here a correlation with the van der Waals volume exists:

$$\{V_c(298)\} = 1.435\{V_W\} \quad [3]$$

For semicrystalline polymers the molar unit volume can be calculated by linear interpolation between V_a and V_c, if the degree of crystallinity is known. From this V_{sc} value the density can directly be found: $\rho = \{M\}/\{V\}$.

4. *Verification.* The mean deviation between the $\{V\}$ values calculated by means of the technique of additive group contribution and the experimental values from the literature is about 1%.

5. *Thermal expansion and* $\{V\}$ *values at arbitrary temperature.* In order to calculate the $\{V\}$ values at arbitrary temperatures the molar thermal expansion $\{E\}$ must be known; this quantity is related to the thermal expansion coëfficient α $(=(V^{-1}(\partial V/\partial T)_p))$ in the following way:

$$\{E\} = \frac{\alpha\{M\}}{\rho} = \alpha\{V\}$$

$\{E\}$ proved to be numerically related to the van der Waals volume

$$\{E_1\} \approx \{E_r\} \approx 10\{V_W\} \times 10^{-4} \text{ cm}^3/\text{K} \cdot \text{mol}$$

$$\{E_g\} \approx \{E_c\} \approx 4.5\{V_W\} \times 10^{-4} \text{ cm}^3/\text{K} \cdot \text{mol}$$

Combining these values with those of the molar unit volumes (see above) gives the following expressions of the V's as a function of T [3]:

$$\{V_r\}(T) = \{V_l\}(T) \approx \{V_a\}(298) + \{E_r\}(T - 298) \approx \{V_W\}(1.30 + 10 \times 10^{-4}T)$$

$$\{V_g\}(T) \approx \{V_a\}(298) + \{E_g\}(T - 298) \approx \{V_W\}(1.44 + 4.5 \times 10^{-4}T)$$

$$\{V_c\}(T) \approx \{V_c\}(298) + \{E_c\}(T - 298) \approx \{V_W\}(1.30 + 4.5 \times 10^{-4}T)$$

This equation also enables us to calculate the volume expansion $\Delta(V_m)$ during melting [3]:

$$\frac{\Delta\{V_m\}}{\{V_W\}} = 0.165\left(\frac{T_m}{298}\right)$$

ADDITIVE MOLAR FUNCTIONS CONNECTED WITH DIFFERENT FORMS OF INTERNAL ENERGY

Internal Energy of Polymers

The additive molar functions connected with internal energy have their base in the thermodynamic equations of the molar enthalpy and the molar entropy, respectively:

$$H(T) = H(\mathrm{O}) + \int_0^T C_p\, dT + \sum_{\mathrm{tr}} \Delta H_{\mathrm{tr}}$$

$$S(T) = S(\mathrm{O}) + \int_0^T \frac{C_p}{T}\, dt + \sum_{\mathrm{tr}} \Delta S_{\mathrm{tr}}$$

where C_p is the molar heat capacity, ΔH_{tr} is the molar enthalpy of first order phase transitions, and ΔS_{tr} is the molar entropy of the same.

The Molar Heat Capacity Function $\{C_p\}$

$$AMF\ 7 \qquad \{C_p\} = \left\{\sum_i N_i C_{p,i}\right\} = \{M\}c_p$$

The molar heat capacity function in this form was proposed by Satoh [34], who provided a rather complete list of increments for solid organic substances for three temperatures (200, 300, and 400 K). Later, a similar list for organic liquids at 298 K was given by Shaw [35]. When applied on polymers, the given group contributions lead to $\{C_p\}$ values which are in fair agreement with the available experimental literature data on solid (glassy *and* semicrystalline) and liquid (molten and rubbery) polymers. So two sets of increments must be used in the system of additive group contributions: $\{C_{p,s}\}$ and $\{C_{p,l}\}$. Table 2 contains their values at 298 K.

Remarks.

1. *Specific heat at arbitrary temperature.* For a number of polymers the full course of the specific heat as a function of temperature has been determined. In particular, Wunderlich and his school [36,37] did very important work in this field.

Examination of the available data showed that, for the polymers investigated, the molar heat capacity of polymeric solids and liquids above 150 K is a practically linear function of the temperature. The following rules may be used:

$$\{C_{p,s}\}(T) = \{C_{p,s}\}(298)[1 + 3.0 \times 10^{-3}(T - 298)]$$
$$\{C_{p,l}\}(T) = \{C_{p,l}\}(298)[1 + 1.2 \times 10^{-3}(T - 298)]$$

2. *Molar heat at constant volume.* If values for C_v of a polymer are needed they may be calculated from C_p by means of the thermodynamic equation

$$C_p - C_v = \frac{TV\alpha^2}{\kappa}$$

where α is the thermal expansion coefficient (for calculation see page 89; Remark 5) and κ = compressibility = reciprocal bulk modulus (for calculation see page 100).

The Molar Melt Entropy Function $\{\Delta S_m\}$

$$AMF\ 8 \qquad \{\Delta S_m\} = \left\{\sum_i N_i\, \Delta S_{m,i}\right\} = \frac{\Delta H_m}{T_m}$$

where ΔH_m is the molar latent heat of melting (or fusion, or crystallization), i.e., the enthalpy difference between the liquid and the solid state at the melting temperature and ΔS_m is the corresponding melting entropy.

Bondi [2] stated that the ratio $\Delta H_m/T_m$ (the melting entropy) shows a fairly regular relationship with the chemical structure, and has additive properties.

Lack of really reliable data on the molar heats of fusion of polymers is the reason that only a small number of increments could be derived. They are given in Table 2. From the additive molar quantity ΔS_m the value of ΔH_m can be estimated by multiplication with T_m.

Cohesive Energy of Polymers

The molar cohesive energy is the increase in energy per mole of substance if *all inter*molecular forces are eliminated, i.e., if the substance is transformed from a condensed state into an ideal solution or an ideal gas. For polymers the gas state does not exist.

The Molar Cohesive Energy Function $\{E_{coh}\}$

$$AMF\ 9 \qquad \{E_{coh}\} = \left\{\sum_i N_i E_{coh,i}\right\} = \{V\} e_{coh}$$

where e_{coh} = cohesive energy density (= E_{coh}/V).

Several authors, e.g., Dunkel [38], Hayes [39], Hoftyzer and Van Krevelen [40], and Fedors [41], have shown that the molar cohesion energy of solid amorphous polymers is an additive function. Table 2 presents the most reliable values. In view of the fact that the basic experimental data show a very large spread, the accuracy of the additive calculation is only about 10%, which is within the limits of accuracy of the experimental data.

ADDITIVE MOLAR FUNCTIONS CONNECTED WITH THE INTERPLAY OF POLYMERS AND LIQUIDS

The Phenomena of Solubility, Wetting, and Repulsion

If polymers and liquids have affinity to each other, they may mix; i.e., the polymer will absorb the liquid and *dissolve*. If they have unsufficient affinity the liquid may either *wet* the surface or (in the other extreme) even *undergo repulsion*. We shall now consider these different possibilities.

The Molar Attraction Function $\{F\}$

$$AMF\ 10 \qquad \{F\} = \left\{\sum_i N_i F_i\right\} = \{V\}\delta$$

where δ = solubility parameter (= $(\{E_{coh}\}/\{V\})^{1/2}$)

This additive molar function was introduced by Small [42]. Hoy [43] propsed group contributions to F, slightly different from those of Small. Table II gives the preferable set of group contributions.

Remarks.

1. *Correlations of AMF 9 and 10.* These are alternative functions. It is clear that they are closely related:

$$e_{coh} = \delta^2 \qquad \text{and} \qquad \{F\} = \{E_{coh}\}^{1/2}\{V\}^{1/2}$$

2. *Solubility of polymers.* The quantities e_{coh} and δ are both used in the theory of solubility, as developed by Hildebrand and his school. According to Hildebrand and Scott [44], the enthalpy of mixing (Δh_M) of two components 1 and 2 is determined by the equation:

$$\Delta h_M = \phi_1\phi_2(\delta_1 - \delta_2)^2$$

where ϕ_1 and ϕ_2 are the volume fractions of the two components. If $\delta_1 = \delta_2$; the enthalpy of mixing is zero. The *gain* in free enthalpy is then at its maximum, since in the equation

$$\Delta G_M = \Delta H_M - T\,\Delta S_M$$

the mixing entropy is always positive. This means that complete solubility can occur if $(\delta_P - \delta_S)^2$ is as small as possible (the indexes P and S refer to polymer and solvent). This is the mathematical expression of the well-known empirical rule that chemical *and* structural similarity favors solubility (*Similia similibus solvuntur*).

3. *The (absolute) value* $|\delta_P - \delta_S|$. The absolute value of this difference plays a part in numerous phenomena, such as

Swelling of polymers by liquids
Crazing and cracking by liquids
Shrinkage of polymer fibers in liquids
Deterioration of mechanical properties in liquids
Induction of crystallization by liquids
Viscosity of solutions (related to the exponent a in the Mark-Houwink equation)
Free energy of polymer solutions (related to the interaction parameter χ)

4. *Relationship to other physical properties.* As a matter of fact, e_{coh} and δ are related to other properties connected with cohesive internal energy, so e_{coh} is proportional to the inner pressure π of the state equation, and is related to the interfacial energy:

$$e_{\text{coh}}\ \text{MJ/m}^3 \approx 1.5\gamma^{3/2}\ \text{mN/m}$$

5. *Refinements in the* $\{E_{coh}\}$ *and* $\{F\}$ *calculations.* Although no rigorous additivity exists in the case of $\{E_{\text{coh}}\}$ and $\{F\}$, a fair estimation of e_{coh} and δ is possible. However, this is only true if no specific forces are active between the structural units of the molecules involved. If one of the substances (polymer or solvent) contains strong polar groups or hydrogen bonding groups, even for $\delta_P = \delta_S$ no dissolution may occur; conversely, if both substances contain these special groups, solubility may be promoted, even if $\delta_P \neq \delta_S$.

A refinement in the calculation of the solubility parameter is based on the splitting of the cohesive energy into three terms, corresponding to the three types of interaction forces: *dispersion* or *van der Waals–London* forces, *polar* or *Keesom-Debye* forces, and *hydrogen bond* or *Pauling* forces:

$$E_{\text{coh}} = E_d + E_p + E_h$$

$$\delta^2 = \delta_d^2 + \delta_p^2 + \delta_h^2$$

For details we refer to [3].

Important work in this field was done by Beerbower et al. [45], Crowley et al. [46], and especially by Hansen [47].

Surface Energy and Interfacial Energy

Surface energy is a direct manifestation of the intermolecular forces. The molecules at the surface of a liquid or a solid are influenced by unbalanced molecular forces and so possess additional energy, in contrast with the molecules inside the liquid or solid.

In liquids the surface energy manifests itself as a force which tends to reduce the surface area to a minimum. The specific free surface energy of liquids is called *surface tension* (γ).

The Molar Surface Energy Function or Parachor $\{P_s\}$

AMF 11 $$\{P_s\} = \left\{\sum_i N_i P_{s,i}\right\} = \{V\}\gamma^{1/4}$$

where γ is the surface tension (of the liquid). This function (Parachor) was introduced by Sugden [48], who gave a set of increments for liquids. This set was later somewhat improved and supplemented by Mumford and Phillips [49] and Quayle [50].

The $\{P_s\}$ function is valid for liquids (and polymer melts), but is also applicable for amorphous solid polymers. In the latter case the original increments of Sugden proved to give the best correspondence with the experimental values. The Parachor is nearly independent of the temperature.

Remarks.

1. *Surface energy of solid polymers.* In solids the surface energy can only be measured by means of indirect methods. One method is the extrapolation of the surface tension of melts. A second method is the measurement of the *contact angle*, ϑ, and use of the equation

$$\gamma_s \text{ and } \gamma_l \frac{(1 + \cos\vartheta)^2}{4\phi^2}, \quad \text{where} \quad \phi = \frac{4[\{V_s\}\{V_l\}]^{1/3}}{[\{V_s\}^{1/3} + \{V_l\}^{1/3}]^2}$$

A third method was developed by Zisman [51], who introduced the concept of "critical surface tension of wetting" (γ_{cr}). In a plot of cos ϑ versus γ_l, the γ_{cr} is the intercept with the horizontal line cos $\vartheta = 1$. A liquid with a γ_l less than γ_{cr} will spread on the surface. It was proved that $\gamma_{cr} = \gamma_s$. If no measurements of cos ϑ are available, division of the calculated Parachor of the liquid by the molar volume of the solid gives a fair approximation of $\gamma_s^{1/4}$ of amorphous solid polymers.

2. *Spreading tendency of a liquid on a solid.* This phenomenon is determined by the ratio γ_s/γ_l and is characterized by the contact angle ϑ. The values γ_s/γ_l and $\cos\vartheta$ are related by the equation of Girifalco and Good [52]:

$$1 + \cos\vartheta \approx 2\phi\left(\frac{\gamma_s}{\gamma_l}\right)^{1/2} \quad \text{where} \quad \phi \approx \frac{4(V_s V_l)^{1/3}}{(V_s^{1/3} + V_l^{1/3})^2} \approx 1$$

If $\gamma_s/\gamma_l > 1$, the liquid will not spread on the surface, if the ratio < 1, it will spread.

3. *Interfacial tension.* The interfacial tension between two liquids can be estimated from the equation

$$\gamma_{12} \approx (\gamma_1^{1/2} - \gamma_2^{1/2})^2$$

The same equation is valid in the case of a liquid (1) on a polymer (2).

4. *Stability of adhesive joints in liquids.* If an interface $s_1 - s_2$ is immersed in a liquid L, spontaneous separation will occur if

$$\gamma_{s_1 s_2} > \gamma_{Ls_1} + \gamma_{Ls_2}$$

5. *Effect of hydrogen bonds.* The equations mentioned above are strictly valid only for substances without hydrogen bonds, as was demonstrated by Fowkes [53] and Owens [54]. Refinements in the calculations are possible again by splitting the surface energy in separate terms:

$$\gamma = \gamma_d + \gamma_h$$

where d and h refer to the dispersive and the hydrogen bond forces, respectively.

ADDITIVE MOLAR FUNCTION CONNECTED WITH THE MICROPORE STRUCTURE OF POLYMERS

Permeation

Knowledge of the permeation of gases and vapors is very important in the application of polymers as protective coatings and packaging films. The main practical property is the *permeability*; its fundamental equation and definition is

$$P = SD \tag{1}$$

where S = solubility and D = diffusivity. These two quantities are (in the case of simple gases) typical Arrhenius functions of the temperature:

$$S(T) = S_0 \exp\left(-\frac{\Delta H_s}{RT}\right) \tag{2}$$

$$D(T) = D_0 \exp\left(-\frac{E_D}{RT}\right) \tag{3}$$

where ΔH_s = molar heat of sorption and E_D = activation energy of diffusion. So

$$P(T) = S_0 D_0 \exp\left(-\frac{\Delta H_s + E_D}{RT}\right) \tag{4}$$

$$P(298) = S_0 D_0 \exp\left(-\frac{\Delta H_s + E_D}{298\ R}\right) \tag{5}$$

The literature contains much data on S, D, and P, from which information on S_0, D_0, ΔH_s, and E_D can be calculated for a considerable number of combinations of polymers and gases. The problem is to find the relationships between these quantities and the nature (structure) of the polymers and gases. It can be expected that the polymer-gas interaction plays a role. It is, e.g., probable that ΔH_s depends on a force constant of the gas (possibly the *Lennard-Jones potential* (ϵ/k)) or the related critical temperature) and on the gas-polymer interaction (possibly the solubility parameter). It is also probable that the activation energy of diffusion is determined by the kinetic *diameter of the gas molecule* (σ) and by the cohesion density of the polymer. Furthermore, as in many similar rate processes, it may be expected that D_0 will be related to E_D/R and S_0 possibly with $\Delta H_s/R$ (the "compensation effect").

Salame [56] formulated four semiempirical equations for the quantitites S_0, D_0, ΔH_s, and E_D in their functional relationship with a new physical quantity π, for which he coined the name *Permachor*. The four permeation parameters are independent of the concentration of the permeant only in the case of simple gases.

The Specific Permachor as a Characterization Index for the Microporous Structure of Polymeric Materials

Salame's specific permachor π is a typical practical index. In this respect it is comparable with other practical indices such as the Centigrade scale for the common temperature characterization (°C.) and the pH of watery solutions as an index of the acidity.

The specific Permachor π is defined by the equation

$$\pi = -\frac{1}{s}\ln\frac{P}{P^*} = -\frac{2.3}{s}\log\frac{P}{P^*}$$

where P = permeability of a gas (by preference a "standard gas") in an arbitrary polymer

P^* = permeability of the same gas in a chosen "standard polymer"

s = a scaling factor, determined by the choice of the "fixed points" of π

As "standard polymer" Salame selected natural rubber for several reasons. It is a generally known and available polymer with a well-defined chemical composition (poly-*cis*-isoprene), and it is representative of the whole class of elastomers, with a relatively high permeability. This also implies that for natural rubber $\pi = 0$ (by definition !).

As a second fixed point in the π scale, Salame chose one of the most impermeable polymers, widely used for packing purposes for food, namely *polyvinylidene chloride* (semicrystalline and stretched, generally known as Saran). For this material π is, by definition, 100.

The preferable standard gas is ntirogen; it is the main component of air and is highly typical for "normal" gases (in boiling point and molecular diameter).

The value of log $P(298)$ for the nitrogen/natural rubber system is -12 ± 0.3 and that for ntirogen/PVDC is -17 ± 0.3 (in $cm^2s^{-1}Pa^{-1}$). The s value of nitrogen is 0.115. Since in the definition of π a linear relationship between π and log P is implied, a graphical representation of log P versus π is a straight line through the fixed points. For every value of log $P(298)$ a value of π can be read from such a graph.

The proportionality of π to the negative logarithm of the relative permeability is comparable to the pH of a solution as the negative logarithm of the H^+ ion concentration. If the pH of a solution is high, then the acidity is low. In the same way a polymeric film has a low permeability if the π-value is high. (The pH scale also has two fixed points: 0 and 14.)

The Molar Permeability Function: Molar Permachor $\{\Pi\}$

Salame found that the product $\{N\}\pi$ (= Π) is an additive function, the molar Permachor:

$$AMF\ 12 \qquad \{\Pi\} = \left\{\sum_i N_i\Pi_i\right\} = \{N\}\pi$$

where $\{N\}$ = number of characteristic groups per structural unit
Π_i = contribution of the group considered to $\{\Pi\}$

The available group contributions to $\{\Pi\}$ are given in Table 2.

Rearranging AMF 12 gives the value of π:

$$\pi = \frac{\sum_i N_i \Pi_i}{\sum N_i} = \frac{\{\Pi\}}{\{N\}}$$

This leads immediately to the log $P(298)$ value of the polymer considered (for nitrogen as the permeant)

$$\log P(298) = \log P^*(298) - \frac{s\pi}{2{\cdot}3} = -12 - 0.05\pi$$

The value of $P(298)$ (for nitrogen permeation) being known, that for other gases can be estimated by means of a series of multiplication factors, given in the following table:

Relative P Values for Some Gases ($P_{N_2}(298) = 1$)

CO	1.2	He	15
CH_4	3.4	H_2	22.5
O_2	3.8	CO_2	24

Physical Basis of the Permachor

Salame succeeded in giving his starting equation (the definition of π as $\pi = f(-\log P)$) a sound physical basis, showing that it is, in essence, identical to the fundamental equation (5). For this purpose he derived (from the available experimental data) four semiempirical equations, namely, for the four permeation parameters of Eq. (5).

These four equations (where we have chosen cm, J, and Pa as units of length, energy, and pressure) are the following:

Equation		Rubbers (elastomers)	Glassy amorphous polymers
I	E_D/R	$\frac{\sigma_X^2}{\sigma_{N_2}}(3125 + 78\pi)$	$\frac{\sigma_X^2}{\sigma_{N_2}}(2875 + 45\pi)$
II	$\log D_0$	$-4.0 + 10^{-3}\, E_D/R$	$-3.5 + 0.6 \times 10^{-3}\, E_D/R$
III	$\Delta H_S/R$	$1550 - 13.25\, \epsilon/k$	$450 - 13.25\, \epsilon/k$
IV	$\log S_0$	$-5{,}3 - 0.0057\epsilon/k - 0.013\pi$	$-6.5 - 0.0057\epsilon/k - 0.013\pi$

Expressions I–III are analogous to earlier expressions in the literature (Van Amerongen [57], Van Krevelen [3]); Eq. IV is new.

After substitution of these equations in (4) and (5), Salame combined all terms containing π into a product $s\pi$ (resulting in a full expression of s) and all terms not containing π into $A(T)$, identical to $P^*(T)$, so giving the full expression for P^*. Both s and P^* vary with the nature of the gas and with temperature.

Remarks. All π values calculated by means of AMF 12 are valid only for amorphous polymers. For conversion into the values of semicrystalline polymers, Salame gives the following formula:

$$\pi_{sc} \approx \pi_a - 18 \ln a \approx \pi_a - 41.5 \log(1 - x_c)$$

where a = fraction amorphous and x_c = fraction crystalline. Another expression for semicrystalline polymers [58] is

$$S_{sc} = S_a(1 - x_c), \qquad D_{sc} = D_a(1 - x_c),$$

so

$$P_{sc} = S_{sc}D_{sc} = S_aD_a(1 - x_c)^2 = P_a(1 - x_c)^2$$

ADDITIVE MOLAR FUNCTIONS CONNECTED WITH ELASTIC ENERGY (MECHANICAL PROPERTIES OF POLYMERS)

Elastic Quantities of Polymers

Mechanical properties of polymers are of interest in all applications where polymers are used as structural materials. Mechanical behavior involves deformations under influence of applied forces or (constant or variable) force fields. The most characteristic mechanical properties are called moduli: the bulk modulus (B), the shear modulus or rigidity (G), and the tensile modulus (E). They are measures of stiffness. The reciprocal moduli are called compliances: the bulk compliance or compressibility (K), the shear compliance (J), and the tensile compliance (D).

A further important quantity is the Poisson ratio (ν), the ratio of lateral contraction to the simultaneous axial extension (strain). It is a measure for the "fluidness" or "plasticity" of a material. Extremely rigid materials, like diamond, have a Poisson ratio tending to zero, whereas liquids have a value of 1/2. Linear polymers have Poisson ratios between 1/3 and 1/2. In purely elastic materials the moduli and the Poisson ratio are interrelated:

$$E = 2G(1 + \nu) = 3B(1 - 2\nu)$$

This equation implies that at $\nu = 1/2$ (liquid state), E and G, but not B, become zero.

The Molar Elastic Wave Function $\{U\}$

$$\textit{AMF 13} \qquad \{U\} = \{V\}\left[u^2 \frac{1+\nu}{3(1-\nu)}\right]^{1/6} \approx \{V\}u^{1/3}$$

where u is the longitudinal sound velocity.

The additive function $\{U\}$ was introduced by Rao [59] for molecular liquids and extended for solids by Schuyer [60], who found that for all phases the following expression holds:

$$u_{\text{long}}^2 = \frac{B}{\rho}\frac{3(1-\nu)}{1+\nu}$$

For $\{U\}$ we may therefore write

$$\text{AMF 13a} \qquad \{U\} = \left\{\sum_i N_i U_i\right\} \approx \{V\}\left(\frac{B}{\rho}\right)^{1/6}$$

The group contributions for $\{U\}$ are given in Table 2.

Remarks.

1. *Relationship between bulk modulus and Poisson ratio.* Van Krevelen [3] found an interesting empirical relationship between B/ρ and ν:

$$\log\left(\frac{1}{2}\frac{B}{\rho}\right) \approx 2(1-2\nu)$$

(see Fig. 3). This relationship, in combination with AMF 13a and with the earlier mentioned interrelation of the moduli, enables us to make a fairly good estimation of the other moduli (G and E).

2. *Relationship between bulk modulus and thermal properties.* According to the *rule of Grüneisen-Tobolsky* [61,64], the following expression is valid for molecular crystals:

$$B\ \text{GN/m}^2 \approx 8.0\,\frac{\{E_{\text{subl}}\}}{\{V\}}\,\frac{J/\text{mol}}{m^3/\text{mol}}$$

For polymers (per structural unit) the analogous expression is

$$B\ \text{GN/m}^2 \approx 8.0\,\frac{\{E_{\text{coh}}\} + x_c\{\Delta H_m\}}{\{V\}}$$

Since $\{\Delta H_m\}$ can be calculated from $\{\Delta S_m\} = \{\Delta H_m/T_m\}$, it is possible to estimate B from two additive molar functions, $\{E_{\text{coh}}\}$ and $\{\Delta S_m\}$, whose increments are known (see Table 2).

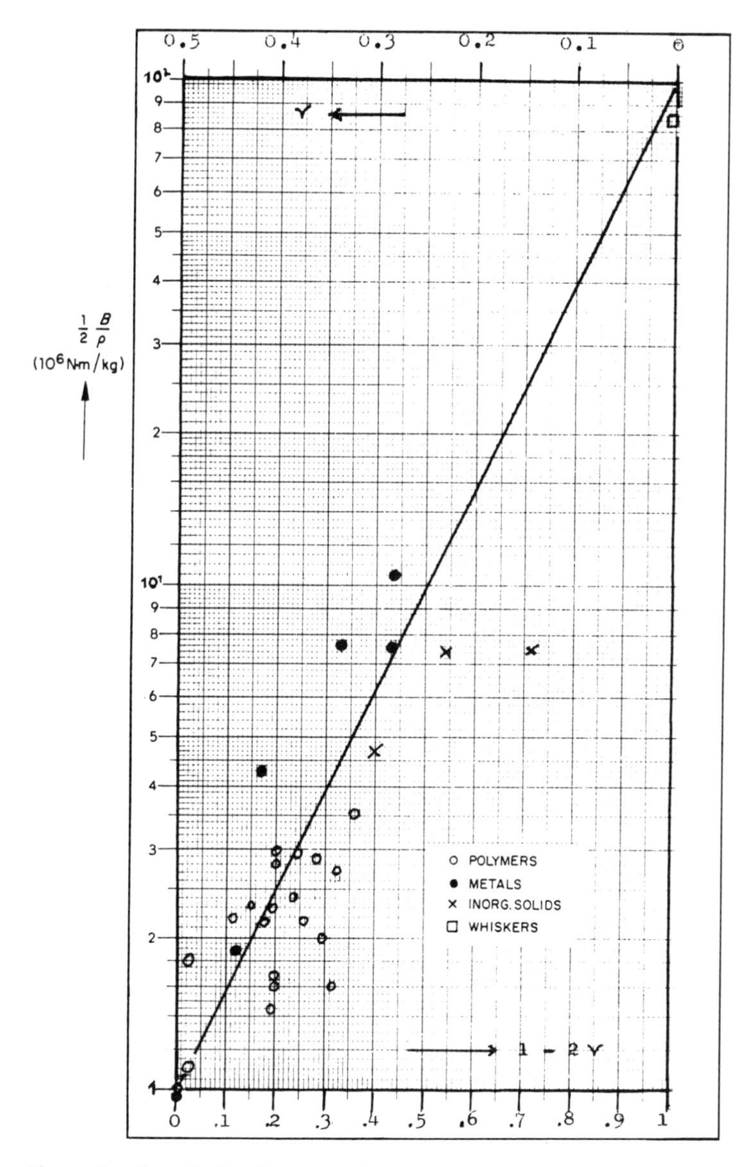

Figure 3 Correlation between the specific bulk modulus and the Poisson ratio of different materials.

3. *Relationship between shear modulus and glass-transition temperature.* A direct correlation between the shear modulus G and the two main transition temperatures was found by Van Krevelen [3]:

for $T_g < 298$ $\quad G_g(298) \approx 10^6\ \mathrm{N/m^2}$

for $T_g > 298$

$$\begin{cases} G_g(298) \approx \dfrac{T_g}{T_g/3 + 200} \quad 10^9\ \mathrm{N/m^2} \\ G_c(298) \approx \dfrac{T_m - 298}{100} \quad 10^9\ \mathrm{N/m^2} \\ G_{sc}(298) \approx G_g(298) + x_c^2\,(G_c(298) - G_g(298)) \end{cases}$$

where x_c = degree of crystallinity; the indices c and sc stand for crystalline and semicrystalline.

4. *Temperature dependence of the rigidity.* Van Krevelen and Hoftyzer found another useful empirical relationship for the temperature dependence of the moduli [3]: If information is available on the value of G (or E) at a certain reference temperature T_R (e.g., $T_R = 298$) the following equations are suitable for the temperature dependence of G and E [3]:

for *glassy polymers* and $T < T_g$:

$$\frac{G_g(T)}{G_g(T_R)} = \frac{E_g(T)}{E_g(T_R)} \approx \frac{T_g/T_R + 2}{T_g/T_R + 2\,T/T_R}$$

for *semicrystalline polymers* with $T_g \ll T_R$:

$$\log \frac{G(T)}{G(T_R)} = \log \frac{E(T)}{E(T_R)} \approx -1.15 \times \frac{T_m/T_R - T_m/T}{T_m/T_R - 1}$$

Viscoelastic Behavior

Polymers never behave as purely elastic bodies; they always show a combination of elastic and viscous properties: they are *viscoelastic*. Relaxations of volume, shape, stress, and creep (as a function of time) in solid polymers, and nonNewtonian flow and melt elasticity in liquid polymers, are typical viscoelastic phenomena.

In order to handle the behavior of viscoelastic solids mathematically, the basic equations of the moduli must be written as relationships of complex quantities:

$$E^* = 2G^*(1 + \nu^*) = 3B^*(1 - 2\nu^*)$$

where each of the moduli (M) can be written as

$$M^* = M' + iM''$$

Here M' is called *storage modulus* (= the "real" or purely elastic part of the complex modulus M^*) and M'' is the *loss modulus* (the "imaginary" or purely viscous part of M^*). The latter is directly related to a viscosity, e.g., $G'' = \omega_n$ with ω the frequency or reciprocal time.

The complex modulus is influenced by both temperature and time (the time during which the field of force acts or changes). Temperature and time proved to be "equivalent." A higher temperature is equivalent to a longer time, and vice versa. *Going from very low to very high frequencies has the same effect as going from low to high temperatures.*

The ratio M''/M' is called the *damping* or *dissipation factor*, or the *loss tangent* (tan δ); peaks in tan δ are found when characteristic transition points are passed. In this way one obtains a mechanical "spectrum" of the polymer. In every transition region there is a certain fall of the modulus, accompanied by a peak of the loss tangent.

As a first approximation the time dependence of the *moduli* (M) can be described by the two formulas:

$$M_{\text{rlx}}(t) \approx M_0 t^{-n} \qquad \text{and} \qquad M_{\text{rlx}}(t) \approx M_0 \exp\left(-\frac{t}{\theta_{\text{rlx}}}\right)$$

where $n = (\tan \delta)/(\pi/2) \approx 2/3$ and θ_{rlx} = relaxation time.

The temperature dependence of the moduli is described by the well-known WLF equation [63,64]:

$$-\log \frac{M}{M_R} = \frac{C_1(T - T_R)}{C_2 + (T - T_R)}$$

for temperatures between T_g and $T_g + 50$

Consequently the *time-temperature equivalence (or superposition) principle* can be formulated as a *shift factor* according to the expression

$$\log a_T = \log \frac{t}{t_R} = -\frac{C_1(T - T_R)}{C_2 + (T - T_R)}$$

for amorphous polymers (valid for $T_g < T < T_g + 50$)

The values of C_1 and C_2 are

Chosen reference temperature T_R	C_1	C_2
$T_R = T_g + 50$	8.86	101.6
$T_R = T_g$	17.44	51.6

Ultimate Mechanical Properties

The ultimate mechanical properties are very important in practice. They represent the maximum values which can be reached before collapse (e.g., ultimate strength, strain, torsion, compression, etc.). Also, hardness, scratch resistance, and "critical times" of resistance to continuous or cyclic deformation belong to this group of properties. Theoretical values of these properties are always higher than practical ones, due to the fact that the latter are determined to a high degree by imperfections and surface damage. Therefore only empirical rules are available. Some of these will be mentioned here.

Tensile Strength ($\hat{\sigma}_t$). The theoretical strength of polymers is about one-tenth of the modulus. In fact the actual strength is much lower (10 to 100 times lower!) due to imperfections. For unoriented polymers the following rule may be used as a first approximation [3]:

$$\hat{\sigma}_t \approx 30E^{2/3} \qquad (\hat{\sigma}_t \text{ and } E \text{ in N/m}^2)$$

This relationship is shown in Figure 4. This equation holds for the breaking strength of brittle (amorphous) polymers as well as for the yield strength of ductile (semicrystalline) polymers. In the first class the failure is due to surface imperfections and crack propagation; in the latter class it is due to sliding "flow" of the chains.

Ultimate Tensile Strain ($\hat{\epsilon}_t$). The elongation at failure (in %) may be approximated by the rule

$$\log \hat{\epsilon}_t \approx 4 - 14(1 - 2\nu)$$

Compression Strength ($\hat{\sigma}_c$). Northolt found an interesting relationship between compression strength and T_g [62]:

$$\log \hat{\sigma}_c \text{ [MPa]} \approx 2 \log T_g + 3$$

Strength Ratios. Two practical rules are available for the ratios between compressive and flexural strength and tensile strength [3]:

$$\log \frac{\hat{\sigma}_c}{\sigma_t} \approx 2.5\,(1 - 2\nu) - 0.3 \qquad \text{(see Fig. 5)}$$

$$\frac{\hat{\sigma}_{fl}}{\hat{\sigma}_t} \approx 1.7$$

Hardness. The indentation hardness of a polymer is closely related to Young's modulus, as shown in Figure 6. The drawn line obeys the equation

$$H_p = 10E^{3/4}$$

if both H_p and E are expressed in N/m^2.

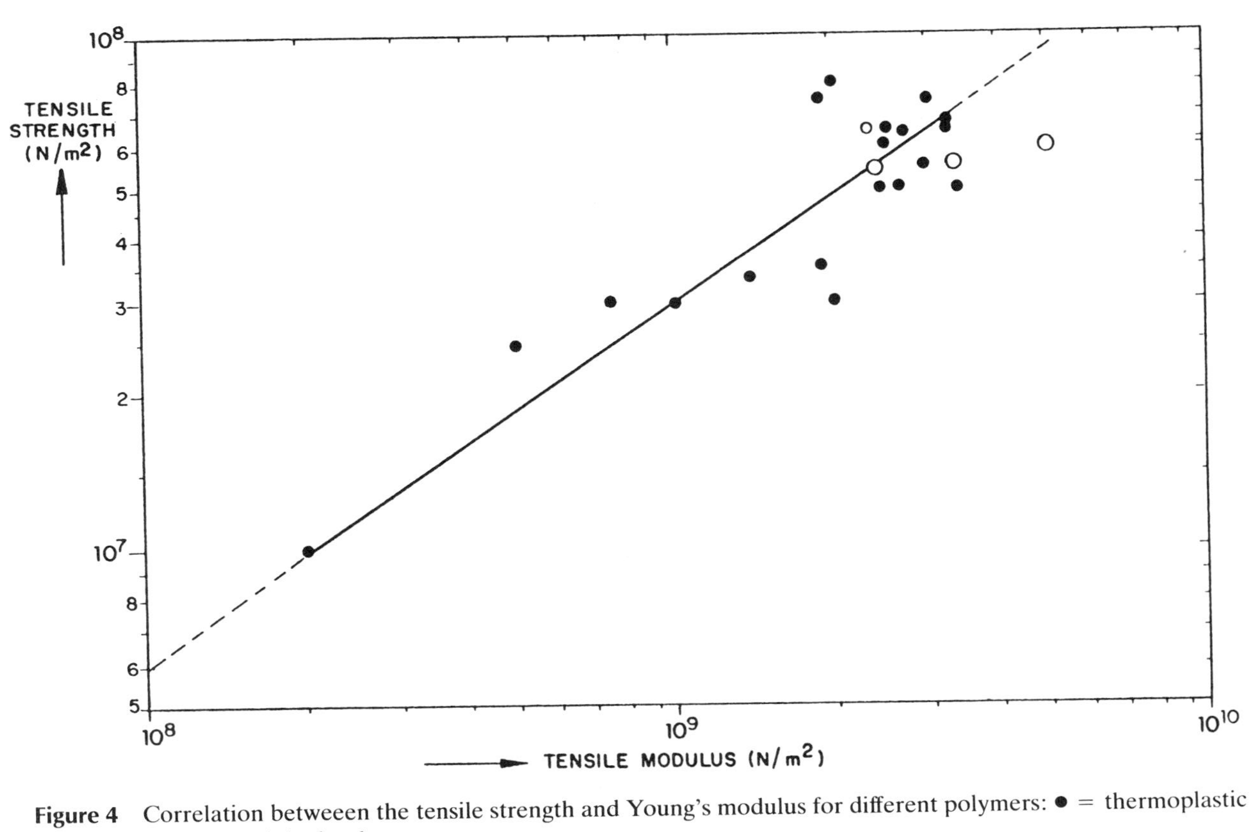

Figure 4 Correlation betweeen the tensile strength and Young's modulus for different polymers: ● = thermoplastic polymers, ○ = cross-linked polymers

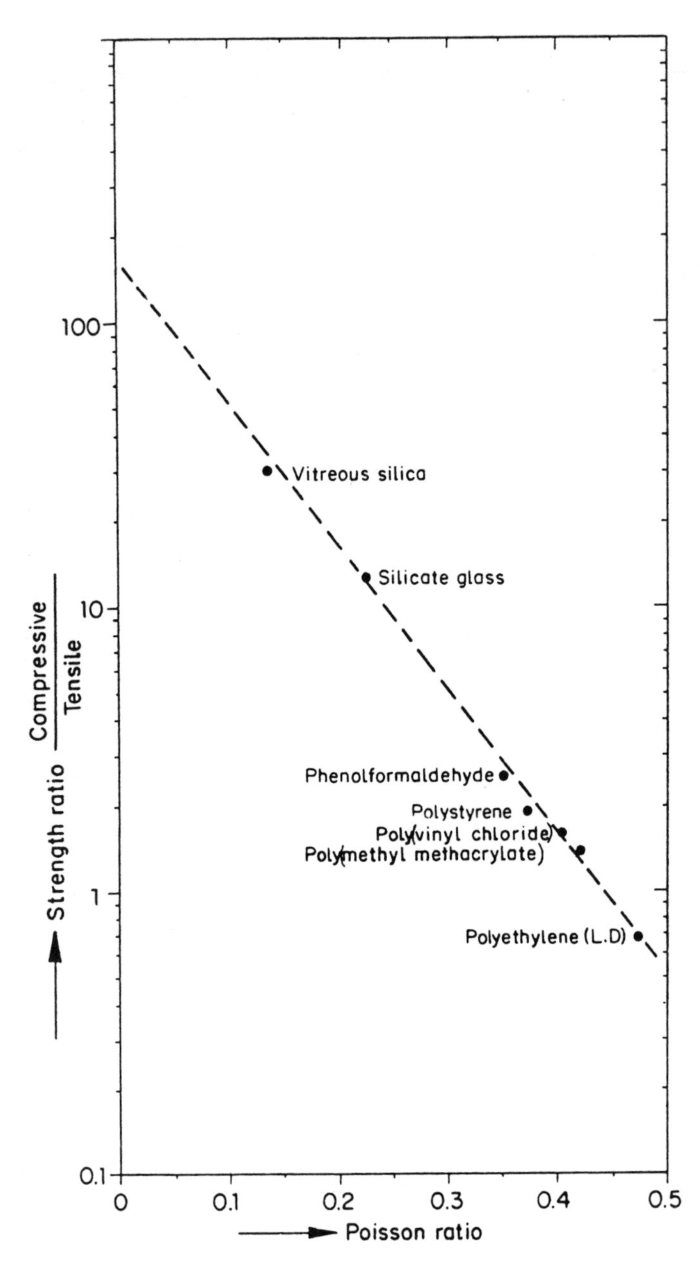

Figure 5 Correlation between the strength ratios (compressive/tensile) and the Poisson ratios of different materials.

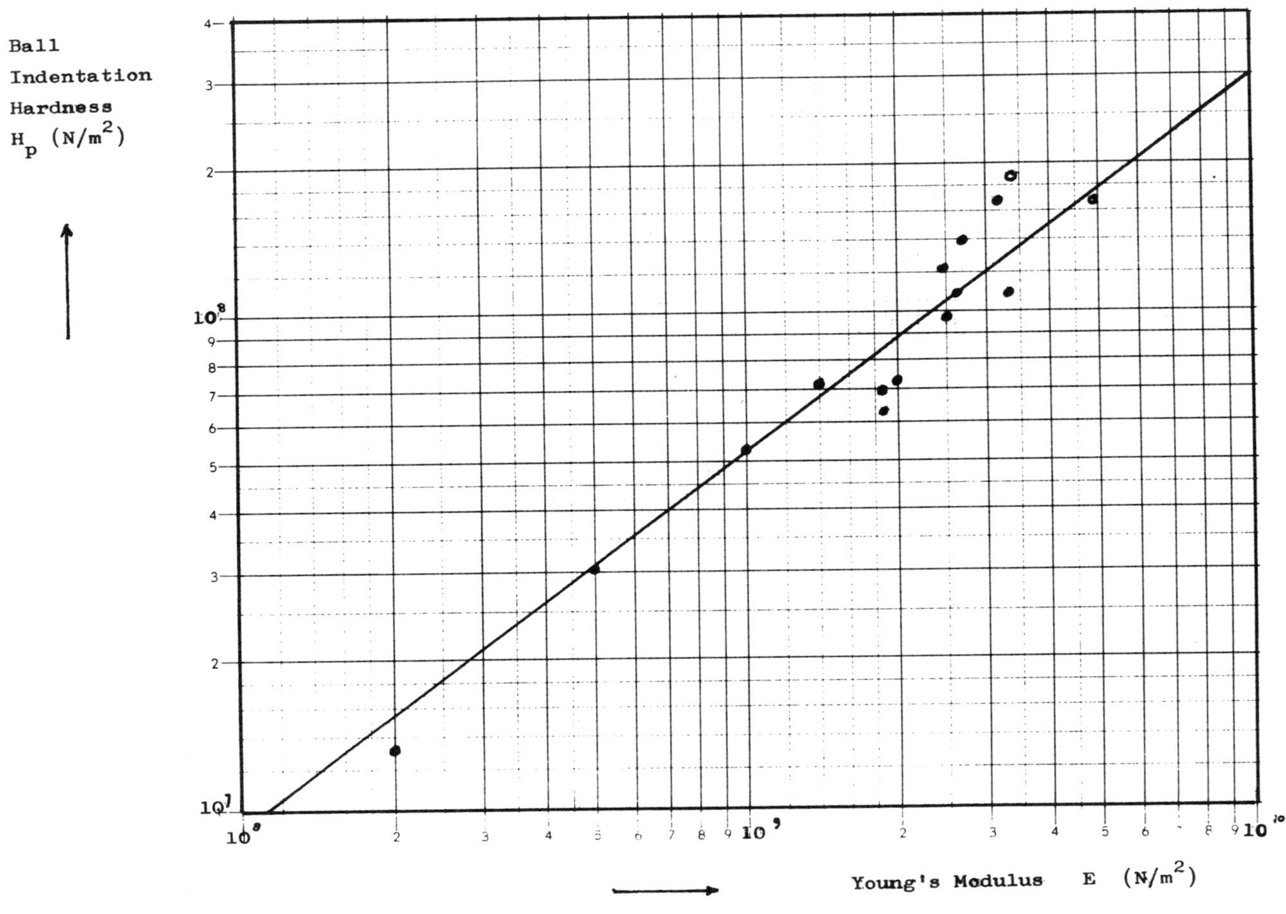

Figure 6 Correlation between the indentation hardness and Young's modulus of different polymers.

ADDITIVE MOLAR FUNCTIONS CONNECTED WITH MOLECULAR MOBILITY (VISCOMETRIC AND RHEOLOGICAL PROPERTIES OF POLYMERS)

The Molar Intrinsic Viscosity Function $\{\mathcal{J}\}$

$$AMF\ 14 \qquad \{\mathcal{J}\} = \left\{\sum_i N_i \mathcal{J}_i\right\} = K_\theta^{1/2}\{M\} - 4.2\,\{Z\}$$

The term K_θ is related to the *limiting viscosity number* $[\eta]$ (formerly called *intrinsic viscosity*) under *θ-conditions* of very diluted solutions, i.e., where the polymer molecules are in their *unperturbed state*. In these conditions the following equation is valid:

$$[\eta]_\theta = K_\theta \overline{M}_v^{1/2}$$

This is the famous *Mark-Houwink equation for θ-conditions*, connecting the limiting viscosity number with the average molecular mass (the "viscosity-averaged" M, which is nearly equal to the "weight-averaged" mass $\overline{M}_w$).

The additive molar function $\{\mathcal{J}\}$ was introduced by Van Krevelen and Hoftyzer [3,65]. The group contributions are given in Table 2.

Remarks.

1. *The physicochemical quantity* K_θ [66,67]. The quantity K_θ is extremely important for theoretical treatment of polymer solutions and mobility of polymer molecules. First of all it is directly connected to the *dimensions of the polymer coil* in its *unperturbed state* (dilute solutions and melts):

$$K_\theta = 6N_A\left(\frac{R_{G.\theta}}{M^{1/2}}\right)^3$$

where N_A = Avogadro number = $6.025 \times 10^{23}\ \text{mol}^{-1}$

$R_{G.\theta}$ = *radius of gyration* of the coil

In common diluted solutions the coil will expand beyond its unperturbed state due to interaction with the solvent:

$$R_g = R_{G.\theta}\alpha$$

where α is the linear expansion coefficient, a function of the nature of the solution. Second, K_θ is related to the critical molecular mass $\overline{M}_{cr}$ (see page 77), according to a rule found by Van Krevelen [3]:

$$K_\theta M_{cr}^{1/2} = 0.013\ \text{m}^3/\text{kg}$$

Third, K_θ is one of the principal determining factors of the general Mark-Houwink equation for arbitrary polymer solutions:

$$[\eta] = KM^a$$

where K is proportional (inter alia) to K_θ.

The Mark-Houwink Equation [68]

The general Mark-Houwink equation reads

$$[\eta] = K\overline{M}_v^a$$

where $[\eta] = \lim_{c\to 0}\left(\frac{n_{spec}}{c}\right) = \lim_{c\to 0}\left(\frac{\eta - \eta S}{\eta_S}\right)\frac{1}{c}$

K = Mark-Houwink "constant"

a = Mark-Houwink "exponent"

n_S = viscosity of the solvent

Both the exponent a and the coefficient K of the Mark-Houwink equation are correlated to the chemical structure and can be predicted from it (albeit indirectly), as will be shown.

The M-H Exponent a

The value of exponent a depends on the interaction of polymer and solvent, characterized by the difference of the solubility parameters of polymer (δ_p) and solvent (δ_S). The following rules have been found [3]:

for $|\delta_P - \delta_S| > 3, \quad a = 1/2$

for $|\delta_P - \delta_S| \leq 3, \quad a = 0.80 - 0.10(|\delta_P - \delta_S|)$

For theta-solutions a is always equal to 1/2. Since the δ_P value can be calculated from AMF 10 (and also that of δ_S), the correlation of a with the structure is laid.

Note. The M-H exponent a is also correlated to the expansion coefficient α (see page 110) and with the Huggins constant k_H (see page 110)

The M-H Coëfficient K

Theoretical treatment leads to the expression

$$[\eta] = [\eta]_\theta\alpha^3 = K_\theta\overline{M}_v^{1/2}\alpha^3$$

with α = (linear) expansion coefficient of the coil. This equation is universally valid, e.g., for the critical molecular mass:

$$[\eta]_{cr} = [\eta]_{\theta.cr}\alpha_{cr}^3 = K_\theta\overline{M}_{cr}^{1/2}\alpha_{cr}^3$$

This must be equal to the Mark-Houwink equation for the critical molecular mass:

$$[\eta]_{cr} = K\overline{M}_{cr}^{a}$$

so that

$$K\overline{M}_{cr}^{a} = K_{\theta}\overline{M}_{cr}^{1/2}\alpha_{cr}^{3}$$

or

$$K = \frac{K_{\theta}\alpha_{cr}^{3}}{\overline{M}_{cr}^{a-1/2}} \qquad [3]$$

Of the four variables in this equation, three can be calculated as described earlier (K_θ from AMF 13, $\overline{M}_{cr}$), and the exponent a from the equation on page 109). The fourth variable α_{cr} proved to be related to the exponent a, according to a semiempirical formula found by Hoftyzer and Van Krevelen [3]:

$$\log \alpha_{cr}^{3} \approx 13(a - 1/2)^{3}$$

So the M-H coefficient is fully and quantitatively related to the chemical structure of the polymer by means of the additive molar relationships (10 and 13).

The Viscosity of Dilute Polymer Solutions

For dilute polymer solutions Huggins [69] formulated the correlation

$$n_{sp} = [\eta]c + k_H[\eta]^2c^2$$

so that one obtains for the solution viscosity

$$\eta = \eta_S(1 + [\eta]c + k_H[\eta]^2c^2)$$

The solution viscosity is thus determined by $[\eta]$, n_S, c, and k_H, where k_H is the Huggins constant.

Van Krevelen [3] found the following relationship between k_H and the M-H exponent a:

$$k_H + a = 1.1$$

so that η of the solution can be calculated.

The Molar Meltiviscosity-temperature Function $\{H_\eta\}$

$$AMF\ 15 \qquad \{H_\eta\} = \sum_i N_i H_{\eta,i} = \{M\}(E_{\eta,\infty})^{1/2}$$

where $E_{\eta,\infty}$ is the *activation energy of viscous flow* at high temperature (i.e., well above $T = 2T_g$) and at low shear rate (i.e., under *Newtonian conditions*). This function was introduced by Van Krevelen and Hoftyzer [70], who demonstrated that it determines a universal dimensionless master graph from which the melt viscosity and the viscosity of concentrated solutions can be calculated (Fig. 7(b)).

Viscosity of Polymer Melts

The viscosity of polymer melts depends on three parameters: the average *molecular mass*, the *temperature* and the *shear rate*. At low shear rates the viscosity is *Newtonian* ($\eta_N(T,M)$).

1. The dependence of the average molecular mass is according to the, more or less, general rule [3, p. 336]

$$\log \eta \sim 3.4 \log M$$

or

$$\log \frac{\eta_N}{\eta_{N,\mathrm{cr}}} = 3.4 \log \frac{M}{M_{\mathrm{cr}}}$$

(This means that polymers with $M = M_{\mathrm{cr}}$ form a kind of reference.)

2. The influence of the temperature is more complicated. In the temperature region between T_g and $T_g + 100$ the well-known Vogel-WLF equation [63] holds for amorphous polymers (WLF refers to Williams, Landell, and Ferry [63,64]):

$$\log \frac{\eta_N(\mathrm{T})}{\eta_N(T_g)} = \log \frac{\eta_{\mathrm{N,cr}}(\mathrm{T})}{\eta_{\mathrm{N,cr}}(T_g)} = -\frac{17.4(T - T_g)}{51.6 + (T - T_g)}$$

where

$$\eta_{N,\mathrm{cr}}(T_g) \approx 10^{12}\ \mathrm{N \cdot s/m^2}$$

$$\eta N(T_g) = 10^{12} + 3.4 \log \frac{M}{M_{\mathrm{cr}}}$$

For semicrystalline polymers and also for amorphous polymers at temperatures $> 2T_g$ an Arrhenius type equation gives a much better approximation to the experimental data:

$$\log \eta_N(\mathrm{T}) = \log \eta_{N,\infty} - \frac{E_\eta}{2.3T}$$

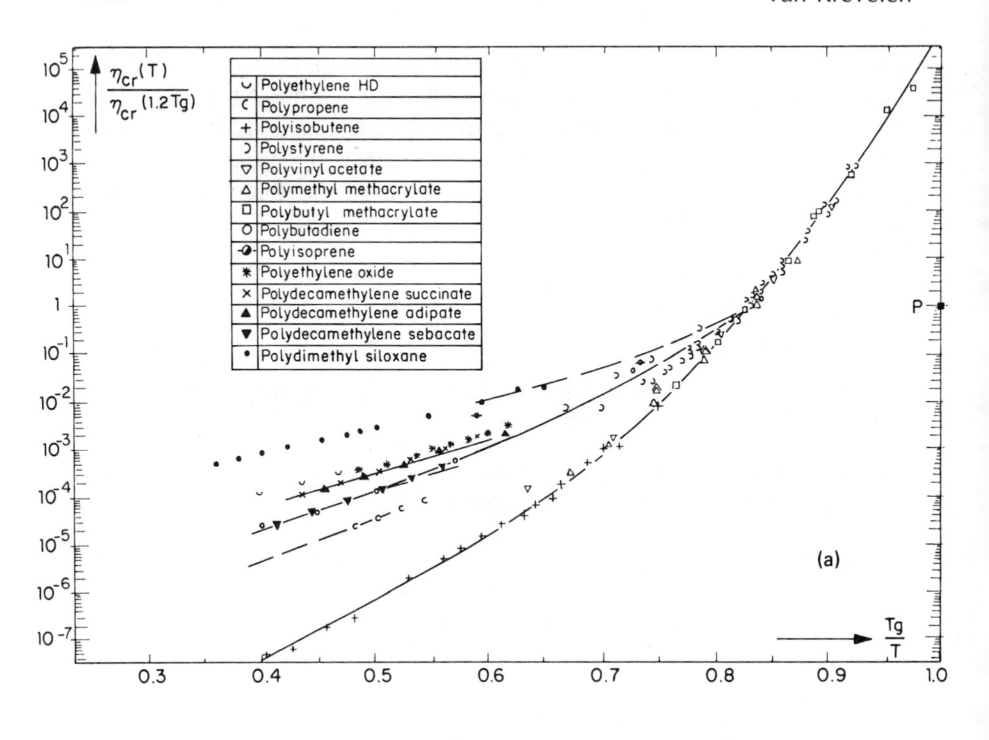
$\eta_{cr}(T)$ / $\eta_{cr}(1.2\,Tg)$
Polyethylene HD
Polypropene
Polyisobutene
Polystyrene
Polyvinyl acetate
Polymethyl methacrylate
Polybutyl methacrylate
Polybutadiene
Polyisoprene
Polyethylene oxide
Polydecamethylene succinate
Polydecamethylene adipate
Polydecamethylene sebacate
Polydimethyl siloxane
P
(a)
Tg/T
10^5 10^4 10^3 10^2 10^1 1 10^{-1} 10^{-2} 10^{-3} 10^{-4} 10^{-5} 10^{-6} 10^{-7}
0.3 0.4 0.5 0.6 0.7 0.8 0.9 1.0

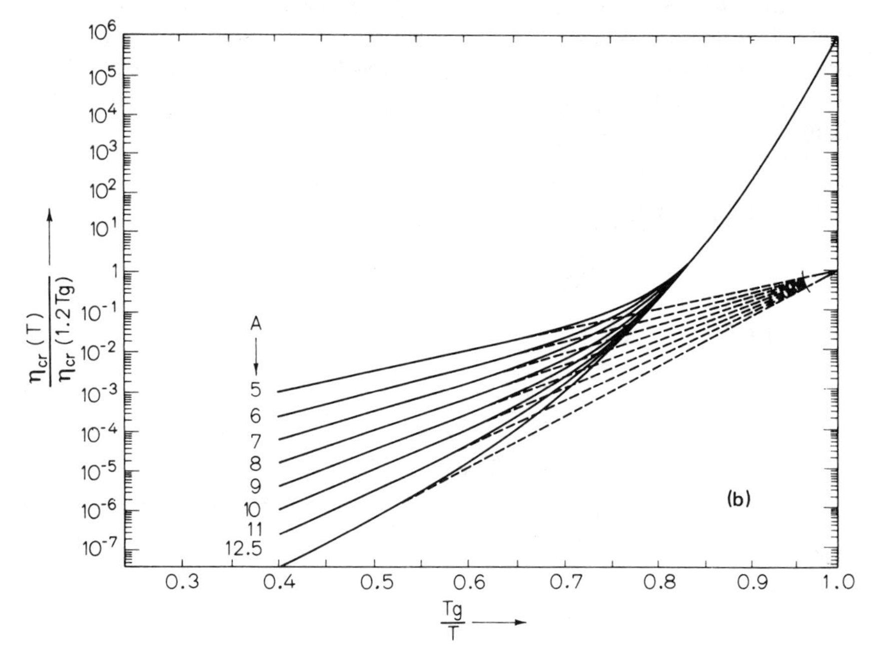
$\eta_{cr}(T)$ / $\eta_{cr}(1.2\,Tg)$
A
5
6
7
8
9
10
11
12.5
(b)
Tg/T
10^6 10^5 10^4 10^3 10^2 10^1 1 10^{-1} 10^{-2} 10^{-3} 10^{-4} 10^{-5} 10^{-6} 10^{-7}
0.3 0.4 0.5 0.6 0.7 0.8 0.9 1.0

E_η has a characteristic value depending on the chemical structure of the polymer; it can be calculated from the additive molar function $\{H_\eta\}$ (AMF 15).

3. Van Krevelen and Hoftyzer [70] derived a dimensionless master diagram which correlates all available experimental data on melt viscosities for the whole temperature scale of the melt. In this diagram all η_N-T_g/T curves coincide in the WLF region, whereas for temperatures $>2T_g$ they form a bundle of diverging straight lines whose slope is determined by E_η (Fig. 7(a) and 7(b)). The full expression in the Arrhenius region ($T > 2T_g$) is

$$\log \eta_N(T) = \frac{E_\eta}{2.3RT_g}\left(1 - \frac{T}{623}\right) + 3.4 \log \frac{M}{M_{cr}} - 1.4$$

4. For *high shear rates* the viscosity decreases, due to molecular orientation of the melt. This increase may amount to several decades; thanks to this phenomenon polymer melts are spinnable. The new parameter that describes the shear rate is the dimensionless group of Weissenberg, $\dot{\gamma}\,\theta_M$, where

$$\dot{\gamma} = \text{shear rate}$$

$$\theta_M = \frac{3}{5}\frac{\eta_N M_w}{\delta RT} Q = \text{characteristic deformation time}$$

$$Q = \frac{m_w}{M_n} \qquad \text{(see page 000)}$$

The viscosity of the melt thus becomes [3]

$$\text{for } \dot{\gamma}\,\theta_M < 1 \qquad \log \eta(T) = \log \eta_N(T)$$

$$\text{for } \dot{\gamma}\,\theta_M > 10^2 \qquad \log \eta(T) = \log \eta_N(T) - 0.75 \log \dot{\gamma}\theta_M + 0.5$$

The region $1 < \dot{\gamma}\theta_M < 10^2$ is a transition region.

5. *Extensional viscosity of polymer melts* (λ). The extensional viscosity is the ratio

$$\frac{\text{tensile stress } \sigma}{\text{rate of extension } \dot{\epsilon}}$$

Figure 7 Graphical correlation between the logarithm of the reduced viscosity ($\eta_{red} = \eta_{cr}(T)/\eta_{cr}\,(1.2T_g)$) and the reduced reciprocal temperature ($1/T_{red} = T_g/T$) for a series of widely different polymers. η_{cr} is the viscosity, extrapolated to the critical molecular mass M_{cr} (see page 76): (a) experimental data; (b) dimensionless master graph. The parameter A is proportional to the activation energy of viscous flow at sufficiently high temperature (the Arrhenius region: $A = E/2.3RT_g$).

Under Newtonian conditions a simple relationship exists between λ_N and η_N, namely $\lambda_N = 3\eta_N$.

Measurements of rheological quantities in tensile deformation of polymer melts are extremely difficult; they require the development of very special apparatus and techniques (see, e.g., Meissner [72]). At low extensions the values of λ/λ_N as a function of θ_M/t are the same as those of η/η_N as a function of $\dot{\gamma}\theta_M$; with increasing extension, however, the extensional viscosity increases considerably, due to molecular orientation. In contradistinction to the shear viscosity, the extensional viscosity not only depends on the rate of deformation but also on the *amount of deformation*. The experimental data are too scarce to provide general relationships.

Viscosities of Concentrated Polymer Solutions

Solutions with a polymer concentration larger than 5% are considered "concentrated." Their behavior is much more complicated than that of dilute solutions: an extra parameter plays an important part, the *concentration*. Very concentrated solutions behave as diluted melts. This implies first of all a dilution effect and furthermore a lowering of the T_g of the system ("plasticizer effect" of the solvent). The viscosity of a concentrated solution at a constant temperature obeys a power law:

$$\eta \approx Kc^{\alpha}\overline{M}^{\beta} \qquad \text{where } \alpha = 5.1 \quad \text{and} \quad \beta = 3.4 \quad \text{(as in melts)}$$

so that

$$\eta \approx (c^{1.5}\overline{M})^{3.4} \approx (c\overline{M}^{0.67})^{5.1}$$

The first expression is similar to that of the melts, with the product $c^{1.5}\overline{M}$ in the same role as $\overline{M}$ in the melt viscosity. Here also a critical value of $\overline{M}c^{1.5}$ exists, below which the exponent 3.4 changes to 1.

A calculation method for estimating the viscosity of concentrated solution was developed by Hoftyzer and Van Krevelen [3,73]. For details refer to the original paper.

ADDITIVE MOLAR FUNCTIONS CONNECTED WITH ELECTROMAGNETIC PHENOMENA

The Molar Dielectric Polarization Function $\{P\}$

$$AMF\ 16 \qquad \{P\} = \left\{\sum_i N_i P_i\right\} = \left\{V\frac{\epsilon - 1}{\epsilon + 2}\right\}$$

This function was introduced by Lorentz [7].

Remarks.

1. The *dielectric constant* or *electric inductive capacity* (ϵ) is one of the most important properties of diëlectric substances. It is directly related to the capacity of an electric condensor. It is also related to the electrical resistivity (reciprocal conductivity):

$$\log R\ [\Omega \cdot \text{cm}] = 23 - 2\epsilon$$

A qualitative relationship exists between δ and the *static electrification*: the sequence of polymers according to their electric charging behavior (the *triboelectric series*) is the same as the sequence of the dielectric constants [76].

2. ϵ is also related to the *cohesive energy density*:

$$\epsilon = \frac{1}{7} e_{\text{coh}}^{1/2} \qquad (e_{\text{coh}} \text{ in J cm}^{-3})$$

3. The dielectric behavior of a polymer in an oscillating electric field is similar to its dynamic-mechanical behavior under oscillating stresses. In both cases the delay between changes in the force field and changes in its effect (polarization) can be expressed as a *loss angle* (δ); tan δ is again called *damping*.

In analogy to the mechnical moduli also ϵ can be expressed as a complex quantity:

$$\epsilon^* = \epsilon' - i\epsilon''$$

where ϵ' = *real* dielectric constant
ϵ'' = dielectric absorptivity

The Molar Optical Refraction Function {*R*}

AMF 17 $\quad \{R_{GD}\} = \left\{\sum_i N_i R_{GD,i}\right\} = \{V\}(n - 1)$

AMF 17a $\quad \{R_{LL}\} = \left\{\sum_i N_i R_{LL,i}\right\} = \{V\}\dfrac{n^2 - 1}{n^2 + 2}$

AMF 17b $\quad \{R_{Lo}\} = \left\{\sum_i N_i R_{Lo,i}\right\} = \{V\}(N^{2/3} - 1)$

Function AMF 17 was introduced by Gladstone and Dale (1858) [6]. Lorentz [7] and, independently, Lorenz [8] derived AMF 17a from the electromagnetic theory of light. Looyenga [10] proved 17b to be an almost exact approximation to 17a and much easier to use. For polymers AMF 17 is preferred because of its simplicity, its temperature independence, and the availability of a detailed list of group contributions for polymers (Goedhart [74]). The increments are given in Table 2.

Remarks.

1. In the electromagnetic theory of light the complex dielectric constant is equal to the square of the complex refractive index:

$$\epsilon^* = (n^*)^2 = \epsilon' - i\epsilon''$$

$$\epsilon' = n^2 - n^2K^2$$

$$\epsilon'' = 2n^2K^2$$

Here K is the well-known *absorption coëfficient*; as a function of the wavelength, it is the characteristic "fingerprint" of a substance.

2. Related to the refractive index (n) is the *specific refraction index increment*, dn/dc, important in light scattering studies. As has been shown by Goedhart [74], it is possible to estimate dn/dc rather accurately from the simple equation

$$\frac{dn}{dc} = \frac{n_P - n_S}{\rho_P}$$

where n_S is the refractive index of the solvent, and n_P and δ_P are the refractive index and the density of the polymer. Since the latter quantitites can be calculated from $\{M\}$, $\{V\}$, and $\{R\}$, dn/dc can be estimated.

3. Another quantity related to n is the *light reflectance r*. The reflectance on a boundary plane between two media is a function of the refractive indices of these media. If the light strikes the boundary plane perpendicularly, the reflectance will be given by the Fresnel-Beer relationship [75]

$$r = \frac{(n_2 - n_1)^2 + (n_2K_2)^2}{(n_2 + n_1)^2 + (n_2K_2)^2} \approx \left(\frac{n_2 - n_1}{n_2 + n_1}\right)^2$$

where the indices 1 and 2 denote the two media; in this case 1 = 1 air, 2 = polymer.

The Molar Magnetic Susceptibility Function $\{X\}$

$$AMF\ 18 \qquad \{X\} = \sum_i N_i X_i = \{M\}\chi$$

This function was introduced by Henricksen (1888) and Pascal [79,80]. Important work in this field has been done by Dorfmann [77] and Haberditzl [78]. Group contributions can be found in Table 2.

ADDITIVE MOLAR FUNCTIONS CONNECTED WITH THE THERMAL STABILITY AND DECOMPOSITION OF POLYMERS

The Molar Free Enthalpy of Formation Function $\{\Delta G_f^\circ\}$

$$AMF\ 19 \qquad \{\Delta G_f^\circ\} = \left\{\sum_i N_i \, \Delta G_{f,i}^\circ\right\} = \Delta H_f - T \, \Delta S_f^\circ$$

The free enthalpy of formation (formerly called free energy of formation) is a measure of the relative thermal stability of a chemical compound compared with its elements. The more negative the value of ΔG_f°, the more stable is the compound. The value of ΔG_f° of the elements is zero by definition. The degree symbol (°) represents the standard state, namely, the free molecule. The first introduction of $\{\Delta G_f^\circ\}$ as an additive molar quantity was by Anderson et al. [81], followed by Bremner and Thomas [82], Souders et al. [83], and Franklin [84]. The most elaborate system of increments was developed by Van Krevelen and Chermin [85]. A comprehensive survey was given by Janz [86].

Remarks.

1. ΔG_f° is a nearly linear function of the temperature in the interval 300–600 K,:

$$\{\Delta G_f^\circ\} = \{A\} + \{B\}T$$

According to Ulich [87], $A \approx \Delta H_f^\circ$ (298) and B $\approx \Delta S_f^\circ$ (298).

2. Both A and B are additive:

$$\{A\} = \left\{\sum N_i A_i\right\} \qquad \text{and} \qquad \{B\} = \left\{\sum_i N_i B_i\right\}$$

The increments of A [kJ/mol] and B [kJ/mol · K] are given in Table 2.

3. As mentioned, all values are standardized for the ideal state of the free molecule (fugacity = 1). If the reactants/reaction products are not in the standard state, corrections have to be made to obtain the values under "real" conditions. For comparisons of stability, however, the standard state may be used.

4. For any chemical reaction $\Sigma_n\ n_R R \ 1\ \Sigma_n\ n_P P$, the total change in free enthalpy can be calculated as the difference in free enthalpy of the products (P) formed and the reactants (R) considered:

$$\Delta G_f^\circ \text{ (reaction)} = \sum n_P\{\Delta G_{f,P}^\circ\} - \sum n_R\{\Delta G_{f,R}^\circ\}$$

5. In many reactions a simple gas is formed as one of the reaction products, so that for the calculation it is necessary to know its free enthalpy of formation.

For a number of frequently occurring reactions where simple gases are formed, the following ΔG_f° values may be used:

Reaction	Gas formed	ΔG_f° of gas [kJ/mol]
Dehydrogenation	H_2	0
Dehydration	H_2O	$-243 + 0.0482T$
Demethanation	CH_4	$-79 + 0.0925T$
Decarbonylation	CO	$-111 - 0.090T$
Decarboxylation	CO_2	$-395 - 0.002T$
Deammoniation	NH_3	$-48 + 0.017T$
Dehydrochlorination	HCl	$-93 - 0.009T$
Dehydrofluorination	HF	$-270 - 0.006T$

The Molar Thermal Decomposition Function $\{Y_{d,1/2}\}$

$$AMF\ 20 \qquad \{Y_{d,1/2}\} = \left\{\sum_i N_i Y_{d,1/2,i}\right\} = \{M\}T_{d,1/2}$$

where $T_{d,1/2}$ means the temperature at which the polymer has lost half of its potential "volatiles" during heating under standardized conditions (in vacuo, during 30 min heating). The quantity $T_{d,1/2}$ was introduced by Madorsky and Straus in their pioneering work [88,89], and was used by Korshak and his school [90]. The function $Y_{d,1/2}$ is introduced in analogy with Y_g [3]. The group contributions are given in Table 2.

Remarks. The thermal decomposition process (also called *pyrolysis*) is characterized by four main parameters:

1. The *temperature of initial decomposition* ($T_{d,0}$)
2. The *temperature of "half decomposition"* ($T_{d,1/2}$)
3. The temperature of *maximum rate of decomposition* (loss in weight) at a standard rate of heating, e.g., 2°C/min ($T_{d,\max}$)
4. The *average energy of activation* ($E_{d,\mathrm{act}}$), if the thermal decomposition is considered as a quasi-first-order process (with the fractional loss in weight as variable)

These four parameters are *interrelated*:

$$T_{d,0} \approx 0.9T_{d,1/2}$$

$$T_{d,\max} \approx T_{d,1/2}$$

$$E_{d,\mathrm{act}}\ [\mathrm{kJ}] \approx C_d T_{d,1/2}\ \mathrm{K}$$

The value of C_d is 0.4 if the main chain consists of carbon atoms; heteroatoms lower the value of C_d. In the case of nitrogen atoms in the main chain it is 0.28; with oxygen in the backbone it may be as low as 0.2.

The Molar Char Forming Tendency Function $\{C_{FT}\}$

$$AMF\ 21 \qquad \{C_{FT}\} = \sum_i N_i C_{FT,i} = \{M\}\left(\frac{CR}{12}\cdot\frac{1}{100}\right)$$

where CR is the char (coke) residue in percent of the original weight of the polymer after heating to a standard end temperature (normally 900°C). So, according to the definition, C_{FT} is the amount of char, expressed as carbon equivalents, per structural unit of the polymer.

The $\{C_{FT}\}$ function was introduced by Van Krevelen [91]. The set of increments was derived from the results of pyrolysis experiments with about 100 polymers. It is given in Table 2. It is restricted to nonhalogen-containing polymers.

From AMF 21 the char residue on pyrolysis can be estimated for arbitrary (halogen-free) polymers.

The average deviation in the experimental values for the polymers investigated is 3.5%.

Remarks.

1. The C_{FT} function is based on the assumption that each of the structural groups of the polymer contributes to the char in a characteristic way. Aliphatic hydrocarbon groups do not contribute to the char, but are completely volatilized; they may even have a negative contribution, especially if they are rich in hydrogen. In this case they supply hydrogen to decomposing aromatic groups and so enhance the overall volatilization. The reason for that is the well-known hydrogen shift in disproportionation reactions. Aliphatic groups containing oxygen or halogen will contribute to the char, since their hydrogen is consumed for water and hydrohalogenic acid formation.
2. The $\{C_{FT}\}_i$ increments are not valid for polymers which contain halogens, since the latter are "built in" soot formers (flame retardants) by secondary reactions.

Flammability of Polymers; The Oxygen Index (O.I.)

The $\{C_{FT}\}$ function is also important for the estimation of another useful property of polymers: their flame resistance, or its reverse, flammability. Van Krevelen [91, 92] found a close relationship between the amount of char residue on pyrolysis and the *oxygen index*, a parameter introduced by Fenimore and Martin [93]. The O.I. is defined as the minimum fraction of oxygen in a test atmosphere which just supports the combustion (after ignition); the test is carried out under standardized conditions. The correlation reads

$$\text{O.I.} \times 100 \approx 17.5 + 0.4\text{CR}$$

Like the $\{C_{TF}\}$ function it can only be used for polymers that have no halogens. A polymer is considered non-flammable if its O.I. value is 0.295 or higher.

For polymers with a certain halogen content a direct correlation exists between the oxygen index and the following *composition parameter* P(E.C.) [92]:

$$P(\text{E.C.}) = \frac{\text{H}}{\text{C}} - 0.65\left(\frac{\text{F}}{\text{C}}\right)^{1/3} - 1.1\left(\frac{\text{Cl}}{\text{C}}\right)^{1/3}$$

where H/C, etc., are atomic ratios of the *e*lemental *c*omposition.

The correlation with the O.I. is the following:

If $P(\text{E.C.}) \geq 1$, O.I. ≈ 0.175.

If $P(\text{E.C.}) \leq 1$, O.I. $\approx 0.60 - 0.425P(\text{E.C.})$.

REFERENCES

1. R. C. Reid, J. M. Prausnitz, and Th. K. Sherwood, *The Properties of Gases and Liquids*, 1st ed., 3rd ed., McGraw-Hill, New York (1958, 1977).
2. A. Bondi, *Physical Properties of Molecular Crystals, Liquids and Glasses*, Wiley-Interscience, New York (1968).
3. D. W. Van Krevelen, *Properties of Polymers; Their Estimation and Correlation with Chemical Structure*, 1st ed., 2nd ed., Elsevier, Amsterdam (1972, 1976).
4. A. A. Askadskii and Y. I. Matveyev, *Chemical Structure and Physical Properties of Polymers*, Moscow (1983). (in Russian) Askadskii published an extract of this book, entitled *Prediction of Physical Properties of Polymers* in the *Polymer Yearbook* IV, pp. 93–147 (R. A. Pethrick and G. E. Zaikov, eds.), Harwood Academic, London (1987).
5. K. Van Nes and W. A. Van Westen, *Aspects of the Constitution of Mineral Oils*, Elsevier, Amsterdam, (1951), D. W. Van Krevelen, *Coal; Typology, Chemistry, Physics and Constitution* Elsevier, Amsterdam (1961); 2nd printing (1981).
6. J. H. Gladstone, and T. P. Dale *Trans. Roy. Soc. (London)*, *A148*, 887 (1858); *A153*, 317 (1863).
7. H. A. Lorentz, *Ann. Phys.*, *9*, 641 (1880).
8. L. V. Lorenz, *Ann. Phys.*, *11*, 70 (1880).
9. A. Vogel, *Chem. & Ind.*, *1950*, 358; *1951*, 376; *1952*, 514; *1953*, 19; *1954*, 1045.
10. H. Looyenga, *Mol. Phys.*, *9*, 501 (1965); *J. Pol. Sci. Pol. Phys. Ed.*, *11*, 1331 (1973).
11. H. G. Weyland, P. J. Hoftyzer, and D. W. Van Krevelen, *Polymer*, *11* 79 (1970).
12. G. L. Slonimskii, A. A. Askadskii, and A. I. Kitaigorodski, *Vysokomolekularnie Soyedinenia*, *12*, 494 (1970).

13. R. A. Pethrick, in *Polymer Yearbook*, Vol. 3, p. 129. (R. A. Pethrick and G. E. Zaikov, eds.), Harwood, London (1986).
14. D. W. Van Krevelen, D. J. Goedhart, and P. J. Hoftyzer, *Polymer*, *18*, 750 (1977).
15. J. M. Barton, and W. A. Lee, *Polymer*, *9*, 602 (1968).
16. U. T. Kreibich, and H. Batzer, *Angew. Makromol. Chem.*, *83*, 57 (1979); *105*, 113 (1982).
17. R. Becker, u. Faserforschung *Textiltechn.*, *29*, 361 (1973).
18. R. A. Hayes, *J. Appl. Polym. Sci.*, *5*, 318 (1961).
19. W. E. Wolstenholme, *Polym. Eng. Sci.*, *8*, 142 (1968).
20. A. J. Staverman, *Rheologica Acta*, *5*, 283 (1966).
21. H. Breuer, and G. Rehage, *Kolloid Z.*, *216/17*, 1159 (1967).
22. R. F. Boyer, *Macromolecules*, *6*, 288 (1973).
23. R. F. Boyer, *2nd Int. Conf. Phys. Chem.*, *Paris*, June 6 (1952).
24. R. G. Beaman, *J. Polym. Sci.*, *9*, 472 (1953).
25. C. W. Bunn, (Chap. 12,) in *Fibres from Synthetic Polymers*, (R. Hill, ed.), Elsevier, Amsterdam (1953).
26. W. A. Lee and G. J. Knight, *Brit. Polymer*, *J.*, *2*, 73 (1970).
27. R. F. Boyer, *J. Polym. Sci. Symp.* No. *50*, 189 (1975); *Polymer Yearbook*, Vol. 2, pp. 233–243 (R. A. Pethrick, ed.), London (1985).
28. J. Traube, *Ber. Deutsch. Chem. Ges.*, *28*, 2722 (1895).
29. H. G. Davis, and S. Gottlieb, *Fuel*, *42*, 37 (1963).
30. E. K. Harrison, *Fuel*, *44*, 339 (1965); *45*, 397 (1966).
31. A. E. Rheineck, and K. F. Linn, *J. Paint Techn.*, *40*, 611 (1968).
32. D. W. Van Krevelen, and P. J. Hoftyzer, *J. Appl. Polym. Sci.*, *13*, 871 (1969).
33. R. F. Fedors, *Polym. Eng. Sci.*, *14*, 147, 472 (1974).
34. S. Satoh, *J. Sci. Res. Inst. Tokyo*, *43*, 79 (1948).
35. R. Shaw, *J. Chem. Eng. Data*, *14*, 461 (1969).
36. B. Wunderlich, *J. Phys. Chem.*, *64*, 1062, (1960).
37. B. Wunderlich, and L. D. Jones, *J. Macromol. Sci. Phys. B*, *3*, 67 (1969).
38. M. Dunkel, *Z. Phys. Chem. A*, *138*, 42 (1928).
39. R. A. Hayes, *J. Appl. Polym. Sci.*, *5*, 318 (1961).
40. P. J. Hoftyzer, and D. W. Van Krevelen, *JUPAC Pol. Conf.* 1970, Leyden, Book of Prepr. IIIa, 15.
41. R. F. Fedors, *Polym. Eng. Sci.*, *14*, 147, 472 (1974).
42. P. A. Small, *J. Appl. Chem.*, *3*, 71 (1953).
43. K. L. Hoy, *J. Paint Techn.*, *42*, 76 (1970).
44. J. H. Hildebrand, and R. L. Scott, *The Solubility of Non-Electrolytes*, 1st ed., 3rd ed., Van Nostrand Rinehold, New York (1936, 1949).
45. A. Beerbower, L. A. Kaye, and D. A. Pattison, *Chem. Eng.*, Dec. 18, 118 (1967).
46. J. D. Cromley, G. S. Teague, J. W. Lowe, *J. Paint Techn.*, *38*, 269 (1966), *39*, 19 (1967).
47. C. M. Hansen, *J. Paint Techn.*, *39*, 104, 511 (1967); *Ind. Eng. Chem. Prod. Res. Dev.*, *8*, 2 (1969).

48. S. Sugden, *J. Chem. Soc.*, *125*, 1177 (1924); *The Parachor and Valency*, Routledge, London (1930).
49. S. A. Mumford, and J. W. C. Phillips, *J. Chem. Soc.*, *130*, 2122 (1929).
50. O. R. Quayle, *Chem. Rev.*, *53*, 439 (1953).
51. W. A. Zisman, in *Contact Angle, Wettability and Adhesion*, Adv. Chem. Ser., Vol. 43, pp. 1–51, Am. Chem. Soc. (1964).
52. L. A. Girifalco, and R. J. Good, *J. Phys. Chem.*, *61*, 904 (1957); *62*, 1418 (1958); *64*, 561 (1960).
53. F. M. Fowkes, in *Contact Angle, Wettability and Adhesion*, Adv. Chem. Ser., Vol. 43, pp. 108–110, Am. Chem. Soc. (1964).
54. D. K. Owens, *J. Appl. Polym. Sci.*, *14*, 1725 (1970).
56. M. Salame, *Polym. Eng and Sci.*, *26*, 1543 (1986).
57. G. J. Van Amerongen, *J. Polym. Sci.*, *2*, 381 (1947); *5*, 307 (1950); *Rubber Chem. Techn. 37*, 1065 (1964).
58. V. Stannett, in *Diffusion in Polymers*, Chap. 2 (J. Crank, and G. S. Parks, eds.), Academic Press, New York (1968).
59. R. Rao, *Indian J. Phys.*, *14*, 109 (1940); *J. Chem. Phys.*, *9*, 682 (1941).
60. J. Schuyer, *Nature*, *181*, 1394 (1958); *J. Polym. Sci.*, *36*, 475 (1959).
61. E. Grüneisen in *Handbuch der Physik*, Vol. 10, Springer, Berlin (1926).
62. M. G. Northolt, *J. Mater. Sci.*, *16*, 2025 (1981).
63. M. L. Williams, R. F. Landel, and J. D. Ferry, *J. Am. Chem. Soc.*, *77*, 3701 (1955).
64. A. V. Tobolski, *Properties and Structure of Polymers*, Wiley, New York (1960).
65. D. W. Van Krevelen, and P. J. Hoftyzer, *J. Appl. Polym. Sci.*, *11*, 2189 (1967).
66. H. G. Elias, *Makromoleküle*, Hüthig and Wepf, Basel (1971).
67. M. Kurata, and W. H. Stockmeyer, *Fortschr. Hochpolym. Forschung*, *3*, 196 (1963).
68. D. W. Van Krevelen, and P. J. Hoftyzer, *J. Appl. Polym. Sci.*, *10*, 1331 (1966); *11*, 1409 (1967); *11*, 2189 (1967).
69. M. L. Huggins, *J. Am. Chem. Soc.*, *64*, 2716 (1942).
70. D. W. Van Krevelen, and P. J. Hoftyzer, *Z. Makromol. Chem.*, *52*, 101 (1976).
71. J. D. Ferry, *Viscoelastic Properties of Polymers*, 1st ed., Wiley, New York (1961).
72. J. Meissner, *Trans. Soc. Rheol.*, *16*, 405 (1972).
73. P. J. Hoftyzer, and D. W. Van Krevelen, *Z. Makromol. Chem.*, *54*, 1 (1976).
74. D. J. Goedhart, *Comm. Gel Perm. Chromatogr.*, Intern. Seminar, Monaco, Oct. 12–15 (1969).
75. R. W. Pohl, *Einführung in die Optik*, Berlin (1943).
76. A. Coehn, *Ann. Phys.*, *64*, 217 (1898).
77. J. G. Dorfmann, *Diamagnetismus und Chemische Bindung*, Deutsch, Frankfurt (1964).
78. W. Haberditzl, *Magnetochemie*, Akademie Verlag, Berlin (1968).
79. P. Pascal, *Compt. Rend.*, *147*, 56 (1908).
80. P. Pascal, A. Pacault, and J. Hoarau, *Compt. Rend.*, *233*, 1078 (1951).
P. Pascal, F. Galais, and J. F. Labarre, *Compt. Rend.*, *252*, 18, 2644 (1961).

81. J. W. Anderson, G. H. Beyer, and K. M. Watson, *Nat. Petr. News*, *36R*, 76 (1944).
82. J. G. M. Bremner and G. D. Thomas, *Trans. Farad. Soc.*, *44*, 20 (1948).
83. M. Souders, C. S. Matthews, and C. O. Hurd, *Ind. Eng. Chem.*, *41*, 1037 (1949).
84. J. L. Franklin, *Ind. Eng. Chem.*, *41*, 1070 (1949).
85. D. W. Van Krevelen, and H. A. G. Chermin, *Chem. Eng. Sci.*, *1*, 66 (1951); 1, 238 (1952).
86. G. J. Janz, *Estimation of Thermodynamic Properties of Organic Compounds*, Academic Press, New York (1958).
87. H. Ulich, *Chemische Thermodynamik*, Leipzig, (1930).
88. S. L. Madorsky, and S. Strauss, *J. Res. Natl. Bur. Standards*, *53*, 361 (1954); *55*, 223 (1955); *63A*, 261 (1959).
89. S. L. Madorski, *Thermal Degradation of Organic Polymers*, Interscience, New York (1964).
90. V. V. Korshak, *The Chemical Structure and Thermal Characteristics of Polymers*, Transl. Israel Progr. Sci. Transl. (1971).
91. D. W. Van Krevelen, *Polymer*, *16*, 615 (1975).
92. D. W. Van Krevelen, *J. Appl. Polym. Sci.*, Appl. Polym. Symp., *31*, 269–292 (1977).
93. C. P. Fenimore, and F. J. Martin, *Combustion and Flame 10*, 135 (1966).

Handbooks and Encyclopedias

1. J. Brandrup, and F. H. Immergut, eds., *Polymer Handbook*, 2nd ed., 3rd ed. Wiley-Interscience, New York (1975, 1989).
2. H. F. Mark, N. M. Bikales, and C. G. Overberger, eds., *Encyclopedia of Polymer Science and Engineering*, 2nd ed. Wiley, New York (1985).
3. M. B. Bever, ed.-in-chief, *Encyclopedia of Materials Science and Engineering*, Pergamon Press, Oxford (1986/1987).

part two

FORCE-FIELD TECHNIQUES

2

Force-Field Techniques and Their Use in Estimating the Conformational Stability of Polymers

William J. Welsh
University of Missouri-St. Louis
St. Louis, Missouri

INTRODUCTION

As pointed out several years ago, the structure, energy, and chemical behavior of molecules are all interrelated [1]. To be sure, polymers pose no exception to this corollary. Beyond the specification of chemical structure, it is apparent that a molecule's conformation and structural topology often play a crucial role in dictating its physical and chemical properties. One could argue that this is even more so the case for polymers.

Polymer chains have access to a virtually unlimited number of conformational arrangements, or "spatial configurations," achieved by different rotations about the single bond contained in the chain backbone. Indeed, so many of the unique macroscopic properties associated with polymers are derived in part from the conformational variability of their composite chains. A list of their conformation-dependent properties would include rubber elasticity, amorphous-crystalline morphology, glass-transition temperature, many electrical and optical properties, dipole moment, and diverse mechanical properties, to name but a few [2].

To choose one example, rubber elasticity is best understood in terms of long-chain molecules responding to an imposed force by converting from random, relatively compact, configurations to ones having higher spatial extension along the axis of deformation. Upon removal of the force, the elastomer returns to its original shape since this restores the entropy to its original, and considerably larger, value. Thus the mechanism for reversibility of the deformation is based primarily on considerations of entropy rather than enthalpy and is derived from differences in conformational versatility between the undeformed and deformed polymer chain [3].

With respect to the conformational analysis of polymers, several notable differences exist between synthetic polymers and natural polymers. One important difference is that natural polymers such as proteins have definite molecular formulae and weights as a consequence of the fixed number and sequence of monomer units constituting each molecule of a given type. In contrast, synthetic polymers are in general polydisperse; i.e., a sample of a given polymer will be comprised of a distribution of chain lengths commensurate with the varying numbers of repeat units in the chain. Another difference which bears consequences for conformational analysis is that synthetic polymers can experience cross-linking and chain branching to a degree beyond that normally found in most natural polymers. Third, whereas hydrogen bonding normally plays a dominant role in the structure and conformation of natural polymers, it is either absent or of relatively minor significance in most synthetic polymers.

Configuration-dependent properties of polymers, such as rubber elasticity mentioned above, are obviously of the greatest interest and importance in the characterization and utilization of polymeric materials. The theoretical interpretation, understanding, and prediction of these properties are conveniently provided by a formalism known as the rotational isomeric state (RIS) approximation [2]. According to RIS theory, each rotatable skeletal bond is assumed to occur in one of a small number of discrete torsional domains or "rotational states." In most chain molecules, these states are located at or near the conformations designated *anti* or *trans* (t), *gauche positive* (g^+), and *gauche negative* (g^-), which by convention are assigned the rotational angles 0°, 120°, and −120°, respectively. (Unfortunately, the convention adopted by most other chemistry subdisciplines is just the converse, viz., *trans* is 180° with *gauche* states at 60° and −60°.) A spatial configuration is then specified by the particular sequence of conformations or rotational states adopted by the skeletal bonds of the chain. The choice of rotational state for a given skeletal bond depends on the relative energies of the rotational states available to that bond. Likewise, the average configuration of the chains and their configuration-dependent properties, which are calculable from the equations of statistical mechanics, depend on the relative energies of these rotational states taken collectively [2].

One of the principal utilities of force-field methods in the conformational analysis of polymers has been to provide estimates of these relative energies which can thereupon serve as input in the application of RIS theory to compute values of the polymer's configuration-dependent properties [2,4]. In return, comparisons of calculated and corresponding experimental values of a configuration-dependent property provide a sensitive test of the force-field in general and its parameterization in particular. While the methods and applications of RIS theory will be elaborated on later, the foregoing discussion highlights the close interplay between the conformational analysis of polymers as deduced by theoretical force-field methods [5] and the delineation of a polymer's configuration-dependent properties obtainable from RIS methods.

The theory and practice of conformational analysis deals with delineating and hopefully quantifying the conformational profile of a particular molecule, whether polymeric or otherwise [1,6]. A frequent objective is identification of the various conformational isomers (conformers), including their relative energies (conformation energies), and the energetic barriers (rotational barrier heights) between them. Achieving such a seemingly modest goal is nonetheless often beset with an assortment of obstacles. Foremost among these, a molecule or polymer segment will possess an often large array of torsional degrees of freedom roughly commensurate to its number of single, and presumably rotatable, bonds. (In most but not all applications, bonds of bond order appreciably above 1 are assumed nonrotatable.) The resulting multidimensional conformational energy surface typically exhibits a multitude of local energy minima in addition to a global energy minimum. This phenomenon is commonly referred to as the "multiminima problem," and as a consequence can give rise to at least several if not a profusion of energetically stable conformers [7]. These difficulties are magnified should it be necessary to scan the conformational energy terrain for occurrence of rotational barriers and saddle regions. In this case, domains of both high and low conformational energy would need to be explored.

The use of theoretical methods to study the structure, energy, and conformation of molecules provides in principle a straightforward, reliable, and accessible avenue for many investigators. The list of methods can be divided into two general categories: (1) classical mechanical or "empirical" and (2) quantum mechanical. The latter topic will be covered in Part V of this book. It will suffice to note here that the advent of high-speed computers and supercomputers, replete with large memory and storage capacities, has expanded the utility of quantum mechanical techniques deeper into the realm of conformational analysis. The prospects will continue to heighten with the availability and increased rigor of the so-called semiempirical techniques such as AM1 [8], with the development and codifying of promising new *ab initio* methods such as density functional theory [9], and through the adaptation and evolution of computer operating systems and processors designed to maximize rigor and CPU time efficiency. (In this context, we are reminded that the supercomputers of today are the mere computers of tomorrow.)

Notwithstanding the enlarged role ensured for quantum mechanical methods in conformational analysis, classical force-field methods are almost guaranteed to maintain, if not advance, their position as the alternative of choice. A primary reason for this relates to the point mentioned above, namely, the expanse of conformational dimensionality enjoyed by many nonpolymeric and most polymeric molecules and the associated multi-minima problem. In practical terms, the long CPU turnaround times characteristic of quantum mechanical methods will remain intolerable for the foreseeable future and thus limit their widespread use in applications of conformational analysis for all but the most simple or rotationally inflexible of molecules. A simple example can illustrate this point. The conformational analysis of one rotational (i.e., torsion) angle taken through its full range of values from 0° to 360° in increments of 20° would require 18 separate energy minimizations (invoking geometry optimization) to calculate the structure and associated energy. For a concerted analysis of more than a single torsion angle, the number of minimizations jumps to 324 for two torsions, 5832 for three torsions, and so on. Clearly, the demands imposed on contemporary computer resources by employing rigorous quantum mechanical methods for such tasks quickly outstrip the computer budgets and the patience of most investigators. If not by virtue of its relative mathematical simplicity alone, the classical force-field method is now and will likely remain the leading theoretical technique amenable to multidimensional conformational analysis in computational chemistry [5,10–12].

The emphasis of this chapter is on the force-field methods employed in the conformational analysis of synthetic polymers. One principal aim is to describe in sufficient detail the potential functions which constitute the typical force fields applied to polymers. Consequently, this will not be an exhaustive recapitulation of the results obtained from innumerable applications. Rather we will cite examples of applications to illustrate the power and utility of the general procedures commonly employed. In particular, the latter part of this chapter will examine examples of the use of force-field techniques in providing the geometrical, energetic, and conformational information useful as input for application of RIS theory in calculations of the configuration-dependent properties of polymers [2]. The important area of biological macromolecules is outside of the scope of this book; hence the interested reader is referred to several excellent recent publications covering this topic [13–15].

By today's standards, the force-field hamiltonians employed in earlier applications of conformational analysis on polymers are noted for their simplicity, notably in terms of the number and mathematical rigor of their individual interaction terms [2,6,16–18]. Of course, this simplicity was a deliberate expediency conceived in recognition of the fact that the chief limiting factor was computational resources. Inasmuch as these early force fields are the progenitors of current schemes, we will cover their development and utilization in some detail. Some features of more contemporary force fields [10–12,19] are then introduced as extensions of these earlier basic forms. The tacit assertion subsumed herein is that it is instructive if not imperative for the reader to view the subject of force fields, including those employed in polymer conformational analysis, from the context of their historical underpinnings rather than merely as immutable creations of recent vintage.

Not surprisingly, the explosion in accessible computer power in recent years has augured a concomitant increase in complexity of the force-field hamiltonians [10–12,19]. The force-field protocols found in many of the present-day "canned" molecular modeling packages will commonly contain hamiltonians comprised of more realistic and mathematically more rigorous potential energy functions. Some of these modern force fields feature Morse equations to describe bond stretching as well as a growing assortment of "cross terms" (e.g., stretch-bend) for fine-tuning. One of the aims of these embellishments is to yield calculated vibrational frequencies within a 30–40 cm^{-1} tolerance of the experimental values [19]. Inclusion of explicit terms for hydrogen bonding interactions as well as attempts to treat dielectric effects more plausibly have added further complexity to the hamiltonian. Likewise, the complete specification of the so-called force-field parameters found in modern force-field programs now requires long inventory lists.

One gratifying observation is that the operative force fields remain substantially invariant whether applied to nonpolymeric or polymeric molecules [2,6,10]. Thus we can and shall describe force-field methods with a fair degree of generality. The principal distinction between the two in this regard is the existence of long-range interactions in the polymer molecule [2]. However, primarily for the sake of computational efficiency, it is common practice in conformational analysis to represent a given chain by means of model compounds or sometimes molecular segments. An example of this is illustrated in Figure 1. In general, these model compounds should be constructed sufficiently long (in terms of polymer repeat units) so as to incorporate most if not all of the pertinent features of the polymeric chain

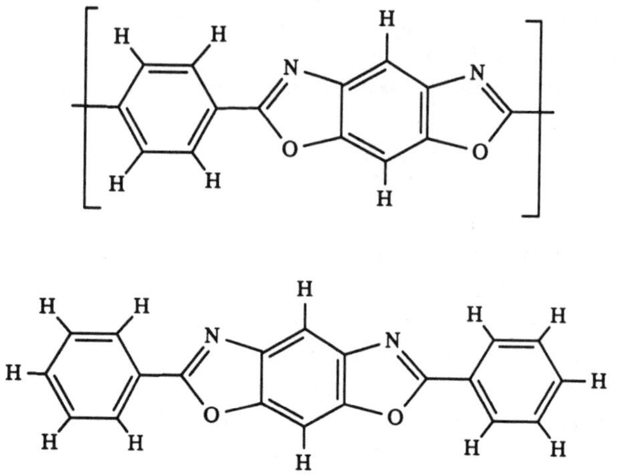

Figure 1 The repeat unit for the polymer *cis*-poly(benzobisoxazole) (PBO) along with a typical model compound.

and to curtail the "end effects." Conversely, the size of the model compound is limited largely by the capacity of one's computer resources and by the desire to maintain computational tractability. In practice, the typical model compound or molecular segment culminating from these opposing constraints contains no more atoms or internal degrees of freedom than most nonpolymeric molecules subjected to conformational analysis. This subsequent conformity in size and structure between nonpolymeric molecules and polymer model compounds is reflected in the virtual indistinguishability of the force fields applied to polymers and nonpolymers.

DESCRIPTION OF THE FORCE-FIELD METHOD FOR CALCULATING CONFORMATIONAL ENERGIES OF POLYMERS

Force-field methods originated with the pioneering work of Westheimer [20] and were later implemented and expanded by Hendrickson [21] and by Wiberg [22]. The basic concept was to relate some measure of the energy of a molecule to its geometric variables which are assumed to be readily calculable or observable parameters. Gradually, improvements in implementation led to the generation of computer programs capable of automatically altering the molecular geometry subject to a specified force field in pursuit of a nearby energy minimum. The force-field computer algorithms of today largely represent refinements of this paradigm.

We should begin by defining the term *force field* with regard to conformational analysis. In the present context, we regard a force field as a potential energy surface represented by a closed set of analytical potential energy functions. In principle, this potential energy surface is an outcome of the Born-Oppenheimer approximation which allows us to separate nuclear and electronic motion. However, the simplification implicit in force-field methods is that the electrons adjust optimally and instantaneously to nuclear movement and hence need not be considered explicitly [10]. The classical force field thus obviates the need for quantum mechanical derivation of electronic motion. Force-field methods in effect adopt a mechanical view of molecules. Stated simplistically, molecules are composed of atoms represented as spheres which in turn are connected by bonds represented as springs. This "ball and spring" perspective is classical mechanical rather than quantum mechanical, as connoted by the familar synonym "molecular mechanics" bestowed on force-field methods [10–12].

Each of the potential energy functions which define a given force field contains adjustable parameters for describing this surface. These parameters are then optimized by a best-fit process of calculated and experimental properties for a preselected basis set of test molecules. The data set of molecular properties often includes experimental structural geometries, crystal structure data, thermodynamic data such as heats of formation and of sublimation, dipole moments, and spectroscopic data such as vibrational force constants and torsional barrier heights [10,23]. The results of numerous rigorous semiempirical and rigorous *ab initio* molecular orbital calculations have contributed more in recent years to force-field parameter development [10–12,23]. In fact, the use of quantum mechanical calculations to derive force-field parameters represents another example of the complementarity of the two methodologies. The operational assumption underlying force-field methods is that the particular force field and its associated parameter set are transferable to molecules structurally similar to, but separate from, the test molecules consitituting the basis set [1,10].

Excellent treatments of the theory and development of force-field methods for the conformational analysis of molecules are presented elsewhere [1,10–12,19]. We will thus start with a discussion of some typical force-field equations employed frequently in earlier studies on polymers [2,6,16–18]. The force-field equation is composed of a sum of energy terms. Each term is an analytical expression representing a particular internal motion or interaction that contributes to the potential energy associated with a given structure or conformation. As depicted in Eq. (1), these earlier force fields as applied to synthetic polymers typically included only three terms: (1) a van der Waals term E_{vdW} for nonbonded interactions between all the pairs of atoms not covalently bonded to each other, (2) an electrostatic term E_{elec} for electrostatic interactions between permanent dipoles, and (3) a torsional term E_{tor} for correct portrayal of bond rotation potentials beyond that depicted by the former two terms alone [2,6,16–18]:

$$E_{conf} = \Sigma E_{vdW} + \Sigma E_{elec} + \Sigma E_{tor} \quad (1)$$

As discussed by Flory [2], the van der Waals interactions between substituent atoms on either side of the bond in question are alone not sufficient for depicting rotational barriers. While inclusion of electrostatic interactions provides a definite improvement, disparities persist between experimentally derived information on rotational barriers and calculated barriers derived from fitting the E_{vdW} and E_{elec} terms alone. The E_{tor} term was thus introduced in part as an empirical correction required so that the calculated barriers agreed with the experimentally determined barriers. Its theoretical basis has been ascribed to exchange interactions of electrons in bonds adjacent to the bond undergoing rotation [16,24–26].

The sum expressed in Eq. (1) is conspicuous for its absence of additional terms for such contributions as bond stretching, bond-angle bending, and the assortment of cross terms commonplace in today's force fields [10–12, 19]. This "fixed valence-geometry" picture assumes that contributions from these terms are independent of conformation and thus make a constant increment to the total energy. Hence, they need not be considered so long as relative and not absolute conformational energies are of interest. We shall proceed by discussing each term on the right-hand side of Eq. (1).

The van der Waals Term E_{vdW}

Both the basis and analytical forms of equations for describing the nonbonded van der Waals–type interactions find their origins in the theoretical analysis of intermolecular forces [27,28]. In general, the total potential energy of an N-particle system, in the absence of external force fields, can be given by

$$E_{tot} = \Sigma E_{i,j} + \Sigma E_{i,j,k} + \Sigma E_{i,j,k,l} + \cdots \quad (2)$$

where the sum is taken over all atoms in the assembly. The sum includes contributions from all possible pairs of interactions plus additional higher-order terms representing simultaneous 3-body interactions, 4-body interactions, and beyond. These additional *many-body* terms arise due to the polarizability of the electronic charge cloud about each atom [6,27,28].

While the subject of many-body effects remains one of considerable interest, their contributions are as yet not well understood but assumed to be small in terms of their influence on conformational behavior. Thus, most of the familiar force fields omit explicit inclusion of the higher-order terms. This omission is likely a mere mathematical convenience to maintain computational simplicity. In any case, many-body effects are embodied implicitly in the pair (i.e., 2-body) potential terms whose empirical parameters are derived from experimental data which by definition incorporate many-body effects [27,28]. As a consequence, the sole surviving 2-body term is often described as a *pseudo*-pair potential.

The pair potential $u(r)$ has the general shape illustrated in Figure 2. For small values of r, the potential energy is very large corresponding to the interpenetration of the electron clouds of the two interacting atoms. As r increases, $u(r)$ drops sharply, crosses the r axis at $r = \sigma$ and becomes negative, passes through a minimum at $r = r_m$ with value $-\epsilon$, and then asymptotically approaches zero.

The unique form of this potential arises from the combination of two operative effects. Specifically, at short distances strong coulombic repulsions arise from the close nuclear-nuclear interactions coupled with the overlap of the electron clouds between the two atoms. These repulsions dominate the attractive coulombic interactions between the nuclear core and electrons on the two different atoms. At distances near and beyond r_m, the dominant effect is an attractive one arising from the so-called London dispersion forces, the form of which can be obtained from second-order perturbation theory [27,28]. Specifically, by expanding the electrical potential energy in a multipole series, one obtains

$$u(r) = -c_6 r^{-6} - c_8 r^{-8} - c_{10} r^{-10} - \cdots \tag{3}$$

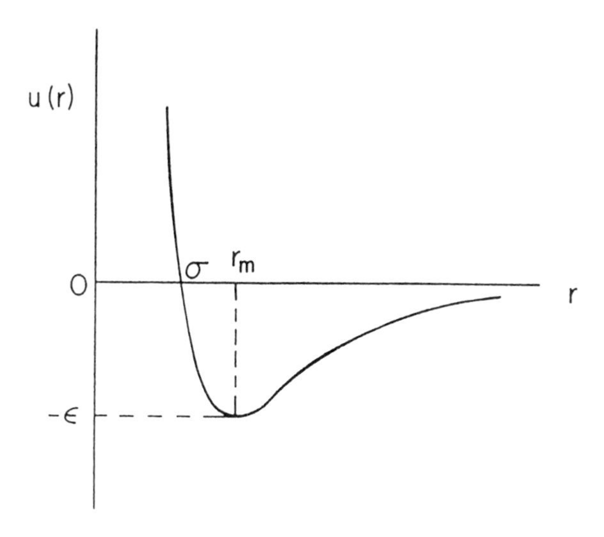

Figure 2 The general shape of the pair potential $u(r)$.

The three terms shown above represent, respectively, interactions due to the induced dipole, induced quadrupole, and induced octupole associated with the London dispersion forces. For large values of r, the induced dipole interaction term ($-c_6 r^{-6}$) predominates. Thus in most conventional force fields only this term is retained, while the coefficient c_6 can be adjusted to compensate in part for the neglect of the higher-order expansion terms in Eq. (3).

Over the years, many analytical functional forms for $u(r)$ have been proposed [6,27,28]. These functional forms are designed to have the general shape shown in Figure 2 and contain two or more parameters which are determined by a comparison of theoretical and experimental physical properties. At least in the case of these earlier force fields, an attempt was made to depict a reasonable facsimile of the actual pair potential without sacrificing mathematical convenience. For the conformational analysis of polymeric and nonpolymeric molecules alike, two of the most popular functions have been the *Lennard-Jones* (LJ) *n-m potential* and the *Buckingham* "exp-6" *potential.* While other functional forms are found in popular force fields [10–12,19], the LJ and exp-6 potentials are considered prototypical.

The LJ n-m potential expresses $u(r)$ as the sum of two terms: a negative "attractive" term proportional to r^{-m} and a positive "repulsive" term proportional to r^{-n} with $n > m > 0$ [29,30]. The equation appears in two equivalent forms:

$$u(r) = \frac{n\epsilon}{n-m}\left(\frac{n}{m}\right)^{m/(n-m)}\left[\left(\frac{\sigma}{r}\right)^{n} - \left(\frac{\sigma}{r}\right)^{m}\right]$$

$$u(r) = \frac{n\epsilon}{n-m}\left[\frac{m}{n}\left(\frac{r_m}{r}\right)^{n} - \left(\frac{r_m}{r}\right)^{m}\right] \tag{4}$$

where ϵ, σ, and r_m have the significance denoted in Figure 2. Inasmuch as the asymptotic behavior of $u(r)$ is of the form $-c_6 r^{-6}$, m in the LJ n-m potential is usually taken to be 6. There is no equivalent theoretical basis for the value of n in the repulsive term. In fact, the repulsive part of the curve decreases roughly as an exponential in the separation, not as r^n [27,28]. However, values for n in the range 9 to 14 are acceptable provided that values for ϵ and σ are appropriately chosen based on comparison with experimental data such as the second virial coefficient. When n is chosen as 12, the LJ potential assumes a particularly simple form

$$u(r) = 4\epsilon\left[\left(\frac{\sigma}{r}\right)^{12} - \left(\frac{\sigma}{r}\right)^{6}\right]$$

$$= \epsilon\left[\left(\frac{r_m}{r}\right)^{12} - 2\left(\frac{r_m}{r}\right)^{6}\right] \tag{5}$$

where r_m and σ are related by $r_m = 2^{1/6}\sigma$. Equation (5) is the well-known LJ 12-6 potential which has been reviewed extensively elsewhere [27].

For applications in force-field calculations, Eq. (5) has been simplified further to [6,17,18]

$$u(r) = \frac{A_{ij}}{r_{ij}^{12}} - \frac{C_{ij}}{r_{ij}^{6}} \tag{6}$$

where A_{ij} and C_{ij} are adjustable parameters and i and j are indices for the interacting atoms. The parameters C_{ij} characterizing the attractions are evaluated from the effective number of electrons N_{eff} and the atomic polarizabilities α by application of the Slater-Kirkwood equation [26,31,32]

$$C_{ij} = \frac{(3/2)e(h/2\pi)m^{1/2}\alpha_i\alpha_j}{(\alpha_i/N_i)^{1/2} + (\alpha_j/N_j)^{1/2}} \tag{7}$$

where α_i and α_j are the atomic polarizabilities taken from Ketelaar [33], e and m are the electronic charge and mass, respectively, and N_i and N_j are the "effective" values of N for atoms i and j taken from Scott and Scheraga [26]. A short list of values of α, N_{eff}, and r_m is given in Table 1.

The common basis for evaluating the repulsive parameters A_{ij} is to require that the complete potential function, or some component thereof, conforms to experimentally derived criteria. Values of A_{ij} can be calculated by imposing the condition that the potential function has a minimum at a distance r_m, which is usually taken as the sum of the van der Waals radii r_{vdW} obtained from crystal structure data. A major source of values for r_{vdW} has been the tabulations of Bondi [34] based on x-ray diffraction data. By

Table 1 Selected Values[a] of α,[b] N_{eff},[c] and r_m[d]

Interaction	$10^{24}\alpha$[b]	N_{eff}[c]	r_m[d]
C···C[e]	0.93	5.0	3.4
C···C[f]	1.23	5.0	3.7
N···N	1.15	6.0	3.1
O···O	0.59	7.0	3.0
S···S	3.04	14.8	3.6
H···H	0.42	0.9	2.4

[a]Taken from Ref. 48.
[b]Polarizability, in cm^3.
[c]Effective number of electrons.
[d]Sum of van der Waals radii for two interacting atoms, in Å.
[e]Aliphatic carbon atoms.
[f]Aromatic carbon atoms.

applying the condition $[du_{LJ}(r)/dr]_{r=rm} = 0$ to Eq. (6), the relation $A_{ij} = (1/2)C_{ij}r_m^6$ [6] is obtained from which values of A_{ij} are calculated [6,18].

A brief list of values for A_{ij} and C_{ij} for some common interatomic interactions is given in Table 2. In some applications, the values adopted for r_{vdW} have been "augmented" by 0.1 Å, as suggested by Flory [2] and by Allinger et al. [35]. This augmentation is introduced to compensate for the absence in most conformational energy calculations of long-range attractive forces that are operative in the crystal.

For applications on organic polymers, accurate experimental conformational energies and geometries available for model compounds such as the alkanes, alkenes, and cycloalkanes have been used to parameterize the potential function [2,17,36]. Using this scheme, potential parameters are deliberately chosen so as to reproduce the known energetic barriers to rotation for representative molecules or the differences in energy for various stable conformers, e.g., the energy for the *gauche* conformations versus the alternative *trans* conformation [2].

The analogous Buckingham exp-6 potential function assumes the form [2,6,16,24–26,36]

$$E_{vdW} = A_{ij}\exp(-B_{ij}r_{ij}) - \frac{C_{ij}}{r_{ij}^6} \tag{8}$$

in which the r^{-12} function of the LJ 12-6 potential has been replaced by an exponential decay term. Whereas the LJ 12-6 equation is a two-parameter function, the exp-6 equation is a three-parameter function. Values for C_{ij} are obtained as described above for the LJ potential. Values of B for like atom pairs (i.e., B_{ii} and B_{jj}) are usually obtained directly by interpolating from a plot given by Scott and Scheraga [26] of B versus atomic

Table 2 Selected Values[a] of the LJ 12-6 Potential Parameters A and C

Interaction	$10^{-3}A$[b]	C[b]
C···C	908.	363.0
H···H	10.37	47.1
Si···Si	4590.	3060.
C···H	86.1	127.0
Si···H	1970.	371.4
Si···C	2100.	1050.

[a]Taken from Ref. 54.
[b]Units are such as to give energy in kcal mol^{-1} for r in Å.

Table 3 Selected Values[a] of the Buckingham Potential Parameters A, B, and C.

Interaction	$10^{-3}A^{b}$	B^{b}	C^{b}
C···C[c]	541.4	4.59	363.0
C···C[d]	1820	4.59	556.7
N···N	393.2	4.59	547.3
O···O	135.8	4.59	217.2
S···S	906.3	3.90	3688
H···H	7.323	4.54	47.1

[a]Taken from Ref. 48.
[b]Units are such as to give energy in kcal mol^{-1} for r in Å.
[c]Aliphatic carbon atoms.
[d]Aromatic carbon atoms.

number Z for the inert-gas elements. Values of B for an unlike atom pair (i.e., B_{ij}) are obtained by taking the geometric mean of B_{ii} and B_{jj}, i.e., $B_{ij} = (B_{ii}B_{jj})^{1/2}$. Corresponding values of the parameter A_{ij} are once again determined by minimizing Eq. (8) at $r = r_m$. A brief set of values of the Buckingham potential parameters A, B, and C for some common interatomic interactions is given in Table 3.

Compared with the LJ potential's r^{-m} term, the Buckingham potential's exponential term is, in principle, a better portrayal of the true repulsive interaction. However, in practice it provides no worthwhile advantages and some disadvantages [17,18]. Notably, the Buckingham function introduces a third parameter unnecessarily into the force-field equation. Of more serious concern, it contains a spurious maximum in the range $0 < r < \sigma$ and unrealistically (and catastrophically) plummets to increasingly negative energies as r approaches zero [27]. This latter deficiency can cause problems in conformational analysis, particularly when one is dealing with sterically or conformationally congested polymers. As a remedy in some applications, simple spline functions have been joined to the Buckingham potential in the problem region so as to extrapolate correctly to high energies as r approaches zero [37].

The Electrostatic Term E_{elec}

The electrostatic interactions between noncharged polar groups within (and between) individual polymer molecules make a separate contribution to the conformational energy and are, in general, conformation dependent.

In classical electrostatics, the electrostatic energy E_{elec} associated with these dipole-dipole terms is given by the well-known Jeans' formula [6,10].

$$E_{dipole} = \frac{\mu_i \mu_j}{D r_{ij}^3} (\cos \chi - 3 \cos \alpha_i \cos \alpha_j) \tag{9}$$

where D is the effective dielectric constant, χ is the angle between the two dipoles μ_i and μ_j, and the α's are the angles that the dipoles form with the vector connecting the two. However, most force fields applied to polymers have adopted the mathematically and physically more simple Coulomb's law formula $E_{Coul} = kq_iq_j/Dr_{ij}$, where q_i and q_j are the partial atomic charges separated by a distance r_{ij}, D is again the effective dielectric constant, and k is a conversion factor commonly set equal to 332.072 in order to produce E_{Coul} in kcal mol^{-1} when r_{ij} is in Å and q is in units of fraction of an elementary electronic charge. Electrostatic contributions to the conformational energy are then estimated using Coulomb's law by assigning atomic partial charges to each atom in the (model) molecule under consideration, or at least to each atom involved in a polar bond [18]. In turn, the atomic partial charges q^+ and q^- are obtained from values of the appropriate bond dipole moment by applying $q = \mu/l$, where μ in this case is the bond's dipole moment and l is its length. Bond dipole moments can be estimated from the observed dipole moment of a simple molecule containing that bond dipole by applying elementary geometrical arguments to resolve the molecular dipole moment vector into its component bond dipoles. This method for obtaining suitable atomic partial charges has been, and still remains, popular in applications of force-field methods to polymers [38–44]. A standard source for experimental dipole moments has been the compilations of McClellan [45].

More recently, some studies have employed atomic partial charges derived from quantum mechanical calculations carried out on model compounds [46–49]. A cautionary note in this regard is that the different quantum mechanical methods currently available (e.g., CNDO [50], MINDO/3 [51], MNDO [52], and AM1 [8], *ab initio* [53]) can be expected to yield different, and sometimes qualitatively irreconcilable, charge distributions. Of a more general nature, another potential problem involves the question whether the charge distribution is invariant, or nearly so, to conformational changes. While the assumption of conformationally independent charge distributions is commonplace, it may falter particularly in delocalized systems. These quandaries can pose difficulties if a particular conformation-dependent property of interest is sensitive to the electrostatic potential energy.

With regard to choosing a suitable value for the effective dielectric constant D, one sees a fair degree of variation among various studies of polymers. In studies of organic polymers, most investigators choose values ranging from 1.0 to 5.0 debyes (D). The *in vacuo* value of 1.0 D is prescribed in cases where the solvent (i.e., the dielectric) molecules are considered explicitly or where one merely wishes to maximize the electrostatic contributions to the conformational energy [47]. Of course, the proper choice for D is critical in applications where E_{elec} makes a significant contribution to E_{conf}; fortunately, this is rarely the case for synthetic polymers. Flory and his coworkers [2,38] among others [6,17,37,40,42–44,46–49,54,55] have adopted values in the 2.0–4.0 D range most frequently.

Of final note, calculated values of the atomic partial charges are sensitive to one's choice of values for the geometrical parameters (i.e., bond lengths and bond angles). Most often, workers have used structural data derived from quantum mechanical calculations or from experimental studies performed on model compounds. Another source of structural information is available from comprehensive tables, most notably the earlier compendium of Sutton [56] and the more recent work of Allen et al. [57].

The Torsional Term E_{tor}

The change in potential energy associated with a torsional rotation about the C—C bond in a simple molecule like CH_3—CH_3 is depicted in Figure 3. The potential energy profile is periodic and threefold symmetric, a con-

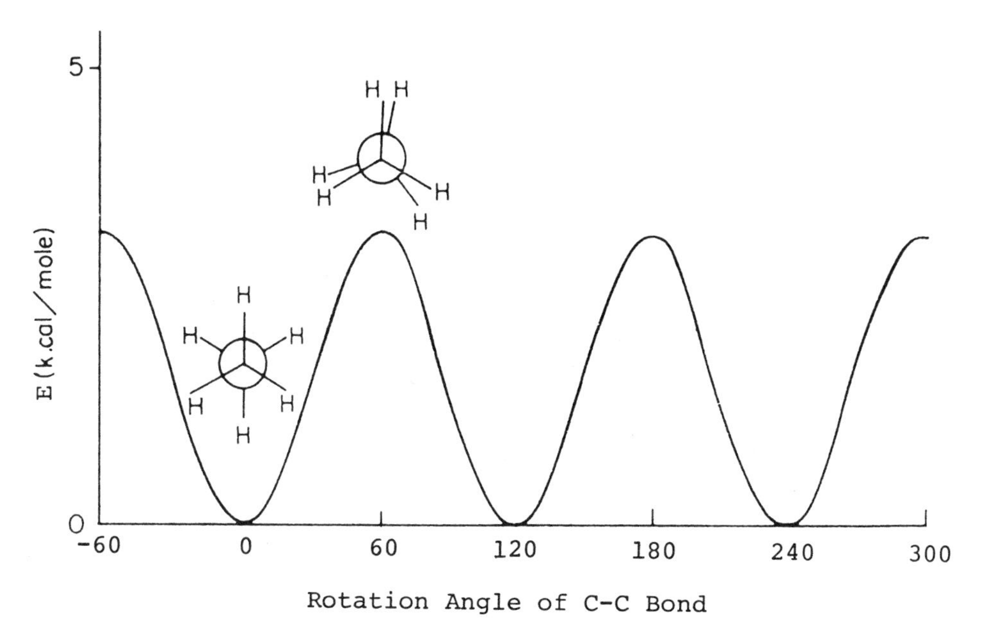

Figure 3 Plot of E_{tor} versus rotation about the C—C bond in ethane.

sequence of the C_{3v} symmetry of each methyl group. The periodic fluctuation in energy has been ascribed [2,6,10,16,26] to two contributing factors: (1) steric contributions arising from the substituent H···H interactions between the respective groups; these interactions are most favorable (i.e., of lowest energy) in the staggered conformations while least favorable in the eclipsed conformations; (2) quantum mechanical contributions arising from the non-single-bond character of the orbitals constituting the C—C bond.

It has become common practice to account for the steric interactions in the E_{vdW} term, thus it would be redundant to include them again in E_{tor}. Most force fields, both past and present, accordingly regard the E_{tor} term as a correction factor to bring rotational barriers based on force-field calculations, inclusive only of steric (and, if applicable, electrostatic) interactions, into agreement with the actual rotational barrier obtained from experiment or from quantum mechanical calculations. Stated in operational terms, the torsional barrier height parameter E^0_{tor} defining the amplitude of the E_{tor} term is what remains after subtracting the applicable steric and electrostatic energies due to substituent interactions from the actual rotational barrier. Inasmuch as E_{tor} represents a measure of a bond's "natural" resistance to rotation, it often carries the descriptor *intrinsic* torsional energy [2,6,10,16,26].

A particularly simple form adopted for E_{tor} in many applications on polymers is [2,6,10]

$$E_{tor} = \frac{E^0_{tor}}{2}(1 - \cos n\phi) \tag{10}$$

where ϕ is the angle of rotation from the staggered, or *trans*, form, n is the periodicity (i.e., 1, 2, 3) of the function, and E^0_{tor} is defined as above.

A typical force field applied to polymers will contain a separate E_{tor} term for each individual bond whose rotation would engender unique conformations. That is to say, terms for such bonds as C—H and C═O would be meaningless and are omitted. On the other hand, terms are often included for multiple bonds (partially double, double, and triple) since fluctuations away from the planar conformations are not uncommon.

In some of today's force fields [10,19], the expression for E_{tor} is written as a Fourier series expansion by analogy to the formalism used in vibrational spectroscopy [58]. Terminated after three terms, the expansion takes the form

$$E_{tor} = \frac{E^0_1}{2}(1 + \cos \phi) + \frac{E^0_2}{2}(1 - \cos 2\phi) + \frac{E^0_3}{2}(1 + \cos 3\phi) \tag{11}$$

where E_1^0, E_2^0, and E_3^0 are respectively the first, second, and third torsional constants. The physical picture attributed to these torsional terms is that the onefold term is a dipole-dipole interaction, the twofold term is from conjugative effects, and the threefold term is steric [10,59]. In theory, some of these terms would be expected to vanish (i.e., $E_n^0 = 0$) for symmetry reasons. However, better results are obtainable by judiciously choosing a set of values of E_1^0, E_2^0, and E_3^0 more as empirical fitting parameters for the torsional potential [10,50].

Today's force fields have also adopted the practice of decomposing the value of E_{tor} for a bond into individual contributions from all the various combinations of four-atom sequences which bracket this bond in a particular molecule [10,19]. For example, the E_{tor} term in $H_3C—CH_3$ would be decomposed into nine separate "H—C—C—H" contributions. Likewise, the torsional potential energy for the central C—C bond in $H_3C—CH_2—CH_2—CH_3$ is computed as the sum of four H—C—C—H terms, two each of H—C—C—C and C—C—C—H terms, and one C—C—C—C term. The advantage of this scheme is its ability to tailor a torsional potential to the unique nature of the rotatable bond inclusive of its full complement of specific substituents. Using this system, the torsional potential for $H_3C—CH_2F$ differs from $H_3C—CH_3$ by the replacement of three of the "H—C—C—H" terms by "H—C—C—F" terms. Of course, this enhanced opportunity for finetuning is afforded at the expense of complicating the force field and enlarging considerably the body of E_1^0, E_2^0, and E_3^0 torsional parameters.

Specific values of torsional force-field parameters have been obtained by variations of two different approaches. The most straightforward and physically meaningful way has been to estimate E_{tor}^0 from values of the actual torsional barrier height determined by experiment or by quantum mechanical calculations carried out on small molecule analogues. Adopting the simple analytical form given in Eq. (10), this process involves choosing a value for E_{tor}^0 which reproduces the actual torsional barrier of a small model molecule when Eq. (10) is combined with the nonbonded energies (E_{vdW} and E_{elec}) evaluated by the force field. More precisely, one first calculates the difference in nonbonded energies between the least and most favored conformations [e.g., $E_{nonbonded}$(eclipsed) − $E_{nonbonded}$(staggered)]. This difference is then subtracted from the actual torsional barrier to obtain E_{tor}^0. This first approach has remained popular in numerous polymer applications [40,41,46,49].

The second approach employs more elaborate fitting procedures which are meant to optimize the force-field parameters to attain a best fit between results calculated by the particular force-field equation and tabulated data of experimental and quantum mechanical results for a large number of model molecules [10,19,60]. This approach is indicated when more complicated forms of E_{tor} such as Eq. (11) are used, although initial guesses of E_1^0, E_2^0, and E_3^0 are often obtained by the process described in the first approach. A disadvantage of the "best-fit" approach is that it tends to mix the separate contributions in the force-field expression. As a result, a physical interpretation of individual energy contributions (E_{vdW}, E_{tor}, E_{elec}, etc.) as separate components of the total conformational energy E_{conf} becomes ambiguous. Furthermore, the force-field parameters are now correlated which greatly limits the interchangeability of a given parameter between different force fields.

MORE RECENT EXTENSIONS OF FORCE FIELDS

The evolutionary development of force fields, along with the increased availability of fast and powerful computers, has given rise to more elaborate forms for delineation of the potential energy surface. For example, a contemporary force-field equation might include explicit terms for bond stretching and bond-angle bending. In addition, today many force fields feature a growing number of crossterms [10,19]. A typical expression may take the form

$$E_{conf} = \sum E_{bond} + \sum E_{angle} + \sum E_{tor} + \sum E_{vdW} + \sum E_{elec} + \text{crossterms} \tag{12}$$

The crossterms have been added to force fields in large part to enable calculations of vibrational frequencies within 30–40 cm^{-1} of experimental values. Their inclusion is not considered essential for assessments of coformational energies, hence the reader is referred to other sources for further discussion of this topic [10,19]. However, terms for both bond-length and bond-angle deformations have become standard fare in today's force fields including those applied to polymers [10,19,41,61–64].

Bond Stretching and Bond-Angle Bending

The potential energy terms for these represent the energy expended in deforming a bond or a bond angle from its relaxed or equilibrium value. As a first approximation, a harmonic function obeying Hooke's law is used for each [10,19]:

$$E_{bond} = \frac{1}{2} K_R (R - R_0)^2 \tag{13}$$

$$E_{angle} = \frac{1}{2} K_\theta (\theta - \theta_0)^2 \tag{14}$$

An example of these functions is illustrated in Figure 4. Their harmonic nature accounts for equal costs in energy for compression and for expansion. The empirical parameters include the idealized or undeformed values R_0 for bond lengths and θ_0 for bond angles, and the so-called stretching and bending force constants K_R and K_θ. It is considerably easier in terms of energy costs to deform a bond angle than a bond, hence $K_R \gg K_\theta$. (As a rule of thumb, bond-angle deformation potentials are about 10 times "softer" than bond-stretching potentials.)

The true potential energy curve (Fig. 4) for these motions is anharmonic and asymmetric about R_0 and θ_0, respectively. The discrepancy between the real curve and the harmonic Hooke's law function widens at larger deformations. It can be seen that a realistic bond stretching function will approach zero energy (i.e., bond dissociation) asymptotically at large R, whereas the Hooke's law function incorrectly approaches infinity at large R reflecting an ever-increasing restoring force. The functional representation can be improved by adding cubic and even quartic terms to the quadratic term. Instead, the Hooke's law function can be replaced entirely by a Morse potential form [10,19,65]. Still, in most polymer-related applications severe deviations from the undeformed state are uncommon, hence the simple quadratic functions are usually adequate. Values for R_0 and θ_0 are typically obtained from experimental data or from rigorous quantum mechanical calculations. However, one should be aware of the potential pitfalls in assuming a priori transferability of R_0, θ_0, K_R, and K_θ from these sources into force fields or from one force field to another [10].

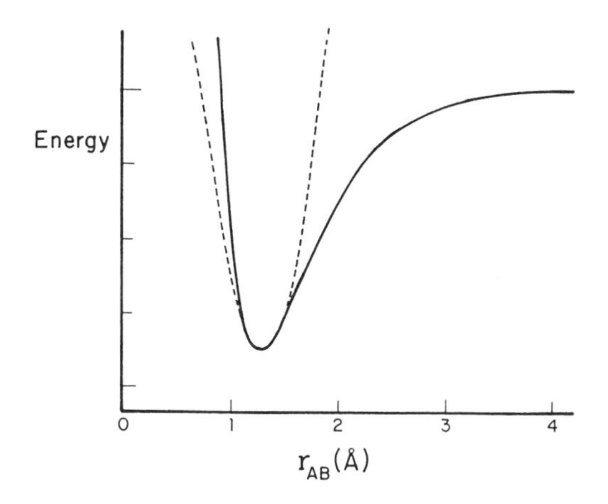

Figure 4 Sketch of the bond deformation energy E_{bond} versus bond length A—B. The dashed line is a harmonic function given by Hooke's law; the solid line is the actual shape as portrayed by a Morse function.

ENERGY MINIMIZATION

The routine use of computational procedures to achieve geometry optimization via energy minimization is a fairly recent phenomenon in applications to polymers. Indeed, full geometry optimization is absent in the preponderance of conformational energy calculations on polymers published up to the early or mid-1980s. Even a cursory scanning of the recent polymer literature familiar to the author reveals assorted examples of force-field methods applied to polymers in the absence of full geometry optimization [44,49,54,66–69]. In contrast, use of full geometry optimization in force-field applications seems to have gained popularity faster among practitioners in other areas such as biological molecules [13–15].

The reason for the slower growth among polymer scientists may stem from the fact that their chief interest in conducting these calculations has often been to locate and quantify entire conformational domains rather than isolated energy minima associated with low-energy regions [2,6,18,42,50,69–71]. The conformational data so derived was more amenable to calculations of the configuration-dependent properties of a polymer by means of RIS methods [2]. Given the meager computer resources available to polymer scientists of yesterday, energy minimization at each backbone rotational state was not conducive to the generation of conformational energy maps composed of finely spaced grids needed to identify the entire energy surface within the entire range of accessible conformations. In applications of RIS theory, it has been shown that identification of rotational isomeric states with discrete energy minima, without regard for the relative size and shape of the low-energy domains, can lead to large inaccuracies when treating the configurational statistics of a chain molecule [2,71–73]. In any case, calculated values of configuration-dependent properties were deemed (either implicitly or explicitly) rather insensitive to whether full geometry optimization was employed or not.

Another primary interest of polymer scientists has been to investigate the preferred conformations of a chain molecule, and not so much to explore those conformational regions such as barriers and saddle points not readily accessible to the polymer chain. Consequently, the necessity for geometry optimization was regarded as less essential in many polymer applications. The issue is somewhat moot today since the savings in CPU time once realized by omitting geometry optimization are negligible given the superior speed of modern computers. Authoritative discussions of the common geometry optimization techniques employed in force-field calculations are presented elsewhere for the interested reader [10–15].

EXAMPLES OF APPLICATIONS

The volume of high-quality work encompassing this chapter's subject area is substantial. In the preceding, I have mentioned only a small fraction of the many contributions present in the literature, and it is beyond the scope of the present discussion to attempt a survey or even a summary. In the following, I will briefly discuss a small sample of three representative studies spanning the past 15 years of the polymer literature. By my choosing a diversity of articles in terms of their publication date, it is intended to dramatize the fact that a wealth of relevant work has been sustained over a significant period of time.

Conformational Analysis of Polypropylene

Our first example is a 1975 publication of Suter and Flory [71] dealing with the conformational analysis and configurational statistics of polypropylene. Conformational energies were calculated for a dyad (dimer) of the polypropylene chain using force-field methods, and the conformational energies so derived were instrumental as input for a theoretical analysis of these chains' configurational statistical properties based on RIS theory [2]. The force field included a threefold intrinsic torsional term for each C—C bond and a nonbonded term represented by an all-atom LJ 12-6 potential for interactions between atoms separated by more than two bonds.

In order to first estimate the effect of variations in bond angles on the conformational characteristics, the conformational energy was calculated in regions of favorable conformations as a function of bond-angle deformations. Suter and Flory found that the precise locations of the conformational minima vary somewhat with their choice of bond angle. However, *averages* taken over low-energy conformational domains were less sensitive to the bond angles. This finding justified their adoption of fixed bond angles for the proceeding calculations.

In considering both meso (isotactic) and racemic (syndiotactic) dyads of polypropylene, Suter and Flory treated the conformational energy as dependent upon the backbone torsions ϕ_i and ϕ_{i+1} and the pendant methyl rotational angles χ_{i-1} and χ_{i+1}. However, satisfactory results were obtained by partially decoupling the separate contributions from backbone and pendant group torsions. The resulting conformational energy was then simplified as $E(\phi_i, \phi_{i+1})$, which in turn was minimized for each pair of ϕ_i, ϕ_{i+1} values with respect to the pendant-group torsions χ_{i-1} and χ_{i+1} varied by applying an iterative routine. For all low-energy conformations, the pendant groups were found to assume the staggered orientations or nearly so.

The results were presented as scans of $E(\phi_i, \phi_{i+1})$ versus ϕ_i and ϕ_{i+1} for both the meso and racemic dyads. After identifying distinct minima from the scans, Suter and Flory isolated specific low-energy domains of configuration space (ϕ_i, ϕ_{i+1}) by circumscribing regions within 6 kcal mol^{-1} of the local energy minima. The partition function z, average energy $\langle E \rangle$, and average rotation angles $\langle \phi_i \rangle$ and $\langle \phi_{i+1} \rangle$ were then calculated for eleven distinct domains by numerical integration of the familiar statistical mechanical equations

$$
\begin{aligned}
z &= \sum\sum \exp\left(\frac{-E}{RT}\right) d\phi_i \, d\phi_{i+1} \\
\langle E \rangle &= z^{-1} \sum\sum E \exp\left(\frac{-E}{RT}\right) d\phi_i \, d\phi_{i+1} \\
\langle \phi \rangle &= z^{-1} \sum\sum \phi \exp\left(\frac{-E}{RT}\right) d\phi_i \, d\phi_{i+1}
\end{aligned}
\tag{15}
$$

A significant conclusion is that these averages, rather than the values of E and ϕ associated with individual minima, provide a superior basis for allocating the configurational domains to rotational states within the RIS scheme. Based on extended calculations of the variations in z and $\langle E \rangle$ over the full range of pendant-group rotations χ_{i-1} and χ_{i+1}, it was concluded that the assumption of partial decoupling of the backbone and pendant-group rotations is acceptable in this application.

The above analysis yielded a rotational isomeric scheme based on five rotational states as opposed to the more typical three rotational states. These were identified as t, t^*, g^*, g, and g^-, corresponding to 15, 50, 70, 105, and $-115°$, respectively. The appropriate statistical weight matrices were constructed for both the meso and racemic dyads, and the statistical weight ξ for a pair of states was given in the form of a pre-exponential factor ξ_0 multiplied by the Boltzmann-weighted energy E_ξ

$$
\xi = \xi_0 \exp\left(\frac{-E_\xi}{RT}\right) \tag{16}
$$

A self-consistent set of values for the statistical weight parameters was derived by solving a set of linear equations based on the partition function z_ξ. The corresponding values of ξ_0 and E_ξ provide a measure of the size and shape as well as the average energy of each domain.

The characteristic ratio $C_n = \langle r^2 \rangle_0 / nl^2$, where $\langle r^2 \rangle_0$ is the "unperturbed" chain dimensions and n is the number of backbone bonds each of length l, was calculated by application of standard RIS methods within the five-state scheme [2,71]. (The chain dimensions or mean-square end-to-end distance $\langle r^2 \rangle$ is described as "unperturbed," as denoted by the subscript zero above, to mean that its calculation is based on neglect of the long-range excluded volume effects [2]). Values of C_n were determined versus temperature and as a function of the fraction ω_m of meso dyads. The values of C_n decreased monotonically as ω_m increased, ranging from about 13.8 for $\omega_m = 0$ to about 5.5 for $\omega_m = 1$. The dependence of C_n on stereoregularity also decreased as ω_m increased. These results were largely consistent with the findings of earlier calculations based on a three-state model [74–76]. Calculated values of C_n and its temperature coefficient $-d \ln\langle r^2\rangle_0/dT$ were in reasonably good agreement with published experimental values for different chain tacticities.

Conformational and Configurational Analysis of Polysilanes

The polysilanes [—SiRR′—] and polygermanes [—GeRR′—], where R and R′ represent various organic groups, are members of a family of polymers known as *sigma conjugated* [77,78]. Much of the intense scientific interest in the polysilanes stems from the strong temperature dependency of their electronic spectra for some members of this series. This temperature sensitivity has been interpreted in terms of the conformational preferences of the backbone Si—Si bonds.

In a 1986 paper, Welsh, Debolt, and Mark [54] compared three force fields with respect to their calculated conformational energies of model systems for the two simplest polysilanes [—SiH_2—] and [—$Si(CH_3)_2$—]. The earlier force-field calculations of Damewood and West [79] employed the MM2 program [80] with full geometry relaxation at specific rotational states and were thus abbreviated FR for "full relaxation." Welsh, Debolt, and Mark [54] assumed instead a rigid structural geometry, but generated full scans of the conformational energy E versus the backbone torsion angles ϕ_i and ϕ_{i+1}. In one set of calculations referred to as NR for "no relaxation," no geometry relaxation was selected. In a second set of calculations denoted as PR for "partial relaxation," optimization of the methyls' rotational angles in the [—$Si(CH_3)_2$—] model system was carried out so as to minimize the sidechain torsional energy for a given backbone conformation. The conformational energies so derived were to serve as input for calculations of the chains' dimensions based on RIS theory [2].

An example of the chain segments considered for these calculations is shown in Figure 5. For the NR calculations, the nonbonded interactions were described by an exp-6 function with the A, B, and C parameters derived as described earlier in this chapter. For the PR calculations, the nonbonded energies were calculated by an LJ 12-6 function with parameters A' and C' similarly derived. In these calculations, the electrostatic interactions were considered negligible and omitted.

Welsh, Debolt, and Mark [54] stressed that important differences are apparent between the polysilanes and analogous alkanes in terms of struc-

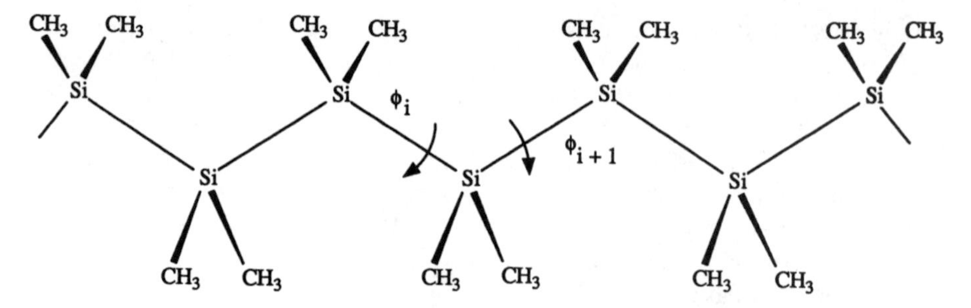

Figure 5 Illustration of a typical polysilane chain segment.

tural geometry and applicable force fields, and these differences are manifested in their respective conformational properties. In particular, the bond lengths for the Si—Si and Si—C bonds are 2.34 Å and 1.87 Å, respectively, compared with 1.53 Å for the C—C bond. Furthermore, the potential energy minimum for the Si···Si nonbonded interactions found in the polysilanes is roughly four times deeper and located 0.5 Å more distant than that for the corresponding C···C interactions in their carbon-based counterparts. Moreover, the intrinsic torsional barrier E^0_{tor} was set at only 1.2 kcal mol^{-1} for the Si—Si bond, compared with 2.8 kcal mol^{-1} typically employed for the C—C bond [2]. As a result of these differences, the polysilanes are significantly more flexible conformationally than the analogous alkanes, both in terms of thermodynamic flexibility (larger low-energy domains) and kinetic flexibility (lower barriers between minima).

Calculations of the conformational energy $E(\phi_i, \phi_{i+1})$ over the full range of torsion angles ϕ_i and ϕ_{i+1} were used to produce conformational energy maps and to evaluate values of the conformational partition function z, the average energy $\langle E \rangle$, and the average torsional angle $\langle \phi \rangle$ associated with each low-energy domain.

A standard three-state (i.e., t, g^+, g^-) model was adopted within the RIS framework [2]. In the case of the NR and PR calculations, the statistical weight parameters for each rotational state were derived directly from an analysis of relative values of z in a manner similar to that employed by Suter and Flory [71] as described above. An important advantage is gained by comparing z values rather than merely energy minima when evaluating the statistical weights or, indeed, any aspect of relative conformational stability. As alluded to earlier, the value of z takes into account the size and shape of a domain (i.e., the "entropy factor") as well as its energy. As such, it represents a physically more realistic measure of relative conformational stability. However, proper evaluation of conformational partition functions requires quantitative investigation of the entirety of configurational space scanned using a fine grid. The literature contains numerous examples of conformational energy studies, particularly those employing energy minimization techniques, in which predictions of relative conformational stability are based solely on comparisons of energy minima without due consideration of domain size. Such an analysis invites the risk of serious error [71]. This is particularly so in the case of conformationally flexible molecules, such as the polysilanes and polygermanes, where domain sizes around energy minima may be substantial. In fact, it is not uncommon for predictions of relative conformational stability based solely on energy minima to be contradistinctive to those based on conformational partition functions [54].

In the absence of conformational maps from the FR calculations [79] to afford evaluation of z, values of the statistical weight parameters based on the FR calculations were taken as simple Boltzmann factors in the energy for the various energy minima. This procedure is tantamount to assuming equal domain sizes circumscribing the various energy minima. Characteristic ratios C_n based on the results of the FR, PR, and NR calculations were then determined by standard RIS methods [2].

In the case of the [—SiH_2—] segment, nearly all regions of configurational space were within 2 kcal mol^{-1} of the global minima. This chain was thus predicted to be highly flexible, in contrast to the analogous polyethylene chain for which large regions of prohibitively high energy are found [2,36]. Based on values of the energy minima obtained from both the NR and FR calculations, the preference of rotational states followed the order $GG > TG > GG^- > TT$. However, based on values of z derived from the NR calculations the order rearranged to $TG > TT > GG > GG^-$. It should be noted that all four rotational states fell within a range of 0.7 kcal mol^{-1}. Even the $G^{+/-}G^{-/+}$ states, which for the n-alkanes and many organic polymers give rise to the energetically prohibitive "pentane effect," [2] are readily accessible to the [—SiH_2—] chain. These results indicate that the [—SiH_2—] chain is conformationally similar to the fictitious "freely rotating chain" frequently alluded to in treatments on RIS theory [2].

With regard to the [—$Si(CH_3)_2$—] chain, the three force-field methods produced qualitatively different results. The FR calculations of Damewood and West [79] indicated a preference for the GG state, with the TT and TG states higher by about 0.9 kcal mol^{-1}. According to the NR calculations, the TG and TT states are preferred and nearly isoenergetic, while the GG states were found higher by about 4 kcal mol^{-1}. Seemingly splitting the difference between the NR and FR results, the PR calculations found the TG and GG states favored and nearly isoenergetic while the TT state was higher by about 0.4 kcal mol^{-1}. All three force-field methods found the GG^- states prohibitively high in energy; thus, even for the polysilanes (as found for polyethylene and the n-alkanes [2,36]) the "pentane effect" re-emerges once the steric bulk of the pendant groups becomes significant.

The generation of conformational energy maps afforded by the NR and PR calculations on the [—$Si(CH_3)_2$—] chain segment again permitted evaluation of the partition function z associated with each rotational state. Interestingly, based on values of z both methods predict the order of preference as $TT > TG >> GG >> GG^-$. In summary, the results seem to suggest that the TT, TG, and GG states are all accessible energetically, yet the TT state and possibly the TG states are more probable statistically (i.e., entropically) based on their greater domain size. Above all, the analysis of these polysilanes by Welsh, Debolt, and Mark [54] dramatizes, as did the earlier paper by Suter and Flory [71], the importance of considering both domain sizes and relative energies, rather than energy minima alone, when comparing the conformational stability of rotational states.

Values of C_n were determined by standard RIS procedures [2] using conformational energies obtained from the foregoing force-field calculations. Both the FR and NR methods yielded $C_n = 4.0$ at 25°C for the [—SiH_2—] chain. The low value reflects the high degree of conformational flexibility of these chains predicted by the force-field calculations. Comparative values for polyethylene and poly(dimethylsiloxane), two flexible polymers of note, are 6.7 [2,36] and 6.4 [2,81], respectively.

Values of C_n at 25°C for [—$Si(CH_3)_2$—] were 15, 13, and 12 based on the NR, PR, and FR methods, respectively. It was thus concluded that [—$Si(CH_3)_2$—] is more extended than [—SiH_2—], as expected considering the increased steric demands in the former. The closeness in values of C_n is surprising given the appreciable qualitative differences in conformational preferences predicted by the three force-field methods. The unexpected similarity was rationalized on the basis of compensating factors. Specifically, the inclusion of sequences such as ...$TTTT$... and ...$TGTG$... suggested from the NR results leads to higher chain dimensions than the corresponding sequences ...$GGGG$... suggested from the FR results. However, the NR results also suggest the appearance of sequences such as ...$TGTG^-$... and ...$TGTTG^-$... which would tend (in a compensatory manner) to foreshorten the chain dimensions.

The Conformational Features of Polycarbonates

In a 1989 paper [69], Sundararajan undertook a comparison of the conformational characteristics of a series of polycarbonates differing by the substituent at C_α. This work represented an extension of the large body of earlier theoretical and experimental studies [70,82–97] carried out on the polycarbonates. The familiar polymer bisphenol A polycarbonate

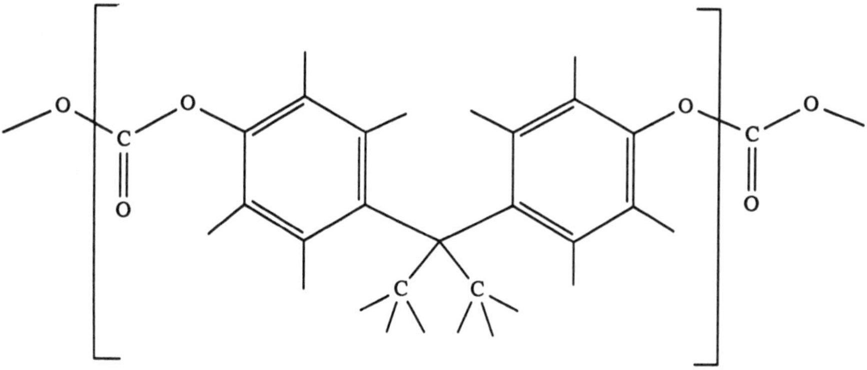

Figure 6 Sketch of the prototypical polycarbonate BPABC.

(BPAPC) is illustrated in Figure 6. The primary purpose of Sundararajan's analysis was to assess the effect of the substituents on the rotational flexibility of the backbone phenylene groups and to evaluate in each case the chain dimensions. The C_α substitutions considered by Sundararajan were (1) CH_3,CH_3 (corresponding to BPAPC); (2) H,H; (3) H,CH_3; (4) H,Ph (Ph=phenyl); (5) CH_3,Ph; (6) Ph,Ph; (7) Cy (Cy=cyclohexyl); and (8) =CCl_2, all of which were represented by model compounds.

The force field consisted of an LJ 12-6 potential to describe the nonbonded interactions, while the torsional contributions associated with rotation about the C_α—Ph bonds were deliberately omitted as negligibly small (about 1 kcal mol^{-1}). However, a quadratic torsional potential with E_{tor}° = 15 kcal mol^{-1} was included for the rotation about the C_α=CCl_2 bond.

To obtain a quantitative measure of relative conformational flexibility across this series of polycarbonates, values of the partition function z for each were evaluated. These were presented as Z, which included rotation of the backbone phenylene groups but not the C_α substituents, and Z', which included rotation of both the backbone and substituent groups.

In the case of CH_3,CH_3 substitution (corresponding to BPAPC), the conformational energy map showed a continuous range of conformations within the 2–3 kcal mol^{-1} contours. This feature was noted earlier both by Tonelli [70] and by Bicerano and Clark [93]. The preferred conformation found was that in which the two phenylene groups are twisted (like a propeller) in opposite directions by about 40–60°. Conversely, conformations in which the phenylene groups become coplanar are highly disfavored due to the strong steric repulsion between the ortho hydrogens. The energy needed to assume a "butterfly" conformation is high at about 7.5–9.0 kcal mol^{-1}, however Sundararajan pointed out that complete π

flips of one phenylene can be achieved at much lower energy cost (2–3 kcal mol^{-1}) if the other phenylene is permitted to adjust rotationally in a concerted manner. These conclusions are consistent with the calculations of Bicerano and Clark [93] and of Jones et al. [97] using semiempirical molecular orbital methods. Calculated values of Z, Z', and C_n are 179, 411, and 1.79, the latter value being in good agreement with the experimental values averaging about 1.9 [98–100].

For the case of H,H substitution, Sundararajan found isolated domains of low energy in contrast to the continuous domains found for the case of CH_3,CH_3 substitution. The presence of these localized domains suggested that the phenylenes can engage in broad oscillatory motions about their conformational minima, but complete π flips (even concerted ones) are energetically prohibitive. Thus, even though the H,H substituents are sterically less bulky than the CH_3,CH_3 substituents, the conformational profile of the former case is more restrictive. For the case of H,H substitution, the values of Z and Z' correspondingly dropped (both to 84) while C_n (1.8) was affected only slightly.

The results found for the case of H,CH_3 substitution resembled those found for the case of H,H substitution, particularly in terms of the presence of localized rather than continuous low-energy domains. However, these localized domains were more contracted, thus the values of Z and Z' (41 and 75, respectively) were appropriately smaller.

Across the series of substitutions H,Ph, to CH_3,Ph, to Ph,Ph, the conformational energy profiles were found to become more restrictive. For the case of H,Ph substitution, the low-energy domains were again continuous, suggesting the possibility of concerted π flips by the backbone phenylenes. The conformational domains were small and again localized in the case of CH_3,Ph substitution, suggesting the possibility of low-amplitude oscillations.

For the case of substitution by Cy in its preferred chair form, Sundararajan found the full range of backbone conformations accessible (<3.5 kcal mol^{-1}) to one phenylene as long as the other phenylene assumed a "butterfly" orientation. For the case of $=CCl_2$ substitution, the conformational energy map looked surprisingly similar to that found for the case of H,H substitution. The localized conformational domains suggested the possibility of oscillations by the backbone phenylenes but of very low amplitude. This theoretical prediction is consistent with NMR evidence [97] indicating that the phenylene motion in this case is more restricted compared to that in BPAPC.

Sundararajan's study demonstrates convincingly that a polymer's freedom of rotation depends heavily on the nature of the substituent, and moreover that a decrease in the steric bulk of a substituent is not necessarily conmmensurate with increased conformational flexibility or versatility of the polymer chain. That the conformational flexibility (and the value of Z or Z') for the case of CH_3,CH_3 substitution is actually greater than that for the case of H,H substitution is a case in point. Sundararajan [69] concluded with the observation that, in light of their limited conformational freedom, the polycarbonates cannot be considered as freely rotating chains. Still, the unique conformational symmetry of this series of polymers permitted their treatment as such in previous studies [70,82]. Indeed, for this series under study the values of C_n calculated by Sundararajan all fell within the range 1.73–1.80; these are barely higher than the value of 1.736 corresponding to the freely rotating chain.

CONCLUDING REMARKS

Force-field methods undoubtedly began in part to bridge the gap between the need for a facile yet versatile theoretical method suitable for quantitative conformational analysis and the scarcity of computer resources and complexity of quantum mechanical methods to serve in this capacity. Despite the recent dawning of vectorized and massively parallel supercomputers, it is not likely that force-field methods will gradually fade into obscurity. The reasons for this assertion are manifold. First of all, force-field techniques are so thoroughly entrenched as almost routine tools in computational chemistry [1,5,10]. Their applications now span virtually all branches of chemical science. Moreover, compared with that for quantum mechanical methods, the hamiltonian associated with force field methods is more easy to comprehend and interpret physically. Additionally, force fields are now widely employed to generate the potential energy surface required in applications of classical molecular dynamics (MD) and Monte Carlo (MC) simulations on macromolecular systems [64]. In fact, virtually all of the popular turn-key molecular modeling computer packages offer at least one "canned" force-field procedure with provisions for conducting energy minimizations and a variety of MD simulation procedures (all pleasingly animated and analyzed graphically on computer graphics monitors if desired), and the appearance of complementary MC algorithms is at hand [64].

Force-field methods thus serve as indispensable tools for conducting conformational analysis on macromolecular systems and for simulating real-time dynamic behavior, both of which are presently beyond the reach of quantum mechanical analysis. In such applications, quantum mechanical calculations still can be used to supply many of the input parameters constituting the classical force field. In this manner, the force-field and quantum mechanical methods can work in tandem and indeed are often applied in such a manner. Above all, the ubiquity achieved by force-field methods gives testimony to their general reliability for interpreting and even predicting the physical and chemical properties of molecular systems of all types [1,5,10–12].

ACKNOWLEDGMENTS

The author wishes to acknowledge the funding agencies who supported the author's work cited in this article, including the Air Force Office of Scientific Research, The Petroleum Research Fund (administered by the donors of the American Chemical Society), The Plastics Institute of America, and The Research Corporation. The author also wishes to thank Professor James E. Mark for sharing our mutual research interest in this field.

REFERENCES

1. J. E. Williams, P. J. Stang, and P. v. R. Schleyer, in *Annual Review of Physical Chemistry*, Vol. 19, pp. 531–558 (H. Eyring, C. J. Christensen, and H. S. Johnston, eds.), Annual Reviews, Palo Alto, CA (1968).
2. P. J. Flory, *Statistical Mechanics of Chain Molecules*, Interscience, New York (1969).
3. J. E. Mark, *Accts. Chem. Res.*, *12*, 49 (1979).
4. M. V. Volkenstein, *Configurational Statistics of Polymeric Chains*, Interscience, New York (1963); T. M. Birshtein and O. B. Ptitsyn, *Conformations of Macromolecules*, Science Publishing House, Moscow (1964).
5. A list of review articles on the force-field method would include, in addition to references 1 and 10, the following: (a) E. M. Engler, J. D. Andose, P. v. R. Schleyer, *J. Am. Chem. Soc.*, *95*, 8005 (1973); (b) C. Altona and D. H. Faber, *Fortschr. Chem. Forsch.*, *45*, 1 (1974); (c) J. D. Dunitz and H.-B. Bürgi, *MTP Int. Rev. Sci.*: *Org. Chem.*, *81* (1976); (d) N. L. Allinger, *Adv. Phys. Org. Chem.*, *13*, 1 (1976); (e) O. Ermer, *Struct. Bonding (Berlin)*, *27*, 161 (1976); (f) D. N. White, *J. Comput. Chem.*, *1*, 225 (1977); (g) A. I. Kitaigorodsky, *Chem. Soc. Rev.*, *7*, 133 (1978); (h) K. Mislow, D. A. Dougherty, and W. D. Hounshell, *Bull. Soc. Chim. Belg.*, *87*, 555 (1978).
6. A. J. Hopfinger, *Conformational Properties of Macromolecules*, Academic Press, New York (1973).
7. G. Chang, W. C. Guida, and W. C. Still, *J. Am. Chem. Soc.*, *111*, 4379 (1989).
8. M. J. S. Dewar, E. G. Zoebisch, E. F. Healy, and J. J. P. Stewart, *J. Am. Chem. Soc.*, *107*, 3902 (1985).
9. R. G. Parr and W. Yang, *Density Functional Theory of Atoms and Molecules*, Oxford Press, New York (1989).
10. U. Burkert and N. L. Allinger, *Molecular Mechanics*, ACS Monograph 177, American Chemical Society, Washington, DC (1982).

11. T. Clark, *A Handbook of Computational Chemistry*, Wiley-Interscience, Wiley, New York (1985).
12. D. M. Hirst, *A Computational Approach to Chemistry*, Blackwell Scientific, London (1990).
13. J. A. McCammon and S. C. Harvey, *Dynamics of Proteins and Nucleic Acids*, Cambridge University Press, Cambridge (1987).
14. C. L. Brooks III, M. Karplus, and B. M. Pettitt, *Proteins: A Theoretical Perspective of Dynamics, Structure, and Thermodynamics*, Wiley, New York (1988).
15. W. F. van Gunsteren and P. K. Weiner, eds., *Computer Simulation of Biomolecular Systems*, ESCOM Science, Leiden, The Netherlands (1989).
16. R. A. Scott and H. A. Scheraga, *J. Chem. Phys.*, *44*, 3054 (1966).
17. D. A. Brant, A. E. Tonelli, and P. J. Flory, *Macromolecules*, *2*, 228 (1969).
18. W. G. Miller, D. A. Brant, and P. J. Flory, *J. Mol. Biol.*, *23*, 67 (1967).
19. N. L. Allinger, Y. H. Yuh, and J.-H. Lii, *J. Am. Chem. Soc.*, *111*, 8551 (1989); J.-H. Lii and N. L. Allinger, *J. Am. Chem. Soc.*, *111*, 8566, 8576 (1989).
20. F. H. Westheimer, in *Steric Effects in Organic Chemistry*, Chap. 12, (M. S. Newman, ed.), Wiley, New York (1956).
21. J. B. Hendrickson, *J. Am. Chem. Soc.*, *83*, 4537 (1961); *84*, 3355 (1962); *86*, 4854 (1964).
22. K. B. Wiberg, *J. Am. Chem. Soc.*, *87*, 1070 (1965).
23. W. J. Orville-Thomas, ed., *Internal Rotation in Molecules*, Wiley, London (1974).
24. L. Pauling, *Proc. Nat. Acad. Sci.*, *44*, 211 (1958).
25. L. Pauling, *The Nature of the Chemical Bond*, 3rd ed., p. 130, Cornell University Press, Ithaca, NY (1960).
26. R. H. Scott and H. A. Scheraga, *J. Chem. Phys.*, *42*, 2209 (1965).
27. J. O. Hirschfelder, C. F. Curtiss, and R. B. Bird, *Molecular Theory of Gases and Liquids*, Wiley, New York (1954).
28. G. C. Maitland, M. Rigby, E. B. Smith, W. A. Wakeman, *Intermolecular Forces* (R. Breslow, J. B. Goodenough, J. Halpern, and J. S. Rowlinson, eds.), Clarendon Press, Oxford (1987).
29. G. Mie, *Ann. Phys.*, *11*, 657 (1903).
30. J. F. Lennard-Jones, *Proc. Roy. Soc.* (London), *Ser. A*, *106*, 463 (1924).
31. K. S. Pitzer, *Adv. Chem. Phys.*, *2*, 59 (1959).
32. J. C. Slater and J. G. Kirkwood, *Phys. Rev.*, *37*, 682 (1931).
33. J. Ketelaar, *Chemical Constitution*, Elsevier, p. 91, Amsterdam (1953).
34. A. Bondi, *J. Phys. Chem.*, *68*, 441 (1964).
35. N. L. Allinger, M. A. Miller, F. A. Catledge, J. A. Hirsch, *J. Am. Chem. Soc.*, *89*, 4345 (1967).
36. A. Abe, R. L. Jernigan, and P. J. Flory, *J. Am. Chem. Soc.*, *88*, 631 (1966).
37. W. J. Welsh, D. Bhaumik, and J. E. Mark, *J. Macromol. Sci.-Phys.*, *B20(1)*, 59 (1981).
38. A. Abe and J. E. Mark, *J. Am. Chem. Soc.*, *98*, 6468 (1976).
39. R. Sundararajan and P. J. Flory, *J. Am. Chem. Soc.*, *96*, 5025 (1974).
40. A. Abe, *Macromolecules*, *13*, 541 (1980).

41. G. D. Smith and R. H. Boyd, *Macromolecules*, *23*, 1527 (1990).
42. P. R. Sundararajan, *Macromolecules*, *11*, 256 (1978).
43. J. E. Mark and C. Sutton, *J. Am. Chem. Soc.*, *95*, 1083 (1972).
44. E. Riande, J. G. de la Campa, D. D. Schlereth, J. de Abajo, and J. Guzman, *Macromolecules*, *20*, 1641 (1987).
45. A. L. McClellan, *Tables of Experimental Dipole Moments*, vol. I, W. H. Freeman, San Francisco, CA (1963); vol. II, Rahara Enterprises, El Cerrito, CA (1974).
46. W. J. Welsh, D. Bhaumik, and J. E. Mark, *Macromolecules*, *14*, 947 (1981).
47. D. Bhaumik and J. E. Mark, *Macromolecules*, *14*, 162 (1981).
48. D. Bhaumik, W. J. Welsh, H. H. Jaffe, and J. E. Mark, *Macromolecules*, *14*, 951 (1981).
49. W. J. Welsh, in *Current Topics in Polymer Science*, Vol 1, pp. 217–234 (R. M. Ottenbrite, L. A. Utracki, and S. Inoue, eds.), Hanser (1987).
50. U. W. Suter, *J. Am. Chem. Soc.*, *101*, 6481 (1979).
51. M. J. S. Dewar, R. C. Bingham, and D. H. Lo, *J. Am. Chem. Soc.*, *97*, 1285 (1975).
52. M. J. S. Dewar and W. Thiel, *J. Am. Chem. Soc.*, *99*, 4899 (1977).
53. W. J. Hehre, L. Radom, P. v. R. Schleyer, and J. A. Pople, *Ab Initio Molecular Orbital Theory*, Wiley-Interscience, New York (1986).
54. W. J. Welsh, L. Debolt, and J. E. Mark, *Macromolecules*, *19*, 2978 (1986).
55. A. Abe, *J. Am. Chem. Soc.*, *106*, 14 (1984).
56. L. E. Sutton, *Tables of Interatomic Distances and Configuration in Molecules and Ions*, Chemical Society Special Publication Nos. 11, 18, Chemical Society, London (1958), (1965).
57. F. H. Allen, O. Kennard, and D. G. Watson, L. Brammer, and A. G. Orpen, *J. Chem. Soc. Perkin Trans. II*, *S1* (1987).
58. E. B. Wilson, Jr., J. C. Decius, and P. C. Cross, *Molecular Vibrations*, Dover, New York (1955).
59. L. Radom, W. J. Hehre, and J. A. Pople, *J. Am. Chem. Soc.*, *94*, 2371 (1970).
60. A. J. Hopfinger and P. A. Pearlstein, *J. Comput. Chem.*, *5*, 486 (1984).
61. J. Bicerano, *Macromolecules*, *22*, 1408 (1989).
62. D. W. Noid, G. A. Pfeffer, S. Z. D. Cheng, and B. Wunderlich, *Macromolecules*, *21*, 3482 (1988).
63. D. T. Baldwin, W. L. Mattice, and R. D. Gandour, *J. Comput. Chem.*, *5*, 241 (1984).
64. A number of relevant papers were presented at the Symposium on Computer Simulation of Polymers during the National Meeting of the American Chemical Society, Miami Beach, FL, September, 1989; Polym. Prepr., Div. of Polymer Chemistry, Inc., pp. 1–99, American Chemical Society, Washington, DC, *30*(2) (1989).
65. P. Morse, *Phys. Rev.*, *34*, 57 (1929).
66. R. Napolitano, *Macromolecules*, *21*, 622 (1988).
67. A. Tonelli, *Macromolecules*, *15*, 290 (1982).
68. B. L. Farmer, J. F. Rabolt, and R. D. Miller, *Macromolecules*, *20*, 1169 (1987).

69. P. R. Sundararajan, *Macromolecules*, *22*, 2149 (1989).
70. A. E. Tonelli, *Macromolecules*, *5*, 503, 558 (1972).
71. U. W. Suter and P. J. Flory, *Macromolecules*, *8*, 765 (1975).
72. P. J. Flory, *J. Polym. Sci., Polym. Phys. Ed.*, *11*, 621 (1973).
73. P. J. Flory, *Macromolecules*, *7*, 381 (1974).
74. U. Biskup and H. J. Cantow, *Macromolecules*, *5*, 546 (1972).
75. Y. Fujiwara and P. J. Flory, *Macromolecules*, *3*, 280 (1970).
76. A. Abe, *Polym. J.*, *1*, 232 (1970).
77. For recent reviews, see (a) J. M. Zeigler, *Synthetic Metals*, *28*, C581 (1989); (b) R. D. Miller and J. Michl, *Chem. Rev.*, *89*, 1359 (1989).
78. A number of relevant papers were presented at the Symposium on Sigma Conjugated Polymers held during the National Meeting of the American Chemical Society, Washington, DC, August 1990; Polym. Prepr., Div. of Polymer Chem., Inc., pp. 221–305, American Chemical Society, Washington, DC, *31*(2) (1990).
79. J. R. Damewood, Jr. and R. West, *Macromolecules*, *18*, 159 (1985).
80. N. L. Allinger, *J. Am. Chem. Soc.*, *99*, 8127 (1977).
81. V. Crescenzi and P. J. Flory, *J. Am. Chem. Soc.*, *86*, 141 (1964).
82. A. D. Williams and P. J. Flory, *J. Polym. Sci., Polym. Phys. Ed.*, *6*, 1945 (1968).
83. B. Erman, D. C. Marvin, P. A. Irvine, and P. J. Flory, *Macromolecules*, *15*, 664 (1982).
84. P. R. Sundararajan, *Can. J. Chem.*, *63*, 103 (1985).
85. P. R. Sundararajan, *Macromolecules*, *20*, 68 (1987).
86. S. Perez and R. P. Scaringe, *Macromolecules*, *20*, 1534 (1987).
87. B. C. Laskowski, D. Y. Yoon, D. McLean, and R. L. Jaffe, *Macromolecules*, *21*, 1629 (1988).
88. A. A. Jones, in *High Resolution NMR Spectroscopy of Synthetic Polymers in Bulk*, Chap. 7 (and references cited therein), (R. A. Komoroski, ed.), VCH, Deerfield Beach, FL (1986).
89. P. Tekely and E. Turska, *Polymer*, *24*, 667 (1983).
90. M. A. Mora, M. Rubio, and C. A. Cruz-Ramos, *J. Polym. Sci., Polym. Phys. Ed.*, *24*, 239 (1986).
91. R. A. Davenport and A. J. Manuel, *Polymer*, *18*, 557 (1977).
92. J. T. Bendler, *Ann. N.Y. Acad. Sci.*, *371*, 299 (1981).
93. J. Bicerano and H. A. Clark, *Macromolecules*, *21*, 585, 597 (1988).
94. P. Tekely, *Macromolecules*, *19*, 2544 (1986).
95. D. Perchak, J. Skolnick, and R. Yaris, *Macromolecules*, *20*, 121 (1987).
96. A. F. Yee and S. A. Smith, *Macromolecules*, *14*, 54 (1981).
97. A. A. Jones, J. F. O'Gara, P. T. Inglefield, J. T. Bendler, A. F. Yee, and K. L. Ngai, *Macromolecules*, *16*, 658 (1983).
98. W. Gawrisch, M. G. Brereton, and E. W. Fischer, *Polym. Bull. (Berlin)*, *4*, 687 (1981).
99. D. Y. Yoon and P. J. Flory, *Polym. Bull. (Berlin)*, *4*, 693 (1981).
100. G. V. Reddy, M. Bohdanecky, D. Staszewska, and L. Huppenthal, *Polymer*, *29*, 1894 (1988).

3

Polydimethylsiloxane: Conformational Analysis and Configurational Properties

Stelian Grigoras
Dow Corning Company
Midland, Michigan

INTRODUCTION

One of the most important structural features of a polymer is chain flexibility of the backbone. A promising tool for the prediction of chain flexibility, based solely on molecular structure, is given by conformational analysis. Generally, conformational analysis identifies the stable isomeric states for polymer chains and the energy barriers between them. Conformational analysis, in the case of polymers, can become a cumbersome task, because the architecture of polymer chains allows for a large number of degrees of freedom which must be studied simultaneously. To obtain meaningful information, the analysis must be simplified, and only the most significant conformational elements should be studied. The rotational isomeric states (RIS) model [1,2] is one of the ways to analyze the conformational states of polymers but, more importantly, provide data which can be related to measurable properties of polymers.

Polysiloxanes constitute an excellent family of polymers to be studied in terms of flexibility, because they are known to have a high degree of conformational freedom. Furthermore, this class of polymers becomes more relevant considering the fact that this is a typical example of an inorganic polymer with properties and performances which cannot be attained by organic polymers. The conformation of polydimethylsiloxane (PDMS) is different from that of other commodity polymers like polyethylene, for example, and so are its properties. Previous RIS studies [3–7] of PDMS attempted to describe the conformation of the backbone of this polymer by using an analogy with the conformational states of polyethylene (PE), i.e., a three-rotational-states model consisting of the *trans* state and two *gauche* states with *trans* state being the lowest energy state. Recently, it has been shown that the conformation of PDMS is different from that of polyethylene [8–10]. The most stable conformation of PDMS is described by the *cis-trans* sequence of states in contrast to PE, where the *trans-trans* conformation is the most stable. This result was obtained from ab initio molecular orbitals (MO) calculations [10] and the study was extended to larger molecular structures using molecular mechanics potentials derived from fitting the ab initio MO conformational analysis. It has been shown that the electrostatic potential, in addition to steric potential, plays a major role in determining the *cis-trans* sequential conformation as being the most stable [10]. This conformation, *cis-trans*, was suggested in 1985, by Tsvankin et al. [11] for crystalline polydiethysiloxane based on x-ray scattering data. Furthermore, a very recent study [12] reports molecular mechanics and molecular dynamics results in agreement with the newly proposed conformation for PDMS. This recent study, elegantly shows the conformational differences between PDMS and PE chains, and using molecular dynamics simulations relates the probability distribution of isomeric states to the statistical behavior of the chain and to the macrocyclization equilibrium constants.

This paper reports the results of a conformational study of PDMS and describes the isomeric states of the backbone chains in terms of the RIS model. The conformational analysis provides the data on the isomeric states which can be used directly to calculate the configurational properties of this polymer.

COMPUTATIONAL METHODS

Molecular Geometry

The conformational analysis, reported herein is based on the variation of nonbonding interactions for torsional deformations. The valence geometry has been maintained constant, including the Si—O—Si bond angle, but the conformational analysis has been repeated for different values of this bond angle. All bond lengths and all other angles were maintained constant at values which are listed in Table 1. The CH_3 group has been approximated

Table 1 Valence Geometry of Oligomers Included in the Present Study

Si—O	1.645 Å	Si—O—Si	149.5°
Si—Me	1.897 Å	O—Si—O	109.3°
O—H	0.965 Å	O—Si—Me	108.8°
		Si—O—H	112.5°

[a]Si—O—Si bond angles studied: 145°, 147°, 149.5°, 152°.

as a united atom. The nonbonding molecular mechanics potentials used to characterize the conformation of PDMS chains are described below.

Nonbonding Potentials

Previously [10], a set of molecular mechanics potentials was generated and compared with the ab initio MO results for organosilicon compounds. The nonbonding interactions include two terms: steric and electrostatic.

The steric potential is calculated by using a Hill function with van der Waals radii approximately 20% larger than the original values [13]. The electrostatic potential is based on net atomic charges calculated by using molecular orbital method from a Mullikan population analysis. The net atomic charges are calculated with the extended Huckel theory method [14]. A dielectric constant of 8.7 offers the best fit for the electrostatic interactions. The dielectric constant used in the actual context has a role of adjusting factor for the net atomic charges, as they were calculated, rather than being considered having a physical meaning. The parameters used for the steric potentials and the net atomic charges for electrostatic potentials are listed in Table 2.

Overall, this approach provides good agreement with the ab initio MO results of nonbonding interactions without the use of additional empirical torsional potentials. Also, it provides the capability to analyze independently the contribution of steric versus electrostatic potentials to specific conformations.

Table 2 Parameters for Steric Potentials and Net Atomic Charges

Atom	r^*	ε	Charge (e.u.)
Si	2.70	0.036	+1.8363
O	2.088	0.036	−1.1976
Me	2.65	0.08	−0.3263
H (at OH)	1.80	0.0494	+0.6043
O (at OH)	2.088	0.036	−1.2723

Scanning with Full Torsional Relaxation

The present conformational analysis includes a systematic torsional deformation of one or two consecutive bonds and the full torsional relaxation of the other bonds in the system. The results were obtained with CHEMLAB II molecular modeling package [15] and with an original program, SCAN, which allowed for full torsional relaxation of the adjacent bonds.

Torsional relaxation of adjacent bonds was carried as follows. The torsional angle corresponding to one bond, which was assigned as the principal bond, was scanned systematically. In this process, the principal bond is fixed at a specific torsional angle. Then the nearest bonds are scanned successively in increments of 5° for a complete rotation. The torsional angle corresponding to the lowest energy becomes the new value for the corresponding bond, and the next adjacent bond is scanned. This process is continued sequentially for all other bonds included in the molecular structure. To increase the probability of determining the global minimum, this sequential scanning is performed twice. To refine the search for global minimum the sequential scanning is repeated two more times with the starting angle at −30° and smaller increment, of 2°, to the final point at 30° from the value corresponding to the previously determined most stable conformation. The new lowest energy conformation replaces the previous most stable conformation. One more scan is performed from −10 to +10° in increments of 1° and again the lowest energy conformation replaces the previous one. The energy corresponding to this final conformation is recorded as the value associated with the torsional angle of the main bond. Then everything is repeated for a new value of the torsional angle of the main bond.

The advantage of the sequential versus a simultaneous scanning of all bonds is the shorter computer time required for the calculations. This advantage becomes significant for molecular systems with a large number of torsional degrees of freedom. Both methods have similar accuracies in predicting the optimum conformation. The sequential scanning used to search for minimum energy conformations is more efficient than using regular optimization methods, because the bonds are scanned for a complete torsional rotation, therefore the chances to stop the minimization process in a local minimum are lower. However, sequential scanning cannot guarantee that the global minimum is achieved, instead of a local minimum, particularly if the absolute minimum has a very steep variation as a function of the corresponding torsional angles. To reduce the chances of stopping the search at a local minimum, in each case, several starting points and various torsional sequences were used for the relaxation of the bonds, and the conformation corresponding to the lowest energy was recorded.

CONFORMATIONAL ANALYSIS

Bending Flexibility of Si—O—Si Bond Angle

Polysiloxanes show two different levels of flexibility in terms of molecular structure: torsional flexibility, which is common to the majority of organic polymers, and bending flexibility. The torsional flexibility is defined as the ability of atoms to rotate around chemical bonds. Torsional flexibility does not necessarily involve distortion of the valence geometry, i.e., the bond lengths and bond angles of the chemically bonded atoms may be considered unchanged during the torsional rotation, as a first approximation. Bending distorsion generally occurs in cases where the steric repulsion between nonbonded atoms is very large for energetically unfavorable torsional angles.

The bending flexibility is also present in most organic polymers, with an angular amplitude of several degrees, but it is more prominent in polysiloxanes. For a siloxane backbone two different bond angles are formed: Si—O—Si and O—Si—O. The Si—O—Si bond angle is very flexible. This angle has a value between 140° and 180° and a small barrier of linearization (0.3 kcal/mol) [16]. In contrast, the O—Si—O is a rigid bond angle with values between 102 ° and 112° depending on the nature of the two substituents on Si [10].

The nature of the unusual flexibility of Si—O—Si bond angle has been discussed previously [17]. It has been shown, using ab initio MO calculations, that the electronic charge localized at lone pairs on oxygen is transferred to the covalent bonding region between Si and O when a correct basis set [18] is used to calculate the charge density. The lone pairs at oxygen are diminished and the sp^3 hybridization geometry is altered. The consequence is a wider and flexible Si—O—Si bond angle. The electronic charge transferred from the oxygen's lone pairs creates a charge excess in the covalent region between silicon and oxygen. This excess of electronic charge shortens the Si—O bond. An alternative explanation [19], based on frontier orbital analysis, shows that the unusual flexibility of Si—O—Si angle may be attributed to the high energy of the π (SiR_3) group orbitals and poor mixing of π (SiR_3) orbitals with the oxygen lone pairs and to some mixing of the lone-pair orbitals with the relatively low-lying π^* (SiR_3) group orbitals.

Also, it has been shown [17] that the variation of Si—O—Si bond angle significantly alters the values of the potential barriers between conformational minima and the relative energy of these minima. The relaxation of the adjacent bonds has an important effect on bond torsions. It modifies the potential barriers, as well as the energy and the position of the minima. By using the nonbonding potentials, which will be described in the following section, the torsional barriers in hexamethyldisiloxane are estimated in excellent agreement with the experimental values: 0.6 kcal/mol, and 1.6 kcal/mol for the torsions around Si—O and Si—C bonds, respectively [17].

Molecular Structures and Conventions

The strategy used to analyze the conformation of polymers is based on studying small molecular samples, which include the repeating units of the polymer of interest. The effects of chain folding are not taken in consideration at this level of theory, but the results given by this approach provide the starting point and the needed parameters for the study of long polymeric chains with Monte Carlo atomistic approach or molecular dynamics methods. To evaluate the importance of selection of the molecular sample upon the conformational analysis results, several molecular structures have been considered in the present study:

(a) $Me_3Si{-}O\underset{1}{\text{——}}SiMe_2\underset{2}{\text{——}}O{-}SiMe_3$

(b) $HO{-}SiMe_2\underset{1}{\text{——}}O\underset{2}{\text{——}}SiMe_2{-}OH$

(c) $Me_3Si{-}O\underset{1}{\text{——}}SiMe_2\underset{2}{\text{——}}O\underset{3}{\text{——}}SiMe_2\underset{4}{\text{——}}O{-}SiMe_3$

(d) $HO{-}SiMe_2\underset{1}{\text{——}}O\underset{2}{\text{——}}SiMe_2\underset{3}{\text{——}}O\underset{4}{\text{——}}SiMe_2{-}OH$

(e) $Me_3Si{-}O\underset{1}{\text{——}}SiMe_2\underset{2}{\text{——}}O\underset{3}{\text{——}}SiMe_2\underset{4}{\text{——}}$
$O\underset{5}{\text{——}}SiMe_2\underset{6}{\text{——}}O{-}SiMe_3$

The bonds which were scanned systematically are numbered. Each molecular sample has been analyzed, from a torsional deformation point of view, as a two-dimensional conformational map for sequential torsional angles. These angles have been scanned systematically for complete bond rotations of 360° with increment of 10°. The other dihedral angles have been relaxed in order to achieve the conformation with the lowest energy corresponding to each particular pair of fixed torsional angles which were scanned systematically.

The notation for the isomeric states which will be used in this paper is presented below. The torsional angle of 0° corresponds to *cis* (c) conformation, and 180° to *trans* (t). A torsional angle of 160° corresponds to a quasi-*trans* (t) conformation, which will be abbreviated as t^+ and $-160°$ as t. Following this convention g^+ corresponds to a torsional angle of 60° (*gauche*) state or in the vicinity of this value (an angle of 70°, for example, would also be labeled as g^+). Correspondingly, an isomeric state defined by a torsional angle of $-60°$ would be labeled as g^-.

Results and Discussion

The study of PDMS requires the analysis of two sequences of atoms: Si—O—Si—O—Si, and O—Si—O—Si—O.

Figures 1 and 2 show the conformational maps of these two sequences which belong to different molecular structures, and which provide the best description for PDMS chains, between the siloxane molecules considered in the present study. Figure 1 shows the three-dimensional (3-D) maps,

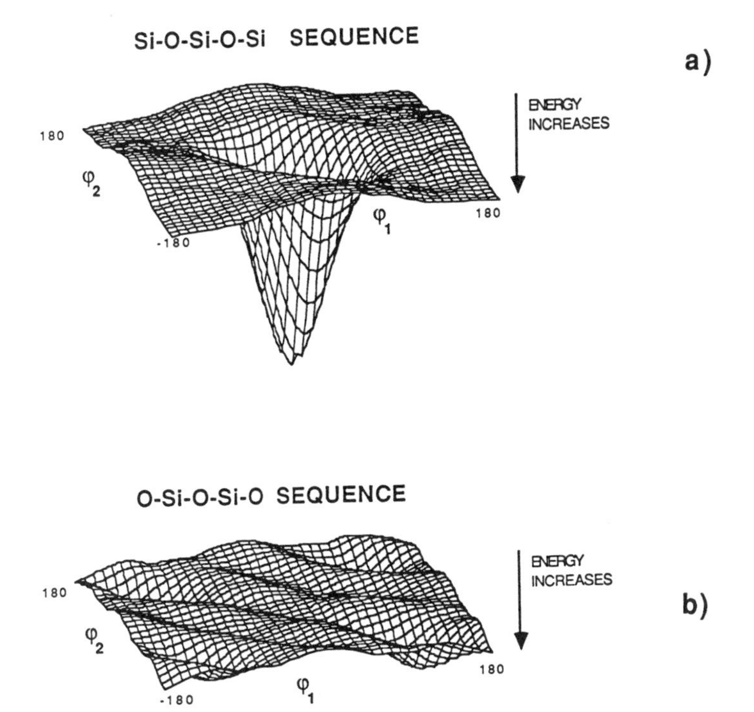

Figure 1 The 3-D conformational maps. (a) The Si—O—Si—O—Si sequence as part of molecular structure **e,** bond i corresponds to O—Si and bond $i+1$ to Si—O. (b) The O—Si—O—Si—O sequence as part of molecular structure **c,** bond i corresponds to Si—O and bond $i+1$ to O—Si.

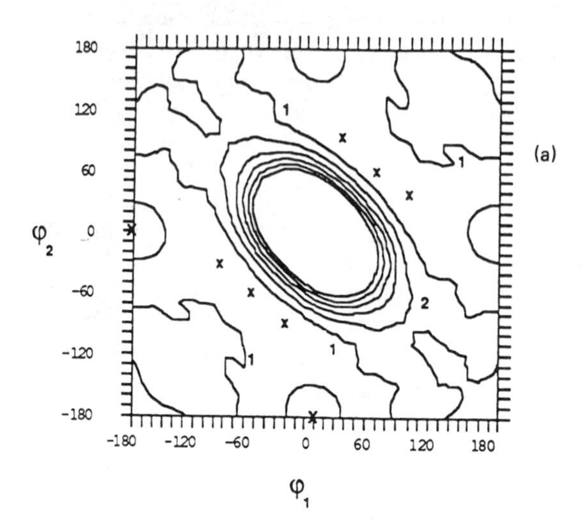

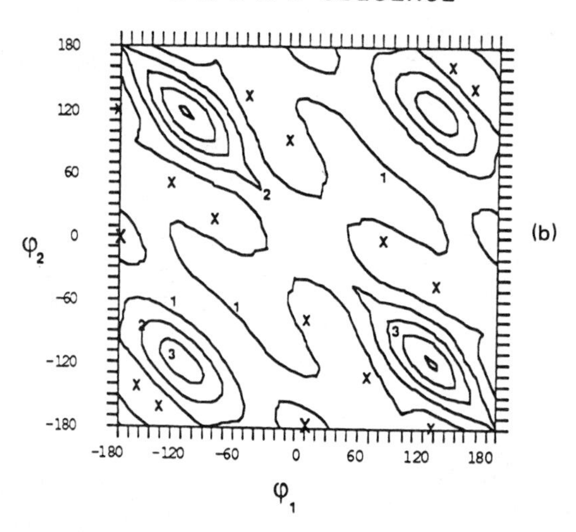

Figure 2 The isoenergy conformational maps. (a) The Si—O—Si—O—Si sequence as part of molecular structure **e,** bond i corresponds to O—Si and bond $i + 1$ to Si—O. (b) The O—Si—O—Si—O sequence as part of molecular structure **c,** bond i corresponds to Si—O and bond $i + 1$ to O—Si. The x labels indicate minima on the map, the numbers indicate the energy (kcal/mol) of the corresponding isoenergy curve.

and Figure 2 the isoenergy contour lines for these two sequences. These conformational maps have a diagonal symmetry and they indicate important differences between these two sequences. These differences will be discussed below.

The first sequence to be discussed in detail is Si—O—Si—O—Si. This sequence as part of molecular structure **e,** with torsional rotations around bonds 3 and 4 provided the results shown in Figures 1(a) and 2(a). The 3-D map of this sequence shows that the lowest energy minima correspond to the pairs of rotational angles (0,180), (180,0) around O—Si and Si—O bonds, respectively. Two local minima are located at (−90, −30) and (−30, −90) having energy 0.5 kcal/mol higher than that of the absolute minima, and two other minima at (−40, −130) and (−130, −40) with energy 0.6 kcal/mol higher than that of the absolute minima. These four local minima are replicated in the upper diagonal part of the conformational map. All local minima are located in + + or − − quadrants, and none in + − or − +, which means that g^+g^- or g^-g^+ sequences are not energetically favored. The *cis-cis* sequence has very high energy, 25 kcal/mol, relative to the absolute minima. *Trans-trans* conformation does not correspond to an energy minimum; therefore it cannot be a stable conformational state. Its energy is 2.75 kcal/mol above the energy of the absolute minima.

The other sequence to be discussed is, O—Si—O—Si—O, is part of structure **c** where the rotations were done around bonds 2 and 3. Figure 1(b) shows for this sequence features which are different than the previous sequence. The diagonal symmetry can be clearly observed in this figure and in Figure 2(b). There are six diagonal low-energy regions oriented symmetrically and parallel to the diagonal line defined by the pairs of angles (180, −180) and (−180,180), with a general aspect of parallel folds. Hence, the low-energy values are distributed on six folds, and the difference in energy between minima and maxima is much lower (less than 6 kcal/mol) than in the case of the previous sequence.

The symmetry of these conformational maps make suitable the analysis of minima along the diagonals on which they are aligned. The six folds are distributed symmetrically along the main diagonal, and for this reason only three regions, which are located in the lower part of the map, will be compared. These three folds can be followed by changing both torsional angles by 10° along the diagonal connecting the following pairs of torsional angles:

(i) (120, −180) and (−180,120)
(ii) (0, −180) and (−180,0)
(iii) (−120, −180) and (−180, −120)

The first region has the lowest energy starting on the map at the pair of angles 120 and −180° and it ends at the point defined by the angles −180° and 120°. Along this line two local minima occur at (−160, −140) and (−140, −160) having an energy slightly higher, 0.5 kcal/mol, than the lowest minimum located at (0, −180). The energy has a variation of 0.2 kcal/mol along this region. The second diagonal region starts at the point defined by 0° and −180° torsional angles and ends at the point (−180,0), points which define the absolute minima. The highest energy in this region is 0.6 kcal/mol at (−90, −90). The third diagonal region starts at the conformation defined by 120°, −180° and it ends at −180°,120°. The variation of the energy in this region is also small (0.5 kcal/mol) with shallow minima. The lowest energy local minima occur at −90,20 and −20,90° having an energy which is 0.6 kcal/mol above the energy of the absolute minima.

The Effect of Molecular Environment

A comparison of the conformational maps obtained from the analysis of the five molecular structures would allow us to identify the effect of different substituents at the terminal atoms of each sequence upon the location of isomeric states and upon the height of potential barriers between these states. In addition, such comparison would allow us to select which structure is relevant for studying the conformation of longer polymeric chains. In the previous selection it has been shown that important differences occur between the Si—O—Si—O—Si and O—Si—O—Si—O bond sequences, and in this section each of these two sequences will be compared in the considered structures.

The first sequence, Si—O—Si—O—Si, in all molecular structures, indicate that *cis-cis* conformation is energetically highly unfavorable and it is located at the top of a potential barrier. The height of this barrier depends upon the steric repulsion between the substituents at the silicon atoms located at the end of the sequence. The highest barrier has a value of 38.5 kcal/mol in the structure **a** which has three methyl groups at each terminal silicon atom. This barrier becomes smaller, 30 kcal/mol, when two methyl groups and siloxane group are at one terminal silicon atom and three methyl groups at the other terminal silicon atom as in structure **c** when bonds 1 and 2, or 3 and 4 are rotated. This barrier becomes 24.4 kcal/mol when the terminal atoms in the sequence have two methyl groups and one siloxane group, as in structure **e** when bonds 3 and 4 are rotated. If the two siloxane groups are replaced by hydroxyl groups as in structure **d,** the energy barrier decreases to a value of 8.7 kcal/mol.

The lowest energy minima on the conformational maps of Si—O—

Si—O—Si sequences correspond to *cis-trans*, or quasi *cis-trans* conformation. In structure **a** the absolute minima occur at ct^+, ct^- states, where t^+ means 160°. The difference in energy between ct^+ and *ct* is 0.1 kcal/mol in this structure.

The *trans-trans* conformation is an energy maximum. The difference in energy between this conformation and the *cis-trans* conformation is 3.2 kcal/mol with the exception of structure **d** which has bonds 2 and 3 rotated, and structure **e,** with bonds 3 and 4 rotated successively. These two structures have one oxygen atom and two methyl groups at each terminal silicon in the sequence. The difference in energy between *tt* and *ct* states in these two cases is 2.7 kcal/mol. The energy of the *tt* conformational sequence is lower in these two cases because the terminal silicon atoms have two methyl groups instead of three, which is the case with the other structures. When the third methyl group at a terminal silicon is replaced by hydroxyl or siloxane groups with negatively charged oxygens, a change in balance between the electrostatic and steric interactions occurs and this structural modification lowers the relative energy of the *tt* conformation.

The comparison of conformational maps for the Si—O—Si—O—Si sequence in various molecular structures indicates that the sequence of bonds 3 and 4 in structure **e** can be considered as the most representative for a PDMS chain. This study indicates that conformational results are distorted if the terminal silicon atoms have as substituents hydroxyl groups, e.g., structure **d,** or trimethyl groups, e.g., structure **a,** instead of two methyl groups and one siloxane group.

The second sequence which will be discussed is O—Si—O—Si—O. The diagonal symmetry is present in all conformational maps analyzed in this study. The orientation of this symmetry on conformational maps indicates that g^+g^- sequences are energetically unfavored. The rotational states in the vicinity of *cis-cis* are energetically allowed. Surprisingly, in structure **b,** the most stable conformations correspond to c^+c^+ and c^-c^- sequences with torsional angles of (20,20) and $(-20, -20)$, respectively. Comparing the energy profile along fold **i** it can be observed, in Figure 3, that the energy change follows a curve with a concave shape for the structures ending with siloxane (**c** and **e**), and a convex shape for the structures ending with hydroxyl groups (**b** and **d**). The energy variation from minimum to maximum value along the curve is only 0.2–0.3 kcal/mol. The lowest energies along these lines have values of 0.2–0.5 kcal/mol above the energy of the absolute minima for **c** and **e** structures, and larger values 1.2–1.8 kcal/mol for **b** and **d** structures which contain hydroxyl groups.

Figure 4 (a) plots the energy change along the second fold, **ii,** for bonds 2 and 3 in structure **c** which is terminated with siloxane groups, and Figure 4(b) shows the same plot for structure **b** terminated with hydroxyl groups.

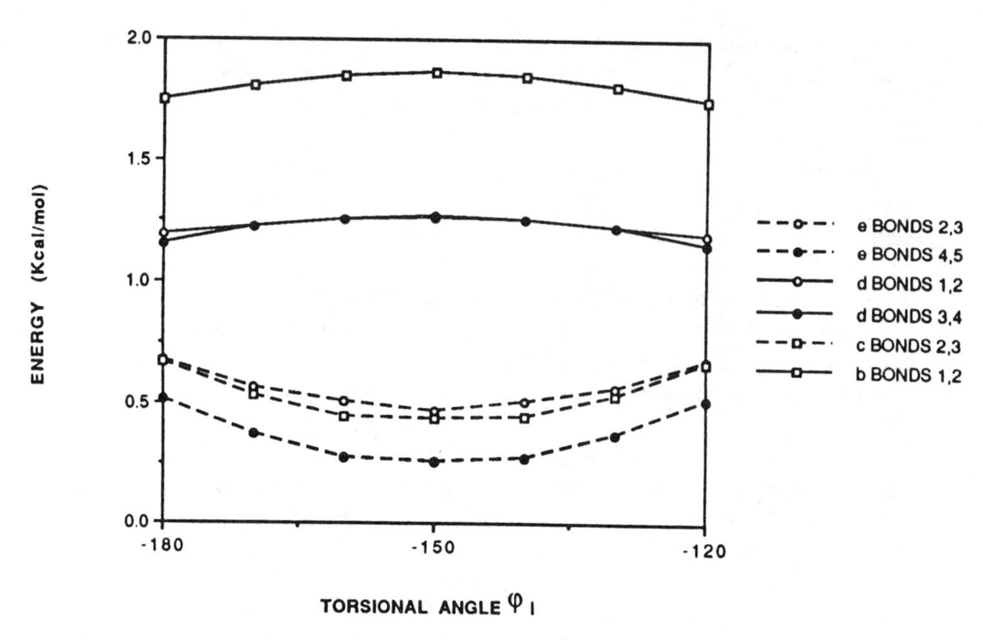

Figure 3 Torsional energy plot for the O—Si—O—Si—O sequence along fold i: from the starting point (−180, −120) both torsional angles have been changed simultaneously by 10°. The ordinate shows only the values of bond *i*.

The curves have different profiles, and they indicate that the siloxane-terminated structures favor the *cis-trans* sequence and the hydroxy-terminated structures favor the c^-t^- conformation, with the energy variation along this fold of 0.9 kcal/mol for structure **c** and 0.35 kcal/mol for structure **b.** The corresponding curves for the other compounds are similar to these shown in Figure 4. The energy change along each curve is less than 1 kcal/mol. The maximum energy on each curve occurs for the pairs of torsional angles (−90, −90), and the energy minima have different locations for these two molecular environments: (0, −180) and (−180,0) for structures terminated with siloxane groups, and (−40, −140) and (−140, −40) for structures terminated with hydroxyl groups.

Figure 5 plots the energy change along the third fold, **iii,** for the same bonds as discussed for the second fold. The energy variation is 0.6 kcal/mol for structure **c** which is terminated with siloxane groups, and 1.5 kcal/mol for structure **b,** terminated with hydroxyl groups. Structure **c**, shows four energy minima, the lowest being located at (−180,120) and (120, −180), and in structure **b** the minima correspond to the *cis-gauche* conformation.

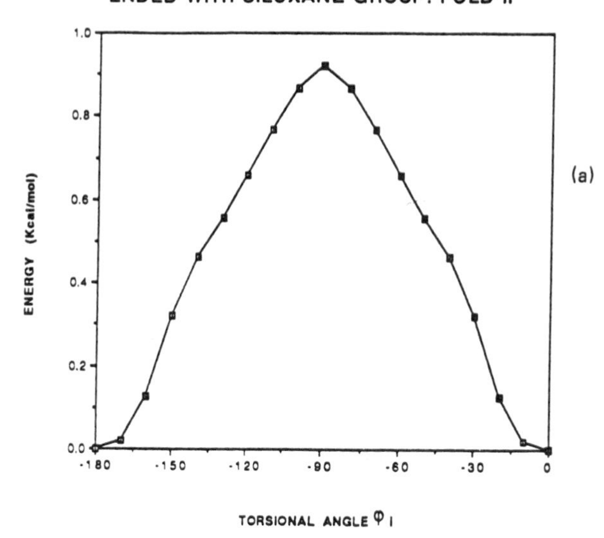

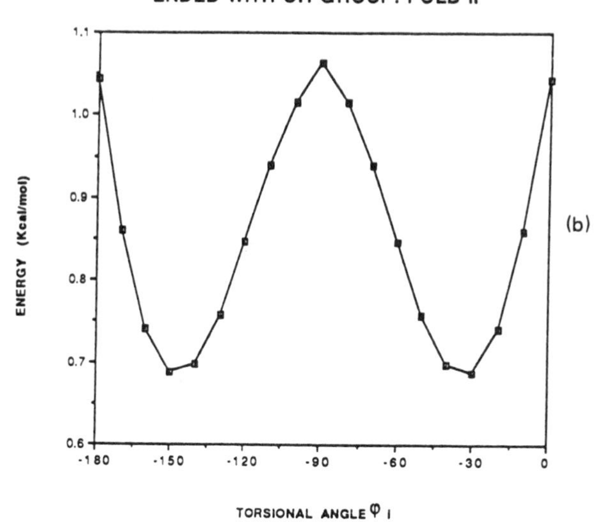

Figure 4 Torsional energy plot for the O—Si—O—Si—O sequence along fold ii: from the starting point (−180,0) both torsional angles have been changed simultaneously by 10°. The ordinate shows only the values of bond *i*.

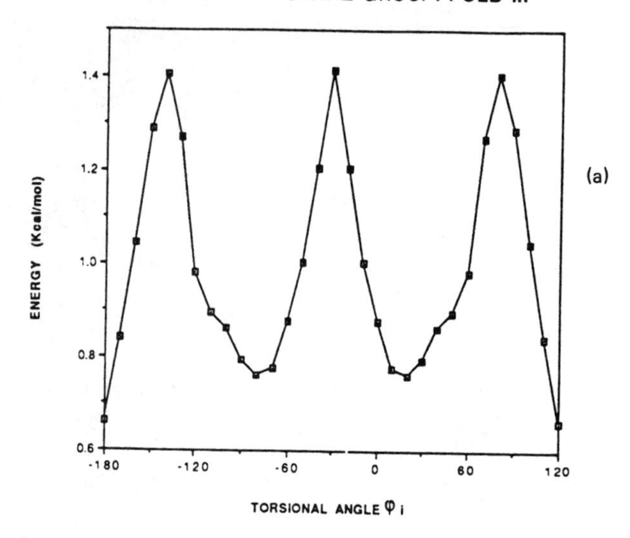

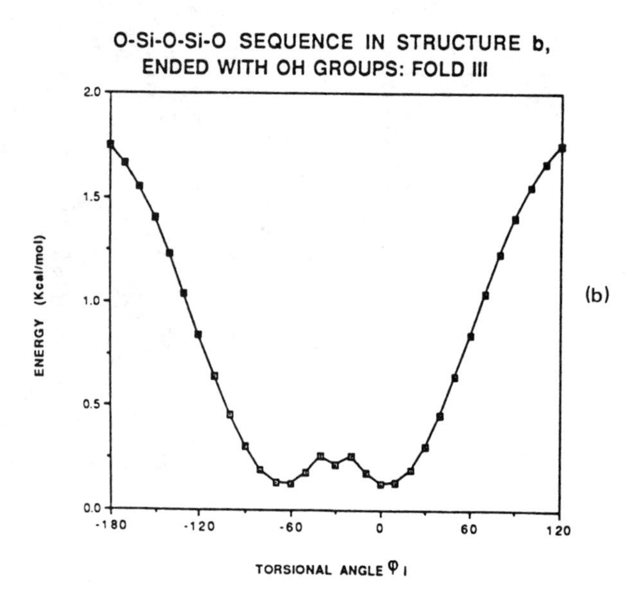

Figure 5 Torsional energy plot for the O—Si—O—Si—O sequence along fold iii: from the starting point (−180,120) both torsional angles have been changed simultaneously by 10°. The ordinate shows only the values of bond *i*.

These comparisons of conformational maps for the O—Si—O—Si—O sequence in various molecular structures indicate that the sequence of bonds 2 and 3 in molecule **c** can be considered as typical for a PDMS chain. Using hydroxyl functionality at terminal silicon atoms diminishes the steric repulsions and increases the electrostatic interactions, which results in overestimation of the nonbonding attraction interactions.

This study shows significant differences between Si—O—Si—O—Si and O—Si—O—Si—O sequences and indicates that the pertinent bonds in the molecular structures considered are structure **e,** with rotations around bonds 3 and 4, and structure **c,** with rotations around bonds 2 and 3, respectively. The conformational results of these two cases will be discussed for different values of the Si—O—Si bond angle.

Effect of the Si—O—Si Bond Angle

The previous conformational analysis was done by assuming the same valence geometry with a value of 149.5° for the Si—O—Si bond-angle value, which has been suggested experimentally [16] and theoretically for this bond [18]. The flexibility of the Si—O—Si bond angle, or, in other words, the small differences of energy for deviations of this angle from the equilibrium value, involves a large degree of uncertainty when this angle is considered for the polysiloxane polymeric backbone. This uncertainty requires a comparison of conformational results for several values of the Si—O—Si bond angle. The sequences Si—O—Si—O—Si and O—Si—O—Si—O have been considered in the most relevant structures identified above for PDMS. Conformational analysis has been performed for the following values of the Si—O—Si bond angle: 145., 147., 149.5, and 152.

The results of this analysis will be discussed below for the two sequences. In all cases, both sequences have the most stable isomeric state at the *cis-trans* conformation, and the conformational maps obtained with different values for the Si—O—Si bond angle are similar to those shown in Figures 1 and 2. The differences between the samples with various Si—O—Si bond angles will be discussed for several slices from the conformational maps. A slice from a conformational map is defined by the variation of energy as a function of the torsional angle of one bond, when the torsional angle corresponding to the other bond, represented on the conformational map, is held constant.

Significant differences occur for the sequence Si—O—Si—O—Si which has been analyzed in structure **e** with torsional rotations around bonds 3 and 4. These rotations will be assigned as torsional angles φ_i and φ_{i+1}, respectively.

Figure 6(a) shows a slice through the conformational map where the

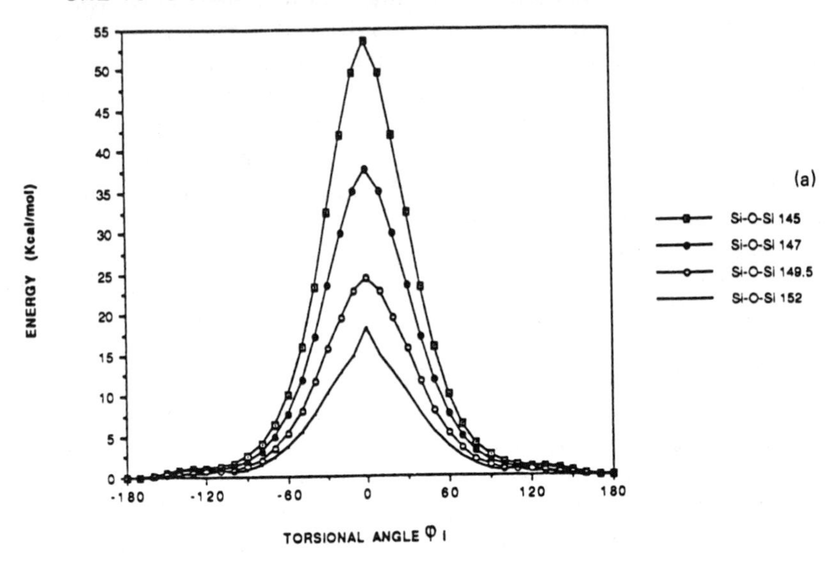

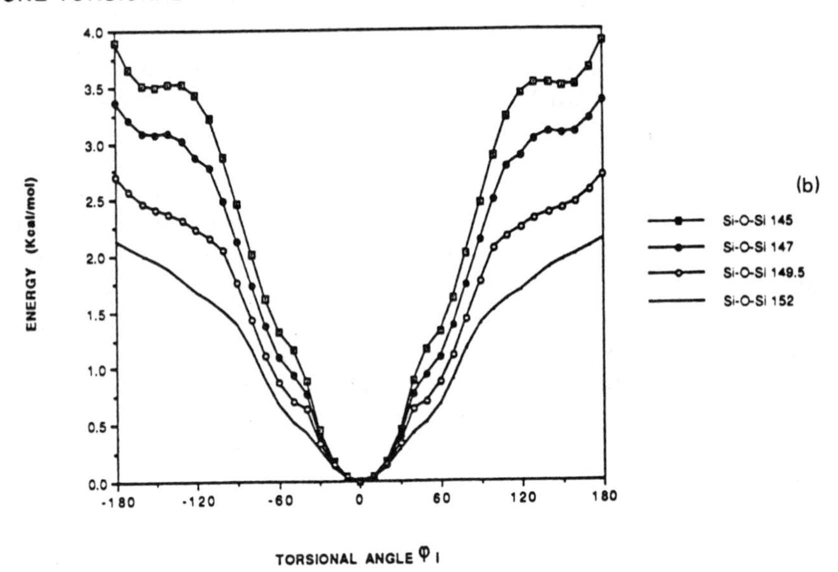

Figure 6 Slices through the conformational map of the Si—O—Si—O—Si sequence.

torsional angle φ_{i+1} has been held constant at 0° and the other angle has been varied from −180° to 180°. The energy barrier corresponding to the *cis-cis* conformation is increasing dramatically for smaller values of the Si—O—Si bond angle. In contrast, Figure 6(b), which shows the slice where the torsional angle φ_{i+1} has been held constant at 180°, indicates an increase of the energy barrier at the *trans-trans* conformation by an order of magnitude less than the previous barrier.

The same slices through the corresponding conformational maps for the sequence O—Si—O—Si—O are shown in Figure 7. This figure shows that when the Si—O—Si bond angle decreases, the potential barriers become higher and the relative energy of the local minima becomes higher compared to the energy of the absolute minima. Unexpectedly, *cis-cis* conformation is not extremely unfavorable. These features are clearly illustrated in Figure 7(b). This figure shows a potential barrier of 2.0 kcal/mol for *cis-cis* relative to *cis-trans* conformation for the Si—O—Si bond angle of 149.5°.

An important difference, which will become more relevant in the next section, is that some local minima have different locations on the conformational map for different values of the Si—O—Si bond angle. For the sequence Si—O—Si—O—Si, when the Si—O—Si bond angle is 145° or 147°, Tables 3 and 4 show that there are two sets of local minima. The first set includes the minima located at (−90, −30), (−30, −90) and the symmetrical positive values for the same pairs, φ_i and φ_{i+1}, of angles. The second set is located at (−60, −60), and its symmetrical minimum at (60,60). For the Si—O—Si bond angles of 149.2° and 152°, Tables 5 and 6 show that the second set of local minima is located at (−80, −130) and (−130, −80), and to the corresponding positive values of these torsional angles. The location and the relative energy of various local minima, when the Si—O—Si bond angle is varied, will be discussed further in the next section, where the rotational isomeric states will be used to estimate the configurational properties of PDMS.

Overall this conformational analysis shows that to study the flexibility of PDMS, two sequences must be considered: Si—O—Si—O—Si and O—Si—O—Si—O as the central portions of structures **e** and **c**, respectively. They have different conformational features, but also have the common elements: (a) the most stable conformation is *cis-trans*, and (b) they have a diagonal symmetry. The second sequence shows higher flexibility, in terms of lower potential barriers and more local minima. The potential barriers have a height between 2.5 and 3 kcal/mol. Litvinov et al. [20,21] measured a torsional barrier of 2.75 kcal/mol for the segmental motion in PDMS above T_g. At this time we cannot specify which energy barrier is measured experimentally, but most probably the measured value repre-

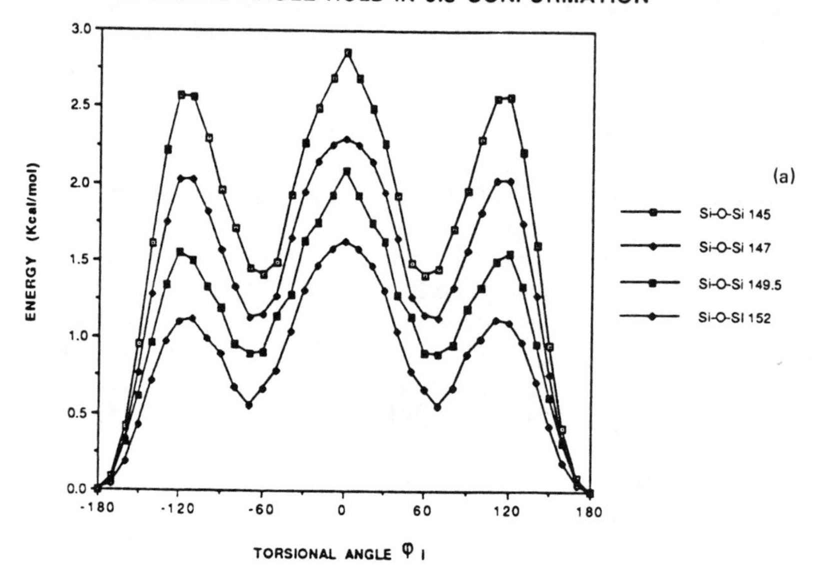

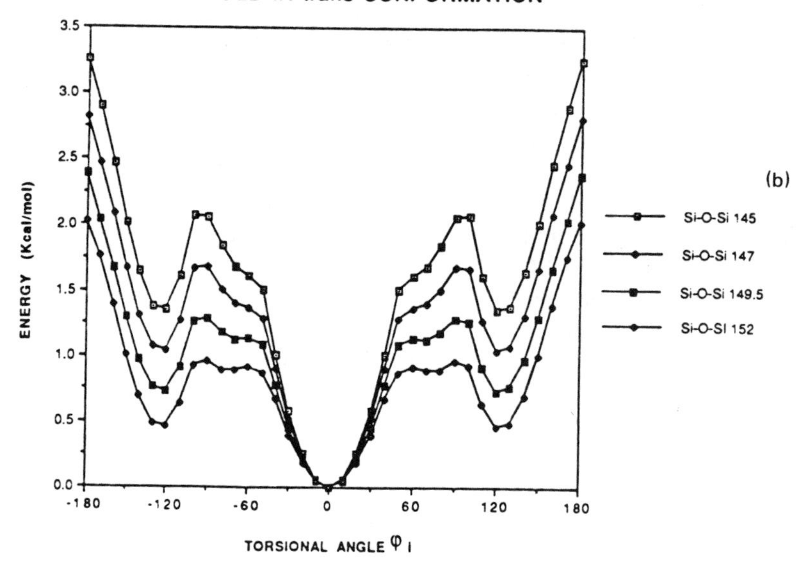

Figure 7 Slices through the conformational map of the O—Si—O—Si—O sequence.

Table 3 Relative Energy Minima and Partition Functions From Conformational Maps, With Si—O—Si Bond Angle Fixed at 145°

Sequence	Rel. Energy (kcal/mol)	φ_1	φ_2	Z at 300K	Z at 400K
O—Si—O—Si—O	0.000	0	180	1.000	1.000
	0.000	180	0	1.000	1.000
*	0.978	−160	−150	0.109	0.170
**	1.058	−130	50	0.179	0.270
**	1.145	−30	100	0.129	0.197
*	1.342	−90	−70	0.082	0.138
	1.354	120	180	0.052	0.090
	1.354	180	120	0.052	0.090
Si—O—Si—O—Si	0.000	0	180	1.000	1.000
	0.000	180	0	1.000	1.000
*	1.094	−90	−30	0.088	0.137
***	1.236	60	60	0.193	0.295

Note: a complete list of minima would include symmetrical combinations of pairs of angles:
* −a, −b; a, b; −b, −a; b, a
** −a, b; a, −b; −b, a; b, −a
*** a, a; −a, −a
The energies and statistical weights for such pairs are the same.

Table 4 Relative Energy Minima and Partition Functions From Conformational Maps, With Si—O—Si Bond Angle Fixed At 147°

Sequence	Rel. Energy (kcal/mol)	φ_1	φ_2	Z at 300 K	Z at 400 K
O—Si—O—Si—O	0.000	0	180	1.000	1.000
	0.000	180	0	1.000	1.000
*	0.726	−160	−150	0.167	0.236
**	0.865	−130	50	0.247	0.342
**	0.880	−20	90	0.193	0.268
	1.046	180	120	0.098	0.151
	1.046	120	180	0.098	0.151
Si—O—Si—O—Si	0.000	0	180	1.000	1.000
	0.000	180	0	1.000	1.000
*	0.809	−90	−30	0.062	0.087
***	0.988	60	60	0.276	0.385

See Table 3 for meaning of asterisks.

Table 5 Relative Energy Minima and Statistical Weights From Conformational Maps, with Si—O—Si Bond Angle Fixed at 149.5°

Sequence	Rel. Energy (kcal/mol)	φ_1	φ_2	Z at 300 K	Z at 400 K
O—Si—O—Si—O	0.000	0	180	1.000	1.000
	0.000	180	0	1.000	1.000
*	0.510	−160	−140	0.236	0.296
**	0.617	−20	90	0.290	0.361
**	0.694	−120	40	0.210	0.266
	0.729	180	120	0.166	0.225
	0.729	120	180	0.166	0.225
**	0.817	−160	60	0.178	0.225
Si—O—Si—O—Si	0.000	0	180	1.000	1.000
	0.000	180	0	1.000	1.000
*	0.524	−90	−30	0.138	0.172
*	0.615	−40	−130	0.130	0.162

See Table 3 for meaning of asterisks.

Table 6 Relative Energy Minima and Partition Functions From the Conformational Maps, with Si—O—Si Bond Angle Fixed at 152°

Sequence	Rel. Energy (kcal/mol)	φ_1	φ_2	Z at 300 K	Z at 400 K
O—Si—O—Si—O	0.000	0	180	1.000	1.000
	0.000	180	0	1.000	1.000
*	0.281	−160	−150	0.337	0.401
**	0.405	−20	90	0.412	0.478
	0.466	180	120	0.244	0.302
	0.466	120	180	0.244	0.302
**	0.615	−160	60	0.225	0.265
Si—O—Si—O—Si	0.000	0	180	1.000	1.000
	0.000	180	0	1.000	1.000
*	0.303	−90	−30	0.263	0.312
*	0.917	−80	−140	0.070	0.105

See Table 3 for meaning of asterisks.

sents an average of the barriers which have been shown to occur in this study.

A different experiment, based on measuring the changes in depolarization ratio of Raman bands in PDMS as a function of temperature [22], provides a value of 0.82 kcal/mol for the potential barrier between the most populated states in this polymer. This work has shown that the most populated states are *trans* and *cis*. The potential barrier measured by this experiment can be described theoretically by scanning systematically only one bond. The results show similar plots for any single bond in the considered sequences. Figure 8 shows the energy variation when one bond is rotated, considering several values of the Si—O—Si bond angle. This figure shows that the *trans* conformation is isoenergetic with the *cis* conformation when one bond is scanned systematically and the other bonds in the system are allowed to relax. The potential barrier between these two states varies in height between 0.5 kcal/mol and 1.3 kcal/mol when the Si—O—Si bond angle is varied from 152° to 145°, respectively. A barrier in the range of the experimental value of 0.82 kcal/mol corresponds to the siloxane bond

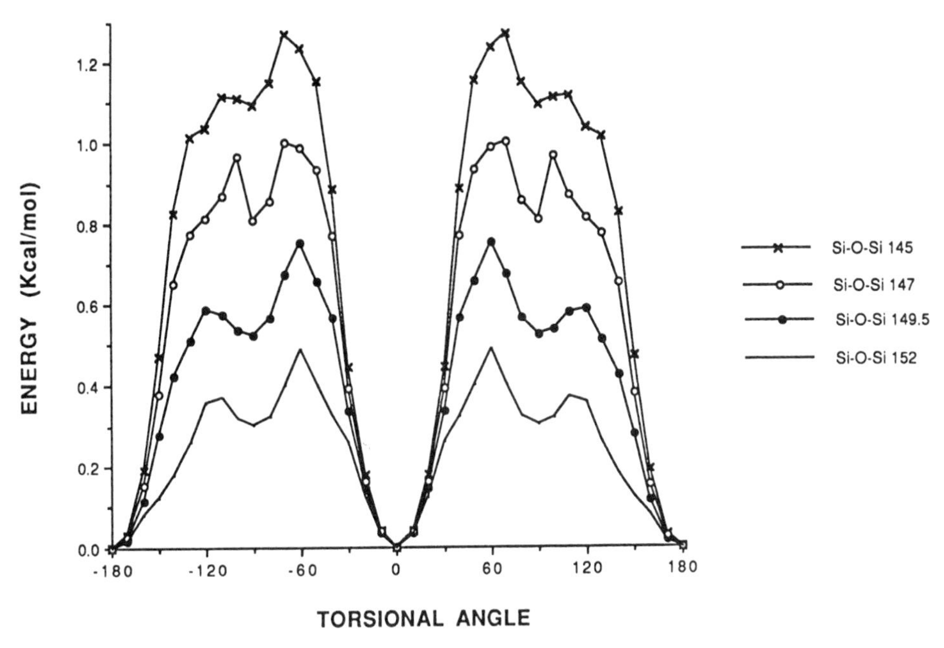

Figure 8 Torsional energy for one-bond rotation in PDMS. All other bonds are allowed to relax.

angle between 147° and 150°. Figure 8 shows that when one bond is scanned, the *gauche* conformation corresponds to the top of the potential barrier, and a local minimum occurs for torsional angles of ±90°.

ROTATIONAL ISOMERIC STATES

Conformational analysis can provide information on stable isomeric states and can describe completely the chain flexibility. This is the key element required to establish the conceptual bridge between the polymer structure and its physical properties. The results provided by conformational analysis were used according to the RIS model to calculate configurational properties of PDMS.

The rotational isomeric states model is a theoretical approach based on statistical methods which succeeds in using the information provided by conformational analysis of small molecular samples and determining the conformational properties of long polymeric chains. A permanent challenge for this model is to obtain calculated values in agreement with the experimental data for PDMS. Using the present conformational results in this model, a reevaluation of the characteristic ratio will be discussed and the temperature dependence of the polymer dimension will be analyzed by taking into consideration the bending flexibility of the siloxane bond.

Computation of the Partition Functions from Conformational Maps

It has been shown [23] that conformational analysis can be used to estimate the partition function z_ζ for each rotational state ζ according to the equation

$$z_\zeta = \iint \exp\left(\frac{-E}{RT}\right) d\varphi_i \, d\varphi_{i+1} \qquad (1)$$

The rotational isomeric state ζ is defined by a pair of torsional angles, φ_i and φ_{i+1}, on the conformational map, which corresponds to a minimum energy. The integration is performed over the domain of torsional angles which define the potential well, starting from the angles corresponding to the energy minimum and ending at the top of the potential barrier. Due to the asymmetric nature of conformational maps, for a two-dimensional map, each potential well was defined by four paraboloidal segments. One segment is determined by the second-order polynomial coefficients of a least-squares curve fit to the energy values as a function of the variation of one torsional angle in one direction, when the other torsional angle is maintained at the value corresponding to the lowest energy of that particular rotational isomeric state ζ. The second segment is determined in the same way as the first, maintaining the same torsional angle constant and varying the other angle in the opposite direction to the previous segment. The following paraboloidal segments are defined in the same way as the first two segments, but the torsional angle which was maintained constant is then varied and the other is kept at the value which corresponds to the energy minimum. This process is illustrated in Figure 9. The corresponding

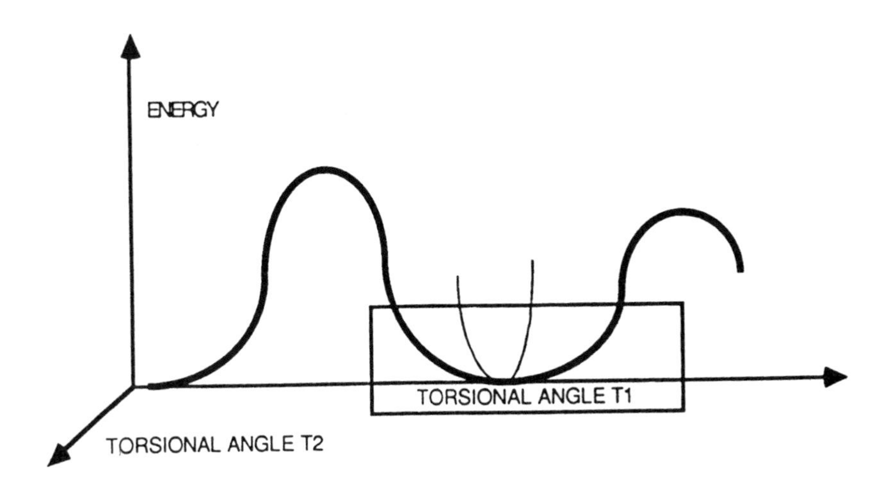

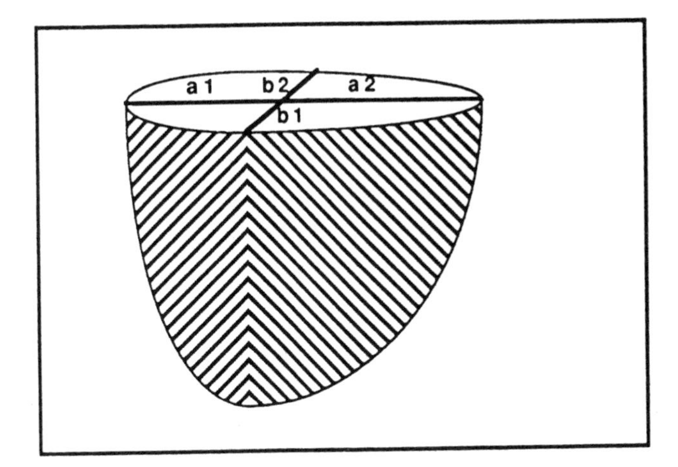

Figure 9 Computational procedure for partition function of an isomeric state from a conformational map.

integrals in Eq. (1) were calculated as the sum of the volumes of the four segments by using a Monte Carlo method [24] with 30,000 random numbers per segment.

Computation of Configurational Properties with the RIS Model

Using the procedure outlined in the previous section, it is possible to estimate the partition function for any isomeric state directly from a conformational map. The partition function of each state, divided by the partition function corresponding to the absolute minimum, is the statistical weight of that particular isomeric state, and they can be used to estimate the elements of *U* matrix in the RIS model. The *U* matrix is a representation of the isomeric states from the conformational map.

The procedure used in this work to estimate configurational properties of PDMS was developed by Suter et al. and is described in detail in Refs. 23 and 25. This procedure allows for the direct use of conformational results in a flexible isomeric state scheme.

From the above discussion it is clear that the O—Si—O—Si—O sequence is more flexible than the Si—O—Si—O—Si sequence of bonds. It has been shown that both sequences of bonds show a diagonal symmetry and that the first sequence is characterized by parallel folds which contain the conformational minima.

Tables 3–6 list the pairs of torsional angles φ_i and φ_{i+1} defining the energy minima, the corresponding relative energy, and the statistical weight as calculated from the conformational maps corresponding to each sequence of five atoms discussed above, maintaining the Si—O—Si bond angle at the fixed values: 145°, 147°, 149.5°, and 152°. The statistical weights have been calculated according to the method described above for temperatures 300 and 400°K. The lowest energy minima are located at the same pair of torsional angles: 0°, 180° which correspond to the *cis-trans* conformation. Note that φ_i and φ_{i+1} for some local minima differ when the Si—O—Si bond angle varies.

First, the sequence O—Si—O—Si—O will be discussed. The local minima which are common to all calculated conformational maps are defined by the following pairs of torsional angles: (−160;−150), (−20, or −30; 90, or 100), and (180,120) and their corresponding symmetrical values. Their energy, relative to the energy of the absolute minimum, is less than 1 kcal/mol and decreases for increasing values of the Si—O—Si bond angle. The statistical weights of these minima increase when the same bond angle increases. Table 3, which lists the minima for the Si—O—Si bond angle fixed at 145°, shows a minimum at (−90,−70), which is not present for other values of the Si—O—Si bond angle. Another minimum located at (−130,50) has large statistical weight and can be found only for the lower values of the Si—O—Si bond angle: 145° and 147°. Tables 5 and 6 list a different minimum which occurs only at larger values of the Si—O—Si bond angle: 149.5° and 152°. This minimun is located at (−160,60) and corresponds to the t^-g^+ conformation.

The Si—O—Si—O—Si sequence contains one local minimum present in all cases where the Si—O—Si bond angle has been studied. This common local minimum, located at (−90, −30), has features similar to the common minima in the previous sequence: the statistical weights increase and the energy decreases when the Si—O—Si bond angle increases. The next local minima with higher energy, in Tables 3 and 4, are located at (−60, −60) and (60,60) corresponding to the g^+g^+ and g^-g^- conformation. These minima occur only at smaller values of the Si—O—Si bond angle, 145° and 147°, and for the larger values the location of the next local minimum is changed. If the Si—O—Si bond angle is 149.5°, as shown in Table 5, this local minimum is located at (−40, −130) and at the corresponding symmetrical pairs of torsional angles. For a more extended Si—O—Si bond angle, at 152°, this local minimum is at (−80, −140) and at the corresponding symmetrical pairs as indicated on Table 6.

The data listed in Tables 3–6 have been used to estimate two configurational properties of PDMS: the end-to-end distance and the characteristic ratio. These calculations were performed using a program provided by Professor Ulrich Suter. The algorithm is discussed in detail in Ref. 25. The valence geometry used to obtain the configurational properties were the same as listed in Table 1. The statistical weights calculated from the conformational maps of the O—Si—O—Si—O sequence were used in the first-order interactions matrix U'' for different siloxane bond angles. This sequence has been selected to be used in the U'' matrix because it is the most flexible. A RIS model with 12 states has been used. The rotational isomeric states were located at the following torsional angles: 180, 150, 120, 90, 60, 30, 0, −30, −60, −90, −120, and −150. The statistical weights included in this matrix can be retrieved from Tables 3–6. If the torsional angles of the isomeric states did not coincide with the angles which define the U'' matrix, the nearest value in the matrix has been considered.

The results provided by the conformational analysis of the Si—O—Si—O—Si sequence were used to estimate the U' matrix, but in an approximate way. The angles which position the isomeric states for the second-order interactions matrix U' were the same as for U''. The elements of the U' matrix were considered all ones or zeros, as they would correspond to a low or high energy on the conformational map of this sequence.

The U' matrix has been considered the same for all cases.
The U' matrix used is

	180	150	120	90	60	30	0	−30	−60	−90	−120	−150
180	0	0	0	1	1	1	0	1	1	1	0	0
150	0	0	1	1	1	1	1	1	1	1	0	0
120	0	1	1	1	1	1	1	1	1	0	0	0
90	1	1	1	1	1	1	1	1	0	0	0	1
60	1	1	1	1	1	1	1	0	0	0	1	1
30	1	1	1	1	1	0	0	0	0	1	1	1
0	1	1	1	1	1	0	0	0	1	1	1	1
30	1	1	1	1	0	0	0	0	1	1	1	1
60	1	1	1	0	0	0	1	1	1	1	1	1
90	1	1	0	0	0	1	1	1	1	1	1	1
120	0	0	0	0	1	1	1	1	1	1	1	1
150	0	0	0	1	1	1	1	1	1	1	1	0

Table 7 lists the results obtained from the RIS model for the end-to-end distance and the characteristic ratio at 300 and 400 K, using each set of statistical weights listed in Tables 3–6 for the four values of the Si—O—Si bond angle considered in this study. The characteristic ratio is defined as

$$C_\infty = \frac{\langle r^2 \rangle_0}{nl^2} \tag{2}$$

where r is the end-to-end distance of a polymer chain formed by n bonds and l is the bond length.

A comparison of the experimental value for the characteristic ratio of 6.3–7.7 [26] with the calculated values at 300 K for the Si—O—Si bond angle larger or equal to 147° shows good agreement.

In the case of disiloxane, it has been shown [18] that the equilibrium value for the Si—O—Si bond angle is 149.5° and that the bending deformational curve is assymetrical: the deformational bending energy rises fast

Table 7 Configurational Properties of PDMS for Different Values of Si—O—Si Bond Angle at 300 and 400 K for Chains with 8192 Units

	300		400	
Si—O—Si	$\langle r^2 \rangle$	C_∞	$\langle r^2 \rangle$	C_∞
145	261,000	5.89	237,900	5.36
147	283,500	6.39	260,600	5.88
149.5	334,800	7.55	317,000	7.15
152	312,700	7.05	303,400	6.84

for lower values of the Si—O—Si bond angle, and this deformational energy rises slow when this bond angle tends toward higher values. The assymetric deformational energy curve would lead toward higher values of the Si—O—Si bond angle by raising the temperature. This feature is expected to occur in case of disiloxane as well as in the case of PDMS. The Si—O—Si bond angle is not available from experimental measurements for this polymer, but such data are available for model compounds. This bond angle has a value of 148° ± 3° in hexamethydisiloxane as measured by electron diffraction [27]. In cyclic dimethylsiloxanes this angle is 144.8° (D4), 146.5° (D5), and 149.6° (D6) [28], where D*n* indicates a cyclic dimethyl siloxane containing *n* Si—O groups.

A significant success of the RIS model is the ability to estimate correctly the temperature dependence of unperturbed dimensions of polymer chains, when the backbone of such polymers are characterized by rigid valence geometry. This model shows that such polymers have a negative temperature coefficient, which is defined as

$$d\,\frac{\ln \langle r^2\rangle_0}{dT} \tag{3}$$

The success of this model to explain this property is based on the fact that usually *trans* conformation, which is the most extended although the most stable, and the more compact conformations, as *gauche*, would have a higher energy and consequently a lower statistical weight. However, at higher temperatures the more compact isomeric states, which have higher energies, become more populated by more bonds and the polymers would become shorter. This is not the case with PDMS, which is characterized by a positive temperature coefficient of +0.0007/deg [29]. This work, at the present stage, does not intend to provide a quantitative estimation of the temperature coefficient, but indicates that considering the flexibility of the Si—O—Si bond angle it is possible to account for this "anomaly" of PDMS.

These results suggest a possible way to explain the positive temperature coefficient considering that the Si—O—Si bond angle becomes wider at higher temperatures. Table 7 shows the values of $\langle r^2\rangle_0$ as calculated using the present method at 300 and 400K for several values of the Si—O—Si bond angle: 145°, 147°, 149.5°, and 152°. These data indicate that, for fixed values of the Si—O—Si bond angle, the rise of temperature decreases the dimension of the unperturbed chains in all cases, if the Si—O—Si bond angle does not increase when the temperature rises. This would imply a negative temperature coefficient in all cases. However, by superimposing the variation of the Si—O—Si bond angle from 147° to 149.5° with the temperature variation from 300 to 400K, Table 7 indicates that the characteristic ratio increases from 6.39 (Si—O—Si bond angle 147° at 300K) to 7.15 (Si—O—Si bond angle 149.5° at 400°K).

These results show that the bending flexibility interferes with torsional flexibility, and these two types of flexibilities should not be separated. The present theory does not allow for taking into consideration this complex case, but a future study using molecular dynamics simulations would allow this problem to be overcome.

CONCLUSIONS

It has been shown that the torsional relaxation method can provide an accurate and detailed description of the conformation of PDMS chains which is in good agreement with the available experimental values. The selection of molecular samples considered to be representative for a polymeric chain is important, and this work has shown that $Me_3Si—O—(SiMe_2—O)_n—SiMe_3$, where $n = 2,3$, constitute suitable structures to allow for studying PDMS. The difference between the two sequences of five atoms belonging to the siloxane backbone, Si—O—Si—O—Si and O—Si—O—Si—O, are significant, and the second sequence shows a higher flexibility than the first. The configurational properties of chains can be estimated directly from conformational analysis results, and this approach relieves the RIS model of some adjustable parameters. The bending flexibility of the siloxane bond plays an important role in temperature dependence of configurational properties, and this role has been explained qualitatively in this chapter.

ACKNOWLEDGMENT

The author is grateful to Professor U. W. Suter for providing his program to estimate the configurational properties and for helpful discussions, and to Professor W. L. Mattice for providing a preprint of his most recent work. Special thanks to Dr. G. Lie and Dr. T. H. Lane for helpful discussions. The author would also like to thank the Dow Corning MIS Departments, both in the United States and Europe, for providing the computing resources for this work.

REFERENCES

1. P. J. Flory, *Statistical Mechanics of Chain Molecules*, Interscience, New York (1969).
2. M. V. Volkenstein, *Configurational Statistics of Polymeric Chains*, Interscience, New York (1963).

3. P. J. Flory, V. Crescenzi, and J. E. Mark, *J. Am. Chem Soc.*, *86*, 146 (1964).
4. J. B. Carmichael, *J. Polym. Sci. Part A*, *3*, 4279 (1965).
5. W. L. Mattice, *Macromolecules*, *11*, 517 (1987).
6. S. Bruckner and L. Malpezzi, *Makromol. Chem.*, *183*, 2033 (1982).
7. G. R. Mitchell and A. Odajima, *Polym. J.*, *16*, 351 (1984).
8. Gas Research Institute Report No. 5082-260-066 (1986).
9. S. Grigoras, *Polym. Preprints*, *28*, 2, 405 (1987).
10. S. Grigoras and T. H. Lane, *J. Comput. Chem.*, *9*, 25 (1988).
11. D. Ya. Tsvankin, V. S. Papkov, V. P. Zhokov, Yu. K. Godovsky, V. S. Svistunov, and A. A. Zhdanov, *J. Pol. Sci.: Pol. Chem. Ed.*, *23*, 1043 (1985).
12. I. Bahar, I. Zuniga, R. Dodge, and W. L. Mattice, *Macromolecules*, *24*, 2986 (1991).
13. N. L. Allinger and Y. H. Yuh, Quantum Chemistry Program Exchange, Bloomington, IN, Program #395.
14. J. Howell, A. Rossi, D. Wallace, K. Haraki, and R. Hoffmann, Quantum Chemistry Program Exchange, Bloomington, IN, Program #469.
15. CHEMLAB II, Molecular Design Ltd., San Leandro, CA.
16. J. R. Durig, M. J. Flanagen, and V. F. Kalasinsky, *J. Chem Phys.*, *66*, 275 (1977).
17. S. Grigoras and T. H. Lane, in *Silicon-Based Polymer Science: A Comprehensive Resource*, p. 125 (J. W. Zeigler and F. W. G. Fearon, eds.), Advances in Chemistry Series, vol. 224, Am. Chem. Soc., Washington, DC (1990).
18. S. Grigoras and T. H. Lane, *J. Comput. Chem*, *8*, 84 (1987).
19. S. Shambayaty, J. F. Blake, S. G. Wierschke, W. L. Jorgensen, and S. L. Schreiber, *J. Am. Chem. Soc*, *112*, 697 (1990).
20. V. M. Litvinov, B. D. Lavrukhin, and A. A. Zhdanov, *Polym. Sci. USSR*, *27*, 2777 (1985).
21. V. M. Litvinov, B. D. Lavrukhin, and A. A. Zhdanov, *Polym. Sci. USSR*, *27*, 2786 (1985).
22. J. Maxfield and I. W. Shepherd, *Chem. Phys.*, *2*, 433 (1973).
23. U. W. Suter and P. J. Flory, *Macromolecules*, *8*, 765 (1975).
24. N. V. Kopchenova and I. A. Maron, *Computational Mathematics*, p. 189, Mir, Moscow (1975) (English transl.).
25. U. W. Suter, E. Saiz, and P. J. Flory, *Macromolecules*, *26*, 1327 (1983).
26. V. Crescenzi and P. J. Flory, *J. Am. Chem. Soc.*, *86*, 141 (1964).
27. B. Csakvari, Zs. Wagner, P. Gomory, and F. C. Mijlhoff, *J. Organomet. Chem*, *107*, 287 (1976).
28. H. Oberhammer, W. Zeil, and G. J. Fogarasi, *J. Mol. Struct.*, *8*, 413 (1971).
29. J. E. Mark and P. J. Flory, *J. Am. Chem. Soc.*, *86*, 138 (1964).

4

Mechanics of Strophons in Glassy Amorphous Polymers: A Unified View of Stiffness, Yielding, and Crazing Behavior

Robert R. Luise
E. I. DuPont de Nemours and Company, Wilmington, Delaware

Ioannis V. Yannas
Massachusetts Institute of Technology, Cambridge, Massachusetts

INTRODUCTION

The relation between mechanical behavior and molecular structure of glassy amorphous polymers has been pursued by several authors [1–18]. Important advances in understanding have been made during the past few years, and it is gradually becoming clear how the specific features of the chemical repeat unit conspire to determine the glassy modulus and its dependence on temperature, the glass-transition temperature, the distinction between yielding and crazing, the craze stress, and the yield stress. The nature of the chemical unit affects both the stiffness of the isolated chain and the magnitude of the intermolecular interactions among chain segments in neighboring chains. However, the relative importance of these interactions in the mechanical behavior of glassy amorphous polymers has been disputed.

In this chapter we summarize the elements of a theoretical treatment which leads to a self-consistent calculation of the relative importance of configurational (intramolecular) and intermolecular (chain-chain) interactions on the mechanical behavior of six glassy, amorphous polymers. The theory is based on the postulate that the thermal and mechanical behavior of isotropic, glassy amorphous polymers can be predicted by assuming that stress is transferred at the molecular level via the interaction of strophons. The strophon is defined as the irreducible unit of deformation in isotropic, glassy amorphous polymers, a skeletal sequence of at least three virtual bonds which deforms only by restricted rotation around the backbone bonds and interacts with a strophon in a neighboring chain by a Lennard-Jones potential [4,5,7]. Two distinct deformation mechanisms have been identified [10,11]. In the first, strophon motion or rotation is resisted primarily by intermolecular (chain-chain) barriers, and stress transfer accordingly occurs across chains. In the other, strophon motion is resisted principally by configurational (intramolecular) barriers, and stress transfer occurs along the chain backbone. The theory and the relevant experimental evidence point at the unique importance of the dimensionless "locking factor," r/a_0, where r is the radius of the circle described by rotation of the characteristic virtual bond around the chain backbone bond (the moment arm of the rotating strophon) and a_0 is the van der Waals distance between two interacting strophons. The magnitude of r/a_0 determines the relative importance of intermolecular versus configurational interactions for a given repeat unit [10,11].

Previously, a simple strophon model was used to calculate the modulus and yield stress of the polycarbonate of bisphenol A at 0 K [4,5]. Subsequently [10,11], the model was modified to include strophon pair interactions and extended to the calculation of the modulus at two temperatures, 0 K and 298 K, and two pressures, 0 and 34.56 Pa, for a number of other glassy polymers, including poly(2,6-dimethyl-*p*-phenylene oxide) (PPO), poly[*N*,*N'*-(*p*,*p'*-oxydiphenylene pyromellitimide)] (KH), polystyrene (PST), poly(vinyl chloride) (PVC), and poly(methyl methacrylate) (PMMA). In elevated temperature/pressure calculations, volumetric expansion/compression was assumed to be the sole degree of freedom. The calculations suggested that, while stress transfer is primarily intermolecular in the cases of nonvinyl polymers, configurational stress transfer is of equal importance in the case of the vinyls. The distinction was largely attributed to the smaller values of r/a_0 characteristic of vinyl polymers [10,11]. These observations were more recently extended to interpret the yielding/crazing behavior of the same polymers at large deformation [12]. Intermolecular force calculations suggested that the vinyl polymers with preference to craze must overcome much lower force barriers than nonvinyl polymers which show a preference to yield. The analysis also explained the enhanced tendency to craze with increasing temperature and reduced pressure as a reduction of the same intermolecular force barriers.

In this work, we extend modulus calculations through the glass transition for the same six polymers, utilizing the strophon pair model. We also account for the first time for the temperature dependence of torsional oscillations in the form of a simple single-node potential, and consider it as an added degree of freedom. Extension of modulus calculations through the glass temperature allows prediction of the relative importance of the intermolecular and configurational contributions through the entire region of glassy behavior. This leads to a new interpretation of the principal mechanism for stress transfer in the glassy polymer near T_g, and to prediction of a critical interstrophon separation distance at which the polymer loses its glassy status (devitrification) while being heated. Study of the interstrophon distance at the yield modulus reveals an interesting similarity between T_g and the onset of yielding. An analysis of yielding and crazing is then pursued, and certain important similarities and differences of strophon interactions which characterize these two phenomena are pointed out. The result is a simple but unified view of several aspects of mechanical behavior for glassy amorphous polymers, both at small and large strains.

ELEMENTS OF STROPHON MECHANICS

The principles of strophon mechanics have been described in detail elsewhere, including an analysis of geometric compatibility, equilibrium of forces, and a constitutive relation for a strophon [4,5,10,11]. The essential elements of the calculation of the glassy modulus are reviewed below, together with additions which include a simplified procedure for obtaining potential parameters and an extension of modulus-temperature calculations to the glass transition which incorporates effects of restricted rotation. Calculations of the force required to separate one strophon from another by simple translation and calculations of the force to induce strophon rotation as well as the yield stress are presented in later sections.

The theory introduces the strophon (Gr., *strophi*, turn) as the irreducible molecular unit of deformation in glassy amorphous polymers, i.e., a short skeletal sequence of the main chain, comprising at least three virtual bonds, which is allowed to deform only by rotations about skeletal bonds. Measurements of the infrared dichroism in isotropic, glassy polycarbonate, in the study of absorption bands which have transition moment vectors unambiguously related to the orientation of chain backbone bonds, are consistent with the conclusion that significant rotation around backbone bonds occurs at temperatures as low as 126°C below T_g at a tensile strain exceeding 0.006 [19,20]. In the theoretical treatment the macromolecule is modeled in the manner often employed in treatments of the statistics of chain configuration, i.e., as a sequence of rigid segments (virtual bonds), related to but distinct from the covalent backbone bonds of the chemical repeat unit, which are connected by "hinges" or backbone bonds around which rotation is relatively free. In order to satisfy geometric compatibility with the rest of the chain, a reference rigid segment can become displaced by an externally applied force only when three to five sequential segments participate in cooperative motions (rotations) with the reference segment [10,11]. Thus, an increment in chain distortion is defined in a geometrically compatible manner as a cooperative change in configuration occurring by rotation of between one and three strophons.

It is postulated, therefore, that deformation at the molecular scale is propagated by configurational changes, i.e., by strophon motion. Such motion takes place by cooperative configurational distortions which can be simply visualized to occur along two main pathways. In the first case, strophons may move principally *along* the chain backbone, eventually lengthening the end-to-end distance of the chain, somewhat in the manner of a wrinkle which is propagated from one end of a carpet to the other. Alternately, strophons may move principally *across* chains, eventually propagating the configurational distortion of a few segments in a reference chain over several adjacent chains ("hopping"). Alternative use of the two pathways can lead to elements of macroscopic strain, with strophons switching from a configurational path along a given chain to an intermolecular path over several neighboring chains, and back to a configurational path along another chain, eventually tracing a random walk through a macroscopic, isotropic specimen [10,11].

Strophon rotation is simultaneously resisted by configurational (intramolecular) and intermolecular (chain-chain) energy barriers which are represented by potential energy functions. The two barriers can be modeled as in Figure 1, which shows two strophons, *ABCD* and *EFGH*, belonging to neighboring chains. Rotation of strophon *ABCD* around the z axis is resisted by the configurational barrier $U(\phi)$ [10,11]:

$$U(\phi) = \frac{U_0}{2}(1 - \cos n(\phi - \phi_0)) \quad (1)$$

where U_0 is the height of the rotational barrier and ϕ is the angle of rotation. Here, ϕ_0 is a reference angle which defines the conformation, so that $\phi = \phi_0$, $\phi = \phi_0 + \pi$ for the *trans* and *cis* forms, respectively, of a strophon with twofold ($n = 2$) rotational symmetry (applicable to polycarbonate and the other two nonvinyl polymers examined). Threefold rotational symmetry ($n = 3$) has been assumed for vinyl polymers. Rotation around the x axis is also resisted by an intermolecular barrier modeled as the conven-

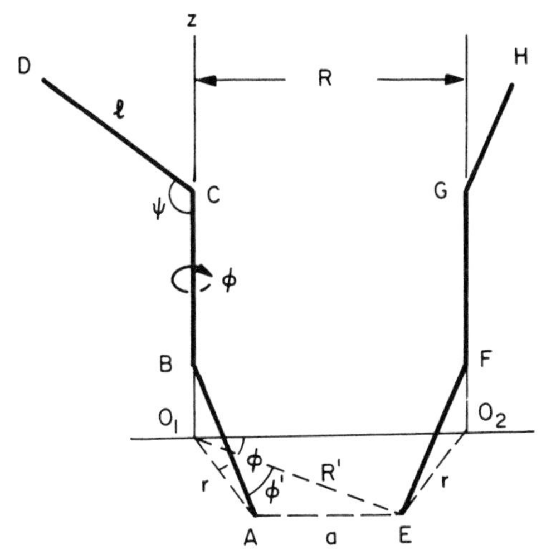

Figure 1 Two three-segment sequences (strophons), $ABCD$ and $EFGH$, belonging to neighboring chains. Rotation of strophon $ABCD$ around the z-axis by an angle ϕ is resisted both by a configurational (intramolecular) and by a chain-chain (intermolecular) barrier. The length of each rigid segment is l while the stereochemical angle is ψ. The axes of rotation of the two strophons are separated by a mean distance R. The intermolecular interaction which resists rotation of a strophon $ABCD$ is considered to be centered at points A and E, separated by a distance a. The regular trapezoid O_1AEO_2 lies in a plane perpendicular to the z-axis (see text for discussion of model validity).

tional 6-12 Lennard-Jones potential of interaction, $V(a)$, between points A and E of Figure 1 [10,11]:

$$V(a) = V_0 \left[\left(\frac{a_0}{a} \right)^{12} - 2 \left(\frac{a_0}{a} \right)^{6} \right] \tag{2}$$

where a is the distance between the strophon point centers of interaction at A and E, a_0 is the distance at 0 K, or van der Waals distance, and V_0 is the potential energy at 0 K, i.e., the distance at $a = a_0$. The total strain energy, $W(\phi,a)$, is written as

$$W(\phi,a) = U(\phi) + V(a) \tag{3}$$

We note that Figure 1 is a special case of relative orientation of a strophon pair in which central segments BC and FG are parallel. Calculation shows that the choice of relative orientation does not affect the major conclusions. This is not unexpected, since both the moment arm, or swing radius, r for strophon rotation and the interstrophon distance a in Figure 1, which dominate Young's modulus, are unaffected by the spatial orientation of the strophons.

To calculate the stress acting on the strophon, the latter is inscribed in a homogeneous, elastic right circular cylinder (Fig. 2), and an expression is obtained for the externally applied moment M_t and the twist angle ϕ:

$$dM_t = \frac{\pi r^4 G_s}{2h} d\phi \tag{4}$$

where $r = l \sin \psi$ (Fig. 1), $h = l - 2l \cos \psi$ (Fig. 2), and G_s is the shear rigidity of the cylindrical element. For pure torsional deformation this leads to an expression for G_s. Finally, the rigorous continuum-mechanical treatment of Christensen and Waals [21] is used to calculate Young's modulus

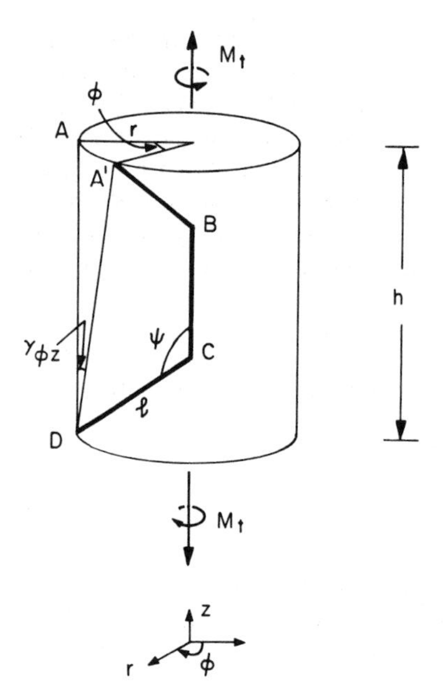

Figure 2 Representation of a three-segment sequence (strophon) $ABCD$ as a mechanically equivalent solid. The latter is a homogeneous, isotropic, linearly elastic cylindrical element which deforms in pure torsion, as does also the inscribed strophon, by the twist angle ϕ when loaded by a twisting moment M_t. The cylinder and the strophon inscribed in it have the same shear rigidity G_s. The cylinder dimensions are $r = l \sin \psi$ and $h = l - 2l \cos \psi$. A cylindrical coordinate system $r\phi z$ is used.

of the macroscopic, isotropic specimen ($E = (8/15)G_s$) in terms of the strophon interaction parameters [10,11]:

$$E = \frac{32h}{15\pi r^4}\left(\frac{d^2U}{d\phi^2} + \frac{d^2V}{d\phi^2}\right) \tag{5}$$

The complete expression for Young's modulus is derived from Eqs. (1), (2), and (5):

$$E = \frac{32h}{15\pi r^4}\left\{\frac{n^2}{2} U_0 \cos n(\phi - \phi_0) + \frac{12V_0 r^2 \sin^2\phi}{a_0^2} \times \left[13\left(\frac{a_0}{a}\right)^{14} - 7\left(\frac{a_0}{a}\right)^{8}\right] - \frac{12V_0 r\cos\phi}{a_0}\left[\left(\frac{a_0}{a}\right)^{13} - \left(\frac{a_0}{a}\right)^{7}\right]\right\} \tag{6}$$

where the quantities r, a, and ϕ are defined in Figure 1. The first term accounts for the configurational contribution to the modulus (E_{conf}), while the second and third terms represent the intermolecular contribution (E_{inter}).

TEMPERATURE DEPENDENCE OF THE GLASSY MODULUS

Equation (6) has formed the basis for calculation of the temperature and pressure dependence of Young's modulus, as previously reported for several vinyl and nonvinyl polymers [10,11]. Strophon geometry for these polymers is reproduced in Tables 1 and 2.

The basic assumptions in previous calculations were (a) linear volumetric expansion (contraction) with temperature (pressure) as determined by the coefficients of thermal expansion (compressibility), which affects the interstrophon distance a and the interchain distance R, (b) rotation which retains its 0 K level of restriction independently of changes in temperature or pressure so that the rotation angle ϕ remains constant ($= \phi_0$, the value at 0 K). These assumptions are depicted in Figure 3, which shows the dimensional changes with temperature of the regular trapezoid O_1O_2AE that forms the projection of the strophon pair (Fig. 1) in the xy plane. For calculation of the temperature dependence the interstrophon distance, $a = a(T)$, and the interchain distance, $R = R(T)$, were related as follows:

$$a(T) = R(T) - 2r\cos\phi_0 \tag{7}$$

where $R(T)$ was estimated from the molar glassy volume $V(T)$ of a spherically symmetric repeat unit:

$$R(T) = 2\left[\frac{3}{4\pi N} V_g(T)\right]^{1/3} \tag{8}$$

where N is Avogadro's number.

Table 1 Structure Parameters of Strophons (not to scale)

Polycarbonate of bisphenol A (PC)	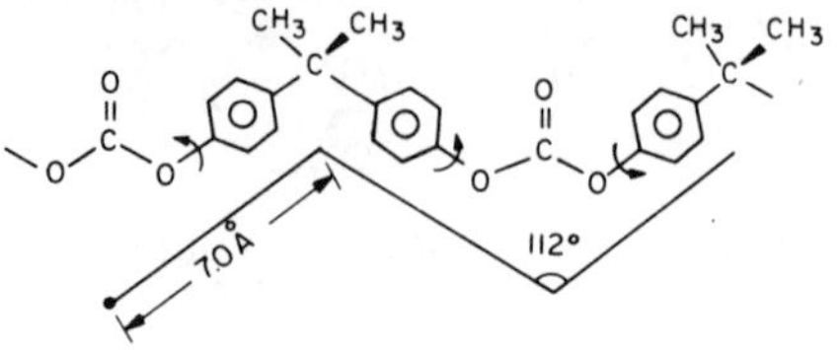 7.0 Å; 112°
Poly(2,6-dimethyl *p*-phenylene oxide) (PPO)	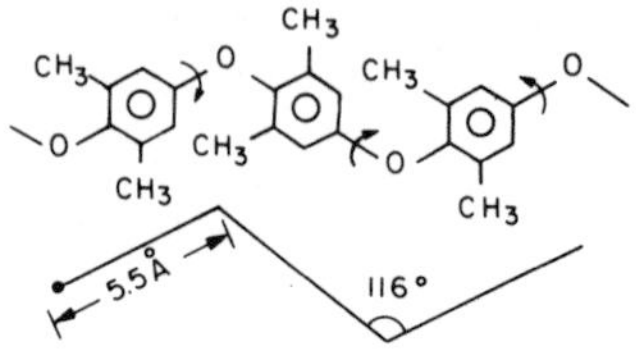 5.5 Å; 116°
Poly[*N*,*N*′-(*p*,*p*′ oxydiphenylene, pyromellitimide)] (KH)	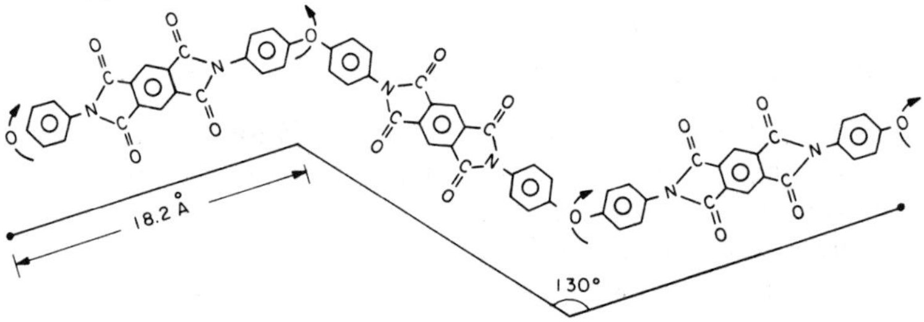 18.2 Å; 130°
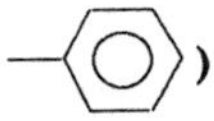 Polystyrene (x = —⟨phenyl⟩) (PST) Poly(vinyl chloride) (x = Cl) (PVC) Poly(methylmethacrylate) (x = $-COOCH_3$) (PMMA)	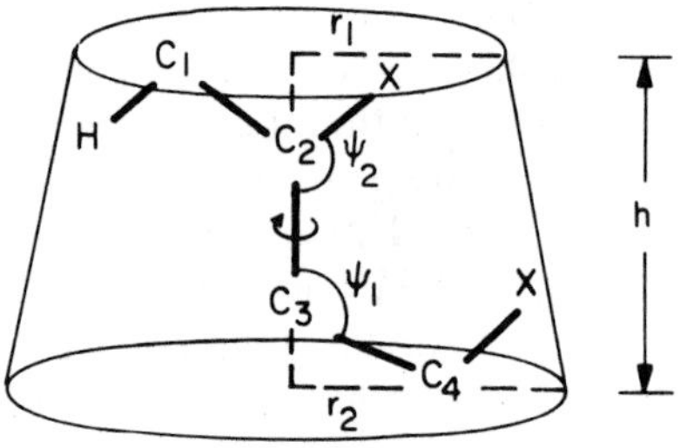 r_1, C_1, C_2, C_3, C_4, ψ_1, ψ_2, r_2, h, H, X

In previous work, estimates of geometric and potential parameters for several types of vinyl and nonvinyl polymers (see below) led to estimates of the glassy modulus at 0 K and 298 K, and at elevated pressure, which were in reasonable agreement with experiment for the six polymers [10,11]. The calculations further showed that the intermolecular contribution to the modulus prevailed over the configurational contribution in the cases

Table 2 Structural Parameters of Strophons[a]

	ψ (deg)	l (Å)	h (Å)	r (Å)			
Nonvinyl polymers (see Fig. 2 for nomenclature)							
PC	112	7.0	12.3[b]	6.5[c]			
PPO	116	5.5	10.3[b]	4.9[c]			
KH	130	18.2	41.6[b]	13.8[c]			
	ψ_1 (deg)	ψ_2 (deg)	l (Å)	h (Å)	r_1 (Å)	r_2 (Å)	$(r_1 + r_2/2$
Vinyl Polymers (see Table 1 for nomenclature)							
PST	114	112	1.53	3.7[d]	4.0[e]	4.5[f]	4.25[g]
PVC	112	112	1.53	2.8[b]	2.15[e]	2.7[f]	2.4[g]
PMMA	122	109.5	1.53	4.7[d]	2.9[e]	3.6[f]	3.3[g]

[a]See Table 2, Ref. 11 for sources.
[b]Calculated from $h = l - 2l \cos \psi$ (Fig. 2).
[c]Calculated from $r = l \sin \psi$ (Fig. 1).
[d]Calculated from $h = l - l_1 \cos \psi_1 - l_2 \cos \psi_2$; l_1, l_2 = $C_3 - C_4 - X$, $C_2 - X$ projections (Table 1).
[e]Based on $C_2 - X$ projection for PST, PMMA and on C_2—C_1—H projection for PVC (Table 1).
[f]Table 1.
[g]Used as the value for r in Eq. (5).

of the nonvinyl polymers due to the larger moment arm, r, of the latter (Fig. 1 and Table 1) [10,11] which, in turn, led to large values for r/a_0 for these polymers.

The construction of intermolecular and configurational interaction parameters for the strophon repeat unit has been described elsewhere [10,11]. Briefly, in the configurational case, the barrier to rotation is taken as the sum

$$U = U_0^* + \Delta U \tag{9}$$

where U_0^* = torsional energy barrier ($\cong$3 kcal mole^{-1} for most single-bond rotations) and ΔU represents the contribution due to nonbonded interactions [11]. Values of ΔU were estimated from energy contour maps referenced previously [10,11].

Previous estimates of the Lennard-Jones intermolecular parameters, V_0 and a_0, were taken from Bondi's zero-point enthalpy (H_0^0) and van der Waals volume (V_W), respectively, as listed in Van Krevelen [22], assuming hexagonal packing of quasi-lattice sites and spherical symmetry. Hence, $V_0 = (1/6)H_0^0$ and $a_0 = 2(3V_w/4\pi N)^{1/3}$. The latter values were estimated from the atomic contributions of a single virtual bond strophon segment, as listed in van Krevelen [22]. In the present work we depart slightly from this protocol in order to simplify extension of these calculations to other

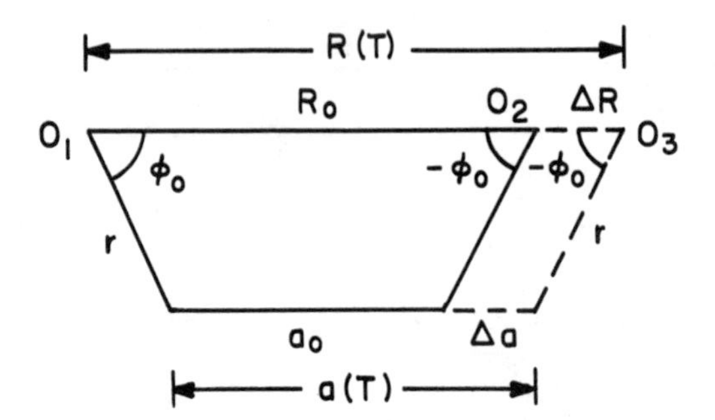

Figure 3 An increase in temperature from 0 K to $T < T_g$ is assumed to increase 0 K value of the mean distance of chain separation R_0 to $R(T)$ while leaving the angle ϕ_0 unchanged. In the process, the Lennard-Jones parameter a_0 increases to $a(T)$.

polymers as well as to be consistent with the calculation of $R(T)$, which has derived from values for the molar volume of a repeat unit (as listed also in van Krevelen [22]). Since in most cases (except for polycarbonate), the molar repeat unit is identical to the rigid segment, the results are essentially unchanged. A summary of potential parameters is shown in Table 3.

Calculation of the modulus-temperature curve through the glass transition has been carried out as previously done at lower temperatures using Eq. (6). A provision for hindered rotation, however, has been added to account for increased torsional vibrations at elevated temperature, as has been recognized by other authors [23]. For simplicity, a simple single-node

Table 3 Potential Parameters of Strophons

Polymer	U_0[c] (kcal mole^{-1})	V_0 (kcal mole^{-1})	V_g (0 K) (cc mole^{-1})	R_0[b] (Å)	V_W (cc mole^{-1})	a_0[a] (Å)
PC	6	2.7	193.6	8.50	136.7	7.57
PPO	6	2.6	102.6	6.88	72.5	6.13
KH	12	6.1	269.5	9.50	190.3	8.46
PST	5	1.9	90.4	6.60	63.9	5.88
PVC	3.5	0.7	41.2	5.08	29.1	4.52
PMMA	9	1.4	78.2	6.29	55.2	5.60

[a] $a_0 = 2(3V_w/(4\pi N))^{1/3}$
[b] $R_0 = 2(3V_g(0\text{ K})/(4\pi N))^{1/3}$ where $V_W = V_g$ (298 K)/1.55 (Ref. 22).
[c] See Ref. 10 for sources.

potential has been utilized:

$$U(\phi) = \frac{U_0}{2}(1 - \cos\phi) \tag{10}$$

The latter has been used elsewhere for vinyl polymers [23] and is a good approximation for twofold symmetric strophons, if it is recognized that there is a finite energy difference between *trans* and *gauche* states (e.g., polycarbonate), i.e., $\overline{\cos\phi} \neq 0$ [24].

The configurational contribution to the modulus (first term of Eq. (6)) becomes

$$E_{\text{conf}}(T) = \frac{32h}{15\pi r^4} \cdot \frac{n^2}{2} U_0(\overline{\cos\phi''(T)}) \tag{11}$$

where $\phi''(T) = \phi(T) - \phi_0$. The limit of $kT \ll U_0$ (valid near T_g) yields a simple, well-known expression [23]:

$$(\overline{\cos\phi''(T)}) \cong 1 - \frac{kT}{U_0} \tag{12}$$

and

$$E_{\text{conf}}(T) = \frac{32h}{15\pi r^4} \cdot \frac{n^2}{2} U_0 \left(1 - \frac{kT}{U_0}\right) \tag{13}$$

Equation (13) reduces at $T = 0$ to the expression for $E_{\text{conf}}(T)$ used in previous work [10,11].

It can be further shown [23] that

$$\cos\phi(T) = \cos\phi_0 \left(1 - \frac{kT}{U_0}\right) \tag{14}$$

The latter relation affects $a(T)$, the interstrophon separation distance, as follows:

$$a(T) = R(T) - 2r\cos\phi(T) \tag{15}$$

where $\cos\phi(T)$ is given by Eq. (14). Thus, in the new calculation $a(T)$ increases with temperature by (a) volume expansion, through $R(T)$, and (b) hindered rotation. This calculation effectively "couples" configurational and intermolecular motions.

The dimensional changes of the trapezoid for the strophon pair including rotational effects are shown in Figure 4. For thermal expansion, i.e., $R(T)$, we use Eq. (8) and the Simha-Boyer relation [22] for the expansion coef-

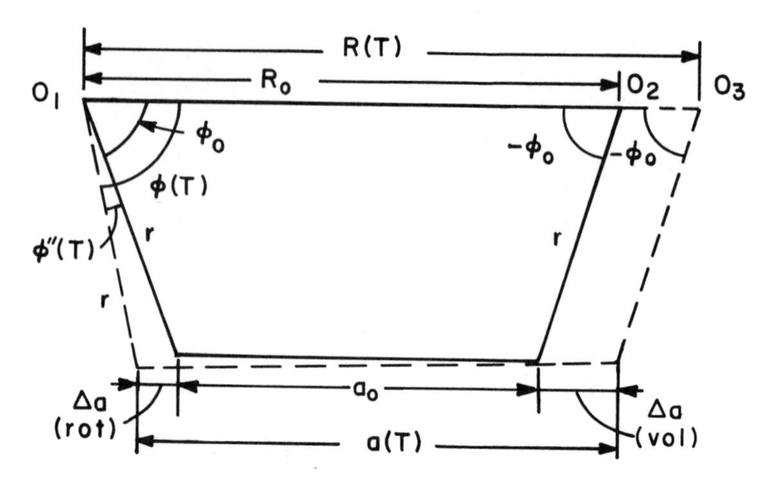

Figure 4 Strophon model for temperature dependence of volume and restricted rotation.

ficient, $E_g = 4.5 \times 10^{-4}\ V_W\ (T < T_g)$, and, therefore,

$$V_g(T) = V_g(298\ \mathrm{K}) + E_g(T - 298\ \mathrm{K}) \tag{16}$$

CHARACTERISTICS OF THE GLASSY STRUCTURE NEAR T_g

Modulus-temperature curves have been calculated for the six amorphous polymers of interest and are compared in Figures 5 and 6 with experimental data. These curves have been based on Eqs. (6), (13), (14), (15), and the data in Tables 1–3. The contributions to the modulus due to configurational interactions alone, E_{conf}, as well as contributions due to intermolecular interactions alone, E_{inter}, are presented separately. Since the two barriers to deformation operate in parallel as the strophon rotates, the two contributions are generally additive and their sum is the total barrier to strophon rotation. Accordingly, the modulus of the polymer equals the sum of these two contributions [10,11]:

$$E = E_{\mathrm{inter}} + E_{\mathrm{conf}} \tag{17}$$

Also shown in Figures 5 and 6 are curves showing the effect of hindered rotation. The correction of the intermolecular contribution for hindered rotation is indicated in Figures 5 and 6 by "inter + rot" (calculated using the second and third terms of Eq. (6), together with Eq. (15)) whereas the analogous correction for the configurational contribution is indicated by "conf + rot" (calculated using the first term of Eq. (6) plus the r.h.s. of

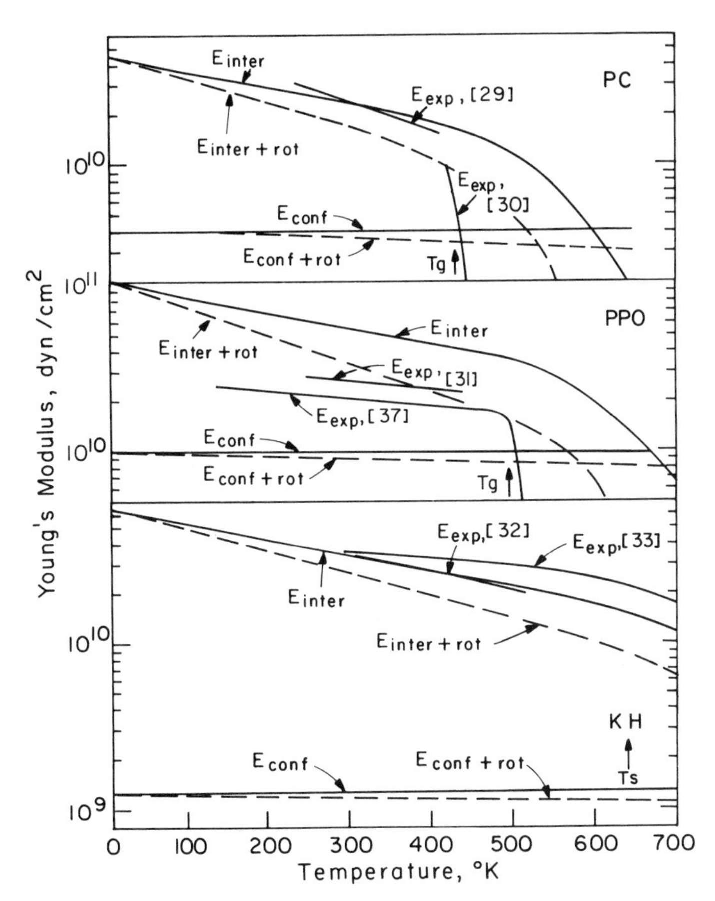

Figure 5 Calculated versus experimental moduli versus temperature for indicated polymers. T_g is indicated for PC and PPO while T_s (softening point) is indicated for KH polyimide.

Eq. (13)). Curves labeled E_{conf} were calculated using the first term of Eq. (6), together with Eq. (7).

Several direct conclusions emerge from these calculations. First, the magnitude of the experimental glassy modulus and the general shape of the experimental curve are generally predicted by the intermolecular modulus (E_{inter}) derived from the Lennard-Jones potential. The drop in the calculated modulus with temperature is, however, much more diffuse than experiment, dropping steeply at 100–200°C above T_g. Second, the effect of adding restricted torsional rotation to the intermolecular contribution is to reduce E_{inter} more sharply with temperature (through Eq. (15)); how-

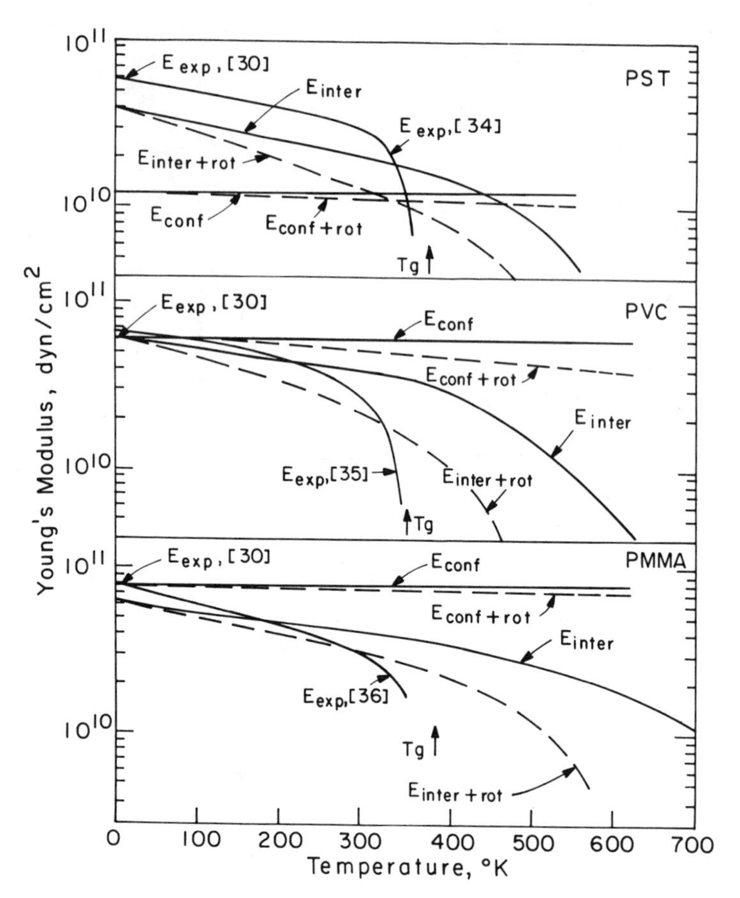

Figure 6 Calculated versus experimental moduli versus temperature for indicated polymers.

ever, the calculated drop near T_g is still more diffuse than the observed values. We attribute the diffuseness of the calculated modulus drop near T_g to the assumption that the strophon pair is isolated rather than participating in cooperative motions in the macroscopic network. This interpretation of the discrepancy between calculated and experimental values is somewhat analogous mathematically to the view of a very diffuse cooperative configurational transformation (e.g., a particularly diffuse Zimm-Bragg transition involving a rather large value of the nucleation perameter σ) [25]. Third, the temperature behavior of the configurational modulus, corrected for hindered rotation, is relatively flat through T_g, since kT_g/U_0 amounts to only 0.10–0.20 for the six polymers studied. Below T_g, the magnitude of E_{conf} is comparable with experiment (and with E_{inter}) only in the cases of poly(vinyl chloride) and poly(methyl methacrylate); in the remainder of the examples E_{conf} is generally much less than the experimental values (Figs. 5 and 6).

In previous calculations considerably below T_g [10,11] it was observed that nonvinyl polymers with large values of r/a_0 yielded relatively low values of the configurational modulus (E_{conf}), suggesting an intermolecular, across-chain mechanism for stress transfer (hopping). In contrast, vinyl polymers with small values of r/a_0 yielded configurational and intermolecular moduli of comparable magnitude, suggesting equal importance of stress transfer along and across the chain, respectively, for these polymers. However, the new modulus calculations suggest that, at least near T_g, the intermolecular "hopping" mechanism is the dominant mechanism of stress transfer at small strains for both classes of polymers.

The most interesting finding from these calculations near T_g is that the ratio of the interstrophon distance at T_g to that at 0 K appears to be constant for the six polymers studied, taking the value (Table 4):

$$a(T_g)/a_0 = 1.07 \pm 0.01 \tag{18}$$

This value is significantly lower than the value $a(T_g)/a_0 = 1.109$, which holds at $E_{inter} = 0$. The latter, higher ratio is obtained when the third (dominant) term in Eq. (6) is set equal to zero. The implication of Eq. (18) is that the same critical interstrophon separation is required for loss of the glassy structure (devitrification) with each of the six chemically distinct polymers.

Table 4 Interstrophon Distances at $T = T_g$ and at $\sigma = \sigma_y$[a]

Polymer	T_g (K)	$a(T_g)/a_0$	T_1 (K)[b]	a_y/a_0
PC	423	1.066	380	1.074
PPO	482	1.066	520	1.073
KH	658	1.082	600	1.083
PST	368	1.068	350	1.061
PVC	373	1.067	385	1.071
PMMA	378	1.053	500	1.089
Average		1.067 ± 0.008		1.075 ± 0.009

[a]Distances calculated from $E_{inter+rot}$ (see Figs. 5 and 6).
[b]Temperature at which $E_{inter} = 9.0 \times 10^9$ dyne/cm^2. The latter is equal to the experimental ratio of yield stress to yield strain (secant modulus at yield) for several glassy amorphous polymers [26].

In summary, the combined findings from Figures 5 and 6 suggest that, for the six polymers studied, the glassy structure near T_g transfers stress principally by intermolecular stophon motion. Observations of polycarbonate chain backbone motion by infrared dichroism below T_g [19,20] cannot, therefore, be interpreted as being resisted by intrinsic chain stiffness; instead, the resistance to such motion appears to arise almost entirely from strophon interaction across chains. In addition, the empirical findings summarized in Eq. (18) suggests that the six polymers share a common interstrophon distance at T_g.

INTERSTROPHON DISTANCE AT THE YIELD MODULUS

Literature data on shear banding (yielding) of a number of glassy polymers have previously shown that the secant modulus at yield (ratio of yield stress to yield strain, E_y) is about $(9 \pm 1) \times 10^9$ dyne/cm^2 for all polymers included in the survey [26]. To test this preliminary observation, we performed detailed measurements in simple tension with polycarbonate (T_g = 149°C) in the range 24–142°C [27]. These measurements have shown that yielding is initiated when the ratio of yield stress to yield strain (yield secant modulus) takes the value $(9.5 \pm 0.3) \times 10^9$ dyn/cm^2 independently of temperature, as shown in Table 5. Since the observed secant modulus values at yield are very close to modulus values near T_g for the six polymers (Figs. 5 and 6), the interstrophon distance at yield would appear to be comparable to that near T_g. A closer look at that distance is, therefore, warranted.

Compiled in Table 4 are calculated interstrophon distances at the temperature T_1, where T_1 is the experimental temperature at which $E_{\text{inter}} = 9.0 \times 10^9$ dyne/cm^2. The latter is equal to the observed value of the secant modulus at the onset of yielding for several polymers [26]. These distances were calculated using both the intermolecular and the rotational contribution to the modulus, since the combination of these two thermal effects gives modulus-temperature curves which approximate experimental values most closely (Figs. 5 and 6). Since these distances have been calculated under conditions where yielding has been just initiated, the interstrophon distance calculated in this manner is labeled a_y below.

The resulting ratio of the interstrophon distance at the yield modulus, a_y, to the 0 K distance is (Table 4)

$$\frac{a_y}{a_0} = 1.075 \pm 0.01 \tag{19}$$

which is roughly constant for the polymers of interest and is very close to the corresponding value at T_g (= 1.07 ± 0.01, Table 4). A critical inter-

Table 5 Secant Modulus of Polycarbonate[a] at Yield [27]

T (°C)	Yield stress (dyne/cm^2) $\times 10^{-8}$ ($\pm$1.5%)	Yield strain (%) $\times 10^2$ ($\pm$3%)	Secant modulus[b] (dyne/cm^2) $\times 10^9$ ($\pm$3%)
23	6.8	7.4	9.2
40	6.35	6.75	9.4
50	6.0	6.15	9.8
60	5.8	5.95	9.7
65	5.6	5.7	9.8
70	5.5	5.7	9.6
75	5.4	5.7	9.5
80	5.2	5.35	9.7
90	4.95	4.9	10.1
100	4.4	4.8	9.2
110	4.1	4.3	9.5
120	3.5	3.7	9.5
130	3.15	3.2	9.8
142	2.1	2.5	8.4

[a]Amorphous polycarbonate of bisphenol A studied in simple tension [27].
[b]Yield stress/yield strain. Average over the temperature range 23–142°C: $(9.5 \pm 0.3) \times 10^9$ dyne/cm^2.

strophon distance for yielding which is very close to the separation distance at T_g has been apparently identified. This finding suggests that yielding and devitrification can be interpreted on a common basis using strophon mechanics as critical transitions occurring at an equivalent dilatational level. This level is crossed either by subjecting the polymer to a critical stress (σ_y) or by heating above a critical temperature (T_g).

CRAZING AND THE FORCE REQUIRED TO SEPARATE TWO STROPHONS

The differences in crazing behavior among polymeric glasses can be considered as a difference in the difficulty with which two segments in neighboring chains can be separated from each other by a distance large enough to induce void formation and thereby initiate crazing. Translation of a reference strophon away from a near neighbor may proceed, thereby generating a microvoid, until further motion is opposed by repulsive intermolecular forces between the reference strophon and a third close neighbor which lies in the path of the translating strophon.

The force which induces strophon translation, F_{inter}, must overcome the intermolecular energy barrier (Eq. (2)) and is given as follows [12]:

$$F_{\text{inter}} = \frac{dV}{da} = 12V_0\left(\frac{a_0}{a}\right)^6\left[1 - \left(\frac{a_0}{a}\right)^6\right]a^{-1} \tag{20}$$

The force is evaluated at the "distance of closest approach," $a = R' - r$, obtained by ϕ' rotation of the moment arm, r, of the left strophon, centered at O_1 in Figure 1, until coincident with the R' direction. Values of F_{inter} calculated at 0 K, 298 K, and T_g are shown in Table 6 and plotted graphically versus r/a_0 in Figure 7. Values at elevated temperature have been calculated with and without the hindered rotation correction.

The results in Table 6 show generally a decrease in F_{inter} with increasing temperature and decreasing magnitude of r/a_0. Clearly, the tendency to craze increases with decreasing values of F_{inter}. At T_g, for example, values of F_{inter} for polycarbonate and for polyphenylene oxide approach that for polystyrene at 0 K. Since it is known that polystyrene crazes over a wide temperature range, while polycarbonate and polyphenylene oxide craze appreciably only near T_g [38], it is possible to estimate a threshold barrier for the transition from yielding to crazing, F_{inter} ~7000 kcal mol^{-1} Å, from Table 6 and Figure 7. Thus, the transition from yielding to crazing occurs at the critical value $r/a_0 \cong 0.8$. The use of calculated F_{inter} values to predict the transition is illustrated in Figure 8.

Similarly, the transition from crazing to yielding found for polystyrene induced by a hydrostatic pressure of about 0.7 GPa [39] has been shown by a similar calculation to correspond to a threshhold value of the intermolecular barrier of similar magnitude [12]. It is likely that the striking transition from yielding to crazing which is induced by uniaxial tensile load

Table 6 Value of F_{inter}[a] Versus Temperature

		F_{inter} ($\times 10^3$ kcal mol^{-1} Å^{-1})			
		298 K		T_g	
Polymer	0°	Vol.	Vol./Rot.	Vol.	Vol./Rot.
PC	34.4	17.6	9.2	9.2	6.8
PPO	32.0	9.9	6.8	7.6	6.0
KH	25.7 × 10³	16.0 × 10³	10.1 × 10³	6.5 × 10³	5.5 × 10³
PST	7.1	4.0	3.3	2.9	2.6
PVC	0.33	0.24	0.19	0.20	0.16
PMMA	1.27	0.72	0.65	0.60	0.54

[a] F_{inter} is the force to translate the strophon. See Eq. (20).

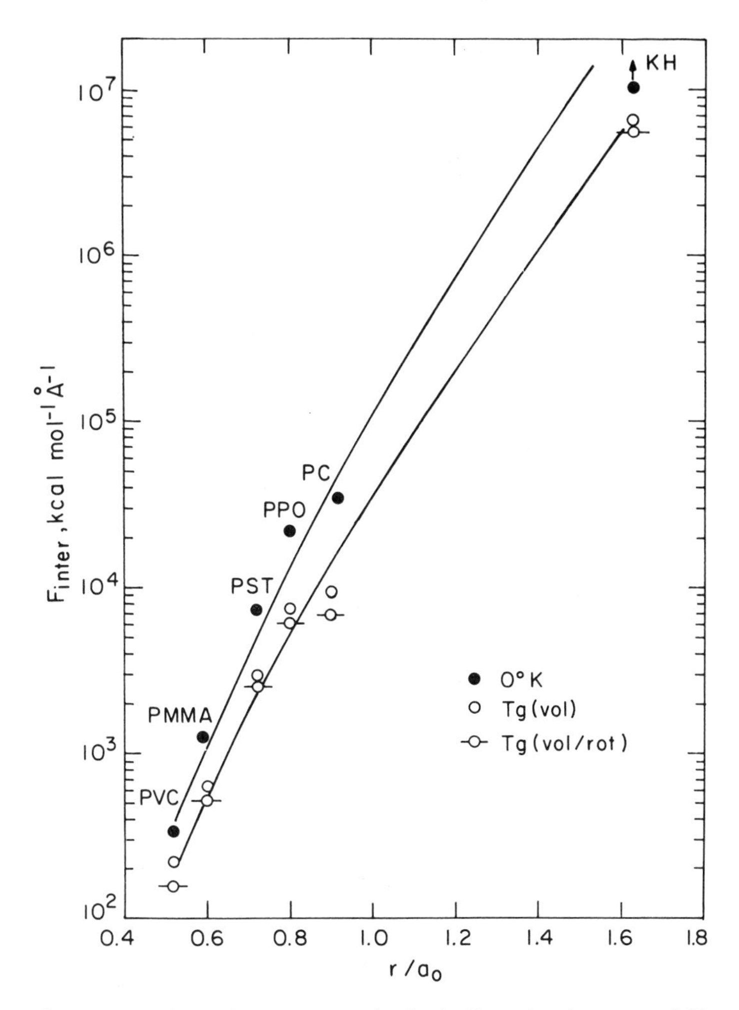

Figure 7 Values of F_{inter} versus r/a_0 for indicated polymers at 0 K and T_g.

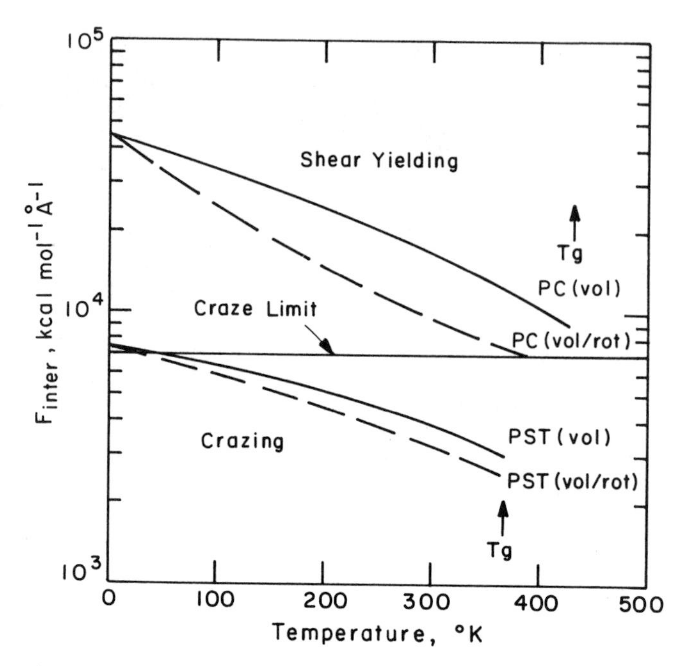

Figure 8 Values of F_{inter} versus temperature for polycarbonate and polystyrene.

isothermally in amorphous poly(ethylene terephthalate) below T_g [40] can be explained similarly.

THE YIELD STRESS AND THE FORCE REQUIRED TO ROTATE A STROPHON

Crazing and shear banding appear, by all accounts, to recruit quite different molecular mechanisms. In fact, it has been noted that shear bands stop craze growth [38]. Crazing is initiated by microvoid formation and appears, according to the above calculations, to depend strongly on chain separation (simulated above by strophon translation). Shear banding, by contrast, is the manifestation of very high deformation which scales to the dimension of the macroscopic specimen (e.g., shear banding across the thickness of a thin, flat specimen subjected to plane stress) without formation of microvoids. When shear banding is accompanied by drawing (e.g., tensile yielding), the polymer is subjected to strains of order 1.0 and very high chain orientations can be unambiguously demonstrated [19]. Accordingly, shear banding will be modeled at the molecular level as the conversion of a highly coiled chain to a highly stretched one. The force necessary to bring about such a transformation can be readily calculated as the force to rotate a strophon against generally two simultaneous barriers, an intermolecular and a configurational one.

Strophon mechanics has been previously used to estimate the yield stress of polycarbonate at 0 K [5]. The force required to rotate a strophon in polycarbonate at that temperature was found to be primarily intermolecular. Yielding was postulated [5] to occur at the point of maximum intermolecular force, i.e., at $dV/d\phi = 0$ in Eq. (2), which occurs at $a_y/a_0 = 1.10$. The force may be expressed as

$$F_{\text{inter},\phi} = \frac{1}{r}\frac{dV}{d\phi} = \frac{-12V_0 \sin\phi}{a_0}\left[\left(\frac{a_0}{a_y}\right)^{13} - \left(\frac{a_0}{a_y}\right)^{7}\right] \tag{21}$$

and may be related to the macroscopic yield stress σ_y, as follows [5]:

$$\sigma_y = \frac{2}{3}\sigma_s = \frac{2}{3}\left(\frac{2F_{\text{inter},\phi}}{\pi r^2}\right) \tag{22}$$

where σ_s is the torsional stress on the strophon cylinder (Fig. 2).

Utilizing the potential parameters from Table 3 and the critical interstrophon distance $a_y\ (0\ \text{K})/a_0 \cong a(T_g)/a_0 = 1.07$ in place of $a_y/a_0 = 1.109$, which was used previously [5], we have calculated the yield stress at 0 K using Eqs. (21) and (22). The results shown in Table 7 indicate that the intermolecular contribution underestimates the extrapolated experimental values, particularly in the case of the vinyls.

Values of $F_{\text{inter},\phi}$ are listed in Table 8. The latter values (force to rotate the strophon) are considerably lower than F_{inter} values of Table 6 (force to translate strophon), mainly due to the difference in separation distance used in each case (values in Table 6 were calculated at the distance of closest approach). However, the nonvinyls show consistently higher values in both calculations.

A configurational contribution to the yield stress may be estimated from the force, F_{conf}, which can be obtained from Eq. (1) as follows:

$$F_{\text{conf},\phi} = \frac{1}{r}\frac{dV}{d\phi} = \frac{V_0}{2r}\, n \sin n(\phi - \phi_0) \tag{23}$$

The maximum force, $F_{\text{conf-max}}$, occurs at $\phi - \phi_0 = 45°$ and $n = 2$ for nonvinyls, whereas $\phi - \phi_0 = 30°$ and $n = 3$ for vinyl polymers. The results shown in Table 8 indicate larger values of $F_{\text{conf-max}}$ for vinyls relative to nonvinyls. This leads, with use of Eq. (22), to a large yield stress contribution at 0 K in better agreement with the experimental values in Table 7.

The calculations may be extended to elevated temperature by applying

Table 7 Comparison of Calculated and Experimental Yield Stress, σ_y ($\times 10^8$ dyne/cm^{-2})

Polymer	0 K				298 K	
	Exp.	Calc.			Exp.	Calc.
		Inter	Conf	Tot		Inter
PC	14–20[a], 26–30[c]	5.7	3.9	9.6	6.2[b], 8.4–8.8[d]	2.8
PPO	—	11.8	9.0	20.8	8.0[c], 10.3[d]	7.4
KH	—	2.5	0.8	3.3	(10.3)[d,e]	2.0
PST	29–30[f]	12.1	17.4	29.5	4.8[b], 6.0–9.6[d]	6.0
PVC	—	19.5	67.8	87.2	6.9–11.0	7.7
PMMA	73–87[f]	15.5	66.7	82.2	9-0–13.1[d]	7.7

[a]Extrapolated data from Ref. 5.
[b]Tension, Ref. 20, Fig. 6.8.1.
[c]Tension/compression, Ref. 20, Table 6.4.1, by extrapolation.
[d]Flexure, Ref. 21.
[e]Value for unspecified polyimide.
[f]Compression, Ref. 20, Table 6.4.1, by extrapolation.

a linear correction for the increase in interstrophon distance, $a(T)$, due to thermal expansion, so that the separation component for yielding becomes

$$\frac{a_y(T)}{a_0} = \frac{a(T_g)}{a_0} - \left(\frac{a(T) - a_0}{a_0}\right) \tag{24}$$

The latter reduces to 1.07 (equal to $a(T_g)/a_0$) at 0 K and to 1.0 at T_g, where $\sigma_y = 0$. This calculation provides a simple explanation for the well-

Table 8 Values of Rotational Force, F_ϕ (kcal mole^{-1} Å^{-1})

Polymer	$F_{\text{conf-max}}$[a]	F_{inter} (0 K)[b]	F_{inter} (298 K)[b]
PC	0.92	1.36	0.91
PPO	1.22	1.60	1.00
KH	0.87	2.69	2.15
PST	1.77	1.23	0.79
PVC	2.19	0.63	0.25
PMMA	4.09	0.95	0.47

[a]Eq. (23).
[b]Eq. (21).

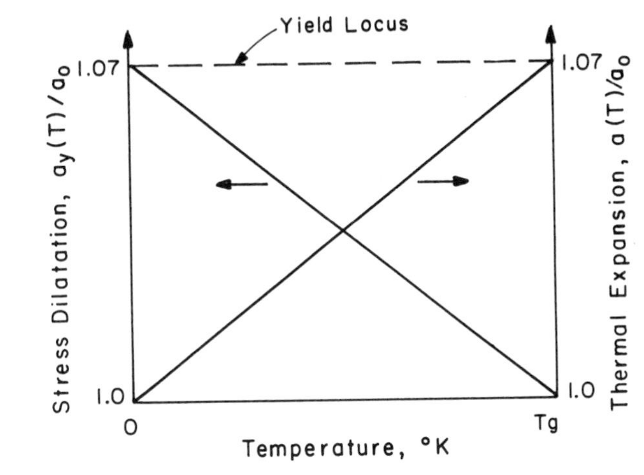

Figure 9 Critical separation distance ratios for yielding, $a_y(T)/a_0$, and devitrification $a(T)/a_0$, shown schematically versus temperature.

known phenomenon of the decrease of σ_y to zero as the glassy polymer is heated to T_g. This is further illustrated in Fig. 9. Since $a(298\ \text{K})/a_0 \cong 1.05$ is essentially constant, $a_y(298\ \text{K})/a_0$ is very close to 1.02 for all six polymers.

Estimates of F_{inter} and σ_y at 298 K using Eqs. (21), (22), and (24) are shown in Tables 7 and 8. Agreement with experiment in the case of σ_y is improved relative to the case at 0 K, suggesting increasing importance of the intermolecular contribution at elevated temperature.

YIELDING, CRAZING, AND CHAIN ENTANGLEMENTS

The force calculations summarized in Table 8 and the yield stress results in Table 7 suggest an interpretation of the distinction between crazing and yielding as follows. Crazing appears to be determined by the size of the intermolecular force barrier, i.e., interchain separation by translation which requires comparatively small rotation angle (ϕ) displacement. Accordingly, vinyl polymers which have a propensity to craze have comparatively small barriers relative to nonvinyls and $F_{\text{inter}} < F_{\text{conf}}$. Yielding, on the other hand, appears to be determined in part by the size of the configurational force barrier and hence requires much larger displacement angles: nonvinyl polymers which tend to yield have comparatively low configurational force barriers relative to vinyl polymers and $F_{\text{conf}} < F_{\text{inter}}$ at 0 K. The latter results mainly from the relatively larger moment arm, r, of the nonvinyl polymers. Chain straightening, which characterizes yielding, is therefore easier than chain separation (crazing) with nonvinyl polymers, whereas the reverse appears to be true with vinyls. At high temperatures (e.g., near T_g), interchain separation becomes sufficient for crazing to occur in nonvinyl polymers (e.g., as in polycarbonate). Predictions reached by strophon-mechanical calculations as above are in agreement with the extensive data of Wellinghoff and Baer [38], as well as the results of a limited study by Yannas [40], that document the difference between Type I polymers (vinyls), which generally craze rather than yield, and Type II polymers (arylene polymers with flexible oxygen linkages, such as PC and PPO), which craze only near T_g.

This analysis leads to an alternative topology for chain-chain entanglements. Rather than being necessarily considered as chains looping round each other or as knots, entanglements are presented here as a cluster of "locked" strophons [5] which are prevented from entering into localized deformation by strong intermolecular interactions leading to high values of F_{inter}. This reasoning clearly predicts that entanglements as described here would be more likely to occur with polymers which show high values of r/a_0 and is, therefore, consistent with the detailed observation by Donald and Kramer [28,41] of higher entanglement densities in nonvinyl polymers which form shear bands in preference to crazing. Strophon mechanics lead, therfore, to the view that neighboring segments in nonvinyl polymers simply engage each other much more readily than vinyl polymers due to large values of r/a_0. These "crowded glasses" transfer stresses primarily by intermolecular interactions, and the result is an efficient scale-up of the deformation leading to development of shear bands across the macroscopic specimen rather than formation of microscopic voids consistent with crazing.

ALTERATIVE REPRESENTATIONS OF GLASSY STRUCTURE AND MECHANICAL BEHAVIOR

Previously, Andrews and Kimmel [42] proposed that glassy polyacrylonitrile is an assembly of chains held together by van der Waals bonds and dipole-dipole interactions and that devitrification results from loosening of these bonds. This hypothesis [42] contradicts the often-held view that the glass transition occurs when the intrinsic chain stiffness is overcome by thermal energy. Equation (18) summarizes the finding that in all six polymers studied in our work the same interstrophon distance is reached upon being heated to T_g. Our finding supports the view that T_g occurs when intermolecular bond (secondary bonds, including various London dispersion interactions) are "dissociated" by thermal expansion. These secondary bonds, rather than intrinsic chain stiffness, are therefore considered primarily responsible for the stiffness of glassy polymers near T_g.

Andrews [43] also first proposed the hypothesis that the effect of stress on glassy amorphous polymers is to induce formation of nuclei in which the intermolecular bonding has been erased, and proposed that yielding can thereby be regarded as a stress-induced glass transition [44]. Our finding, Eqs. (18) and (19), that all six polymers studied show the same interstrophon distance at yield as they show at T_g, provides strong support for Andrews's hypothesis and extends it by providing a common dilatational mechanism for the "dissociation" of intermolecular (secondary) bonds which occurs when each of these molecularly equivalent phenomena occur.

A study of plastic deformation of a series of aromatic polyimides by Argon and Bessonov [8,9] showed that plastic deformation became less local, and the thermally activated complex grew in size, as the spacing between natural hinges on the polyimide chains increased with increasing chemical complexity. The theoretical underpinning of this study was Argon's theory [3] in which molecular kinking was modeled by a pair of wedge disclination loops. The conclusions of Argon and Bessonov [8,9] that deformation becomes less local as the size of the rigid backbone unit increases are consistent with our strophon mechanical calculations in this study, which show that polymers which exhibit values of the locking factor r/a_0 higher than about 0.8 undergo shear banding rather than crazing.

Our calculations provide a simple interpretation of the yield-to-craze transition observed by Yannas [40] with amorphous poly(ethylene terephthalate) near T_g, as well as the insightful correlation by Wellinghoff and Baer [38] of the propensity toward shear banding shown by polyarylenes relative to the crazing tendencies of vinyl polymers.

The detailed study of entanglements by electron microscopy in crazing by Donald and Kramer [28,41] has led them to the conclusion that polymers which show the highest entanglement density form shear bands rather than craze. Our calculations provide a quantitative criterion for the yield-to-craze transition in terms of the critical intermolecular force to translate a strophon, 7000 kcal mol^{-1} Å^{-1}, which in turn corresponds to the value $r/a_0 \cong 0.8$. Although we do not dispute the notion that looping or knotting of chains is certainly a form of chain entanglement, we question the need to invoke a highly specific topological model to explain the electron microscopic observations of Donald and Kramer [28,41]. We propose instead that the microscopic data [28, 41] can be explained in terms of *nonlooped* sequences of neighboring segments which interfere by local potentials [5]. The traditional model of chain entanglement as a loop should, however, be anticipated to be critically important in polymer liquids (e.g., melts).

Theodorou and Suter [13,14] developed a detailed atomistic model of atactic, glassy polypropylene. In this model bond lengths and bond angles were fixed, the intermolecular interaction was modeled by the Lennard-Jones potential and the barrier to backbone skeletal torsion was modeled by a threefold potential [13,14]. A series of energy minimization steps was performed in order to generate a dense glassy model, and the temperature was introduced by choice of the density. A thermodynamic analysis of deformation showed that entropic contributions to the elastic response can be neglected in polymeric glasses. Elastic constants were predicted closely by the model [13,14]. It was concluded that deformation in the glassy state is accompanied by concerted displacement of chain segments approximately 10 bonds long [13,14]. A detailed model has recently been constructed for polycarbonate by Hutnik, Argon, and Suter [18] in an effort to predict aspects of the inelastic behavior of this polymer. Mott, Argon, and Suter [17] have reported preliminary studies of plastic deformation in polypropylene by the same method. Sylvester, Yip, and Argon [16] are currently extending this molecular dynamics approach to the study of structural differences between the liquid and glassy states of atactic polypropylene. It remains to be seen to what extent these detailed atomistic studies will address the issue of chain stiffness versus intermolecular interactions which plays such a critically predictive role in strophon mechanical analyses in terms of the dimensionless number r/a_0. It is also interesting to anticipate the possibility of a detailed theoretical test of the simple prediction, arrived at here by calculation, that yielding and devitrification are dilatationally equivalent states.

In conclusion, we suggest that the approach of strophon mechanics [4] described in this chapter provides an alternative method for viewing devitrification, crazing, and yielding of glassy amorphous polymers which, in spite of its simplicity, appears to interpret, predict, and unify for the first time a large number of important observations.

REFERENCES

1. P. Bowden, *Polymer, 9,* 449 (1969).
2. R. N. Howard and J. R. McCallum, *Polymer, 12,* 189 (1971).
3. A. S. Argon, *Phil. Mag., 28,* 839 (1973).
4. I. V. Yannas, *Polym. Prepr. Am. Chem. Soc., 14(2)*, 802 (1973).
5. I. V. Yannas, in *Proceedings of IUPAC International Symposium on Macromolecules*, p. 265 (E. B. Mano, ed.), Elsevier, Amsterdam (1975).
6. L. Holliday, in *Structure and Properties of Amorphous Polymers*, Chap. 7 (J. M. Ward, ed.), Wiley, New York (1975).
7. I. V. Yannas and A. C. Lunn, *Polym. Prepr. Am. Chem. Soc., 16(2)*, 564 (1975).
8. A. S. Argon and M. I. Bessonov, *Phil. Mag., 35*, 917 (1977).
9. A. S. Argon and M. I. Bessonov, *Polymer Eng. Sci., 17*, 174 (1977).
10. I. V. Yannas and R. R. Luise, *J. Macromol. Sci-Phys., B21(3)*, 327 (1982).

11. I. V. Yannas and R. R. Luise, in *The Strength and Stiffness of Polymers*, Chap. 6 (R. S. Porter, ed.), Dekker, New York, (1983).
12. G. A. Kardomateas and I. V. Yannas, *Phil. Mag.*, *A52*, 39–50 (1985).
13. D. N. Theodorou and U. W. Suter, *Macromolecules 18*, 1476 (1985).
14. D. N. Theodorou and U. W. Suter, *Macromolecules 19,* 379 (1986).
15. D. G. Gilbert, M. F. Ashby, and P. W. Beaumont, *J. Mat. Sci.*, *21*, 3194 (1986).
16. M. F. Sylvester, S. Yip, and A. S. Argon, *Polym. Prepr. Am. Chem. Soc. Polym. Chem. Div.*, *30(2)*, 34 (1989).
17. P. H. Mott, A. S. Argon, and U. W. Suter, *Polym. Prepr. Am. Chem. Soc.*, *30(2)*, 34 (1989).
18. M. Hutnik, A. S. Argon, and U. W. Suter, *Polym. Prepr. Am. Chem. Soc.*, *30(2)*, 36 (1989).
19. A. C. Lunn and I. V. Yannas, *J. Poly. Sci.-Polym. Phys.*, *10*, 2189 (1972).
20. J-F. Jansson and I. V. Yannas, *J. Polym. Sci.-Polym. Phys.*, *15*, 2103 (1977).
21. R. M. Christensen and F. M. Waals, *J. Composite Mater.*, *6*, 518 (1972).
22. D. W. Von Krevelen, *Properties of Polymers*, pp. 55, 62, 70, 74, 132–133, Elsevier, New York (1976).
23. M. V. Volkenstein, *Configurational Statistics of Polymer Chains*, pp. 130–135, Interscience, New York (1963).
24. W. H. Stockmayer, private communication.
25. P. J. Flory, *Statistical Mechanics of Chain Molecules*, pp. 286–296, Interscience, New York (1969).
26. I. V. Yannas, *Abs. 4th IUPAC Microsymposium*, p. J1, Prague, Sept. 1–4, (1969).
27. H-H. Sung and I. V. Yannas, unpublished data.
28. A. M. Donald and E. J. Kramer, *J. Polym. Sci.*, *20*, 899 (1982).
29. H. Schnell, *Chemistry and Physics of Polycarbonates*, p. 142, Interscience, New York (1964).
30. L. E. Nielson, *Mechanical Properties of Polymers*, pp. 187, 207, Reinhold, New York (1962).
31. H. F. Mark, ed., *Encyclopedia of Polymer Science and Technology*, vol. 10, p. 102, Wiley, New York (1969).
32. H. F. Mark, ed., *Encyclopedia of Polymer Science and Technology*, vol. 10, p. 268, Wiley, New York (1969).
33. S. L. Cooper, A. D. Mair, and A. V. Tobolsky, *Textile Res. J.*, 1110 (December 1965).
34. J. A. Brydson, *Plastics Materials*, p. 360, Butterworths, London (1975).
35. J. Brandrup and E. H. Immergut, eds., *Polymer Handbook*, p. V-46, Wiley, New York (1976).
36. H. Ruff and R. Scott, eds., *Handbook of Common Polymers*, p. 93, CRC Press (Butterworths) London (1971).
37. R. Kambour, personal communication.
38. S. T. Wellinghoff and E. Baer, *J. Appl. Polym. Sci.*, *22*, 2025–2045 (1978).
39. R. P. Kambour, *J. Polym. Sci.*, *D7*, 1 (1973).

40. I. V. Yannas, *Science*, *166*, 227–228 (1969).
41. A. M. Donald and E. J. Kramer, *J. Mater. Sci.*, *17*, 1871 (1982).
42. R. D. Andrews and R. M. Kimmel, *Polym. Lett.*, *3*, 167 (1965).
43. R. D. Andrews, *Polym. Prepr. Am. Chem. Soc. Polym. Chem. Div.*, *10(2)*, 1110 (1969).
44. R. D. Andrews and Y. Kazama, *J. Appl. Phys.*, *38*, 4118 (1967).

part three

THERMODYNAMIC AND STATISTICAL MECHANICAL TECHNIQUES

5

Monte Carlo and Molecular Dynamics Simulations of Amorphous Polymers

Kurt Binder
Johannes-Gutenberg-Universität, Mainz, Germany

INTRODUCTION: WHAT PROBLEMS CAN MONTE CARLO AND MOLECULAR DYNAMICS SIMULATIONS DEAL WITH?

Computer simulation has been a prominent tool of research in polymer science for a long time: in particular, Monte Carlo studies of self-avoiding walks on lattices have played a pivotal role in elucidating excluded-volume effects on the configurational statistics of long flexible polymer chains in dilute solution [1–4]. This application is outside of consideration here, where we rather are concerned with simulations *of dense macromolecular materials, such as melts and blends*. In this chapter, much attention is paid to static and dynamic properties of polymers in fluid phases. Understanding these properties is useful also for the understanding of amorphous polymers, which are produced by cooling from the respective fluid phase through the glass transition temperature T_g of the considered material. Thus, certain structural characteristics of the solid disordered polymeric material are just frozen in from the fluid phase. Also residual local motions of monomers are still possible in the amorphous state and can be understood along similar lines.

In most applications of the Monte Carlo and molecular dynamics methods one does not take into account the chemical structure of the polymers forming the considered material in a realistic way; rather one works with simplified models, which often even represent a whole class of materials. Since this approach thus involves taking properties at the truly molecular level into account only in a crude, approximate fashion, the question should be answered first: why is such an approach useful?

The answer is that such simulations are particularly useful to check the validity of theoretical concepts. Of course, one could also try to check the validity of an analytical theory by directly comparing it to experimental data. But often this comparison is rather inconclusive, since the theory involves two types of approximations: (a) A theoretical model needs to be formulated, where one restricts attention to those degrees of freedom only which one believes to be relevant. The price paid for this reduction in complexity of the problem is the introduction of suitable parameters, which then are not known explicitly a priori. Rather they are estimated from a fit of experimental data to the theoretical formulae. (b) Due to the complexity of many-body problems, usually it is not possible to derive mathematically rigorous expressions from the considered model; rather crude approximations are required, whose accuracy often is very uncertain.

These facts hamper the comparison between theory and experiment. Due to the unknown parameters contained in a model it often is difficult to judge whether a theory is reliable. *Often a quantitatively inaccurate theory can be fitted to experiment; only the fitted parameters are systematically in error.* Of course, such a feature is disadvantageous if one wishes to make predictions on properties of materials using such a theory, and sometimes large errors can occur.

In this situation, computer simulations help: one can study precisely the same simplified model on which the analytical theory is based, but avoids the mathematical approximations! All parameters occurring in the model can be varied within large limits and thus the reliability of an approximation can be stringently tested. Often hints are obtained on how to improve upon an approximation. A particular merit of the simulations is that they provide information on molecular configuration and their evolution in time on a microscopic scale. This often yields new insights into the nature of physical states and processes. (By "microscopic scale" we mean here the length scale of a few bonds.)

On the other hand, a direct comparison of a simulation to experiment may also be possible and useful: it allows to estimate adjustable parameters of a model designed to describe a real material more reliable than by using approximate theories for that purpose. One can check whether a model is at all adequate to describe a real system. If an analytical theory fails to reproduce experimental data, we often do not know whether it is the model or some approximation in the mathematical treatment of the model which fails.

This chapter thus will illustrate this interplay of computer simulaton with both analytical theory and experiment by several examples, and describe the state of the art of both Monte Carlo and molecular dynamics simulations of dense polymer systems. Before doing that sound background information on both the Monte Carlo technique and the molecular dynamics technique will be given. The last section summarizes our conclusions and gives some hints concerning the directions of future research.

TECHNICAL ASPECTS

In this section, the general type of modeling of polymers as it will be alluded to in this article will be introduced. Some background on computer simulation methods in general will be given first, and then we proceed to aspects which are specific for applications to macromolecules. Both the potential advantages and the limitations of the various variants of the methods discussed will be pointed out. More thorough introductions to the Monte Carlo method in general can be found in Refs. 4–8, and to the molecular dynamics method in Refs. 8–11.

The Monte Carlo Method

The Monte Carlo method addresses the problem of calculating thermal averages along the lines of statistical thermodynamics, by constructing a random walk through the configuration space of the considered model system. The desired physical properties are then obtained as "time-averages" along this stochastic trajectory in phase space. To the extent that the actual time evolution of the considered degrees of freedom can be modelled by a Markov process, one can in fact also obtain information on dynamical (time-displaced) correlation functions along this Markov chain of configurations. This is usually the case when the description considers only a subset of dynamical variables of the system, such that the subset variables are much slower than the rest. If such a separation of time scales does not occur, the Monte Carlo method may still be useful for computing static averages, but for the study of dynamic properties the molecular dynamics method (see the following subsection) must be used.

It is the explicit construction of the Markov chain of random system configurations, where random numbers are used which have given the Monte Carlo method its name [12]. Despite the availability of very fast computers, one still is far from being able to handle the statistical mechanics of dense polymer systems in its full complexity. Thus, rather simplified models of polymers have to be used, where a lot of detailed information on the chemical structure is disregarded. We start by discussing the most common of these models.

Modeling of Polymers for Monte Carlo Simulation

In principle, one may wish to consider systems which contain only one kind of polymer, or systems composed of polymers differing in their chemical nature. Branched polymers as well as linear polymers may be considered, and among the latter flexible polymers as well as stiff chains need to be distinguished. Most of the work that exists so far, however, addresses long flexible macromolecules existing in solution or melt in a randomly coiled state. Modeling of such chains hence will be emphasized here.

First of all, we wish to distinguish off-lattice models from lattice models of macromolecules (Figs. 1, 2). Since the relaxation time needed to equilibriate the configurations of long flexible chains increases so strongly with chain length, it is necessary to perform a sort of coarse-graining along the chain, by which many details of the local chemical structure is lost: the basic unit is *not* the chemical monomer but rather a group of successive monomers, which, at least roughly, has the size of the order of a Kuhn step length already. This basic element mostly is taken of rigid length (Figs. 1a, b, Fig. 2a-c) but sometimes it also may fluctuate in length (Figs. 1(c), 2(d)).

In the model of freely jointed chains (Fig. 1(a)) [13] each polymer hence is modelled by a succession of N rigid bonds of length l jointed together at arbitrary angles. The steps of the Monte Carlo procedure then consist in rotations of beads. For example, bead i is rotated by a randomly selected angle φ_i from its old position to a new position. This attempted move is not always carried out but only with a transition probability $W(\varphi_i)$: e.g., this transition probability is set equal to zero if the attempted move leads to some intersection of bonds, and if it is the dynamics of melts that one wishes to simulate. Forbidding bond intersection then represents the entanglement restriction that in their motion polymer chains cannot cross each other. Of course, if one is just interested only in static thermal equilibrium properties of the system, such a restriction needs not to be made.

The freely jointed chain is surely an extremely unrealistic model of polymers, since chains can come arbitrarily close to each other, while in reality along the backbone of the chain due to the nonzero size of the radii of the atoms forming the backbone (and possible sidegroups attached to it) there is a volume region excluded from occupation by monomers of other chains. The pearl necklace model (Fig. 1(b)) is a crude way of taking such an excluded volume interaction into account. Also the fact that during their motions chains cannot cross each other can be taken into account by

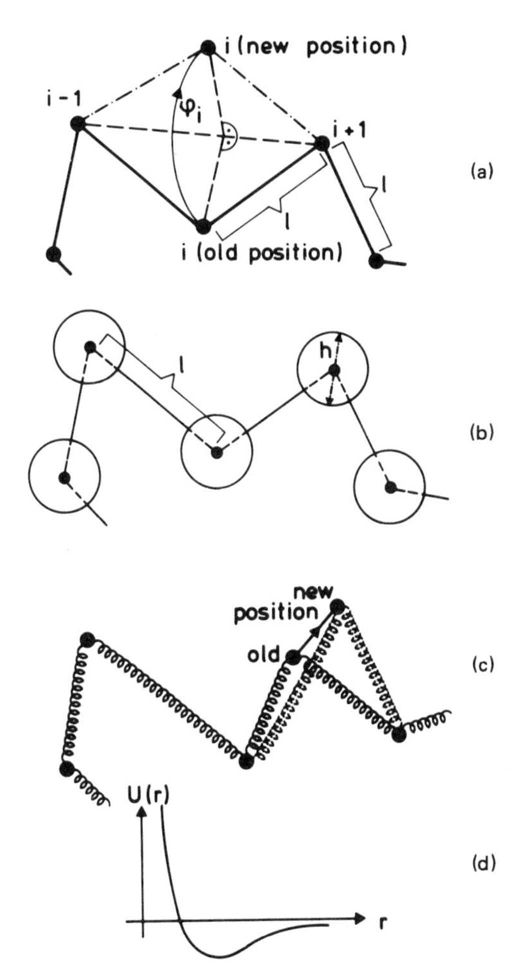

Figure 1 Off-lattice models for polymer chains. In the freely jointed chain (a) rigid links of length l are jointed at beads marked by dots and may there have arbitrary angles with each other. The stochastic chain conformational changes are modelled by random rotation of bonds around the axis connecting the nearest-neighbor beads along the chain, as indicated. The new position of a bead i may be chosen by assigning an angle φ_i chosen randomly from the interval $[-\Delta\varphi, +\Delta\varphi]$ with $\Delta\varphi \leq \pi$. For the simulation of melts, the freely jointed chain is often supplemented by a Lennard-Jones-type potential (d) between any pair of beads [13]. An alternative model is the pearl necklace model (b), where the beads are at the center of hard spheres of diameter h, which must not intersect each other. This model has also proven useful for the simulation of melts [14]. Still another alternative is the bead-spring model (c), which also has been used for melt simulations [15].

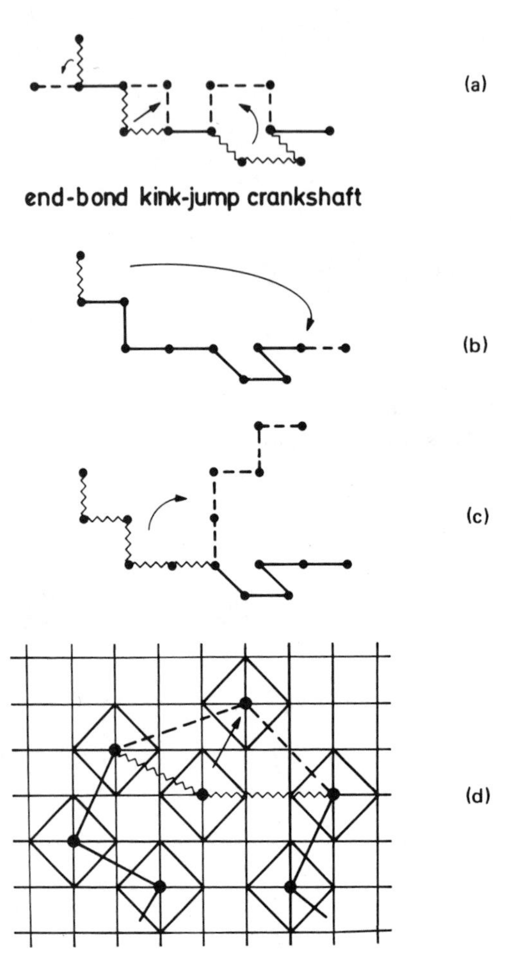

Figure 2 Various examples of dynamic Monte Carlo algorithms for lattice models of polymers. Bonds indicated as wavy lines are moved to new positions, other bonds are not moved. (a) Generalized Verdier-Stockmayer [4,16–18] algorithm on the simple cubic lattice, showing three types of motion: end-bond motion, kink-jump motion, crankshaft motion; (b) slithering-snake (reptation) algorithm [19–21]; (c) "pivot" ("wiggle") algorithm [22–24]; (d) bond length fluctuation algorithm [25,26] on the square lattice. Each monomer blocks four sites around its center from occupation by other monomers. The bond length is allowed to fluctuate from a minimum value $\sqrt{4}$ to a maximum value of $\sqrt{13}$ lattice spacings shown in the example shown. Note that this algorithm also works for branched polymers, star polymers, etc. (From Ref. 4.)

proper choice of the radii [14]. By varying the ratio between the diameter h of the hard spheres attached to the beads and the bond length l, one can vary the effective strength of the excluded volume interaction. The moves which are carried out are still of the type shown in Figure 1(a), but any motions which would lead to overlap of the hard spheres are forbidden. It is possible, however, to choose $l/2 < h < l$, which implies that successive spheres along the chain are allowed to overlap; still this is a valid definition of the volume region occupied by the atoms forming a chain.

While the pearl necklace model is strictly athermal, it is also possible to work with a potential energy which is finite, such as the Lennard-Jones potential

$$U(\mathbf{r}_i,\mathbf{r}_j) = 4\epsilon\left[\left(\frac{\sigma}{|\mathbf{r}_i - \mathbf{r}_j|}\right)^{12} - \left(\frac{\sigma}{|\mathbf{r}_i - \mathbf{r}_j|}\right)^{6}\right] \tag{1}$$

which is qualitatively sketched in Figure 1(d). While ϵ characterizes the strength of the potential, σ characterizes its range (the minimum occurs for a distance $|\mathbf{r}_i - \mathbf{r}_j| = 2^{1/6}\sigma$). The first term on the right side of Eq. (1) thus accounts for the repulsive excluded volume interaction at short distances between the beads at sites $\mathbf{r}_i$, $\mathbf{r}_j$. The potential Eq. (1) is applied between all possible pairs of beads. Now when one carries out a trial motion as shown in Figure 1(a), one has to compute the energy change δU associated with this move, and carry out the motion with probability $w = \exp(-\delta U/k_BT)$ only, if $\delta U > 0$. As will be justified below, this algorithm ensures that one generates configurations which have the right Boltzmann weight $\exp\{-\Sigma U(\mathbf{r}_i,\mathbf{r}_j)/k_BT\}$, with k_B being Boltzmann's constant and T the absolute temperature. Carrying out the motion with probability w, means in practice that one draws a random number ζ uniformly distributed between zero and unity: if $w \geq \zeta$, the trial motion is actually carried out but otherwise it is rejected. Then the old configuration needs to be counted once more for the averaging.

Still another alternative off-lattice model is the bead-spring model of Figure 1(c). There, two neighboring beads are no longer a fixed distance apart, but only on average they have the distance l, maintained via a harmonic potential $U_{el}(\mathbf{r}_i - \mathbf{r}_{i-1}) \equiv k[(\mathbf{r}_i - \mathbf{r}_{i-1})^2 - l^2]$, where k is some suitably chosen spring constant. It is also possible to work with anharmonic spring potentials, of course; to ensure a finite maximum extension l_{max} of a link one may choose, for instance, a potential $U_{el}(\mathbf{r}_i, \mathbf{r}_{i-1}) = k\{-l^2_{max} \ln[1 - (\mathbf{r}_i - \mathbf{r}_{i-1})^2/l^2_{max}] - l^2\}$. In this model the motion of a single bead can be a displacement in an arbitrary direction, but the change of the elastic energy U_{el} needs to be included in the computation of the transition probability W. Just as the freely jointed chain, the bead-spring model also needs to be complemented by a potential of the type of Fig 1(d) to account for excluded volume interactions at short distances. We note that this bead-spring model is also a suitable model for molecular dynamics calculations of polymers [27]; see page 239.

After this discussion of a few variants of off-lattice polymer models we now also discuss some lattice models-again we aim not at all at a complete and exhaustive description of all models existing in the literature, but focus on a few cases of widespread use. A simple very common lattice model of polymers is the self-avoiding random walk of N steps on a lattice (Figs. 2(a)–(c)). The beads are now situated at lattice sites, the bonds usually have the length of the lattice spacing so that nearest-neighbor beads along the chemical sequence of the chain are also nearest neighbors at the lattice. Of course, models with larger step size (=bond length) also have occasionally been used. Since each lattice site can at most be occupied by one bead, the walk cannot intersect itself (or other walks, respectively), and thus an excluded volume interaction is already automatically included. Clearly, while the arbitrary angles between successive links of a chain are reasonable if we interpret these links as the Kuhn effective segments formed from groups of several monomers along the chemical sequence of the chain, the rigid angles (0°, 90° at the square and simple cubic lattice) between the successive bonds of the lattice model are a further idealization. But there are reasons to believe that for the large scale properties of polymers this does not matter. Only prefactors in asymptotic laws but not the laws themselves should be affected by the lattice structure; thus lattice models nevertheless are very widely used. A reason for this is that for many simulation purposes one can devise much more efficient computer programs using lattice rather than off-lattice models [4]. A further motivation is that many analytical theories are formulated in the framework of a lattice model, e.g., the well-known Flory-Huggins model [28] of polymer mixtures. Use of the same lattice model in a simulation allows to immediately perform a stringent test of such theories [18,29–31].

If we adopt such a lattice model, then there are general possibilities to generate new chain configurations, starting from some given initial configuration. One rather common technique is the generalized Verdier-Stockmayer [16–18] algorithm (Fig. 2(a)). In the original work [16,17], randomly a bead was selected: if it is at a chain end, the link is rotated to a randomly chosen new direction (if this does not violate the excluded volume constraint, otherwise the move is of course rejected). If the randomly selected bead is in the chain interior, a motion is attempted only if the two bonds jointed together at this bead form a right angle: then one can attempt to carry out a move which we call kink-jump (Fig. 2(a)); again the move is accomplished only if the new lattice site to which the bead is moved was still empty.

Later it was realized [32,33] that this algorithm in the form described so far is very inefficient. This is due to the fact that kink-jumps do not create any new "bond vector" connecting two neighboring bonds, but rather only exchange two neighboring bond vectors. New bond vectors only are created at the ends and then diffuse slowly towards the interior of the chain. Such an algorithm fails manifestly for ring polymers. Only a small subset of the configuration space of a ring polymer would be sampled, the algorithm would be strongly nonergodic. Thus, one has to allow for additional motions, such as the 90° crankshaft motion, Figure 2(a), which creates new bond vectors. In this form the algorithm already is widely used; see, e.g., [18,29–31]. Even for this algorithm, *ergodicity* is a problem [34]: there are certain "forbidden configurations" of the walk which cannot be reached from other configurations and also could not relax by the motions shown in Figure 2(a), if one would generate these "forbidden configurations" by some other method. There is ample evidence, however, that this nonergodicity does not matter in practive [29]. The statistical weight of these locked-in configurations is negligibly small for chain lengths of practical interest, as the comparison of results obtained with different simulation techniques shows [4].

Such an alternative is the "slithering snake"—or "reptation"—algorithm [19–21] (Fig. 2(b)). There, one randomly chooses one end of the chain and tries to remove it from there and add it at the other chain end in a random direction. Again the move is carried out only if it does not violate excluded volume constraints.

An advantage of this algorithm in comparison to the previous one is the fact that it relaxes distinctly faster [4,35]. A disadvantage, of course, is that it cannot be interpreted as representing motins of the real chain, unlike the kink-jump algorithm of Figure 2(a): the random local conformation changes induced by random friction forces acting from its environment on a chain. Thus, the kink-jump algorithm leads to a Rouse-like chain relaxation [36], as demonstrated in [37]. In that respect, the relaxation of the model in Figure 2(a) corresponds to the relaxation found for the off-lattice algorithms of Figure 1 [13,38]. Also the "reptation algorithm" of Figure 2(b) suffers from the problem that locked-in configurations exist which can neither relax nor be reached; these configurations are not the same ones as for the algorithm of Figure 2(a), of course [34]. Again it is believed that this problem can be forgotten in practice. Now an algorithm which would not suffer from ergodicity problems is the "pivot" algorithm [22–24] depicted in Figure 2(c). There a bead of the chain is randomly chosen and then the part of the chain to the right of this bead one attempts to rotate (like a rigid object) in a new direction. Of course, the new configuration is accepted only if it does not violate the excluded volume constraint. Otherwise this trial motion is rejected, and another trial is made. While this algorithm is very efficient for the simulation of isolated chains [24], it presumable is not useful for the simulation of dense polymer systems, and thus will not be considered further in this chapter. For a description of several other "single-chain" algorithms, we refer to the literature [2–4].

At this point, we also mention more complicated algorithms where an effective monomer does not occupy a single lattice site only but a whole group of neighboring lattice sites [25,26]. Since then the bond connecting two neighboring beads typically should be much larger than the lattice spacing, one need not choose the bond length constant, but one can work with a fluctuating bond length (Fig. 2(d)). A particular merit of this algorithm is that it is also useful for simulating branched polymers. A branch point of a star polymer or a branch point in a polymer network according to this algorithm can easily move, while neither of the algorithms in Figures 2(a)–(c) would work then. Furthermore, it is likely that there are no ergodicity problems, if one chooses the range over which the bond lengths fluctuates properly. In one elementary step, a monomer is moved by one lattice spacing only.

An important condition which all valid Monte Carlo algorithms must fulfill is the detailed balance principle. For the athernal Monte Carlo problems modeled by Figures 1(a), 1(b), or 2(a)–(d), this simply means that the probability to carry out the considered move must be equal to the probability to carry out the inverse motion. While this is straightforwardly satisfied in the case of Figure 1(a), care must be exerted that no bias is introduced when one works with a combination of several types of motion such as in the case of Figure 2(a).

Of course, it also is possible to introduce various energy parameters into lattice simulations such as shown in Figure 1(a)–(d), and thus simulate thermal rather than athermal problems. For example, for the simulation of a polymer blend one imagines that there are two types (A, B) of chains, and the interaction energies ϵ_{AA}, ϵ_{AB}, or ϵ_{BB} are won, if a nearest-neighbor pair of sites is taken by two AA-monomers, by an AB (or BA) pair of monomers, or a BB pair [18,29–31]. Alternatively, one may also work with an energy parameter ϵ controlling chain stiffness: it may cost an energy ϵ if two bonds at a bead form a 90° angle, while it costs no energy if the two bonds continue in the same direction [39,40]. In all such thermal problems the energy change occurring in an attempted move enters the transition probability W exactly in the same way as described for the off-lattice case.

While in all simulation algorithms described so far the chain length N (number of bonds) is held fixed, it sometimes is useful to consider simulations where this is not the case, and one allows the chain length to fluctuate as well. In grand-canonical Monte Carlo methods for single chains

[41–43], one allows for the addition or annihilation of bonds by a random procedure (Fig. 3(a)). The acceptance of these trials is controlled by the chemical potential μ leading to an average chainlength $\langle N \rangle$. In practice, one can proceed by selecting a bond at random and trying to displace it at random by one lattice unit to one of the $2(d\text{-}1)$ directions orthogonal to the bond. If this attempt to move would violate the excluded volume restriction, it would, of course, be rejected. If it complies with the self-avoiding walk condition, the chain length changes either by $\Delta N = \pm 2$ or it does not change ($\Delta N = 0$). In order to satisfy the detailed balance condition with the chosen fugacity per bond β, the transition probabilities must satisfy the relation $W(\Delta N)/W(-\Delta N) = \beta^{2\Delta N}$. We do not go into the details of this method and related approaches [44,35] here, since we are not aware of any applications to dense polymer systems yet.

Techniques that can be used for very dense polymer systems are more important for our purposes, such as the techniques illustrated in Figures 3(b)–(d). Note that all the techniques shown in Figure 2 require vacant sites, and thus the acceptance rate for any attempted motions goes to zero as the vacancy concentration $\phi_v \to 0$. In fact, as discussed in the third section ("Simulation of the Glass Transition"), a freezing into a glassy state may already occur at a small but nonzero concentration of vacancy already. But for $\phi_v = 0$, no motion whatsoever would be allowed for the algorithms of Figure 2. To overcome this problem, several techniques have been proposed: the chain-breaking methods due to Olaj and Lantschbauer [45] and Mansfield [46] (Fig. 3(b), (d)) accept some polydispersity in the system; the so-called COMO ("collective motion") algorithm of Pakula and Geyler [47,48] (Fig. 3(c)) allows simulation of strictly monodisperse multichain systems at $\phi_v = 0$, but since its accuracy is somewhat doubtful [4] it will not be considered further. Since the width of the chainlength distribution can be controlled in various ways [45,49] and can be made very small, we feel that the polydispersity of the model is not a real disadvantage. Remember that experimental polymeric materials anyhow always are somewhat polydisperse. Moreover, it has been shown that for small widths of the chain length distribution the chain linear dimensions are hardly dependent on this width, in the method used in [45].

References [45 and 46] include motions of the "bond flip" -type, where a pair of parallel bonds is rotated if the bonds belong to neighboring chains. If the two bonds belong to the same chain, this move is forbidden since it would lead to formation of a ring polymer. This motion does not involve the chain ends of either chain. The second move common to both refs. [45,46] is the so-called end attack (Fig. 3(d)): From a site of a chain that is nearest neighbor to a chain end of another chain a bond is rotated, so that the previous chain end is now an inner bead of one of the chains, and

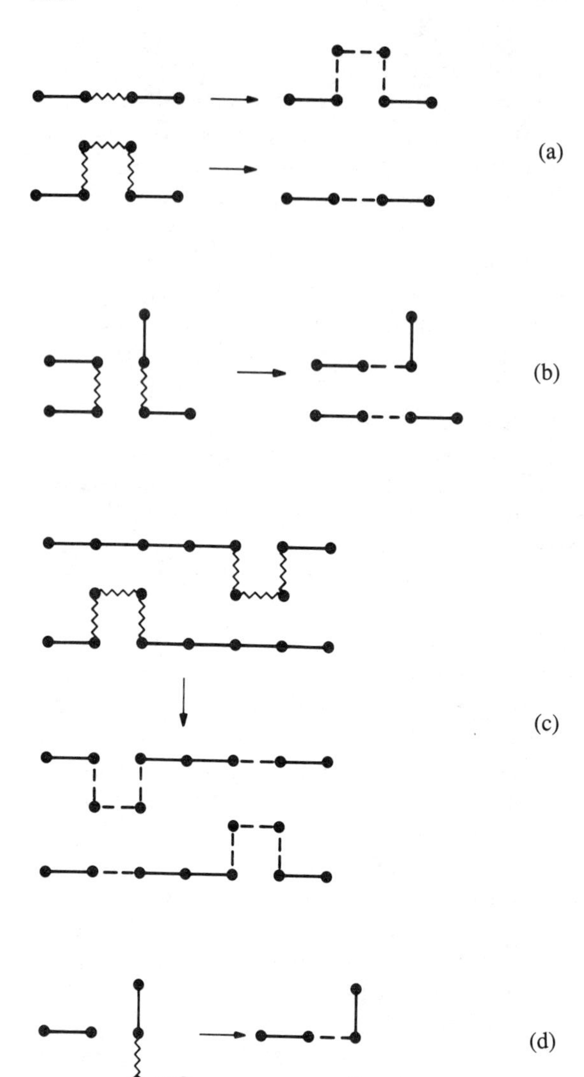

Figure 3 Dynamic Monte-Carlo algorithms involving creation or annihilation of bonds [41–43] (a); or exchange between pairs of parallel bonds which reorient to form a different connection bewteen the two parts of the chain (or two different chains, respectively) [45,46] (b); exchange between crankshaft configurations between two different parts of a chain (or two different chains, respectively) [47,48] (c); rotation of a bond to form a different connection between two parts of a chain (or two different chains, respectively) [45,46] (d). (From Ref. 4.)

the site from which the bond moved away becomes a chain end. Mansfield [46] also allows intramolecular end attacks (the "backbite move"), if it does not lead to ring formation. On the other hand, Ref. 45 includes another reaction between two adjacent chain ends: The chains may "dimerize" at this bond and subsequently are subject to a chain scission at a statistically chosen bond. This reaction ensures a rapid migration of end segments.

If no further restriction is introduced, the resulting distribution of chain lengths would be rather broad. Mansfield [46] therefore forbids any reactions, by which chains shorter than some given length N_{min} would be created. Madden [49], working at vacancy densities $\phi_v \approx 0.1$ to $\phi_v \approx 0.2$ and using a mixture of "chain-breaking" methods and the reptation method, introduced both an upper and lower cutoff, so that all chain lengths lie in the interval $N_{min} \leq N \leq N_{max}$. However, it turns out that the chainlength distribution generated by this method is rather close to a rectangular distribution, which is rather different from the experimentally relevant distributions (which are often well approximated by a log-normal distribution [50].

Olaj and Lantschbauer [45] control the degree of polydispersity in a very elegant way using the idea of an importance sampling Monte Carlo method. Suppose a pair of chains is considered for a reaction and ΔN_{old} is the absolute value of the difference of their chain lengths. If this difference would be ΔN_{new} after the attempted reaction, the trial reaction always is accepted if $\Delta N_{new} \leq \Delta N_{old}$. On the other hand, if $\Delta N_{new} > \Delta N_{old}$, the new configuration is accepted with probability of $\exp[-f(\Delta N_{new} - \Delta N_{old})]$ only, where the parameter f controls the width of the distribution; otherwise the trial configuration is rejected, and the old configuration is retained. Choosing $f = 10/N$ yields a distribution of similar shape as in experiment and a polydispersity index N_w/N_n (weight average/number average) of about 1.02, i.e., monodisperse for practical purposes.

Clearly, all the polymer models considered in Figures 1–3 are extremely simplified, and yield a coarse-grained picture of an amorphous polymer only (see Fig. 4 for an example), where a correlation between the polymer properties and details of its chemical structure are hardly possible. A more realistic atomistic modeling of polymers, based for instance on the rotational isomeric state model [51], certainly is possible [52,53], but will not be considered here, since so far this approach is practically restricted to static nonthermal properties (it is essentially a zero-temperature Monte Carlo method). Once more we emphasize our opinion that the simple models of Figures 1–3 can yield many useful results. This is true even for the most simple version, the freely jointed chain (Fig. 1(a)) without any interaction, which recently was used to estimate the partition coefficient of polymers in pores [54].

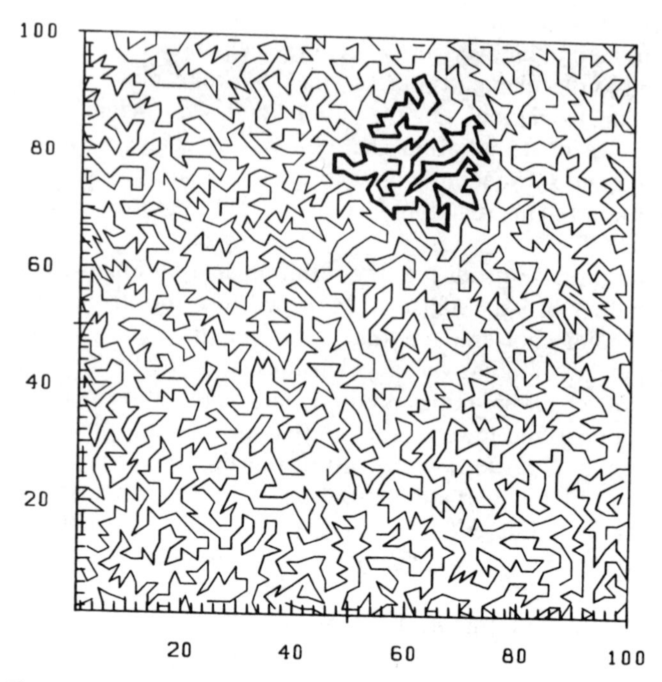

Figure 4 Snapshot picture of a two-dimensional melt, consisting of chains with $N = 50$ on a 100×100 square lattice, with a vacancy concentration $\phi_v = 0.2$. Chain configurations were generated with the bond fluctuation method, as explained in Figure 2(d). (From Ref. 26.)

Brief Review of Importance Sampling Monte Carlo Method

In this subsection, we sketch the background on the Monte Carlo method as it is used in statistical mechanics in general [4–8]. These "dynamic" Monte Carlo methods (which can be contrasted with "static," "simple sampling" methods [1–4] useful for single-chain problems) are based on a stochastic Markov process, where subsequent configurations $\mathbf{x}_\gamma$ of the system are generated from the previous configuration $\{\mathbf{x} \to \mathbf{x}' \to \mathbf{x}'' \to \ldots\}$ with some transition probability $w(\mathbf{x} \to \mathbf{x}')$. For the examples shown in Figures 1–4, a configuration $\mathbf{x}$ is specified by giving the coordinates of all beads of all chains. Examples for the meaning of the move $\mathbf{x} \to \mathbf{x}'$ are also shown in Figures 1(a), 1(c), 2(a)–(d), 3(a)–(d). The choice of the transition probability for these moves is not unique: we only require the principle of detailed balance with the equilibrium distribution $P_{eq}(\mathbf{x})$,

$$P_{eq}(\mathbf{x})w(\mathbf{x} \to \mathbf{x}') = P_{eq}(\mathbf{x}')w(\mathbf{x}' \to \mathbf{x}), \tag{2}$$

with $P_{eq}(\mathbf{x})$ the canonic distribution of the system, expressed in terms of the system Hamiltonian $\mathcal{H}(\mathbf{x})$,

$$P_{\text{eq}}(\mathbf{x}) = \frac{\exp\{-\mathcal{H}(\mathbf{x})/k_B T\}}{Z}, \quad Z = \operatorname{Tr} \exp\left\{-\frac{\mathcal{H}(\mathbf{x})}{k_B T}\right\}. \tag{3}$$

Of course, Eq. (2) contains as a special case also athermal hard-core problems (cf. Fig. 1(b)), where the potential energy U is infinite for the "forbidden" configurations, and zero otherwise: then the temperature T does not enter explicitly, since $P_{\text{eq}}(\mathbf{x}) = 0$ if $U = \infty$ and $P_{\text{eq}}(\mathbf{x}) = 1/Z$ otherwise. Equations (2 and 3) lead to the requirement

$$\frac{w(\mathbf{x} \to \mathbf{x}')}{w(\mathbf{x}' \to \mathbf{x})} = \exp\left\{-\frac{\delta\mathcal{H}}{k_B T}\right\}, \tag{4}$$

with $\delta\mathcal{H} = \mathcal{H}(\mathbf{x}') - \mathcal{H}(\mathbf{x})$ being the energy change produced by the move. Following Metropolis et al. [12], one can satisfy Eq. (4) by taking $w(\mathbf{x} \to \mathbf{x}') = \exp(-\delta\mathcal{H}/k_B T)$ if $\delta\mathcal{H} > 0$ and $w(\mathbf{x} \to \mathbf{x}') = 1$ otherwise: this is the algorithm which was already mentioned after Eq. (1).

In the limit where the number of configurations M generated in this way tends to infinity, the states $\mathbf{x}$ generated with this procedure are distributed proportional to the equilibrium distribution $P_{\text{eq}}(\mathbf{x})$ (provided there is no problem with the ergodicity of the algorithm). Then the canonical average of any observable $A(\mathbf{x})$ is approximated by the simple arithmetic average which can also be interpreted as a time average [4–8]:

$$\langle A \rangle \approx \overline{A} = \frac{1}{M - M_0} \sum_{\gamma = M_0 + 1}^{M} A(\mathbf{x}_\gamma) = \frac{1}{t - t_0} \int_{t_0}^{t} dt' \, A(t'). \tag{5}$$

In Eq. (5) it is anticipated that the first M_0 of the M generated configurations are omitted from the averaging, since the starting configuration usually is not representative for the equilibrium distribution (e.g., one may start with completely stretched chains which first must "equilibrate" to form random coils). We associate a (pseudo)-time variable $t' \equiv \gamma/\tilde{N}$ with the label γ of successively generated configurations, where usually the time unit is defined by requiring that $\tilde{N}$ is the total number of monomers in the system ($\tilde{N}$ = chain length N times number of chains n). Thus, M_0 and M correspond to times $t_0 = M_0/\tilde{N}$ and $t = m/\tilde{N}$, respectively. The Monte Carlo procedure thus is interpreted as numerical realization of a Markov process described by a master equation for the probability $P(\mathbf{x},t)$ that a configuration $\mathbf{x}$ occurs at time t:

$$\frac{d}{dt} P(\mathbf{x},t) = -\sum_{\mathbf{x}'} w(\mathbf{x} \to \mathbf{x}')P(\mathbf{x},t) + \sum_{\mathbf{x}'} w(\mathbf{x}' \to \mathbf{x})P(\mathbf{x}',t). \tag{6}$$

Obviously, the detailed balancing principle, Eq.(2), guarantees that $P_{\text{eq}}(\mathbf{x})$

(Eq. (3)) is the steady-state solution of the master equation, Eq. (6). If all states are mutually accessible, $P(\mathbf{x},t)$ must relax toward $P_{eq}(\mathbf{x})$ as $t \to \infty$, irrespective of the initial configuration $\mathbf{x}_1$.

This dynamic interpretation of Monte Carlo averaging is useful for two purposes:

1. The dynamics is modeled as a Brownian motion in a heat bath, which stochastically induces local conformational changes. These processes are qualitatively represented by moves such as indicated in Eqs. (1a), (2a), or (2d), respectively. It is believed that the Rouse model accounts for the dynamics of not too long real chains in concentrated solutions and melts rather well [55].

2. In order to obtain reliable results, one must know how large the number M_0 of configurations which are initially omitted has to be to ensure proper equilibration. In addition, one needs to know how many configurations $M - M_0$ need to be kept for the average Eq. (5) in order to reach a desired size of the statistical error. In a simple sampling procedure [4] where the various states generated are completely uncorrelated, the error δA of A based on $n >> 1$ observations $A_\gamma \equiv A(\mathbf{x}_\gamma)$ is

$$\langle(\delta A)^2\rangle \approx \frac{\langle A^2\rangle - \langle A\rangle^2}{n}. \tag{7}$$

This formula does *not* hold for the importance sampling scheme described here, because subsequent configurations are highly correlated with each other. In the dynamic interpretation, this correlation is simply expressed in terms of linear (l) and nonlinear (nl) relaxation times of a variable $A(t)$ defined by the stochastic model, Eq. (6):

$$\tau_A^{(l)} \equiv \int_0^\infty \phi_A^{(l)}(t)\,\mathrm{d}t, \qquad \phi_A^{(l)}(t) \equiv \frac{\langle A(0)A(t)\rangle - \langle A\rangle^2}{\langle A^2\rangle - \langle A\rangle^2}, \tag{8}$$

$$\tau^{(\mathrm{nl})} \equiv \int_A^{(\mathrm{nl})} (t)\,\mathrm{d}t, \qquad \phi_A^{(\mathrm{nl})}(t) \equiv \frac{\langle A(t)\rangle - \langle A(\infty)\rangle}{\langle A(0)\rangle - \langle A(\infty)\rangle}; \tag{9}$$

here $\phi_A^{(l)}(t)$ and $\phi_A^{(\mathrm{nl})}(t)$ are termed "linear (nonlinear) relaxation functions." Now the condition that the average Eq. (5) includes only well-equilibrated configurations simply reads

$$t_0 >> \tau_A^{(\mathrm{nl})}. \tag{10}$$

The error estimate Eq. (7) gets replaced by

$$\langle(\delta A)^2\rangle \approx \frac{1}{n}\left[\langle A^2\rangle - \langle A\rangle^2\right]\left\{1 + 2\left(\frac{\tau_A^{(l)}}{\delta t}\right)\right\}, \qquad t_n \equiv n\delta t >> \tau_A^{(l)}, \tag{11}$$

where $t_n = \mathrm{t} - t_0$ in Eq. (5) and δt is the time interval between successive observations used for the average Eq. (5).

Note that the conditions $t_0 >> \tau_A^{(\mathrm{nl})}$, $t - t_0 >> \tau_A^{(\mathrm{l})}$ must hold for all quantities $\{A\}$ that are calculated, and hence it is important to focus on the slowest relaxing quantity, for which the times $\tau_A^{(\mathrm{l})}$, $\tau_A^{(\mathrm{nl})}$ are largest. This matter of time correlations clearly is rather important: already for simple random walks (excluded volume interactions being absent or screened out) the Rouse model [36,55] implies that the single-chain relaxation time τ_N grows with N as

$$\tau_N = \hat{\tau} N^2, \tag{12}$$

where $\hat{\tau}$ is also rather large in a dense system if time is measured in attempted motions per monomer: most motions will be forbidden due to excluded volume or entanglement constraints. From the point of view of making τ_N as small as possible, the reptation algorithm (Fig. 2(b)) seems preferable, since it is believed to satisfy a law [4,35] $\tau_N \sim N$, but the price which is paid is that this is not at all a microscopically realistic description of the chain dynamics. This reptation algorithm must not be confused with the true reptative diffusion motion of chains in the "tubes" provided by their surrondings, where the relaxation time behaves theoretically [55] as $\tau_N \sim N^3$ and experimentally [56] even as $\tau_N \sim N^{3.4}$, respectively. Using the reptation picture of a chain in a tube, each attempted move now corresponds to a time of order N^2 of the real reptating chain, yielding the time scaling quoted above.

Special Problems: Sampling of Chemical Potential, Grand-Canonical Techniques, and Others

In a multichain system where all chains have the same length N we can define a chemical potential μ which is the thermodynamic variable conjugate to the number n of chains in our system. In fact, it is desirable to obtain the equation of state $n = n(T,\mu)$ of the polymer melt (or polymer solution, respectively) from the simulation.

In principle, one could try to perform the simulation at fixed μ, T rather than keeping n, T fixed: n would then be statistically fluctuating, since the Monte Carlo algorithm then must contain moves where chains are randomly removed from the system or added to the system, in accord with the prescribed transition probability which depends on μ. However, since any chain that is inserted into the system as a whole must obey the excluded volume constraint with all the other chains present in the system, this method can work only in dilute solution where the polymer concentration is sufficiently small.

A variant of this method which suffers from the same problem but nevertheless has been applied is the test chain insertion method of Okamoto et al. [57–64]. It is essentially based on the same idea as Widom's [65] particle insertion method for ordinary fluids. This method samples the chemical potential in a calculation performed in the canonical ensemble (n fixed) by sampling the expectation value of the acceptance rate of a move where a test chain would be inserted into the system somewhere. In contrast to truly grandcanonical simulations, the test chain insertion is never carried out, even if the move is "accepted." The chemical potential then is obtained as

$$\mu = -k_B T \ln \left\langle \exp\left(\frac{-U_t}{k_B T}\right)\right\rangle_n . \tag{13}$$

Here U_t is the energy of the inserted chain, and $\langle \cdots \rangle_n$ means the ensemble average over the n-chain system.

While this method is exact, but works for rather short chain lengths only, a reasonalby accurate approximate method seems the idea of "scanning future steps" due to Meirovitch [66]. Since it has been applied for studying properties of semidilute solutions only [67–69], it will not be discussed here, however.

An interesting alternative exact method was recently proposed by Dickman [70,71], relating the pressure of the system to the segment density at a repulsive wall. While usually in simulations one considers a d-dimensional cubic box with all linear dimensions equal to L and periodic boundary conditions, in this method one applies a lattice of length L in d-1 dimensions and of length H in the remaining direction, with which one associates the coordinate x. There is an infinite repulsive potential at $x = 0$ and $x = H + 1$, while in the other directions periodic boundary conditions apply. The partition function of n N-mers on the lattice then is $Z(N,n,L,H) = (n)^{-1}\Sigma \exp(-U/k_B T)$, where the sum runs over all chain configurations on the lattice, and the potential U incorporates restrictions which define then chain structure, prohibit overlaps, etc. While for a model in continuous space the pressure is

$$\Pi^* = \frac{\partial}{\partial V}[\ln Z(N,n,L,H)] = L^{-(d-1)}\frac{\partial}{\partial H}[\ln Z(N,n,L,H)], \tag{14}$$

the analogous expression for the lattice model is

$$\Pi^* = L^{-(d-1)}[\ln Z(N,n,L,H) - \ln Z(N,n,L,H-1)]. \tag{15}$$

The difference of free energies required in this equation now is evaluated by associating with each occupied site in the plane a factor λ, $0 < \lambda < 1$;

it may be viewed as being due to an additional finite repulsive potential next to the wall. Denoting the number of occupied sites in the plane $x = h$ as $N(H)$ the partition function becomes

$$Z(N,n,L,H,\lambda) = (n!)^{-1}\Sigma \exp\left(\frac{-U}{k_B T}\right)\lambda^{N(H)}, \tag{16}$$

with

$$Z(N,n,L,H,0) = Z(N,n,L,H-1), \qquad Z(N,n,L,H,1) = Z(N,n,L,H). \tag{17}$$

This yields

$$\Pi^* = L^{-(d-1)} \int_0^1 \left(\frac{\partial \ln Z}{\partial \lambda}\right) d\lambda = \int_0^1 \frac{d\lambda}{\lambda}\, \varphi_H(\lambda), \qquad \varphi_H \equiv N(H)L^{d-1}. \tag{18}$$

Thus, one must carry out simulations for several values of λ to sample $\varphi_H(\lambda)$, the fraction of occupied sites in the plane $x = H$, in order to perform the integration in Eq. (18) numerically [70,71].

A truly grand-canonical simulation can be performed for symmetrical binary polymer mixtures (in the presence of vacancies) [18,29–31]. There the quantity of interest is the difference in the number of A-chains n_A and the number of B-chains n_B, or the relative volume fractions $\phi = n_A/(n_A + n_B)$, $1 - \phi = n_B\,(n_A + n_B)$ of the two components. The conjugate thermodynamic variable is the chemical potential difference $\Delta\mu = \mu_A - \mu_B$ between the species. Holding $\Delta\mu$ fixed the total number of chains $n = n_A + n_B$ does not change but one must allow for moves where an A-chain transforms into a B-chain, and vice versa. If the mixture is symmetric ($N_A = N_B$), the excluded volume interaction is not at all affected by a move where a chain changes its identity but precisely keeps its configuration. Only the finite energies between neighboring AA, AB, BB pairs of monomers ($\epsilon_{AA},\epsilon_{AB},\epsilon_{BB}$) need to be taken into account in the transition probabilities, as well as $\Delta\mu$.

The Molecular Dynamics Method

In the molecular dynamics method [8–11,72] one attempts to simply solve Newton's law for the classical equation of motion of the many-body system

$$m \frac{d^2}{dt^2}\mathbf{r}_i(t) = -\nabla_i U(\{\mathbf{r}_j\}), \tag{19}$$

where m is the mass of the ith monomer, $\mathbf{r}_i(t)$ its position in space, and the right side describes the force acting on it due to the interaction potential $U(\{\mathbf{r}_j\})$ between all the particles. Involving the ergodicity principle of statistical mechanics, any average properties of the chains may be calculated as time averages along the trajectories $\mathbf{X}(t)$ in phase space generated by the solution of Eq. (19).

In practice the differential equation Eq. (19) is transformed into a difference equation employing a finite time step Δt. There are various schemes available in the literature [8–11] which also thoroughly discuss the resulting discretization error. While Eq. (19) implies a strict conservation of total energy and hence an averaging carried out in the microcanonical ensemble, energy is no longer strictly conserved in the actural schemes employing finite time steps Δt. This problem is particularly cumbersome for simulations of macromolecules, of course, since due to the slowness of the relaxation of these objects one wants to follow the system over a much larger time interval than one would want to do in a simulation of a simple liquid [8–11]. Thus, standard molecular dynamics simulations are mostly restricted to studies of rather short chains surrounded by solvent molecules [73–76], apart from recent work by Rigby and Roe [77,78], described in detail on pages 275–278. Rigby and Roe [77,78] apply a microscopically realistic model of alkane chain molecules, apart from the fact that they use a spring constant for bond length stretching weakened by a factor of about 7 in comparison to the realistic value. This was necessary in order to arrive at timescales for the various local processes which are of the same order of magnitude, since the fastest mode dictates the choice of time step. Attempts to simulate realistic models of long polymer chains using realistic spring constants for bond length stretching [79] fail to equilibrate even isolated chain configurations. This failure is expected, of course, since for real polymers time constants for bond length vibrations are much smaller than for bond reorientations.

A technique better suited for the simulation of dense systems is obtained when one augments Eq. (19) by a friction term $\Gamma\, d\mathbf{r}_i(t)/dt$ and a random force $W_i(t)$ [27],

$$m \frac{d^2}{dt^2} \mathbf{r}_i(t) = -\nabla_i U(\{\mathbf{r}_j\}) - \Gamma \frac{d\mathbf{r}_i}{dt} + W_i(t), \tag{20a}$$

where the random force satisfies a fluctuation-dissipation relation

$$\langle W_i(tW_j(t')\rangle = \delta_{ij}\delta(t - t')6k_BT\Gamma. \tag{20b}$$

While in the molecular dynamics method in strict sense (Eq.(19)) the energy is conserved and the conjugate variable, the temperature, is fluctuating and only implicitly follows as an output of the simulation for a given initial value of the energy, Eqs. (20a) and (20b) define a simulation at constant temperature. Of course, by choosing the friction coefficient Γ very small the dynamics described by Eqs (20a) and (20b) approximate the dynamics described by Eq. (19) rather closely [80]. This is not really desirable in the case of polymers, for which models such as shown in Figure 1(d), are not microscopically realistic, and then there is no point in attempting an accurate modeling of the dynamics of such a model on short time scales. Choosing Γ in the range $0.5\tau^{-1} \leq \Gamma \leq 1.5\tau^{-1}$ [27], the ballistic motion of a monomer is restricted to a range between one and four bond lengths in a dilute solution case. If we choose Γ very large in Eq. (20), the inertial force $m(d^2/dt^2)\mathbf{r}_i(t)$ in Eq. (20) may be neglected altogether and one arrives at the so-called Brownian dynamics method [8], which also is a useful technique of polymer simulation [15,81–85]. The resulting Langevin equation to which Eq. (20) then reduces is equivalent to a Smoluchowski equation for the probability distribution of a configuration $\Psi(\{\mathbf{r}_i\},t)$ of the chains,

$$\frac{d\Psi(\{\mathbf{r}_i\},t)}{dt} = \mathrm{D} \sum_{j=1}^{Nn} \nabla_j \cdot \left[\nabla_j \Psi(\{\mathbf{r}_i\}) + \frac{\Psi(\{\mathbf{r}_i\},t)\nabla_j U}{k_B T} \right], \tag{21}$$

where D is the diffusion constant of a monomer and U the total intermolecular potential. The sum runs over all monomers. Since this equation is a kind of diffusion equation, it can be solved by simulating an equivalent random walk problem again by a Monte Carlo technique [15,81,82], and no explicit numerical time integration as done for solving Eqs. (19) or (20a) and (20b) is needed. For solving Eqs. (19), (20a) and (20b), one can, for instance, apply standard molecular dynamics techniques, such as the Verlet algorithm or predictor-corrector techniques [27,8–11]. Typically, one applies time steps of the order $\Delta t = 0.002\tau$ up to 0.008τ, with τ being the characteristic time for a Lennard-Jones problem (Eq. (1)),

$$\tau = \sigma \left(\frac{m}{\epsilon}\right)^{1/2}. \tag{22}$$

Since τ is of the order of 10^{-12} s and the accessible number of time steps is at most about 10^7 [86], the physical time scales that can be described by Molecular Dynamics methods are fairly short, $t \lesssim 10^{-8}$ s if fairly realistic microscopic models are simulated. For "coarse-grained" polymers the accessible time scale is $t \lesssim 10^{-5}$ s [86,87]. Nevertheless, this yields useful insight in the behavior of melts [86,87] and information on the slowing down near the glass transition [77,78], as will be described below.

We conclude this section by quoting some technical details of the simulation of polymer liquids and glass [77], as an example. There, the Verlet algorithm [88,89] was used, where positions $\mathbf{r}_i$ and velocities $\mathbf{v}_i$ are updated as follows:

$$\mathbf{r}_i(t + \Delta t) = -\mathbf{r}_i(t - \Delta t) + 2\mathbf{r}_i(t) + (\Delta t)^2\mathbf{F}_i(t), \tag{23a}$$

$$\mathbf{v}_i(t) = \frac{\mathbf{r}_i(t + \Delta t) - \mathbf{r}_i(t - \Delta t)}{2\,\Delta t}, \tag{23b}$$

$\mathbf{F}_i$ being the force on the ith segment (no random force was used in Ref. 77). Equation (23) shows that knowledge of the configuration at times t and $t - \Delta t$ needs to be kept to calculate the configuration of the system at time $t + \Delta t$.

Typical speeds obtainable at CRAY XMP computers are 0.02 s per single integration steps for systems containing 500 segments for the rather complicated model of Ref. 77, while in favorable cases [86,87] the speed even is considerably faster. For taking advantage of the vector processing ability of the computer, it is necessary to evaluate the nonlocal interaction (described by a truncated Lennard-Jones potential) with a neighbor list-technique [90,91], with updating of the list every 20 time steps. Codes imposing the periodic boundary constraints also are vectorized [91].

SELECTED APPLICATIONS

As has been emphasized in the introductory section, the intent of this article is to provide the reader with some background on computer simulation techniques, and to show how these techniques are applied in a useful way, by discussing some characteristic examples of investigations. On the other hand, no complete coverage of the field is intended, and hence a lot of related work is not treated at all. Clearly, the selection of applications presented next is highly subjective, and perhaps reflects more the topics with which the author is most familiar, rather than implying that topics not covered would be not important.

Statistics and Dynamics of Polymer Melts

Chain Configurations in Dense Polymer Systems

In dense systems of long flexible polymer chains such as concentrated solutions, melts, and amorphous solids one expects the chain configurations to resemble Gaussian coils, as far as long-distance properties are concerned [28, 92]. Excluded volume interactions are screened out, over all distances except extremely short ones. Thus, we expect asymptotic laws for the mean square end-to-end distance $\langle R^2 \rangle$ and mean square gyration radius $\langle R^2_{gyr} \rangle$ in agreement with gaussian statistics. That is, for a chain of N links of length

l we should have

$$\langle R^2 \rangle = c_\infty l^2 N, \ \langle R^2_{\text{gyr}} \rangle = \frac{\langle R^2 \rangle}{6} \qquad \text{for } N >> 1, \tag{24}$$

where c_∞ is a constant that depends on the detailed chemical structure of the macromolecule [51], and is not of great interest in the context of this article. Furthermore, the coherent structure factor $S_N(\mathbf{q})$ of a chain, which is defined as follows:

$$S_N(\mathbf{q}) = \sum_{i=1}^{N} \sum_{j=1}^{N} \left\langle \frac{\exp[\mathbf{iq} \cdot (\mathbf{r}_i - \mathbf{r}_j)]}{N} \right\rangle, \tag{25}$$

can be described simply by the Debye function $f_D(x)$; i.e.,

$$S_N(q) = N f_D(q^2 \langle R^2_{\text{gyr}} \rangle), \qquad f_D(x) = \frac{2}{x} \left\{ 1 - \frac{e^{-x} - 1}{x} \right\}. \tag{26}$$

Equations (24)–(26) are also well established from small-angle neutron scattering experiments.

The tasks to be treated by computer simulations hence are the following: (a) one wishes to confirm that the models under study are consistent with the behavior described in Eqs. (24)–(26), since this is both theoretically predicted and experimentally observed and hence a necessary condition that our models are sensible is that they reproduce this well-established behavior. (b) Since Eqs. (24), (26) are asymptotic laws, which need to hold only in the limit $N \rightarrow \infty$, it is interesting over which range of N deviations from these laws are still significant for short chains. (c) When a concentrated solution is more and more diluted with solvent, a crossover is predicted [92] in a different regime, the so-called semidiluted concentration regime, where a correlation length $\xi >> l$ appears: chains are swollen over distances less than ξ but obey Gaussian statistics over larger distances.

As an example, Figure 5 shows the data of Olaj and Lantschbauer [45] for a lattice completely filled with chains, and Figure 6 shows the data of Kremer et al. [86] for their off-lattice model of a polymer melt. In both cases it is seen that the results are nicely consistent with the expected laws, Eq. (24). The lattice data, however, show the asymptotic behavior only for relatively long chains, indicating that on short-length scales the excluded volume interaction still gives rise to a "persistence length" distinctly larger than the lattice spacing l. In this respect, the approach of off-lattice data to the thermodynamic limit seems faster. Figure 7 finally shows results for the single-chain structure factor $S_N(\mathbf{q})$, obtained from a tetrahedral lattice

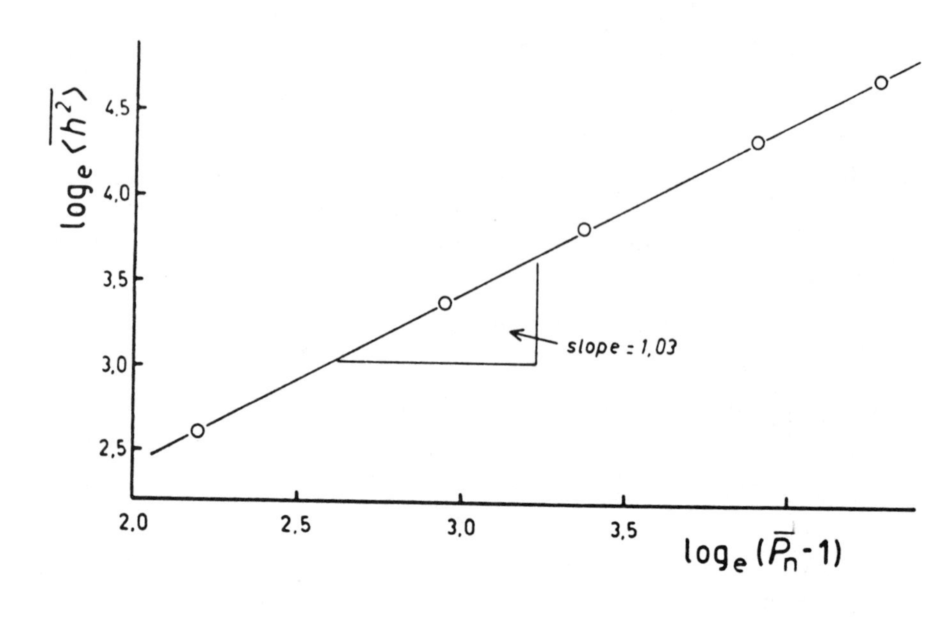

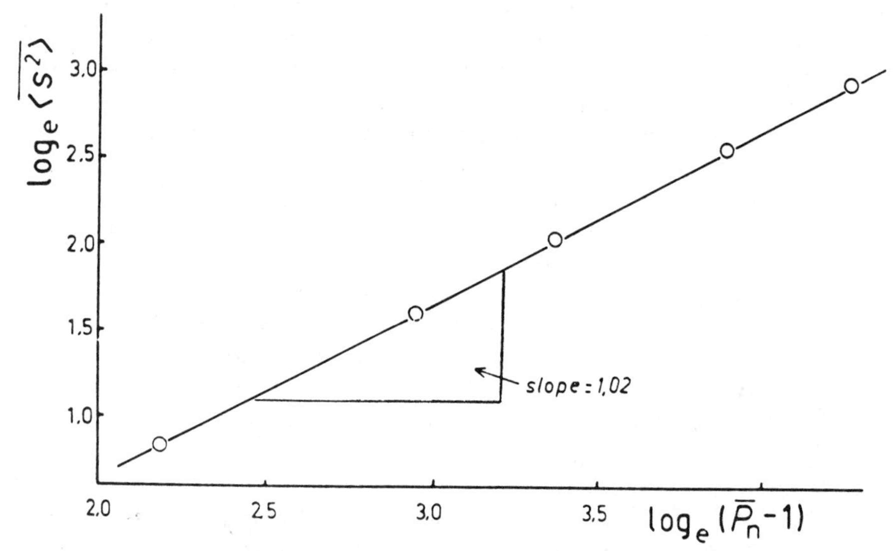

Figure 5 Averaged mean-square end-to-end distance, (upper part) and normalized gyration radius square (lower part) plotted versus number average chain length P_n. Data are obtained for a simple cubic lattice of dimensions 30^3 completely filled with chains [45]: chain configurations were sampled with the "chain-breaking" method. (From Ref. 45.)

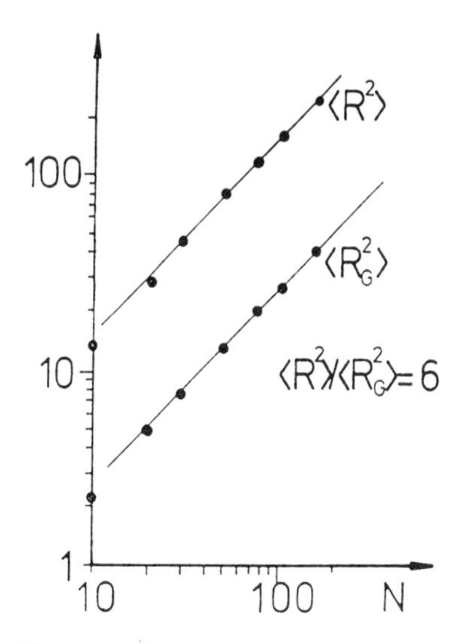

Figure 6 Log-log plot of the mean-square end-to-end distance $\langle R^2 \rangle$ and the gyration radius square $\langle R^2_{\mathrm{gyr}} \rangle$ versus N, for the off-lattice bead-spring model of Figure 1(d) with the Lennard-Jones interaction Eq. (1) truncated at a distance 2.5σ. These data were obtained from the molecular dynamics method using Eqs. (2) and (21), using a box containing typically 20 chains at a density $\rho\sigma = 0.85$ and at the temperature $k_B T/\epsilon = 1.0$. (From Ref. 86.)

model simulation, with a volume fraction $\phi_v = 0.66$ of vacancies, for a chin length of $N = 200$ [37]. It is clear that due to this large vacancy content the excluded volume interaction is relevant at short-length scales (i.e., large values of q), and thus the behavior $S_N(q) \sim q^{-2}$ as implied by the Debye function (Eq. (26)) can be seen over an intermediate regime of concentrations only (on the other hand, we must also have $q^2\langle R^2_{\mathrm{gyr}} \rangle > 1$).

The effect of increasing dilution (note that in the context of a lattice model lattice sites not taken by monomers may either be interpreted as "vacancies" ["free volume"] V or as solvent molecules) on chain dimensions has been studied in many investigations [29,61,62,93–95]. Figure 8 presents an example. While in the dilute case $\langle r^2 \rangle \sim \langle R^2_{\mathrm{gyr}} \rangle \sim N^{2\nu}$, where the exponent $\nu \approx 1.18$ [92], the radii for concentrated solution first increase with N stronger than linear, and then crossover to the behavior consistent with Eq. (24); cf. the leveling-off of the curves in Figure 8 with increasing chain length. Since this crossover occurs the later the larger the vacancy concentration ϕ_v is, a strong increase of the constant c_∞ with increasing vacancy content is implied. It turns out that this crossover behavior can be understood in terms of a simple scaling theory, which should hold in

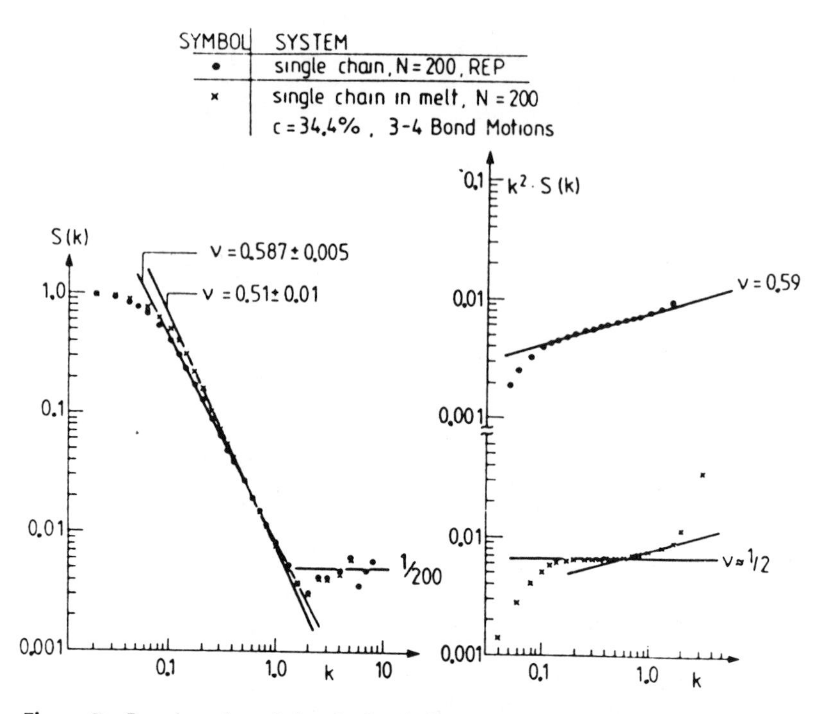

Figure 7 Log-log plot of the single-chain structure factor $S_N(k)$ versus wave number k for chains of length $N = 200$ on the tetrahedral lattice. Both the results for a single chain and for a "melt" at concentration c = 34.4% of monomers are shown. The right replots the data as $k^2S(k)$ vs k. (From Ref. 37.)

the semidilute concentration range only:

$$\langle R^2 \rangle = N^{2\nu} l^2 f\{(1 - \phi_v)^{3\nu-1} N\}, \tag{27}$$

where the scaling function $f(\mathfrak{z})$ must have the limits $f(\mathfrak{z} << 1) \rightarrow$ const and $f(\mathfrak{z} >> 1) \sim \mathfrak{z}^{1-2\nu}$, in order to yield Eq. (24) in this limit. Although in principle Eq. (27) should hold for semidilute solutions only, where $1 - \phi_v$ is small, e.g., $1 - \phi_v < 0.1$, Figure 9 shows that Eq. (27) is a reasonable approximation over the entire concentration range, including concentrated solutions [96]. Similar data for various two- and three-dimensional lattices can be found in [93] and the concentration dependence for off-lattice chains is studied in [97].

A scaling form similar to Eq. (27) also holds for the pressure and related quantities such as the second virial coefficient, and this behavior has been checked by various simulations [67–69,71,93]. Here we are more interested in the behavior in the concentrated solution regime. Figure 10 compares recent simulation results [71] for the compressibility factor $Z = N\Pi^*/$

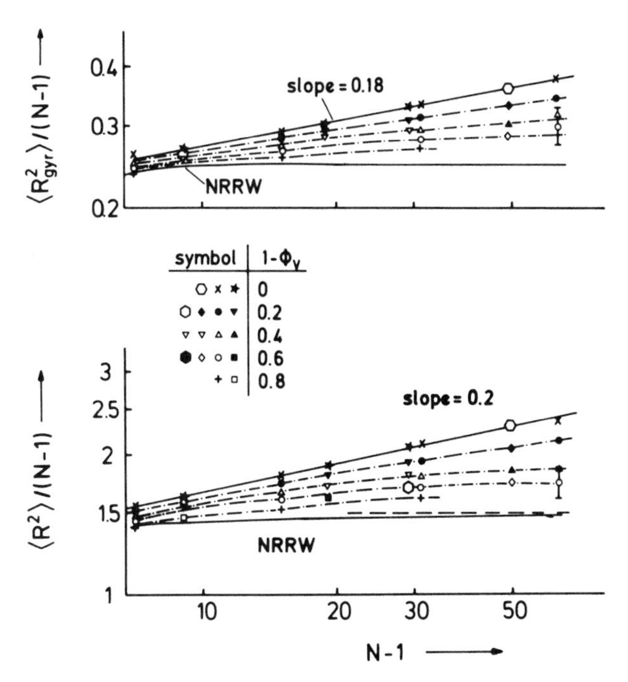

Figure 8 Log-log plot of the normalized gyration radius square $\langle R^2_{gyr}\rangle/(N - 1)$, upper part, and the normalized squared end-to-end distance $\langle R^2\rangle/(N - 1)$, lower part, versus the number of bonds $N - 1$ per chain, for various concentrations of vacancies ϕ_v as indicated in the figure, for the simple cubic lattice. Data for $N =$ 10, 20, 30 are due to De Vos and Bellemans [94,95] and Okamoto [62], while data for $N = 50$ are due to Olaj and Lantschbauer [45] (at $(1 - \phi_v) = 0, 0.4, 0.6$) and Okamoto [62]. All other data points are obtained by Sariban and Binder [29]. Dash-dotted curves are guides to the eye only. Full curves represent the simple power law $\langle R^2\rangle/(N - 1) \sim (N - 1)^{2\nu-1}$ (for $1 - \phi_v = 0$, the choices of the exponent $2\nu - 1$ are indicated in the figure) and the exact result for the nonreversal random walk (NRRW), respectively. Statistical errors are always smaller than the size of the symbols, except for the point $N = 64$, $1 - \phi_v = 0.6$, where the error bar is given. (From Ref. 29.)

ϕ, where $\phi = 1 - \phi_v$ is the polymer concentration with predictions of various equations of state: the expressions are due to Flory [86],

$$Z_F = \frac{1 - N[\ln(1 - \phi) + \phi]}{\phi}, \tag{28a}$$

Flory and Huggins [98,99], z being the coordination number of the lattice,

$$Z_{FH} = Z_F + \frac{Nz}{2\phi}\left\{\ln\left[1 - \frac{2\phi}{z}\left(1 - \frac{1}{N}\right)\right] + \frac{2\phi}{z}\left(1 - \frac{1}{N}\right)\right\}, \tag{28b}$$

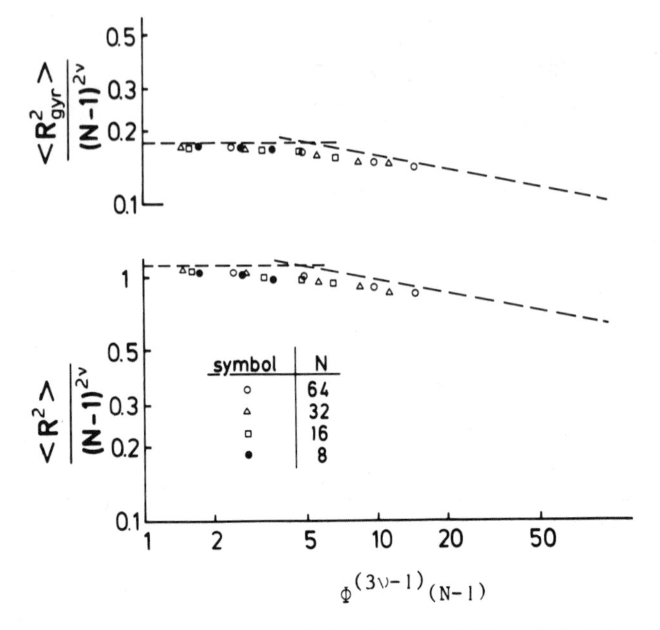

Figure 9 Log-log plot of $\langle R^2\rangle/N^{2\nu}$ versus $N(1 - \phi_v)^{3\nu-1}$ for chains of lengths N = 8, 16, 32, and 64 and various vacancy concentrations on the simple cubic lattice. (From Ref. 96.)

and to Bawendi and Freed [100];

$$Z_{BF} = Z_F - \phi\left\{\frac{(N-1)^2}{zN} + \frac{3(N-1)^4 - 8(N-1)^2 - 4(N-1) + 1}{z^2N^3}\right\} - \phi^2\left\{\frac{-20(N-1)^4 + 4(N-1)^3 + 24(N-1)^2}{3z^2N^3}\right\} - \phi^3\left\{\frac{6(N-1)^4}{z^2N^3}\right\}. \tag{29}$$

It is seen that while the Flory approximation is unsatisfactory, both the Flory-Huggins approximation and the Bawendi-Freed approximation describe the equation of state well at high volume fractions ϕ. At small volume fractions, however, neither of these approximations is very accurate, as expected, since there one has entered a semidilute concentration regime, where a scaling description analogous to Eq. (27) is needed.

Dynamics of Melts: Rouse Model versus Reptation

In this subsection we discuss some of the simulations [13–15,37,86,101–109] that have been performed to test the tube model of reptative motion

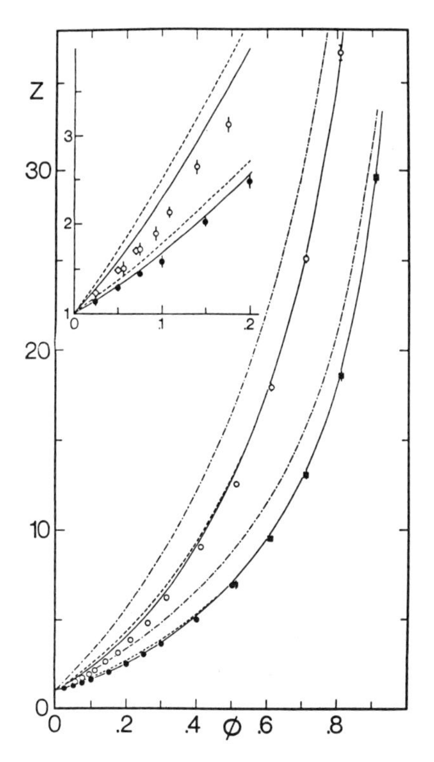

Figure 10 Compressibility factor Z plotted versus volume fraction, for simple cubic lattice chains of length $N = 20$ (filled symbols) and $N = 40$ (open symbols), compared to various theories: the Flory theory (dash-dotted curves) [98], Flory Huggins-theory (broken curves) [99], and the Bawendi-Freed theory (full curves) [100], respectively. Circles represent data obtained from the repulsive wall method, while squares or diamonds are obtained from the test-chain insertion method. (From Ref. 71.)

of polymers, proposed by de Gennes [92,110–112], Edwards [55,113,114], and Doi [55,114]. According to this model, entanglements of a chain with other chains have the effect of forming a "tube" in which each chain diffuses predominantly along its own contour. However, neither is the precise microscopic meaning of entanglements clear, nor can the tube diameter be theoretically be predicted on the basis of microscopic model parameters or basic chain properties. In addition, refinements of the tube model clearly are needed in polydisperse systems, in polymer mixtures, for block copolymers, etc. This field still is an area of active research, and simulations are hoped to yield a refined insight into the mechanisms which need to be modeled.

The first attempt to simulate the dynamics of melts [13,101,102] used

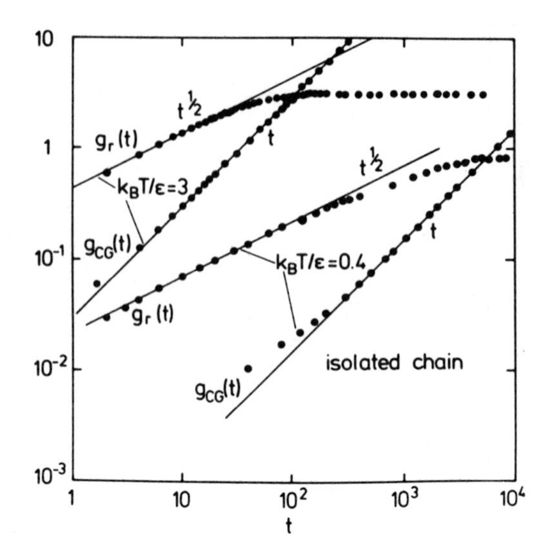

Figure 11 Log-log plot of average mean-square displacement $g_r(t)$ of a bead relative to the center of gravity of the chain and of the average mean-square displacement $g_{CG}(t)$ of the center of gravity of the chain versus time (in units of Monte Carlo steps per bead). Two temperatures are shown. Straight lines indicated $t^{1/2}$ and t power laws, respectively. (From Ref. 13.)

a model of freely jointed chains (Fig. 1(a)) where chains consisted of $N = 16$ monomers only, which interact with a Lennard-Jones potential (Fig. 1(d), Eq. (1)) where $\sigma = 0.4l$ was chosen and the temperature was set at $k_BT/\epsilon = 3$. Of course, a chain length $N = 16$ is extremely short, and thus one needed to study the dynamics of single isolated chains first, to show that the model in fact exhibits the expected Rouse [36] behavior. Figures 11, 12 show that both the mean-square displacement $g_r(t)$ of beads in the center-of-gravity system of the chain

$$g_r(t) \equiv \langle\{\mathbf{r}_i(t) - \mathbf{r}_{CG}(t) - [\mathbf{r}_i(0) - \mathbf{r}_{CG}(0)]\}^2\rangle \tag{30}$$

as well as the logarithm of the incoherent scattering function $S_{\text{inc}}(\mathbf{q},t)$,

$$S_{\text{inc}}(\mathbf{q},t) = \frac{1}{N}\sum_i \langle\exp\{i\mathbf{q}\cdot[\mathbf{r}_i(t) - \mathbf{r}_i(0)]\}\rangle \tag{31}$$

exhibit rather nicely the expected square root of time behavior [115],

$$g_r(t) \sim t^{1/2}, \qquad \ln[S_{\text{inc}}(\mathbf{q},t)] \sim q^2t^{1/2}. \tag{32}$$

The mean-square displacement of the center of gravity of the chain

$$g_{CG}(t) \equiv \langle[\mathbf{r}_{CG}(t) - \mathbf{r}_{CG}(0)]^2\rangle \tag{33}$$

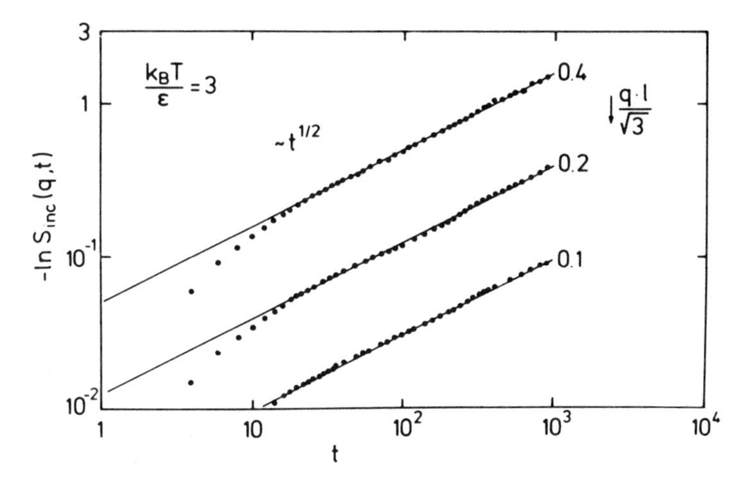

Figure 12 Log-log plot of the logarithm of the incoherent scattering function $S_{inc}(q,t)$ versus time (in units of Monte Carlo steps per bead) and $k_BT/\epsilon = 3$. Three wavevectors q are shown for a single off-lattice freely jointed chain with $N = 16$ beads (Fig. 1 a). Straight lines indicate the expected $t^{1/2}$ behavior. (From Ref. 13.)

exhibits the simple diffusive behavior $g_{CG}(t) = 6D_Nt$, where $D_N \sim 1/N$ is the diffusion constant of the chains, over the entire time range accessible to the simulation, as expected.

Turning now to the simulation of a dense multichain system, again for $N = 16$, one finds that the chain motions are slowed down by an order of magnitude, but a description in terms of the Rouse model [36] still applies. Figure 13 includes the normalized coherent structure factor $S_{coh}(\mathbf{q},t)$,

$$S_{coh}(\mathbf{q},t) = \frac{1}{N^2}\sum_{i,j} \langle \exp\{\mathbf{iq} \cdot \mathbf{r}_i(t) - \mathbf{r}_j(0)]\}\rangle, \tag{34}$$

since this quantity is experimentally accessible via neutron spin echo methods [101,102]. Thus Figure 13 shows no evidence for reptation.

Of course, it is not clear that one should expect any such evidence from this simulation, since the chains are so short. But, surprisingly, a very different conclusion is reached if after some time one artificially freezes in all chains in the system apart from one, which still can move in a "network" of frozen obstacles formed by all the surrounding chains. In this simulation the predictions of the reptation model [110] for an intermediate time regime,

$$g_r \sim ld_T\left(\frac{t}{\tau_1}\right)^{1/4}, \quad g_{CG}(t) \sim \left(\frac{t}{\tau_1}\right)^{1/2}, \qquad \tau_1\left(\frac{d_T}{l}\right)^4 \le t \le \tau_1 N^2, \tag{35}$$

where d_T is an effective tube diameter and τ_1 an effective monomer reori-

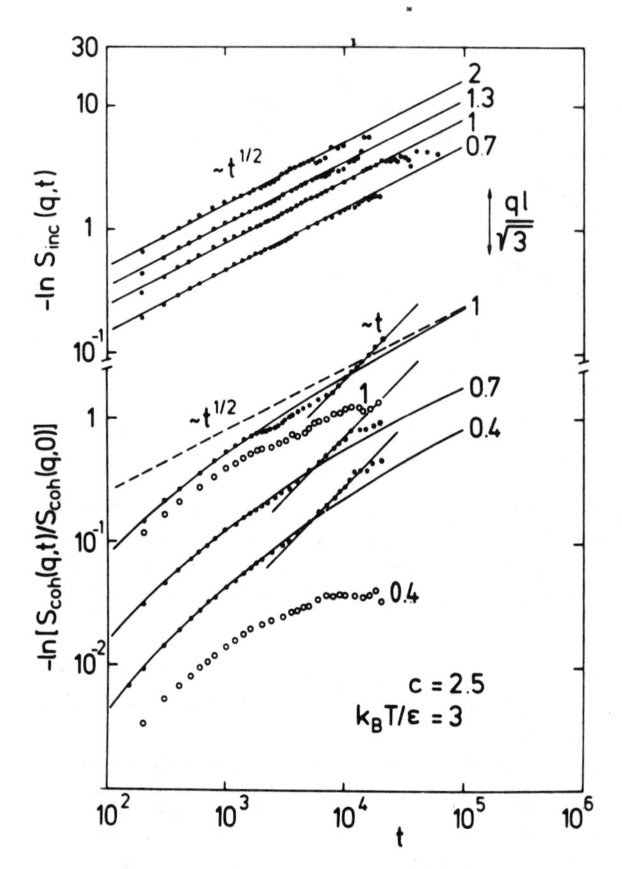

Figure 13 Log-log plot of $-\ln S_{inc}(q,t)$ {upper part} and of $-\ln[S_{coh}(q,t)/S_{coh}(q,0)]$ (lower part) versus time, for a melt at temperature $k_BT/\epsilon = 3$ and "concentration" $c \equiv Nn/(L/l)^3 = 2.5$, where $n = 10$ chains in a box with linear dimensions $L = 4l$ and periodic boundary conditions were simulated. Several wave vectors are shown as indicated. Time is measured in units Monte Carlo steps per bead. Straight lines indicate the $t^{1/2}$ and t^1 laws, while full curves in the lower part are the Rouse model predictions for the normalized coherent scattering function, Eq. (34), with the time-scale parameter suitably adjusted. Open symbols show corresponding data if the center of gravity motion would be subtracted. (From Ref. 13.)

entation time, are verified rather nicely (Fig. 14). Thus, some features of reptation are seen easily for chains moving in randomly frozen networks of obstacles, and this is consistent with work where chains moving in a regular network of obstacles were simulated [116,117]. These results [13] and related simulations for a tetrahedral lattice model of a polymer melt at $\phi_v = 0.67$ [37] for much longer chains ($N = 200$) show that the tube

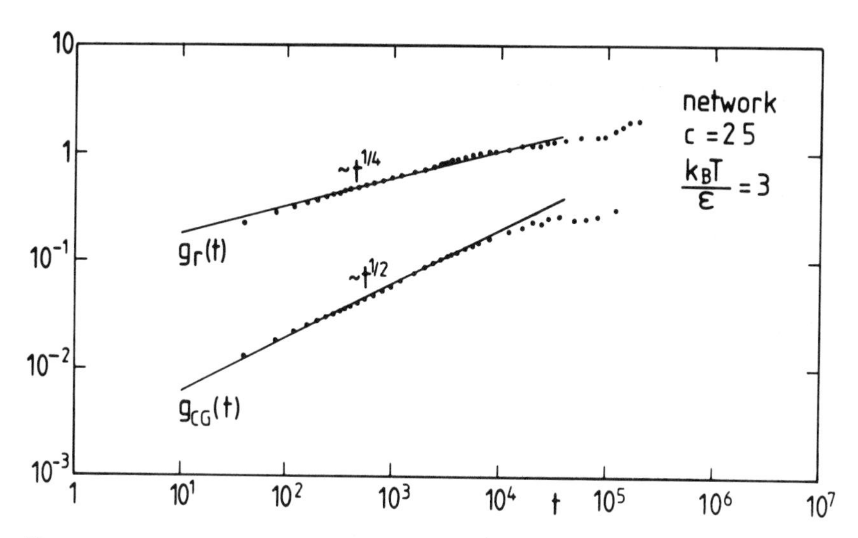

Figure 14 Log-log plot of average mean square displacement $g_r(t)$ of a bead of one mobile chain in a "network" of frozen-in chains (at a configuration corresponding to thermal equilibrium at $k_BT/\epsilon = 3$, $c = 2.5$, cf. Fig. 13) and of the center-of-gravity mean-square displacement $g_{cq}(t)$ versus time. Straight lines indicate the reptation model predictions (Eq. (35)). (From Ref. 13.)

diameter in a melt cannot be a quantity determined exclusively by geometrical constraints describing the entanglement topology, as it was occasionally suggested [118].

Of course, one still can argue that for a mobile environment the tube is not a rigid but an elastic object and due to a kind of Debye-Waller factor the effective tube diameter is much larger than for a corresponding immobile environment. Such a hypothesis clearly could account for the work of Refs. 13 and 37. After all, these early simulations showed Rouse model behavior only, thus implying that diffusion constant D_N and largest relaxation time τ_N (which is experimentally accessible from melt viscosities [56]) scale as

$$D_N \sim N^{-1}, \quad \tau_N \sim N^2 \qquad \text{(Rouse model [36,55])} \tag{36a}$$

while the experimental data imply [55,56]

$$D_N \sim N^{-2}, \quad \tau_N \sim N^{3.4}, \qquad N >> N_e, \tag{36b}$$

if N exceeds the "average chain length between two entanglements" N_e, which is related to the tube diameter d_T via $N_e \sim (d_T/l)^2$.

Rather extensive lattice simulations due to Kolinski et al. [107–109] find systematic deviations from the Rouse model when the volume fraction ϕ

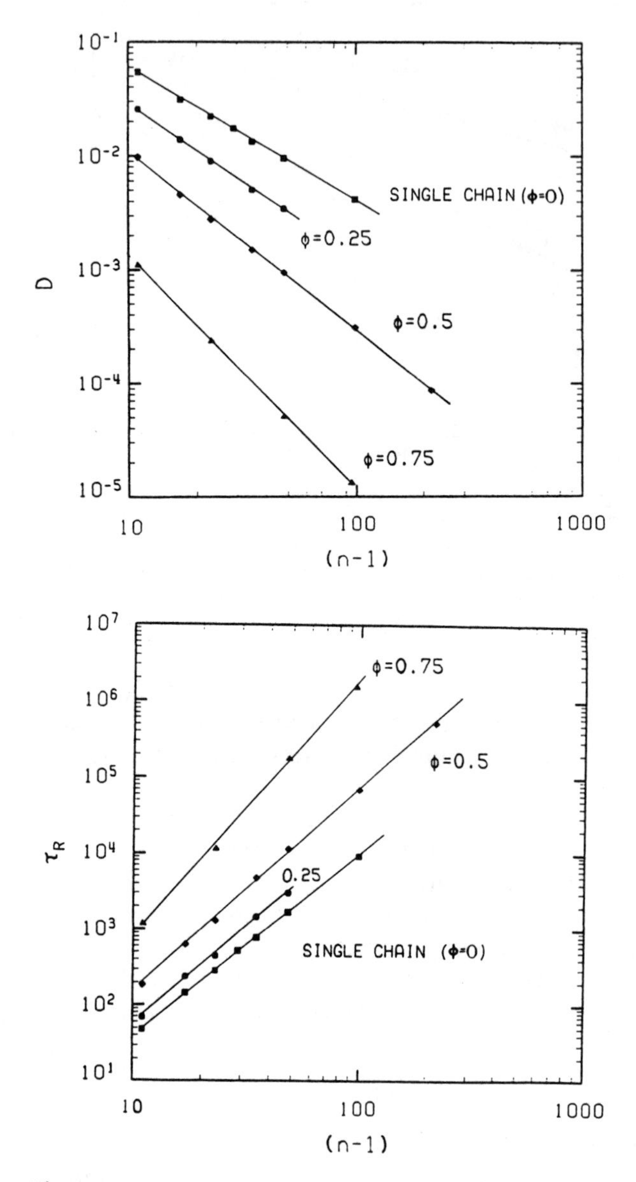

Figure 15 Log-log plot of polymer self-diffusion constant D_N versus the number of bonds (a) and relaxation time τ_R of the end-to-end distance (b), for the simple cubic lattice and various values of the monomer volume fraction ϕ, as indicated. (From Ref. 107.)

of lattice sites taken by monomers gets larger than about one-half (Fig. 15). While the single-chain results $D_N \sim N^{-1.15}$ and $\tau_N \sim N^{2.35}$ are close to the results expected for the Rouse model in the case of strong excluded volume interactions [92],

$$D_N \sim N^{-1}, \quad \tau_N \sim N^{1+2\nu} \approx N^{2.18} \qquad (\nu \approx 0.59) \tag{37}$$

it seems that with increasing ϕ the exponents steadily increase from the values of Eq. (37) towards the values of Eq. (36b), which roughly are reached for $\phi \sim 0.75$, or even larger values at values of ϕ near the critical concentration for glass formation, which is estimated at [107] $\phi_c \approx 0.92$ for the simple cubic lattice. However, even for $\phi = 0.75$ the mean-square displacements do not satisfy Eq. (35): rather one finds only a small reduction of the exponents in comparison to the Rouse model predictions, $g_r \sim t^{0.46}$ instead of $g_r \sim t^{0.5}$ (Eq. (32)), $g_{CG} \sim t^{0.89}$ instead of $g_{CG} \sim t^{1.0}$. At the same time, a direct inspection of coarse-grained chain configurations reveals that the chain has as much tendency to move perpendicular to its contour as to move along its contour, as the reptation model would suggest. The slowing down of the chain dynamics seen in these simulations, hence, is attributed to a highly cooperative motion among the chains. Subsequent

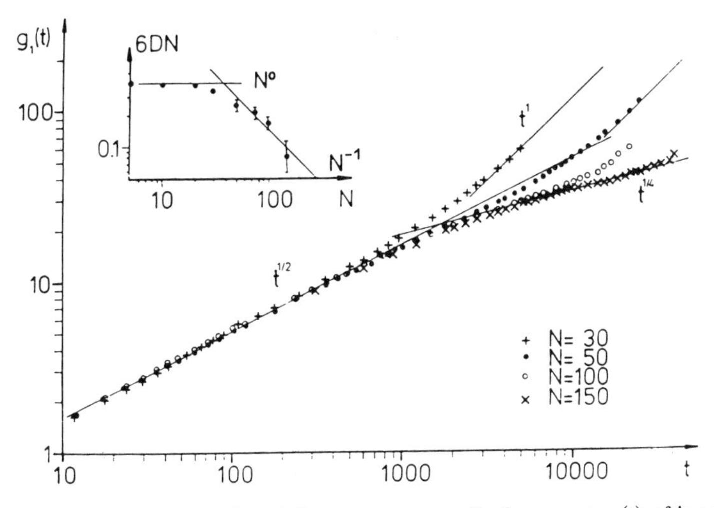

Figure 16 Log-log plot of the mean square displacement $g_1(t)$ of inner monomers $\{g_1(t) = (1/5) \sum_{i=N/2-2}^{i=N/2+2} \langle [r_i(t) - r_i(0)]^2 \rangle\}$ versus time {measured in units of $\tau = \sigma(m/\epsilon)^{1/2}$ where σ, ϵ are the parameters of the Lennard-Jones potential acting between the beads in the bead-spring model simulated with molecular dynamics techniques}. Inset shows log-log plot of the normalized chain diffusion constant versus N. Straight lines indicate behaviors $DN \sim N^0$ or N^{-1}, respectively. (From Ref. 86.)

work at $\phi = 0.5$ where chain lengths up to $N = 800$ were explored [108] seemed to confirm this interpretation. However, it is possible that in these simulations the crossover from Rouse to reptation dynamics and the slowing down due to the glass transition as $\phi \rightarrow \phi_c$ are interwoven in a complicated way, and thus it is not completely clear that typical melt behavior could actually be studied in this work. In addition, the possibility is not ruled out that the lattice models with just very few elementary kinds of motion (Fig. 1) are too artificial to describe the dynamics of strongly entangled chains in melts reliably.

In this controversial situation the off-lattice molecular dynamics work of Kremer et al. [86] has been most helpful. Figure 16 shows that for long chains ($N \geq 100$) the chains diffuse in accord with the reptation model (Eq. (36b)), while for short chains the Rouse-like behavior $D_N N \sim$ const. (Eq. (37)) is seen. But the crucial point is that for the long chains the mean-square displacement of inner monomers is nicely consistent with Eq. (35). And the visual inspection of coarse-grained projections of the chain configurations [87] gives compelling direct evidence that the chains move in effective tubes.

To further compare these results with experiments, one needs to map the monomers of the simulations [86,87] onto chemical species. Here, this is done for two examples, namely polydimethylsiloxane (PDMS) at room temperature and polystyrene (PS) at about 210°C. Since an analysis of the Rouse mode spectrum τ_p in the simulation reveals a crossover from $\tau_p \sim (N/p)^2$ to $\tau_p \sim (N/p)^3$ for $N_e = N/p \approx 35$, one can map N_e to the entanglement molecular weight M_e, which is $M_e = 18{,}000$ (PS) and 9000 (PDMS), respectively [119]. The longest chains in Figure 16 then correspond to molecular weights $M = 77{,}000$ (PS) and 38,000 (PDMS), respectively. Using then the relation $\langle R^2 \rangle = c_\infty \langle l^2 \rangle N$ [51] where $(\langle l^2 \rangle c_\infty)^{1/2} = 6$ Å (PDMS) and 7 Å (PS), the tube diameters become, using the relation [55] $d_T^2 = 0.8 \langle R_{N_e}^2 \rangle$, d_T (PS) ≈ 82 Å and $d_T \approx 60$ Å (PDMS). These large numbers (which are consistent with the previous estimates [119,55,56], of course) again indicate that the chains of previous off-lattice simulations [13,101,102] were too short to show any sign of reptation. Thus, we feel that with the work of Kremer et al. [86,87] this problem is settled, but again it should be clear that there is no hope to see reptation in simulations of chemically more realistic models (such as those of Rigby and Roe [77], for instance, where $N \leq 50$): with chemical monomer molecular weights of 104 (PS) and 72 (PDMS), one would have values $N_e \approx 173$ (PS) and $N_e \approx 125$ (PDMS): thus chainlengths $N >> N_e$ are completely out of reach for chemically realistic polymer models.

Polymer Blends

In this section we discuss the modeling of phase separation in binary fluid mixtures of long flexible macromolecules, such as polystyrene (PS) and polymethylvinylether (PMVE) [120–123]. The temperature-concentration phase diagrams of such mixtures exhibit a coexistence curve ("binodal") separating the one-phase region from the two-phase region. This coexistence curve ends in a critical point.

There has been a lot of interest in such partially miscible polymer blends recently: even though most polymers are immiscible, unmixing proceeds rather slowly [121,122], and quenching a mixture, which is unmixed only on microscopic scales but still homogeneous on macroscopic scales, to temperatures below its glass transition may produce amorphous polymeric materials with interesting properties for various applications [121], similarly as metallic alloys often have more suitable mechanical properties as the pure metals [124]. Thus there is great interest in a better understanding of the molecular origins of polymer compatibility or incompatibility.

The traditional approach to describe such phenomena is the Flory-Huggins theory [28,98,99,125]. Phase separation occurs since the entropy of mixing is overruled by energetic effects driving the two partners (which we denote as A, B in the following) apart. These effects are represented by one effective parameter, χ. This parameter is experimentally accessible from the small-angle scattering of light, x-rays, or neutrons, since (according to the random-phase approximation [92]) the intensity $I(\mathbf{q})$ recorded under scattering vector $\mathbf{q}$ is written as

$$[I(\mathbf{q})]^{-1} \sim \frac{4}{S_T^{\mathrm{coll}}(\mathbf{q})} = (\phi_A N_A)^{-1} + (\phi_B N_B)^{-1} - 2\chi + 2q^2 \left[\frac{\langle (R_{\mathrm{gyr}}^{A})^2 \rangle}{\phi_A N_A} + \frac{\langle (R_{\mathrm{gyr}}^{B})^2 \rangle}{\phi_B N_B} \right] \tag{38}$$

Here, ϕ_A, ϕ_B are the volume fractions taken by the effective segments of the two polymers; N_A, N_B are the number of segments per chain; $\langle (R_{\mathrm{gyr}}^{A})^2 \rangle$, $\langle (R_{\mathrm{gyr}}^{B})^2 \rangle$ are the mean square gyration radii. The angle brackets $\langle \cdots \rangle$ denote a configurational and thermal average.

All these quantities can be found from independent measurements. Hence a fit of $I(\mathbf{q})$ to Eq. (38) yields χ [126,127]. However, typically [126] one finds that for each temperature T and volume fraction ϕ_A a different χ-parameter $\chi(T,\phi_A)$ results. The theory thus does not contain only one adjustable parameter, but rather an undetermined function of two variables. Of course, the feasibility of the fit hence is *not* a sensitive test of the basic theoretical model. A more stringent test can be performed by Monte Carlo methods as will be described next [18,29–31].

In the Flory-Huggins lattice model [28] every chain is represented by a (self-avoiding) random walk on the lattice, and every site is taken by one effective segment of a chain. Here we generalize the original model slightly, by including a third component which can be interpreted either as "free volume" ("vacancies" V) or a low-molecular-weight solvent. This is necessary in order that the chain configurations can relax via the motions shown in Figure 2(a), and also is of physical interest. We denote the volume fraction of the vacancies by ϕ_v.

The energy parameters causing the unmixing are attractive energies ϵ_{AA}, ϵ_{BB} which are won if two nearest-neighbor sites are taken by effective segments of the same kind, or a repulsive energy ϵ_{AB} between unlike neighbors. In the mean-field approximation [28,92], the χ-parameter mentioned above then is related to these energies only in the particular combination ϵ,

$$\epsilon \equiv \epsilon_{AB} - \frac{\epsilon_{AA} + \epsilon_{BB}}{2}, \qquad \chi = \frac{z\epsilon}{k_B T}, \tag{39}$$

when z is the coordination number of the lattice ($z = 6$) for the simple cubic lattice studied by Sariban and the present author [18,29–31]. Neglecting also the fact that chains must not intersect themselves or other chains, Flory [28] gets for the excess free energy of the mixtures

$$\frac{F}{k_B T} = \frac{\phi_A \ln \phi_A}{N_A} + \frac{\phi_B \ln \phi_B}{N_B} + \phi_v \ln \phi_v + \chi \phi_A \phi_B \tag{40}$$

from which one predicts that unmixing occurs if χ exceeds a critical value χ_c, which for a symmetric mixture simply is

$$\chi_c = \frac{z\epsilon}{k_B T_c^{FH}} = \frac{2}{N(1 - \phi_v)}. \tag{41}$$

Of course, this result is based on many approximations—neglect of excluded volume, of correlations in site occupancies, etc.; in addition, Eq. (39) contains a significant overcount of the number of neighbors a monomer can interact with. An inner segment of an A chain can have at most $z - 2$ rather than z B neighbors.

In a Monte Carlo simulation, precisely this Flory-Huggins model as specified above can be studied, avoiding any of the above approximations; i.e., one carries out essentially an exact statistical-mechanical calculation of the model. The limitations are only statistical errors, which can be made smaller by increasing the amount of computer time, and finite lattice size effects which need not be disregarded in the simulation study of phase transitions [5–8,128], but are well controlled by the application of finite size scaling theories [128,129]. Thus, we fully may profit from the advantage that no adjustable parameter whatsoever enters the comparison between theory and simulation!

Figure 17 anticipates some of the main results, comparing the actually

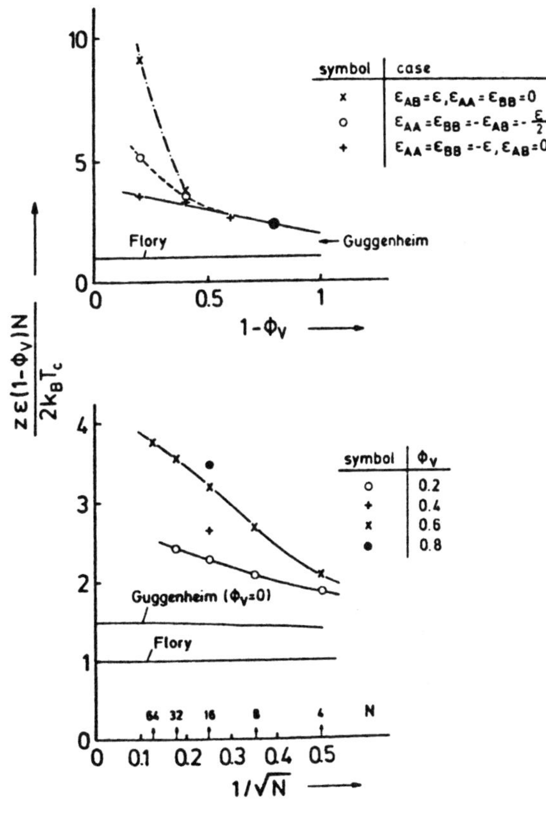

Figure 17 Ratio between the critical temperature of the Flory-Huggins approximation, Eq. (41), $T_c^{MF} = z\epsilon N(1 - \sigma_v)/(2k_B)$, and the critical temperature T_c obtained from Monte Carlo simulation. Upper part plots this ratio for $N = 16$ versus polymer concentration $1 - \phi_v$, for three different choices of the interactions ϵ_{AA}, ϵ_{BB}, ϵ_{AB} (for $1 - \phi_v = 0.8$ these three choices yield an identical result, indicated by a black dot). Lower part plots the same ratio vs the inverse square root of the chain length, for various vacancy concentrations as shown in the figure. (From Ref. 29.)

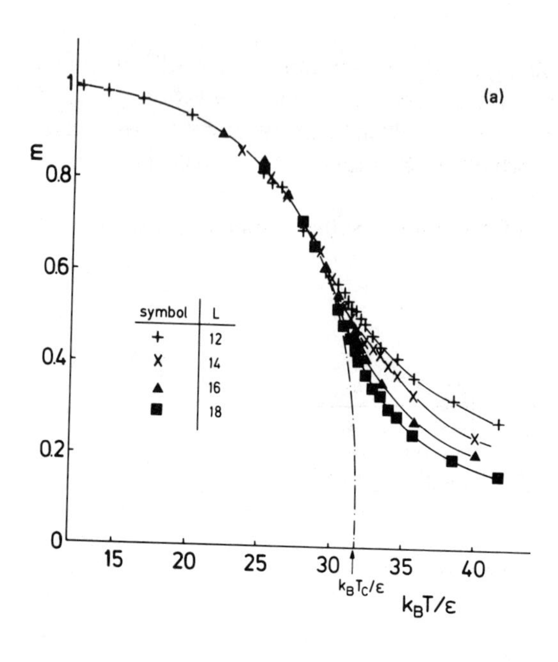
(a)
m
1
0.8
0.6
0.4
0.2
0
15
20
25
30
35
40
k_BT_c/ε
k_BT/ε
symbol L
12
14
16
18

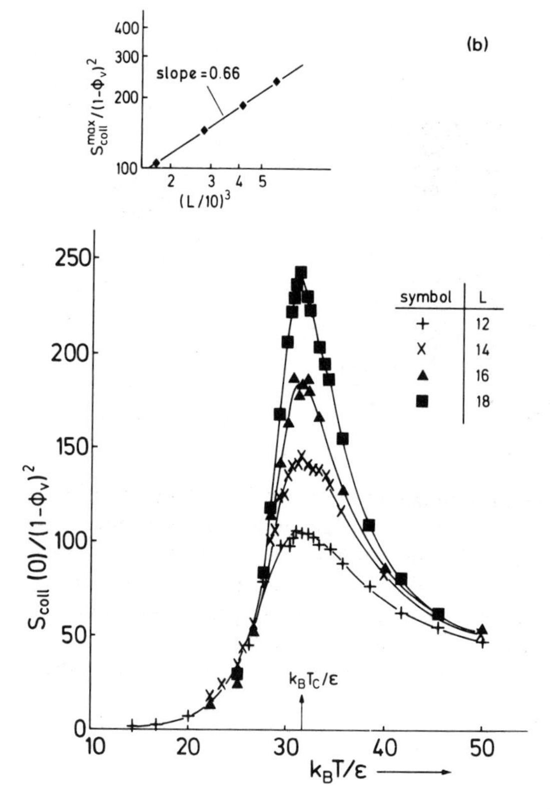
(b)
$S_{coll}^{max}/(1-\Phi_v)^2$
400
300
200
100
slope = 0.66
2
3
4
5
$(L/10)^3$
$S_{coll}(0)/(1-\Phi_v)^2$
250
200
150
100
50
0
10
20
30
40
50
k_BT_c/ε
k_BT/ε
symbol L
12
14
16
18

observed critical temperature T_c of the simulation to the prediction, Eq. (41): it is seen that the latter equation is very inaccurate. Even for $\phi_v \to 0$ there is a discrepancy by about a factor of 2 (the Guggenheim [130] approximation then is somewhat better, though not exact). This discrepancy increases with increasing vacancy content and with increasing chain length. Also the "one-energy-parameter approximation" breaks down for $\phi_v \geq 0.6$: *T_c does not only depend on* ϵ (Eq. (39)) but on ϵ_{AB}, ϵ_{AA}, ϵ_{BB} separately.

Before the physical interpretation of these discrepancies is discussed, a few comments are given on how the results of Figure 17 are derived. Chain configurations are relaxed by various stochastic local motions, see Figure 2(a), which correspond to the dynamics described by the Rouse model [36, 55]. In addition, the grand-canonical simulation technique described on page 239 is used, where the chemical potential difference $\Delta\mu$ between the two species is held fixed, rather than ϕ_A/ϕ_B. Since from time to time A-chains transform into B-chains only the total number $n = n_A + n_B$ of the two types of chains is fixed, while $\Delta n = n_A - n_B$ fluctuates. Since for a symmetric mixture the binodal occurs for $\Delta\mu = 0$, the "order parameter" m of the unmixing transition can simply be obtained from sampling Δn for $\Delta\mu = 0$:

$$m \equiv \frac{\phi_A^{\text{coex}(2)} - \phi_A^{\text{coex}(1)}}{\phi_A + \phi_B} = \langle |M| \rangle \equiv \frac{\langle |\Delta n| \rangle}{n} \tag{42}$$

Also the collective structure factor S_T^{coll} $(q = 0)$ can be obtained simply from sampling the fluctuation of Δn or $M = \Delta n/n$, respectively:

$$S_T^{\text{coll}}(0) = (n_A + n_B)N(1 - \phi_v)[\langle M^2 \rangle - \langle |M| \rangle^2], \quad \Delta\mu = 0, \quad T < T_c, \tag{43a}$$

$$S_T^{\text{coll}}(0) = (n_A + n_B)N(1 - \phi_v)\langle M^2 \rangle, \quad \Delta\mu = 0, \quad T > T_c. \tag{43b}$$

Figure 18 shows typical results obtained for m and $S_T^{\text{coll}}(0)$ (only data referring to Eq. (43a) is shown comparing results for different linear dimensions L in order to check for finite size effects [18,128]. While for T distinctly

---

Figure 18 Order parameter m (a) and structure factor $S_{\text{coll}}(0)/(1 - \phi_v)^2$ (b) plotted versus temperature for $\phi_v = 0.2$, $N = 32$, $\epsilon_{AB} = 0$, $\epsilon_{AA} = \epsilon_{BB} = -\epsilon$, and various lattice sizes. Insert in (b) shows the increase in the maximum value of $S_{\text{coll}}(0)$ at T_c with L on a log-log plot. Critical temperature T_c and order parameter m for $L \to \infty$ (dash-dotted curve) was obtained from extrapolations using finite size scaling theory, as described in the text. (From Ref. 18.)

less than T_c the Monte Carlo data are fairly independent of lattice size, in the critical region finite size effects are clearly apparent. In particular, m has pronounced finite size tails above T_c. In order to locate T_c, both extrapolation of peak heights of $S_T^{\text{coll}}(0)$ {Fig. 18b} and specific heat C (Fig. 19(a)) can be used, as well as using the intersection temperature of the cumulant U_L (Fig. 19(b)),

$$U_L = 1 - \frac{\langle M^4 \rangle}{3\langle M^2 \rangle^2}. \tag{44}$$

The justification for these procedures to locate T_c can be derived from a discussion of the probability distribution $P_L(M)$ for the order parameter in a finite d-dimensional box of volume L^d [128,132]. This probability distribution (for $\Delta\mu = 0$) changes from a single Gaussian far above T_c to a double Gaussian far below T_c, Figure 20, and satisfies near T_c a finite size scaling hypothesis [132]

$$P_L(M) = L^{\beta/\nu} \tilde{P}\left[ML^{\beta/\nu}, \frac{L}{\xi}\right], \qquad L \to \infty, \tag{45}$$

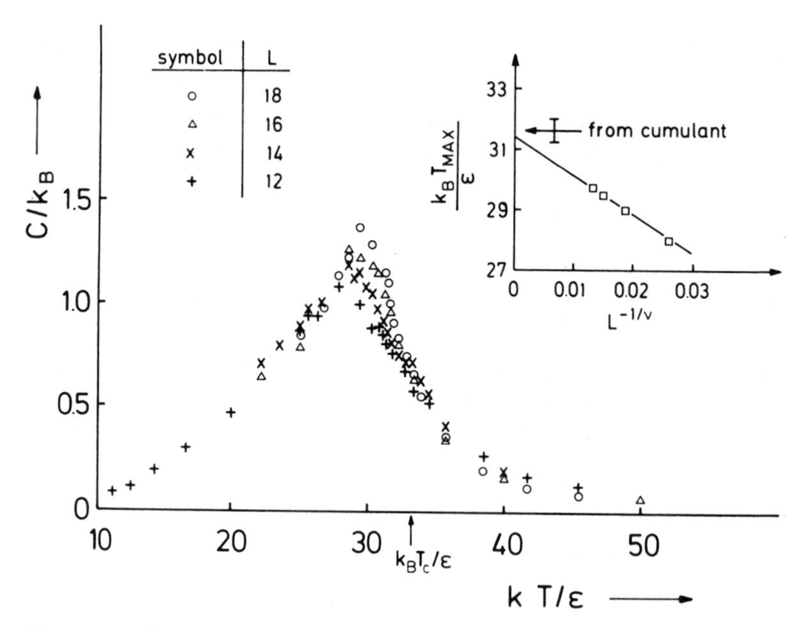

Figure 19 (a) Specific heat per chain C/k_B plotted vs temperature k_BT/ϵ for $N = 32$, $\phi_v = 0.2$, $\epsilon_{AB} = 0$, $\epsilon_{AA} = \epsilon_{BB} = -\epsilon$, and various lattice sizes. Insert shows that extrapolation of temperatures of the specific heat maxima yields T_c consistent with the cumulant intersection point.

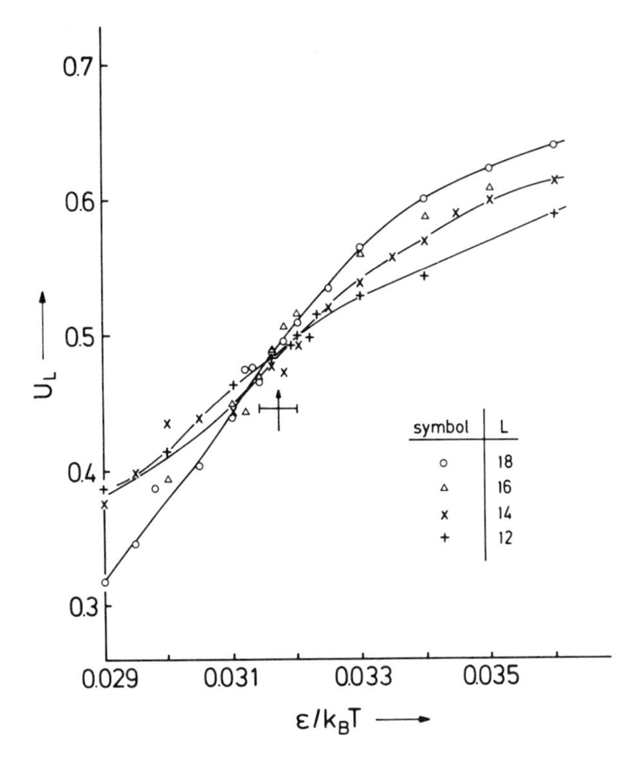

Figure 19 (b) Cumulant U_L plotted versus inverse temperature ϵ/k_BT for the same case and various values of L as shown in the figure. Arrow with error bar indicates the observed estimate for T_c and its uncertainty. (From Ref. 18.)

where ξ is the correlation length of order parameter fluctuations in the system, $\tilde{P}$ is the so-called scaling function, and β, ν are the critical exponents of order parameter and correlation length, respectively [133],

$$\langle |M| \rangle_{L\to\infty} \sim \left(\frac{1 - T}{T_c}\right)^{\beta}, \qquad T \leq T_c, \tag{46a}$$

$$\xi \sim \left|\frac{1 - T}{T_c}\right|^{-\nu}, \qquad T \to T_c. \tag{46b}$$

The finite size scaling hypothesis, Eq. (45), expresses the rule that $P_L(M)$ depends on the three variables L, M, and $1 - T/T_c$ not in the most general form, but rather via two scaling variables $ML^{\beta/\nu}$, L/ξ. From Eq. (45) one readily derives finite size scaling expressions for m, $S_T^{\text{coll}}(0)$ and U_L, namely

$$m = L^{-\beta/\nu}\tilde{M}\left(\frac{L}{\xi}\right), \qquad L \to \infty, \quad \xi \to \infty, \tag{47a}$$

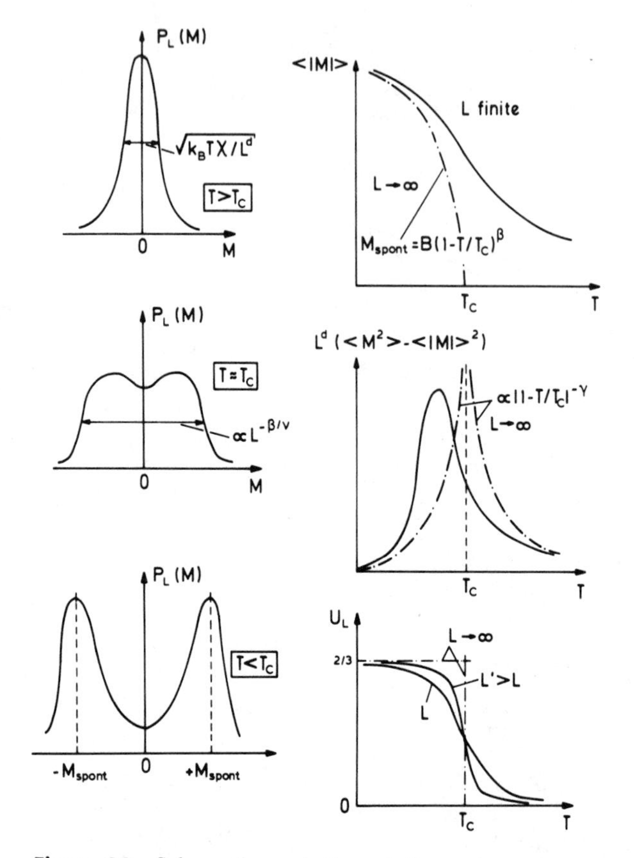

Figure 20 Schematic evolution of the order parameter probability distribution $P_L(M)$ from $T < T_c$ (lower part) to $T > T_c$ (upper part), and corresponding temperature dependence or order parameter $\Psi = \langle |M| \rangle$, collective structure factor at zero wave vector $k_B T \chi' = L^d (\langle M^2 \rangle - \langle |M| \rangle^2)$, and the reduced fourth-order cumulant $U_L = 1 - \langle M^4 \rangle_L / (3 \langle M^2 \rangle^2)$. Dash-dotted curves indicate the singular variation which results in the thermodynamic limit, $L \to \infty$. (from Ref. 131.)

$$S_T^{\text{coll}}(0) = L^{\gamma/\nu} \tilde{\chi} \left(\frac{L}{\xi} \right), \qquad L \to \infty, \quad \xi \to \infty, \tag{47b}$$

$$U_L = \tilde{U} \left(\frac{L}{\xi} \right), \qquad L \to \infty, \quad \xi \to \infty. \tag{47c}$$

Here the scaling functions $\tilde{M}$, $\tilde{\chi}$, and $\tilde{U}$ derive from suitable moments $\langle x^k \rangle \equiv \int dx\, x^k \tilde{P}(x, L/\xi)$, using the definitions Eqs. (42), (43), and (44), with $k = 1,2$, and 4. Eq. (47c) provides a justification for the cumulant

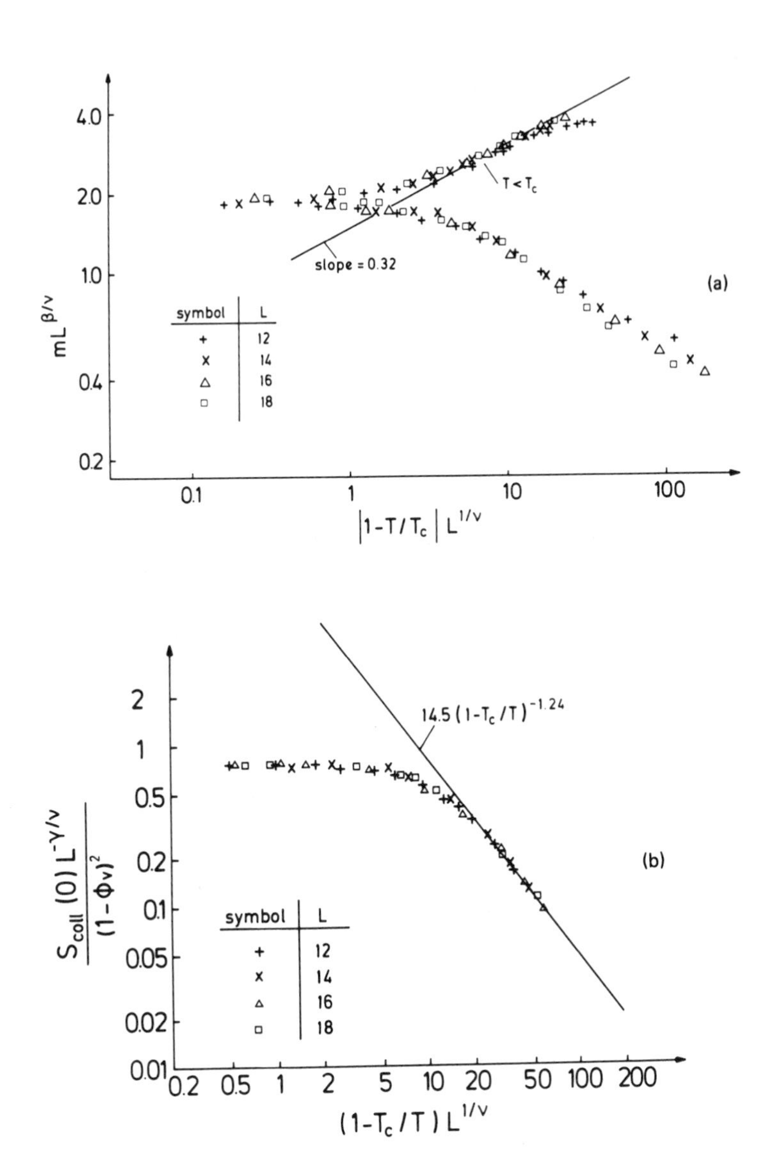

Figure 21 Log-log plot of $S_T^{coll}(0)L^{-\gamma/\nu}(1 - \phi_v)^2$ versus $(1 - T_c/T)L^{1/\gamma}$ (lower part) and of $mL_{\beta/\nu}$ versus $|1 - T/T_c|L^{1/\nu}$ (upper part), for $\phi_v = 0.2$ and $N = 32$. Ising model exponents (Eq. (49)) are used. Straight lines indicate the asymptotic power laws, Eq. (46a) and Eq. (48), including the prefactors. The straight line in (a) corresponds to the dash-dotted line in Figure 18(a). (From Ref. 18.)

intersection method: approaching T_c at fixed L all U_L take the same value $\tilde{U}(0)$. In Eq. (47b), we have defined the critical exponent γ $(= d\nu - 2\beta$ [134]) of the collective structure factor,

$$S_T^{\text{coll}}(0) \sim \left|\frac{1 - T}{T_c}\right|^{-\gamma}, \qquad T \to T_c. \tag{48}$$

Since peaks of $S_T^{\text{coll}}(0)$ as well as of the specific heat C correspond to a case L/ξ = const., or $L^{-1/\nu}|1 - T/T_c|$ = const., it follows that peak positions should be extrapolated to $L \to \infty$ linearly versus $L^{-1/\nu}$, as done in Figure 19(a). Throughout our analysis we have found that the critical exponents of the three-dimensional ising model [133,134], namely [135]

$$\beta \approx 0.32, \qquad \nu \approx 0.63, \qquad \gamma \approx 1.24 \tag{49}$$

provide an appropriate description of our model. This is seen in the insert at Figure 18(b), where one sees direct evidence for the relation $S_{\text{coll}}^{\text{max}} \sim L^{\gamma/\nu} \sim (L^3)^{0.66}$, and in Figure 21, where relations (47a) and (47b) are checked (further examples for this finite size scaling behavior for other values of N and ϕ_v see Refs. 18 and 29).

The Ising critical behavior, Eq. (49), is not in accord with the Flory Huggins [28] approximation, which rather yields "mean-field" exponents [133]

$$\beta = 1/2, \qquad \nu = 1/2, \qquad \gamma = 1. \tag{50}$$

Now from the universality principle [134] one expects that very close to T_c all binary mixtures should show the Ising exponents. On the other hand, a Ginzburg criterion [92,122,136] implies that this non-mean-field critical behavior for very long polymer chains is indeed only visible very close to T_c, while further away from T_c the data should be compatible with Eq. (50). The chains of the simulations described here are rather short [29], and thus Eq. (50) is not at all a good approximation then.

The non-mean-field critical behavior {Eq. (49)} has several observable effects: (a) the binodal near T_c is much flatter than the meanfield parabola; see Figure 22(b). And while $[S_T^{\text{coll}}(0)]^{-1}$ plotted versus temperature should vanish linearly at T_c if data in the one-phase region at the critical volume fraction are used, an exponent $\gamma > 1$ implies that $[S_T^{\text{coll}}(0)]^{-1}$ vanishes at T_c according to a curved function, the slope getting zero. This behavior is seen in both simulation (Fig. 23(a)) and recent experiments (Fig. 23(b)) [137].

Of course, the depression of T_c due to critical fluctuations can only account for part of the discrepancy noted between the Monte Carlo simulation and Flory-Huggins theory. Most of the discrepancy actually is due to the overcount in the number of contacts, Figure 24. The total number

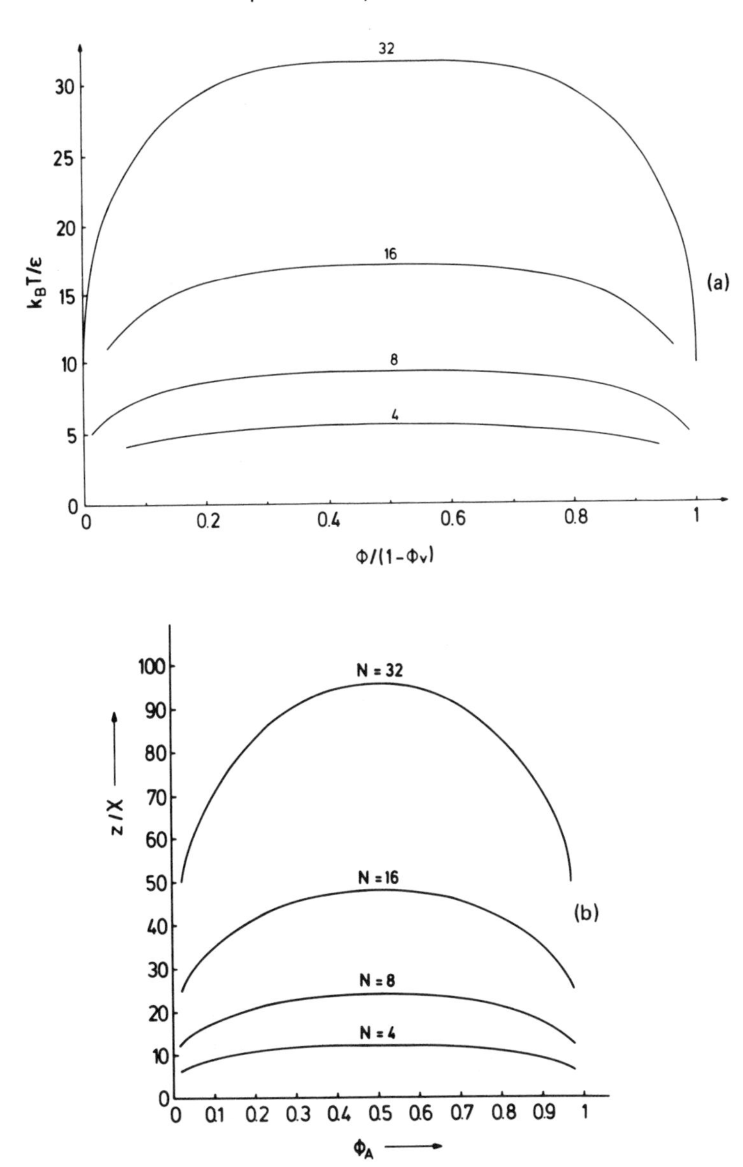

Figure 22 (a) Coexistence curves for four different chain lengths (N = 4, 8, 18, and 32) resulting from the Monte Carlo calculation for the model with ϕ_v = 0.2, $\epsilon_{AB} = 0$, $\epsilon_{AA} = \epsilon_{BB} = -\epsilon$. (b) Coexistence curves resulting from the Flory-Huggins approximation for the same chain lengths, but ϕ_v = 0. Note that the case with $\phi_v \neq 0$ simply follows via replacing ϕ_A by $\phi_A/(1 - \phi_v)$ and z/χ by $(z/\chi)(1 - \phi_v)$. (From Ref. 18.)

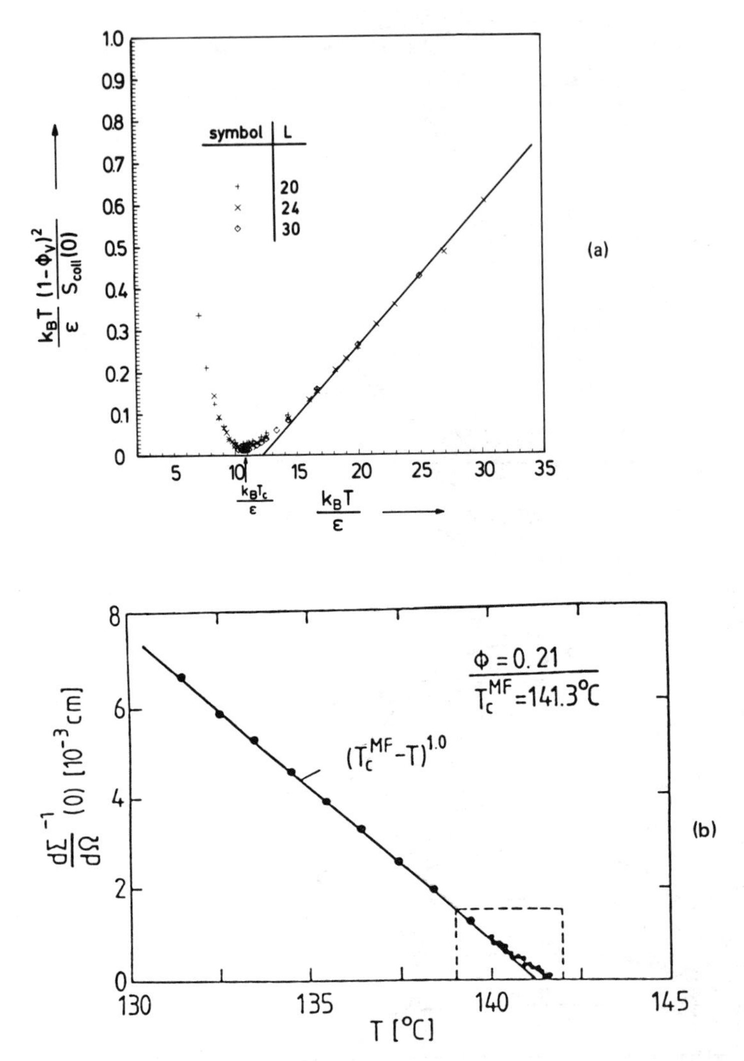

Figure 23 (a) Inverse collective structure factor plotted versus temperature, for the model with $\phi_v = 0.6$, $N = 16$, $\epsilon_{AB} = 0$, $\epsilon_{AA} = 0$, $\epsilon_{AA} = \epsilon_{BB} = -\epsilon$. (From Ref. 29.) (b) Inverse small-angle neutron scattering intensity for mixtures of polystyrene ($M_w = 232{,}000$) and polyvinylmethylether ($M_w = 89{,}000$) at the critical concentration $\phi = 0.21$, plotted versus temperature. Note that the slope of this plot is negative rather than positive as in Figure 11, since the system has a lower consolute critical point rather than the upper critical consolute point of the present model calculations (Fig. 22). The region inside the dashed straight lines is consistent with Eqs. (48) and (49). (From Ref. 137.)

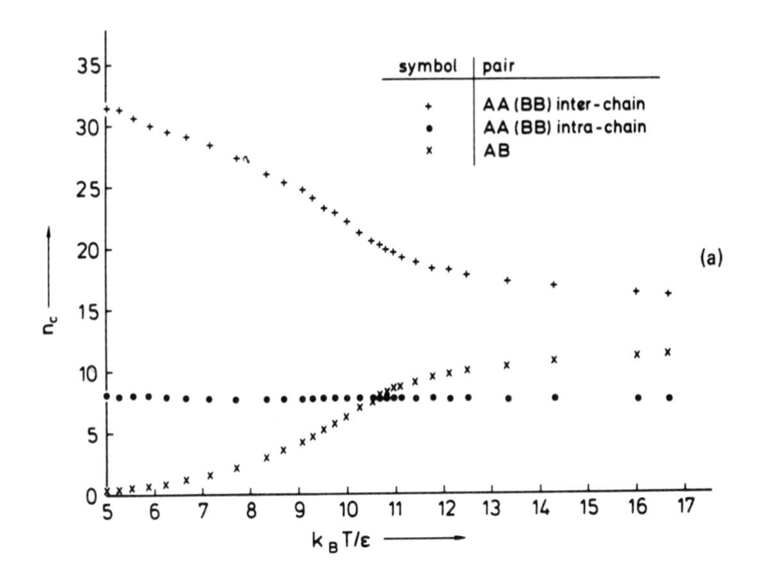

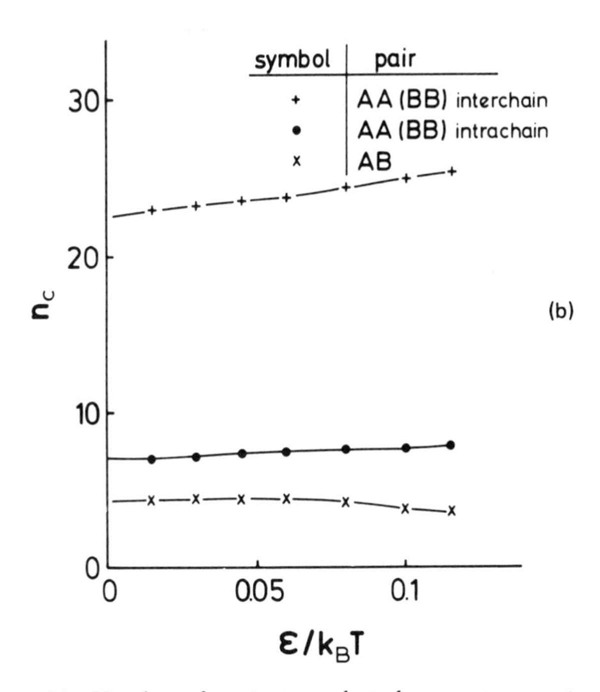

Figure 24 Number of contacts n_c plotted versus temperature (a) or inverse temperature (b) for the case $\phi_v = 0.6$, $N = 32$, $\epsilon_{AB} = 0$, and $\epsilon_{AA} = \epsilon_{BB} = -\epsilon$. Case (a) refers to the grand-canonical situation, where $\Delta\mu = 0$; i.e., $\phi_A = \phi_B = \phi_{crit} = (1 - \phi_v)/2$ for $T > T_c$, while for $T < T_c$ ($k_BT_c/\epsilon \approx 10.9$ here) $\phi_A = \phi_A^{coex(1)}$. Case (b) refers to $\phi_A/(1 - \phi_v) = 0.9$, $\phi_B/(1 - \phi_v) = 0.1$, i.e., distinctly off-critical volume fractions. (From Ref. 30.)

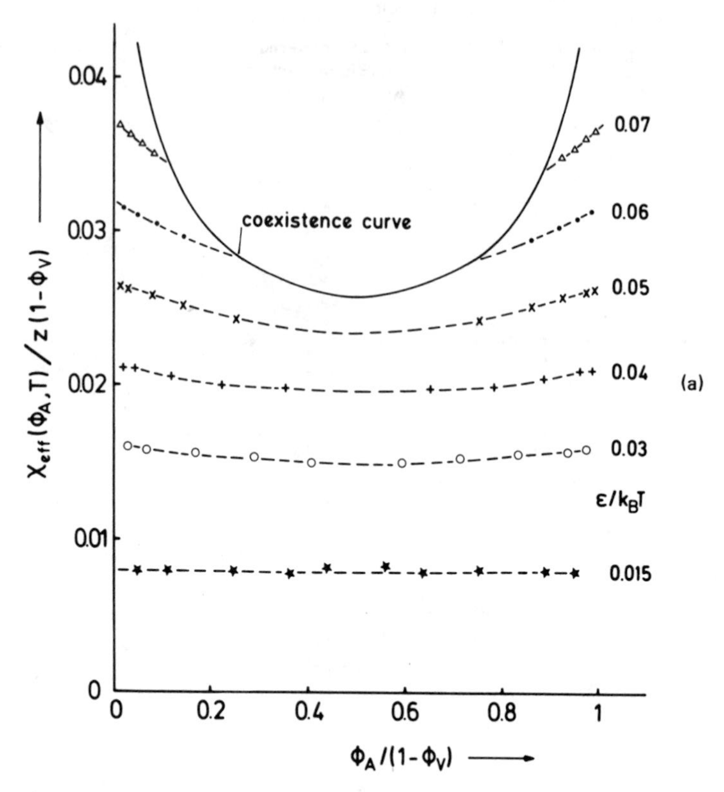

Figure 25 (a) Effective Flory-Huggins parameter $\chi_{\text{eff}}(\phi_A, T)$ plotted versus relative concentration $\phi_A/(1 - \phi_v)$ for several temperatures and the case $\phi_v = 0.6$, $N = 32$, $\epsilon_{AB} = 0$, $\epsilon_{AA} = \epsilon_{BB} = -\epsilon$. Broken curves are guides to the eye only. (From Ref. 29.) (b) Effective Flory-Huggins parameter $\chi_{\text{eff}}(\phi, T)$ plotted versus volume fraction of deuterated polyethylethylene in a *P*-polyethylethylene/*d*-polyethylethylene mixture. Filled symbols refer to $N_w^p = 1710$, open symbols to $N_w^p = 818$, while $N_w^d = 1330$. (From Ref. 127.)

n_c^{tot} of contacts that a chain makes with monomers of other chains is much less than predicted and also weakly depends on temperature. For the example shown in Figure 24, one finds $n_c^{\text{tot}} = n_c\{\text{AA(BB) interchain}\} + n_c(\text{AB}) \approx 32$ for low T in a state at the binodal, $n_c^{\text{tot}} \approx 28.5$ for a mixture at the critical point, and $n_c^{\text{tot}} \approx 27$ at high temperatures. Within statistical errors these numbers do not depend on volume fraction. Of course, the individual numbers $n_c\{\text{AA(BB) interchain}\}$ and $n_c(\text{AB})$ do depend on volume fraction, $n_c(\text{AB}) \to 0$ as $\phi_A \to 0$ or $\phi_B \to 0$. Thus the strong decrease of $n_c(\text{AB})$ in Figure 24(a) simply reflects the decrease in the number of AB pairs as one follows the binodal.

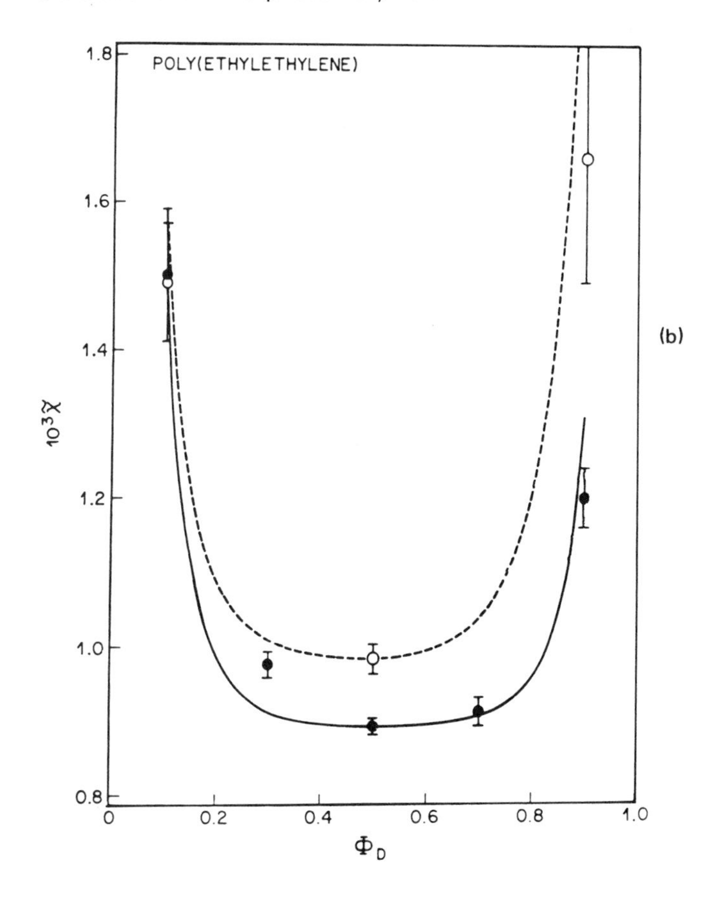

There occurs a nonneglible number of AA(BB) intrachain contacts ($n_c^{\text{self}} \approx 7$ to 8 for the example shown), which obviously cannot contribute to a phase separation between A-chains and B-chains. Of course, these are neglected both in the Flory approximation (which would predict $n_c^{\text{tot}} = zN(1 - \phi_v) \approx 77$) and in the Guggenheim approximation (which would predict $n_c = [(z - 2)N + 2](1 - \phi_v) \approx 54$). Both the selfcontacts and the further reduction of any contacts due to excluded volume, which is locally already rather effective at $\phi_v = 0.6$, are responsible for the low value of the actual n_c^{tot}, and hence T_c is small (since $T_c \sim n_c^{\text{tot}}$) in comparison with the theoretical prediction.

The consistency of this interpretation is corroborated by a study of the temperature and concentration dependence of an effective Flory-Huggins parameter $\chi^{\text{eff}}(\phi_A, T)$ extracted from the simulation. This parameter results from a fit of the Flory-Huggins equation of state to the Monte Carlo data,

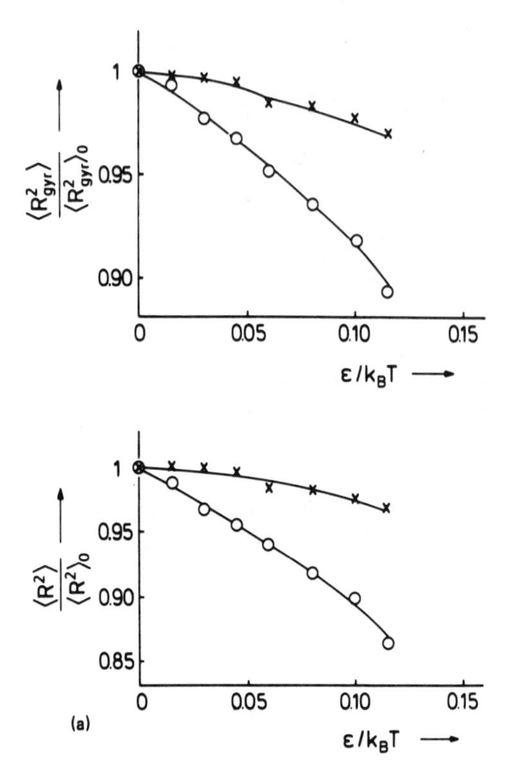

Figure 26 (a) Meansquare gyration radius (upper part) and meansquare end-to-end distance (lower part) at constant $\phi_A/(1 - \phi_v) = 0.9$ plotted versus inverse temperature, for $N = 32$, $\phi_v = 0.6$, $\epsilon_{AB} = 0$, $\epsilon_{AA} = \epsilon_{BB} = -\epsilon$. The crosses refer to the majority component (A), the circles to the minority component (B). The largest value of ϵ/k_BT corresponds to a point at the binodal. All radii are normalized by their values at $\epsilon/k_BT = 0$. (From Ref. 29.) (b) Mean-square end-to-end distance normalized by the chain length (upper part) and mean-square gyration radius normalized by the chain length (lower part) on a log-log plot versus chain length for one minority (B) chain among A-chains and vacancies ($\phi_v = 0.6$, $\epsilon_{AB} = 0$, $\epsilon_{AA} = \epsilon_{BB} = -\epsilon$). Various temperatures are shown as indicated. Dashed straight lines show the asymptotic slope for collapsed chains. (From Ref. 138.) (c) Same as (b) but replotted in scaled form versus $N\epsilon/k_BT$. Three values of N are included, $N = 16$ (squares), $N = 32$ (triangles), and $N = 64$ (circles). (From Ref. 138.)

similarly as experimental data are fitted to Eq. (38). If Eqs. (39)–(41) were true we would expect $\chi_{\text{eff}}(\phi_A,T)/z(1 - \phi_v) = \epsilon/k_BT$ while the actual result is much lower (Fig. 25(a)). For small ϵ/k_BT ($T >> T_c$) this reduction is just a constant factor, namely n_c^{tot} (actual)/n_c^{tot} (Flory Huggins). Near T_c there occurs an additional spurious dependence on volume fraction. Per-

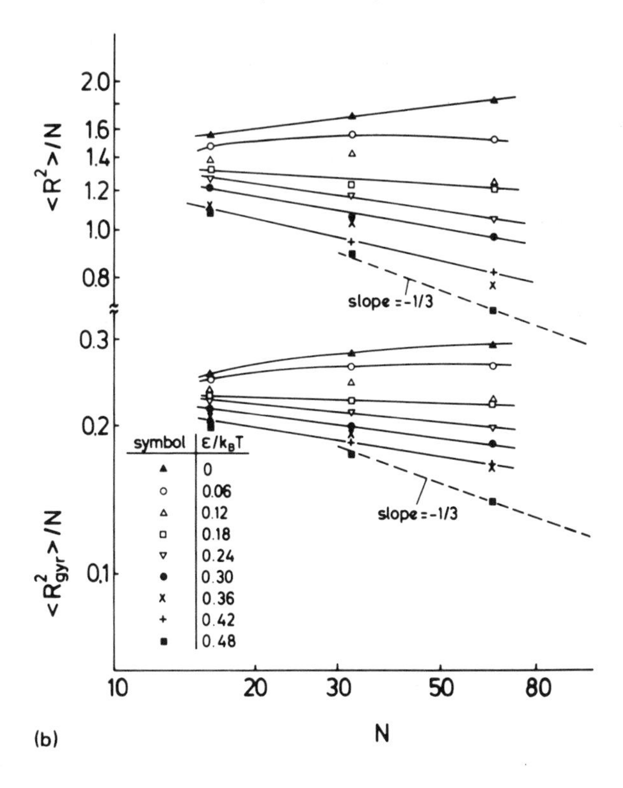
$<R^2>/N$
2.0
1.6
1.4
1.2
1.0
0.8
0.3
0.2
0.1
$<R^2_{gyr}>/N$
slope = -1/3
slope = -1/3
symbol
ε/k_BT
0
0.06
0.12
0.18
0.24
0.30
0.36
0.42
0.48
10
20
30
50
80
N
(b)

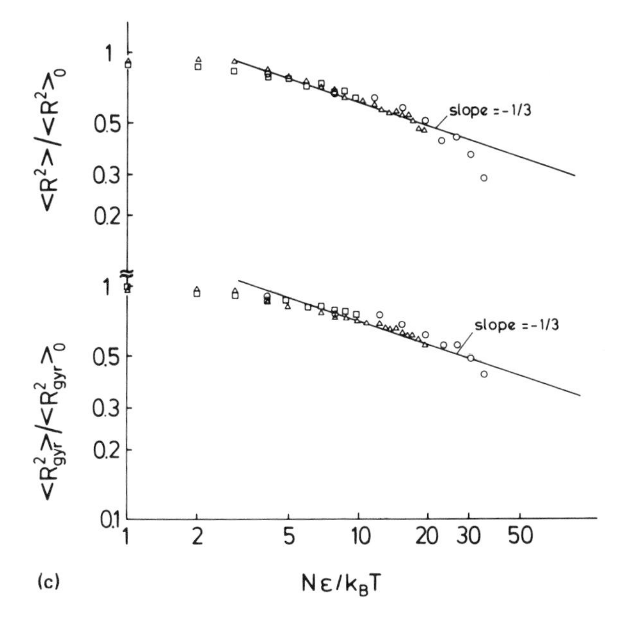
$<R^2>/<R^2>_0$
$<R^2_{gyr}>/<R^2_{gyr}>_0$
1
0.5
0.3
0.2
0.1
slope = -1/3
slope = -1/3
1
2
5
10
20
30
50
$N\varepsilon/k_BT$
(c)

haps there is a connection between this observation and the volume fraction dependence of $\chi_{\mathrm{eff}}(\phi,T)$ seen in mixtures of protonated and deuterated polyethylethylene (Fig. 25(b)) [127].

The weak temperature dependence in the number of contacts (Fig. 24) is accompanied also by a small contraction in the chain radii, which is also an effect not predicted by the Flory-Huggins approximation. For asymmetric volume fractions, this effect is more pronounced for the minority component (Fig. 26(a)). Particularly dramatic effects occur if one species is rather dilute (Fig. 26(b)): then a gradual transition to a collapsed state

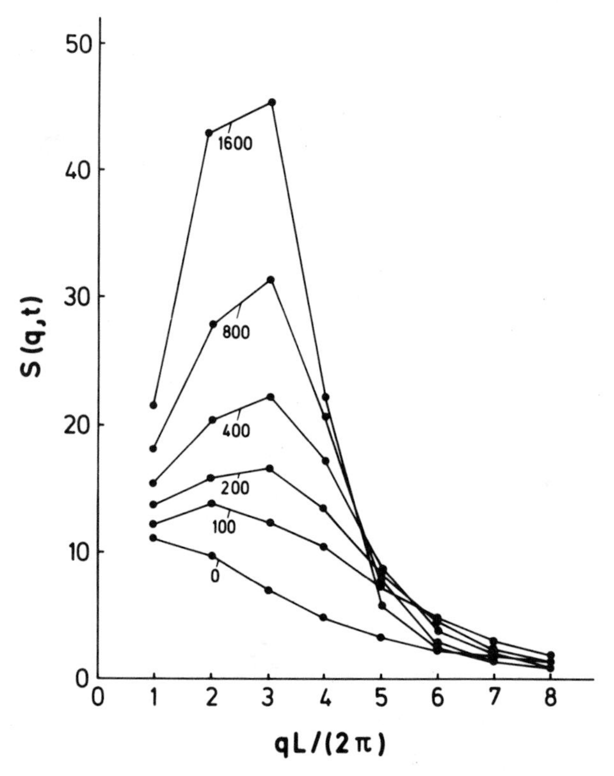

Figure 27 Collective structure function $S(q,t)$ plotted versus q at various times t after a quench from infinite temperature to $\epsilon/k_BT = 0.6$, for $N = 32$, $\phi_v = 0.6$, $\epsilon_{AB} = 0$, $\epsilon_{AA} = \epsilon_{BB} = -\epsilon$. Different curves are labeled by the time t (measured in units of attempted moves per monomer). Note that S(q,t) is defined only for the discrete values $q_\gamma = \gamma(2\pi/L)$, $\gamma = 1.2, \ldots$, characterized by the dots ($L = 40$ in this simulation). For clarity, these dots are connected by straight lines in the figure. (From Ref. 139.)

occurs [138], where the chain gyration radii behave as $\sqrt{\langle R^2_{\text{gyr}} \rangle} \sim N^{1/3}$ [92]. Figure 26(c) shows that the ratio $\langle R^2_{\text{gyr}} \rangle_T / \langle R^2_{\text{gyr}} \rangle_{T=\infty}$ is a function of the product $N\epsilon/k_BT$ or, alternatively, χ/χ_c only [138]. As a result, one predicts such a chain collapse if a chain is dissolved in a melt of other chains in a nearly incompatible blend: forming a lot of self-contacts due to the collapse is then the most favorable situation.

Finally, we draw attention to studies of the dynamics of phase separation [21,139]. By a rapid temperature quench one may bring a system which is in a homogeneous state in thermal equilibrium into the two-phase region. Now long-wavelength concentration inhomogeneities are unstable, and the system starts to unmix via the so-called "spinodal decomposition" mechanism [120–123,140,141]. This shows up in a characteristic time dependence of the collective scattering function $S(q,t)$: a peak grows at a characteristic wave number q_m. As time proceeds, the peak position moves to smaller values, since the precipitated structures coarsen. Figure 27 shows a simulation of this process [139]. One may also simulate interdiffusion of chains in a blend in thermal equilibrium [142].

Simulation of the Glass Transition

In a first study of a dense off-lattice polymer system [13], it was found that either at rather low temperature or at high density the chains were essentially frozen in; i.e. even over a very long time only mean-square displacements much less than the square of a bond length occurred. This was interpreted as vibrational motion in an amorphous glasslike structure [13].

Since the model used in Ref. 13 (the freely jointed chain, Fig. 1(a), with Lennard-Jones interaction, Fig. 1(d) is not microscopically realistic, the detailed behavior found there cannot directly be compared to experiment. In contrast, Rigby and Roe [77,78] simulated a model to mimic polyethylene by molecular dynamics methods, as described earlier. Measuring the pressure from the virial theorem as usual [8–11] for a variety of densities as a function of temperature, one can replot these original data in the form of a density-temperature diagram (Fig. 28(a) [77,143]. These data were obtained simulating a cubic box with periodic boundary conditions, containing 500 segments (for chainlengths $N = 10$ and 20) or 2000 segments (for chain length 50, respectively). The time step Δt of the molecular dynamics algorithm, Eq. (23), was set equal to $0.005\tau = 10^{-14}$ s, with τ being defined in Eq. (22). Each simulation was run for a duration of 50 or 60 τ (i.e., 100 to 120 ps.). Figure 28(a) shows that there are two linear regions, separated by a rather well-defined kink, which is interpreted as the glass transition. This interpretation is consistent with a behavior of the self-diffusion coefficient (Fig. 28(b), which is found to essentially vanish

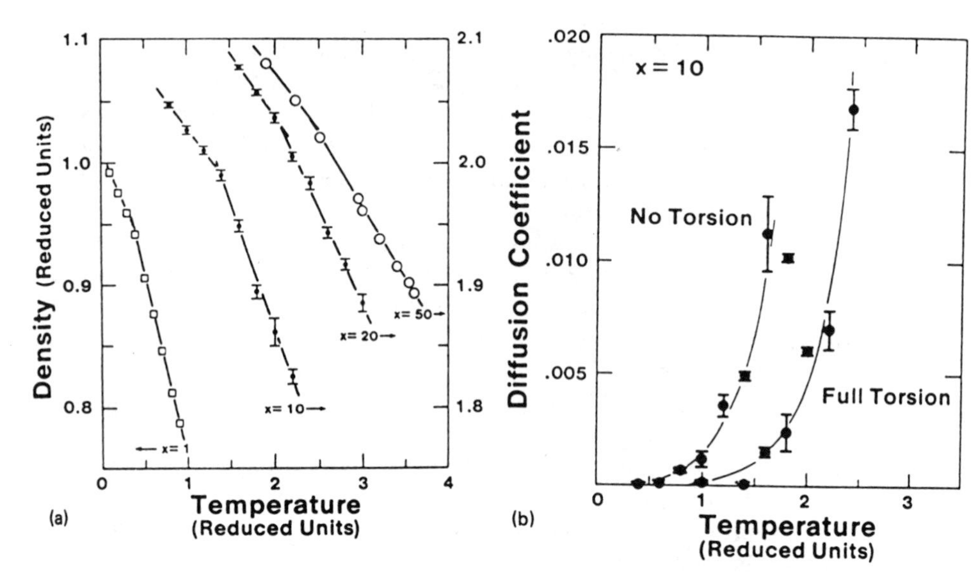

Figure 28 (a) Plot of density in reduced units {the unit of density is $\rho^* = 1/\sigma^3$ where $\sigma = 0.38$ nm is the characteristic length of the Lennard-Jones potential, Eq. (1), between the segments} versus temperature in reduced units (the unit of temperature is $\epsilon/k_B = 60.1$ K, ϵ being the characteristic energy in the Lennard-Jones potential, Eq. (1)). Curves for chain lengths (denoted by x in this figure) 10, 20, and 50 are shown, and data for a simple fluid with Lennard-Jones interaction [143] (denoted by $x = 1$) are included for comparison. (b) Self-diffusion coefficient in reduced units plotted versus temperature in reduced units for a chainlength of 10 and a reduced pressure of $p = 3$ (units of pressure is ϵ/σ^3). Upper curve refers to a simulation where the torsion potential was arbitrarily switched off. (From Ref. 77)

as the temperature is lowered; although the estimation of the diffusion coefficient near the temperatures where the density exhibits the kink is very difficult and subject to large errors, since one may not have reached the asymptotic regime of times yet, it is clear that there is a pronounced slowing down of the dynamics.

In very recent further work [78] the short-range order and orientational correlations among the C—C bonds between neighboring subchains were investigated for this model. Intermolecular radial distribution functions $g_k(r)$ were obtained for subchains consisting of k units within chains having 20 units in total. Figure 29 shows the effect of varying temperature at constant pressure $p = 3$ (in reduced units), for a system containing in total 2000 CH_2 units [78]. In the example shown in Figure 29, the glass transition

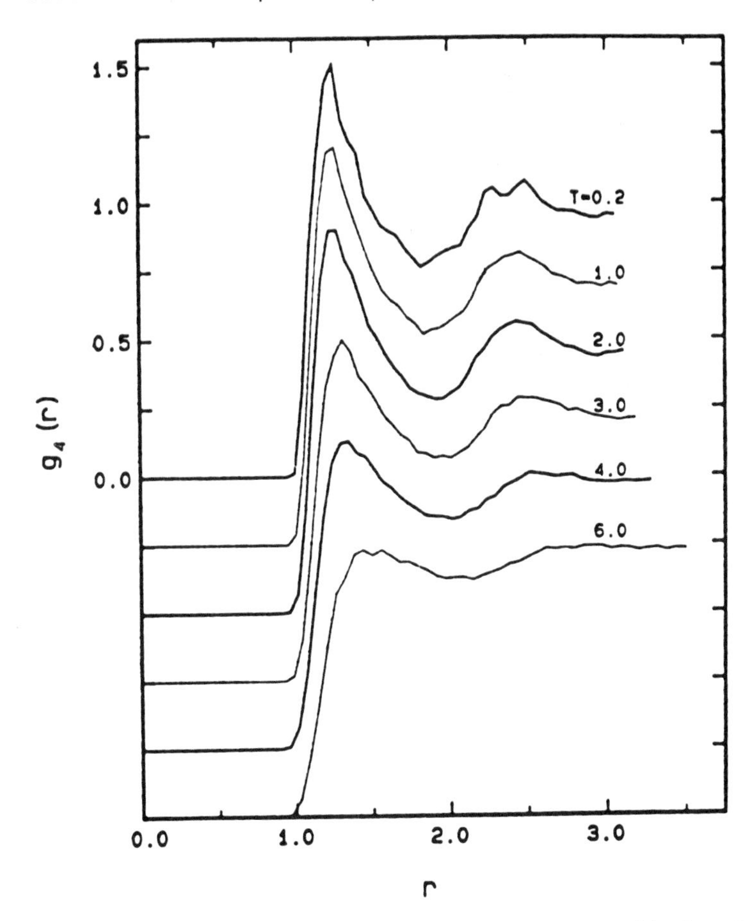

Figure 29 Intermolecular radial distribution function g_4(r) plotted versus r (in units of σ) at various temperatures (in reduced units, cf. Fig. 28) as indicated in the figure. Curves for $T \leq 4$ were obtained by stepwise cooling at constant pressure of a system first equilibrated at $T = 4.0$. Note that the ordinate zero is displaced by 0.1 for each curve for the sake of clarity. (From Ref. 78.)

is estimated to occur at about $T \approx 1.96$. It is seen that $g_4(r)$ does not show any abrupt change there. The splitting of the second peak in the radial distribution function, which sometimes [144] is taken as a characteristic signature of glass states in general, occurs in the present model for the lowest temperature only.

The behavior of the orientational correlations, which get distinctly stronger as T is lowered, suggest that the glass transition competes with a transition to a nematic phase with liquid crystalline-long range order. In our view, much larger systems and, at the same time, much longer computing times, would be required to allow reliable estimates of the temperature where this isotropic-nematic transition would occur in thermal equilibrium.

Although the study of this model for polymethylene [77,78] clearly allows a variety of interesting conclusions to be drawn, the complicated competition between the glass transition and liquid crystalline order makes again the availability of a simpler model desirable, if one wishes to elucidate the behavior of the glass transition. Again one feels that maybe the simplest models are lattice models, similar to some of the basic theoretical approaches [145]. So far only rather preliminary studies of the glass transition in lattice models are available [146,147]. On the tetrahedral lattice for

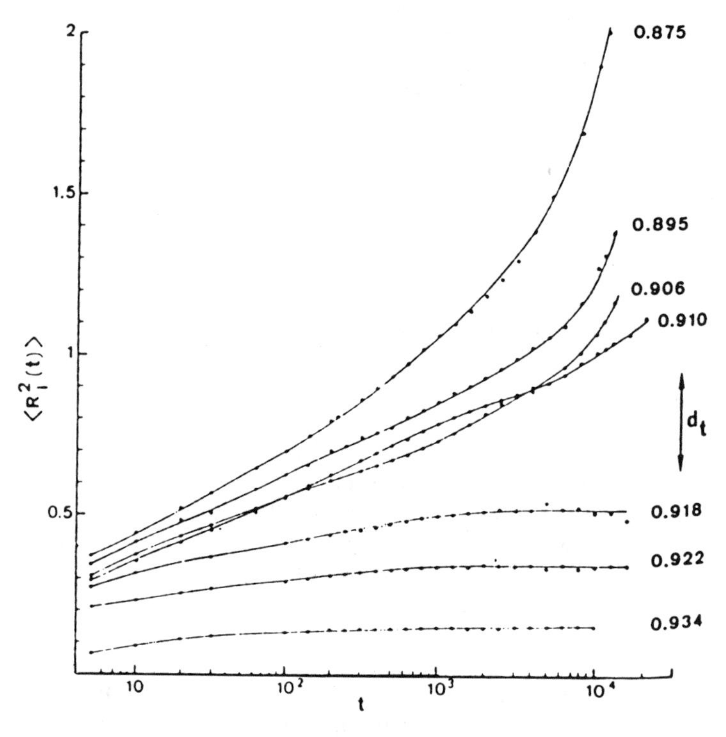

Figure 30 Mean-square displacements of polymer beads plotted versus the logarithm of time (measured in the unit of attempted Monte Carlo steps per bead), for a tetrahedral lattice of linear dimension $L = 16$ and periodic boundary conditions, at various values of total bead density (denoted as d_t in the figure). (From Ref. 146).

chain lengths $N = 32$ a glass transition is found for systems having total bead densities of about 0.92; see Figure 30. Of course, in this case the preparation of an initial configuration at these high densities is rather nontrivial. In Ref. 146 the chains are introduced one by one inside the box by the slithering snake technique [20], which is also used for equilibration, while the dynamics of diffusion near the glass transition is then studied by a combination of 3-bond motions and 4-bond motions [148]. In order to be able to vary the density in very small steps, one chain was chosen to have a smaller number of beads. Also the effect of adding free beads was studied. An advantage of such lattice models is that "free volume" can be characterized straightforwardly and hence contact to corresponding analytical theories of the glass transition [149] can be made. More work, however, is still necessary to judge the reliability of the various theories [145,149]. A study of this problem is in progress [147].

Branched Polymers (stars, gels, rubber, etc.)

The computer simulation of branched polymers is just at its beginning. The molecular dynamics simulation technique [150] and the bond fluctuation method for the Monte Carlo simulation of lattice models [26] has been extended to the study of single-star polymers. So far, the focus of this work has been on checking scaling ideas on the static and dynamic properties of star polymers in dilute solution, and thus this work is outside the scope of the present review. We expect, however, that these techniques will soon be extended to the study of melts of stars and other dense branched polymers. (For a study of star polymer reptation in a regular entanglement network, see [151].)

Also we shall not consider any work on the sol-gel transition [152,153], where computer simulations have concentrated on the concept of percolation models [154] and related approaches.

But clearly, the simulational modeling of gels and rubbers will be a very promising field for future work. As an example of the type of questions that can be asked we mention work on the effect of uniaxial extension of an elastomer on junction fluctuations [155]. The entangled micronetwork there was simulated putting three chains of 41 beads each on a cubic lattice, with one trifunctional junction point fluctuating in position, and three fixed-chain ends. The coordinates of these chain ends were $(0, \lambda\sqrt{2N/3}, 0)$, $(\sqrt{N/3}\lambda, -\lambda\sqrt{N/6}, 0)$, and $(9\sqrt{N/2}\lambda, -\lambda\sqrt{N/6}, 0)$, where λ is the uniaxial extension ratio. Thus, the starting positions for the free fixed-chain end in the unstrained, unentangled state were determined by assuming Gaussian chain statistics, and furthermore, it is assumed that the fixed-chain ends and the entanglement net deform affinely when a uniaxial strain

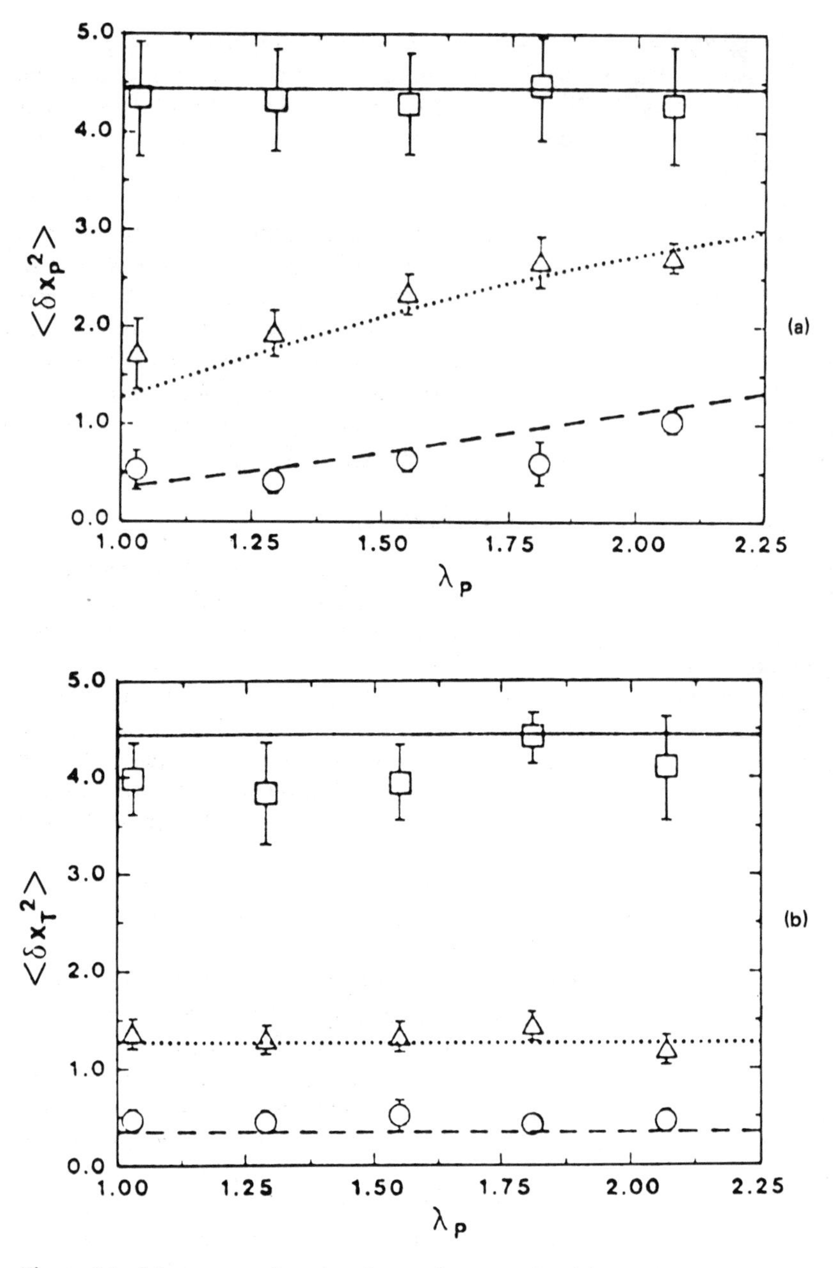

Figure 31 Meansquare junction fluctuations parallel (a) and perpendicular (b) to the uniaxial extension plotted versus extension ratio for simulation runs with different mesh spacings c of the entanglement lattice: $c = 2$ (squares), $c = 4$ (triangles) and $c = \infty$ [unentangled state] (squares). Lines are theoretical prediction for a

is applied. The entanglement net is represented by a regular rigid obstacle net, similar as done by Edwards and coworkers [116,151,156], which does not overlap the chain lattice, so that there is no excluded volume at all, irrespective of the mesh spacing of this obstacle lattice. Also Verdier-Stockmayer type [16] motions which allow bead overlap are used, and thus it is possible to move the junction point without fluctuation in bond lengths.

Figure 31 shows that the results for the mean square junction fluctuations are in reasonable agreement with the Flory [157] theory of constrained junction motion. However, it is clear that this simulation involves several rather special assumptions, and should be considered only as a first step. For an alternative attempt to simulate elastomers see [158].

Polymers at Interfaces

If polymers are confined between two media with which the polymer film is completely incompatible, or if polymers are adsorbed very strongly at a wall, one may model the polymer configuration as essentially two-dimensional. Thus chains in two-dimensional melts are under investigation [26]; see Fig. 4. Apart from this application, the effect of reduced dimensionality often is of interest as a means of checking theoretical predictions. In this spirit, also a computer simulation of spinodal decomposition in polymer films has been performed [21]. While the behavior of the collective structure factor is similar to the three-dimensional case, described on page 274 the two-dimensional case allows to study the coarsening of the phase-separated structure by direct inspection of the polymer configurations (Fig. 32).

Here we are more interested in cases where the polymers are not confined to a strictly two-dimensional configuration but rather may extend in the direction normal to the interface too. While initially most work has focused on properties of isolated single chains [159–165] which is outside of our scope here, recent work has considered density profiles of dense polymer systems near walls [166], density profiles of concentrated polymer layers adsorbed at a solid/solvent interface [167], and the structure of the grafted polymer brush [168]. For example, Figure 33 [166] shows that a polymer model consisting of freely joint tangent hard spheres exhibits for very short chains ($N = 4$) pronounced density oscillations near a hard wall, similarly as it occurs for hard-sphere monatomic fluids [169], while for long

---

phantom network (full curve) and for the Flory theory [157], where the adjustable constant κ measuring the severity of the entanglement constraints on the network points was taken to be $\kappa = 2.5$ (dotted curves) and $\kappa = 12$ (broken curves), respectively. (From Ref. 157.)

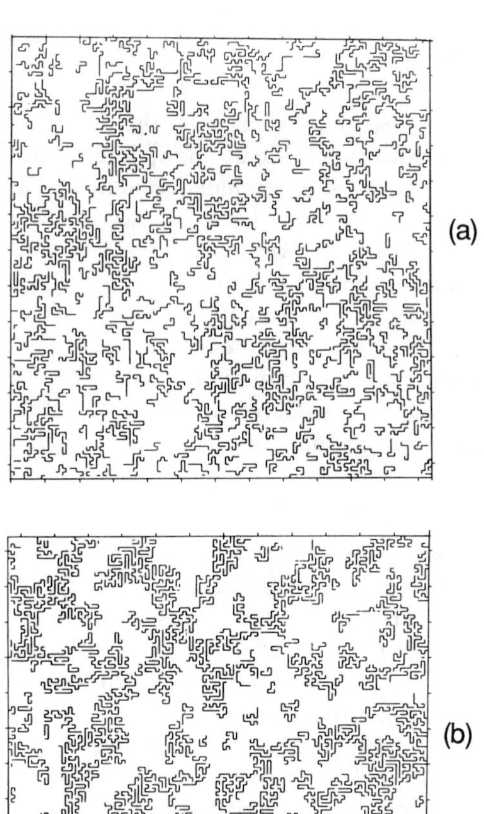
(a)

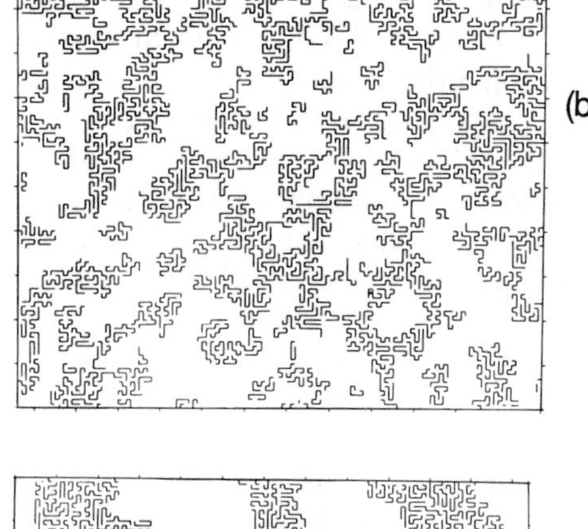
(b)

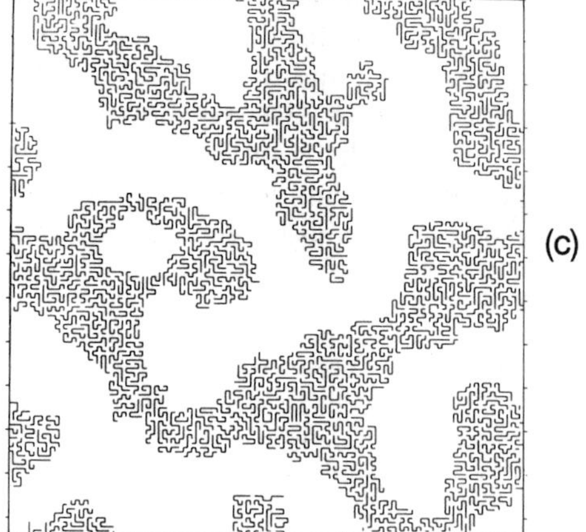
(c)

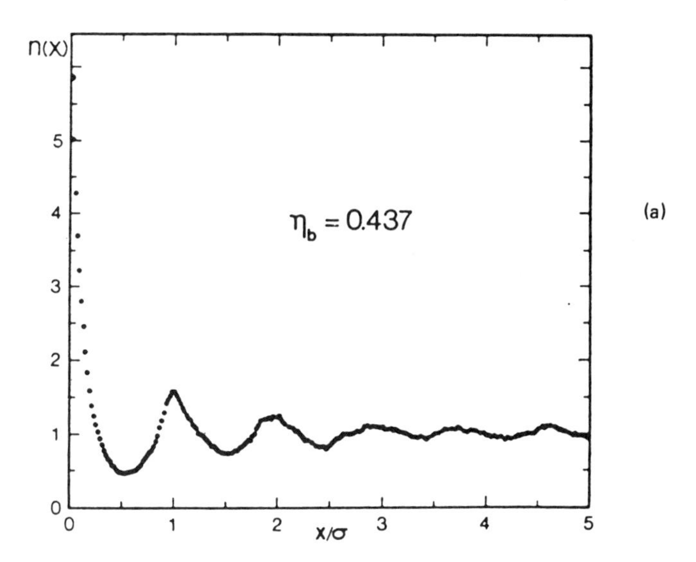

Figure 33 Normalized segment density profile $n(x)$ versus distance from the hard wall, from simulation of the hard-chain fluid with $N = 4$ spheres of a diameter σ at a bulk volume fraction of $\eta_b = 0.437$. (From Ref. 166.)

chains one expects a depletion region near the wall. No detailed analytical theory predicting this behavior seems to be available yet. On the other hand, the simulated density profiles near solid/solvent interfaces [167] can, at least roughly, be accounted by an extension of the Flory-Huggins theory [98,99] to inhomogeneous systems [170].

The work on end-grafted polymer chains at a wall [168] where the molecular dynamics techniques of Grest and Kremer [27,86,87,150] are extended to interface problems is particularly interesting. Figure 34 shows typical configurations at three surface coverages. Data of this sort are used to calculate the density profile and the probability distribution to find the other end of a grafted chain a distance z from the wall [168]. The results

---

Figure 32 Snapshot pictures of a square lattice of 123 × 123 sites with periodic boundary conditions, containing chains (A,B) of length $N_A = N_B = N = 10$, volume fractions $\phi_A = \phi_B = 0.488$ and empty sites at concentration $\phi_v = 0.024$. Only the 738 A chains are shown, B-chains as well as vacancies are left white. Case (a) refers to "infinite temperature," i.e., a noninteracting case which is the initial state for a quench to $k_B T/\epsilon = 1.0$, ϵ being the energy which is won if two neighboring lattice sites are taken by monomers of the same kind. Cases (b) and (c) refer to times $t = 3750$ and $t = 44{,}500$ Monte Carlo steps per bead after the quench, applying the "slithering snake" algorithm [20]. (From Ref. 21.)

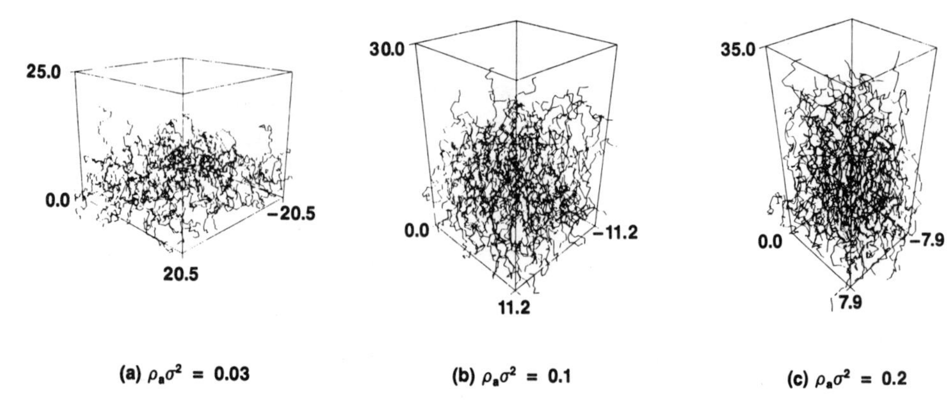

Figure 34 Snapshot pictures of a system with 50 polymers of length $n = 50$ interacting with the potential Eq. (1) at three surface coverages: $\rho_a\sigma^2 = 0.03$ (a), $\rho_a\sigma^2 = 0.1$ (b), and $\rho_a\sigma^2 = 0.2$ (c). Each polymer is grafted at one end randomly onto the lower surface $z = 0$ and in addition experiences a potential of the form $U^{\text{wall}}(z) = \epsilon(\sigma/z)^{12} + Az + B$, for $z < \sigma/2$, to ensure that the monomers do not cross the grafting surface. The constants A and B were chosen such that both $U^{\text{wall}}(z)$ and $\partial U^{\text{wall}}/\partial z$ vanish for $z = \sigma/2$. The simulations were carried out at $k_B T/\epsilon = 1.2$, using periodic boundary conditions in the horizontal plane. (From Ref. 168.)

agree well with a self-consistent field theory [171] while too simplified scaling ideas [172] are ruled out.

Orientational Order in Dense Polymer Systems

The behavior of semiflexible chain molecules at high densities is a long-standing problem [173–175]. Flory [173] suggested that such systems at low temperature should have an orientationally ordered phase, where the rod-like chains are parallel oriented to each other even in the absence of attractive interactions between different chains. This ordering should result simply from the impossibility of packing rods on a lattice to high density in a disordered configuration. However, more refined theories [174,175] have questioned the validity of this approach.

This problem has been studied by simulations of lattice models in both two and three dimensions [39,40,176–179,60]. It turns out that for this problem a careful analysis of finite size effects is extremely crucial: while one observes for the choice $L = N$ for $L \to \infty$ a first-order transition to a well-oriented single-domain state [39], for $L >> N$ the system rather attains a disordered multidomain configuration at low temperatures (Fig. 35), and long-range orientational order is lacking [177–179]. In this model, an energy

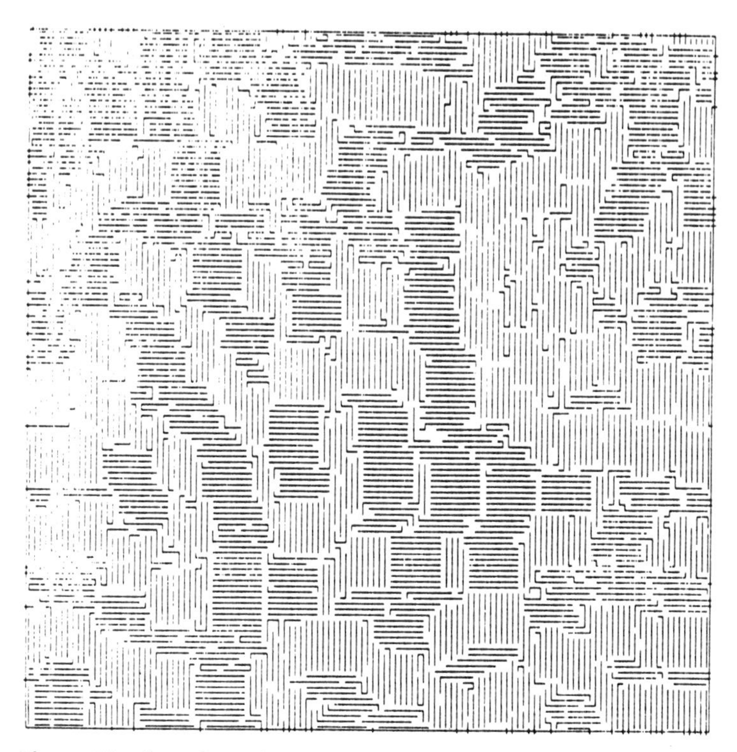

Figure 35 Snapshot picture of a 123 × 123 square lattice with 1476 chains of length $N = 10$, simulated with a modified "slithering snake" algorithm [176], at a temperature $k_B T/\epsilon = 0.5$ where ϵ is the energy which is won if two successive bonds are colinear. (From Ref. 177.)

ϵ is won if "*trans* bonds" occur (i.e., two successive steps of the self-avoiding walk on the lattice go into the same direction), while no energy is won for "*gauche* bonds" (i.e., two successive steps of the self-avoiding walk go into different directions). Thus, the intermolecular energy per bond E is a measure of the "configurational" order parameter, with $0 < E < 1$. In Figure 35, $E = 0.948$ so the configurational order is nearly perfect, although there is very little orientational order ($s = 2\langle f_y \rangle - 1 = -0.00015$ here, with f_y being the fraction of lattice steps in the y direction). It appears that the correlation length for orientational order is of the same order of magnitude as the chain length, however, which means that real high-molecular-weight systems for many practical purposes would not differ from truly long-range ordered systems. Figure 36 shows that the behavior of the configurational order is rather gradual, and can approximately be described by the transition of a single "unperturbed chain" [173], i.e., a nonreversal

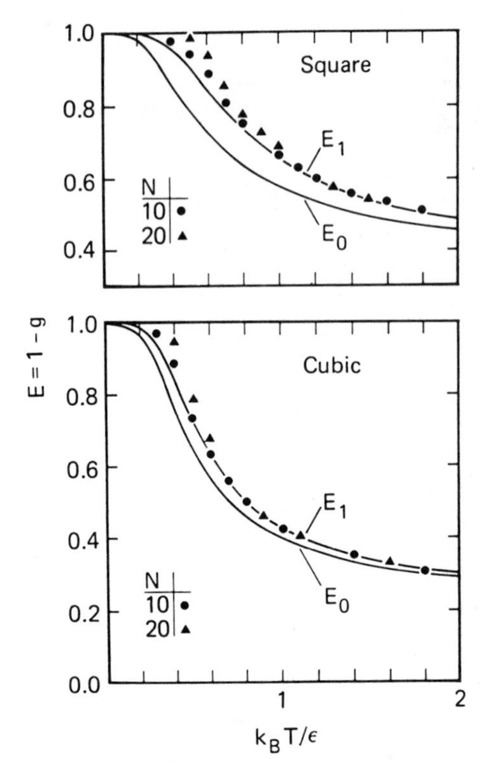

Figure 36 Configurational order parameters E versus temperature for square lattice models (upper part) and cubic lattice models (lower part). Simulated systems contained 1476 chains of length 10 on a 123 × 123 lattice, 756 chains of length 20 on a 126 × 126 lattice, 2883 chains of length 10 on a 31 × 31 × 31 lattice, or 3528 chains of length $N = 20$ on a 42 × 42 × 42 lattice, respectively. E_0 is the approximation quoted in Eq. (51) while E_1 is an improved approximation. (From Ref. 177.)

walk

$$E_o = 1 - \left\{1 + (z - 2)^{-1} \exp\left(\frac{\epsilon}{k_B T}\right)\right\}^{-1}. \tag{51}$$

Including then a local orientation-dependent interaction of van der Waals type, a second-order phase transition to a long-range ordered phase is observed on two-dimensional lattices [176,178] and a first-order transition on three-dimensional lattices [178,179,60].

Orientational short-range order has also been seen in recent molecular dynamics simulations of polymethylene [78]. There, the short-range order is that of a nematic liquid crystal, however, while the present lattice model rather develops full crystalline order up to a length scale N. Presumably off-lattice models will be crucial to allow for both liquid/liquid crystalline and crystalline phases, but this is probably beyond the present computer capabilities. Similarly, a realistic treatment of interfaces between amorphous and cyrstalline polymer lamellae is desirable but very difficult. So far simulations have mainly addressed questions such as to what extent adjacent reentry of chains in a crystalline lamella occur, applying simple lattice models [180].

The question of orientational order comes up also in the study of lipid monolayer models. While work with molecular dynamics [181] and off-lattice Monte Carlo [182] methods tried to model the systems rather realistically, but was restricted to extremely small systems, there has been extensive recent simulation work, e.g., [183], studying cooperative phenomena in lattice models of these systems which is reviewed elsewhere [184].

SUMMARY AND OUTLOOK

In this article, we have reviewed the models for dense polymer systems, which have been simulated by Monte Carlo and molecular dynamics methods. We have briefly described technical aspects of these models, mentioning both their advantages and their limitations. Applications on the statics and dynamics of polymer melts as well as on the thermodynamics of polymer mixtures are described in some detail, while other topics (elastomers, polymers at walls, melting of polymers) are mentioned only rather briefly.

On various occasions we have emphasized the point that it is not yet possible to push these calculations to full molecular detail: polymers are both very large and very slow objects, and taking into account all degrees of freedom it would not be possible to equilibrate the configuration of the polymer, sampling a sufficiently large "cross section" of its phase space. For example, in the molecular dynamics approach one encounters the problem that the characteristic time for a bond length vibration is orders of magnitude smaller than the time taken for a local reorientation of a C—C bond, which needs to overcome a potential energy barrier in order to jump from, say, the *trans* configuration to a *gauche* configuration, or vice versa. Thus, in order to make the approach tractable at all, the recipe is either to work with rigid bonds (this idea was followed already in very early work on alkane simulation [185]), or to make the bonds artificially soft, in order to slow down the bond length vibrations [77,78]. Even then it is necessary to work with a time step of $t = 10^{-14}$ s, for a small number of chains with at most $N = 50$ monomers! As a consequence, a full equilibration of a polymer melt even of such short chains is not possible, since already in the Rouse model regime the relaxation time is $\tau_1 N^2$, with τ_1 the monomer reorientation time: although it is not known what τ_1 is for the artificial model of [77,78] (for real polymers it is at least of the order of 10^{-11} s), certainly we must have $\Delta t << \tau_1$ for a valid numerical integration algorithm, and hence it follows that Rigby and Roe [77,78] could not simulate their model up to the Rouse time of their model for $N = 50$. Of course, the real dynamical response of polymer melts is dominated by entanglement effects, and hence chain lengths much larger than N_e, the entanglement polymerization index which is between 100 and 200, need to be simulated, such as $N \approx 10^3$. Since in this regime the relaxation time is $\tau_{rep} \approx \tau_1 N_e^{-1} N^3 \approx 10^{-4}$ s or larger, one would have to follow the simulation over at least 10^{10} integration steps, which is many orders of magnitude more than is possible today.

Thus, the way to proceed is to sacrifice some of the chemical-molecular details by simulating *coarse-grained* polymer models, where the basic effective unit is one "Kuhn segment" formed by several chemical monomers already, and the time scale for the vibrations in the length of this segment can be adjusted to the time scale of its orientational relaxation [27, 86]. Since this simpler model could be simulated over up to 10^7 time steps, and due to this rescaling, the effective $N_e^{(\mathrm{eff})}$ is smaller (one finds $N_e^{(\mathrm{eff})} \approx 35$ Kuhn segments corresponds to the actual N_c), chain lengths $N \approx 150$ suffice to study the dynamics of entangled melts (page 255). We have also mentioned some earlier attempts choosing different models to emphasize the point that the art of inventing the properly simplified model, which still contains the essential physics of the problem but at the same time is computationally tractable, is entirely nontrivial.

An even higher level of idealization, in our view, is necessary if one studies phase transitions in dense polymer systems, such a unmixing of blends, or the onset of cyrstalline or liquid crystalline order (pages 257–287). For such problems, useful work has been done only for very simplified lattice models. The emphasis of this work has usually been to test simplified analytical theories which have used precisely the same lattice model, such as theories of Flory and Huggins [98,99] and Flory [173]. Limitations of approximations used by these theories then are clearly recognized, and ideas for their improvement are gained. This line of research contributes to a molecular level understanding only indirectly. Recall that concentration and temperature dependencies of effective parameters such as the Flory-Huggins χ-parameter to some extent are often phenomenologically linked to molecular origin, while the simulations show that part of these dependencies is spurious, resulting solely from too simplified statistical mechanics. Many of the features seen in the simulations go beyond the analytical theories and can be used to interpret phenomena in real systems qualitatively. Thus, also the Monte Carlo simulation of lattice models is a very useful tool for understanding real polymers. Note that results such as the simulated time-dependent structure factor $S(q,t)$ for spinodal decomposition (Fig. 27) looks very similar to corresponding experimental data, but there the physical time scale of the process is of the order of 10^2 s, and thus one is even more remote from the possibility of microscopically realistic simulations.

These remarks carry over to more complicated problems, such as elastomers—where the average distance between chemical crosslinks introduces an additional large length scale—adsorbed polymers on walls and other interfacial problems, such as the interface between amorphous and crystalline polymers. Again very simplified models are only accessible to the computer simulation, but this nevertheless allows a useful and significant test of the corresponding theories. With the continuous increase in computer power, very useful further work should be possible in the future: in these fields, the modeling by simulations only just has begun!

ACKNOWLEDGMENTS

Part of the work described here is based on a longstanding and fruitful colloboration with A. Baumgärtner, K. Kremer, and A. Sariban. It is a great pleasure to thank them. This work is supported in part by the Deutsche Forschungsgemeinschaft under Grant No. SFB 262/D1, by the Materialwissenschaftliches Forschungszentrum of the University of Mainz, and by the Fonds der Chemischen Industrie im VCI, Frankfurt.

REFERENCES

1. F. T. Wall, S. Windwer, and P. J. Gans, in *Methods in Computational Physics*, vol 1 (B. Alder, S. Fernbach, and M. Rotenberg, eds.), Academic Press, New York (1963).
2. S. Windwer, in *Marcov Chains and Monte Carlo Calculations in Polymer Science* (G. C. Cowry, ed.) Marcel Dekker, New York (1970).
3. A. Baumgärtner, in *Applications of the Monte Carlo Method in Statistical Physics*, Chap. 5 (K. Binder, ed.), Springer, Berlin and New York, (1984).
4. K. Kremer and K. Binder, *Comput. Phys. Rept.*, *7*, 259 (1988).
5. K. Binder (ed.), *Monte Carlo Methods in Statistical Physics*, Springer, Berlin and New York (1979).
6. K. Binder (ed.), *Applications of the Monte Carlo Method in Statistical Physics*, Springer, Berlin and New York (1984).

7. K. Binder and D. W. Heermann, *Monte Carlo Simulation in Statistical Physics: Introduction*, Springer, Berlin and New York (1988).
8. D. W. Heermann, *Introduction to Computer Simulation Methods in Theoretical Physics* Springer, Berlin and New York (1986).
9. M. P. Allen and D. J. Tilesley, *Computer Simulations of Liquids* Clarendon Press, Oxford (1987); G. Cicotti and W. G. Hoover (eds.), *Molecular Dynamics Simulations of Statistical—Mechanical Systems*, North-Holland, Amsterdam (1986).
10. G. Ciccotti, D. Frenkel, and I. R. McDonald (eds.), *Simulation of Liquids and Solids* (Molecular Dynamics and Monte Carlo Methods), North-Holland, Amsterdam (1987).
11. W. G. Hoover, *Molecular Dynamics*, Springer, Berlin and New York (1987); R. W. Hockney and J. W. Eastwood, *Computer Simulation Using Particles*, Adam Hilger, Bristol (1988).
12. N. Metropolis, A. W. Rosenbluth, M. N. Rosenbluth, A. N. Teller, and E. Teller, *J. Chem. Phys.*, *21*, 1087 (1953).
13. A. Baumgärtner and K. Binder, *J. Chem. Phys.*, *75*, 2994 (1981).
14. A. Baumgärtner, *Ann. Rev. Phys. Chem.*, *35*, 419 (1984).
15. M. Bishop, D. Ceperley, H. L. Frisch, and M. H. Kalos, *J. Chem. Phys.*, *76*, 1557 (1982).
16. P. H. Verdier and W. H. Stockmayer, *J. Chem. Phys.*, *36*, 227 (1962).
17. P. H. Verdier, *J. Chem. Phys.*, *45*, 2122 (1966).
18. A. Sariban and K. Binder, *J. Chem. Phys.*, *86*, 5859 (1987).
19. A. K. Kron, *Polymer Sci. USSR*, *7*, 1361 (1965).
20. F. T. Wall and F. Mandel, *J. Chem. Phys.*, 63, 4592 (1975).
21. A. Baumgärtner and D. W. Heermann, *Polymer*, *27*, 1777 (1986).
22. M. Lal, *Mol. Phys. 17*, 57 (1969).
23. B. Mac Donald, N. Jan, D. L. Hunter, and M. O. Steinitz, *J. Phys. A*, *18*, 2627 (1985).
24. N. Madras and A. D. Sokal, *J. Stat Phys.*, *50*, 109 (1988).
25. I. Carmesin and K. Kremer, *Macromolecules*, *21*, 2819 (1988).
26. I. Carmesin and K. Kremer, in *Polymer Motion in Dense Systems*, p. 214 (D. Richter and T. Springer, eds.) Springer, Berlin and New York (1988).
27. G. S. Grest and K. Kremer, *Phys. Rev. A*, *33*, 3628 (1986).
28. P. J. Flory, *Principles of Polymer Chemistry*, Cornell University Press, Ithaca, NY (1953).
29. A. Sariban and K. Binder, *Macromolecules*, *21*, 711 (1988).
30. A. Sariban and K. Binder, *Colloid Polym. Sci.*, *266*, 389 (1988).
31. A. Sariban and K. Binder, *Colloid Polym. Sci.*, *267*, 469 (1989).
32. H. J. Hilhorst and J. M. Deutch, *J. Chem. Phys.*, *63*, 5153 (1975); H. Boots and J. M. Deutch, *J. Chem. Phys.*, *67*, 4608 (1977).
33. M. T. Gurler, C. C. Crabb, D. M. Dahlin, and J. Kovac, *Macromolecules*, *16*, 398 (1983); C. Stokeley, C. C. Crabb, and J. Kovac, *Macromolecules*, *19*, 860 (1986).
34. N. Madras and A. D. Sokal, *J. Stat. Phys.*, *47*, 573 (1987).

35. A. Beretti and A. D. Sokal, *J. Stat. Phys.*, *40*, 483 (1985).
36. P. E. Rouse, *J. Chem. Phys.*, *21*, 127 (1953).
37. K. Kremer, *Macromolecules*, *16*, 1632 (1983).
38. A. Baumgärtner and K. Binder, *J. Chem. Phys.*, *71*, 2541 (1979).
39. A. Baumgärtner and D. Y. Yoon, *J. Chem. Phys.*, 79, 521 (1983).
40. D. Y. Yoon and A. Baumgärtner, *Macromolecules*, *17*, 2864 (1984).
41. B. Berg and D. Förster, *Phys. Lett. B*, *106*, 323 (1981).
42. C. Aragão de Carvalho and S. Caracciolo, *J. Phys. (Paris)*, *44*, 323 (1983).
43. C. Aragão de Carvalho, S. Caracciolo, and J. Fröhlich, *Nucl. Phys. B*, *215* [FS7], 209 (1983).
44. S. Redner and P. J. Reynolds, *J. Phys. A*, *14*, 2679 (1981).
45. O. F. Olaj and W. Lantschbauer, *Makromol. Chem. Rapid Commun.*, *3*, 847 (1982).
46. M. L. Mansfield, *J. Chem. Phys.*, *77*, 1554 (1982).
47. T. Pakula and S. Geyler, *Macromolecules*, *20*, 2909 (1987); T. Pakula, *Macromolecules*, *20*, 679 (1987).
48. T. Pakula, *Coll Polym. Sci.*, *75*, 171 (1987).
49. W. G. Madden, *J. Chem. Phys.*, *88*, 3934 (1988); *87*, 1405 (1987).
50. H. G. Elias, *Makromoleküle: Struktur-Eigenschaften-Synthesen-Stoffe*, *Technologie*, Hüthig and Wepf, Basel (1981).
51. P. J. Flory, *Statistical Mechanics of Chain Molecules*, Interscience, New York (1966).
52. D. N. Theodorou and U. W. Suter, *Macromolecules*, *19*, 379 (1986).
53. D. N. Theodorou and U. W. Suter, *Macromolecules*, *18*, 1467 (1985).
54. M. G. Davidson, U. W. Suter, and W. M. Deen, *Macromolecules*, *20*, 1141 (1987).
55. M. Doi and S. F. Edwards, *The Theory of Polymer Dynamics*, Clarendon Press, Oxford (1986).
56. J. D. Ferry, *Viscoelastic Properties of Polymers*, Wiley, New York (1980).
57. A. Bellemans and E. De Vos, *J. Polym. Sci. Symp.* No. *42*, 1195 (1973).
58. H. Okamoto and A. Bellemans, *J. Phys. Soc. Japan* *42*, 955 (1979).
59. H. Okamoto, K. Itoh, and T. Araki, *J. Chem. Phys.*, *78*, 975 (1983).
60. R. Dickman and C. K. Hall, *J. Chem. Phys.*, 85, 3023 (1986).
61. H. Okamoto, *J. Chem. Phys.*, *64*, 2686 (1975); *70*, 1690 (1979).
62. H. Okamoto, *J. Chem. Phys.*, *79*, 3976 (1983).
63. H. Okamoto, *J. Chem. Phys.*, *83*, 2587 (1985).
64. H. Okamoto, *J. Chem. Phys.*, *88*, 5095 (1988).
65. B. Widom, *J. Chem. Phys.*, *39*, 2808 (1963).
66. H. Meirovitch, *J. Phys. A*, *15*, L735 (1982).
67. H. C. Öttinger, *Macromolecules*, *18*, 93 (1985).
68. H. C. Öttinger, *Macromolecules, 18*, 1348 (1985).
69. P. G. Khalatur, S. G. Pletneva, and Y. G. Papulov, *J. Chem. Phys.*, *83*, 97 (1984).
70. R. Dickman, *J. Chem. Phys.*, *86*, 2246 (1987).
71. A. Hertanto and R. Dickman, *J. Chem. Phys*, *89*, 7577 (1988).

72. B. J. Alder and T. E. Wainwright, *J. Chem. Phys.*, *33*, 1439 (1960).
73. J. P. Ryckaert and A. Bellemans, *Chem. Phys. Lett.*, *30*, 123 (1975); *Faraday Discuss.*, *66*, 95 (1978).
74. J. H. R. Clarke and D. Brown, *Mol. Phys.*, *58*, 815 (1986).
75. M. Bishop, M. K. Kalos, and H. L. Frisch, *J. Chem. Phys.*, *70*, 1299 (1979); *79*, 3500 (1983).
76. Yu. Yu. Gotlib, *Macromolecules*, 13, 602 (1980).
77. D. Rigby and R. J. Roe, *J. Chem. Phys.*, *87*, 7285 (1987).
78. D. Rigby and R. J. Roe, *J. Chem. Phys.*, *89*, 5280 (1988).
79. B. Jung, and B. L. Schürmann, *Macromolecules*, *22*, 477 (1989).
80. T. Schneider and E. Stoll, *Phys. Rev.*, *B17*, 1302 (1978).
81. D. Ceperley, M. H. Kalos, and J. L. Lebowitz, *Phys. Rev. Lett.*, 41, 313 (1978).
82. D. Ceperley, M. H. Kalos, and J. L. Lebowitz, *Macromolecules*, *14*, 1472 (1981).
83. T. A. Weber, *J. Chem. Phys.*, *69*, 2347 (1978); *70*, 4277 (1979).
84. T. A. Weber and A. Helfand, *J. Chem. Phys.*, *71*, 4760 (1979); *87*, 2881 (1983).
85. E. Helfand, Z. Wasserman, and T. Weber, *Macromolecules*, *13*, 526 (1980).
86. K. Kremer, G. S. Grest, and I. Carmesin, *Phys. Rev. Lett.*, *61*, 566 (1988).
87. K. Kremer, and G. S. Grest, *J. Chem. Phys.*, 92, 5057 (1990).
88. L. Verlet, *Phys. Rev.*, *159*, 98 (1967).
89. For a comparative discussion of various algorithms for the approximate integration of molecular dynamics equations, see also J. Kushick and B. J. Berne, in *Modern Theoretical Chemistry*, vol. 6, *Statistical Mechanics B*, Chap. 2 (B. J. Berne, ed.), Plenum Press, New York (1977).
90. W. F. van Gunsteren, H. J. C. Berendsen, F. Colonna, D. Perahia, J. P. Hollenberg, and D. Lellouch, *J. Comput. Chem.*, 5, 272 (1984).
91. O. Teleman and B. Jönsson, *J. Comput. Chem.*, *7*, 58 (1986).
92. P. G. de Gennes, *Scaling Concepts in Polymer Physics*, Cornell University Press, Ithaca, NY (1979).
93. M. Bishop, M. H. Kalos, A. D. Sokal, and H. L. Frisch, *J. Chem. Phys.*, *79*, 3496 (1983).
94. E. De Vos and A. Bellemans, *Macromolecules*, *7*, 812 (1974).
95. E. De Vos and A. Bellemans, *Macromolecules*, *8*, 651 (1975).
96. A. Sariban and K. Binder, unpublished.
97. M. Bishop, D. Ceperley, H. L. Frisch, and M. H. Kalos, *J. Chem. Phys.*, *72*, 3228 (1980).
98. P. J. Flory, *J. Chem. Phys.*, *9*, 660 (1941).
99. M. L. Huggins, *J. Chem. Phys.*, *9*, 440 (1941).
100. M. G. Bawendi and K. F. Freed, *J. Chem. Phys.*, *88*, 2741 (1988).
101. D. Richter, A. Baumgärtner, K. Binder, B. Ewen, and J. B. Hayter, *Phys. Rev. Lett.*, *47*, 109 (1981).
102. D. Richter, A. Baumgärtner, K. Binder, B. Ewen, and J. B. Hayter, *Phys. Rev. Lett.*, *48*, 1695 (1982).
103. K. Kremer and K. Binder, *J. Chem. Phys.*, *81*, 6381 (1984).

104. A. Baumgärtner, *Faraday Symp. Chem. Soc.*, *18*, 221 (1983).
105. C. C. Crabb and J. Kovac, *Macromolecules*, *18*, 1430 (1985).
106. J. M. Deutsch, *Phys. Rev. Lett.*, *54*, 56 (1985); *J. Phys. (Paris)*, *48*, 141 (1987).
107. A. Kolinski, J. Skolnick, and R. Yaris, *J. Chem. Phys.*, *86*, 1567 (1987).
108. A. Kolinski, J. Skolnick, and R. Yaris, *J. Chem. Phys.*, *86*, 7164 (1987).
109. A. Kolinski, J. Skolnick, and R. Yaris, *J. Chem. Phys.*, *86*, 7174 (1987).
110. P. G. de Gennes, *J. Chem. Phys.*, *55*, 572 (1971).
111. P. G. de Gennes, *Macromolecules*, *9*, 587 (1976).
112. P. G. de Gennes, *J. Chem. Phys.*, *72*, 4756 (1980).
113. S. F. Edwards, *Proc. Phys. Soc.*, *91*, 513 (1967).
114. M. Doi and S. F. Edwards, *J. Chem. Soc. Faraday Trans.* 2, *74*, 1789, 1802, 1818 (1978).
115. Actually for an isolated chain in a good solvent, where the chain radii behave as $\langle R^2 \rangle \propto \langle R^2_{gyr} \rangle \propto N^{2\nu}$, with $\nu \approx 0.59$ due to excluded volume effects, dynamic scaling implies [103] $g_r(t) \propto t^{1/(1+1/2\nu)} \approx t^{0.54}$ rather than $t^{1/2}$. The data of Ref. 13 do not extend over sufficient regimes of times to distinguish between a $t^{0.5}$ law and a $t^{0.54}$ law, however. Evidence for the $t^{0.54}$ law has recently been presented in Ref. 27, and for $t^{0.6}$ in $d = 2$ in dimensions [25].
116. K. E. Evans and S. F. Edwards, *J. Chem. Soc., Faraday Trans.* 2, *77*, 1891 (1981).
117. J. Naghizadeh and S. Kovac, *J. Chem. Phys.*, *84*, 3559 (1986).
118. S. F. Edwards and J. W. V. Grant, *J. Phys. A*, *6*, 1169, 1186 (1973).
119. W. W. Graessley, *J. Polym. Sci. Polym. Ed.*, *18*, 27 (1980).
120. T. Nose, *Phase Transitions*, *8*, 245 (1987).
121. T. Hashimoto, in *Current Topics in Polymer Science*, vol. 2 (R. M. Ottenbrite, L. A. Utrachi, and S. Inoue, eds.) p. 299 Hanser, New York (1987).
122. K. Binder, *Colloid Polym. Sci.*, *265*, 273 (1987).
123. T. Hashimoto, in *Dynamics of Ordering Processes in Condensed Matter*, p. 421 S. Komura and H. Furukawa, eds.), Plenum, New York (1988).
124. A. Gonis and L. M. Stocks (eds.), *Alloy Phase Stability*, Kluwer, Dordrecht (1989).
125. R. Konigsveld, A. Kleintjens, and E. Nies, *Croat. Chim. Acta*, *60*, 53 (1987).
126. C. Herkt-Maetzky and J. Schelten, *Phys. Rev. Lett.*, *51*, 896 (1983).
127. F. S. Bates, M. Muthukumar, G. D. Wignall, and L. J. Fetters, *J. Chem. Phys.*, *89*, 535 (1988).
128. K. Binder, *Ferroelectrics*, *73*, 43 (1987).
129. M. N. Barber, in *Phase Transitions and Critical Phenomena*, vol. 8, Chap. 2 (C. Domb and J. L. Lebowitz, eds.), Academic, New York (1983).
130. E. A. Guggenheim, *Proc. Roy. Soc. (London) A*, *183*, 201, 231 (1945).
131. K. Binder, *Colloid Polym. Sci.*, *266*, 871 (1988).
132. K. Binder, *Z. Phys. B*, *43*, 119 (1981).
133. H. E. Stanley, *An Introduction to Phase Transitions and Critical Phenomena*, Oxford University Press, Oxford (1971).
134. M. E. Fisher, *Rev. Mod. Phys.*, *46*, 597 (1974).
135. J. C. Le Guillou and J. Zinn-Justin, *Phys. Rev. B*, *21*, 3976 (1980).

136. P. G. de Gennes, *J. Phys. Lett. (Paris)*, *38*, L44 (1977); J. F. Joanny, *J. Phys. A*, *11*, L117 (1978).
137. D. Schwahn, K. Mortensen, H. Yee-Madeira, *Phys. Rev. Lett.*, *58*, 1544 (1987).
138. A. Sariban and K. Binder, *Makromol. Chem.*, *189*, 2357 (1988).
139. A. Sariban and K. Binder, *Polym. Comm.*, *30*, 205 (1989).
140. P. G. de Gennes, *J. Chem. Phys.*, *72*, 4756 (1980); P. Pincus, *J. Chem. Phys.*, *75*, 1996 (1981); K. Binder, *J. Chem. Phys.*, *79*, 6387 (1983).
141. K. Binder, in Ref. 124.
142. W. Jilge, I. Carmesin, K. Kremer, and K. Binder, *Macromolecules*, *23*, 5001 (1990).
143. J. R. Fox and H. C. Andersen, *Ann. N.Y. Acad. Sci.*, *371*, 123 (1981).
144. Y. Waseda, *The Structure of Non-Crystalline Materials, Liquids and Amorphous Solids*, McGraw-Hill, New York (1980).
145. J. H. Gibbs and E. di Marzio, *J. Chem. Phys.*, *28*, 373 (1958).
146. R. D. de la Batie, J.-L. Viovy, and L. Monnerie, *J. Chem. Phys.*, *81*, 657 (1984).
147. H.-P. Wittmann, K. Kremer, and K. Binder, *J. Chem. Phys.* (in press).
148. F. Geny and L. Monnerie, *J. Polym. Sci. Polym. Phys. Ed.*, *17*, 131 (1979).
149. M. H. Cohen and G. S. Grest, *Phys. Rev. B*, *20*, 1077 (1979).
150. G. S. Grest, K. Kremer, and T. A. Witten, *Macromolecules*, *20*, 1376 (1987); G. S. Grest, K. Kremer, S. T. Milner, and T. A. Witten, *Macromolecules*, *22*, 1904 (1989).
151. R. J. Needs and S. F. Edwards, *Macromolecules*, *16*, 1492 (1983).
152. D. Stauffer, A. Coniglio, and M. Adam, *Adv. Polym. Sci.*, *44*, 103 (1982).
153. F. Family and D. P. Landau (eds.), *Kinetics of Aggregation and Gelation*, North-Holland, Amsterdam (1984).
154. D. Stauffer, *An Introduction to Percolation Theory*, Taylor and Francis, London (1985).
155. D. B. Adolf and J. G. Curro, *Macromolecules*, *20*, 1646 (1987).
156. For simulations of chains diffusing around irregularly spaced obstacles see K. Kremer, *Z. Phys. B*, *45*, 149 (1981); A. Baumgärtner and M. Muthukumar, *J. Chem. Phys.*, *87*, 3082 (1987).
157. P. J. Flory, *J. Chem. Phys.*, *66*, 5720 (1977).
158. Y.-K. Leung and B. E. Eichinger, *J. Chem. Phys.*, *80*, 3877, 3885 (1984); N. A. Neuburger and B. E. Eichinger, *J. Chem. Phys.*, *83*, 884 (1985).
159. F. L. Mc Crackin, *J. Chem. Phys.*, *47*, 1980 (1967); F. T. Wall, F. Mandel, and J. C. Chin, *J. Chem. Phys.*, *65*, 2231 (1976).
160. I. Webman, J. L. Lebowitz, and M. H. Kalos, *J. Phys. (Paris)*, *41*, 579 (1980).
161. A. T. Clark and M. Lal, *J. Chem. Soc. Faraday Trans.* 2, *77*, 981 (1981).
162. E. Eisenriegler, K. Kremer, and K. Binder, *J. Chem. Phys.*, *77*, 6292 (1982).
163. K. Kremer, *J. Chem. Phys.*, *83*, 5882 (1985).
164. F. van Dieren and K. Kremer, *Europhys. Lett.*, *4*, 569 (1987).
165. S. Lione and H. Meirovitch, *J. Chem. Phys.*, *88*, 4498 (1988); H. Meirovitch and S. Lione, *J. Chem. Phys.*, *88*, 4507 (1988).

166. R. Dickman and C. K. Hall, *J. Chem. Phys.*, *89*, 3168 (1988).
167. T. Cosgrove, T. Heath, B. van Lent, F. Leermakers, and J. Scheutjens, *Macromolecules*, *20*, 1692 (1987).
168. M. Murat and G. S. Grest, *Macromolecules*, *22*, 4054 (1989).
169. J. K. Percus, *J. Stat. Phys. 15*, 423 (1976); J. R. Hendersen and F. van Swol, *Mol. Phys.*, *51*, 991 (1984), and references therein.
170. J. M. H. M. Scheutjens and G. J. Fleer, *J. Phys. Chem.*, *83*, 1619 (1979); *84*, 178 (1980).
171. S. T. Milner, T. A. Witten, and M. E. Cates, *Europhys. Lett.*, *5*, 413 (1988); *Macrmolecules*, *27*, 2610 (1988).
172. S. Alexander, *J. Phys. (Paris)*, *38*, 983 (1977).
173. P. J. Flory, *Proc. Roy. Soc. (London) A*, *234*, 60 (1956).
174. J. F. Nagle, *Proc. Roy. Soc. (London) A*, *337*, 569 (1974).
175. P. D. Gujrati, *J. Stat Phys.*, *28*, 441 (1982).
176. A. Baumgärtner, *J. Chem. Phys.*, *81*, 484 (1984).
177. A. Baumgärtner, *J. Phys. A*, *17*, L971 (1984).
178. A. Baumgärtner, *J. Phys. (Paris) Lett.*, *46*, L-659 (1985).
179. A. Kolinski, J. Skolnick, and R. Yaris, *Macromolecules*, *19*, 2550 (1986).
180. D. Y. Yoon and P. J. Flory, *Polymer*, *18*, 509 (1977); *Faraday Disc. Chem. Soc.*, *68*, 288 (1979).
181. R. M. J. Cotterill, *Biochim. Biophys. Acta*, *433*, 264 (1976); S. Toxvaerd, *J. Chem. Phys. Phys.*, *67*, 2056 (1977); S. H. Northrup and M. S. Curvin, *J. Phys. Chem.*, *89*, 4707 (1985); P. van der Ploeg and H. J. C. Berendsen, *J. Chem. Phys.*, *76*, 3271 (1982); *Mol. Phys.*, *49*, 233 (1983); J. P. Bareman, G. Cardini, and M. L. Klein, *Phys. Rev. Lett.*, *60*, 2152 (1988); H. J. C. Berendsen, in Ref. 9, p. 496.
182. H. L. Scott, *Biochim. Biophys. Acta*, *469*, 264 (1977); V. Buscio and M. Vacatello, *Mol. Cryst. Liq. Cryst.*, *97*, 195 (1983); D. P. Fraser, R. W. Chantrell, D. Melville, and D. J. Tildesley, *Liquid Crystals*, *3*, 423 (1988).
183. O. G. Mouritsen and M. J. Zuckermann, *Chem. Phys. Lett.*, *135*, 294 (1987); *Eur. Biophys., J.*, *12*, 75 (1985); *15*, 77 (1987); L. Cruzeiro-Hansen and O. G. Mouritsen, *Biochim. Biophys. Acta*, *944*, 63 (1988).
184. O. G. Mouritsen, in *Molecular Description of Biological Membrane Components by Computer Aided Conformational Analysis* (R. Brasseur, ed.), CRC Press, Boca Raton, FL (1989).
185. J. P. Ryckert, G. Cicotti, and H. J. C. Berendsen, *J. Comp. Phys.*, *25*, 327 (1977); W. F. van Gunsteren and H. J. C. Berendsen, *Mol. Phys.*, *34*, 1311 (1977).

6

Free-Volume Theory and Its Application to Polymer Relaxation in the Glassy State

Richard E. Robertson
The University of Michigan
Ann Arbor, Michigan

INTRODUCTION

The concept of free volume and its association with movement is ancient. Indeed, it is as old as the atomic theory itself and is in part responsible for it.

The doctrine of what was afterwards known as the "elements"—fire, air, earth, and water—had already been discarded (but not for the last time) by the earliest Greek philosophers, the Milesians, including the person traditionally thought of as being the first philosopher, Thales (c. 585 B.C.), and by Pythagoras (c. 530 B.C.) and his school. Rather, these early philosophers regarded all of the material world as being states of a single substance, a substance that presented different apperances according as it was more or less rarefied or condensed [1]. These early philosophers had formed the concept of an eternal matter out of which all things are produced and into which all things return. Having assembled the two great concepts of matter and form, the next problem to be faced was that of movement.

At first, movement had simply been taken for granted. But when the new concept of eternal matter began to be taken seriously, difficulties appeared. If reality were regarded as continuous, there seemed no possibility for anything else, not even for empty space. Empty space could only be identified with the unreal, and it was easy to show that the unreal could not exist. But without empty space, motion seemed impossible, and the world of which we suppose ourselves to be aware, with the movement of air, of water, and of ourselves, must be an illusion. This was the position taken by Parmenides of Elea (c. 475 B.C.). Parmenides had begun as a Pythagorean and was led to apply the rigorous method of reasoning introduced into geometry with such success by the Pythagoreans to the question that had occupied the Milesians, the nature of the world. Parmenides cleared up the ambiguity, not by affirming the existence of empty space, but by denying its possibility. The concept of matter that had been elaborated by his predecessors was shown to lead to the conclusion that reality was continuous, finite, and spherical. There could be nothing outside it and no empty space within it; it was a One. For such a reality, motion was impossible, and the world of the senses was therefore an illusion [2]. The case was further reinforced by Zeno of Elea with arguments and paradoxes of a mathematical nature, including the famous paradox of Achilles and the tortoise. (Achilles can never overtake the tortoise because every time he reaches the point where the tortoise had been, the tortoise has moved on.)

It was the rescue of "reality" that the doctrine of atomism was introduced. The first hint was given by another philosopher of Elea, Melissus (c. 444 B.C.). Melissus still followed the reasoning of Parmenides, but went beyond him and argued that Being is infinite in extent and that it is incorporeal, because otherwise it would have parts, implying plurality. Further, if there were plurality, then all would have to be as unchanging as the One. It was now a short but enormous step that was taken by Leucippus of Miletus (c. 440 B.C.). He asserted that empty space exists just as much as Being and that matter consists of minute particles that do indeed resemble the One in being indivisible (atoma), indestructible, and qualitatively neutral. The atoms were thought to differ from one another only in shape and orientation, and different rearrangements of them produced the effect of changeable qualities such as color, heat, hardness, etc. [3]. This atomist system was later promulgated by the prolific writer Democritus (c. 430 B.C.), to whom we often ascribe the origination of the theory instead of to Leucippus.

A succinct summary of the compelling relationship between movement, atoms, and free volume (the vacuity in condensed matter) was given by the poet Lucretius. Lucretius, though writing in first-century B.C. Rome, recorded with fidelity the thought of Epicurus (341–271 B.C.) in physics. The Epicurean system still vividly alive in first-century B.C. Rome had changed little from that of the time of Epicurus [4].

Yet everything is not held close and packed everywhere in one solid mass, for there is void in things: this knowledge will be useful to you in many matters, and will not allow you to wander in doubt and always to be at a loss as regards the universe. . . . Therefore there is intangible space, void, emptiness. But if there were none, things could not in any way move; for that which is the province of body, to prevent and to obstruct, would at all times be present to all things; therefore nothing would be able to move forward, since nothing could begin to give place. But as it is, we discern before our eyes, throughout seas and lands and the heights of heaven, many things moving in many ways and various manners, which, if there were no void, would not so much lack altogether their restless motion, as never would have been in any way produced at all, since matter would have been everywhere quiescent packed in one solid mass. . . .

And here in this matter I am driven to forestall what is imagined by some,* lest it should lead you also away from the truth. They say that water yields to the pressure of scaly creatures and opens liquid ways, because fish leave room behind them for the yielding waves to run together. Also other things are able to move in and out and to change place, although all is full. You must know that this has been accepted for wholly false reasons. For to where, I ask, will the scaly fish be able to move forward, unless the water shall give way? Into what place, again, will the water be able to move back, when the fish is unable to go? Either, then, all bodies must be deprived of movement, or we must say that void is intermingled in things, as a result of which, each thing may begin to move. . . .

Therefore, however you may demur by making many objections, you must confess, nevertheless, that there is void in things. Many another proof besides I can mention to scrape together credit for my doctrines. But for a keen-scented mind, these little tracks are enough to enable you to recognize the others for yourself. For as hounds very often find by their scent the leaf-hidden resting-place of the mountain-ranging quarry, when once they have hit upon certain traces of its path, so will you be able for yourself to see one thing after another in such matters as these, and to penetrate all unseen hiding-places, and draw forth from them the truth [5].

FREE VOLUME

The concept of free volume for explaining motion in condensed phases is no less compelling for us than it was for the ancient philosophers. Where we differ is our modern attempt to quantify the relationship between free volume and motion.

---

* These included Plato and Aristotle. With the exclusive adoption in the West of the thought of the biologist, Aristotle, advanced Greek thought in physics was not recovered for many centuries.

The principal problem in quantifying the relationship between motion and free volume is the quantification of free volume. Free volume is not directly measurable but must be deduced from other measurements, such as that of the total volume. Early experimental results suggested a simple relationship between free volume and total volume. The viscosity or fluidity (the reciprocal of viscosity) of simple van der Waals liquids was found to be essentially independent of temperature when held at constant volume [6]. Assuming this to be true (and we will come back to this) and that the viscosity reflects the magnitude of the free volume, this suggests that the free volume remains constant at constant volume. This means that the total volume minus the free volume, the "occupied" volume, is independent of temperature and that it is just that, the volume occupied by the molecules. This simple decomposition of the total volume V into a constant occupied volume V_0 and the free volume f suggests that the free volume can be obtained from

$$f = V - V_0 \tag{1}$$

Allowing for the possibility that the occupied volume increases slightly with increasing temperature, as vibrational motion of increasing amplitude within the molecule causes it to enlarge, this decomposition of the total volume is suggested to be that indicated in Figure 1.

On the question of what to use for the magnitude of the occupied volume V_0, Doolittle and Doolittle [7], for example, cite three possible definitions

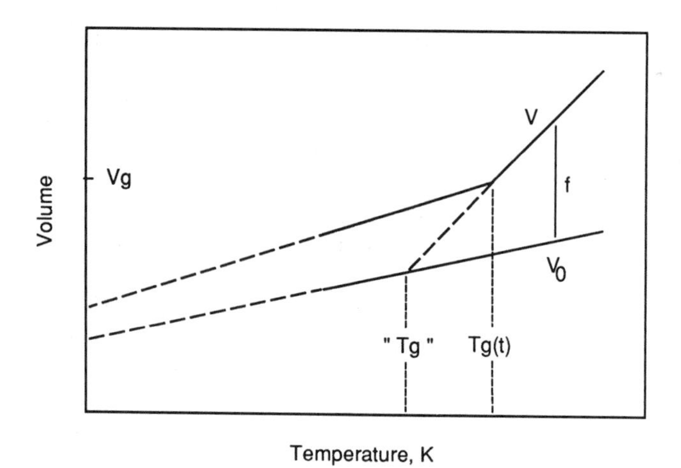

Figure 1 Volume, occupied volume, and free-volume versus temperature. $T_g(t)$ is the usual glass-transition temperature, and "T_g" is the presumed glass transition temperature after an infinite time. (After Ref. 54.)

of "occupied" volume that were suggested by Bondi [8] and Hildebrand [9]: (1) the volume calculated from the van der Waals dimensions, (2) the crystalline volume at 0 K, and (3) the total volume minus the "fluctuation volume." The latter is the volume swept out by the center of gravity of the molecules as a result of thermal motion. Even among these three definitions, the occupied volume still can be defined in various ways, however. For example, the occupied volume used by Doolittle was the volume of the liquid extrapolated *without change of phase* to 0 K [10].

Let us return to examine more closely the constancy of the viscosity at constant volume that had been suggested by the early experiments. We can consider the data of Bridgman [11] for example. Figure 2 shows the viscosity versus density for two simple van der Waals liquids, carbon disulfide and diethyl ether. Data is shown for 30° and 75°C. The relative viscosity is the measured viscosity divided by the viscosity at 30°C and 1 bar pressure. The viscosity versus pressure for these two liquid is shown in Figure 3. At low density (low pressure), the viscosities nearly, but do not quite, superimpose at constant density. At higher densities and pressures, the viscosities depend increasingly strongly on temperature; indeed, the dependence on density is nearly lost as the temperature dependence of viscosity at constant volume (in Fig. 2) approaches that at constant pressure (in Fig. 3). These experiments under elevated pressure show also that the occupied volume is not a fixed, hard sphere-like quantity. Under the pressure of 12,000 kg/cm^2 ($\approx$1200 MPa), the total volumes of both carbon disulfide and diethyl ether have been made to decrease by roughly 30% from that at 1 bar pressure. This is considerably larger than the usual magnitude of the free volume. The dependence of the occupied volume on pressure and on temperature severely limit the utility of the simple decomposition of volume given by Eq. (1).

There have been a number of attempts to measure the free volume in glassy polymers directly. Small-angle x-ray scattering has been used by Roe et al. to determine density fluctuations and then to obtain a measure of free volume size distributions [12,13]. From density fluctuation data and a simple model of a random distribution of free-volume sizes, Roe and Song have estimated the weight-average size of the holes in poly(methyl methacrylate) to have a volume of 0.105 nm^3, for example [13].

Positronium annihilation studies provides estimates of hole sizes from the lifetimes of positronium [14]. Specifically, the lifetime of *ortho*-positronium increases when the dimensions of free-volume holes increase. Using calibration curves to relate free-volume sizes to *o*-positronium lifetimes in simple van der Waals–bonded molecules [15], Jean et al. estimated the hole sizes to vary with temperature from 0.025 to 0.222 nm^3 in amine-cured epoxy resins [16]. More recently, the effect of pressure on hole size

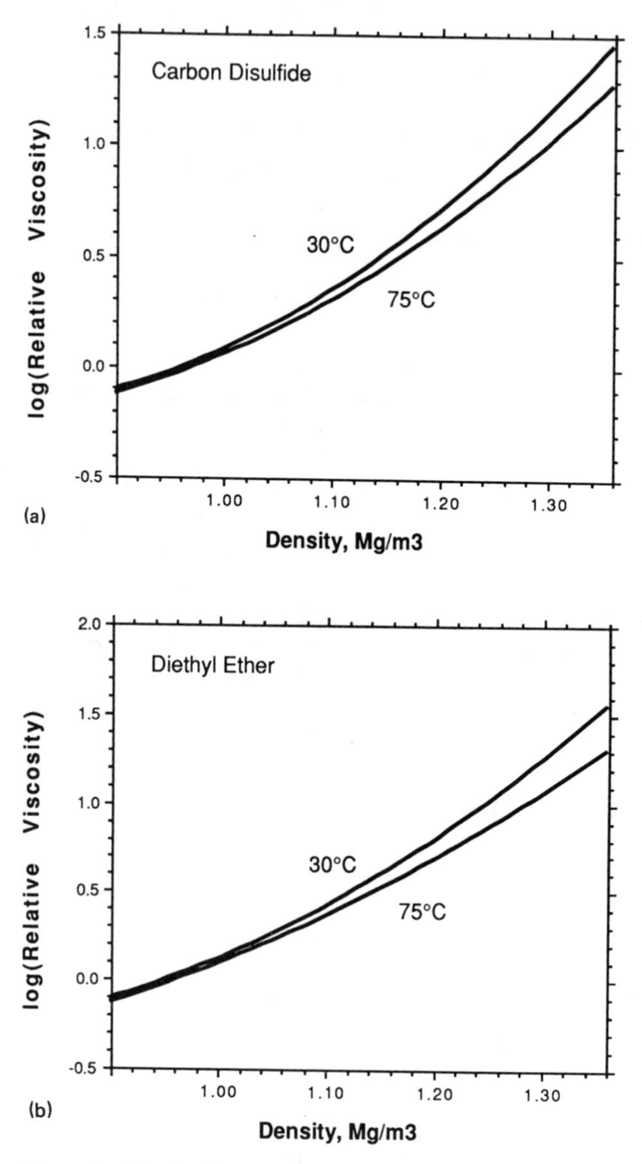

Figure 2 The log(relative viscosity) versus density at 30° and 75°C for (a) carbon disulfide and (b) diethyl ether. (Data from Ref. 11.)

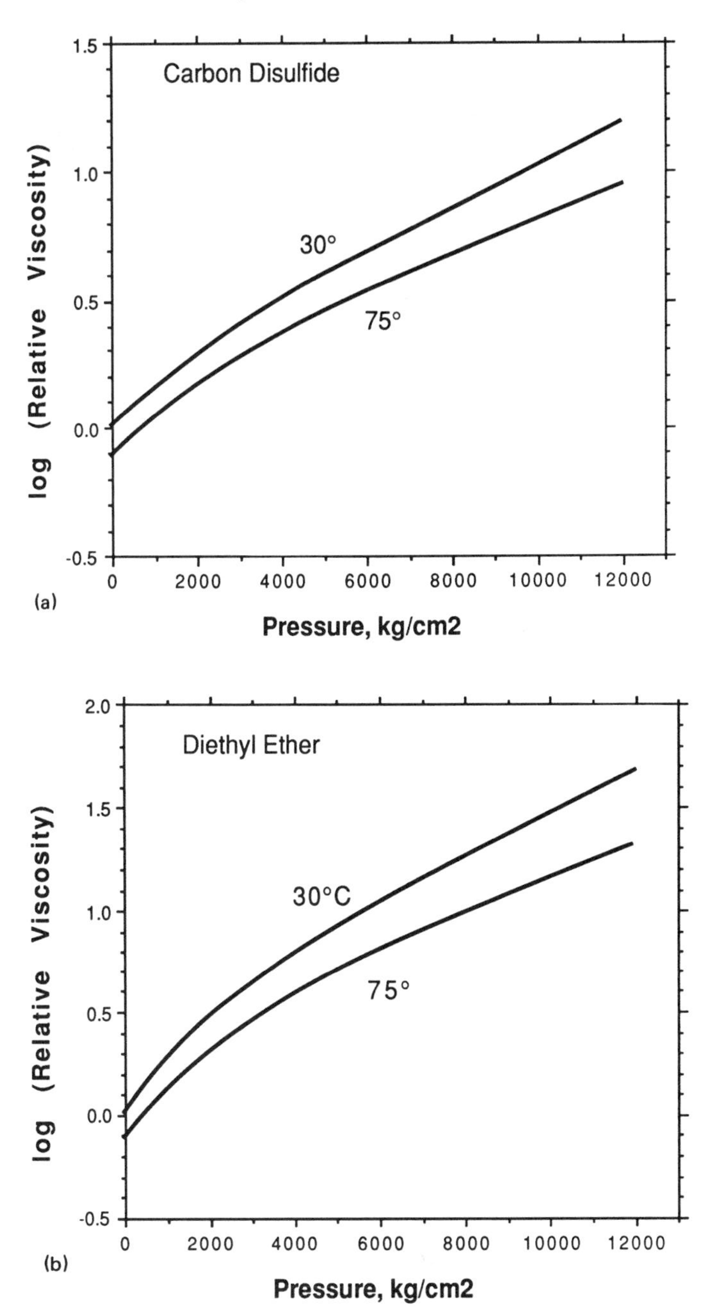

Figure 3 The log(relative viscosity) versus pressure at 30° and 75°C for (a) carbon disulfide and (b) diethyl ether. (Data from Ref. 11.)

in one of these resins was found to change at 100°C from 0.135 to 0.045 nm^3 as the pressure was increased to nearly 400 MPa [17].

For thermoplastic glassy polymers, Malhortra and Pethrick followed the analysis of Ujihara et al. [18] and estimated the hole sizes to be between 0.020 and 0.070 nm^3 in polycarbonate, polysulfone, and polyether sulfone [19]. Poly(vinyl acetate) has been studied extensively by Kobayashi et al. [20]. The lifetime of *o*-positronium and the concentration of *o*-positronium relative to that of free positrons and *p*-positronium were measured in PVAc over a large temperature range that included both the liquid and the glass. The lifetimes indicated an average hole size that ranged from 0.26 nm at 30° to 0.32 nm at 75°C (the hole volumes ranged from 0.078 to 0.134 nm^3). The concentration of *o*-positronium was related to the number of holes, and the product of the concentration and the average hole size was shown to be commensurate with the free-volume fraction computed from the Simha-Somcynsky theory (to be described in the next section).

Small-angle x-ray scattering and positronium annihilation spectroscopy are basically static measures of free volume. Although they give hole sizes directly, they do not relate this to mobility. A technique that does probe the mobility of segments and can be related to free volume hole sizes is photoisomerization.

By using a series of photoisomerizable molecules of different size, Victor and Torkelson were able to determine the average hole size as well as the distribution of hole sizes in glassy polystyrene [21]. The hole size needed for the photoisomerization of each probe was computed from a molecular model of it, and the fraction of molecules of each probe type that were adjacent to holes of the requisite size was inferred from the maximum, long-time extent of photoisomerization induced. The average hole size in unaged polystyrene at 25°C was found to be about 0.27 nm^3 with over 90% of the holes larger than about 0.10–0.12 nm^3 and few larger than 0.40 nm^3. Lamarre and Sung had demonstrated earlier that photochromic labels attached to polymer molecules also could be used to probe the free volume [22]. By attaching the probe molecules to the polymer chain at specific locations, the free volume in those locations could be determined. Although specific hole sizes were not obtained, the relative sizes of the holes in different locations were found. In glassy polystyrene, the hole sizes were shown to be largest where free, unattached probe molecules resided, smaller near probes attached to the polymer chain ends, smaller yet near probes attached to the side of the chain, near the chain center, and smallest for a probe polymerized into the chain backbone [23,24]. In dilute solution, only a small difference in the photoisomerization behavior was found at different sites of the chain.

The quantification of free volume is simplest when the moving objects

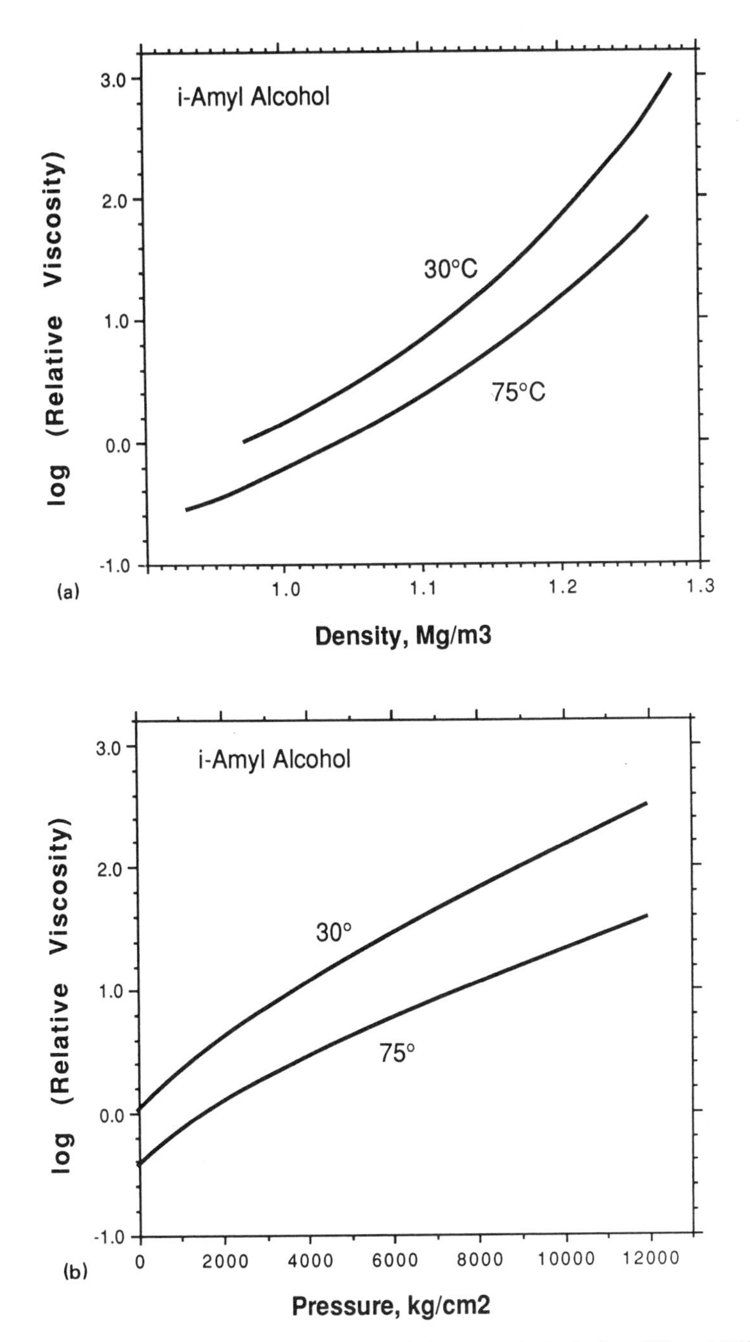

Figure 4 The log(relative viscosity) for *i*-amyl alcohol at 30° and 75°C: (a) versus density, (b) versus pressure. (Data from Ref. 11.)

are round and smooth and they move in a symmetrical force field. This nearly describes the van der Waals liquids carbon disulfide and diethyl ether. It has been approximately true also for the polymers mentioned, which are fairly flexible and, except for their one-dimensional connectivity, interact with their neighbors through nondirectional van der Waals forces. For structured liquids, where the interaction between neighbors is strongly directional, the quantification of free volume is expected to be most difficult. Consider, for example, the hydrogen-bonded liquid, *i*-amyl alcohol. The viscosities of *i*-amyl alcohol versus density and versus pressure are shown in Figure 4 [11]. Little reduction in temperature dependence is gained even at low pressures by plotting the viscosity versus density. The effect of thermal agitation swamps the effect of free volume. This problem can arise as well with stiff, irregularly shaped molecules, like stiff polymers. The free volume required for the rearrangement of a stiff sequence of segments is so large that the rate-determining step is usually the flexing of the chain instead. Macroscopically, one is confronted with this while trying to gather together the dead sticks blown off birch, elm, maple, and other trees following the storms of winter. The pieces are quite rigidly packed when the density of packing is still very low, and this occurs when the sticks are aligned to fit together as well as possible. The tightness is due, of course, to the rigid angularity and branching of the sticks, and it points up how the free volume must be gauged to the size of the moving unit and the degrees of freedom within this unit. As the temperature increases, the surmounting of rotational barriers by thermal agitation effectively shortens the stiff segment sequences.

A difficulty with using any of the above measures of free volume is their general lack of internal consistency. To avoid this difficulty, the free volume derived from the Simha-Somcynsky theory, which is taken up in the next section, will be used for calculations of the relaxation. We should remember, however, that although the viscosity may depend on free volume, and even depend principally on free volume, it need not depend exclusively on free volume.

SIMHA-SOMCYNSKY FREE VOLUME

Although liquids have well-defined thermodynamic properties, which glasses do not, even they are considerably more difficult to describe statistical mechanically than either gases or crystalline solids. As a result, statistical mechanical theories for liquids have usually been formulated in either a solidlike or a gaslike approximation. The solidlike approximation is usually expressed in terms of a cell or lattice model. The gaslike approximation involves models for pair correlation functions. The latter approach, from which the entropy and free energy can be calculated, is considered to be the more fundamental for liquids of small molecules [25].

Despite their faults, cell models retain several advantages [26]. The principal advantage being that of a simple visualization of the liquid structure and local molecular dynamics. Also, the major problem with cell models, that of predicitng the communal entropy, does not enter into the computation of the equation of state for the volume, energy, or enthalpy. For polymer molecules, the effect of a fixed cell or lattice structure on the communal entropy is less severe because only the molecule as a whole, rather than the individual segments, is able to range over the entire volume. Finally, the equations of state given by cell models can be very good. In particular, the Simha-Somcynsky equation of state [27] has been extremely successful for PVT computations [28].

Equations of state based at least in part on cell or lattice ideas have been derived also by Flory et al. [29], Sanchez and Lacombe [30], and others [31].

Simha-Somcynsky Equation of State [27]

The Simha-Somcynsky equation of state is

$$\frac{\tilde{p}\tilde{V}}{\tilde{T}} = [1 - y(2^{1/2}y\tilde{V})^{-1/3}]^{-1} + \frac{y}{\tilde{T}}[2.002(y\tilde{V})^{-4} - 2.409(y\tilde{V})^{-2}], \tag{2}$$

where $\tilde{p}$, $\tilde{V}$, and $\tilde{T}$ are reduced variables, $\tilde{p} = p/p^*$, $\tilde{V} = V/V^*$, $\tilde{T} = T/T^*$, and p^*, V^*, and T^* are the characteristic scaling parameters for each polymer material. Although the characteristic scaling parameters have statistical mechanical significance, as will be seen below, they are usually deduced by fitting p-V-T data. y is the fractional occupancy of the cells. The *fractional free volume* (the "free volume") is defined by

$$f = 1 - y. \tag{3}$$

This is, in effect, the *excess* free volume, the free volume beyond that for full occupancy of the cells. At equilibrium the free volume is not an independent variable, and its value can be determined by a minimization of the free energy, which gives the additional equation

$$1 + y^{-1}\ln(1 - y) = \left[y(\sqrt{2}y\tilde{V})^{-1/3} - \frac{1}{3}\right][1 - y(\sqrt{2}y\tilde{V})^{-1/3}]^{-1} + \frac{y}{6\tilde{T}}[2.409(y\tilde{V})^{-2} - 3.033(y\tilde{V})^{-4}]. \tag{4}$$

When the polymer is not at equilibrium, Eq. (4) is no longer valid, and a different condition is needed to reduce the number of variables in Eq. (2).

Sometimes the volume V can be estimated or determined by experiment. For example, when the temperature or pressure is instantaneously stepped to a new value, the polymer behaves like a glass, and the volume changes accordingly. The new volume is easily estimated, and this value may be used in Eq. (2) to determine an approximate y and thence an approximate free volume f. On the other hand, during structural recovery, the dynamical theory of free volume described below can yield the time-dependent free volume f. Then it is an approximate volume V that may be computed from Eq. (2) using this value for the free volume.

Brief Description of the Derivation

Attempts at formulating a polymer liquid equation of state via the cell model without the specific assumption of vacancies have been relatively unsuccessful. It was only with the inclusion of empty cells by Simha and Somcynsky that a general and useful equation of state was obtained. Assuming that the number N of s-mer molecules occupy the fraction y of the total available sites or cells, the partition function can be written as

$$Z = g(N,y)[v_f(V,y)]^{cN} \exp\left[\frac{-E_0(V,y)}{kT}\right], \tag{5}$$

where g represents the combinatory factor arising from the mixing of molecules and unoccupied sites, v_f is the cell theory free-volume function, and E_0 is the potential energy of the chain with all segments placed in their rest positions. Each cell is occupied by a single segment, and a segment is defined as a grouping of three of the total number of $3c$ external degrees of freedom attributed to the chain.

For molecules and holes placed on a lattice, the volume-dependent part of g is given by

$$g(N,y) \propto y^{-N}(1 - y)^{-Ns(1-y)/y}. \tag{6}$$

The free volume internal to the occupied sites v_f, obtained by averaging linearly over single modes of motion, equals

$$\frac{v_f}{v^*} = [y(\tilde{\omega}^{1/3} - 2^{-1/6}) + (1 - y)\tilde{\omega}^{1/3}]^3, \tag{7}$$

where $v^* = V^*/Ns$ is the characteristic volume per segment, $\tilde{\omega}$ is the reduced segmental cell volume,

$$\tilde{\omega} = \frac{yV}{v^*Ns} = \frac{yv}{v^*},$$

V is the measured volume of the N molecules, and V^* is the corresponding characteristic scaling volume. The potential energy of the system, assuming a Lennard-Jones interaction between nonbonded segments, is given by

$$2E_0 = yNqx\epsilon^*(1.011\tilde{\omega}^{-4} - 2.409\tilde{\omega}^{-2}), \tag{8}$$

where qx is the number of nearest-neighbor sites per chain, equal to $s(x - 2) + 2$, with x being the coordination number, and ϵ^* is the characteristic attraction energy per segment.

The Helmholtz free energy is then

$$F = -kT \ln Z \tag{9}$$

and the equation of state, Eq. (2), which relates p, V, and T, is obtained from the relationship between pressure and Helmholtz free energy: $p = -(\partial F/\partial V)_T$. The characteristic scaling parameters are given by $p^* = qx\epsilon^*/sv^*$, $V^* = Nsv^*$, and $T^* = qx\epsilon^*/ck$.

The subsidiary equation needed to fix the free volume f or occupancy y, Eq. (3), is obtained from $(\partial F/\partial y)_{T,V} = 0$. This yields

$$\frac{s}{3c}\left[\frac{s-1}{s} + y^{-1}\ln(1-y)\right] = \left[y(\sqrt{2}y\tilde{V})^{-1/3} - \frac{1}{3}\right] \times [1 - y(\sqrt{2}y\tilde{V})^{-1/3}]^{-1} + \frac{y}{6\tilde{T}}[2.409(y\tilde{V})^{-2} - 3.033(y\tilde{V})^{-4}] \tag{10}$$

Assuming that each repeating unit or monomer has three external degrees of freedom, then $s/3c = 1$. Therefore, for a high-molecular-weight polymer, s is very large and Eq. (10) becomes that given by Eq. (4).

Mean Free Volume

Calculation of the Mean Free Volume

The mean free volume is obtained from Eq. (3) after the occupancy is obtained. In general, the occupancy y and volume V are both obtained by solving Eqs. (2) and (4) together, assuming that p and T are known but not V. Because of the nonlinearity of these equations, however, they cannot be solved without iteration. In the FORTRAN program given below, Eqs. (2) and (4) were put in the following forms for iteration:

$$y = (y\tilde{V})^2\{\tilde{p}\tilde{V} - \tilde{T}[1 - y(\sqrt{2}y\tilde{V})^{-1/3}]^{-1}\} \times [2.022(y\tilde{V})^{-2} - 2.409]^{-1} \tag{2a}$$

$$(y\tilde{V})^2 = 3.033\left(2.409 - \left[\frac{6\tilde{T}(y\tilde{V})^2}{y}\right]\left\{1 + y^{-1}\ln(1-y) - \left[y(\sqrt{2}y\tilde{V})^{-1/3} - \frac{1}{3}\right]\left[1 - y(\sqrt{2}y\tilde{V})^{-1/3}\right]^{-1}\right\}\right)^{-1} \tag{4a}$$

The volume occurs only in the product, $y\tilde{V}$. Hence, a value of y is assumed, and Eq. (4a) is iterated for $y\tilde{V}$ (from which $\tilde{V}$ can be obtained). The same value of y and the value of $y\tilde{V}$ deduced from Eq. (4a) are then put into Eq. (2a) to find a new value of y. This process is repeated until the value of y obtained from Eq. (2a) is the same as that assumed for the last iteration of Eq. (4a). In the following program, this cycle is performed by the subroutine that finds the zero of the quantity $y' - y''$, where y'' is the value of y assumed for Eq. (4a) and y' is the value subsequently obtained from Eq. (2a).

In the following program, VOUT and FOUT are the values of V and f obtained from the program as output for input values of T and P $= p$. Other symbols are $p^* =$ PSTAR, etc., and $(y\tilde{V})^2 =$ YVV2. The occupancy, y, is expected to be slightly below one and is often around 0.9, and the program searches in a range, from YBEGIN to YEND, around this value. Since values very close to $y = 1$ cause numerical problems, the first part of the program is meant to avoid this region. The subroutine ZERO searches for a zero in the function G in the range of y from YBEGIN to YEND, and the value of y for which G is zero is given as the output value of YEND. The IMSL routine ZBRENT can be used for ZERO, for example. Any error or problem in finding the zero of G is returned in the variable IER, and the calculation can be stopped at this point.

```
      SUBROUTINE YvsT (T, P, VOUT, FOUT)
C
C     COMPUTES y VS T/T* AND P/P* FROM SIMHA-SOMCYNSKY
       EQUATIONS
C
      COMMON/STAR/PSTAR,VSTAR,TSTAR
      COMMON/VST/PP,VV,TT,YVV2
      EXTERNAL G
C
      TT = T/TSTAR
      PP = P/PSTAR
      YBEGIN = 0.99
      YEND = 0.5
 10   CONTINUE
      GX1 = G(YBEGIN)
      IF (GX1 .LT. 0) THEN
        YBEGIN = YBEGIN - 0.005
        IF (YBEGIN .LT. YEND) GO TO 20
          GO TO 10
      ENDIF
 20   CONTINUE
C
```

```
      CALL ZERO(G, YBEGIN, YEND, IER)
C
      IF (IER .GT. 100) THEN
        PRINT*,' FOR YVST, T =',T,'& P =',P,', IER =',IER
        PAUSE
      ENDIF
C
      Y = YEND
      VOUT = VSTAR*SQRT(YVV2)/Y
      FOUT = 1. - Y
C
      RETURN
      END
C
C
      REAL FUNCTION G(Y)
C
      COMMON/VST/PP,VV,TT,YVV2
      YVV2 = 0.9
C
      DO 10 I = 1,15
        YVV21 = YVV2
        YV3 = Y/((2.*YVV2)**(1./6.))
        YVV2 = 3.033/(2.409 - (YVV2/Y)
  1   *6.*TT*(1. + (1./Y)*ALOG(1. - Y)
  2   - (YV3 - 1./3.)/(1. - YV3)))
      IF (ABS((YVV2 - YVV21)/YVV2) .LE. 1.E-5) GOTO 20
 10   CONTINUE
      PRINT*,'FOR Y=',Y,',YVV2 NOT CONVERGING'
      PAUSE
 20   CONTINUE
      VV = SQRT(YVV2)/Y
      YV3 = Y/((2.*YVV2)**(1./6.))
      Y1 = YVV2*(PP*VV - TT/(1. - YV3))/(2.022/YVV2 - 2.409)
      G = Y1 - Y
C
      RETURN
      END
```

The above calculation is for the equilibrium liquid. When the system is not in equilibrium, Eqs. (4) and (4a) cannot be used, and the occupancy must be determined in some other way. Often the volume can be found independently of the equation of state, and then an approximate occupancy can be obtained from Eqs. (2) and (2a). This procedure is followed in the following FORTRAN program, and it is especially useful for quenched glasses, where for constant pressure, the volume can be obtained from

$$V = V_g[1 + \alpha_g(T - T_g)], \tag{11}$$

where V_g is the volume at the glass-transition temperature T_g, and α_g is the thermal expansion coefficient of the glass. Because Eqs. (2) and (2a) were derived under the assumption of equilibrium, they cannot strictly be used away from equilibrium. But McKinney and Simha have shown that the error in following this procedure is small [32].

In the following program, T, p = P, and V = VIN are input to the program and f = FOUT is the output.

```
      SUBROUTINE GYVST(T, P, VIN, FOUT)
C
C     COMPUTES y VS T/T* AND P/P* FROM SIMHA-SOMCYNSKY
       EQUATIONS
C     FOR GLASS, WHERE V = Vg*[1 + ag(T - Tg)]
C
         COMMON/STAR/PSTAR,VSTAR,TSTAR
         COMMON/VST/PP,VV,TT,YVV2
         EXTERNAL GG
C
         VV = VIN/VSTAR
         TT = T/TSTAR
         PP = P/PSTAR
C
C     SQRT((2.022/2.409) + 0.0001) = 0.91622
C
         YBEGIN = (1./VV) * 0.91622
         YEND = 0.999
C
         CALL ZERO(GG, YBEGIN, YEND, IER)
C
         IF (IER .GT. 100) THEN
           PRINT*,' FOR GYVST, T =',T,'& P =',P,', IER =',IER
           PAUSE
         ENDIF
C
         Y=YEND
         FOUT = 1. - Y
C
         RETURN
         END
```

```
C
C
      REAL FUNCTION GG(Y)
C
      COMMON/VST/PP, VV, TT, YVV2
C
      YVV2 = (Y*VV)**2
      YV3 = Y/((2.*YVV2)**(1./6.))
      Y1 = YVV2 * (PP*VV - TT/(1. - YV3)) / (2.022/YVV2 -
        2.409)
      GG = Y1 - Y
C
      RETURN
      END
```

An Example. Let us consider the free volume for both the liquid and glass versus temperature for polystyrene. The characteristic parameters for polystyrene are given by Quach and Simha [33]: $p^* = 7453$ bar, $V^* = 0.9598$ cm^3/g, $T^* = 12{,}680$ K. When used for a constant pressure of 1 bar in the first FORTRAN program given above, which solves Eqs. (2a) and (4a) together, these parameters give the free volume for liquid polystyrene that is shown in Figure 5.

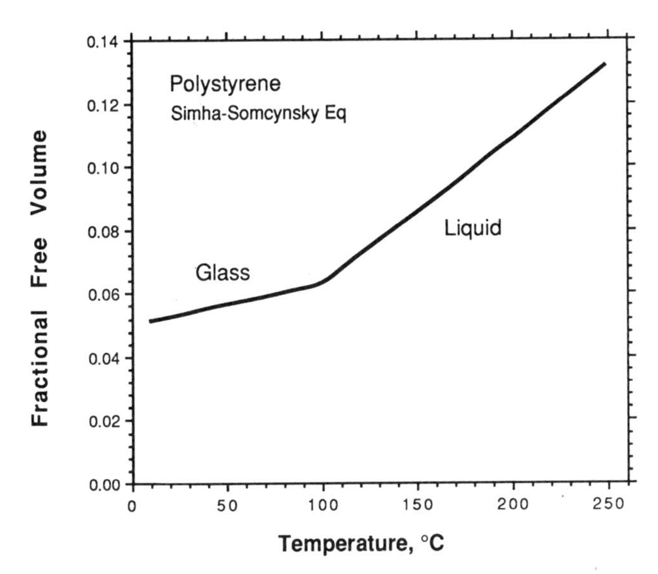

Figure 5 Simha-Somcynsky free volume computed for polystyrene at 1 bar pressure. The free volume has been computed for both the liquid (above 100°C) and the glass. Note that the free volume of the glass is not independent of temperature.

Also shown in Figure 5 is the free volume for glassy polystyrene at 1 bar pressure calculated from the second FORTRAN program given above. For this calculation, the material was assumed to have been quenched from the nominal glass-transition temperature of 100°C. The thermal expansion coefficient for glassy polystyrene measured by Oels and Rehage [34] versus pressure is

$$\alpha_{\text{glass}} = 2.24 \times 10^{-4} - 3.71 \times 10^{-8}\, p + 3.61 \times 10^{-12}\, p^2 \ \text{cm}^3/\text{g} \cdot \text{K}$$

where the pressure p has the units of bar. The Simha-Somcynsky theory gives 0.97642 cm^3/g for the liquid volume at 100°C (V_g), and any volume below this temperature can be computed from Eq. (11). Notice in Figure 5 that the free volume in the glass does not remain constant, independent of temperature.

Thermal Fluctuations in Free Volume

Because of thermal fluctuations, the free volume is not everywhere the same. The mean-squared deviation from the mean free volume can be computed from the general statistical mechanical equation for the mean-squared fluctuations in any quantity u about a mean value u_{av} [35]:

$$\langle \delta u^2 \rangle_{\text{av}} = \langle (u - u_{\text{av}})^2 \rangle_{\text{av}} = kT \left(\frac{\partial^2 F}{\partial u^2} \right)_{u = u_{\text{av}}} . \tag{12}$$

From Eqs. (9) and (5) for the Helmholtz free energy and the partition function at fixed V and T,

$$\begin{aligned} \left(\frac{\partial^2 F}{\partial y^2} \right)_{V,T} &= kT \left[\frac{1}{Z^2} \left(\frac{\partial Z}{\partial y} \right)^2_{V,T} - \frac{1}{Z} \left(\frac{\partial^2 Z}{\partial y^2} \right)_{V,T} \right] \\ &= kT \left\{ \left(\frac{1}{g} \frac{\partial g}{\partial y} \right)^2 - \frac{1}{g} \frac{\partial^2 g}{\partial y^2} + cN \left[\left(\frac{1}{v_f} \frac{\partial v_f}{\partial y} \right)^2 - \frac{1}{v_f} \frac{\partial^2 v_f}{\partial y^2} \right] + \frac{1}{kT} \frac{\partial^2 E_0}{\partial y^2} \right\} . \end{aligned} \tag{13}$$

From Eqs. (6)–(8),

$$\begin{aligned} \left(\frac{\partial^2 F}{\partial y^2} \right)_{V,T} = kT \Bigg\{ & \frac{Ns}{y^2} \left[\frac{s-1}{s} + \frac{2}{y} \ln(1 - y) + \frac{1}{1 - y} \right] \\ & + cN \left[\frac{(\sqrt{2} y \tilde{V})^{2/3} - (8/3) y (\sqrt{2} y \tilde{V})^{1/3} + 3y^2}{y^2 [(\sqrt{2} y \tilde{V})^{1/3} - y]^2} \right] \\ & + \frac{Nqx\epsilon^*}{ykT} [6.066 (y\tilde{V})^{-4} - 2.409 (y\tilde{V})^{-2}] \Bigg\}^{-1} . \end{aligned} \tag{14}$$

Dividing each term in Eq. (14) by Ns, the total number of segments, and writing this as N_s, letting $1/s$ approach zero, writing T^* for $qx\epsilon^*/ck$ and $\tilde{T}$ for T/T^*, and setting $s/3c = 1$, this equation becomes [36]

$$N_s\langle\delta f^2\rangle_{av} = \left\{\frac{1}{y^2}\left[1 + \frac{2}{y}\ln(1 - y) + \frac{1}{1 - y}\right] + \frac{1}{3}\frac{(\sqrt{2}y\tilde{V})^{2/3} - (8/3)y(\sqrt{2}y\tilde{V})^{1/3} + 3y^2}{y^2[(\sqrt{2}y\tilde{V})^{1/3} - y]^2} + \frac{1}{3y\tilde{T}}[6.066(y\tilde{V})^{-4} - 2.409\,(y\tilde{V})^{-2}]\right\}^{-1}, \quad (15)$$

where N_s ($= Ns$) is the nominal number of monomers in the regions of interest. Each such region contains all segments involved in a local segmental rearrangement. For vinyl-type polymers, a chain segment of roughly four backbone carbon atoms, or two monomer units, plus 12 nearest neighbors of like size is expected to constitute a segmental rearrangement cell [37]. This would yield $N_s = 26$ monomers. Different values of N_s may be appropriate for other polymers. Also, problems with the numerical solution sometimes require further changes of N_s. For these reasons, N_s is treated as an adjustable parameter in the calculation of relaxation.

Although Eq. (15), like Eq. (2), is derived under the general assumption of equilibrium, the specific assumption that the free energy is a minimum has not been made. Hence, under conditions where the free-volume distribution has properties, including the shape of the distribution, similar to those existing at equilibrium, Eq. (15) can still be used as an approximation. This is expected to be the situation for the glass formed from an equilibrium liquid by the sudden (instantaneous) quenching by change in temperature or pressure [32].

Calculation of the Fluctuations in Free Volume

The root-mean-square fluctuations in free volume, $\langle\delta f^2\rangle_{av}$, can be obtained in a straightforward way with Eq. (15). A FORTRAN program for doing so is given below. As before, the lowercase symbols, like p, y, and f, are capitalized in the program, $p^* =$ PSTAR, etc., $N_s =$ NS, $\langle\delta f^2\rangle_{av} =$ DF2, and $N_s\langle\delta f^2\rangle_{av} =$ NSDF2.

```
      SUBROUTINE DF2vsT (T, P, V, F, NS, DF2)
C
      REAL*4 NS, NSDF2
      COMMON/STAR/PSTAR,VSTAR,TSTAR
C
      VV=V/VSTAR
      TT=T/TSTAR
```

```
      PP=P/PSTAR
      Y = 1. - F
      YVV2 = (Y*VV)**2
C
      T2 = (2.*Y - Y*Y + 2.*(1. - Y)*ALOG(1. - Y))/(Y*Y*(1. -
     Y))
      YV3 = (2.*YVV2)**(1./6.)
      T3 = (YV3*YV3 - (8./3.)*YV3*Y + 3.*Y*Y)/(3.*Y*((YV3 - Y)
     **2))
      T4 = (2.022/(YVV2**2) - 0.803/YVV2)/TT
      NSDF2 = Y/(T2 + T3 + T4)
      DF2 = NSDF2/NS
C
      RETURN
      END
```

An Example. As an example, the fluctuations in free volume versus temperature for polystyrene calculated for both the liquid and the glass are shown in Figure 6. The parameters used for computing the mean free volume for polystyrene in Figure 5 were again used here, and N_s is set equal to 40.

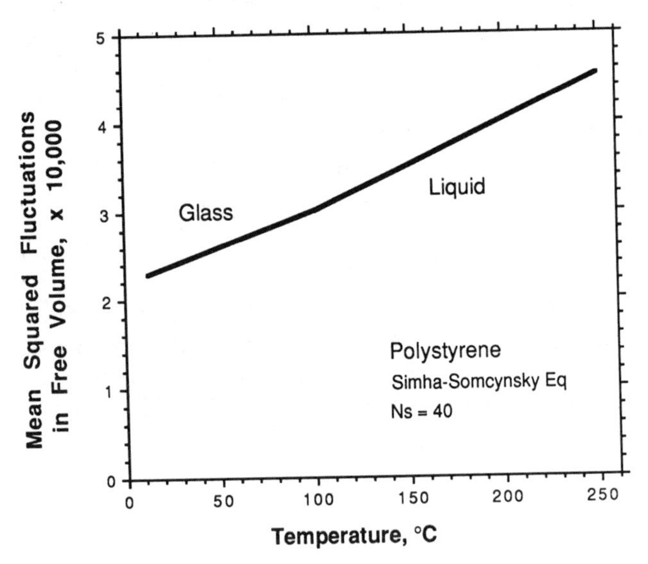

Figure 6 Fluctuations in Simha-Somcynsky free volume computed for polystyrene at 1 bar pressure. The fluctuations have been computed for both the liquid (above 100°C) and the glass.

Free-Volume Distribution

Equations for obtaining the mean free volume and the mean square fluctuations in free volume have been given above. These are the first two moments of a distribution of free volume. When the distribution is approximately symmetrical, as is expected for the equilibrium distribution, the third moment is zero. Thus, a relatively simple distribution can be used to represent the free volume at equilibrium.

There are two well-known distributions for a positive variate (like free volume, for which negative values are undefined) that are defined by the first two moments. These are the gamma distribution for a continuous variate and the binomial dsitribution for a discrete variate. The free volume, being a continuous variate, is most appropriately described by the gamma distribution. But for calculations of relaxation, a finite number of discrete free-volume states is much more convenient, and the binomial distributions also will be considered.

The gamma distribution is given by

$$\xi(f) = \frac{\chi}{\Gamma(\alpha)} (\chi f)^{\alpha-1} e^{-\chi f}, \tag{16}$$

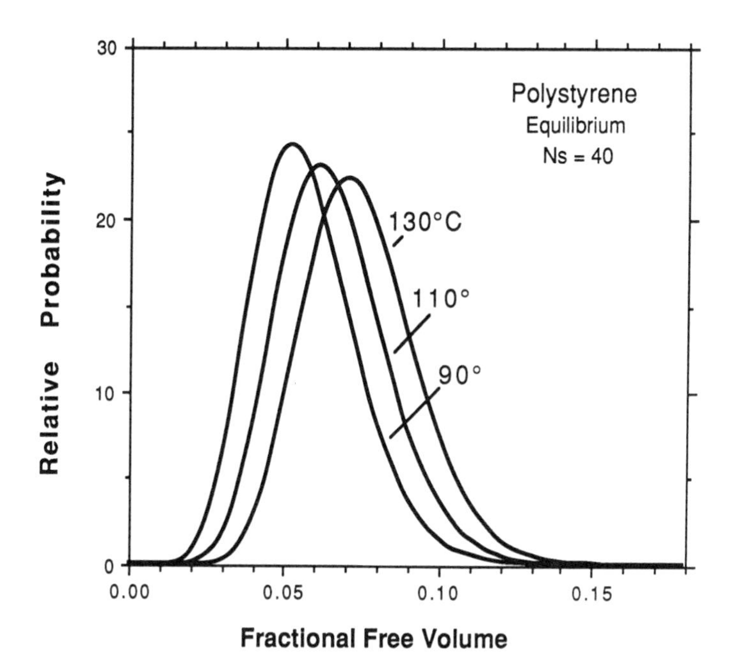

Figure 7 Free-volume distributions for $N_s = 40$ size regions in polystyrene at 90°, 110°, and 130°C, assuming an equilibrium state for each.

where $\xi(f)$ is the relative probability that a given N_s-size region has the free volume f, and α and χ are parameters whose values are determined by the following equations for the mean and mean-squared fluctuations in free volume:

$$\frac{\alpha}{\chi} = f_{av}, \qquad \frac{\alpha}{\chi^2} = \langle \delta f^2 \rangle_{av}. \tag{17}$$

Equation (16) is normalized so that

$$\int_0^\infty \xi(f)\, d\xi = 1. \tag{18}$$

Typical free-volume distributions are shown in Figures 7 and 8. Those in Figure 7 are for $N_s = 40$ size regions in polystyrene for equilibrium states at 90°, 110°, and 130°C. Those in Figure 8 are the corresponding free-volume distributions for the glassy states at 60° and 80°C for glasses quenched from 100°C. (Shown also in Fig. 8 is the distribution for the equilibrium liquid at 100°C.)

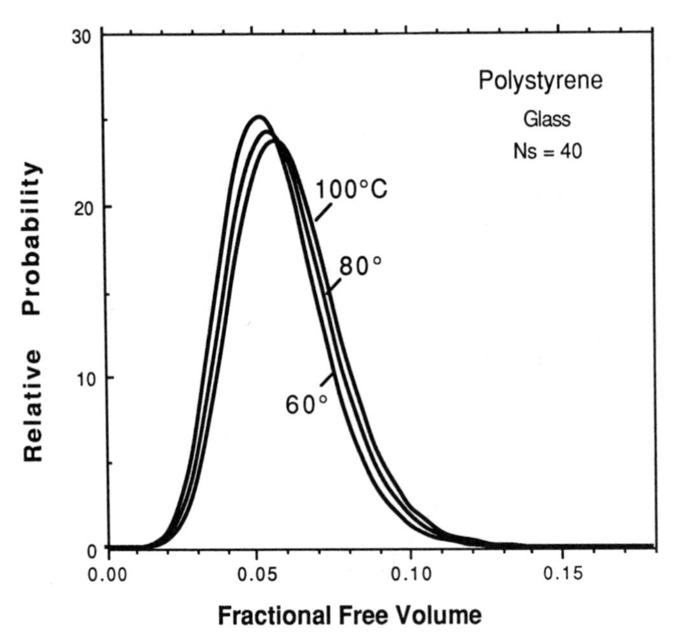

Figure 8 Free-volume distributions for $N_s = 40$ size regions in glassy polystyrene at 60° and 80°C for a glass quenched from 100°C. (Also shown is the distribution for the equilibrium liquid at 100°C.)

The binomial distribution has two parameters: the number of states n and the unit probability p_r. The unit probability p_r and the free-volume increment β are fixed by the mean and variance:

$$(n - 1)p_r\beta = f_{\text{av}},$$
$$(n - 1)p_r(1 - p_r)\beta^2 = \langle \delta f^2 \rangle_{\text{av}}. \tag{19}$$

The state occupancies given by the binomial distribution are

$$\xi_j = \binom{n-1}{j-1} p_r^{j-1}(1 - p_r)^{n-j}, \qquad j = 1, 2, \ldots, n. \tag{20}$$

The occupancies sum to unity:

$$\sum_{j=1}^{n} \xi_j = 1. \tag{21}$$

Typical discrete free-volume distributions using the binomial distribution function are shown in Figures 9 and 10. Like those in Figure 7, those in Figure 9 are for $N_s = 40$ size regions in polystyrene for equilibrium states at 90°, 110°, and 130°C. Those in Figure 10 are the corresponding free-

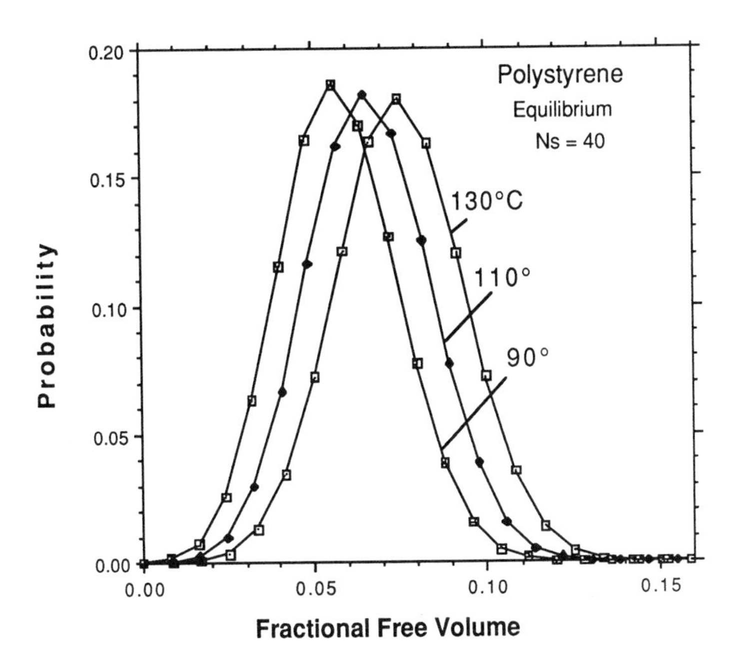

Figure 9 Discrete free-volume distributions for $N_s = 40$ size regions in polystyrene at 90°, 110°, and 130°C, assuming an equilibrium state for each.

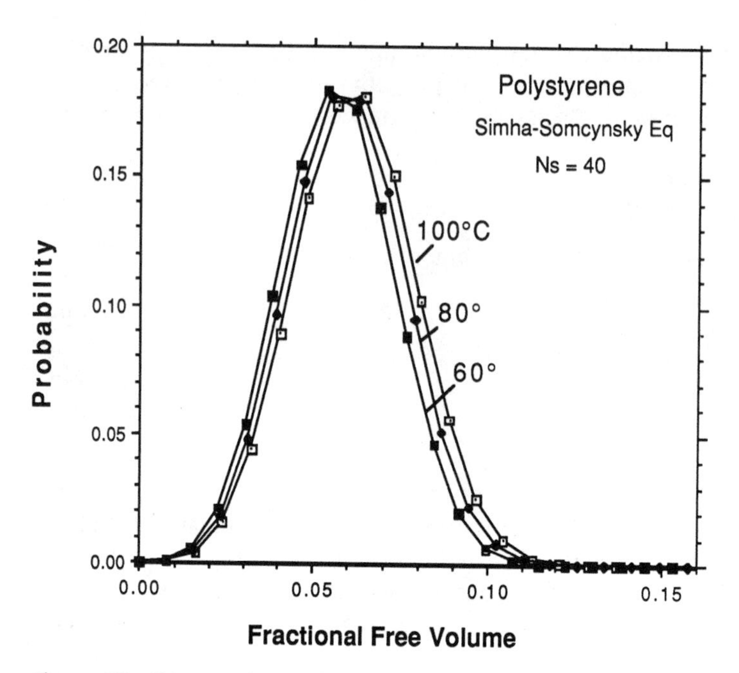

Figure 10 Discrete free-volume distributions for $N_s = 40$ size regions in glassy polystyrene at 60° and 80°C for a glass quenched from 100°C; the curve for the latter is for the equilibrium distribution.

volume distributions for the glassy states at 60° and 80°C for glasses quenched from 100°C; the curve for the latter is an equilibrium distribution. (The points representing the discrete probabilities are connected by lines to aid in distinguishing between the different temperatures. Were the distributions to be exhibited singly, a bar graph might be more revealing.)

KINETIC PROCESSES IN POLYMERS

Although a strictly free-volume theory for rate processes in liquids and glasses is often not completely quantitative, as has been noted, free-volume considerations usually describe a large part of the kinetics. In the following, the Simha-Somcynsky measure of free volume will be applied to kinetic calculations. Fundamental, molecular dynamic calculations of the kinetics of most processes of interest are not presently possible, whether in terms of free volume or not. As a result, the first step in using free volume to compute the kinetics of a process is to establish, usually empirically, a relationship between free volume and a relaxation rate.

The easiest relationship to establish between free volume and a relaxation rate is this: A correspondence is made for the equilibrium liquid between the computed Simha-Somcynsky free volume and the relaxation rate obtained from fluidity or viscosity measurements on the liquid, both at the same temperature and pressure. This correspondence can be derived from fluidity or viscosity measurements often summarized in a moderately simple form, such as by the Arrhenius, Williams-Landel-Ferry (WLF), or Vogel equations. For the temperature region just above the glass transition, where the WLF or Vogel equation applies, the relaxation rate (written as the reciprocal of the relaxation time) can be written at constant pressure as

$$\tau^{-1}(T) = \tau_g^{-1} \exp\left\{2.303c_1 \frac{T - T_g}{c_2 + T - T_g}\right\}, \tag{22}$$

where the subscript g refers to the glass transition, which is taken as the reference temperature [38].

Because of the near linearity between the Simha-Somcynsky free volume f with change in temperature, as seen in Figure 5 for polystyrene, for example, $T - T_g$ may be replaced by $f - f_g$ to good approximation as in the following equation:

$$T - T_g = \frac{(f - f_g)T^*}{f^*}, \tag{23}$$

where f_g is the free volume at the glass transition, and f^* is defined by this equation; in analogy with T^*, f^* may be called the "characteristic free volume." Equation (22) then becomes

$$\tau^{-1}(f) = \tau_g^{-1} \exp\left\{2.303c_1 \frac{f - f_g}{c_2 f^*/T^* + f - f_g}\right\}. \tag{24}$$

One should note that the replacement of T by f in this equation is only that, a simple *replacement*. It has been made simple by the linearity of the Simha-Somcynsky free volume versus temperature. The replacement is unrelated to the correspondence established by Williams, Landel, and Ferry between their equation and the Doolittle equation.

Equation (24) is a general relationship between the relaxation rate and free volume and is as general as the Vogel equation itself. It is assumed to relate average quantities, the average relaxation time of the liquid in terms of the average free volume. To make it more useful, we assume that it applies to local properties. That is, we assume that the local relaxation rate in small regions of the liquid are given by Eq. (24), at least to within a proportionality constant, when the local free volume in those regions is used. Because thermal fluctuations cause the free volume in small regions to differ from the average, the local relaxation rate is likewise expected to differ from the average. It is now a small step to the final assumption, which is that Eq. (24) describes approximately the relaxation rate in any region of interest, wherever the free volume f is defined, irrespective of whether or not the material as a whole is in equilibrium.

For a general kinetic process occurring in a polymer at equilibrium, as in the liquid, the effective relaxation rate is given by

$$\tau_{\text{eff}}^{-1} = \int_0^\infty \tau^{-1}(f)\xi(f)\, df, \tag{25}$$

where $\xi(f)$ is the free-volume distribution function, given by Eq. (16), for example. For a kinetic process occurring in a polymer glass, the effective relaxation rate is assumed to be given by

$$\tau_{\text{eff}}^{-1}(t) = \sum_{i=1}^{n} \tau^{-1}(f_i)w_i(t), \tag{26}$$

where the $w_i(t)$'s are the probabilities of occurrence of the discrete free volumes f_i and are not given by any general equation but must be computed individually. (The free-volume distribution is discretized to reduce the number of probabilities that must be computed.) To use Eq. (25) or (26), the size of the region controlling the proces, N_s, needs to be estimated or chosen. At equilibrium, it is the product $N_s\langle\delta f^2\rangle_{\text{av}}$ that is a constant at a given temperature and pressure. Thus, $\langle\delta f^2\rangle_{\text{av}}$ increases with decrease in N_s according to

$$\langle\delta f^2\rangle_{\text{av}} \sim \frac{\text{const}}{N_s}. \tag{27}$$

For the volume recovery of polystyrene, N_s was estimated to be 40 (monomer units) [39]. This same value was found to be appropriate to describe the photoisomerization of azobenzene moieties attached to and immersed in a polystyrene matrix [24], a summary of which is given at the end of the chapter.

A relaxation problem of considerable challenge, and yet of considerable interest and importance, is that where the state of the material, including the free volume itself, is changing with time. This important problem arises in structural recovery and is described in the next section.

VOLUME RECOVERY

When the temperature of a polymer material is suddenly lowered, for example, there occur instantaneous changes in various properties, like volume and enthalpy. If the final temperature is near or below that of the material's glass transition, there occur in addition delayed changes in properties that are usually in the same direction. The delayed changes are often referred to as "relaxation." The term "relaxation" implies changes from a higher to a lower energy state, as occurs following the sudden lowering of temperature. But a similar behavior can arise if the temperature is suddenly increased from below that of the glass transition. Following the initial, instantaneous increases in volume and enthalpy, there occur the delayed relaxation to *higher* volume and enthalpy. In both cases the material is left in a nonequilibrium state immediately after the changes in temperature, and the delayed changes result from the tendency of the material to seek equilibrium. Thus, the term "recovery" is often used in place of relaxation, to indicate the origin of the process in the recovery of equilibrium.

Delayed changes in properties of glassy polymers occurring after changes in temperature, pressure, or stress sometimes have practical consequences. They also imply structural and dynamic characteristics of the glassy state. For both of these reasons one wants to understand and to be able to predict the kinetics of recovery or relaxation. Studies of the formation of the glass from the liquid state show that the viscosity of polymers and other glass-forming liquids rise very rapidly over a fairly narrow temperature range near the glass transition. This suggests the occurrence of a strong steric or entangling interaction of molecules accompanying the loss of the last bit of free space surrounding them. This space surrounding the molecules is referred to as "free volume."

Review of Theories for Structural Recovery

Although many theories have been suggested, the minimum requirements for a useful recovery theory are (a) that it predict the asymmetry between up and down steps in temperature and (b) that it predict the extremum in volume or enthalpy following the second of a pair of closely timed steps in opposite direction in temperature or pressure. (The contexts of these requirements will be discussed in the last section.) A good summary of theoretical work up to about 1980 has been given by Tant and Wilkes [40].

Models of Moynihan et al., KAHR, and Others

The models discussed here are largely phenomenological. In the model of Moynihan et al. [41], the progress of recovery following a sudden step in temperature at time t_1, say, is described by a nonexponential, nonlinear function, $\phi(t - t_1, t)$, given by

$$\phi(t - t_1, t) = \exp\left[-\left(\int_{t_1}^{t} \frac{dt'}{\tau_0}\right)^{\beta}\right].$$

For the relaxation time τ_0, a function constructed by Narayanaswamy [42] was assumed:

$$\tau_0 = A \exp\left[\frac{x\,\Delta h}{RT} + \frac{(1-x)\,\Delta h}{RT_f}\right].$$

This function depends on both the fictive and real temperatures, T_f and T, and on three adjustable parameters, A, x, and Δh. A procedure to optimize these three parameters and the exponent β in the above integral has been developed by Hodge et al [43] for DSC data, from which good fits have been obtained. Hodge [44] has shown more recently that equations based on the Adam-Gibbs theory [45] can fit the data equally well or slightly better than that based on the Narayanaswamy function, however.

In the KAHR (Kovacs, Aklonis, Hutchinson, and Ramos) model [46,47], the recovery process is divided into a number of subprocesses, each of which represents some (unspecified) internal parameter. Thus, the total deviation from equilibrium, δ, is the sum of the deviation from equilibrium of each of the subprocesses, δ_i. For small deviations from equilibrium, the driving force for recovery is expected to be proportional to the magnitude of the deviation

$$\frac{d\delta_i}{dt} = -\frac{\delta_i}{\tau_i}, \qquad \text{for } i = 1, 2, \ldots, N.$$

The recoveries of the subprocesses are not independent of one another. The relaxation times τ_i are assumed to depend on both the total deviation from equilibrium, δ, and temperature, T. Although any of a number of distributions of τ_i for fixed δ and T produce reasonable fits to the data, a spectrum found sufficient was a two-box-shaped spectrum, with 90% of the spectrum spread over two decades of time and the other 10%, on the short relaxation time side of the first box, spread over three decades.

The KAHR model has been shown to be equivalent to the Moynihan model in the region of the glass transition [46,48]. Recovery in each depends on both the current global volume or enthalpy and a distribution of relaxation times. As mentioned, the models have been extensively applied to data by Hodge [43] and also by others [49]. The values of the parameters in these theories have been found to vary from specimen to specimen in ways that deny a simple physical interpretation, however.

Chow [50] has attempted to derive these models from more physically based concepts. For the division of the recovery process into subprocesses in the KAHR model, for instance, Chow associated a range of hole types. Although the holes differ in energy of formation, each is given the same free volume equivalence.

Further changes in the Moynihan et al.–KAHR models have been suggested by Matsuoka et al. [51]. The latter suggested that the measured dielectric loss spectrum be used for the distribution of relaxation times. For the temperature and structure dependence of the relaxation times, a function that combines both Arrhenius and Vogel-Fulcher-type terms was suggested.

Diffusion Model

Curro, Lagasse, and Simha [52] examined a model in which volume recovery occurs by the diffusion of holes or free volume. The local diffusivity is assumed to be a function of the local free volume, which is derived from the Simha-Somcynsky equation of state [27]. The only adjustable parameter in the theory is a length scale for diffusion. For recovery, the theory is able to describe the asymmetry between upsteps and downsteps in temperature and a "memory effect" following a pair of closely timed temperature steps in opposite directions. The predictions of the diffusion model are essentially equivalent to those of the KAHR model for single temperature steps but underestimates the magnitude of the extremum in the two-step experiment. A model similar to that of Curro et al. was suggested also by Bulatov, Gusev, and Oleinik [53].

Ngai-Rendell Theory

The most detailed fitting of theory to Kovacs's [54] important single temperature-step data on poly(vinyl acetate) is that of Rendell et al. [55]. An excellent fit, probably exact within experimental error, was obtained with Ngai and Rendell's "coupling" model.

This model is based on a general consideration of complex systems involving interactions among elements with a continuum of degrees of coordination and coupling. The two basic equations of the theory, which describe the time development of a macroscopic variable, $\phi(t)$, can be written in the following form if the three parameters, τ_0, ω_c, and n, remain constant, though they are not required to do so by the model:

$$\phi(t) = \phi(0)\exp\left[-\left(\frac{t}{\tau^*}\right)^{1-n}\right], \qquad \text{for } \omega_c t >> 1$$

and

$$\tau^* = [(1 - n)\omega_c^n\tau_0]^{1/(1-n)}.$$

The first equation involves a stretched exponential. This can be expanded as a sum of simple exponentials, suggesting a distribution of relaxation times. But the mathematical intimacy of the first equation with the second, which appears to have received independent experimental verification, is lost by the expansion. The physical origin of the stretched exponential has been discussed also by Palmer et al. [56], who suggest that it arises from a "hierarchical constraint" on the motions of segments or atoms in the glass, a sequencing of motions.

The model's ability to fit a broad range of macroscopic data from complex systems [57] suggests that the mathematical structure of the theory is indeed characteristic of such systems. Also, the three parameters in the model are perhaps optimal for their ability to span disparate systems. Being very general, however, the model is unable to be specific about chemical or structural effects. Of the three parameters, nothing is contained in the model about their values, except that $0 < n < 1$.

RSC Theory

The RSC (Robertson-Simha-Curro) theory [24,36,39,58,59,60], which is the theory to be described at greater length in the following, is based on the local kinetics of polymer segmental rearrangements. The local kinetics are assumed to vary from site to site in the glass because of structural variations caused by thermal fluctuations in the liquid from which the glass was formed. The dependence of recovery on the global structure noted above in connection with the Moynihan and KAHR models arises in the RSC theory from the kinetics of absorption and release of changes in volume by the surroundings. A model for the relaxation of nonpolymeric liquids and glasses was derived from related considerations by Brawer [61].

Free-Volume Theory of Volume Recovery

As described in the previous section, the rate of any kinetic process, including the gain or loss of free volume, is related to the local free volume according to Eq. (24). And, as has also been described, the local free volume usually differs from the mean because of thermal fluctuations, and the magnitude of the fluctuations depends on the size of the region having influence on the process according to Eq. (27). For the processes of volume or enthalpy recovery, regions of two different sizes need to be defined. This is because the region that loses or gains free volume by a single molecular rearrangement step is usually smaller than the region that influences the process. The number of monomer units in the smaller region of volume $\mathcal{V}$, the volume that is changing its free volume, is N_s. The size of the larger region is $z\mathcal{V}$.

Although the fractional free volume existing within any region of interest is a continuous variable, the numerical computation of the kinetics of processes within such regions is more easily performed if the free volume is treated as being discrete. The small regions that change free volume in a single step (the regions $\mathcal{V}$ containing N_s monomer units) will be assumed to contain an amount of free volume that is a multiple of a quantity denoted by β. Thus, the free volume in these regions is one of the amounts, 0, β, $2\beta, \ldots, (j-1)\beta, \ldots, (n-1)\beta$. The index j, which describes the set of free-volume states, runs from 1 to n. The upper limit n is set high enough so that almost no N_s monomer-size cells would have a free volume greater than $(n-1)\beta$. The quantity β is obtained for the binomial distribution by eliminating p_r from Eqs. (19):

$$\beta = \frac{\langle \delta f^2 \rangle_{av}}{f_{av}} + \frac{f_{av}}{n}, \tag{28}$$

where f_{av} and $\langle \delta f^2 \rangle_{av}$ are the mean and mean-squared fluctuations in free volume and n is the number of free-volume states.

Considering all possible distinct, but possibly overlapping, N_s-size region in the glass, the fraction of these that have at time t the free volume $(j-1)\beta$ will be denoted by $w_j(t)$. We will often refer to the n different amounts of free volume as the n free-volume states, or simply, the "states," of the system. The w_j's then are the occupancies of the states of the system, and in the following, we will consider only these states and neglect other properties of the glass. The occupancies have the typical property of probabilities that they sum to unity at all arbitrary times t:

$$\sum_{i=1}^{n} w_i(t) = 1. \tag{29}$$

This follows from the condition that each of all of the N_s-size regions must have one of the n amounts of free volume, or be in one of the n free-volume states, and that this remains true for all times.

The free volume of the larger regions of volume $z\mathcal{V}$, the union of the element of volume $\mathcal{V}$ and its environment, is given by

$$\hat{f}_j = \frac{(j-1)\beta + (z-1)f_{av}}{z}. \tag{30}$$

It is this free volume, $\hat{f}_j$, and not that of the unit, f_j, that controls the kinetics of free-volume change. The free volume of the environment is assumed to be equal to just the average free volume of the entire system because of the considerably larger size of the environment than $\mathcal{V}$. The probability of occurrence of free volume $\hat{f}_j$ at any instant is also $w_j(t)$. When these probabilities are equal to those corresponding to the polymer at equilibrium, we will write them as $w_j = \xi_j$. The use of the symbol w_j instead of ξ_j for the state probabilities is meant to emphasize the fact that the w's usually are not equilibrium occupancies and therefore can change with time even as the temperature and pressure remain constant.

The occupation probabilities, the w_j's, describe the state of the non-equilibrium glass. It is from these that other properties are derived. Thus, the overall fractional free volume at any instant of time is given by

$$f_{av}(t) = \sum_{i=1}^{n} w_i(t) f_i. \tag{31}$$

The change in probability or occupancy of the n free-volume states from an initial time t_0 to the present time t can be described formally by the following set of n equations [62]:

$$w_j(t) = \sum_{i=1}^{n} w_i(t_0) P_{ij}(t - t_0), \qquad j = 1, 2, \ldots, n, \tag{32}$$

where $P_{ij}(t - t_0)$ is a transformation probability—it is the probability that if a given small region has the free volume $(i - 1)\beta$ at time t_0 that it will instead have the free volume $(j - 1)\beta$ at the later time t, that the free-volume state of the region has changed from i to j in the time interval $t - t_0$. An examination of the requirements on the P_{ij}'s will allow us to write a set of differential equations for them from which their time development can be obtained.

Derivation of the State Change Equations

The following are several equations that describe the properties of the P_{ij}'s.

$$\sum_{j=1}^{n} P_{ij}(t'' - t') = 1. \tag{33}$$

The sum of the transition probabilities for an arbitrary time interval must be unity. This is related to the above: all of the regions that were in state i at time t' must be in one of the n states at time t''; i.e., all of the regions must remain accounted for.

$$\lim_{h \to 0} P_{ij}(h) = \delta_{ij}. \tag{34}$$

The transition probabilities are continuous functions of the time interval $h = t'' - t'$, including when h shrinks to zero. This equation expresses the condition that a significant change in free volume or state occupancy occurs only after a finite time interval. (A region is allowed to change its free volume in an instant, but a waiting time is usually required before the change occurs.)

$$P_{ij}(t + h) = \sum_{k=1}^{n} P_{ik}(t) P_{kj}(h). \tag{35}$$

This is the Chapman-Kolmogorov relation, which would follow from Eq. (32) if we considered a time time t' somewhere between t_0 and t. We would then have

$$w_k(t') = \sum_{i=1}^{n} w_i(t_0)P_{ik}(t' - t_0)$$

and

$$w_j(t) = \sum_{k=1}^{n} w_k(t')P_{kj}(t - t').$$

On combining them,

$$w_j(t) = \sum_{k=1}^{n}\sum_{i=1}^{n} w_i(t_0)P_{ik}(t' - t_0)P_{kj}(t - t').$$

The order of the summations can be changed, and w_i, which does not depend on the index k, moved to the left of the summation over k. Then, on comparing with Eq. (32), we see that the Chapman-Kolmogorov relation results.

Because of the continuity of the transition probabilities, mentioned above, a set of differential equations for the transition probabilities can be obtained by differentiating the Chapman-Kolmogorov relation, Eq. (35), with respect to h and evaluating it at $h = 0$:

$$\lim_{h \to 0} \frac{dP_{ij}(t + h)}{dh} = \frac{dP_{ij}(t)}{dt} = \sum_{k=1}^{n} P_{ik}(t)A_{kj}, \tag{36}$$

where

$$A_{kj} = \lim_{h \to 0} \frac{dP_{kj}(h)}{dh}. \tag{37}$$

The number of possible differential equations is the same as the number of possible transition probabilities, n^2. This is obtained by letting each index run from 1 to n. Because the sum of terms on the right-hand side is the same as that resulting from matrix multiplication, we can write the totality of these equations by letting each be a component of the matrix equation

$$\dot{\mathbf{P}}(t) = \mathbf{P}(t)\mathbf{A}, \tag{38}$$

where the time derivative of each component of the **P** matrix is indicated by the dot.

Rates of State Change

By assuming reasonable values for the $P_{kj}(h)$'s for infinitesimal time intervals h, the longer time behavior of the transition probabilities can be obtained from the above differential equation. The values for the P_{kj}'s after infinitesimal time intervals depend on the assumption made concerning how the regions of volume $\mathcal{V}$ change their free volume. One assumption is that the free volume changes in a quasi-continuous manner, that it changes by single units of β at each change. Thus, on changing from state i to j, the free volume would pass through all of the intermediate states between i and j—$i + 1, i + 2, \dots, j - 2, j - 1$, for example. There would usually be a waiting time at each step before the next occurs. This assumption, besides being reasonable, simplifies the description of the $P_{kj}(h)$'s for small h. When the time interval is small enough, there is time, then, for free volume changes to be made only between adjacent states. This can be expressed mathematically by

$$
\begin{aligned}
P_{k,k-1}(h) &= h\lambda_k^- + o(h), \\
P_{k,k+1}(h) &= h\lambda_k^+ + o(h), \\
P_{kk}(h) &= 1 - h(\lambda_k^- + \lambda_k^+) + o(h), \\
P_{kj}(h) &= o(h), \qquad j \neq k - 1, k, k + 1,
\end{aligned} \tag{39}
$$

where λ_k^+ and λ_k^- are the upward and downward transition rates from state k, respectively, and $o(h)$ is a quantity that approaches zero faster than h. From these equations, the A_{jk}'s given by Eqs. (37) are readily found to be

$$
\begin{aligned}
A_{k,k+1} &= \lambda_k^+, \qquad k = 1, 2, \dots, n - 1, \\
A_{kk} &= -(\lambda_k^- + \lambda_k^+), \qquad k = 1, 2, \dots, n, \\
A_{k,k-1} &= \lambda_k^-, \qquad k = 2, 3, \dots, n, \\
A_{kj} &= 0, \qquad j \neq k - 1, k, k + 1.
\end{aligned} \tag{40}
$$

The matrix $\mathbf{A}$ is thus seen to be tridiagonal; i.e., all of the elements are zero except for the diagonal and the subdiagonals on either side of it.

The upward and downward transition rates, λ_k^+ and λ_k^-, are not independent. The state occupancies at equilibrium, ξ_k, are assumed to be unchanging, and the upward and downward transition rates must be consistent with this. Detailed balancing leads to the following:

$$
\xi_k\lambda_k^+ = \xi_{k+1}\lambda_{k+1}^-, \qquad k = 1, 2, \dots, n - 1. \tag{41}
$$

The individual upward and downward rates are expected to be proportional to the relaxation rate given by Eq. (24) with the appropriate free volume given by Eq. (30). Hence, the downward rate from state k ($k = 2, 3, \dots, n$) is written as

$$
\lambda_k^- = R\left(\frac{\xi_{k-1}}{\xi_k}\right)^{1/2} \beta^{-2}\tau_g^{-1} \exp\left\{2.303c_1\left[1 + \frac{c_2}{c_2 + T^*(\hat{f}_k - f_g)/f^*}\right]\right\}, \tag{42}
$$

and the upward rate, from Eq. (41), is given by

$$\lambda_k^+ = \left(\frac{\xi_{k+1}}{\xi_k}\right)\lambda_{k+1}^-, \qquad k = 1, 2, \ldots, n-1. \tag{43}$$

The proportionality constant in Eq. (42) is the triplet of factors R $(\xi_{k-1}/\xi_k)^{1/2}\beta^{-2}$. Here R is a constant parameter that, among other things, ensures the equality of Eq. (25); $(\xi_{k-1}/\xi_k)^{1/2}$ has been included to give the upward and downward rates a similar form; and β^{-2} has been included to remove an implicit dependence of the kinetics on the chosen free-volume unit β.

We might notice that the above definitions for the upward and downward transition rates indicate that **A** is not symmetric. A typical element above the diagonal is $A_{k,k+1} = \lambda_k^+$, and opposite it across the diagonal is $A_{k+1,k} = \lambda_{k+1}^-$. In general, λ_k^+ and λ_{k+1}^- are not equal, unless, according to Eq. (43), $\xi_{k-1} = \xi_k$. We will see below, however, that **A** is readily symmetrized with a similarity transformation using a diagonal matrix involving the ξ_k's along its diagonal.

Solution of the Differential Equations

The differential equations for the transition probabilities become, for $i =$ 1 to n,

$$\begin{aligned}
\frac{dP_{i,1}(t)}{dt} &= P_{i,1}(t)A_{1,1} + P_{i,2}(t)A_{2,1}, \\
\frac{dP_{ij}(t)}{dt} &= P_{i,j-1}(t)A_{j-1,j} + P_{ij}(t)A_{jj} + P_{i,j+1}(t)A_{j+1,j}, \qquad j \neq 1, n, \\
\frac{dP_{i,n}(t)}{dt} &= P_{i,n-1}(t)A_{n-1,n} + P_{i,n}(t)A_{nn}.
\end{aligned} \tag{44}$$

Because the equations are coupled, they must be solved for all of the P_{ij}'s simultaneously. This problem is similar to that of coupled oscillators, such as the atomic vibrational motions of a molecule [63]. As with molecular vibrations [64] the above problem is solved by reforming the variables. The P_{ij}'s are combined into sums such that only a single combined P term appears on the right-hand side of each equation, the same variable as that for which the time derivative appears on the left. The procedure for combining the P's is to find the matrix that diagonalizes the **A** matrix by a similarity transformation, thus converting **A** from a tridiagonal to a diagonal matrix. This is the same as finding the eigenvalues and eigenvectors of **A**, for the eigenvector matrix is the one needed for the similarity transformation. Although the procedure for reforming the variables is the same as that for molecular vibrations, the reformed P's do not have a simple or useful meaning, like the normal vibrational modes of a molecule. Thus, the reformed P's need not be described explicitly.

The desired eigenvalue problem is

$$\mathbf{A}\overline{\mathbf{Q}} = \overline{\mathbf{Q}}\zeta, \tag{45}$$

where ζ, the eigenvalue matrix, is diagonal. (The finding of the eigenvalue and eigenvector matrices, ζ and $\overline{\mathbf{Q}}$, from the matrix $\mathbf{A}$ is done by well-tested computer routines, as is discussed more fully below.) The similarity transformation of $\mathbf{A}$, mentioned above, is obtained by multiplying both sides of Eq. (45) by the inverse of $\overline{\mathbf{Q}}$:

$$\overline{\mathbf{Q}}^{-1}\mathbf{A}\overline{\mathbf{Q}} = \zeta. \tag{46}$$

Multiplying the matrix differential equation for $\mathbf{P}$, Eq. (38), from the right by $\overline{\mathbf{Q}}$ and from the left by the inverse of $\overline{\mathbf{Q}}$ gives

$$\overline{\mathbf{Q}}^{-1}\dot{\mathbf{P}}(t)\overline{\mathbf{Q}} = \overline{\mathbf{Q}}^{-1}\mathbf{P}(t)\overline{\mathbf{Q}}\overline{\mathbf{Q}}^{-1}\mathbf{A}\overline{\mathbf{Q}} = \overline{\mathbf{Q}}^{-1}\mathbf{P}(t)\overline{\mathbf{Q}}\zeta, \tag{47}$$

where use has been made of $\overline{\mathbf{Q}}\overline{\mathbf{Q}}^{-1} = \mathbf{I}$, the unit matrix, inserted between $\mathbf{P}$ and $\mathbf{A}$. The same similarity transformation that diagonalizes $\mathbf{A}$ gives the desired combinations of the P's in the elements of the matrix

$$\Pi(t) = \overline{\mathbf{Q}}^{-1}\mathbf{P}(t)\overline{\mathbf{Q}}. \tag{48}$$

The decoupled differential equations become

$$\dot{\Pi}(t) = \Pi(t)\zeta$$

or

$$\dot{\Pi}_{ij}(t) = \sum_{k=1}^{n} \Pi_{ik}(t)\zeta_{kj} = \Pi_{ij}(t)\zeta_j, \tag{49}$$

where the ζ_j's are the diagonal elements of ζ. The Π_{ij}'s and P_{km}'s are related to each other by

$$\Pi_{ij}(t) = \sum_{k=1}^{n}\sum_{m=1}^{n} (\overline{Q}^{-1})_{ik}P_{km}(t)\overline{Q}_{mj}$$

and

$$P_{ij}(t) = \sum_{k=1}^{n}\sum_{m=1}^{n} \overline{Q}_{ik}\Pi_{km}(t)(\overline{Q}^{-1})_{mj}. \tag{50}$$

The latter equation follows from Eq. (48) by multiplying from the left by $\overline{\mathbf{Q}}$ and from the right by $\overline{\mathbf{Q}}^{-1}$.

The solution of the differential Eqs. (49) is

$$\Pi_{ij}(t - t_0) = \Pi_{ij}(0)\exp\{(t - t_0)\zeta_j\}$$

or

$$P_{ij}(t - t_0) = \sum_{m=1}^{n} \sum_{k=1}^{n} \overline{Q}_{ik}\Pi_{km}(0)(\overline{Q}^{-1})_{mj} \exp\{(t - t_0)\zeta_m\}. \tag{51}$$

This equation can be used to describe the evolution of the free volume over the time interval $t - t_0$. At the beginning of the interval, Eq. (34) indicates that $\mathbf{P}(0)$ and $\Pi(0)$ both are equal to the unit matrix and $\Pi_{ij}(0) = \delta_{ij}$. Therefore, the sum over k has as its sole nonzero term, $\overline{Q}_{im}$, giving

$$P_{ij}(t - t_0) = \sum_{m=1}^{n} \overline{Q}_{im}(\overline{Q}^{-1})_{mj} \exp\{(t - t_0)\zeta_m\}. \tag{52}$$

Using this equation for the P_{ij}'s in Eq. (32), we have for the free-volume state occupancies at time t,

$$w_j(t) = \sum_{m=1}^{n} \sum_{i=1}^{n} w_i(t_0)\overline{Q}_{im}(\overline{Q}^{-1})_{mj} \exp\{(t - t_0)\zeta_m\}. \tag{53}$$

There are two reasons for symmetrizing $\mathbf{A}$ before solving the eigenvalue problem. First, it reduces the amount of computer memory required, and second, most algorithms for solving the eigenvalue problem begin by symmetrizing nonsymmetric matrices anyway. Equation (53) involves three $n \times 1$ matrices, $w_j(t)$, $w_i(t_0)$, and ζ_m, and two $n \times n$ matrices, $\overline{\mathbf{Q}}$ and $\overline{\mathbf{Q}}^{-1}$, few of whose elements are the same. In computer calculations, it is usually necessary to use double precision, at least for the eigenvalue calculation, and it is often convenient to use double precision throughout. Thus, the computer memory required is $2n(3 + 4n)$. The memory required with a symmetrized form for the matrix $\mathbf{A}$ involves a further $n \times 1$ matrix, the diagonal symmetrizing matrix. But the eigenvector matrix resulting from a symmetric matrix has the useful property that the inverse is the same as the transpose. Thus, separate matrices are not needed for the eigenvector matrix and its inverse, and the required memory falls to $4n(2 + n)$. If $n = 60$, for example, the required memory would fall from 29,160 to 14,880 words.

The symmetrizing of the matrix $\mathbf{A}$, because $\mathbf{A}$ is tridiagonal, involves a similarity transformation with a diagonal matrix γ. The elements of γ are $(\xi_k)^{-1/2}$. This is readily shown. A typical off-diagonal element of the symmetrized matrix is

$$(\gamma^{-1}\mathbf{A}\gamma)_{k,k+1} = (\gamma_{kk})^{-1}A_{k,k+1}\gamma_{k+1,k+1} = \left(\frac{\xi_k}{\xi_{k+1}}\right)^{1/2} \lambda_k^+,$$

and the corresponding element of the transposed matrix is

$$(\gamma^{-1}\mathbf{A}\gamma)_{k+1,k} = (\gamma_{k+1,k+1})^{-1}A_{k+1,k}\gamma_{kk} = \left(\frac{\xi_{k+1}}{\xi_k}\right)^{1/2} \lambda_{k+1}^{-}.$$

These are seen to be equal according to Eq. (43), and therefore $\gamma^{-1}\mathbf{A}\gamma$ is symmetric.

The new eigenvalue problem is

$$(\gamma^{-1}\mathbf{A}\gamma)\mathbf{Q} = \mathbf{Q}\zeta, \tag{54}$$

where again the eigenvalue and eigenvector matrices are obtained by computer routines, as is discussed more fully below. Note that because similar matrices have the same eigenvalues, the eigenvalues ζ are unchanged [65]. On multiplying both sides of this equation from the left with γ and comparing the result with Eq. (45), we see that $\overline{\mathbf{Q}} = \gamma\mathbf{Q}$. The inverse of $\overline{\mathbf{Q}}$ is then

$$\overline{\mathbf{Q}}^{-1} = (\gamma\mathbf{Q})^{-1} = \mathbf{Q}^{-1}\gamma^{-1}. \tag{55}$$

The elements of $\overline{\mathbf{Q}}$ and $\overline{\mathbf{Q}}^{-1}$ are simply written, because γ is diagonal:

$$\overline{Q}_{ij} = \gamma_{ii}Q_{ij} = \frac{Q_{ij}}{\sqrt{\xi_i}}$$

and

$$(\overline{Q}^{-1})_{ij} = (Q^{-1})_{ij}\left(\frac{1}{\gamma_{jj}}\right) = \frac{Q_{ji}}{\gamma_{jj}} = \sqrt{\xi_j}\, Q_{ji}. \tag{56}$$

The equation for the evolution of the free volume state occupancy, Eq. (53), may then be written as

$$w_j(t) = \sqrt{\xi_j} \sum_{m=1}^{n} \left(\sum_{i=1}^{n} w_i(t_0) \frac{Q_{im}}{\sqrt{\xi_i}}\right) Q_{jm} \exp\{(t - t_0)\zeta_m\}. \tag{57}$$

Each free-volume state occupancy is seen to be described by a sum of terms, each of which has an exponential factor that depends on one of the eigenvalues ζ_m. The eigenvalues have several important properties [66]. First, because the eigenvalues are derivable from a symmetric matrix, they are all real. Second, because the determinant of $\mathbf{A}$ is zero, one of the eigenvalues is zero. That the determinant of $\mathbf{A}$ is zero can be shown, for example, by subtracting the first column of $\mathbf{A}$ from the second, the second column from the third, and so on until the $(n - 1)$st column is subtracted from the nth. The latter leaves the entire nth column with zeros, which causes the determinant of the matrix to be zero. Third, except for the zero eigenvalue, all of the other eigenvalues are finite and negative. That the eigenvalues are finite and negative can be understood from the fact that the sum of the eigenvalues is equal to the sum of the diagonal elements (the trace) of $\mathbf{A}$, and all of the diagonal elements of $\mathbf{A}$ are negative. The magnitudes of the eigenvalues also are similar to those of the diagonal elements. In summary, the free-volume state occupancies are sums of decaying exponentials plus a steady-state term (the zero eigenvalue).

At very long times, all of the terms in the sums describing the free-volume state occupancies, Eq. (57), approach zero, and only the steady-state term remains. We can see what the occupancies will be then by examining Eq. (57) and the corresponding equation for the transition probabilities

$$P_{ij}(t - t_0) = \sqrt{\xi_j} \sum_{m=1}^{n} Q_{im}\xi_i^{-1/2}Q_{jm} \exp\{(t - t_0)\zeta_m\} \tag{58}$$

obtained from eqs. (52) and (56). When $(t - t_0) \rightarrow \infty$, the only nonzero exponential factor becomes that for $m = n$, which equals unity. (The eigenvalues returned by the eigenvalue problem solver routine are in ascending order, and the algebraically largest eigenvalue is $\zeta_n = 0$.) Using this and the condition on the transition probabilities given by Eq. (33), we obtain

$$\sum_{j=1}^{n} P_{ij}(\infty) = Q_{in}\xi_i^{-1/2} \sum_{j=1}^{n} \sqrt{\xi_j}\, Q_{jn} = 1.$$

Since this is true for all values of i, we see that $Q_{in} \propto \sqrt{\xi_i}$. The proportionality constant is readily found to be unity, so $Q_{in} = \sqrt{\xi_i}$. Using this and Eq. (29) in Eq. (57) gives

$$w_j(\infty) = \sqrt{\xi_j} \left(\sum_{i=1}^{n} w_i(t_0) \frac{Q_{in}}{\sqrt{\xi_i}} \right) Q_{jn} = \xi_j,$$

which is just the result that we expected, that the free-volume state occupancies approach their equilibrium values at long times.

The Eigenvalue Problem

As has been mentioned, the eigenvalue problem is usually solved by computer using standard and proven routines. The routines we have used are from the package called EISPACK, which has been distributed by Argonne National Laboratories.

In the following, the *symmetrized* tridiagonal matrix $\mathbf{A}$ is formed first and then the appropriate EISPACK routines are called. The variables DWNLAM(K) and XI1(K) are λ_k^- and ξ_k, respectively. The other variables are defined in the EISPACK call routine. BANDR, TQLRAT, and TQL2 are the eigenvalue problem solving routines.

```
      REAL*8 XI1(100), DWNLAM(100), A(99,2)
      REAL*8 ZETA(99), Q(99,99), FV1(99), FV2(99)
      .  .  .  .  .  .  .  .  .
      A(1,2) = -DWNLAM(2) * DSQRT(XI1(2)/XI1(1))
      DO 610 J = 2, N - 1
        A(J,2) = -DWNLAM(J) * DSQRT(XI1(J - 1)/XI1(J)) -
          DWNLAM(J + 1) *
    1 DSQRT(XI1(J + 1)/XI1(J))
  610 CONTINUE
      A(N,2) = -DWNLAM(N) * DSQRT(XI1(N - 1)/XI1(N))
C
      DO 620 J = 2,N
        A(J,1) = DWNLAM(J)
  620 CONTINUE
      .  .  .  .  .  .  .  .  .
      CALL RSB(NM, N, MB, A, ZETA, MATZ, Q, FV1, FV2, IERR)
      IF (IERR .NE. 0) THEN
        WRITE (UNIT=6,FMT=310) IERR
  310   FORMAT(' RSB: ERROR VALUE=',I4)
        PAUSE
      END IF
      .  .  .  .  .  .  .  .  .
      SUBROUTINE RSB(NM, N, MB, A, ZETA, MATZ, Q, FV1,
        FV2, IERR)
C
      INTEGER N, MB, NM, IERR, MATZ
      REAL*8 A(NM,MB), ZETA(N), Q(NM,N), FV1(N), FV2(N)
      LOGICAL TF
C
C     This subroutine calls the recommended sequence of subroutines
C       from the eigensystem subroutine package (EISPACK) to find
C       the eigenvalues and eigenvectors (if desired) of a real symmetric
        band matrix.
C
C     On Input:
C     NM must be set to the row dimension of the two-dimensional
C       array parameters as declared in the calling program dimension
        statement;
C
C     N is the order of the matrix A;
```

```
C
C  MB is the half bandwidth of the matrix, defined as the number of
C     adjacent diagonals, including the principal diagonal, required to
C     specify the nonzero portion of the lower triangle of the matrix;
C
C  A contains the lower triangle of the real symmetric band matrix.
C     Its lowest subdiagonal is stored in the last N + 1 - MB
C     positions of the first column, its next subdiagonal in the last
C     N + 2 - MB positions of the second column, further
C     subdiagonals similarly, and finally its principal diagonal in the N
      positions of the last column. Contents of storages not part of
      the matrix are arbitrary;
C
C  MATZ is an integer variable set equal to zero if only eigenvalues
C     are desired; otherwise it is set to any nonzero integer for both
      eigenvalues and eigenvectors.
C
C  On Output:
C
C  ZETA cotains the eigenvalues in ascending order;
C
C  Q contains the eigenvectors if MATZ is not zero:
C
C  IERR is an integer output variable set equal to an error
C     completion code. The normal completion code is zero.
C
C  FV1 and FV2 are temporary storage arrays.
C
---------------------------------------------------------------------
C
      IF(N .LE. NM) GO TO 5
      IERR = 10 * N
      GO TO 50
  5   IF (MB .GT. 0) GO TO 10
      IERR = 12 * N
      GO TO 50
 10   IF (MB .LE. N) GO TO 15
      IERR = 12 * N
      GO TO 50
C
 15   IF (MATZ .NE. 0) GO TO 20
C
C  ===FIND EIGENVALUES ONLY===
```

```
C
      TF=.FALSE.
      CALL BANDR(NM, N, MB, A, ZETA, FV1, FV2, TF, Q)
      CALL TQLRAT(N, ZETA, FV2, IERR)
      GO TO 50
C
C ====FIND BOTH EIGENVALUES AND EIGENVECTORS====
C
 20   TF=.TRUE.
      CALL BANDR(NM, N, MB, A, ZETA, FV1, FV1, TF, Q)
      CALL TQL2(NM, N, ZETA, FV1, Q, IERR)
 50   RETURN
C
C ====LAST CARD OF RSB====
C
      END
```

Free-Volume State Occupancies

Having the eigenvalues, ZETA(K), the eigenvectors, Q(J,K), and the initial state populations, WP(I), the current state populations are readily found according to Eq. (57):

```
      REAL*4 WP(99), W(99), QT(99), QS(99)
      REAL*8 ZETA(99), Q(99,99), FV1(99), XI1(100)
      .   .   .   .   .   .   .   .   .   .   .
      DO 700 M = N, 1, -1
        QT(M) = 0.
        DO 690 I = 1, N
          QT(M) = QT(M) + Q(I,M) * WP(I) / DSQRT(XI1(I))
690   CONTINUE
700   CONTINUE
C
      DO 730 M = 1, N - 1
        QS(M) = QT(M) * EXP(TIME*ZETA(M))
730   CONTINUE
      QS(N) = QT(N)
      DO 750 J = 1, N
        FV1(J) = 0.
        DO 740 M = 1, N
          FV1(J) = FV1(J) + QS(M) * Q(J,M)
740   CONTINUE
      W(J) = FV1(J) * DSQRT(XI1(J))
750   CONTINUE
```

SINGLE TEMPERATURE/PRESSURE STEP

An important experiment for illustrating physical aging in plastics has been the single-temperature-step procedure. A polymer specimen at equilibrium at temperature T_0 is rapidly heated or cooled to the temperature T_1. The physical aging that accompanies the recovery of the specimen toward equilibrium at T_1 can be observed, in the volume, enthalpy, or mechanical behavior. In the following, we will examine the simulation of volume recovery.

In simulating physical aging or recovery experiments, the directly examined and calculated quantity in the simulation is the free-volume distribution, as described above. For volume recovery, we assume that the total volume depends, at least approximately, on the average free volume and not on further details of the distribution. Thus, once we have computed the free-volume distribution or the occupancies $w_i(t)$ of states with free volume $(i - 1)\beta$, $i = 1, 2, \ldots, n$, the quantity of interest is

$$f_{av}(t) = \beta \sum_{i=1}^{n} (i - 1)w_i(t). \tag{59}$$

Because we are using the Simha-Somcynsky free volume, the average free volume from Eq. (59) cannot be used as is the Doolittle free volume, for example, and be added to the "occupied" volume to obtain the total volume. Rather, it must be used in connection with the Simha-Somcynsky equation of state, Eq. (2). That is, the occupancy, $y = 1 - f_{av}(t)$, and the pressure $\tilde{p}$ and temperature $\tilde{T}$, which are presumably known, are used in Eq. (2) to solve for the volume $\tilde{V}$. The Simha-Somcynsky free volume behaves differently from the Doolittle free volume in a number of regards, among them is that it increases faster with increasing temperature than the total volume and the Doolittle free volume does not. That is, the effective occupied volume in the Simha-Somcynsky theory decreases with increasing temperature, rather than remaining constant. This also means that the free volume decreases with falling temperature even in the glass.

To compute the relaxational or recovery properties for a polymer system, there are two categories of decisions to be made. The first is the evaluation of the parameters in the theory, and the second are details of how the calculation is to be performed.

Parameters

The first group of parameters are the Simha-Somcynsky characteristic parameters, p^*, V^*, and T^*. These values are deduced from the fitting of p-V-T data of specific polymer liquids to the Simha-Somcynsky equation of state, and they now have been obtained for a number of polymers by

Simha et al. [67]. The next group of parameters are those for the Vogel equation, τ_g, c_1, and c_2, and the glass-transition temperature T_g. The glass-transition temperature is the reference temperature used in determining c_1 and c_2, and τ_g, or its reciprocal, describes the rate of temperature change under which the glass-transition temperature was measured. These also are widely tabulated [68]; in a pinch, the "universal" values $c_1 = 17.44$, $c_2 = 51.6°C$, and $\tau_g = 1\ h$ (3600 s) can be used. The final group of parameters, which are unique to this calculation, are N_s, z, and R, where N_s is the number of segments filling the minimal volume element $\mathcal{V}$ defining the free volume and for which changes in free volume are registered for any significant change in molecular conformation or molecular packing. The term N_s appears in Eqs. (15) and (27) for the mean-squared fluctuations in free volume, $\langle \delta f^2 \rangle_{av}$. It is from the latter along with the average free volume that the free-volume distribution ξ_i is obtained, defined by Eqs. (20) and (28) for β, and

$$p_r = \left[1 + \frac{(n - 1)\langle \delta f^2 \rangle_{av}}{f_{av}^2}\right]^{-1}.$$

The value of N_s has been estimated to be 26 [69] consisting of a pair of segments and the nearest-neighbor segment pairs surrounding it. Since the theoretical curves are fairly sensitive to values of N_s, small deviations from this estimate often improve fittings to the data. The parameter z defines the size of the environment surrounding the minimal volume element $\mathcal{V}$ that is involved in changes in free volume. (The size of the environment is equal to $(z - 1)\mathcal{V}$.) The term z appears in Eq. (30) for $\hat{f}_j$ and has been estimated to equal 13, arising from the volume element $\mathcal{V}$ and all of the nearest-neighbor volume elements surrounding it. Like the sensitivity to N_s, the theoretical curves are fairly sensitive to values of z, and deviations from 13 are often allowed to fit the data. The R term appears directly in Eq. (42), and its value is chosen to best fit the data. This is done simply by sliding the computed curves parallel with the log (time) axis to obtain the best fit.

Calculational Details

There are two calculational details that need to be mentioned. The first concerns the discretization of the free-volume distribution. The binomial distribution is a convenient discrete replacement for a continuous distribution with a positive variate, like the gamma distribution. Only two parameters, the mean and the mean-square deviation from the mean, are needed for the binomial distribution, and with them can be obtained a fairly symmetrical, bell-shaped distribution. The separation between free-volume states, β, is not an independent parameter with the binomial distribution, however. When the mean or mean-square deviation from the mean change, β changes. That is, the binomial distribution cannot be used with an independent set of free-volume states. An alternative is to define fixed free-volume states and then to define a sufficient number of moments of the distribution beyond the first two (used in the binomial distribution) to obtain an appropriate distribution. Since these higher moments are rarely easy to find with accuracy, we prefer staying with the binomial distribution and to define fixed free-volume states but then to allow as dependent parameters both the number of free-volume states and the mean-square deviation from the mean. The number of states is allowed to vary so as to minimize the difference between the actual and the assumed mean-square deviation from the mean.

To illustrate the above, consider a polymer specimen that is treated as shown in Figure 11. From an equilibrium state at point 0, temperature and volume T_0 and V_0, the specimen is suddenly cooled to point A, temperature and volume T_1 and V_A. Following this, the specimen is allowed to recover its equilibrium properties at point 1, temperature and volume T_1 and V_1.

The free-volume distribution is established at point A. The mean and mean-square deviation from the mean for point A are estimated from Eqs. (2), (3), and (15) using the volume V_A obtained from an equation analogous to Eq. (11). The free-volume separation between adjacent states, β_A, and the distribution are given then by the binomial distribution. (We have used

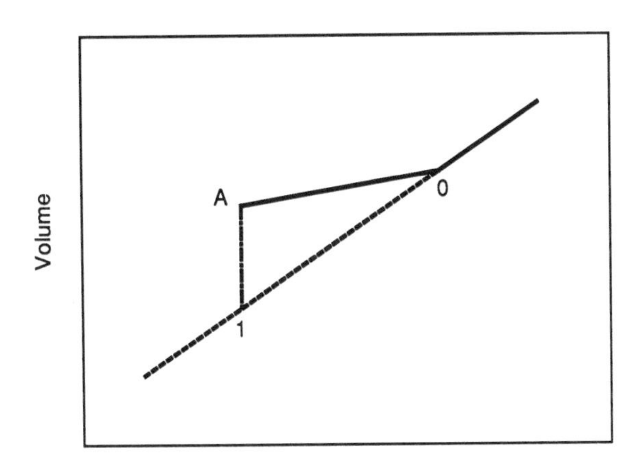

Figure 11 Typical volume recovery experiment. The step from 0 to A is made rapidly, and that from A to 1 occurs slowly at the rate determined by the internal molecular dynamics.

the subscript A here to emphasize that the separation between free-volume states is established at point A.)

As recovery brings the system into equilibrium at point 1, however, the mean and the mean-squared deviation from the mean at this point, $f_{\mathrm{av},1}$ and $\langle \delta f^2 \rangle_{\mathrm{av},1}$ yield a different binomial distribution. The free-volume separation between states for this is given by Eq. (28), and we will denote it by β_1. But rather than letting β deviate in this way from β_A, we allow a variation in the number of states and the mean-square deviation from the mean. The number of states at point 1, n_1, is assumed to be the integer closest to

$$\nu = \frac{f_{\mathrm{av},1}}{\beta_A - \langle \delta f^2 \rangle_{\mathrm{av},1} / f_{\mathrm{av},1}}.$$

This equation is obtained by solving Eq. (28) for n. Because the binomial distributions at both points A and 1 typically have a number of high free-volume states with very low occupancy, we can easily add or drop states on going from n_A to n_1 without significant change to the distribution. To avoid any inconsistency, however, the mean-square deviation from the mean is assumed to be

$$\langle \delta f^2 \rangle_{\mathrm{av},\text{``}1\text{''}} = f_{\mathrm{av},1} \left(\beta_A - \frac{f_{\mathrm{av},1}}{n_1} \right)$$

rather than $\langle \delta f^2 \rangle_{\mathrm{av},1}$.

The second calculational detail to be mentioned is the nonlinearity of recovery. Hence, the differential equations discussed in the section "Free-Volume Theory of Volume Recovery" are approximately correct only during limited time intervals. The reason is that the average free volume, which appears in Eq. (29), is continually changing. Hence, in this approximation, the time interval $(t - t_0)$ in Eq. (57), for example, represents the evolution of the recovery at fixed average free volume. On the question of how often must the average free volume be updated, we have found that it probably is necessary no more often than once each 1°C change in fictive or structural temperature.

An Example: Poly(vinyl acetate) (Refs. 36 and 59). We give here a simulation of the celebrated single-temperature-step data on poly(vinyl acetate) obtained by Kovacs [54]. The simulation will differ in one important respect from that described above. Rather than using the Vogel equation given in Eq. (22) or (24), we will use a modified form given by [59]

$$\tau^{-1} = (\tau_g a_T)^{-1} + (\tau_b b_T)^{-1}, \tag{60}$$

where

$$\log a_T = -c_1 + \frac{c_1 c_2}{c_2 + (f - f_g)T^*/f^*}, \tag{61}$$

$$\log b_T = -C_1 + \frac{C_1 C_2}{C_2 + (f - f_g)T^*/f^*}. \tag{62}$$

Equation (24) is equivalent to having only the first term on the right-hand side of Eq. (60). The change in the rate $(\tau/\tau_g)^{-1}$ induced by including the second terms is shown in Figure 12. Essentially all of the change occurs at temperatures below 30°C (the glass transition temperature is approximately 35°C). The data that defines the curves all occurs above 35°C and are those obtained by Plazek for poly(vinyl acetate) in the "softening region" [70]. Kovacs data was found to be extremely well-fitted with the Vogel equation derived from this data for short times, when the recovery process is dominated by states of high free volume, as is characteristic at temperatures above the glass transition. But at long times, when recovery is dominated by states of low free volume, characteristic of temperatures well below the glass transition, the data was found to be badly fitted. The predicted rate of recovery was much too low. This suggested the extrapolation of $(\tau/\tau_g)^{-1}$

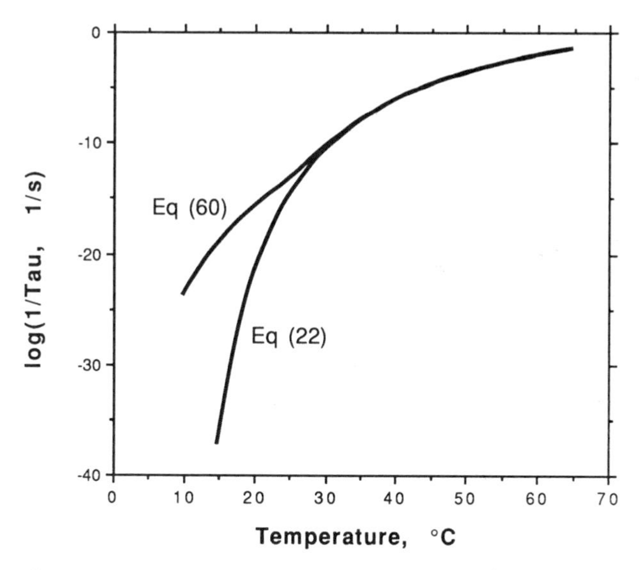

Figure 12 Comparison of the modified relaxation rate for poly(vinyl acetate) given by Eq. (60) with the unmodified rate vs temperature.

Table 1 Parameters for Volume Recovery Kinetics of Poly(vinyl acetate)

Parameter	Description
$p^* = 9380$ bar[a]	Simha-Somcynsky characteristic parameters
$V^* = 814.1$ mm³/g[a]	
$T^* = 9419.$ K[a]	
$c_1 = 12.81$[b]	Time-temperature shift parameters
$c_2 = 28.74$ K[b]	
$C_1 = 11.24$[c]	
$C_2 = 45.96$ K[c]	
$T_g = 308.$ K	Glass-transition temperature
$\tau_g = 1$ h (3600 s)	Nominal relaxation time at T_g
$\tau_b = 36{,}000$ s[c]	Relaxation time of sub-T_g motion at T_g
$N_s = 26$	Number of monomer segments in free-volume transition region
$z = 13$	Size ratio for region controlling free-volume changes
$R = 0.0022$	Translation factor between macroscopic and microscopic processes

[a]J. E. McKinney and R. Simha, *Macromolecules*, *7*, 894 (1974).
[b]D. J. Plazek, *Polym. J.*, *12*, 43 (1980).
[c]R. E. Robertson, *Macromolecules*, *18*, 953 (1985).

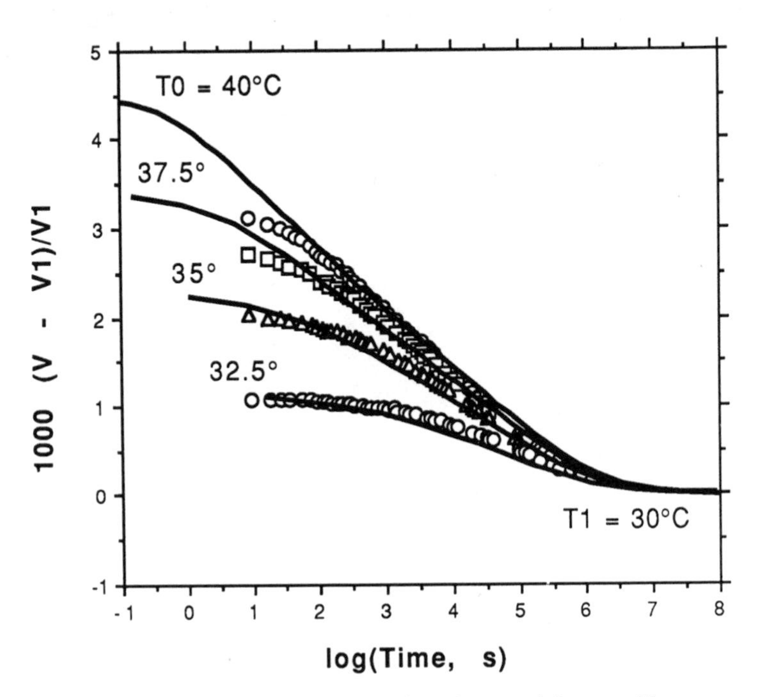

Figure 13 Volume recovery of poly(vinyl acetate) from sudden temperature steps to 30°C from 32.5°, 35°, 37.5°, and 40°C. Comparison of theory and experiment. (Data from Ref. 54.)

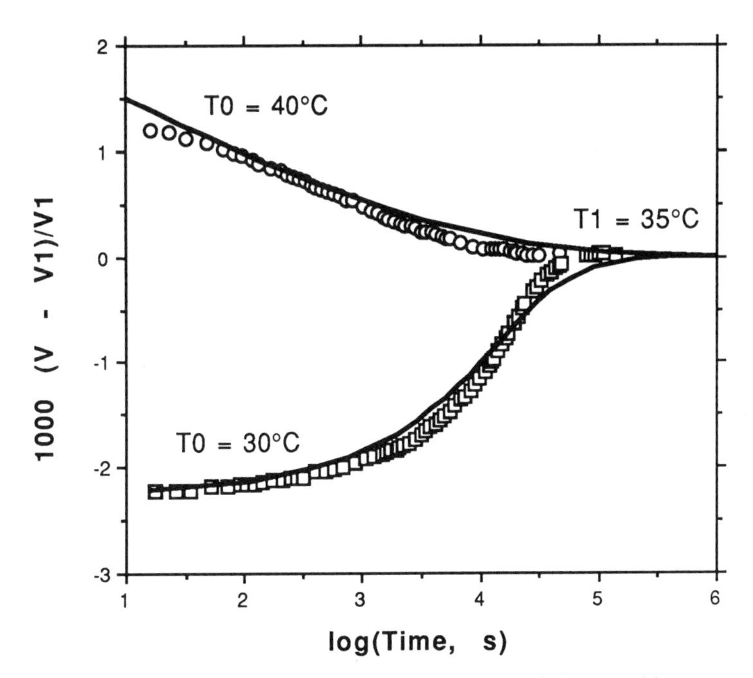

Figure 14 Volume recovery of poly(vinyl acetate) from sudden temperature steps to 35°C from 30° and 40°C. Comparison of theory and experiment. (Data from Ref. 54.)

to low temperature like that in Figure 12. The modification to the normal time-temperature shift parameter was made by adding a term representing sub-T_g motions, which are believed to be affecting low-temperature relaxations.

The various parameters needed for the computation are given in Table 1. A convenient procedure in simulations like this is to make an instantaneous change in the temperature from T_0 to T_1. This simplification causes little problem for upward steps in temperature, but is more problematic for downward steps. This is because in downward steps, considerable relaxation can occur experimentally at the higher temperatures through which the system slowly passes on the way to T_1.

Theoretical rates of volume recovery are compared with experiment in Figures 13–15. Figure 13 shows recovery from sudden steps to 30°C from 32.5°, 35°, 37.5°, and 40°C. Figure 14 shows recovery from sudden steps to 35°C from 30° and 40°C. Figure 15 shows recovery from sudden steps to 40°C from 30°, 32.5°, 35°, and 37.5°C. The data were obtained by Kovacs for a sample of poly(vinyl acetate) that he referred to as PVAc II [54].

A property often of more concern than the volume itself has been the

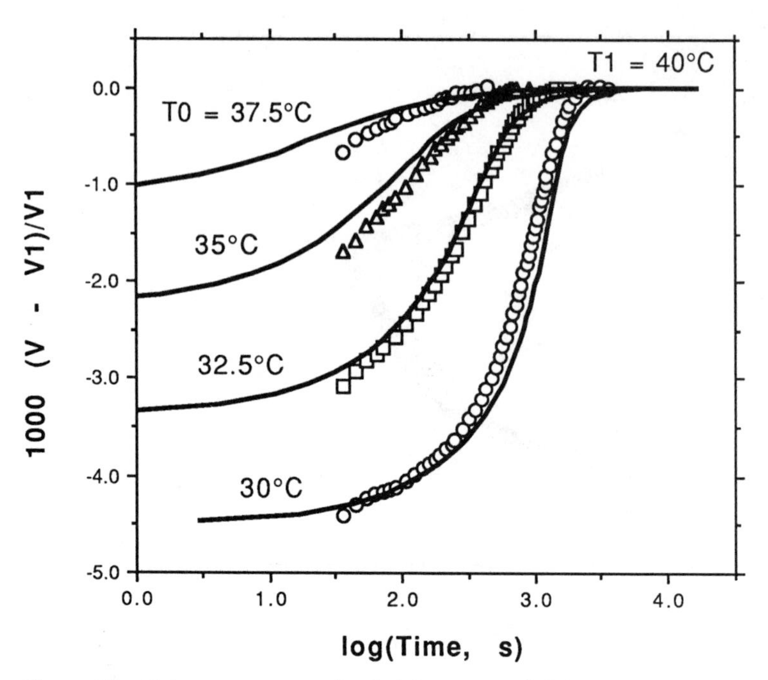

Figure 15 Volume recovery of poly(vinyl acetate) from sudden temperature steps to 40°C from 30°, 32.5°, 35°, and 37.5°C. Comparison of theory and experiment. (Data from Ref. 54.)

effective relaxation time for recovery, τ_{eff}. This is given by

$$\tau_{eff}^{-1} = -\frac{dV}{dt}(V - V_{eq})^{-1} \approx -\frac{df_{av}(t)}{dt}[f_{av}(t) - (f_{av})_{eq}]^{-1}, \tag{63}$$

where the subscript "eq" refers to equilibrium. The second form of the equation arises from the near proportionality between V and f_{av}. f_{av} can be written as

$$f_{av}(t) = \sum_{m=1}^{n} C_m \exp\{(t - t_0)\zeta_m\}, \tag{64}$$

where, from Eqs. (57) and (59),

$$C_m = \beta \sum_{j=1}^{n} (j - 1)\sqrt{\xi_j}\left(\sum_{i=1}^{n} w_i(t_0)\frac{Q_{im}}{\sqrt{\xi_i}}\right) Q_{jm}. \tag{65}$$

Then,

$$\tau_{\mathrm{eff}}^{-1} \approx -\sum_{m=1}^{n-1} \zeta_m C_m \exp\{(t - t_0)\zeta_m\} \times \left[\sum_{m=1}^{n-1} C_m \exp\{(t - t_0)\zeta_m\}\right]^{-1}. \tag{66}$$

In deriving this equation we have made use of the fact that $\zeta_n = 0$ and $C_n = (f_{\mathrm{av}})_{\mathrm{eq}}$. The calculated and experimental effective relaxation times for the volume recovery of poly(vinyl acetate) are compared in Figure 16.

Calculation of Volume from Free Volume

The following routine converts the average free volume $f_{\mathrm{av}}(t)$ = FIN to the total volume $V(t)$ = VOUT. The input variables are T, P, and FIN; the output variable is VOUT.

```
      SUBROUTINE VVST(T, P, VOUT, FIN)
C
      COMMON/STAR/PSTAR, VSTAR, TSTAR
      COMMON/VST/PP, VV, TT, YVV2
```

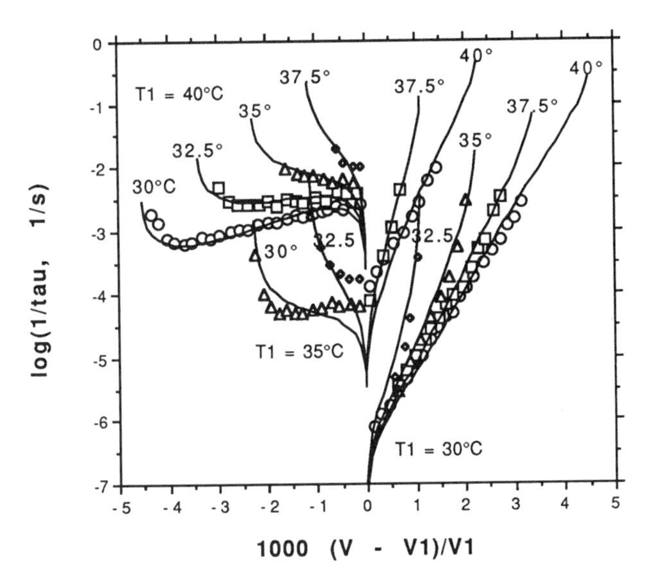

Figure 16 Effective relaxation rates for the volume recovery of poly(vinyl acetate). Comparison of theory and experiment. Attached to each curve is the initial temperature, T_0. (Data from Ref. 54.)

```
C
      TT = T/TSTAR
      PP = P/PSTAR
      Y = 1. - FIN
      YVV2 = 1.
C
      DO 10 I = 1, 6
        YV3 = Y/((2.*YVV2)**(1./6.))
        RYVV2 = 2.409/4.044 + SQRT((2.409/4.044)**2 + (PP*VV -
          TT/(1.
          1 - YV3))/(2.022*Y))
        YVV2 = 1./RYVV2
        VV = SQRT(YVV2)/Y
10    CONTINUE
C
      VOUT=VSTAR*VV
C
      RETURN
      END
```

Multiple Temperature/Pressure Steps [60]

Liquids near the glass transition can exhibit unusual behavior when subjected to a sequence of steps in temperature or pressure. Properties like volume and enthalpy may pass at constant temperature and pressure through extrema. If, for example, the temperature of a liquid at equilibrium at the temperature T_0 is suddenly stepped to T_1 and held there for a relatively short time and then stepped to a third temperature T_2 somewhere between T_0 and T_1 and held at this temperature, the observed volume is likely to pass through either a minimum or a maximum before evolving toward a final, equilibrium value. A maximum occurs when the first step is downward in temperature, and a minimum occurs when the first step is upward. This behavior is sometimes called the "memory effect."

Striking examples of the memory effect were observed in the volume recovery experiments of Kovacs [71,54]. Kovacs found volume maxima for polystyrene [71] and poly(vinyl acetate) [54] from a sequence of temperature steps that began with a downward step. Similar changes in volume have been observed by others in these polymers [72] and in the amorphous part of crystalline polyethylene [73]. Analogous behavior in the dielectric loss of polystyrene blends has also been observed [74]. Memory effects are well known also with inorganic glasses, from both steps in temperature [75] and steps in pressure [76].

This unusual behavior during the recovery of equilibrium is readily explained in terms of a multiplicity of decay or retardation times, such as arises from a multiplicity of local environments. As discussed on page 326 the state of the glass at any instant is assumed to be describable by the temperature, the pressure, and the free volume distribution at that instant. At any given temperature and pressure, the nonequilibrium glass drifts (with a rate that is determined largely by the free-volume distribution) toward the equilibrium state corresponding to that temperature and pressure. Included in the approach toward equilibrium is the free-volume distribution, and this continues until equilibrium is reached. If the temperature or pressure are changed, whether before or after equilibrium is reached, the system immediately begins drifting toward a different equilibrium state and free-volume distribution corresponding to the new temperature and pressure. The rate of change continues to be determined largely by the current free-volume distribution, but at an arbitrary moment this could be very different from a typical equilibrium distribution. Except for the direction in which the system continues to drift, the multiple-temperature- or pressure-step experiment causes little change in the simulation of recovery from that arising from a single temperature or pressure step.

An Example: Poly(vinyl acetate). We continue the simulation of Kovacs's experiments on the volume recovery of poly(vinyl acetate) begun on page 342 with the single-temperature-step experiments and here examine the double-temperature-step experiment. The temperatures and times of the experiments that were simulated are given in Table 2. The results of the computation are shown in Figure 17 with the experimental data of Kovacs [54]. The abscissa is the elapsed time since the last temperature step, namely the step to 303 K. As indicated in Table 2, experiment (1) involves only a single temperature-step, from T_0 to T_2. The other three experiments involve double-steps, with t_1 being the time spent at T_1. The main difference between the real and the simulated experiments is the time required to change temperature. The computer simulation executes an

Table 2 Double-Step Experiments

Experiment	T_0 (K)	T_1 (K)	t_1, (10^5 s)	T_2 (K)
1	313.	—	—	303.
2	313.	283.	5.75	303.
3	313.	288.	5.0	303.
4	313.	298.	3.25	303.

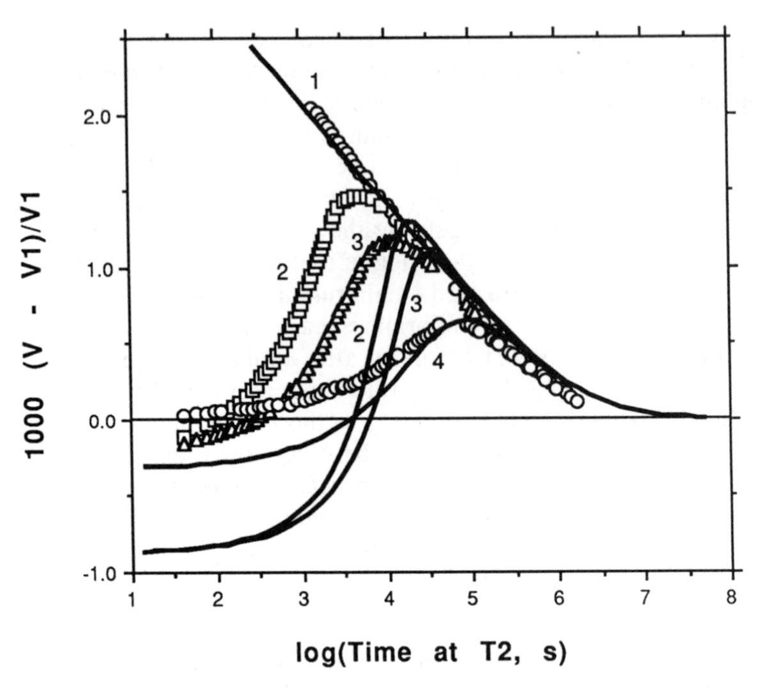

Figure 17 Volume recovery at T_2 (30°C) following the history given in Table 2 for poly(vinyl acetate). Comparison of theory and experiment. (Data from Ref. 54.)

instantaneous temperature change. This difference probably has little effect on the comparison, however.

The computed curves in Figure 17 represent essential features of the data. The general magnitudes of the maxima are well represented, as is the ordering of the curves. Moreover, all of the computed curves are seen to converge at long times, as occurs experimentally. The major discrepancy between theory and experiment in Figure 17 is caused by the magnitudes of the volumes at the beginning of recovery at T_2. The computed volumes have fallen too far during the residence at T_1. As is seen in Figure 17, this is a more serious problem with the lower temperatures T_1. That is, the discrepancy is worse with experiments (2) and (3) than with experiment (4). The problem had not arisen in the previous work with the single-temperature steps because the temperature was never below 30°C. This too rapid fall in volume at T_1 results in two major problems apparent in Figure 17: the times at which the maxima occur after the step to T_2 are delayed, and the heights of the maxima are reduced.

The likely reason for the excessively high rate of recovery predicted at low temperatures is that the individual effects of thermal agitation and structure on the rate have not been properly accounted for. In particular, the theory has not slowed the kinetics in proportion to the diminution in thermal agitation at lower temperatures. The theory being used (the RSC theory) is basically, though not exclusively, a structure or free-volume theory.

By analogy with other molecular processes, if the structure were to remain constant, the recovery rate would be expected to have an Arrhenius form, such as

$$\tau^{-1} = A \exp\left[-\frac{H^{\ddagger}(\theta)}{RT} \right], \tag{67}$$

where $H^{\ddagger}$ is the activation energy for the structure denoted by θ. But the structure of the recovering glass does not remain constant, meaning that $H^{\ddagger}$ changes with time. The changing structure will be assumed to be describable by the single parameter θ, the fictive temperature.

The fictive temperature is that temperature for which the equilibrium liquid has the same structure as the nonequilibrium state of interest. The fictive temperature is obtained by drawing a straight line on a volume-temperature plot parallel with the glass line. The line extends from the point that represents the volume and temperature of the glass of interest to the equilibrium volume-temperature line. The temperature at which this line intersects the equilibrium line is the fictive temperature.

Continuing to use the rate given formally by Eq. (60), the activation enthalpy for a nonequilibrium structure θ is estimated to be [60]

$$H^{\ddagger}(\theta) = R\theta \left\{ \ln(\tau_g^{-1} + \tau_b^{-1}) - \ln[(\tau_g a_T)^{-1} + (\tau_b b_T)^{-1}] + \frac{H_g^{\ddagger}}{RT_g} \right\}, \tag{68}$$

where $H_g^{\ddagger}$ is the activation enthalpy at the glass transition, which must be estimated or otherwise obtained. Note that the common method of taking the derivative of the natural logarithm of eq (60) with respect to $1/T$, with the shift parameters expressed in terms of temperature, does not give the activation enthalpy because the implicit temperature-dependent change in structure is not excluded. The front factor A in Eq. (67) is given by

$$\ln A = \ln(\tau_g^{-1} + \tau_b^{-1}) + \frac{H_g^{\ddagger}}{RT_g}. \tag{69}$$

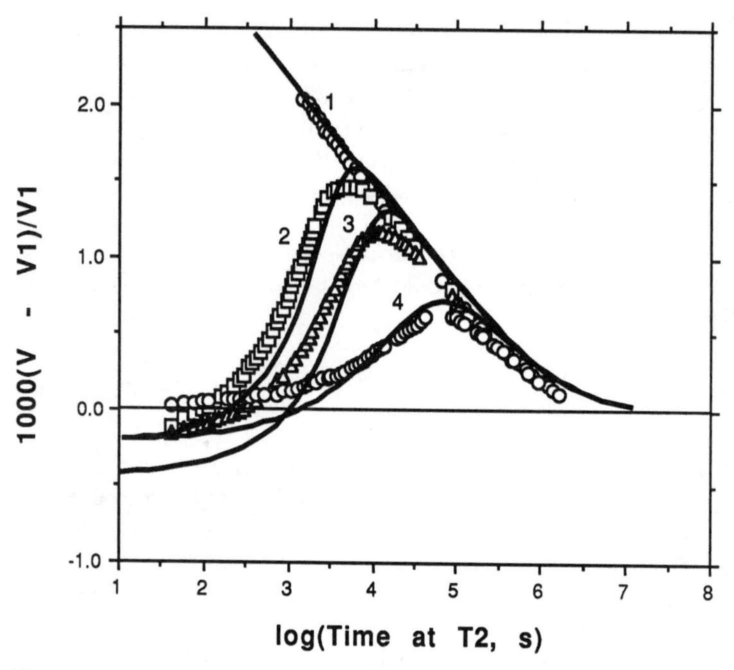

Figure 18 Volume recovery at T_2 (30°C) following the history given in Table 2 for poly(vinyl acetate). Comparison of theory, using Eq. (70) instead of Eq. (60), and experiment. (Data from Ref. 54.)

Combining Eqs. (67) to (69) gives

$$\ln \tau^{-1} = \frac{\theta}{T} \ln[(\tau_g a_T)^{-1} + (\tau_b b_T)^{-1}] + \left(1 - \frac{\theta}{T}\right) \left[\ln(\tau_g^{-1} + \tau_b^{-1}) + \frac{H_g^{\ddagger}}{RT_g}\right]. \quad (70)$$

Note that Eq. (70) reverts to Eq. (60) at equilibrium when $\theta = T$.

A comparison between theory and experiment with $H_g^* = 50$ kJ/mole is shown in Figure 18. The fit is seen to be considerably improved over that in Figure 17. No other changes in the calculations were made, and the values of the parameters N_s, z, and R that had previously been used to fit Kovacs's single step data have been retained. The fit is improved slightly if these parameters are optimzied for Eq. (70) instead of Eq. (60), but this has not been done in Figure 18.

Enthalpy Recovery

Rapid changes in temperature, pressure, or other external variables may leave the system away from equilibrium with respect to enthalpy as well as to volume. The recovery of equilibrium by the enthalpy is expected to be similar to that of volume. In particular, since the free volume is expected to have an enthalpic component, at least this component of the enthalpy is expected to progress toward its equilibrium value just as the free volume progresses toward its equilibrium distribution. For this reason, Jain and Simha, for example, treated the recovery of enthalpy in terms of that of volume, both being functionals of free volume [77]. Although this seems like a reasonable approximation, the volume and enthalpy recoveries are found to be somewhat uncorrelated. This is perhaps best seen in the excess fluctuations (the frozen spatial fluctuations) remaining in the glass as the equilibrium liquid is cooled below the glass transition. The fluctuations are one aspect of the nonequilibrium state. Jäckle has shown that the correlation between the fluctuations in volume and enthalpy can be expressed in terms of the Prigogine-Defay ratio [78]:

$$\frac{\overline{(\Delta H)^2} \cdot \overline{(\Delta V)^2}}{\overline{(\Delta H \cdot \Delta V)}^2} = \frac{\Delta c_p(T_g)\, \Delta \kappa_T(T_g)}{T_g[\Delta \alpha_p(T_g)]^2}, \tag{71}$$

where ΔH and ΔV are the frozen spatial fluctuations in enthalpy and volume, and Δc_p, $\Delta \kappa_T$, and $\Delta \alpha_p$ are the changes at the glass transition of the heat capacity per unit volume at constant presure, the compressibility at constant temperature, and the thermal expansivity at constant pressure, respectively. If the right-hand side of this equation, the reciprocal of the Prigogine-Defay ratio, were equal to unity, the spatial fluctuations in enthalpy and volume would be correlated. But this ratio has been found to exceed one for most systems [79], indicating incomplete correlation between enthalpy and volume.

In spite of the lack of numerical correlation between the rates of volume and enthalpy recovery, there may remain for each a basic correlation with free volume. As with the volume recovery description above, the enthalpy recovery may also be describable by changes in the free-volume distribution, but with different parameters than are used for the volume recovery. This has not yet been explored, however.

General Processes Occurring in the Glass

Two volume elements were defined for recovery. These were the elements of volume $\mathcal{V}$ that define the free volume and the larger elements of volume $z\mathcal{V}$ that control the change in free volume and the process of recovery. For an arbitrary kinetic process occurring in the glass, such as diffusion or a chemical reaction, the controlling volume element is likely to be something different. We can define it to have the volume $x\mathcal{V}$

As was done for recovery, these elements of volume $x\mathcal{V}$ can be defined approximately as having the discrete amounts of free volume given by

$$\hat{f}_{xi} = \frac{f_i + (x - 1)f_{av}(t)}{x} \tag{72}$$

where f_i's are the free volume amounts allowed to the elements of volume $\mathcal{V}$, namely $(i - 1)\beta$. The probability of occurrence of the free volume $\hat{f}_{xi}$ is $w_i(t)$, the same as for f_i, and this approaches ξ_i as equilibrium is approached. Although the values of f_i remain constant, the values of $\hat{f}_{xi}$ do not because f_{av} varies with time. To have a set of time-independent free-volume states for the elements of volume $x\mathcal{V}$, it is convenient to introduce the fixed free-volume states

$$\hat{h}_i = \hat{f}_{xi}(\infty) = \frac{h_i(t) + (x - 1)f_{av}(t)}{x}. \tag{73}$$

A comparison of this equation with Eq. (72) shows that the $h_i(t)$'s are given by

$$h_i(t) = f_i + (x - 1)\,[f_{av}(\infty) - f_{av}(t)]. \tag{74}$$

The probability of occurrence of these states, both the $h_i(t)$'s and $\hat{h}_i$'s, is given by a set of values $u_i(t)$, which are obtained by redistributing the $w_j(t)$'s. That is, when $h_i(t) < f_j < h_{i+1}(t)$, then $w_j(t)$ is distributed between

Table 3 Parameters for Recovery Kinetics of Polystyrene

Parameter	Description
$p^* = 7453$ bar[a]	Simha-Somcynsky characteristic parameters
$V^* = 0.9598$ mm³/g[a]	
$T^* = 12{,}680.$ K[a]	
$c_1 = 13.3$[b]	Time-temperature shift parameters
$c_2 = 47.5$ K[b]	
$T_g = 373$ K	Glass-transition temperature
$\tau_g = 1$ h (3600 s)	Nominal relaxation time at glass transition
$N_s = 40$	Number of segments in recovery region
$z = 12$	Size ratio for regions controlling free-volume change
$R = 5.3$	Translation factor between macroscopic and microscopic processes

[a]A. Quach and R. Simha, *J. Appl. Phys.*, *42*, 4592 (1971).
[b]D. J. Plazek, *J. Phys. Chem.*, *69*, 3480 (1965).

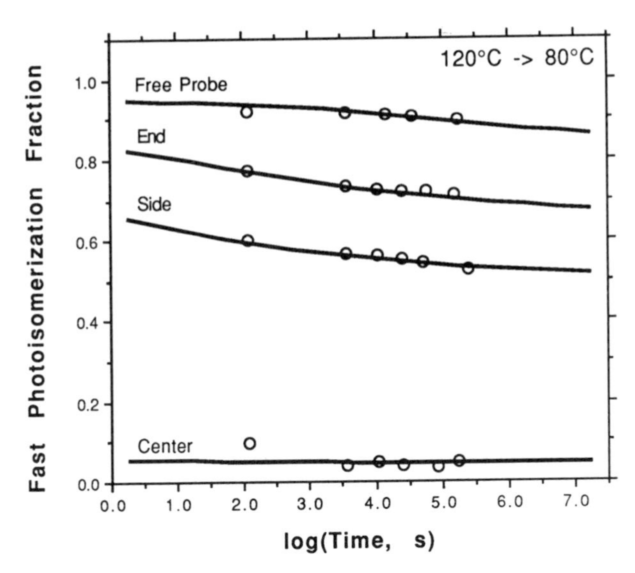

Figure 19 Fast photoisomerization fractions versus annealing time at 80°C following equilibration at 120°C. Comparison of theory and experiment. (After Ref. 24.)

$u_i(t)$ and $u_{i+1}(t)$ according to the relative separation between $h_i(t)$, f_j, and $h_{i+1}(t)$.

Although an arbitrary kinetic process can be expected to require an arbitrary controlling volume element of size $x\mathcal{V}$, it may still occur that the volume element $\mathcal{V}$ is most appropriate. This is what occurred in the following example.

An Example: Photoisomerization in a Polystyrene Matrix [24]. An example of a general kinetic process occurring in a glass is the photoisomerization of azobenzene probes in polystyrene. Azobenzene probes can be attached to polystyrene at different locations as labels as well as be dispersed in the polymer as an unattached small molecule. In this way, the relative freedom of motion or effective hole size at different locations in the glass can be discovered.

While the photoisomerization of the azobenzene probes is characterized by a single rate constant in dilute solution, at least two rates must be assumed for the solid [24]. The fast rate is similar to the rate in dilute solution, and the slow rate is about two orders of magnitude slower. The fraction (α) of the fast process decreases with physical aging but increases with temperature, plasticization, or deformation. Assuming that the faster

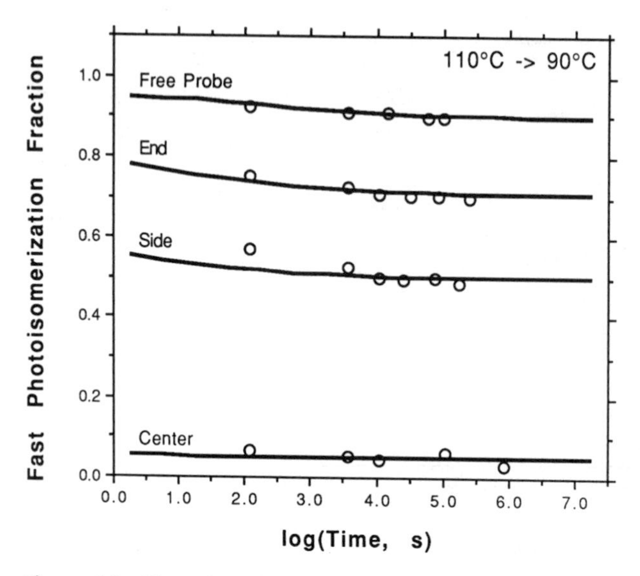

Figure 20 Fast photoisomerization fractions versus annealing time at 90°C following equilibration at 110°C. Comparison of theory and experiment. (After Ref. 24.)

photoisomerization requires a critical free volume in the immediate vicinity of the label, the fast fraction, α, can be interpreted as the cumulative area under the free-volume distribution curve for free volumes above a critical size. With probes at specific sites in the bulk polymer, the fraction of the fast process, α, can be used to indicate the free-volume size distribution in the vicinity of these sites along the polymer chain or in between the chains as a function of aging time.

To be able to predict the kinetics of isomerization, we will want to follow the changes in the population of sites having free volume both above and below some fixed, critical free volume. We will denote this critical free volume by i^*, meaning that it falls between $(i^* - 1)\beta$ and $i^*\beta$. The populations above the critical free volume are given by

$$\alpha = 1 - \sum_{i=1}^{i^*} w_i(t), \tag{75}$$

with the remainder of the population below.

The parameters for the calculation are given in Table 3; the number of free volume states (n) was 16. The parameters are the same as those used previously [39] to fit volume recovery data on polystyrene of Goldbach and Rehage [80]. The regions of interest in the computation are those having the size containing $N_s = 40$ monomer segments. The computation gives the time development of the populations of these regions with 16 amounts of free volume.

The computed fractions of regions having the critical or greater free volume are plotted against aging time in Figures 19–21. Represented in these figures are the temperature steps from 120° to 80°C and from 110° to 90°C and to 80°C, respectively. The data for the rapidly isomerizing fractions have been plotted on the same graphs. The theory gives a family of curves; the particular curves chosen here were those that best represented the critical free volume. The points in Figures 19–21 closest to the ordinate, those just above 100 s, were not aged at the temperatures indicated but were obtained by quenching from the initial temperature to room temperature. Thus, their lack of being fitted by the computed curves, especially in Figure 20, is of little concern.

The computed fit to the data is quite satisfactory. An aspect of the theory being tested here is its uniqueness with respects to other aging theories, namely its reliance on a distribution of free volume, which is assumed to be generated by thermal fluctuations. Thus, with the assumption that the fast and slow photoisomerization processes do in fact arise from different amounts of free volume, then these experiments yield directly the distribution of free volume. Moreover, the distribution it yields seems to be exactly that predicted.

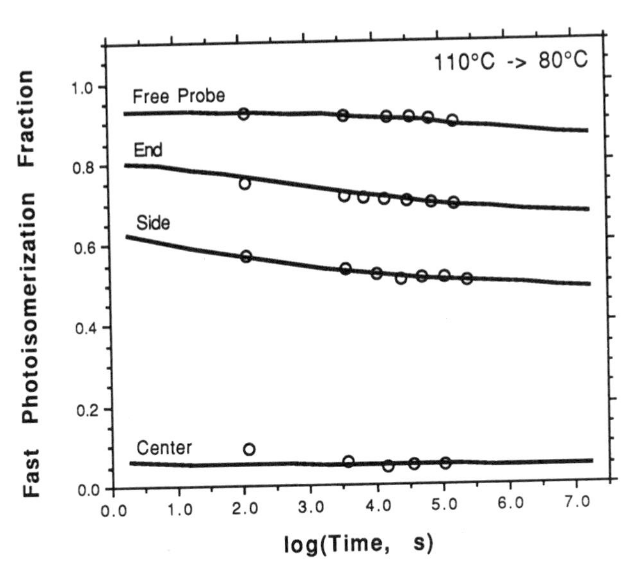

Figure 21 Fast photoisomerization fractions versus annealing time at 80°C following equilibration at 110°C. Comparison of theory and experiment. (After Ref. 24.)

REFERENCES

1. J. Burnet, in *The Legacy of Greece*, pp. 62–64 (R. W. Livingstone, ed.), Clarendon Press, Oxford (1922).
2. Ref. 1, pp. 68–69.
3. M. West, in *The Oxford History of the Classical World*, p. 121, (J. Boardman, J. Griffin, and O. Murray, eds.), Oxford University Press, New York (1986).
4. J. Barnes, in *The Oxford History of the Classical World*, p. 367 (J. Boardman, J. Griffin, and O. Murray, eds.), Oxford University Press, New York (1986).
5. Lucretius, *De Rerum Natura* (W. H. D. Rouse, tr., rev. by M. F. Smith) 2nd ed, pp. 29–35 Harvard Univ. Press, Cambridge, MA (1982).
6. A. J. Batschinski, *Z. Phys. Chem.*, *84*, 644 (1913); R. B. Macleod, *Trans. Faraday Soc.*, *19*, 6 (1923).
7. A. K. Doolittle and D. B. Doolittle, *J. Appl. Phys.*, *28*, 901 (1957).
8. A. Bondi, *J. Phys. Chem.*, *58*, 929 (1954).
9. J. H. Hildebrand and R. L. Scott, *The Solubility of Non-Electrolytes*, pp. 73–75, Van Nostrand Reinhold, New York (1950).
10. A. K. Doolittle, *J. Appl. Phys.*, *22*, 1471 (1951).
11. P. W. Bridgman, *The Physics of High Pressure*, pp. 330–356, Macmillan, New York (1931).
12. R. J. Roe and J. J. Curro, *Macromolecules*, *16*, 428 (1983); J. J. Curro and R. J. Roe, *Polymer*, *25*, 1424 (1984).
13. R. J. Roe and H. H. Song, *Macromolecules*, *18*, 1603 (1985).
14. S. J. Tao, *J. Chem. Phys.*, *56*, 5499 (1972); H. Ache, ed., *Positronium and Muonium Chemistry*, Adv. Chem. Ser. 175, American Chemical Society, Washington, DC (1979); R. W. Siegel, *Ann. Rev. Mater. Sci.*, *10*, 393 (1980).
15. M. Eldrup, in *Positron Annihilation*, p. 753 (P. G. Coleman, S. C. Sharma, and L. M. Diana, eds.), North-Holland, Amsterdam (1982).
16. Y. C. Jean, T. C. Sandrecki, and D. P. Ames, *J. Polym. Sci., Polym. Phys. Ed.*, *24*, 1247 (1986).
17. Y. Y. Wang, H. Nakanishi, Y. C. Jean, and T. C. Sandrecki, to appear.
18. Y. Ujihara, T. Ryuo, Y. Kobayashi, and T. Nomizu, *Appl. Phys.*, *16*, 71 (1978).
19. B. D. Malhortra and R. A. Pethrick, *Eur. Polym. J.*, *19*, 457 (1983); B. D. Malhortra and R. A. Pethrick, *Polymer*, *24*, 165 (1983).
20. Y. Kobayashi, W. Zheng, E. F. Meyer, J. D. McGervey, A. M. Jamieson, and R. Simha, *Macromolecules*, *22*, 2302 (1989).
21. J. G. Victor and J. M. Torkelson, *Macromolecules*, *20*, 2241 (1987).
22. L. Lamarre and C. S. P. Sung, *Macromolecules*, *16*, 1729 (1983).
23. C. S. P. Sung, I. R. Gould, and N. J. Turro, *Macromolecules*, *17*, 1447 (1984).
24. W.-C. Yu, C. S. P. Sung, and R. E. Robertson, *Macromolecules*, *21*, 355 (1988).
25. A. Munster, *Statistical Thermodynamics*, pp. 321–540, Springer-Verlag, Berlin (1974). C. A. Croxton, *Introduction to Liquid State Physics*, Wiley, London (1975).
26. J. A. Barker, *Lattice Theories of the Liquid State*, Macmillan, New York (1963).

27. R. Simha and T. Somcynsky, *Macrmolecules*, *2*, 342 (1969); R. Simha, *Macromolecules*, *10*, 1025 (1977).
28. J. G. Curro, *J. Macromol. Sci.*, *Revs. Macromol. Chem.*, *C11*, 321 (1974).
29. P. J. Flory, R. A. Orwoll, and A. Vrij, *J. Am. Chem. Soc.*, *86*, 3507 (1964); P. J. Flory, *J. Am. Chem. Soc.*, *87*, 1833 (1965); B. E. Eichinger and P. J. Flory, *Trans. Faraday Soc.*, *64*, 2035 (1968).
30. I. C. Sanchez and R. H. Lacombe, *J. Phys. Chem.*, *80*, 2352 (1976); I. C. Sanchez and R. H. Lacombe, *J. Polym. Sci.*, *Polym. Lett. Ed.*, *15*, 71 (1977).
31. T. Nose, *Polym. J.*, *2*, 124 247 (1971); S. Beret and J. M. Prausnitz, *Macromolecules*, *8*, 878 (1975).
32. J. E. McKinney and R. Simha, *Macromolecules*, *9*, 430 (1976); see also the references cited therein.
33. A. Quach and R. Simha, *J. Appl. Phys.*, *42*, 4592 (1971).
34. H. Oels and G. Rehage, *Macromolecules*, *10*, 1036 (1977).
35. R. C. Tolman, *The Principles of Statistical Mechanics*, p. 640, Oxford Unviersity Press, New York (1938).
36. R. E. Robertson, R. Simha, and J. G. Curro, *Macromolecules*, *17*, 911 (1984).
37. L. W. Jelinski, J. J. Dumais, and A. K. Engel, *Macromolecules*, *16*, 492 (1983).
38. M. L. Williams, R. F. Landel, and J. D. Ferry, *J. Am. Chem. Soc.*, *77*, 3701 (1955).
39. R. E. Robertson, R. Simha, and J. G. Curro, *Macromolecules*, *18*, 2239 (1985).
40. M. R. Tant and G. L. Wilkes, *Polym. Eng. Sci.*, *21*, 874 (1981).
41. M. A. DeBolt, A. J. Easteal, P. B. Macedo, and C. T. Moynihan, *J. Am. Ceram. Soc.*, *59*, 16 (1976); C. T. Moynihan, P. B. Macedo, C. J. Montrose, P. K. Gupta, M. A. DeBolt, J. F. Dill, B. E. Dom, P. W. Drake, A. J. Easteal, P. B. Elterman, R. P. Moeller, H. Sasabe, and J. A. Wilder, *Ann. N.Y. Acad. Sci.*, *279*, 15 (1976).
42. R. Gardon and O. S. Narayanaswamy, *J. Am. Ceram. Soc.*, *53*, 380 (1970); O. S. Narayanaswamy, *J. Am. Ceram. Soc.*, *54*, 491 (1971).
43. I. M. Hodge and A. R. Berens, *Macromolecules*, *15*, 762 (1982); I. M. Hodge and G. S. Huvard, *Macromolecules*, *16*, 371 (1983); I. M. Hodge, *Macromolecules*, *16*, 898 (1983).
44. I. M. Hodge, *Macromolecules*, *19*, 936 (1986); I. M. Hodge, *Macrmolecules*, *20*, 2897 (1987).
45. G. Adam and J. H. Gibbs, *J. Chem. Phys.*, *43*, 139 (1965).
46. A. J. Kovacs, J. M. Hutchinson, J. J. Aklonis, in *The Structure of Non-Crystalline Materials*, p. 153 (P. H. Gaskell, ed.), Taylor and Francis, London (1977).
47. A. J. Kovacs, J. M. Hutchinson, J. J. Aklonis, and A. R. Ramos, *J. Polym. Sci.*, *Polym. Phys. Ed.*, *17*, 1097 (1979); A. J. Kovacs, *Ann. N.Y. Acad. Sci.*, *371*, 38 (1981); J. J. Aklonis, *Polym. Eng. Sci.*, *21*, 896 (1981).
48. T. S. Chow and W. M. Prest, *J. Appl. Phys.*, *53*, 6568 (1982); T. S. Chow, *J. Polym. Sci.*, *Polym. Phys. Ed.*, *22*, 699 (1984).
49. W. M. Prest, F. J. Roberts, and I. M. Hodge, *Proc. 12th North American*

Thermal Analysis Conference, Williamsburg, p. 119 (1980); V. P. Privalko, S. S. Demchenko, and Y. S. Lipatov, *Macromolecules*, *19*, 901 1732 (1986).
50. T. S. Chow, *J. Chem. Phys.*, *79*, 4602 (1983); T. S. Chow, *Polym. Eng. Sci.*, *24*, 915 1079 (1984); T. S. Chow, *Macromolecules*, *17*, 2336 (1984).
51. S. Matsuoka, G. Williams, G. E. Johnson, E. W. Anderson, and T. Furukawa, *Macromolecules*, *18*, 2652 (1985).
52. J. G. Curro, R. R. Lagasse, and R. Simha, *Macromolecules*, *15*, 1621 (1982).
53. V. V. Bulatov, A. A. Gusev, and E. F. Oleinik, *Makromol. Chem. Suppl.*, *6*, 305 (1984).
54. A. J. Kovacs, *Fortschr. Hochpolym.-Forsch.*, *3*, 394 (1963).
55. R. W. Rendell, K. L. Ngai, G. R. Fong, and J. J. Aklonis, *Macromolecules*, *20*, 1070 (1987).
56. R. G. Palmer, D. L. Stein, E. Abrahams, and P. W. Anderson, *Phys. Rev. Lett.*, *53*, 958 (1984).
57. J. T. Bendler and K. L. Ngai, *Macromolecules*, *17*, 1174 (1984); D. J. Plazek, K. L. Ngai, and R. W. Rendell, *Polym. Eng. Sci.*, *24*, 1111 (1984); R. W. Rendell and K. L. Ngai, In *Relaxations in Complex Systems*, p. 309 (K. L. Ngai and G. B. Wright, eds.), Government Printing Office, Washington, DC (1985); K. L. Ngai and D. J. Plazek, *J. Polym. Sci., Polym. Phys. Ed.*, *23*, 2159 (1985); G. B. McKenna, K. L. Ngai, and D. J. Plazek, *Polymer*, *26*, 1651 (1985); K. L. Ngai and G. Fytas, *J. Polym. Sci., Polym. Phys. Ed.*, *24*, 1683 (1986); R. W. Rendell, K. L. Ngai, G. R. Fong, A. F. Yee, and R. J. Bankert, *Polym. Eng. Sci.*, *27*, 2 (1987).
58. R. Simha, J. G. Curro, and R. E. Robertson, *Polym. Eng. Sci.*, *24*, 1071 (1984).
59. R. E. Robertson, *Macromolecules*, *18*, 953 (1985).
60. R. E. Robertson, R. Simha, and J. G. Curro, *Macromolecules*, *21*, 3216 (1988).
61. S. A. Brawer, *J. Chem. Phys.*, *81*, 954 (1984).
62. N. G. van Kampen, *Stochastic Processes in Physics and Chemistry*, North-Holland, Amsterdam (1981).
63. M. Golomb and M. Shanks, *Elements of Ordinary Differential Equations*, 2nd ed., Chap. 8, McGraw-Hill, New York (1965).
64. E. B. Wilson, J. C. Decius, and P. C. Cross, *Molecular Vibrations*, Dover, New York (1955).
65. S. Perlis, *Theory of Matrices*, Chap. 9, Addison-Wesley, Reading, MA, (1952).
66. T. M. Apostol, *Calculus*, 2nd ed, vol. 2, Chap. 4, New York, Wiley, (1969).
67. J. E. McKinney and R. Simha, *Macromolecules*, *7*, 894 (1974); O. Olabisi and R. Simha, *Macromolecules*, *8*, 211 (1975); P. Zoller, *J. Polym. Sci., Polym. Phys. Ed.*, *16*, 1491 (1978).
68. J. D. Ferry, *Viscoelastic Properties of Polymers*, 3rd ed., pp. 277–279, 288, Wiley, New York (1980).
69. R. E. Roberston, *J. Polym. Sci., Polym. Phys. Ed.*, *17*, 597 (1979).
70. D. J. Plazek, *Polym. J.*, *12*, 43 (1980).
71. A. J. Kovacs, *J. Polym. Sci.*, *30*, 131 (1958).

72. S. Hozumi, T. Wakabayashi, and K. Sugihara, *Polym. J.*, *1*, 632 (1970); S. Hozumi, *Polym. J.*, *2*, 756 (1971); K. Adachi and T. Kotaka, *Polym. J.*, *12*, 959 (1982).
73. G. T. Davis and R. K. Eby, *J. Appl. Phys.*, *44*, 4274 (1973).
74. K. Pathmanathan, G. P. Johari, J. P. Faivre, and L. Monnerie, *J. Polym. Sci.*, *Polym. Phys. Ed.*, *24*, 1587 (1986).
75. A. M. Kruithof, *Verres Refract.*, *9*, 311 (1955); H. N. Ritland, *J. Am. Ceram. Soc.*, *39*, 403 (1956); S. Spinner and A. Napolitano, *J. Res.*, *Nat. Bureau Stand.*, *70A*, 147 (1966).
76. R. D. Corsaro, *J. Am. Ceram. Soc.*, *59*, 115 (1976).
77. S. C. Jain and R. Simha, *Macromolecules*, *15*, 1522 (1982).
78. J. Jäckle, *J. Chem. Phys.*, *79*, 4463 (1983).
79. P. K. Gupta and C. T. Moynihan, *J. Chem. Phys.*, *65*, 4136 (1976).
80. G. Goldbach and G. Rehage, *J. Polym. Sci.*, *Part C*, *16*, 2289 (1967).

7

Utilization of a Synergistic Combination of Computational Techniques to Study the Transport of Penetrant Molecules Through Plastics

Jozef Bicerano
The Dow Chemical Company
Midland, Michigan

INTRODUCTION

The study of the transport of penetrant molecules through polymers [1,2] is important in many areas of technology. There are two types of industrially important polymeric systems for which such transport phenomena are crucial:

1. *Barrier plastics*, used in food and beverage packaging applications. These materials have high barriers. (*Barrier* is defined as resistance to the permeation of gas and flavor-aroma molecules.)
2. *Separation membranes*, used for the purification of mixtures of gases or liquids flowing through them. These materials have high selectivities and permeabilities that are sufficiently high to allow reasonable recovery of a product of the desired purity. (*Selectivity* is defined as the ratio of the permeabilities of the components of a mixture.)

The understanding of transport properties can be significantly enhanced by using some of the techniques described in this book, in versions adapted specifically to the transport problem, for molecular-level calculations of the structures and properties of polymers and polymer-penetrant systems.

For example, correlations based on group contribution techniques, such as the Permachor method [3–6], are very useful in predicting the permeabilities of penetrant molecules in plastics. The purpose of this chapter is to discuss some recent attempts towards the theoretical and computational modeling of the transport of penetrant molecules in plastics by use of techniques going beyond group contribution calculations.

Barrier plastics will be used as examples. Barrier and selectivity are two sides of the same coin. Both properties are determined by the same types of transport phenomena [7]. Similar techniques can, therefore, also be applied to study separation membranes. The same approximations are valid if the mixture flowing through the membrane is sufficiently dilute that (a) it does not significantly affect the structure and properties of the membrane, and (b) the components of the mixture can be treated as independent penetrants. The same general approaches can also be applied to concentrated mixtures, but only provided that certain simplifying approximations are *not* made.

STRATEGY

A promising combination of techniques to utilize in studying the transport of penetrant molecules in polymers is (a) the free-volume theory of Vrentas and Duda (V&D) [8–15] to provide an overall "global" physical perspective; (b) statistical mechanical models, the best-developed of which are the models of Pace and Datyner (P&D) [16–21], to provide an "intermediate"

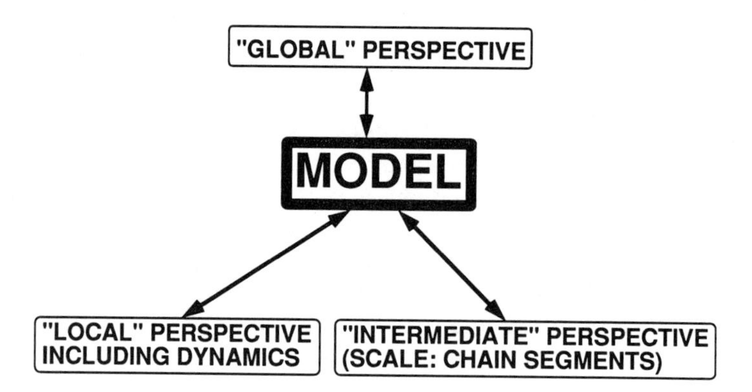

Figure 1 Schematic illustration of how the results of three different types of calculations, each one providing a perspective at a different scale, can be combined synergistically, to construct a unified physical model for the transport of penetrant molecules in plastics.

perspective on the scale of parameters describing short chain segments and their interactions; and (c) study of local unoccupied volume distributions and the dynamics of polymer and penetrant motions to build understanding on a true molecular level, i.e., gain a "local" perspective.

Each perspective can contribute directly to the construction of a unified physical model, as shown schematically in Figure 1. Conversely, as the model is improved by input from any one of these perspectives, the modifications might point out the revisions necessary in the interpretation of phenomena on the scale of the other two perspectives. In other words, the physical model forms the hub of a synergistic interaction between the three perspectives, which constitute a complete and systematic approach to transport phenomena.

Figure 2 shows how glass-transition temperatures (T_g) obtained by dynamic mechanical spectroscopy (DMS), percent crystallinities obtained by wide-angle x-ray scattering (WAXS), experimental diffusion coefficients, and information on tortuosity obtained by studies of morphology, can be useful in applying both the theory of V&D and the model of P&D. The Williams-Landel-Ferry (WLF) parameters [22] c_1^g and c_2^g, which can be

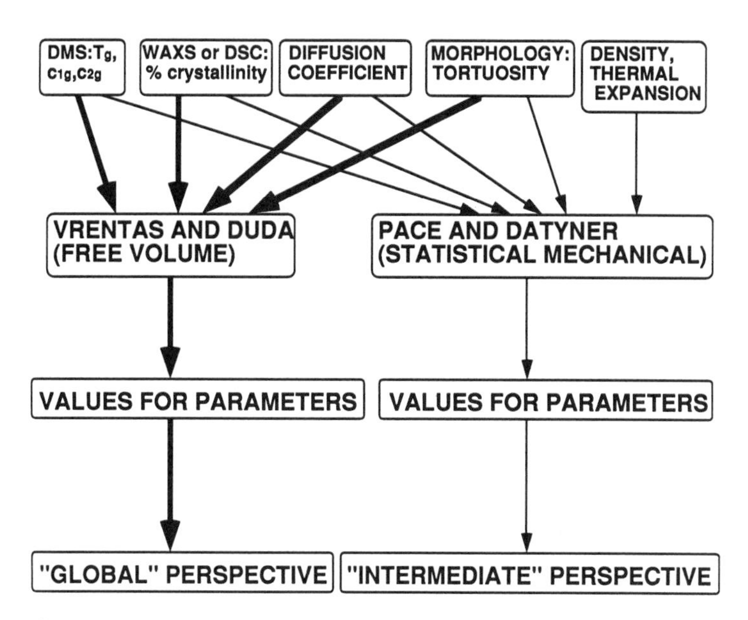

Figure 2 Flowchart summarizing the use of (a) the free-volume theory of Vrentas and Duda to obtain a "global" perspective; and (b) the statistical mechanical model of Pace and Datyner to obtain an "intermediate" perspective on the scale of parameters describing the polymer chain segments.

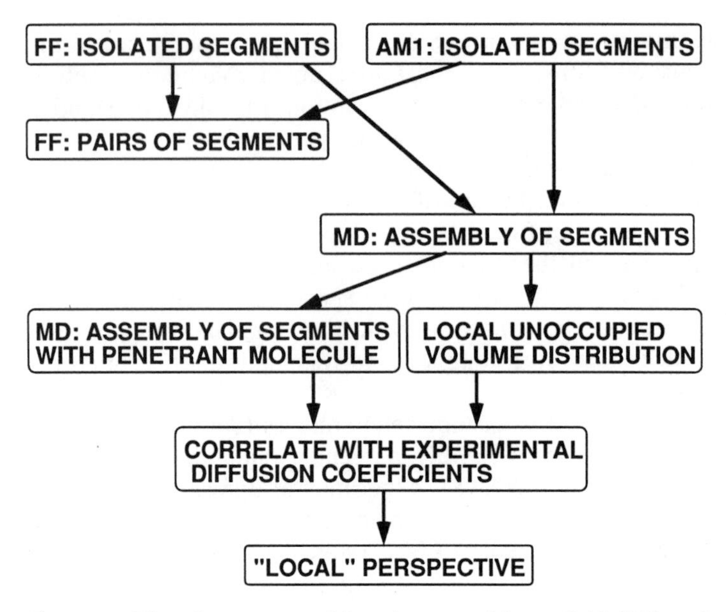

Figure 3 Flowchart summarizing the use of force field (FF) and molecular dynamics (MD) calculations to obtain a "local" perspective including the dynamics of the polymer and the penetrant.

determined by DMS, are needed as additional input for the theory of V&D. Densities and thermal expansion coefficients are needed as additional input for the model of P&D.

Figure 3 shows how a sequence of semiempirical quantum mechanical (QM), force field (FF), and molecular dynamics (MD) calculations can provide a detailed local perspective of the transport phenomena. The information that can be obtained includes the dynamics of the polymer in the absence of the penetrant, the local unoccupied volume distributions, and the dynamics (i.e., trajectory) of the penetrant as it diffuses through the polymer.

POLYMER-PENETRANT SYSTEMS

Sample copolymers of vinylidene chloride (VDC) [23] and vinyl chloride (VC) are being used in the experimental part of this study, which complements the calculations being performed on similar systems by a variety of techniques.

There are two major simplifying features of VDC/VC copolymers:

1. They are excellent barrier materials. Penetrants are likely to be present at concentrations of much less than 2%. Simpler limiting forms of the

V&D and P&D formalisms, for the low-concentration limit, can therefore be used.
2. They are above T_g at room temperature, unless the fraction of VC is high.

The major complicating feature of these copolymers is their semicrystallinity with a possibly substantial crystalline fraction, and all the concomitant structural, morphological, and rheological complexities:

1. The presence of crystallites, which are assumed to be impermeable, causes a fraction of the polymeric volume to become unavailable for transport.
2. The crystallites cause *tortuosity* in the diffusion pathway; i.e., they behave as randomly scattered obstacles which cause the diffusion pathway to become longer than it would have been otherwise. None of the existing simple general expressions for tortuosity form factors is always reliable. [24] There is no clear-cut, foolproof way to extract tortuosity information from morphological studies via techniques such as transmission electron microscopy (TEM) and small-angle x-ray scattering (SAXS) either.
3. The amorphous transitions are attenuated by the presence of crystallites [25], which can cause very significant changes in the rheological behavior. There are, consequently, many serious technical difficulties in the determination of reliable WLF parameters for the amorphous regions of semicrystalline polymers.

Starting with a maximally amorphous VDC/VC sample copolymer (i.e., one of fairly high VC content) can result in major simplifications. Several different compositions were therefore synthesized to allow the complexities introduced by semicrystallinity to become only gradually more important.

COMPARISONS BETWEEN "GLOBAL" FREE VOLUME AND STATISTICAL MECHANICAL MODELS

Expressions for Diffusion Coefficients and Activation Energies

Global parameters, related to the *total* free volume, are defined in the V&D theory [8–15]. This theory can be used to correlate experimental diffusion data by selecting optimum values for its adjustable parameters. Reasonable correlations have been obtained for nonbarrier polymers such as polystyrene (PS) and poly(vinyl acetate) (PVAc). Attempts to use this theory in a fully predictive mode have been less successful.

By contrast, all parameters (whether adjustable or uniquely determined) are expressed in terms of the molecular level structural features in statistical mechanical theories, such as the diffusion model of P&D [16–21]. These parameters are defined by the local structure of, and interactions in, the polymer-penetrant system. Most of them describe features at the scale of polymer chain segments. The trade-off for this apparently more molecular-level perspective is that physical assumptions whose general applicability and validity are questionable have to be made in defining the model.

In the present section, general comparisons will be presented between these two theories. Both theories will be utilized in a correlative mode. The parameters for barrier polymers will be defined for idealized completely amorphous poly(vinylidene chloride) (PVDC) and a VDC/VC copolymer. It will be shown that physically significant qualitative differences exist between the results calculated by the two theories.

The diffusion coefficient D, at the limit of the diffusion of a trace amount of penetrant in a completely amorphous polymer, at temperatures (T) above the T_g of the polymer, is given by the following expression in the theory of V&D.

$$D(\text{V\&D}) = D_{01} \exp\left(\frac{-2.303 c_1^g c_2^g \xi}{c_2^g + T - T_g}\right). \tag{1}$$

In the preliminary calculations on idealized polymers, where the purpose is to obtain "ball park" estimates of major trends, Ferry's "universal" values ($c_1^g = 17.44$ and $c_2^g = 51.6$) [22] will be used for the WLF constants.

The preexponential factor D_{01} and ξ are the only two adjustable parameters in Eq. (1). The parameter ξ is intended to denote the ratio of the critical molar volumes of two "jumping units": penetrant/polymer. In the formalism of V&D, the size of the jumping unit of the polymer is treated as a constant. Consequently, if ξ has been determined for a pair of penetrants in a given polymer, the *ratio* of the ξ's determined for the same pair of penetrants in any other polymer should, in principle, be equal to their *ratio* in the first polymer.

The value of $\xi < 0.8$ for penetrants as large as ethylbenzene in PS. In barrier polymers, the chains are held together very tightly. The free-volume elements (i.e., holes) are therefore likely to be much smaller than in PS. The critical molar volume of a jumping unit of the polymer, needed to fill an average hole, might therefore also be much smaller than in PS. The value of ξ could then become >1, even for small penetrants.

The dependence of D(V&D) on T does *not* have a simple exponential form such as $D = D_0 \exp(-E_a/RT)$. The definition of an apparent activation energy (E_{app}) is, therefore, not as straightforward as with a simple exponential dependence. Energy E_{app} can be defined, however, by the following equation, which would reduce to $E_{app} = E_a$ if D(V&D) had a simple exponential dependence on T:

$$E_{\text{app}} = RT^2\left[\frac{\partial \ln(\text{D})}{\partial T}\right], \tag{2}$$

where ln denotes the natural logarithm function.

The following expression is obtained for E_{app} in the theory of V&D, again at the limit of the diffusion of a trace amount of penetrant in a completely amorphous polymers, with $T > T_g$, by substituting Eq. (1) into Eq. (2):

$$E_{\text{app}}(\text{V\&D}) = \frac{2.303c_1^g c_2^g \xi RT^2}{(c_2^g + T - T_g)^2}. \tag{3}$$

Unlike the situation prevailing with a simple exponential dependence on T, $E_{\text{app}}(\text{V\&D})$ is *not* a constant, but a monotonically decreasing function of T. $E_{\text{app}}(\text{V\&D})$ asymptotically approaches the following limit at high temperatures:

$$E_{\text{app}}(\text{V\&D}, T \to \infty) = 2.303c_1^g c_2^g \xi R. \tag{4}$$

On the other hand, in the model of P&D, for the diffusion of small amounts of "simple" spherical penetrants, such as gas molecules, in "smooth-chained" polymers, such as poly(ethylene terephthalate) (PET) and *cis*-polyisoprene (natural rubber), D is given by the following equation [16]:

$$D(\text{cm}^2\ \text{s}^{-1}) = (9.10 \times 10^{-4})\frac{\overline{L}^2}{\lambda^2}\left(\frac{\varepsilon^*}{\rho^*}\right)^{5/4}\left(\frac{\sqrt{\beta}}{m^*}\right)^{1/2}\frac{d'}{\partial\,\Delta E/\partial d}\,e^{-\Delta E/\text{RT}}, \tag{5}$$

where ΔE denotes the energy required for symmetric chain separation [16]:

$$\begin{aligned}\Delta E = {} & 5.23\beta^{1/4}\left(\frac{\varepsilon^*\rho^*}{\lambda^2}\right)^{3/4}(d')^{-1/4} \\ & \times\left\{0.077\left[\left(\frac{\rho^*}{\rho}\right)^{11}(\rho - 10d') - \rho^*\left(\frac{\rho^*}{\rho^* + d}\right)^{10}\right]\right. \\ & \left. -0.58\left[\left(\frac{\rho^*}{\rho}\right)^{5}(\rho - 4d') - \rho^*\left(\frac{\rho^*}{\rho^* + d}\right)^{4}\right]\right\}^{3/4}.\end{aligned} \tag{6}$$

The parameters entering equations (5) and (6) have the following meanings: (a) d (nm) is the penetrant diameter, (b) β (J · nm/mole) is an effective chain-bending modulus, (c) $\overline{L}$ (nm) is an effective root mean-square jump length for the penetrant, (d) λ (nm) is the mean backbone element separation measured along the chain axis, (e) ρ (nm) is the equilibrium chain separation, (f) ρ^* (nm) is the effective van der Waals diameter of a polymer chain, (g) $d' = (d + \rho^* - \rho)$, (h) ε^* (in J/mole) is the average Lennard-Jones energy parameter of a backbone element, and (i) m^* (daltons) is the average molecular weight per chain element. The reader is referred to the first paper by P&D [16] for more details.

For simple *nonspherical* penetrants, the replacement of the spherical diameter d by an "effective diameter" $d = d_g^2/d_l$ results in good correlations between the P&D model and diffusion data [17]. Here, d_g is the effective "hard sphere diameter" of a gas molecule; and d_l is the longest penetrant dimension, as obtained, for example, from a molecular model. This extension incorporates the fact that the statistically favored mode of motion resulting in the diffusion of a nonspherical penetrant is by moving along its longest axis.

For polymers which possess closely spaced, bulky side groups (such as vinyl and related polymers polypropylene (PP), PVAc, PS, poly(vinyl chloride) (PVC) and poly(ethyl methacrylate) (PEMA)), the replacement of d by $[d\text{-}(\rho^* - \rho_c^*)]$, where ρ_c^* (nm) is the hard cylinder diameter of the chain core, results in good correlations with diffusion data [18]. This extension is tantamount to leaving the parameters describing the polymer unchanged, and treating the extra interchain "space" introduced by chain asymmetry as a reduction of the effective diameter of the penetrant by an equivalent amount. For example, d must be reduced by 0.12 nm in PVC to compute results in good agreement with experimental data [18].

There are two possible ways to define E_{app}(P&D). The first way is to define it in terms of the chain separation energy ΔE, as was done by P&D [16]:

$$E_{app}(\text{P\&D1}) = E(T_0) - \left[\frac{\partial(\Delta E)}{\partial T}\right] T_0, \tag{7}$$

where T_0 is a reference temperature in the range of interest. The second way is to define E_{app}(P&D2) by analogy to the Arrhenius-like equation used in the theory of V&D, substituting Eqs. (5) and (6) into (2), differentiating the resulting expression with respect to T, and assuming (until data become available to support or invalidate this assumption) that the contribution to E_{app}(P&D2) from the possible dependence of L on T is negligible compared to the contribution arising from the dependence of all the other terms on T.

Calculations of Diffusion Coefficients

The following strategy was adopted to determine the general behavior of D and E_{app} as functions of T and penetrant size: (a) obtain D for VC mole fraction (x) values of 0.0, 0.5, and 1.0, via the model of P&D, using Eqs.

(5) and (6), with reasonable values for the parameters, either taken from or extrapolated from experimental data; (b) derive a correlation between the penetrant diameter d used by P&D, and the adjustable parameter ξ used by V&D as an indicator of the size of the penetrant; (c) substitute this correlation into Eq. (1) to express D(V&D) as a function of d; (d) fit the expression for $D(d)$ in the theory of V&D to the expression for $D(d)$ in the model of P&D, to obtain values for D_{01} and ξ; and (e) calculate E_{app}(V&D), E_{app}(P&D1) and E_{app}(P&D2) as functions of d and T.

In applying the model of P&D, β was assumed to have the same value as in PVC (85,000 J · nm/mole). $\overline{L}$ was assumed to behave as described by the empricial correlation derived by P&D from a fit to data [18] collected at 298.15 K:

$$\overline{L} \simeq 0.7 \exp\left(\frac{0.4\,E}{450R}\right) \tag{8}$$

The possible dependence of $\overline{L}$ on T was not taken into account in the calculations at other values of T. In work on actual polymer-penetrant systems, the availability of extensive experimental data on D as a function of d and T will make it possible to treat β and $\overline{L}$ as adjustable parameters and explore their dependence on the polymer composition in more detail.

In order to apply Eqs. (5) and (6), values are also needed for (a) T_g, (b) λ, (c) m^*, (d) the cohesive energy density (δ^2), (e) a reference "amorphous" density, (f) the "glassy" volumetric thermal expansion coefficient below T_g (α_v^g), and (g) the "liquid" volumetric thermal expansion coefficient above T_g (α_v^l).

Most of the values used for the input parameters of PVC are either the same as or similar to those tabulated by P&D [17]. Most of the input values for PVDC were gathered or derived from data provided by Wessling [23]. δ^2 was estimated from the most reliable group contributions tabulated by Van Krevelen [26]. For the copolymer composition with $x = 0.5$, the values used for PVC and PVDC were averaged, except for T_g. See Table 1 for

Table 1 Values of the Input Parameters and Intermediate Quantities Derived from them, in the Application of the Statistical Mechanical Diffusion Model of Pace and Datyner, at T = 298.15 K

x	T_g (K)	λ (nm)	m^* (amu)	δ^2 (J/cc)	Density (g/cc)	$\alpha_v^g \times 10^4$ (K^{-1})	$\alpha_v^l \times 10^4$ (K^{-1})	ρ (nm)	ρ^* (nm)	ϵ^* (J/mol)
0.0	256	0.117	48.5	404	1.78	1.2	5.7	0.623	0.639	520
0.5	291	0.122	39.9	396	1.58	1.7	6.1	0.587	0.602	522
1.0	354	0.127	31.3	389	1.38	2.1	6.4	0.545	0.557	518

The mole fraction of comonomer is denoted by x, so x = 0.0 corresponds to PVDC and x = 1.0 corresponds to PVC.

the values of the input parameters, as well as quantities derived from them that also enter Eqs. (5) and (6).

As can be seen from Table 1, and by comparison with the tabulation of P&D [17], all three derived quantities for amorphous PVDC, which are related to parameters describing chain segments of the polymer, fall into the same ranges as the values of the same parameters in nonbarrier polymers. None of these parameters has a value so significantly different from its values in nonbarrier polymers as to be, by itself, a plausible cause for the much better barrier performance of PVDC. It is their particular combination which results in a low D for PVDC.

The only important parameter whose value for PVDC is very different from its values for the nonbarrier polymers is *not* a molecular level parameter, but a "global" parameter, namely the density, used as an input parameter on which most of the derived parameters [16] depend. The density of PVDC is about 30% higher than the density of any of the homopolymers studied by P&D [17,18]. In contrast, δ^2 is in the same range as in most nonbarrier polymers, and much lower than in PET. On the other hand, brominated polycarbonate has a higher density than PVDC, but a permeability similar to that of PS [27]. Therefore, it is not the density itself, but the *packing efficiency* (i.e., the fraction of the total volume of the polymer occupied by the van der Waals volumes of the atoms, as determined by the shape of the chain contour and the mobility of the chain segments) that makes the main difference. The importance of the density for the set of polymers examined by the model of P&D, both by P&D themselves and in the present work, is mainly its role as an indicator of the packing efficiency. The importance of the packing efficiency in determining the barrier performance of a polymer has also been demonstrated by the work of Lee on the prediction of the gas permeabilities of polymers from specific free-volume considerations [28].

The diffusion coefficients calculated at 298.15 K are depicted as functions of the penetrant diameter d in Figure 4. The curve for PVC with d replaced by $d - 0.12$ corresponds to the results of P&D [18] for PVC, as discussed above. The curves in Figure 4 manifest most of the expected trends: (a) the D's are in the expected ranges, (b) D increases with increasing fraction of VC, and (c) the differences between the D's calculated at different compositions increase with increasing d. A trend which deviates from the expectations is the slight concavity of the curves. There is no evidence in the systematic experimental studies of the d dependence of D [29] for the concavity of D as a function of d. A concave D curve implies an unphysical gradual reduction in the rate at which D approaches zero with increasing d. As d becomes very large, D should probably become a slightly convex function of d because it should rapidly become more difficult to find *any*

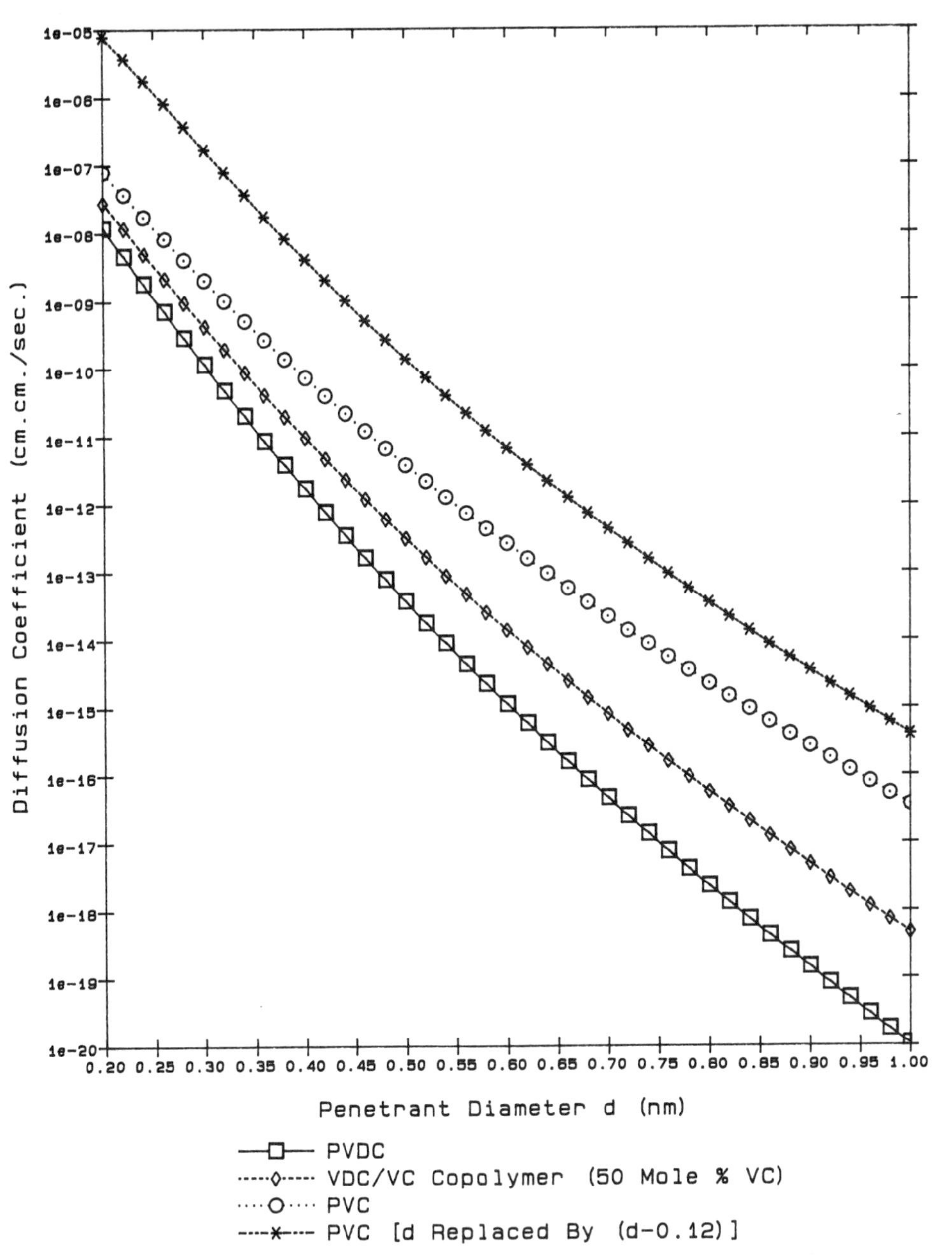

Figure 4 Diffusion coefficients calculated at 298.15 K, by using the statistical mechanical diffusion model of Pace and Datyner for idealized completely amorphous polymers, as functions of the penetrant diameter d.

diffusion pathway, resulting in an acceleration of the rate at which D approaches zero with increasing d.

The penetrant diameter d used in the model of P&D [17] appears to have a relatively unambiguous physical significance. It is an effective hard sphere diameter estimated from viscosity data, whenever necessary modified to correct for the nonsphericity of the shape of the penetrant. Therefore, d has the dimensions of length. On the other hand, the penetrant size parameter ξ used by V&D is somewhat ad hoc. It is described as the ratio of the molar volume of the "jumping unit" of the penetrant to the molar volume of the jumping unit of the polymer, where the latter is a rather ill-defined quantity. Since the molar volume of the jumping unit of the polymer is assumed to be constant, a set of ξ values for a series of penetrants in the same polymer should ideally scale as d^3 if the jumping unit of the penetrant is being described by its molar volume. In practice, ξ is most often used as an adjustable parameter. Its values are estimated by a variety of curve fitting procedures.

Vrentas, Liu, and Duda [10] have provided estimates for ξ for many penetrants in PS; and stated that the values for H_2, CH_4, and C_2H_4 are questionable. The application of correction factors to the values provided for H_2, CH_4, and C_2H_4 might result in more reliable values. These correction factors are equal to the quotient of the value of ξ for ethylbenzene in PS as directly obtained for the PS-ethylbenzene system (0.71) divided by its value extrapolated from results on the penetrant in question (1.0 from H_2, 0.4 from CH_4, and 0.48 from C_2H_4); i.e., 0.71/1.0 for H_2, 0.71/0.4 for CH_4, and 0.71/0.48 for C_2H_4.

Pace and Datyner [17] have provided values of d for many penetrants, nine of which are the same as nine of the penetrants for which ξ values have been provided [10] in PS. The d values provided by P&D [17] and the ξ values provided by Vrentas, Liu, and Duda [10] (where the ξ values for H_2, CH_4, and C_2H_4 have been corrected as described above) are listed in Table 2.

As shown in Figure 5, the correlation between d and ξ is poor, even though there are only nine data points to correlate. The best power law fit is

$$\xi \text{ (in PS)} \simeq 0.63\, d^{1.24}, \tag{9}$$

which is far from the cubic (d^3) correlation that a true volume parameter would be expected to show with a length parameter. Dimensional correlations between such different types of empirically determined molecular size parameters should probably not be taken too literally in any case.

The change of the type of polymer from PS to PVDC only changes the polymer jumping unit size. All penetrant jumping unit sizes remain the

Table 2 Values Provided for the Penetrant Diameter d (nm)[a] and ξ (dimensionless)[b,c]

Penetrant	d	ξ	Penetrant	d	ξ	Penetrant	d	ξ
H_2	0.270	0.09	CH_4	0.410	0.20	C_2H_5OH	0.615	0.30
N_2	0.370	0.17	C_2H_4	0.490	0.27	CH_2Cl_2	0.660	0.35
CO_2	0.370	0.27	CH_3OH	0.520	0.25	$CHCl_3$	0.710	0.46

[a]From Ref. [17].
[b]Defined as the volume of the penetrant jumping unit divided by the volume of the polymer jumping unit (estimated by Vrentas and Duda [10] in PS).
[c]Correction factors (see page 374) have been applied to the values of ξ for H_2, CH_4, and C_2H_4.

same. The value of D(V&D) can therefore be expressed in terms of d by using the correlation given in Eq. (9), and substituting Eq. (10) into Eq. (1):

$$\xi \text{ (in PVDC)} \simeq 0.63cd^{1.24}. \quad (10)$$

The constant c is an adjustable parameter equal to the quotient of the effective polymer jumping unit size in PS divided by the effective polymer jumping unit size in PVDC. In other words, $c > 1$ would imply that a smaller polymer jumping unit is sufficient to fill an average hole in PVDC, and consequently that the size of an average hole in PVDC is smaller than in PS.

There is no systematic correlation between the preexponential factor D_{01} and more direct measures the size of a penetrant, such as its molar volume at 0 K [9]. In the absence of such a correlation, and of suitable experimental data for diffusion in PVDC, D_{01} was treated as a single adjustable parameter. Equations (1) and (10) were used to fit ln[D(V&D,d)] to ln[D(P&D,d)], at 298.15 K, for the idealized completely amorphous PVDC system. The results of this fit are shown in Figure 6. Unlike the curve for D(P&D,d), the curve for D(V&D,d) has the expected slightly convex shape, allowing it to fall off increasingly rapidly with increasing d.

The value of $D_{01} \simeq 5.7 \times 10^{-8}$ cm²/s and $c \simeq 2.252$ are the values of the adjustable parameters which give the best fit. Since $\ln(D_{01}) \simeq -16.7$, the average effective D_{01} calculated for PVDC is much smaller than any of the values calculated in PS. (The lowest $\ln(D_{01})$ listed by VD [9] is -11.4.) It is possible to interpret the average effective D_{01} calculated here as an analogue of $\exp(-0.115\pi)$, where π is the Permachor value [3–5] of the polymer. The average effective D_{01} then becomes an indicator of the intrinsic resistance of the polymer to permeation, independent of the

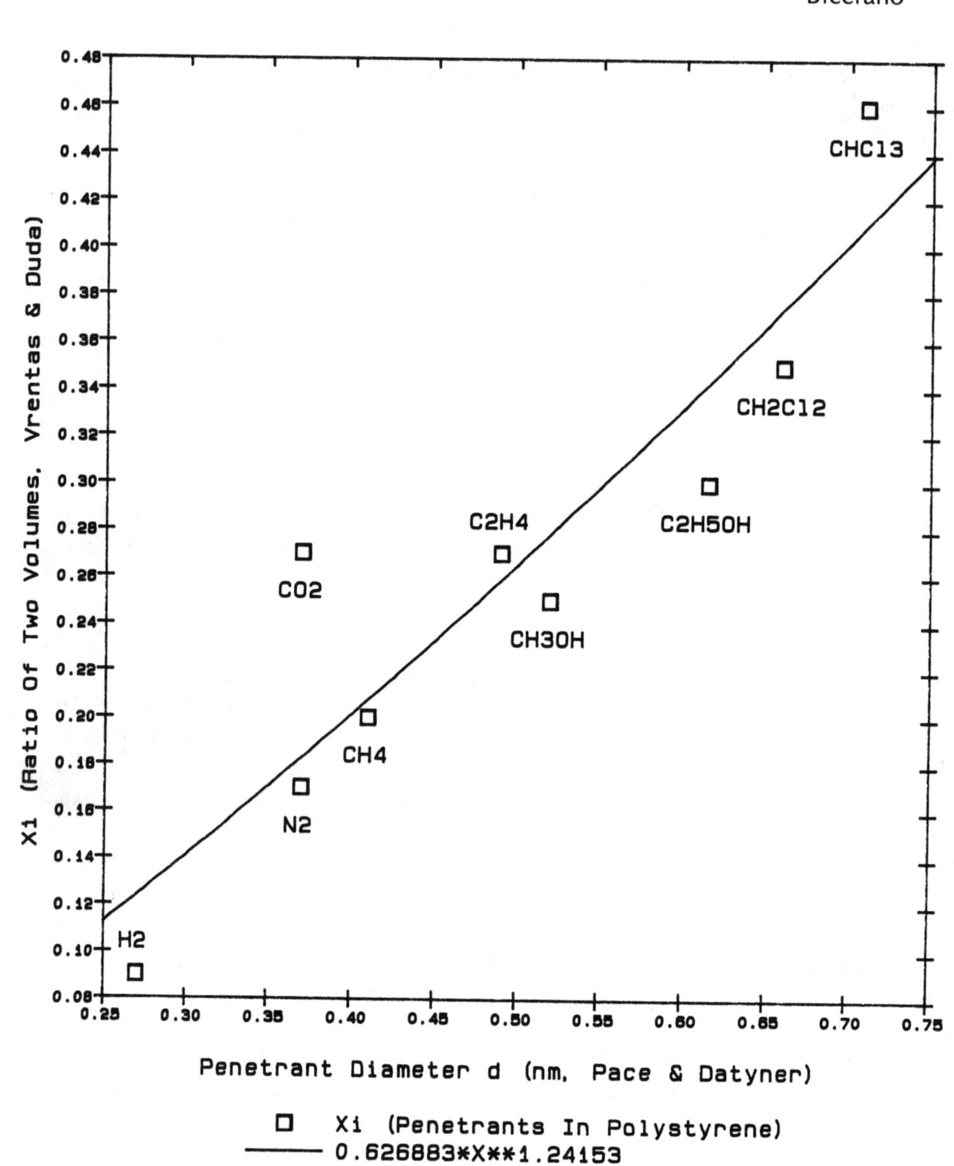

Figure 5 Correlation between penetrant diameter d used in the model of Pace and Datyner and size parameter ξ used in the theory of Vrentas and Duda.

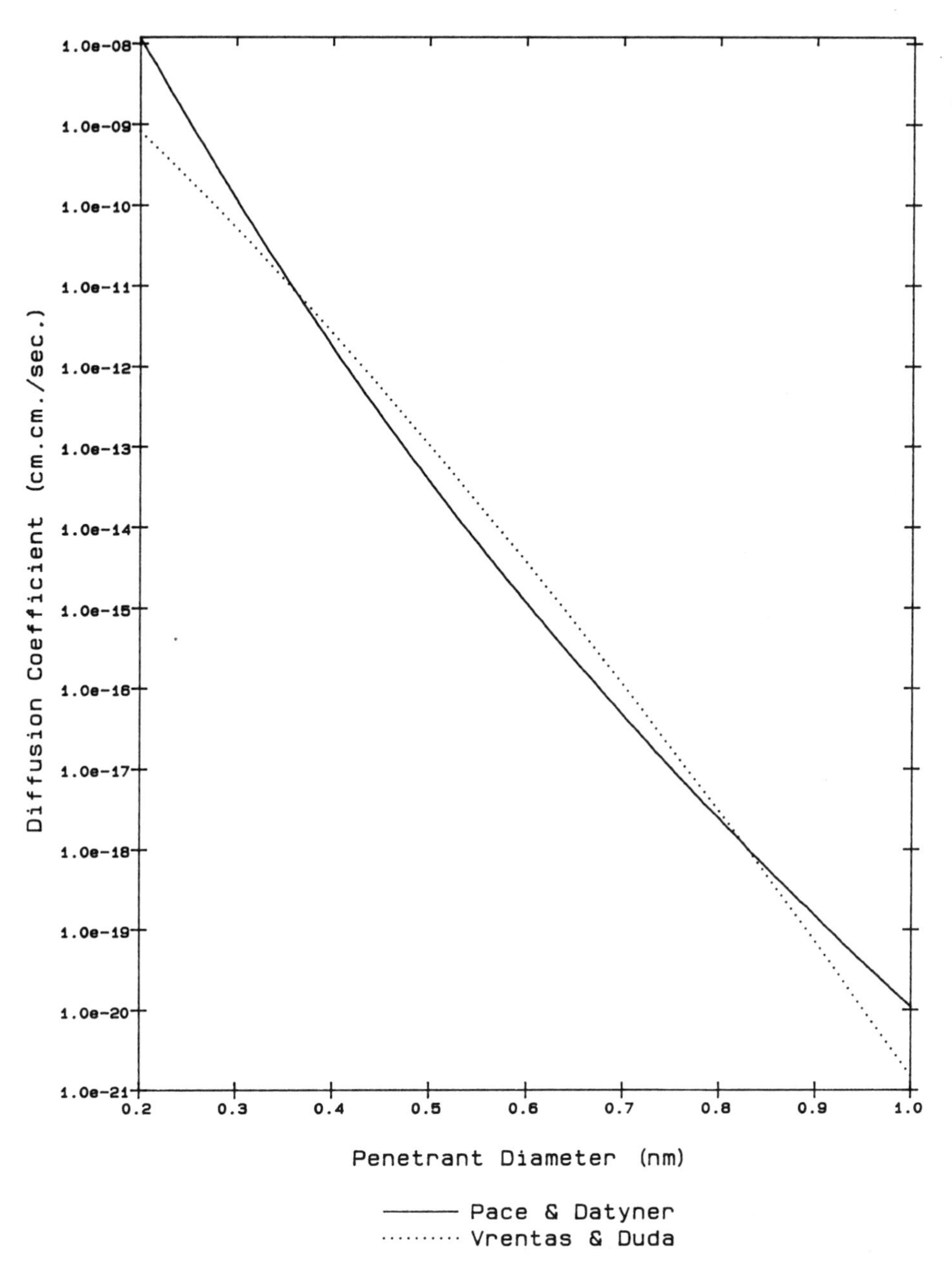

Figure 6 Best fit of the logarithm of the diffusion coefficient D calculated by the theory of Vrentas and Duda to the logarithm of D calculated by the model of Pace and Datyner, at 298.15 K, for an idealized completely amorphous sample of PVDC, as a function of the penetrant diameter d.

penetrant used. Note, also, that $c \simeq 2.252 >> 1.0$. The values of ξ in PVDC are, therefore, larger than those in PS by a factor of 2.252. The average hole size in PVDC is, consequently, much smaller than in PS. The penetrant has much less "space" to diffuse through, and therefore "appears" larger.

The drastic differences of D_{01} and ξ from the values observed in PS, and the observation that density was the only parameter entering the model of P&D with a very different value for PVDC than for *any* of the nonbarrier polymers studied [17,18], indicate that the *packing efficiency* is the most important descriptive *physical* parameter *both* at the "global" and at the "intermediate" scales. The investigation of (a) the local distribution of unoccupied volume and (b) the MD trajectory of the penetrant molecule in the polymer matrix are therefore likely to be very instructive.

It must be kept in mind, however, that *chemical* effects (strong interactions between the penetrant and the polymer), as observed, for example, in (a) the high moisture sensitivity of many polar polymers, (b) the high solubility of some penetrants in some polymers, and (c) the plasticization of the polymer matrix by a penetrant, can cause the behavior of a polymer-penetrant system to become quite different from what would be predicted on the basis of physical effects (such as packing efficiency) alone.

Calculations of Apparent Activation Energies

The Arrhenius-like apparent activation energies E_{app}(V&D) and E_{app}(P&D2), and the alternative apparent activation energy E_{app}(P&D1) calculated in terms of the chain separation energy for the model of P&D, are shown as functions of the penetrant diameter d in Figure 7. E_{app}(V&D) is a function of T whose dependence on T becomes weaker as T increases. Its limiting form as $T \rightarrow \infty$ is given by Eq. (4). By contrast, E_{app}(P&D1) and E_{app}(P&D2) are both almost independent of T. This fundamental qualitative difference in the dependence of E_{app} on T reflects the very different interpretations of the authors of these two models [11,16] of the meaning of E_{app} in a non-Arrhenius system.

E_{app}(V&D) is slightly concave as a function of d, reflecting an acceleration in its rate of increase with d. E_{app}(P&D1) and E_{app}(P&D2) are slightly convex as functions of d, reflecting a deceleration in their rates of increase with increasing d. As with the diffusion coefficients, and for the same reason, the trend in the rate of change of E_{app}(V&D) as a function of d appears more reasonable than the trend in the rate of change of E_{app}(P&D1) and E_{app}(P&D2). In fact, it has been shown [30] that a second power dependence is often found between the Lennard-Jones diameters of penetrant molecules and the activation energy. Such a second power depend-

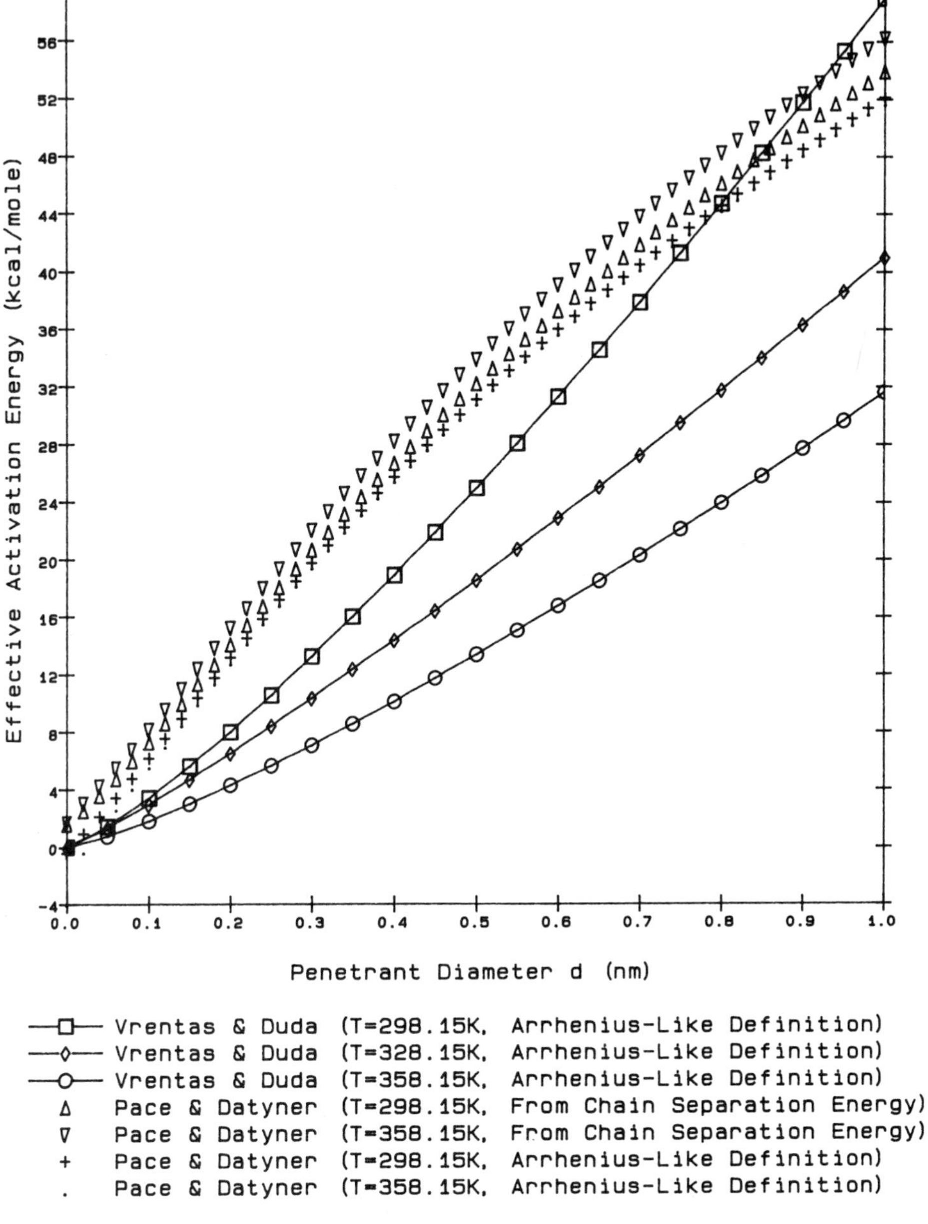

Figure 7 Apparent (or "effective") activation energies as functions of temperature T and penetrant diameter d. See the text for details.

ence, probably related to sensitivity to the cross-sectional area of the penetrant, would obviously be manifested as a concavity in E_{app} when plotted as a function of d with linear x and y axes. E_{app}(V&D) < [E_{app}(P&D1) ≃ E_{app}(P&D2)], except for $d > 0.8$ nm at the lowest value of T (298.15 K) used in the study. When $d = 0.35$ nm (O_2 molecule) and $T = 298.15$ K (room temperature), E_{app}(*V&D*) ≃ 16 kcal/mole, and E_{app}(P&D1) ≃ E_{app}(P&D2) ≃ 24 kcal/mole for this idealized fully amorphous sample of PVDC.

The E_{app} term should extrapolate to zero at some very low positive value of d [31], i.e., at $d = d_0$, where $0.0 < d_0 << 0.22$, and remain zero for all $0.0 \leq d < d_0$. ($d = 0.22$ nm is the diameter of the smallest penetrant (He) used in diffusion studies, and E_{app}(He) > 0.0.) This result is to be expected because an extremely small but still finite-sized penetrant should be able to find a connected pathway of holes, cracks, and crevices large enough to diffuse through. It can be seen from Figure 7 that none of the three types of E_{app} meet this expectation. E_{app}(V&D) becomes zero exactly at $d = 0.0$. E_{app}(P&D1) does not even become zero at $d = 0.0$, unless residual terms adding up to about 1.5 kcal/mole are subtracted out. E_{app}(P&D2) does become zero at $d_0 \simeq 0.02$ nm ≥ 0.0 at the highest value of T used ($T = 358.15$ K), but this result appears to be fortuitous.

SEMIEMPIRICAL CALCULATIONS ON ISOLATED MODEL MOLECULES

Brief Review of Chain Conformations

Standard notation will be used to describe the conformations defined by the values of dihedral angles α about successive C—C bonds in chains of polymers and model molecules containing a backbone of tetravalent carbon atoms: T = *trans* ($\alpha = 180°$), G = *gauche* ($\alpha = 60°$), C = *cis* ($\alpha = 0°$), X = ($180° > \alpha > 0°$, but α not equal to 60°), G′ = −G ($\alpha = -60°$), and X′ = −X.

The letter *Y* will be used to denote a halogen atom, since the more commonly used letter *X* has already been used in describing the conformations. VF, VC, and VB denote vinyl fluoride, vinyl chloride, and vinyl bromide, respectively. In addition to PVDC and VDC/VC copolymers, PVDF and VDF/VF copolymers (Y = F), and PVDB and VDB/VB copolymers (PVDB denotes that Y = Br) will also be studied. Such a study of the entire isoelectronic series (Y = F, Cl, and Br) can provide a more complete understanding of the molecular-level effects determining the barriers of VDC/VC copolymers.

The TTTT (planar zigzag) conformation, which is the preferred crystalline conformation in polyethylene (PE) [32], is not favored in PVDY because of large steric repulsions between halogen atoms. The TGTG′, TXTX′, and TCTC conformations all have less steric repulsion than the TTTT conformation.

PVDF has more than one crystalline form (polymorph). The polymorph of PVDF which most closely resembles the crystalline conformation of PVDC (see below) has a chain conformation of TGTG′ [23].

It has been shown, by ^{13}C [33] and ^{19}F [34] nuclear magnetic resonance (NMR) spectroscopy, that PVDF and other related fluorine-containing polymers have a considerable fraction of defect bonding sites. Here, a head-to-head defect is defined as a defect in which two carbon atoms, both bonded to two hydrogen atoms, are linked to one another along the chain. A tail-to-tail defect is defined as a defect in which two carbon atoms, both bonded to two halogen atoms, are linked to one another along the chain. (To avoid confusion, note that there is no universal convention concerning which type of defect should be called head-to-head, and which type should be called tail-to-tail. A more common practice is to define the head-to-head defect as one with the heavy substituents (the halogen atoms) adjacent to each other. On the other hand, a head-to-head defect was defined as one with the hydrogen atoms adjacent to each other in our previous publications. We will continue to use this alternative convention for the sake of consistency.)

Up to 12% of such defect sites can be found in PVDF, where steric effects and the resonance stabilization of the macroradical growing by polymerization are both much smaller than in PVDC [32]. Farmer et al. have shown that (a) the presence of a sufficient concentration of such defect sites can alter the preferred polymorph of PVDF [35], and (b) the conformation and packing of an alternating copolymer of ethylene and tetrafluoroethylene, which represents a chain of PVDF polymerized entirely as an alternating succession of head-to-head and tail-to-tail defects, are different from those of PVDF [36].

A detailed analysis [37–40], including infrared and Raman spectroscopy and FF calculations, shows that the x-ray data [41] for PVDC can be fitted well with a TXTX′ chain conformation with $\alpha = 32.5°$ for X. PVDB is isomorphous with PVDC [23,32]. Because of the strong kinetic preference for head-to-tail bonding during synthesis, PVDC and PVDB have much fewer head-to-head and tail-to-tail bonding defects than PVDF [32], in spite of the low energies of head-to-head defects (see below).

PVDB has the most sterically hindered vinylidene-type crystal structure for which a TGTG′ or TXTX′ conformation does not have prohibitively large steric repulsions. If the halogen atom is replaced by a methyl group, as in poly(1,1-dimethylene) (commonly referred to as polyisobutylene (PIB)), the crystalline conformation changes [23] to an 8_3 helix, with expanded C—CH_2—C bond angles and a much longer chain repeat distance than in PVDY.

Geometry Optimization

Semiempirical calculations [42] were carried out [43,44] on model molecules simulating isolated chain segments of PVDY and of VDY/VY copolymers. The geometries of the model molecules were optimized [43] by the molecular mechanics 2 (MM2) option in CHEMLAB-II, which is a general-purpose molecular modeling software package developed and owned by Chemlab Incorporated, and marketed by Molecular Design Limited. MM2 is an FF technique for calculating molecular geometries by minimizing the total steric energy E. A lower E implies a thermodynamically more stable molecular structure. The newest version of the molecular mechanics routine in CHEMLAB-II is MMFF; however, MM2 was used in the work being described here.

The simplest types of FF calculations neglect hydrogen bonding and other types of polar interactions. In *isolated* chain segments of PVDY or VDY/VY copolymers, however, the presence of the bulky Y atoms can cause steric repulsions to become the most important factors in determining the molecular geometry. Simple FF calculations can, therefore, give reasonable geometries and relative energies for isolated chain segments of these polymers. On the other hand, as shown for PVDF by Farmer et al. [35], the chain conformation might be affected by packing constraints in a crystalline phase. The results of isolated chain calculations have to be interpreted with some caution.

MM2 was used to optimize the geometries of 12 model molecules of four different types [43]. Each one of the three molecules of any given type contains a different halogen atom (Y = F, Cl, or Br):

1. $C_{20}H_{22}Y_{20}$ with only head-to-tail bonding between successive pairs of monomers. These molecules are oligomers of 10 VDY monomers, terminated by an extra H atom at each end.
2. $C_{20}H_{22}Y_{20}$ with one head-to-head bond (at the tenth and eleventh C atoms).
3. $C_{20}H_{22}Y_{20}$ with one tail-to-tail bond (at the tenth and eleventh C atoms).
4. $C_{20}H_{23}Y_{19}$ with head-to-tail bonding between successive pairs of monomers. These molecules are oligomers of nine VDY and one VY monomers, terminated by an extra H atom at each end. The sixth monomer is the VY unit.

The first type of model molecule formally corresponds to a fully extended chain resembling most closely the crystallographic conformations of PVDF, PVDC and PVDB. All geometry optimizations for this type of model molecule were started from a TXTX′ conformation. No symmetry constraints were imposed on the geometries in any of the optimizations. This conformational search procedure is quite limited, and was *not* intended to be a full conformational analysis. For example, it is less ambitious than the "polymer reduced interaction matrix method" developed by Orchard et al. [45] for the enumeration of the intrachain conformational energetics of very long chains, based on symmetry considerations and molecular mechanics energetics.

Calculations such as those presented in this paper are, nevertheless, useful as starting points for more detailed analysis of chain conformations (i.e., by rotational isomeric state (RIS) theory [46]) or of the physical properties of polymers. The usefulness of RIS calculations for polymers of interest has been demonstrated by the work of Tonelli on the conformational properties of isolated, unperturbed chains of (a) PVDF with head-to-head and tail-to-tail bonding defects [47]; (b) poly(fluoromethylene), poly(trifluoroethylene), and poly(tetrafluoroethylene), again including such defects [48]; and (c) PVB and ethylene-VB copolymers [49].

The molecular geometries resulting from the present calculations can also be used to generate starting coordinates for assemblies of large numbers of chain segments, by using the polymer chain packing routine in CHEMLAB-II. Such assemblies can be used as starting points in (a) studies of chain packing in the desired crystalline phase of the polymer, and (b) MD simulations of dynamic phenomena as functions of time.

The optimized molecular geometries for Y = F and Cl are shown in space-filling illustrations in Figures 8 and 9, respectively [43]. To avoid confusion, note that many of the atoms are hidden from view. The optimized geometries for Y = Br are very similar to those for Y = Cl, and they are therefore not shown. The *E*'s of the four types of model molecules are shown as functions of the halogen atom Y (1 = F, 2 = Cl, and 3 = Br) in Figure 10.

The AM1 (Austin Model 1) Hamiltonian available in Version 3.10 of the MOPAC (Molecular Orbital PACkage) program was used to carry out QM calculations, for the purpose of estimating the partial atomic charges to be used in the calculations on the interactions between pairs of model molecules. $C_{20}H_{22}Y_{20}$ and $C_{20}H_{23}Y_{19}$ were too large to complete the AM1 calculations in a reasonable amount of time. The geometries of the smaller model molecules $C_{12}H_{14}Y_{12}$ and $C_{12}H_{15}Y_{11}$, obtained by truncating the geometries optimized by MM2 for the large model molecules, were therefore used as starting points in the AM1 optimizations [44]. The heats of for-

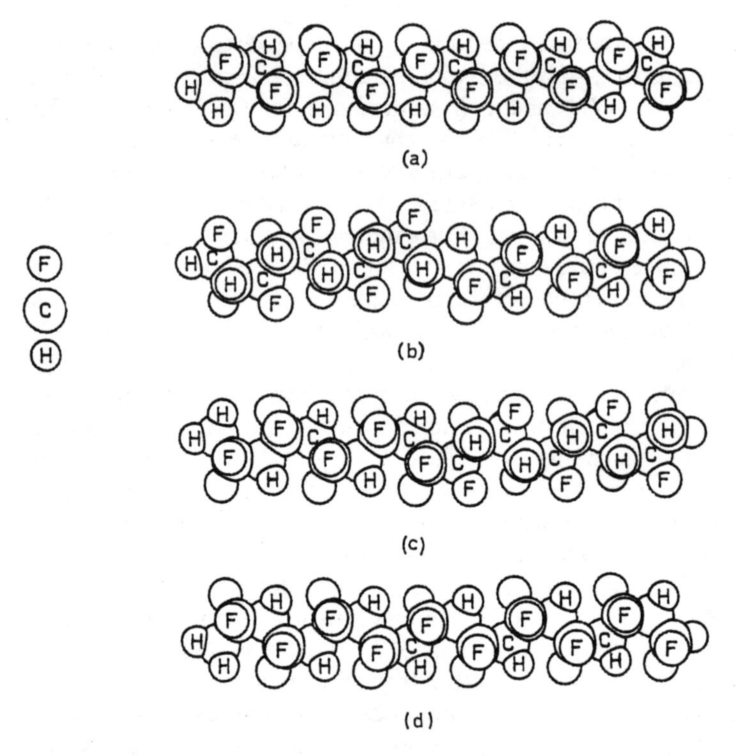

Figure 8 Space-filling illustrations: (a) standard geometry for $C_{20}H_{22}F_{20}$; (b) $C_{20}H_{22}F_{20}$ with a head-to-head bond at the tenth and eleventh C atoms; (c) $C_{20}H_{22}F_{20}$ with a tail-to-tail bond at the tenth and eleventh C atoms; and (d) $C_{20}H_{23}F_{19}$ with the eleventh C atom bonded to both an H and an F.

mation obtained by AM1 are shown, also as functions of the halogen atom Y, in Figure 11 [44].

Optimized Conformations

The optimized geometry of $C_{20}H_{22}F_{20}$ was very close to having TGTG′ symmetry ($\alpha = 60.5 \pm 2.5°$ for G), in agreement with the interpretation of the crystal structure of PVDF. A geometry very similar to TXTX′ was obtained for $C_{20}H_{22}Cl_{20}$ ($\alpha = 48 \pm 3°$ for X). The geometry calculated for $C_{20}H_{22}Br_{20}$ ($\alpha = 46 \pm 2°$ for X) was isomorphous to the geometry computed for $C_{20}H_{22}Cl_{20}$. Of course, each one of these chain segments has more than one stable conformation; however, the many alternative stable (local energy minimum) conformations were not sought in the present calculations.

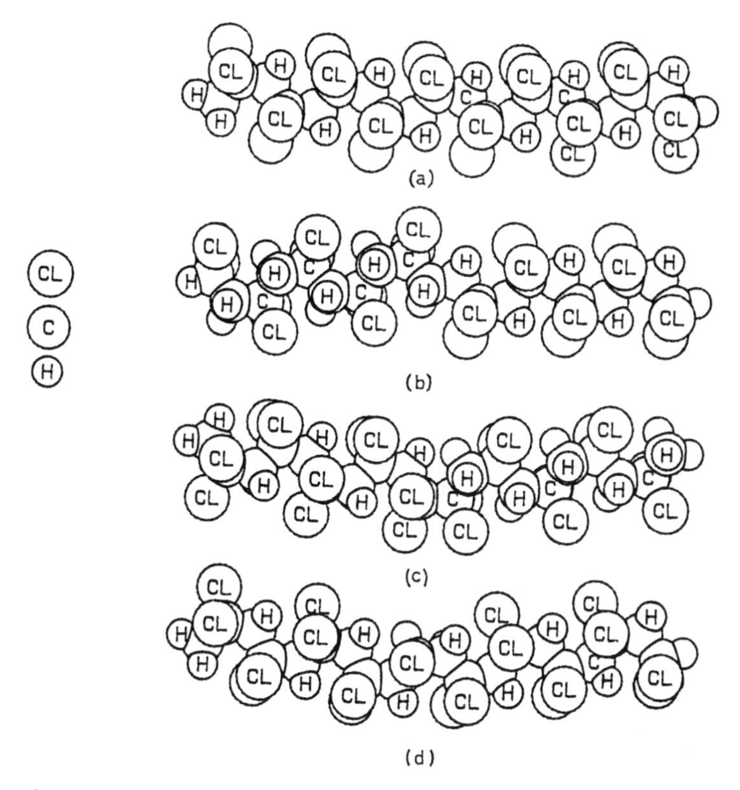

Figure 9 Space-filling illustrations: (a) standard geometry for $C_{20}H_{22}Cl_{20}$; (b) $C_{20}H_{22}Cl_{20}$ with a head-to-head bond at the tenth and eleventh C atoms; (c) $C_{20}H_{22}Cl_{20}$ with a tail-to-tail bond at the tenth and eleventh C atoms; and (d) $C_{20}H_{23}Cl_{19}$ with the eleventh C atom bonded to both an H and a Cl.

These results show that steric energy considerations are crucial factors in determining the chain conformations. Additional evidence of the importance of the steric energy comes from the fact that the study of the conformational characteristics of isolated PIB chains via RIS theory, utilizing results obtained by using an FF containing only steric (Lennard-Jones) energy terms, gives an 8_3 helical conformation with helix parameters in excellent agreement with crystallographic data [50].

Conformations with a head-to-head bonding defect resembled the conformations of the standard structures. The only exception occurred in the vicinity of the two C atoms, tenth and eleventh along the chain, where the defect was incorporated. The bonding around the defect site became *trans*,

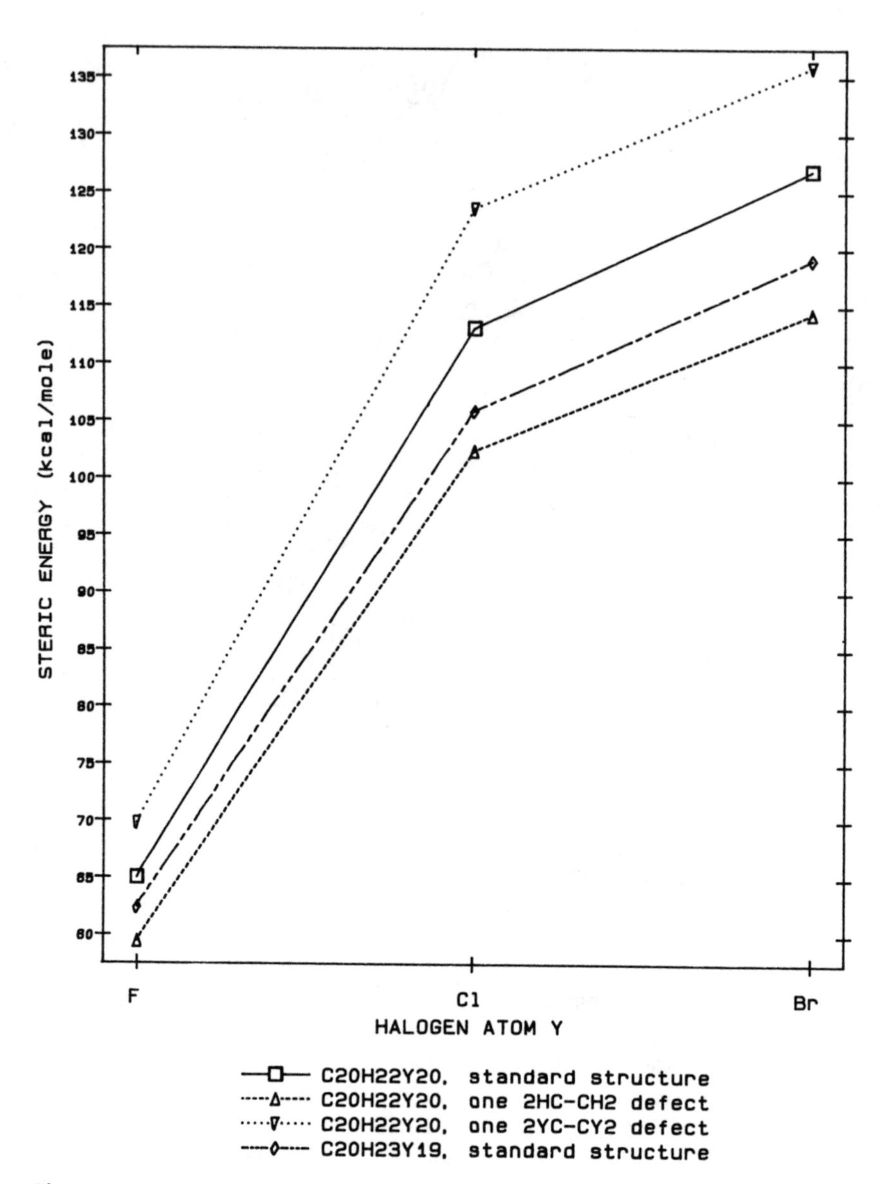

Figure 10 MM2 steric energy as a function of the halogen atom.

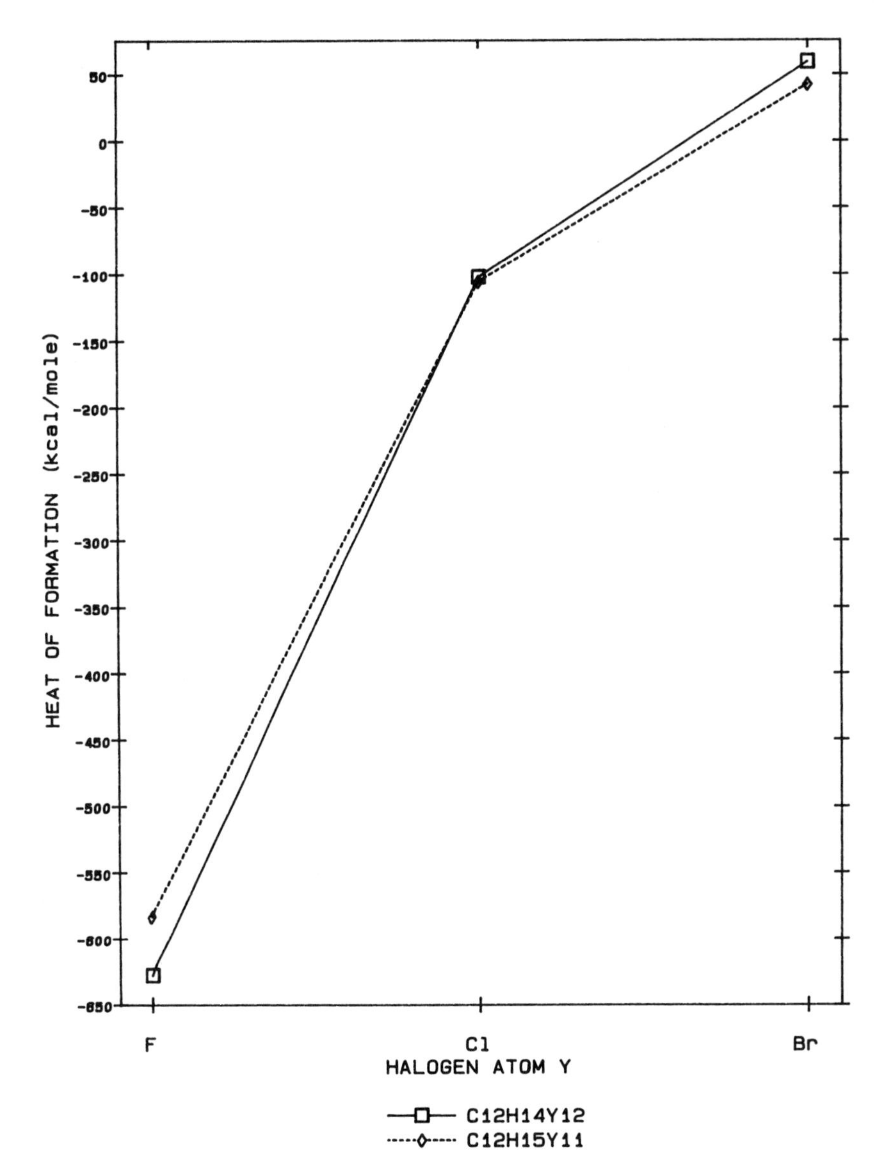

Figure 11 AM1 heat of formation as a function of the halogen atom.

interrupting the TGTG′ or TXTX′ pattern. A succession of three *trans* α's (TTT) resulted, as in the preferred TTTT conformation of PE. The winding pattern then returned to TGTG′ or TXTX′, with an equal likelihood of the "handedness" of the helix being reversed or remaining the same as before the defect. This type of defect might therefore serve as a site where the helix loses all "memory" of its direction of winding, and can either continue in the same direction or reverse directions.

With a tail-to-tail bonding defect, the TGTG′ or TXTX′ winding pattern was retained everywhere along the chain. Around the defect, the α corresponding to X was somewhat larger than the range observed in the standard structures. This defect site α is 70° for Y = F, in comparison with the range of 60.5 ± 2.5°; 68° for Y = Cl, in comparison with the range of 48 ± 3°; and 67° for Y = Br, in comparison with the range of 46 ± 2°. The increase in α at the defect site enables some reduction of the steric repulsions between halogen atoms.

The geometry of each $C_{20}H_{23}Y_{19}$ was very similar to the geometry of the standard structure of $C_{20}H_{22}Y_{20}$. The largest change was observed for Y = Br; however, even with this largest halogen atom, the α corresponding to X, at the eleventh carbon atom along the chain, which is bonded to both an H and a Br, only increases to 52.5°, i.e., very slightly above the range of 46 ± 2° calculated for $C_{20}H_{22}Br_{20}$. Adding a small percentage of VY comonomer should therefore not change the preferred chain conformation. A small number of VY units should be able to enter the crystallites, without distorting the chains away from their preferred conformations. Such defective crystallites may, however, have lower cohesive energies than perfect crystallites of PVDY, as discussed in the following section. As the percentage of VY is increased, the loss of stabilization resulting from the loss of polar interactions, as well as the increasingly imperfect packing, should, at some composition, cause a transition from a predominantly PVDY-like structure with VY defect sites, to a PVY-type structure with VDY defect sites.

Energetics

It can be seen from Figures 10 and 11 that the stability of isolated chain segments decreases in the order Y = F >> Cl > Br. The increases of *E* and of the heat of formation are much larger from Y = F to Y = Cl than from Y = Cl to Y = Br. This result is consistent with the much larger percent differences between the atomic volumes of F and Cl, than the atomic volumes of Cl and Br. The heat of formation is negative for Y = F or Cl, and positive for Y = Br. [A negative (positive) heat of formation indicates that a molecule is thermodynamically stable (unstable) relative to its constituting elements.]

Each structure with a head-to-head bond has a lower *E* than the corresponding standard structure with only head-to-tail bonds. This effect becomes stronger with increasing size of the halogen atom. In amorphous regions, a chain with this type of defect can be expected to have lower thermodynamic energy than the standard structure. In crystalline regions, however, periodicity would be disrupted and the energy would go up. This defect would, therefore, be quite unlikely to be found in crystallites of PVDC and PVDB, because of the large disparity between the sizes of H and of Cl or Br. One of its preferred locations, especially for high percent crystallinity, might be the boundary between a crystallite and the amorphous region surrounding it. After one of these defects is incorporated in a chain, the occurrence of a second such defect in the same chain becomes unlikely. A much less favored tail-to-tail bond is required, to create a new chain end with H attached to the terminal C, and make a second head-to-head bond possible.

The *E* of each defective structure with a tail-to-tail bond is considerably higher than the *E* of the corresponding standard structure. This type of defect is, therefore, thermodynamically disfavored. It is also kinetically much more disfavored than a head-to-head bond, since its formation requires the bonding of two carbon atoms hindered by the bulky Y's.

Replacement of one Y atom by a less bulky H atom results in a decrease in *E*, as shown in Figure 10. This effect becomes more pronounced with increasing size of Y. On the other hand, electronic effects of a quantum mechanical nature, which are accounted for in the QM treatment, stabilize the fully halogenated structure more than the structure with one of the Y atoms replaced by an H atom. Consequently, as shown in Figure 11, the heat of formation of the structure with one of the Y atoms replaced by an H atom does not become distinctly lower except for Y = Br. The electronegativities [51] follow the order F(3.98) > Cl(3.16) > Br(2.96) > C(2.55) > H(2.20. Therefore, Br is the Y atom with the weakest electronic interactions as well as the largest steric repulsions.

SEMIEMPIRICAL CALCULATIONS ON INTERACTING PAIRS OF MODEL MOLECULES

Geometry Optimization

Electrostatic interactions can be very important in interchain interactions. A combination of (a) semiempirical QM calculations and (b) FF calculations utilizing more complex Hamiltonians was therefore utilized. The geometries optimized by MM2 [43] were used for $C_{20}H_{22}Y_{20}$ and $C_{20}H_{23}Y_{19}$. The total intermolecular energy (E_2) of a pair of molecules was calculated [44] as a function of the relative positions of the two molecules. The value of E_2 was taken to be the sum of standard Lennard-Jones 6-12 potential (E_{LJ}) and Coulombic electrostatic potential (E_e) terms; i.e., $E_2 = E_{LJ} + E_e$. Parameter set Z in CHEMLAB-II was used in E_{LJ}. The Mulliken atomic charges calculated by AM1 were used in E_e. The intermolecular geometry optimization was performed [44] by using the PHBIMIN program in CHEMLAB-II to minimize E_2 as a function of the intermolecular configuration of pairs of rigid molecules.

The limitations of such calculations will be summarized, closely following the discussion provided in [44], before discussing the conformations and energetics calculated for interacting pairs of chain segments.

First of all, even when this type of intermolecular geometry optimization is performed repeatedly with the utmost diligence, the intermolecular geometries calculated can only be claimed to probably be close to the true minima, but not guaranteed to be so.

In addition, intermolecular calculations can be heavily dependent upon the selected intrachain conformations, unless intrachain geometries are optimized simultaneously with the interchain geometry. In the present case, the use of rigid intramolecular geometries, which is often a rather crude approximation, was deemed to be adequate at a semiquantitative level because, as discussed on pages 384–388, the TGTG′ conformation calculated for the isolated chain segment of PVDF, and the TXTX′ conformations calculated for PVDC and PVDB, were all in very good agreement with experimental (crystallographic) results, which of course include all the intrachain *and* interchain effects.

Finally, because of differences in the physics of the systems (i.e., because of the many additional interactions in assemblies with more than two chain segments), there can be significant differences between results calculated for only a pair of chain segments and results computed for larger ensembles more representative of the true intermolecular environment in the solid even if the objections described above are overcome. (An example will be given on page 396). Such differences between the physics of chain pairs and of many-chain assemblies clearly show that it is difficult to get reliable chain-packing information from calculations based on only a pair of chains.

Such simple calculations on pairs of chain segments are, therefore, mainly useful as bridges between the single-chain and the many-chain assembly levels of calculation. They provide information on the preferred interactions of two chain segments in the absence of other chain segments. Comparison of their results with the results for isolated chains and multichain assemblies can facilitate the identification of which effects are primarily caused by (a) intrachain factors, (b) the intrinsically preferred patterns of interaction between pairs of chains, and (c) constraints and/or superpositions of effects induced by large-scale packing [44]. The need to identify all these factors as clearly as possible has been shown in studies of the hierarchies of order and disorder in solids [52].

Such many-chain assembly calculations can be carried out by several methods. Calculations on related systems include the work of Farmer et al. [35,36], which was discussed in the previous section, and of Yemni and McCullough [53] on the energetics of orthorhombic-to-monoclinic phase transformations in PE. CHEMLAB-II also contains a routine (CHAINPACK) for packing polymer chains.

A promising new technique for the prediction of polymer crystal structures and properties, by using molecular mechanics and minimizing intermolecular and intramolecular energies simultaneously, has recently been introduced by Boyd's research group and applied to predict structures and properties in reasonable agreement with experimental results for PE and poly(oxymethylene) (POM) [54–56]. Finally, an entirely different approach has been utilized by Blaisdell et al. [57] in their POLY-CRYST program.

The optimized intermolecular configurations of the $C_{20}H_{22}F_{20}$—$C_{20}H_{22}F_{20}$ and $C_{20}H_{22}Cl_{20}$—$C_{20}H_{22}Cl_{20}$ pairs are shown in Figures 12 and 13, respec-

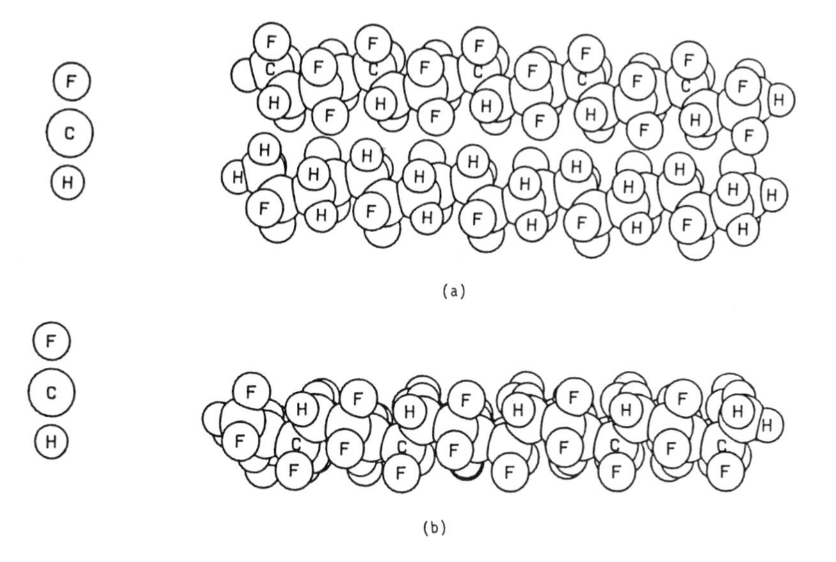

Figure 12 Space-filling illustrations of the $C_{20}H_{22}F_{20}$—$C_{20}H_{22}F_{20}$ pair: (a) a perspective showing both molecules; (b) an alternative perspective with one molecule behind the other, and almost completely eclipsed by it.

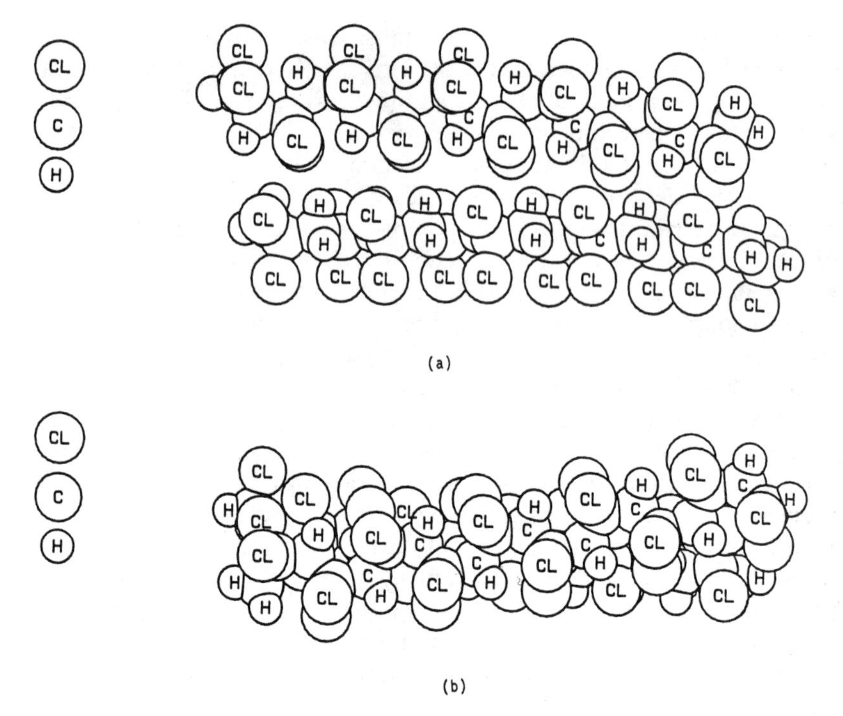

Figure 13 Space-filling illustrations of the $C_{20}H_{22}Cl_{20}$—$C_{20}H_{22}Cl_{20}$ pair: (a) a perspective showing both molecules; (b) an alternative perspective with one molecule behind the other, but only partially eclipsed by it because the two molecules are at an oblique angle.

tively. The intermolecular configurations of (a) $C_{20}H_{22}F_{20}$—$C_{20}H_{23}F_{19}$, which is almost identical to that of $C_{20}H_{22}F_{20}$—$C_{20}H_{22}F_{20}$; and (b) $C_{20}H_{22}Cl_{20}$—$C_{20}H_{23}Cl_{19}$, $C_{20}H_{22}Br_{20}$—$C_{20}H_{22}Br_{20}$, and $C_{20}H_{22}Br_{20}$—$C_{20}H_{23}Br_{19}$, which are almost identical to that of $C_{20}H_{22}Cl_{20}$—$C_{20}H_{22}Cl_{20}$, are not shown. The intermolecular energies are plotted as a function of the period of the halogen atom Y in Figure 14.

Define δE_2 as the change in E_2 resulting from the replacement of $C_{20}H_{22}Y_{20}$ by $C_{20}H_{23}Y_{10}$ as the second molecule in a pair. Here δE_2 is a measure of the energetic effect of replacing 10% of the VDY by VY in one of the two model molecules. Define the structural dissimilarity index σ as the root-mean-square deviation in a geometric match of corresponding $C_{20}H_{22}Y_{20}$ and $C_{20}H_{23}Y_{19}$ molecules. High σ indicates poor match between the two structures. The *magnitude* of δE_2 ($-\delta E_2$ for Y = F and + δE_2 for Y = Cl or Br), is shown as a function of σ in Figure 15.

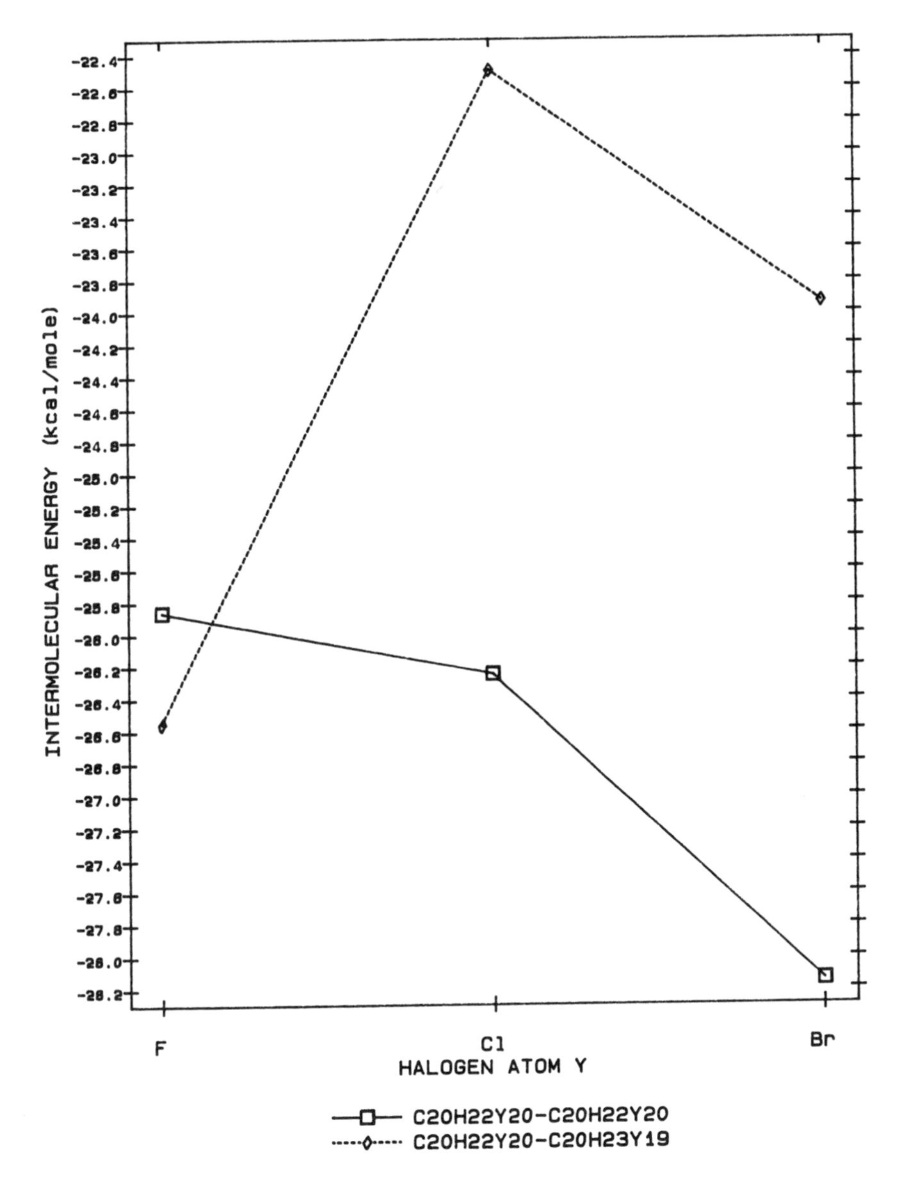

Figure 14 Intermolecular energy as a function of the halogen atom.

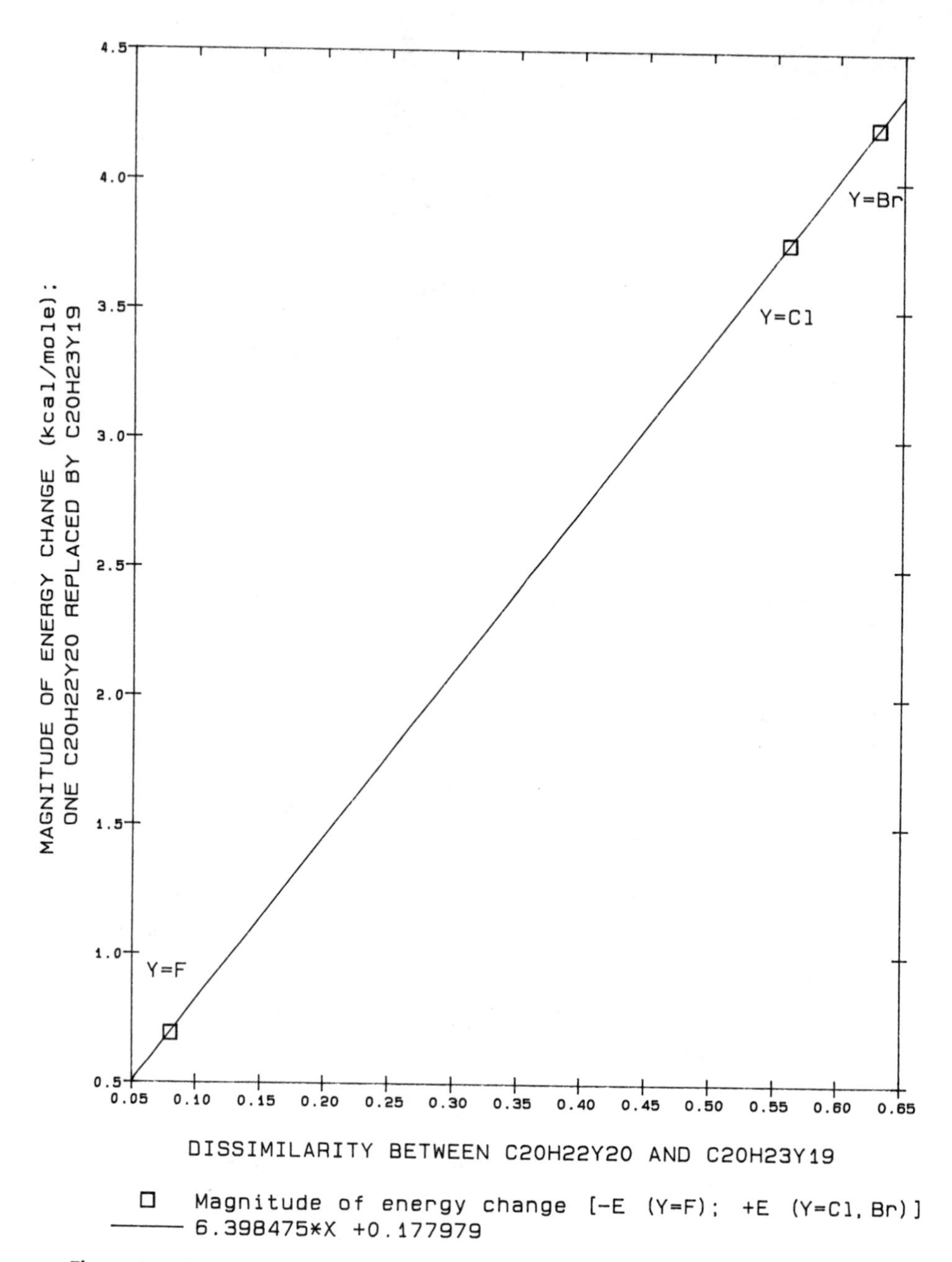

Figure 15 Magnitude of energy change (δE_2) resulting from replacement of one $C_{20}H_{22}Y_{20}$ molecule in a pair by $C_{20}H_{23}Y_{19}$, as a function of dissimilarity (σ) between $C_{20}H_{22}Y_{20}$ and $C_{20}H_{23}Y_{19}$.

Optimized Configurations

The two molecules are antiparallel and perfectly aligned for Y = F. Figure 12(b) shows that the intermolecular configuration can be viewed from a perspective in which one molecule is behind the other, almost completely eclipsed by it. The two molecules are antiparallel but at a slight oblique angle for Y = Cl or Br. Figure 13(b) shows that, in the perspective in which one molecule is behind the other, the molecule in the back is *not* almost completely eclipsed by the one in the front, and a considerable portion of it can still be seen. It is *not* obvious, from simple qualitative considerations, that the use of model molecules much longer than the 20-carbon chains utilized in this work would necessarily result in greater parallelization for Y = Cl or Br. The presence of more than two chains might force additional parallelization, but the energetic cost of forcing pairs of chains into the less favorable fully aligned pairwise configurations could be quite high.

A greater propensity of the chains to parallelize when Y = F than when Y = Cl may favor a higher percent crystallinity in PVDF, but the incorporation of a much larger number of head-to-head bonding defect sites in chains of PVDF [32] may disfavor higher crystallinity in PVDF. Consequently, the crystallinities of PVDF and PVDC are comparable, with PVDC in general being slightly more crystalline than PVDF when prepared in the same manner [58].

The size disparity between Br and H is only slightly larger than the size disparity between Cl and H. Consequently, the crystal structures of PVDC and PVDB can be expected to be much more similar than the crystal structures of PVDC and PVDF. This expectation is in agreement with crystallographic data [41] on oriented fibers.

The replacement of a Y atom by an H atom in a nonterminal site did not cause any significant changes in the intermolecular configurations, just as it did not cause any significant changes in the conformations of isolated chains. Addition of a small percentage of VY comonomer (10% in these calculations) should therefore not cause major changes in the packing patterns.

Energetics

As shown in Figure 14, E_2 was attractive (-26 to -28 kcal/mole) for all three $C_{20}H_{22}Y_{20}$—$C_{20}H_{22}Y_{20}$ pairs. There are twenty monomer units in *two* $C_{20}H_{22}Y_{20}$ molecules. The range of E_2's is therefore equivalent to a stabilization energy of 1.3 to 1.4 kcal/mole per monomer unit.

In order to describe polymer stability quantitatively, it is necessary to perform more detailed calculations, including a calculation of the difference in the heat of formation between the monomer and the reactant. However, the enormous differences between the heats of formation of isolated chain segments of PVDF, PVDC, and PVDB, as calculated by AM1, along with the much smaller intermolecular pair energy which is negative and comparable for each system, makes it possible to draw at least one important semiquantitative conclusion: The instability of PVDB is *not* caused by weak intermolecular binding, but by the instability of its individual chains [44]. This instability is manifested by the positive heat of formation, discussed above, of individual model molecules representing its isolated chain segments. In fact, the *inter*molecular stabilization slightly increases; i.e., E becomes slightly more negative, in the order Y = F < Cl < Br.

The E_2 of the $C_{20}H_{22}Y_{20}$—$C_{20}H_{23}Y_{19}$ pair is slightly *more* negative for Y = F, and considerably *less* negative for Y = Cl or Br, than the E_2 of the corresponding $C_{20}H_{22}Y_{20}$—$C_{20}H_{22}Y_{20}$ pair. As shown in Figure 15, the *magnitude* of δE_2 is proportional to σ, the quantitative index of the exact amount of mismatch between the geometries of corresponding $C_{20}H_{22}Y_{20}$ and C_{20} $H_{23}Y_{19}$ molecules.

The significant loss of stabilization energy caused by the replacement of one Cl or Br atom by an H atom shows that, although the intermolecular packing patterns might not change by the addition of a small percentage of comonomer, the intermolecular stabilization energy (and hence also the cohesive energy of the polymer) may decrease significantly.

The value of E_{LJ} is calculated to be at least 97% of E_2 in every system. All trends would have remained unchanged if E_e had been neglected. The percent contribution of E_e to the total energy should, however, become much larger in periodic crystalline structures, where the summation of Coulombic interactions between atoms with nonzero partial charges located on a periodic lattice should result in a Madelung-type stabilization energy [59]. The results of the crystal-type calculations of Farmer et al. [35] on the polymorphs of PVDF, and of Sorensen et al. [56] on POM, show that this is indeed the case. The present calculations show, therefore, that one fundamental limitation of calculations on pairs of chain segments, when used to model the structures and properties of crystalline polymers with a periodic lattice structure, is that they may grossly underestimate the importance of the electrostatic energy.

On the other hand, in amorphous regions, where there is no periodic lattice, and the interactions are of shorter range and more localized, E_e might remain quite small, albeit still considerably larger than that computed for a pair of chain segments. Finally, it should also be kept in mind that, whether the physical system being studied is an isolated chain, a pair of chain segments, or a crystal, electrostatic effects become increasingly less important, and steric effects become increasingly more important, as (a) the size of the halogen atom increases, and (b) simultaneously its electronegativity decreases, in the series Y = F, Cl, Br. In other words, electrostatic effects will play a less significant effect on PVDC and VDC/VC copolymers, which are the main subjects of this study, than they will in PVDF.

SUMMARY AND CONCLUSIONS

A study is in progress on the industrially important problem of the transport of penetrant molecules in barrier plastics, utilizing techniques at the molecular level. The results obtained so far have been reviewed.

The systematic use of a combination of different types of calculations can provide perspectives at different scales. These perspectives can be used to attempt to construct a unified working model for transport phenomena. Flow charts have been used to summarize how the different types of calculations and experiments fit together as a complete and coherent approach.

Copolymers of vinylidene chloride and vinyl chloride have been used as test cases. Permeation of penetrant molecules is generally believed to occur through the amorphous regions of these semicrystalline polymers. Their study is, however, complicated by the necessity to understand the effects of the presence of the crystallites in addition to the amorphous regions.

General comparisons have been presented between diffusion coefficients and apparent activation energies derivable from the "global" free-volume theory of Vrentas and Duda, and the statistical mechanical model of Pace and Datyner, as an initial step in the utilization of these theories.

The results of semiempirical calculations on isolated chain segments and on interacting pairs of chain segments of PVDY and VDY/VY copolymers (Y = F, Cl, or Br), utilizing both force field and quantum mechanical techniques, have been summarized, and used to draw as many conclusions as possible, from such a limited set of calculations, about chain-packing patterns and stabilities.

The main conclusion of the calculations summarized here is that the *packing efficiency* (as determined by the shape of the chain contour and the mobility of chain segments) is an extremely important physical factor in determining the permeability. This conclusion is also supported by positron annihilation studies of the microstructure of polymers in relation to their diffusional properties. These studies show that the diffusion coefficients of hydrogen and methane in a wide variety of amorphous polymers are determined by the free volume in the disordered regions of the polymer [60]. Finally, a new technique, based on the photoisomerization of photochromic and fluorescent probe molecules, which was recently developed for measuring the distribution of *local* free volume in glassy polymers, appears to be very promising for studying the packing in greater detail [61,62].

ACKNOWLEDGMENT

I thank A. F. Burmester, H. A. Clark, P. T. DeLassus, I. R. Harrison, D. J. Moll, J. K. Rieke, N. G. Rondan, and R. A. Wessling for helpful discussions. In addition, special thanks are due to A. J. Hopfinger, J. S. Vrentas, and R. A. Wessling for their critical reviews of this chapter.

REFERENCES

1. J. Crank and G. S. Park, *Diffusion in Polymers*, Academic Press, New York (1968).
2. V. Stannett, *J. Membrane Sci.*, *3*, 97 (1978).
3. M. Salame, *Polym. Preprints*, *8*, 137 (1967).
4. M. Salame, paper presented at the First International Conference on New Innovations in Packaging Technologies and Markets (November 28–30, 1983).
5. M. Salame, *J. Plastic Film Sheeting*, *2*, 321 (1986).
6. See Chapter 1, where group contribution techniques are reviewed by D. W. Van Krevelen.
7. W. J. Koros, B. J. Story, S. M. Jordan, K. O'Brien, and G. R. Husk, *Polym. Eng. Sci.*, *27*, 603 (1987).
8. *Encyclopedia of Polymer Science and Technology*, the "Diffusional Transport" subsection in the article titled "Diffusion."
9. J. S. Vrentas and J. L. Duda, *J. Appl. Polym. Sci.*, *21*, 1715 (1977).
10. J. S. Vrentas, H. T. Liu, and J. L. Duda, *J. Appl. Polym. Sci.*, *25*, 1297 (1980).
11. S. T. Ju, J. L. Duda, and J. S. Vrentas, *Ind. Eng. Chem. Prod. Res. Dev.*, *20*, 330 (1981).
12. J. S. Vrentas, J. L. Duda, and H.-C. Ling, *J. Polym. Sci.*, *Polym. Phys. Ed.*, *23*, 275 (1985).
13. J. S. Vrentas, J. L. Duda, H.-C. Ling, and A.-C. Hou, *J. Polym. Sci.*, *Polym. Phys. Ed.*, *23*, 289 (1985).
14. J. S. Vrentas, J. L. Duda, and A.-C. Hou, *J. Polym. Sci.*, *Polym. Phys. Ed.*, *23*, 2469 (1985).
15. J. S. Vrentas, J. L. Duda, and A.-C. Hou, *J. Appl. Polym. Sci.*, *33*, 2571 (1987).
16. R. J. Pace and A. Datyner, *J. Polym. Sci.*, *Polym. Phys. Ed.*, *17*, 437 (1979).
17. R. J. Pace and A. Datyner, *J. Polym. Sci.*, *Polym. Phys. Ed.*, *17*, 453 (1979).
18. R. J. Pace and A. Datyner, *J. Polym. Sci.*, *Polym. Phys. Ed.*, *17*, 465 (1979).
19. R. J. Pace and A. Datyner, *J. Polym. Sci.*, *Polym. Phys. Ed.*, *17*, 1675 (1979).

20. R. J. Pace and A. Datyner, *J. Polym. Sci., Polym. Phys. Ed.*, *17*, 1693 (1979).
21. R. J. Pace and A. Datyner, *J. Polym. Sci., Polym. Phys. Ed.*, *18*, 1103 (1980).
22. J. D. Ferry, *Viscoelastic Properties of Polymers*, 2nd ed., Wiley, New York (1970).
23. R. A. Wessling, *Polyvinylidene Chloride*, Gordon and Breach, New York (1977).
24. J. A. Manson and L. H. Sperling, *Polymer Blends and Composites*, p. 410, Plenum Press, New York (1976).
25. T. Alfrey, Jr., and R. F. Boyer, in *Molecular Basis of Transitions and Relaxations*, p. 193 (D. J. Meier, ed.), Gordon and Breach, New York (1978).
26. D. W. Van Krevelen, *Properties of Polymers*, Table 6.2, Elsevier, Amsterdam (1972).
27. N. Muruganandam, W. J. Koros, and D. R. Paul, *J. Polym. Sci., Polym. Phys. Ed.*, *25*, 1999 (1987).
28. W. M. Lee, *Polym. Eng. Sci.*, *20*, 65 (1980).
29. A. R. Berens and H. B. Hopfenberg, *J. Membrane Sci.*, *10*, 283 (1982).
30. S. M. Allen, V. Stannett, and H. B. Hopfenberg, *Polymer*, *22*, 912 (1981). Also see S. M. Aharoni, *J. Appl. Polym. Sci.*, *23*, 223 (1979).
31. This point was brought to the author's attention by D. J. Moll.
32. H.-G. Elias, *Macromolecules*, Vol. 1, *Structure and Properties*, 2nd ed., Plenum Press, New York (1984).
33. A. E. Tonelli, F. C. Schilling, and R. E. Cais, *Macromolecules*, *14*, 560 (1981).
34. A. E. Tonelli, F. C. Schilling, and R. E. Cais, *Macromolecules*, *15*, 849 (1982).
35. B. L. Farmer, A. J. Hopfinger, and J. B. Lando, *J. Appl. Phys.*, *43*, 4293 (1972).
36. B. L. Farmer and J. B. Lando, *J. Macromol. Sci.-Phys.*, *B11*, 89 (1975).
37. M. M. Coleman, M. S. Wu, I. R. Harrison, and P. C. Painter, *J. Macromol. Sci.-Phys.*, *B15*, 463 (1978).
38. M. S. Wu, P. C. Painter, and M. M. Coleman, *Spectrochimica Acta*, *35A*, 823 (1979).
39. M. S. Wu, P. C. Painter, and M. M. Coleman, *J. Polym. Sci., Polym. Phys. Ed.*, *18*, 95 (1980).
40. M. S. Wu, P. C. Painter, and M. M. Coleman, *J. Polym Sci., Polym. Phys. Ed.*, *18*, 111 (1980).
41. S. Narita and K. Okuda, *J. Polym. Sci.*, *38*, 270 (1959).
42. See T. Clark, *A Handbook of Computational Chemistry*, Wiley, New York (1985) for detailed discussions of computational methods.
43. J. Bicerano, *Macromolecules*, *22*, 1408 (1989).
44. J. Bicerano, *Macrmolecules*, *22*, 1413 (1989).
45. B. J. Orchard, S. K. Tripathy, R. A. Pearlstein, and A. J. Hopfinger, *J. Comput. Chem.*, *8*, 28 (1987).
46. P. J. Flory, *Macromolecules*, *7*, 381 (1974).
47. A. E. Tonelli, *Macromolecules*, *9*, 547 (1976).
48. A. E. Tonelli, *Macromolecules*, *13*, 734 (1980).
49. A. E. Tonelli, *Macromolecules*, *15*, 290 (1982).

50. U. W. Suter, E. Saiz, and P. J. Flory, *Macromolecules*, *16*, 1317 (1983).
51. A. L. Allred, *J. Inorg. Nucl. Chem.*, *17*, 215 (1961).
52. J. Bicerano and D. Adler, *Pure Appl. Chem.*, *59*, 101 (1987).
53. T. Yemni and R. L. McCullough, *J. Polym. Sci., Polym. Phys. Ed.*, *11*, 1385 (1973).
54. R. H. Boyd and L. Kesner, *Macromolecules*, *20*, 1802 (1987).
55. R. A. Sorensen, W. B. Liau, and R. H. Boyd, *Macromolecules*, *21*, 194 (1988).
56. R. A. Sorensen, W. B. Liau, L. Kesner, and R. H. Boyd, *Macromolecules*, *21*, 200 (1988).
57. J. J. Blaisdell, W. A. Sokalski, P. C. Hariharan, and J. J. Kaufman, *J. Non-Cryst. Solids*, *75*, 319 (1985).
58. R. A. Wessling and I. R. Harrison, private communication.
59. J. M. Ziman, *Principles of the Theory of Solids*, 2nd ed., Chaps. 2, 4, Cambridge University Press, London (1972).
60. V. V. Volkov, A. V. Gol'danskii, S. G. Dur'garyan, V. A. Onishchuk, V. P. Shantorovich, and Yu. P. Yampol'skii, *Polym. Sci. USSR*, *29*, 217 (1987).
61. J. G. Victor and J. M. Torkelson, *Macromolecules*, *20*, 2241 (1987).
62. J. G. Victor and J. M. Torkelson, *Macromolecules*, *20*, 2951 (1987).

8

Detailed Molecular Structure of a Polar Vinyl Polymer Glass

Peter J. Ludovice
Polygen Corporation
Waltham, Massachusetts

Ulrich W. Suter
Institute für Polymore, ETH
Zürich, Switzerland

INTRODUCTION

Previously, a detailed atomistic molecular mechanics method for polymeric glasses was successfully developed and applied to atactic polypropylene [1]. The atomistic detail on which this model is based makes the investigation of specific interactions in the bulk possible. In order to understand how polar interactions influence the structure of polymer glasses we now attempt to extend this model to include partially charged atoms. Atactic polyvinylchloride (PVC) was chosen for this investigation; it is a polar analogue of atactic polypropylene and allows for comparison between these two systems, which differ essentially only in the nature of interactions between atoms. Eventually we hope to be able to utilize this model to shed some light on the structure of PVC and the interactions that distinguish PVC from other polar polymers.

PREVIOUS WORK

This investigation is a direct extension of previous work [1] in which a static molecular mechanics model to describe polymeric glasses was developed. The method consists of generating an initial guess for the structure of the glass generated in a Metropolis/Monte Carlo fashion from the appropriate rotational isomeric states model, and subsequent minimization of the total potential energy of the resulting structure by a quasi-Newton method [2]. The microstructures consisted of a periodic cube containing a polymer molecule, of degree of polymerization $X = 76$, with fixed bond lengths and bond angles. Atoms interact with a Lennard-Jones potential energy function, and an intrinsic torsional potential acts in the torsional angles of the skeletal carbon-carbon bonds.

After minimization of the total potential energy to a (local) minimum, the resulting microstructures are thought to be representative of real glass configurations that are well relaxed following their initial vitrification. This is in accord with Cohen and Turnbull's view of a glass being in a state of frozen in liquid disorder [3]. In a crystalline material the bulk of the material usually consists of replicas of a single structure represented by the leading term in the partition function, while in an amorphous material, such as a glass, there are many coexisting structures of similar importance with corresponding terms in the partition function of similar magnitude. Therefore many configurations need to be sampled (i.e., an ensemble must be constructed) to obtain structural information representative of a glass.

In the case of polypropylene a sample of 15 microstructures was generated from which material properties were deduced. Properties calculated from these 15 structures of minimum potential energy included values of the Hildebrandt solubility parameter and the elastic constants which compared favorably with experimental results [1,4]. Here we extend this type of model to a polar polymer, PVC.

PVC MODEL PARAMETERS

The modeled chain has the constitution $CH_3—(CHCl—CH_2)_{x-1}—CHCl—CH_3$; the terminal methyl groups are treated as pseudoatoms. Interactions between nonbonded atoms were described with a Lennard-Jones 6-12 potential energy function U^{LJ} and a coulombic interaction energy term U^c. This nonbonded interaction was applied to atoms or atomic groups that are on the same polymer chain but are separated by more than two bonds. The intrinsic torsional potential energy function was identical to that used for polypropylene, as was the method for calculating the Lennard-Jones parameters. Data used to calculate the Lennard-Jones parameters for PVC are contained in Table 1. Also contained in Table 1 are the bond lengths and angles used in the PVC simulation.

The electrostatic interaction energy is modeled as a coulombic inter-

Table 1 PVC Geometry and Potential Parameter Values

	Bond-lengths (Å)	Bond-Angle Supplements (degrees)	
C—C	$l = 1.53$	Intradyad C—C—C	$\theta' = 66$
C—H	$l_H = 1.10$	Interdyad C—C—C	$\theta'' = 68$
C—Cl	$l_{Cl} = 1.79$	Intradyad C—C—H	$\theta'_H = 71$
		Interdyad C—C—Cl	$\theta''_{Cl} = 71$
		Interdyad C—C—H	$\theta''_H = 73.2 = \cos^{-1}\{[(1 - 2\cos\theta'')/3]^{1/2}\}$

i	Atoms or group	r_i^o (Å)[a]	α_i(Å^3)[b]	$N_{e,i}$ [c]
1	H	1.3	0.42	0.9
2	C	1.8	0.93	5.0
3	Cl	1.85	2.28	12.0
4	CH_3	2.0	1.77	7.0

[a]van der Waals radius.
[b]Mean polarizability.
[c]Effective number of electrons.
Source: Ref. 5.

action between atoms carrying point charges, modified by a distance dependent dielectric function $D(r)$:

$$U_{ij}^c(r) = \left(\frac{1}{4\pi D(r)\epsilon_0}\right)\left(\frac{q_i q_j}{r}\right). \tag{1}$$

Here $D(r)$ is the effective permittivity of an electric field in the polymer relative to that in vacuum, which is denoted by ϵ_0. The magnitudes of the point charges on atoms i and j are q_i and q_j, and r is the distance between these charges. Typically, this relative permittivity is taken to assume one of its limiting values; unity for very short separations, or the bulk value of the dielectric constant for large separations. In this study we employ an approximation for $D(r)$ which is based on a function first used by Block and Walker [6]:

$$D(r) = 1, \qquad r < a, \tag{2}$$

$$D(r) = \epsilon_B e^{(-\kappa/r)}, \qquad r > a, \tag{3}$$

$$\kappa = a \log(\epsilon_B), \tag{4}$$

where ϵ_B is the bulk value of the dielectric constant and a is a minimum critical distance. The bulk dielectric constant ϵ_B is taken to be 3.51 for these simulations [7]. The constant κ is defined such that the two portions of the function are joined at $r = a$. Note that the above function produces the correct asymptotic limits for the relative permittivity. Calculations performed by Block and Walker, using this analytical form for $D(r)$, reproduced experimental organic dipole data well [6], but they gave no suggestion as to what the critical distance a should be. Since this distance is the point at which the relative permittivity decreases to unity, its value in vacuo, it should correspond to the point at which no other atoms may fit between the two interacting atoms or atomic groups. A physically reasonable value for this critical value is the sum of the van der Waals radii of the two atoms in question. This approach was employed in the calculations, and an atom pair dependent value of a, a_{ij}, was used for each atom pair. This does present some problems, however, when one attempts to calculate the total potential acting on a given atom due to all surrounding atoms infinitely in all directions, as we will later discuss.

It is important to note that, in the above approach, a form is assumed for the electrostatic potential which is equivalent to Coulomb's law divided by an (ad hoc) effective dielectric function. This is not identical to assuming a form for the actual dielectric function and then solving for the electrostatic potential, which has been done for several dielectric functions for fluids (including the Block-Walker form) [6,8] by solving a Maxwell equation, for the displacement vector **D** in a neutral system [6,9]:

$$\nabla \cdot \mathbf{D} = 0. \tag{5}$$

In principle, this equation could be solved for all the differently charged atoms in a single electrically neutral monomeric unit of PVC. For fluid dielectrics the displacement vector **D** equals the product of the dielectric function with the electric field strength **E** (where **E** is the negative gradient of the electric potential). For solids such as PVC glass this assumption is not necessarily valid, the differential equation (5) represents a difficult problem since any motion of charges changes the mechanical stress conditions of the dielectric and alters its electrical properties as well as causing some mechanical energy change [9]. Therefore, we have chosen here a simple form of the electrostatic interaction rather than solving for it explicity.

The finite size of the periodic system requires that the potential energy functions be limited; in a cubic "box" this range can be no more than half the length of the cube edge. This is accomplished with a quintic spline where the Lennard-Jones potential energy function is limited as in previous work [1]. The potential energy function for the nonbonded *polar* interactions U^{NBP} is similar:

$$U^{NBP}(r_{ij}) = U^c(r_{ij}) \equiv \left(\frac{1}{4\pi D(r_{ij})\epsilon_o}\right)\left(\frac{q_i q_j}{r_{ij}}\right), \qquad r_{ij} < R_1, \tag{6a}$$

$$U^{NBP}(r_{ij}) = \alpha(1 - \xi)^3 \left(\frac{U_1^c}{\alpha} + \left(3\,\frac{U_1^c}{\alpha} + \Delta\,\frac{U_1'^c}{\alpha/\sigma}\right)\xi + \left(6\,\frac{U_1^c}{\alpha} + 3\Delta\,\frac{U_1'^c}{\alpha/\sigma} + \frac{\Delta^2}{2}\,\frac{U_1''^c}{\alpha/\sigma^2}\right)\xi^2\right),$$
$$R_1 < r_{ij} < R, \tag{6b}$$

$$U^{NBP}(r_{ij}) = 0, \qquad R < r_{ij}, \tag{6c}$$

where U_1^c, $U_1'^c$, and $U_1''^c$ represent the value, and the first and second derivatives of the coulombic potential at $r = R_1$, $\Delta = (R - R_1)/\sigma$, and $\xi = (r - R_1)/(R - R_1) = (r/\sigma - R_1/\sigma)/\Delta$. The Lennard-Jones function distance parameter is denoted by σ, and upon replacing the energy parameter $\alpha = q_1 q_2/4\pi\epsilon_0\sigma$ in Eq. (6b) with the Lennard-Jones energy parameter ϵ one obtains the equivalent spline for the Lennard-Jones potential energy function U^{NB} (Eq. (1) of Ref. 1). Note however, that R_1 is different for the Lennard-Jones and the coulombic potential energy function. The value of $R = 2.3\sigma$ for the coulombic potential energy function was chosen to match the Lennard-Jones potential energy function. A value of $R_1 = 0.786\sigma$ was selected and the "critical" radius a of the dielectric function $D(r)$ (see Eqs. (2)–(4)) was taken as the sum of the respective van der Waals radii for the interacting atoms. Figure 1 shows the similarity between the splined potential function and the actual potential function with and without attenuation by the dielectric $D(r)$ in dimensionless units. Both the critical radius and the Lennard-Jones parameter σ are based on the van der Waals radii; $0.7a$ is equal to 0.78σ.

The charges used in the model are constant; a model that includes polarization effects may be more accurate [10,11], but it would be much too computationally demanding for systems to include these effects in polymers of this size. The charges used were deduced from calculations on low-molecular-mass analogues of polyvinylchloride employing semiempirical self-consistent quantum mechanical calculations.

Initial calculations on isopropylchloride and dichloropentane indicated that the methylene hydrogens in PVC may possess nontrivial partial point charges [12]. Therefore, more detailed calculations on the polyvinylchloride trimer 2,4,6-trichloroheptane were carried out using the modified neglect of differential overlap (MNDO) method [13,14] (this method was chosen due to the accuracy with which it reproduces experimental data for many organic compounds [15]). The MNDO calculations were carried out

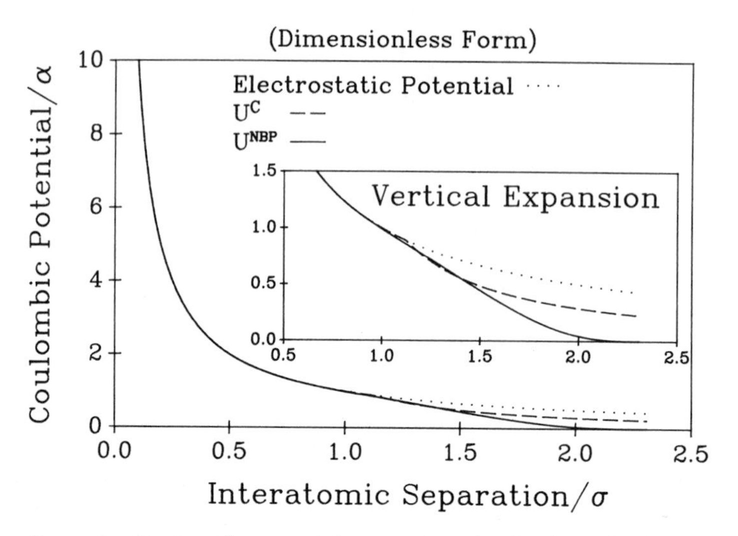

Figure 1 Coulombic potential energy function in dimensionless units. U^c assumes a bulk dielectric constant of 3.51 and a critical radius equal to the sum of the van der Waals radii, U^{NBP} employs the full potential spline range.

on the two most favored conformations of each of the three diastereoisomers of 2,4,6-trichloroheptane, optimizing all geometric parameters (initial guesses of the optimized geometry were made based on known structural data of PVC oligomers [16]). The partial charges used for the calculations are listed in column 3 of Table 2. They are the appropriate arithmetic averages of the charges from the various diastereomers. The methylene

Table 2 Average Partial Charges Deduced from MNDO Calculations[a]

Atom	Average charge	Mean std. deviation	Charge
Methine hydrogen	+0.056	0.004	+0.056
Methine carbon	+0.0897	0.0008	+0.087
Methylene carbon	−0.0162	0.0007	0.000
Methylene hydrogen	+0.035	0.001	+0.035
Chlorine	−0.213	0.002	−0.213
Terminal methyl	—	—	+0.035[b]

[a]Charges were averaged over all atoms in the central methine group and flanking methylene groups for the two most stable conformations of all three stereoisomers of 2,4,6-trichloroheptane; conformations: isotactic (gttg,tgtg), syndiotactic (tttt,ttgg), heterotactic (tttg,ttgt).
[b]Terminal methyl charge is determined by constraining the net charges on the entire polymer to zero.

Table 3 Comparison of Calculated and Experimental Dipoles for 2,4,6-trichloroheptane (in Debyes)

Stereoisomer	Calculated	Boyd & Kesner[a]	Expt[b]
Iso (mm)	2.22[c]	2.12	—
Syndio (rr)	3.85[c]	3.41	4.01[e]
Hetero (rm)	2.66[c]	2.58	2.62[e]
	2.61[d]		2.68[f]

[a]See Ref. 18.
[b]See Ref. 17.
[c]Dipole calculated with *n*-heptane dielectric constant.
[d]Dipole calculated with carbon tetrachloride dielectric constant.
[e]Measured in *n*-heptane.
[f]Measured in carbon tetrachloride.

carbon was assigned a net zero charge because its charge varied between small positive and small negative values over the different diastereomers. It is interesting to note that the charges on the methylene hydrogens are significant relative to the charges on the methine hydrogen and carbon.

In order to test the accuracy of these partial charges they were used to predict molecular dipole moments of the three diastereoisomers of 2,4,6-trichloroheptane. The dipole moments were calculated for the six most energetically preferred conformations of each of the three isomers. A Boltzmann average of these six values was formed using the internal energy calculated from the molecular mechanics model. The dielectric constant of the solvent in the referenced experiment was used as ϵ_B in Eqs. (3) and (4), the potential energy functions were limited in range. The results of these calculations (see Table 3) were compared with dilute solution dipole measurements [17], and with the results of a similar energy minimization carried out by Boyd and Kesner [18]. The model of Boyd and Kesner accounted explicitly for polarization and solvation effects for which the model proposed here does not. Despite this, the proposed model results compare better to the experimental data than the Boyd and Kesner model. The agreement between our calculations and experimental data is good (within less than 6%) indicating that the assigned partial charges are sound.

INITIAL GUESS AND RELAXATION TO MECHANICAL EQUILIBRIUM

PVC microstructures were constructed following the strategy employed for the polypropylene simulations. An initial guess was generated employing a modified Monte Carlo procedure based on a rotational isomeric states model. For the PVC calculations a three-state RIS model [19,20] was substituted for the five-state model [21] used for polypropylene.

The procedure for minimizing the potential energy of the microstructures used in the polypropylene simulations had to be modified somewhat. Even for the apolar structures of polypropylene minimizations require a stagewise approach. However, for PVC more stages were necessary. The Lennard-Jones potential energy function "decays" as $1/r^6$, whereas the coulombic energy function with the distance dependent dielectric diminishes only slightly faster than $1/r$. Hence, this coulombic energy is significantly "felt" by more atoms in the periodic cube, and is effectively of longer interaction range than the Lennard-Jones energy. It is this long range property of the coulombic potential energy function that forces a change in the minimization procedure for polar polymers. Instead of minimizing the potential energy in three stages, as in the case of polypropylene, it was carried out in four stages for PVC.

Only the final step of the minimization employs the aforementioned spline ranges, which are referred to as the "full potential." The first three minimization stages employ a "soft sphere" potential energy function where the spline ranges for the Lennard-Jones and coulombic energies are from 0.94σ to 1.04σ and from 0.337σ to 1.04σ, respectively, the fourth stage, a full potential function with corresponding ranges of 1.45σ to 2.3σ and

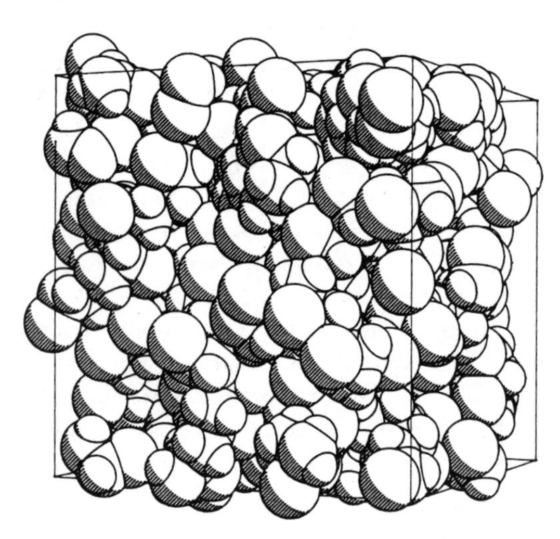

Figure 2 Energy minimized structure of PVC inside periodic box, degree of polymerization 200.

Table 4 Representative Evolution of Computer Relaxations[a]

Minimization step	1	2	3	4
X = 76, cube edge = 17.84 Å				
Starting structure				
energy (kcal/mole)	1.327×10^5	7.716×10^2	4.776×10^3	-6.792×10^1
Ending structure				
energy (kcal/mole)	-1.832×10^{-15}	1.750×10^1	4.204×10^2	-1.722×10^2
iterations	114	1181	1636	869
function evaluations	309	2444	3158	1749
CPU time (hr)	0.263	2.33	3.49	7.31
CPU time per function evaluation (s)	3.0	3.4	3.9	15.0
X = 200, cube edge = 24.64 Å				
Starting structure				
energy (kcal/mole)	6.390×10^{10}	4.087×10^{10}	7.919×10^3	3.253×10^2
Ending structure				
energy (kcal/mole)	4.782×10^1	5.482×10^1	1.416×10^3	-4.168×10^2
iterations	2387	5328	6410	2407
function evaluations	5251	10493	12217	4744
CPU time (hr)	34.32	74.48	95.92	138.91
CPU time per function evaluation (s)	23.5	25.5	28.2	105.4

[a]Calculations were made on an Alliant FX-1 and an Alliant FX-8 with four computational elements. CPU times are from runs on the Alliant FX-1 in full (-byte) precision.

0.786σ to 2.3σ, respectively. The four minimization stages for polar polymers are as follows:

1. Soft sphere potential; atomic radii (hence, σ) and partial charges scaled to half their actual values; no torsional potentials
2. Soft sphere potential; atomic radii and partial charges scaled to 70% of their actual values; full torsional potentials
3. Soft sphere potential; atomic radii and partial charges actual magnitude; full torsional potentials
4. Full potential; atomic radii and partial charges actual magnitude; full torsional potentials

Structures were simulated at a density of 1.39 g/cm^3, with 43% meso diads distributed in a Bernoullian fashion [22]. This density was considered representative of well relaxed PVC at a temperature of 298 K (the temperature used in the initial guess generation). Two different size structures were simulated: an ensemble of 10 structures of degree of polymerization X = 76 (cube edge = 17.84 Å), and a single structure of degree of polymerization X = 200 (cube edge = 24.64 Å). This larger structure, in its periodic cube, is seen in Figure 2. Only one of the larger structures was produced due to the high computational requirements involved (see Table 4), since the necessary computational resources grow approximately as X^2. Most of the structures were calculated on an Alliant FX/1, or an Alliant FX/8 with four computational elements [23] (a net speed-up of approximately three is obtained on four computational elements when explicit concurrency directives are inserted into the code [24]).

TAIL CORRECTIONS

The cohesive energy density U_{coh} is the energy due to only intermolecular forces [25]. This can be estimated by taking the difference between the total interaction energy and that of just the parent chains U_{par}. Energy values computed with the energy functions outlined above are based on range-limited potential energy functions. They must be corrected [1] for the effect of "truncation" by a "tail correction." This tail correction is given by

$$\Delta U_{\text{tails}} = \frac{1}{2} \sum_{\substack{\text{atoms} \\ i}} N_i \sum_{\substack{\text{atoms} \\ j}} \rho_j \left(4\pi \int_{R_{1,ij}}^{R_{ij}} g_{ij}(r)[U_{ij}^{\text{LJ}}(r) - U_{ij}^{\text{NB}}(r)]r^2 \, dr \right.$$
$$\left. + \, 4\pi \int_{R_{ij}}^{\infty} g_{ij}(r) U_{ij}^{\text{LJ}}(r) r^2 \, dr \right)$$

$$+ \frac{1}{2} \sum_{\substack{\text{atoms} \\ i}} N_i \sum_{\substack{\text{atoms} \\ j}} \rho_j \left(4\pi \int_{R_{2,ij}}^{R_{ij}} g_{ij}(r)[U_{ij}^{c}(r) - U_{ij}^{\text{NBP}}(r)]r^2 \, dr \right.$$

$$\left. + 4\pi \int_{R_{ij}}^{\infty} g_{ij}(r) U_{ij}^{c}(r) r^2 \, dr \right) \tag{7}$$

where N_i and ρ_i are the number and number density of groups of species i in the periodic cube. The summation indices refer to the different atomic species: in the first double summation i and j refer to H, C, Cl, and the terminal methyl CH_3; in the second double summation they refer to the differently charged species: H(methylene), H(methine), C(methylene), C(methine), Cl, and CH_3. The atomic pair distribution functions are denoted by $g_{ij}(r)$, and their calculation is discussed later in this paper.

The first and third integrals in Eq. (7) may be evaluated numerically using pair distribution functions obtained from the calculated structures. Due to the periodic nature of the cube, however, the atomic pair distribution functions cannot be calculated for values of r greater than the length of the semidiagonal of the cube R_{sd} [26], and the second and fourth integrals in Eq. (7) cannot be evaluated directly for the range from R_{sd} to infinity. However, if we are to assume that $g_{ij}(r) = 1$ for all values of r greater than some value R^* that is less than the semidiagonal of the cube, the second integral may be evaluated numerically using the pair distribution function from R_{ij} to R^*, and analytically from R^* to infinity [1]. Examination of the atomic pair distribution functions $g_{ij}(r)$ displayed in Figures 5 and 6 confirms that assigning a value of unity to $g_{ij}(r)$ for values of r greater than the cube semidiagonal (15.4 and 21.3 Å for the "small" and "large" PVC simulations respectively) is not unreasonable.

The assumption that $g_{ij}(r) = 1$ for $r > R$ is not sufficient for the convergence of the fourth integral in Eq. (7). However, the local electroneutrality of the PVC chain (e.g., neutrality of the monomeric units) makes this integral convergent (see appendix). This term can be evaluated numerically, but if one further assumes that the critical distance a in the function $D(r)$ is the same for all atomic pairs, it is easily shown that this fourth integral is zero (see appendix). Recall that the a_{ij} values were taken to be the sum of the van der Waals radii for the two interacting atoms. This was "physically realistic" and important in determining the microstructures, but for simplicity we will assume a constant value of a, independent of the interacting species, in evaluating the tail correction.

Assigning a value of unity to $g_{ij}(r)$, for $r > R^*$, as was done above, is tantamount to assuming the microstructures are amorphous on the scale of the periodic cube or larger. However, it would appear that the slow decay of the coulombic energy term may contribute a certain and perhaps significant amount of "correlation" even at relatively large distances. When the total charge of a system is zero (as it is in this case) the sum of the coulombic interactions actually decay as $1/r^3$ for large values of r [27]. Further, when the system is made up of independent dipoles that initially arrange themselves according to a Boltzmann distribution (i.e., in a system in thermodynamic equilibrium, such as a liquid), the sum of the coulombic potential decays as $1/r^6$ [27]. However, the dipoles in the polymeric glasses we attempt to model are highly constrained by the viscosity of the polymeric glass, and probably do not follow a Boltzmann distribution; a coulombic potential energy decay of $1/r^3$ may be more realistic. The presence of some long range order in PVC, that cannot be predicted by the model due to the small size of the periodic cube, may contribute significant error to the calculation of the fourth integral in Eq. (7). In an attempt to quantify this error we will consider another case for estimation of the last term of the coulombic portion of the tail correction (other than the above assumption of total amorphousness). In this other extreme one assumes that liquidlike ordering of the dipoles may occur in a glass. We will consider each charged atom to be surrounded by a sea of randomly distributed monomeric unit dipoles that are oriented according to a Boltzmann energy distribution. In the glassy state the dipoles of the monomeric units are not usually mobile enough to orient themselves according to a Boltzmann distribution at the actual temperature. The actual distribution is closer to the one that occurs at the glass transition temperature. The distribution has relaxed only slightly from this distribution that was "frozen in" during the glass transition, so the structural order in the glass is always less than this hypothetical liquidlike case. Thus, this approach allows us to obtain an upper bound for the magnitude of the fourth integral in Eq. (7), thereby providing a range of possible values for the tail correction integral in PVC. The liquidlike approximation for the fourth integral in Eq. (7) from R^* to infinity is given by (see appendix)

$$\frac{1}{2} \sum_{\substack{\text{atoms} \\ i}} N_i \sum_{\substack{\text{atoms} \\ j}} \rho_j 4\pi \int_{R^*}^{\infty} g_{ij}(r) U_{ij}^{c}(r) r^2 \, dr = \frac{-2\pi X \mu^2}{3kTV(4\pi\epsilon_0\epsilon_B)^2(R - \kappa)} \sum_{\substack{\text{atoms} \\ i}} N_i q_i^2. \quad (8)$$

where π is the effective dipole of a monomeric unit [18]. The degree of polymerization is denoted by X, and R^*, as defined previously, is some distance less than or equal to the semidiagonal of the cube, at which point the pair distribution of the point dipoles is assumed to be unity. The temperature is denoted by T, and k is Boltzmann's constant. All other terms are as described in Eqs. (1) through (7).

RESULTS

Ten PVC structures with degree of polymerization $X = 76$ were simulated as described above, and properties were obtained from averages from this ensemble. Also, a single structure with a degree of polymerization of $X = 200$ was prepared.

Cohesive Energy Density and Solubility Parameter

The solubility parameter may be calculated from the interaction energy in the ensemble of structures [25]. The total potential energy for the 10 smaller structures is (the mean standard deviation is also indicated)

$$U = -157.4\ (\pm 23.1)\ \text{kcal/mole of structures,}$$

and for the single large PVC cube with degree of polymerization $X = 200$:

$$U = -416.8\ \text{kcal/mole of structures.}$$

The values used for R^* were 80% of the length of the continuation vectors of the periodic cubes. These values are slightly less than the cube semidiagonals because the pair distribution functions, $g_{ij}(r)$, are increasingly plagued by numerical error as the value of r approaches the length of the cube semidiagonal. Table 5 contains the results of the tail correction calculations for both the previously discussed approaches, using the possible extreme values for the critical radius a_{ij} in $D(r)$. These were the sum of the van der Waals radii for two hydrogen atoms and two chlorine atoms, respectively. The variation of values on the tail correction due to changes in the critical radius and the model for the distribution of surrounding charges allows us to define a probable range for the tail correction to the ensemble of 10 structures($X = 76$).

$$\Delta U_{\text{tails}} = -87.0\ (\pm 30.0)\ \text{kcal/mole of structures,}$$

and similarly for the single larger PVC structure ($X = 200$):

$$\Delta U_{\text{tails}} = -203.2\ (\pm 66.4)\ \text{kcal/mole of structures.}$$

The total interaction energy of the parent chain, was estimated following the calculation for polypropylene, thus

$$U_{\text{coh}} = U_{\text{par}} - U_{\text{tot}} = U_{\text{par}} - [U + \Delta U_{\text{tails}}]. \tag{9}$$

Table 5 Tail Correction for PVC

Degree of polymerization		
$X = 76$	$X = 200$	Conditions
−57.0	−136.7	$g_{ij}(r)$ from model for $r < R$ $g_{ij}(r) = 1$ for $r > R$ minimum a_{ij} value
−88.3	−221.1	$g_{ij}(r)$ from model for $r < R$ $g_{ij}(r) = 1$ for $r > R$ maximum a_{ij} value
−83.1	−182.4	$g_{ij}(r)$ from model for $r < R$ liquid-like approx. for $r > R$ minimum a_{ij} value
−117.0	−269.6	$g_{ij}(r)$ from model for $r < R$ liquidlike approx. for $r > R$ maximum a_{ij} value
−87 (±30)	−203 (± 66)	Effective range over all of above conditions

The minimum value for the critical radius a_{ij} is the sum of the van der Waals radii for two hydrogen atoms, or 2.60 Å. The maximum value is the sum for two chlorine atoms or 3.296 Å. The values for R are 15.4 and 21.3 for the 76 and 200 degree of polymerization PVC cahins, respectively.

The Hildebrandt solubility parameter δ, defined as the square root of the cohesive energy density,

$$\delta = \left(\frac{U_{\text{coh}}}{V}\right)^{1/2} \tag{10}$$

is obtained from U_{coh} and the density of the system ($\rho = 1.39\ \text{g/cm}^3$). Utilizing the full range of values produced by the different approaches to the tail-correction, we obtain the following ranges for the solubility parameter δ (limits are maximum deviations).

$$\delta = 17.0 \pm (1.2) \times 10^3(\text{J/m}^3)^{1/2} \qquad (10\ \text{structures}, X = 76),$$
$$\delta = 18.8 \pm (0.8) \times 10^3(\text{J/m}^3)^{1/2} \qquad (1\ \text{structure}, X = 200).$$

The large range given for the "smaller" microstructures reflects the variations between the 10 values, while the estimate for the "large" system is derived from a single value, the error range stemming solely from the calculational approach to the tail correction.

The reported range of experimental values for the solubility parameter of commercial PVC is (19.2 to 22.1) × $10^3(\text{J/m}^3)^{1/2}$ [28]. Hence, the theoretical predictions provide reasonable estimates.

Conformation of Individual Chains

The distribution of rotational angles of the two different sized structures is seen in Figure 3. The height of the "Dirac spikes" in this figure is proportional to the a priori probabilities of the states with torsion angles as determined from the RIS model used in the initial guess procedure [19,20]. These a priori probabilities are compared in Table 6 to the corresponding integrated areas in the distribution from the simulation. The overlap of the t and g states, characteristic of poly-α-olefins, is readily observable in Figure 3 [29]. The fact that a three-state RIS model was used in the generation of the PVC initial guess, and that this overlap was still observed in the energy minimized structures indicates that this model is not particularly sensitive to the RIS model used in the initial guess generation procedure. The simulated structures are much richer in the $\bar{g}$ state than the RIS model; in the simulation of polypropylene the same was observed [1].

Comparing the individual chains' radii of gyration with those estimated for unperturbed chains using the RIS model used in the initial guess procedure [19,20] with the generator matrix techniques [30], one observes an unexpectedly small average value for the packed chains (the ratio $\langle s^2 \rangle_{\text{packed}}^{1/2} / \langle s^2 \rangle_0^{1/2}$ is roughly 0.7 (±0.2) at $X = 76$ and $X = 200$). We have traced this phenomenon back to the initial guess generation procedure (see above); during minimization the overall chain dimensions scarcely changed,

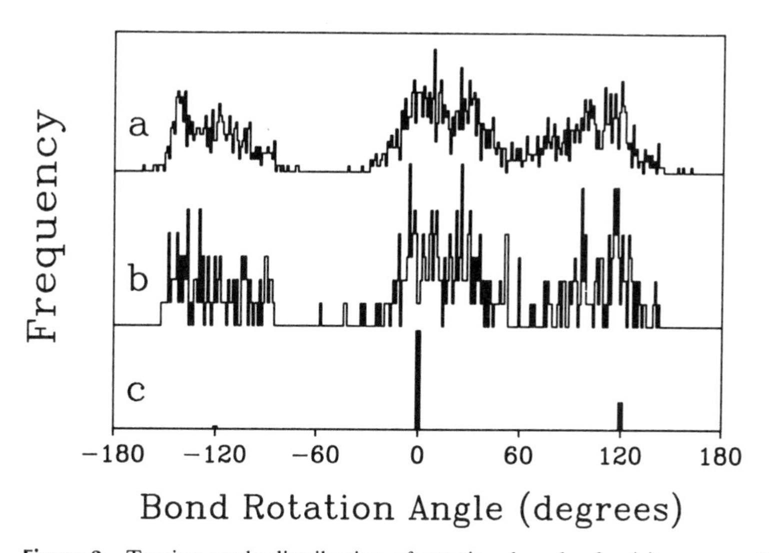

Figure 3 Torsion angle distribution of rotational angles for (a) an ensemble of 10 structures $X = 76$, and (b) a single structure $X = 200$. (c) Corresponding relative RIS a priori probabilities are given as bold vertical lines.

Table 6 Model Torsional Angle Distributions Relative to RIS Predictions for Unperturbed Chains

State	Angle interval	State frequency obtained from 10 model systems ($X = 76$)	State a priori probability from RIS model ($X = 76$)
$\bar{g}$	$-180° < \phi < -60°$	0.264 ± 0.006	0.0191 ± 0.0005
t	$-60° < \phi < 60°$	0.427 ± 0.008	0.7656 ± 0.0006
g	$60° < \phi < 180°$	0.309 ± 0.011	0.2153 ± 0.0006
		State frequency obtained from 1 model system ($X = 200$)	State a priori probability from RIS model ($X = 200$)
$\bar{g}$	$-180° < \phi < -60°$	0.263	0.019
t	$-60° < \phi < 60°$	0.454	0.765
g	$60° < \phi < 180°$	0.283	0.216

although there are significant conformational transitions occurring. It is trivial to screen the initial guess conformations to obtain samples of any desired configurational characteristics, i.e., the chain dimensions of choice. Careful inspection of the set of microstructures generated do not reveal any influence of the chain dimensions on the properties investigated here. We therefore have accepted the initial guess configurations without further selection.

Parenthetically it is interesting to note that the experimental data concerning unperturbed dimensions of PVC span an unusually large range. Many of the commonly accepted values of C_∞ for PVC (6.7) [31] originated from solution measurements in THF and cyclohexanone [32]. Evidence exists to suggest that PVC forms aggregates in these two solvents under various conditions [33], and the low experimental measurement of C_∞, relative to the unperturbed dimensions, may be due the formation of these aggregates. Measurement of C_∞ for PVC in the absence of such aggregation should produce a "true" unperturbed value; for example, at θ conditions in benzyl alcohol a value from 9.6 to 10.8 is obtained for the characteristic ratio [34]. An additional viscometry measurement in benzyl alcohol of a well-characterized sample, with an identical stereochemical distribution as the simulated systems, produced an estimate of 10.9 for C_∞ [35]. The generator matrix calculations using the PVC RIS model, averaged over 100 stereochemical distributions with $m = 43\%$, estimated the characteristic ratio $C_\infty = 11.8$.

Other characteristics of single chain geometry, aside from the unexpectedly small overall dimensions, give "normal" values. For example, the ratios of the average eigenvalues of the radius of gyration tensor $\langle X^2 \rangle$, $\langle Y^2 \rangle$, $\langle Z^2 \rangle$, as well as the shape asphericity $b = \langle X^2 \rangle - (\langle Y^2 \rangle + \langle Z^2 \rangle)/2$, the acylindricity $c = \langle Y^2 \rangle - \langle Z \rangle^2$, and the anisotropy coefficient $\kappa^2 = (b^2 + 3c^2/4)/\langle s^4 \rangle$ [1,36,37] assume values indistinguishable from those for PVC RIS chains, i.e., values computed from Monte Carlo simulations of 2000 chains with the same degree of polymerization and fraction of meso diads [1,37]. These values compared to their RIS counterparts are: $\langle X^2 \rangle / \langle Z^2 \rangle$ is 15.9 ± 3.5 (compared to 18.1), $\langle Y^2 \rangle / \langle Z^2 \rangle$ is 2.9 ± 0.7 (compared to 3.7), b is 0.7 ± 0.2 (compared to 0.69), c is 0.10 ± 0.04 (compared to 0.12), and κ^2 is 0.48 ± 0.04 (comapred to 0.46). In brief, the simulated chains in dense packing assume the same shape as unperturbed chains, however, the dimensions seem to be strongly perturbed.

Linear correlations along the chain can be quantified in the "bond direction correlation function" S given by [38]

$$S = \frac{3\langle \cos^2 \theta \rangle - 1}{2}, \tag{11}$$

where θ is the angle between the directions of parts of polymer chains; the directions are chosen to be represented by chords connecting the midpoints of two adjacent skeletal C—C bonds; i.e., bond chord i is the vector joining either side of carbon i. Also, S as a function of $|i - j|$, the number of backbone bonds separating two given bond chords i and j, is denoted S_{ij}. The averages over all "small" structures as well as those derived from the single "large" structure are only significantly different from zero for $|i - j| < 3$; for chords separated by a large number of bonds $S_{ij} = 0 \pm 0.05$. The fact that this function is not distinguishable from zero, for all but very small values of $|i - j|$, indicates the absence of any orientational correlation *along* the chain. These results are essentially the same as those for polypropylene.

Structure of the Bulk Glass

Analysis of the microstructures generated as described above can yield information on structural order on a length scale smaller than the dimensions of the periodic cube. The single larger PVC microstructure is equally useful in providing statistically significant results as the ensemble of smaller structures because the correlation and distribution functions of interest in the bulk are based on pair correlations, and the number of pair correlations in the single large PVC structure is 69% of the corresponding number in the ensemble of smaller structures. Thus, results from these different-sized structures may be compared directly.

The bond direction correlation function S, described above, may also be calculated for chain directions (bond chords) separated by a given distance r for both a single chain and different images of the same chain; the separation r is measured from the midpoints of the respective bond chords. Values obtained for $S(r)$ are plotted as a dotted line in Figure 4 for the two different-sized structures. $S(r)$ mainly indicates the frequency of the various intramolecular correlations at given distances: positive values indicate a tendency for parallel alignment of the chains, while negative values reflect a tendency toward orthogonality. The positive correlations at very low values of r are due to conformations containing only *trans* states, while the negative correlations originate in conformations that contain some *gauche* states. These curves are very similar to those obtained for polypropylene.

Including only chord pairs between different chains yields the intermolecular form of this function, $S_{\text{inter}}(r)$. It is displayed as the bold lines in Figure 4(a) and 4(b), and is in stark contrast to the values found for polypropylene[1]. A significant positive correlation occurs at an intermolecular separation of about 4.3 Å for the ensemble of smaller structures, and this positive correlation is even stronger in the larger structure, indicating a strong tendency for bond chords on different chains to align parallel when separated by this particular (intermolecular) distance. In polypropylene a strong negative correlation in $S_{\text{inter}}(r)$ occurs at approximately the same intermolecular separation. Thus whatever is responsible for this parrallel alignment is not present in polypropylene.

The results from the two different size PVC microstructures indicate the presence of strong spatial correlations; the fact that they increase with increasing system size suggests that they are of a scale which is larger than at least the smaller of the periodic cubes used in the simulations. This difference comes about because the larger structure is less constrained by its spatial continuity than the smaller system, and hence the larger structure is more "relaxed" than the ensemble of smaller PVC structures. Even though all the microstructures are very well "relaxed" subject to their periodic conditions, they may not be representative of a well-relaxed glass because larger-scale relaxations that can also occur in a real glassy polymer are suppressed in the microsystems. The model reveals the presence of such large-scale correlations by the dependence of the resulting structure on the size of the periodic cube.

Insight into the structural differences between the large and small PVC structures might possibly be gained by examining the atomic pair distribution functions $g_{ij}(r)$. This function is defined such that $(4\pi/V)g_{ij}(r)r^2\,dr$ is equal to the probability of an i atom being separated from a j atom by a distance between r and $r + dr$, where V is the accessible volume. These distribution functions, computed to include only pairs that depend on the

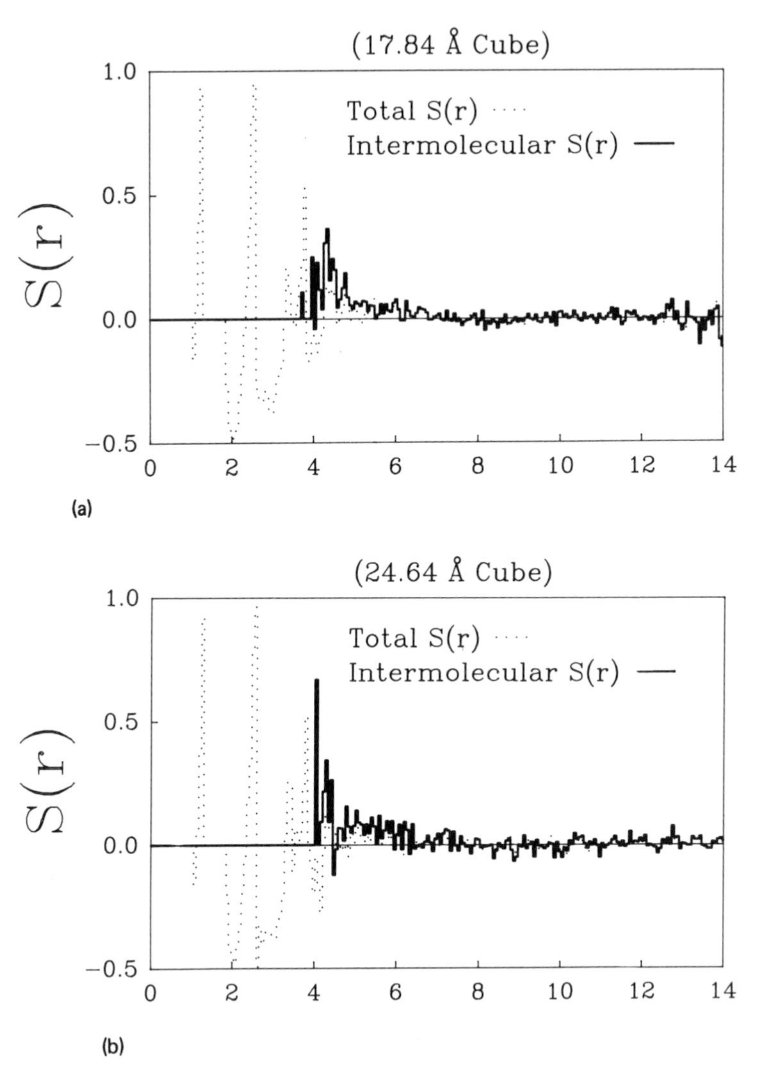

Figure 4 Bond direction correlation function $S(r)$ as a function of separation between midpoints of bond chords for (a) an ensemble of 10 structures $X = 76$ and (b) a single structure $X = 200$. Radial distance, Å.

conformation of the polymer chain, can be seen in Figures 5 and 6 [26]. Both the total and the intermolecular pair distribution functions are not markedly different from the group pair distribution functions computed earlier for polypropylene, despite the markedly different intermolecular chain alignments [1]. This similarity suggests that the atomic pair distri-

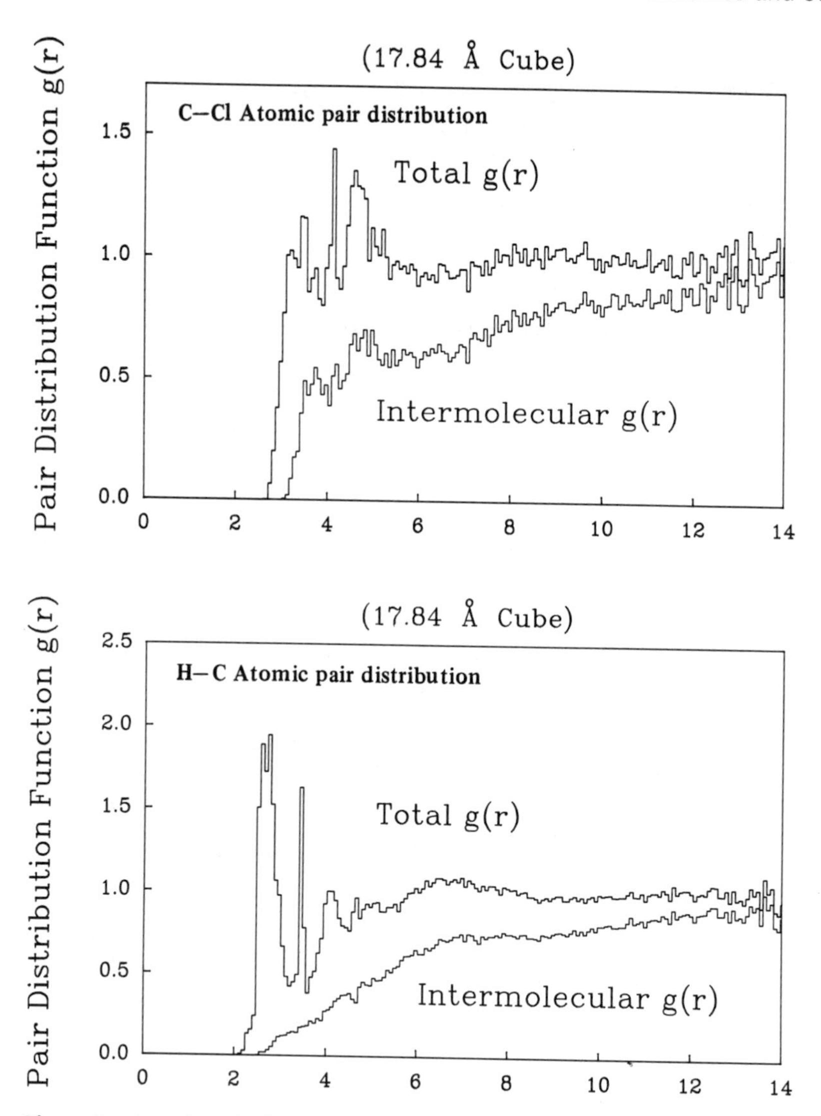

Figure 5 Atomic pair distribution functions $g_{ij}(r)$ for the ensemble of 10 structures with $X = 76$. Radial distance, Å.

butions in both polypropylene and PVC may be determined by the steric hindrance of the polymer chains, and not by any specific polar interaction. The largest intermolecular correlation is between the chlorine atoms and between the chlorine and methine-hydrogen atoms. This might suggest that the interaction between these highly charged species may be important

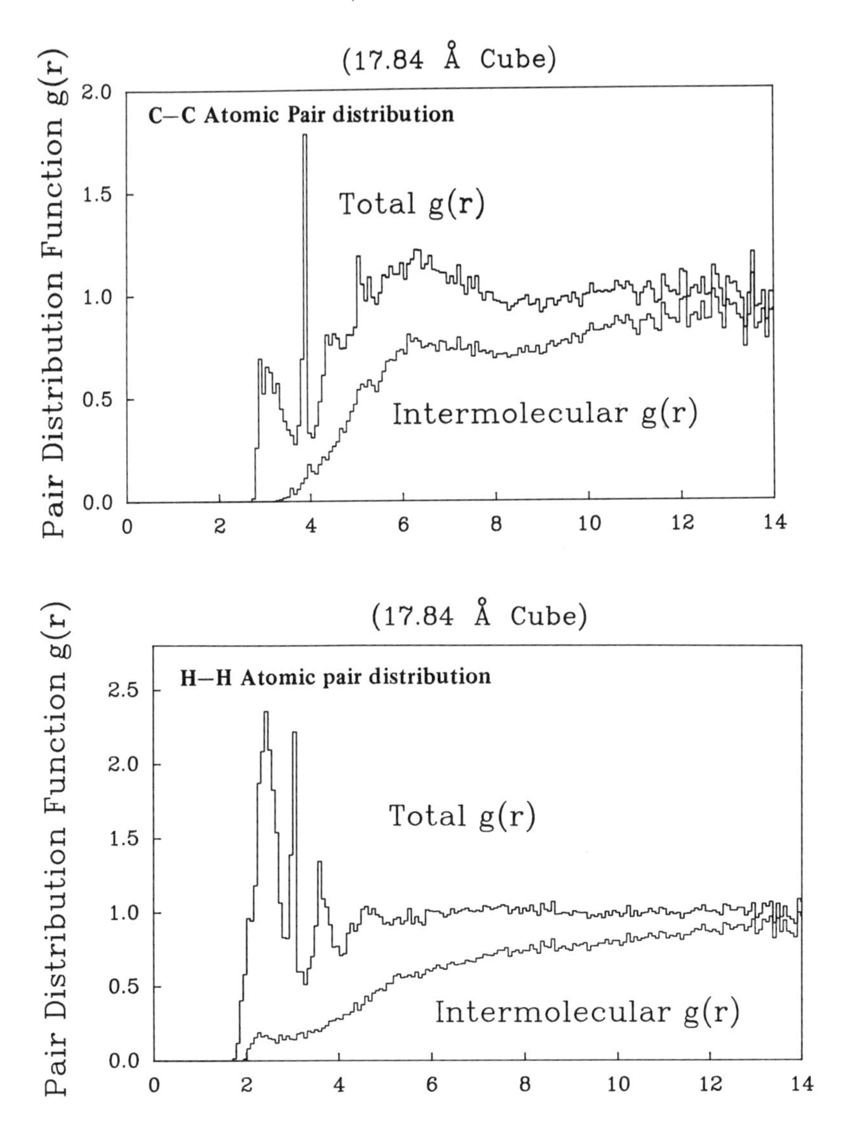

to the formation of the observed parallel alignment of the chains. Although this intermolecular correlation in the atomic pair distribution may be partially caused by the "natural packing," it produces very many attractive interactions between the highly charged chlorine and methine-hydrogen atoms as well as repulsive interactions between chlorine atoms.

If these polar interactions are important, then the observed first-neighbor correlation in the H-Cl pair distribution function may consist of pre-

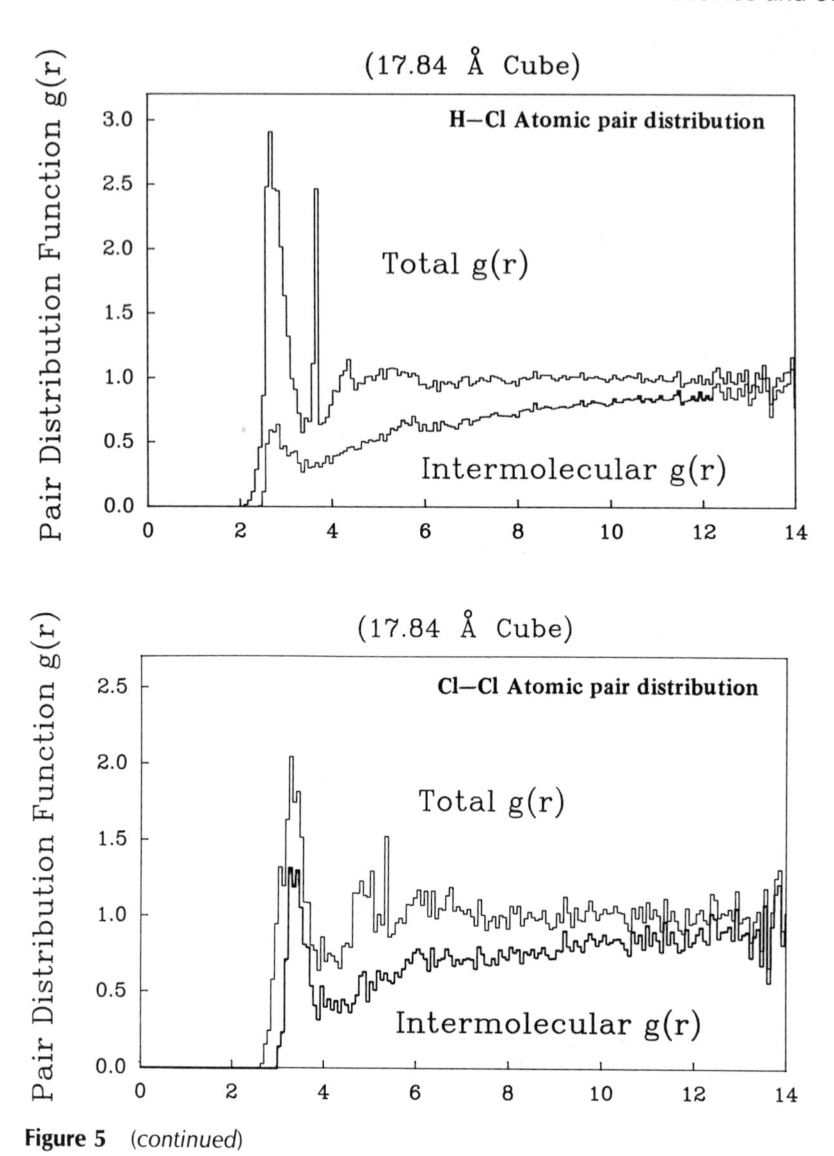

Figure 5 *(continued)*

dominantly methine hydrogen chlorine pairs. Figures 7 and 8 contain the H-Cl atomic pair distribution functions for only the methine and methylene hydrogens respectively. One observes that this first neighbor correlation is more *inter*molecular in nature for the methine hydrogens relative to the methylene hydrogens. This suggests that the relatively large attractive in-

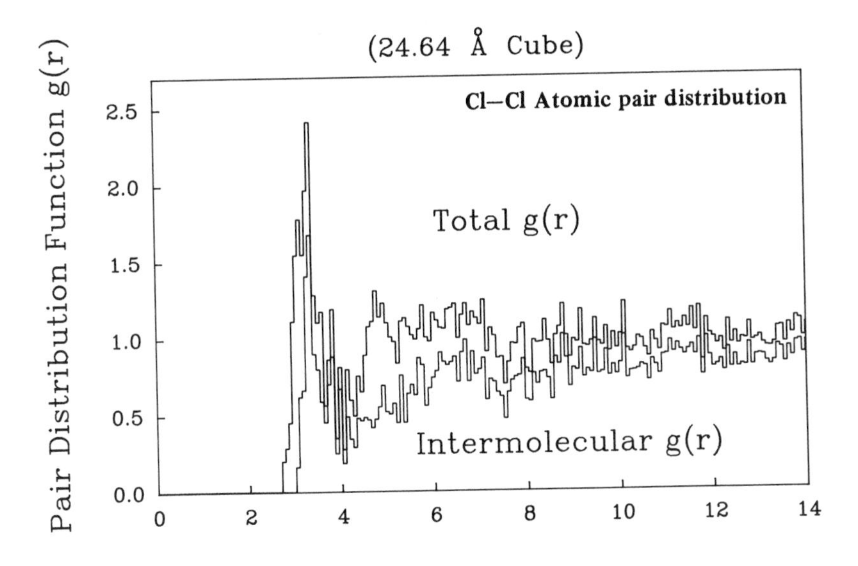

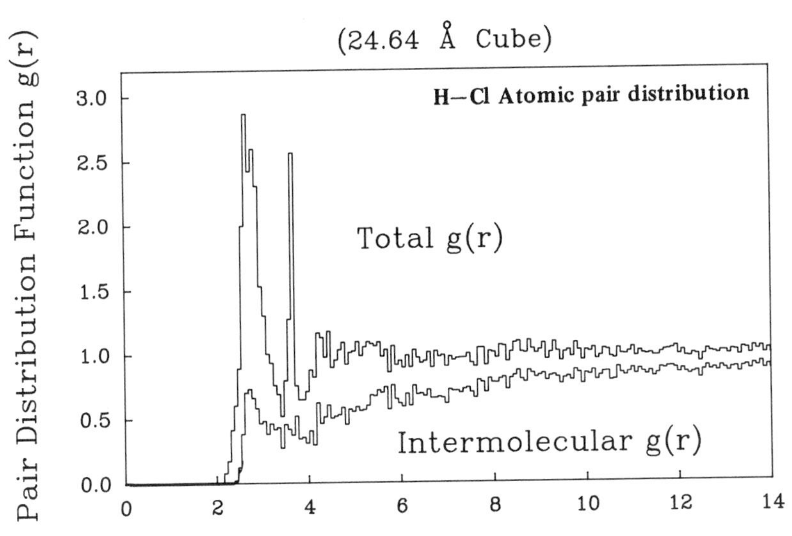

Figure 6 Atomic pair distribution functions $g_{ij}(r)$ for the single structure with $X = 200$. Radial distance, Å.

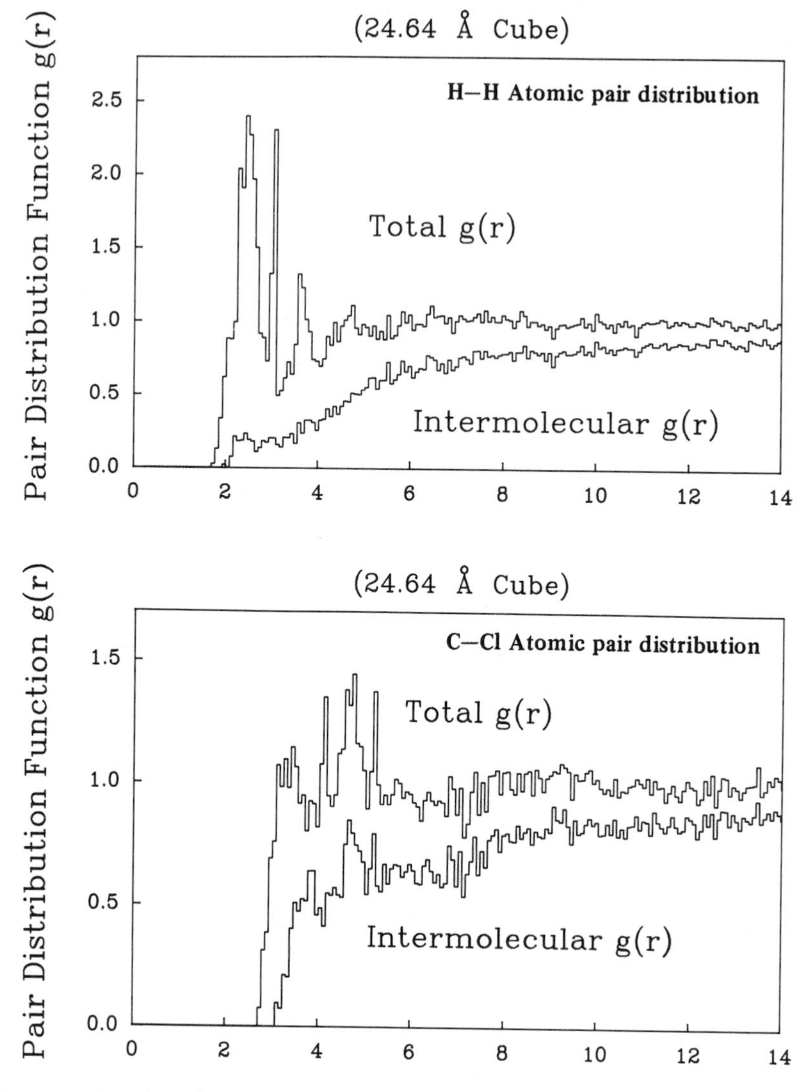

Figure 6 *(continued)*

teraction between the methine hydrogen and chlorine atoms may be significant in the formation of intermolecular structural correlations in PVC.

Additional information on how the chloromethine groups orient relative to each other can be gained by examining a two-dimensional pair distribution function. We base such a distribution on the molecular separation of the chlorine and methine-hydrogen atoms from different polymer chains.

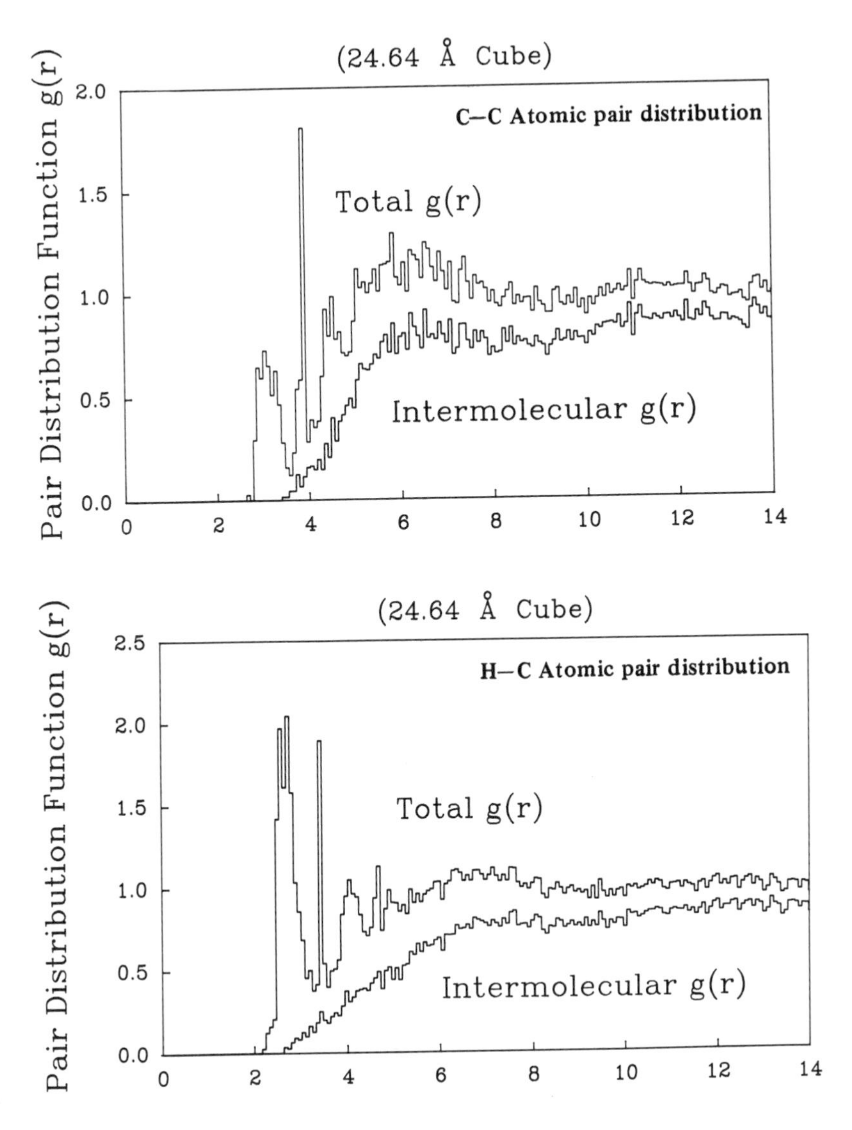

This function $g_{ij}(r_1,r_2)$ is defined such that $16\pi^2 r_1^2 r_2^2 g_{ij}(r_1,r_2)dr_1dr_2/V^2$ is equal to the probability of finding a given pair of methine groups on different chains with *one* of its H-Cl pairs separated by a distance between r_1 and $r_1 + dr_1$, and its second H-Cl pair separated by a distance between r_2 and $r_2 + dr_2$. This function is graphed for distances r_1, $r_2 < 6$ Å, for the two different sized PVC structures in Figure 9. There are two distinct ridges visible, characterized by r_1 = 2.5–3 Å, and r_2 = 2.5–3 Å. The average chlorine methine-hydrogen separation in these structures is 3.34 Å, and

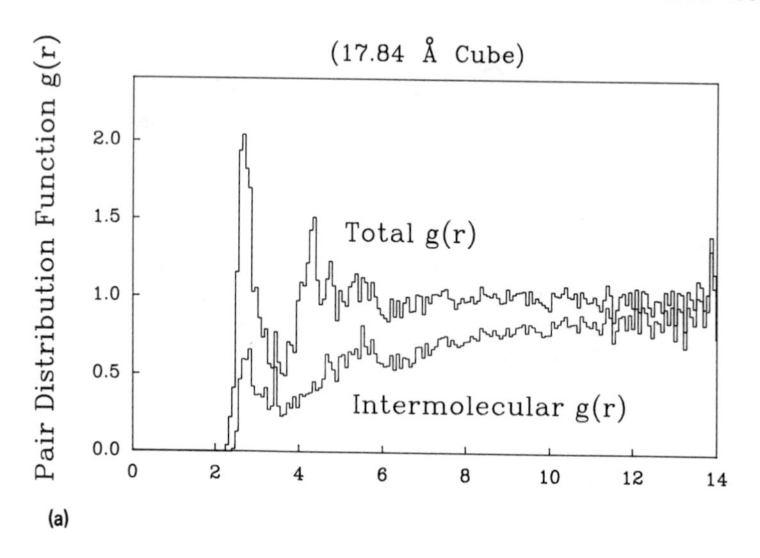

(a)

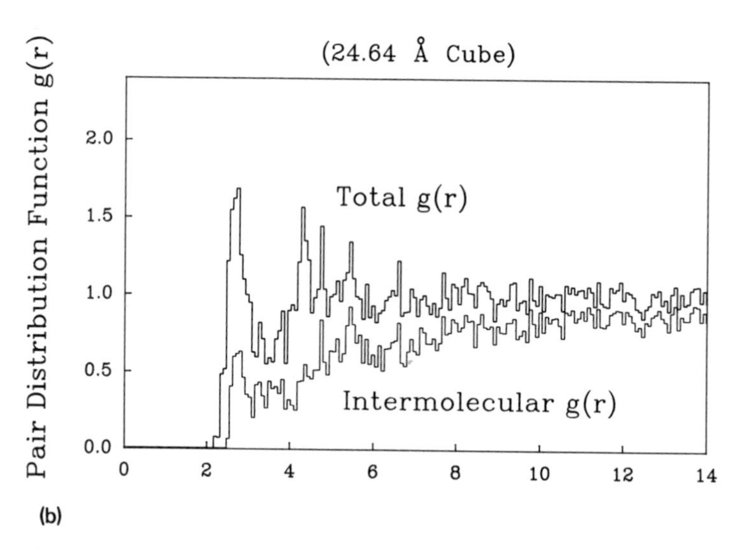

(b)

Figure 7 H-Cl atomic pair distribution functions $g_{ij}(r)$ of the methine hydrogen and chlorine atomic pair for (a) an ensemble of 10 structures $X = 76$, and (b) a single structure $X = 200$. Radial distance, Å.

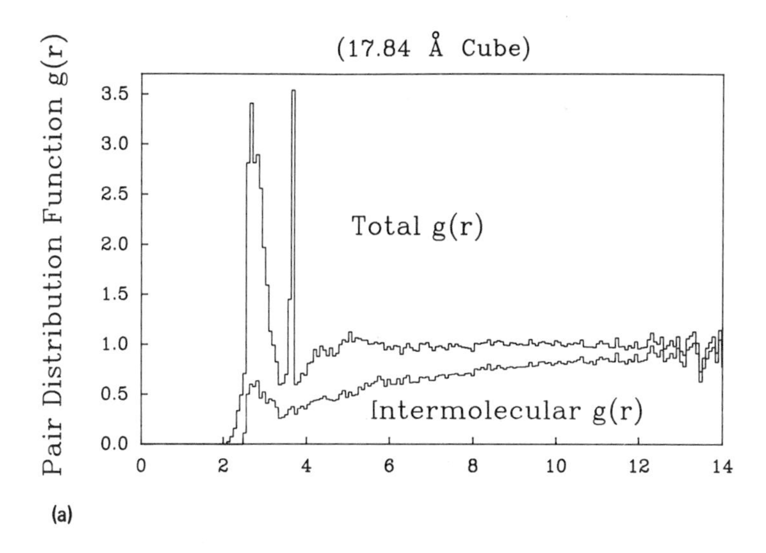

(a)

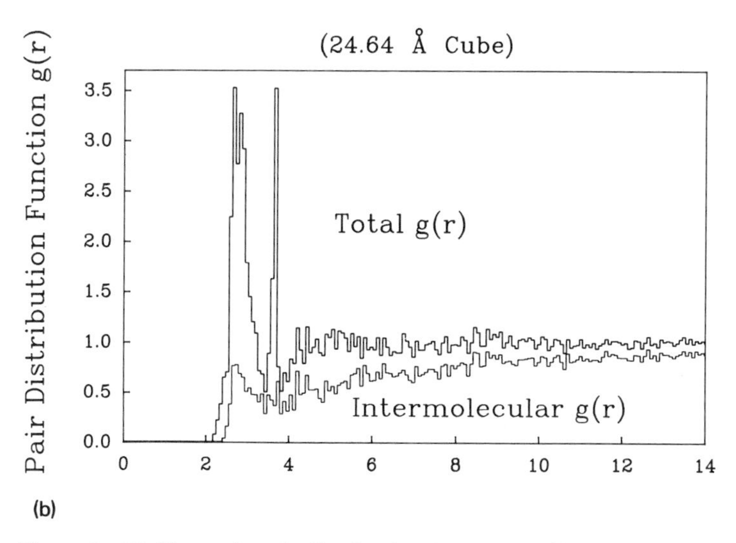

(b)

Figure 8 H-Cl atomic pair distribution functions $g_{ij}(r)$ of the methylene hydrogen and chlorine atomic pair for (a) an ensemble of 10 structures $X = 76$ and (b) a single structure $X = 200$. Radial distance, Å.

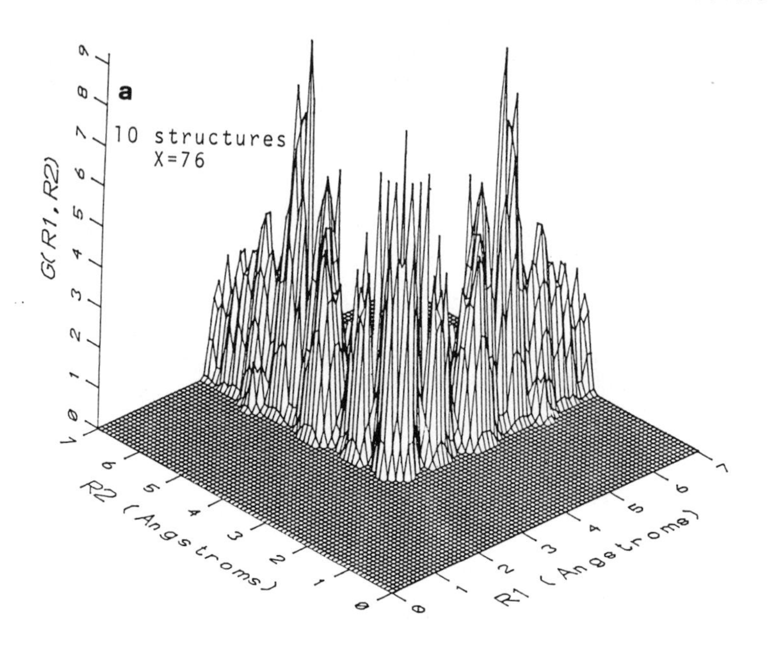

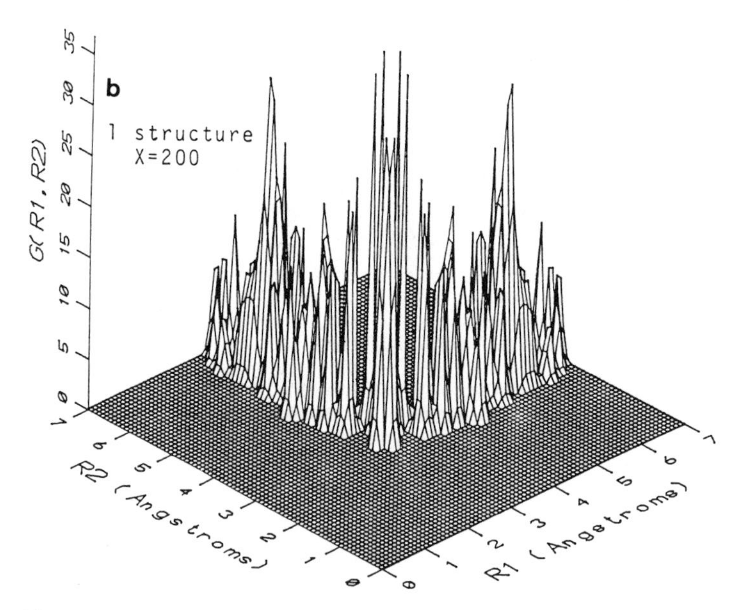

Figure 9 Two-dimensional pair distribution function of methine H-Cl pairs for (a) an ensemble of 10 structures $X = 76$ and (b) a single structure $X = 200$.

$g_{ij}(r_1,r_2)$ values for r_1 or r_2 less than this distance might be thought to represent "attractive" H-Cl interactions between methine groups. The region where r_1 and r_2 are greater than this average distance form a broad group of peaks slightly behind these ridges, and have been suppressed in Figure 9 for clarity. The diagonal group of peaks in this region at $r_1 = r_2 = 2.5$ Å is indicative of a cyclic structure in which methine groups are oriented at 180° relative to each other. In these cyclic structures the chlorine atom from one methine group is adjacent to the hydrogen atom of another methine group, and vice versa; they are refered to as "cyclic dimers" [39]. The largest groups of off-diagonal peaks in this region, at $r_1 = 2.5/r_2 = 5.5$ Å are indicative of a single "hydrogen bonding" interactions between chloromethine groups. These peaks, indicative of a hydrogen bonding phenomenon increase in both height and sharpness from the small to the large PVC microstructures (note different axis scales in Fig. 9). The fact that this increase in the hydrogen bonding accompanies an increase in intermolecular parallel chain alignment suggests that hydrogen bonding may be intimately related to the formation of order in atactic PVC.

A widely accepted theory that addresses the nature of the observed gelation in PVC connects this gelation behavior with a crystallization of portions of the polymer chain that are predominantly syndiotactic. An analysis of the sharp positive peak in the intermolecular $S(r)$ function for the larger PVC structure in Figure 4(b) was carried out. Stereochemical data for the portions of the chain that contribute to this sharp positive peak were assembled in Table 7. Triads are indicative of bond chords that flank a methine carbon, and include the stereochemical configuration of this carbon as well as the adjacent methine carbons on either side. Diads are indicative of bond chords that flank a methylene carbon, and are comprised

Table 7 Ratios of Observed Populations of Interacting Pairs in the Highly Aligned Portions of the PVC Chain to the Bernoullian Populations ($X = 200$)

	Triads			Diads	
	RM	RR	MM	R	M
RM	0.000	1.477	0.000	1.263	1.116
RR		0.000	0.000	1.271	0.000
MM			0.000	1.116	2.960
R				1.448	0.960
M					0.000

of the stereochemical configuration of the two methine carbons on either side of this methylene carbon. The data in Table 7 was normalized for the total number of the particular combinations of diads or triads that would occur in a Bernoullian system with the same diad populations as the modeled systems. Bond chords that were separated by a distance between 4.0 and 4.5 Å, that produced an equivalent $S(r)$ of greater than 0.2 were included. This represents only 18 structures, and this lack of statistical significance makes it difficult to draw any conclusion about the role of stereochemistry in this parallel alignment. The absence of interactions between the isotactic (mm) triad and other triads might indicate a tendency to favor syndiotactic (rr) triads. However, the absence of interactions between syndiotactic triads contradicts this speculation. Despite the allowance for a small sample size, one is hard-pressed to point to a strong correlation between stereochemical distribution and the formation of intermolecular chain alignment. This lack of a strong stereochemical correlation occurs despite the fact that this chain of degree of polymerization 200 contains, purely by chance, a sequence of 21 syndiotactic triads in a row.

The model calculations predict the existence of some vestiges of order in atactic polyvinylchloride, despite its atactic stereochemical distribution. This order takes the form of a tendency toward the intermolecular alignment of polymer chains that appears directly related to a "hydrogen bonding" interaction between the methine hydrogen and chlorine atoms. The subtle nature of this order makes for difficulty in observing it in normal atomic pair distribution functions. Examination of these functions indicate that the atomic spatial distribution is similar to that of atactic polypropylene. The amorphous nature of the modeled polypropylene indicates that these atomic pair distributions reflect little more than a generally amorphous arrangement of atoms. The lack of any distinct difference in these functions between the large and small simulated PVC structures also exhibits their failure to elucidate the exact mechanism of order formation in PVC.

The dependence of the structural properties on the size of the periodic cube indicate that the simulated structures are clearly not well relaxed. The larger of the two simulated PVC structures represents a more relaxed glass because of its inclusion of larger-scale relaxations during the minimization, but both glasses are relaxed on only a rather small scale. Glassy atactic PVC must actually contain large scale correlations (greater than order 20 Å) the accurate prediction of which would require a much larger simulated system. The vestiges of order in the simulated structures may reflect only the onset of order formation in PVC and whether this is related to the actual order is not yet known.

ACKNOWLEDGMENTS

We gratefully acknowledge support by NSF Grant No. DMR-8312694 (Polymers Program) and the Bayer Professorship at the Department of Chemical Engineering at MIT. The assistance of Dr. Kenzabu Tasaki with the charge calculations was also greatly appreciated.

APPENDIX: DERIVATION OF COULOMBIC PORTION OF TAIL CORRECTION INTEGRAL

The fourth integral in the tail correction expression (Eq. (7)) is given by

$$\frac{1}{2}\sum_i N_i \sum_j \rho_j \left(4\pi \int_{R_{2,ij}}^{\infty} g_{ij}(r)U_{ij}^{c}(r)r^2\,dr\right). \tag{A.1}$$

This integral can be evaluated explicitly using the pair distribution functions $g_{ij}(r)$ from $R_{2,ij}$ to the value R^* at which the pair distribution function is no longer known or is approximated by unity. The remaining portion to be analyzed is

$$\frac{1}{2}\sum_i N_i \sum_j \rho_j \left(4\pi \int_{R^*}^{\infty} g_{ij}(r)U_{ij}^{c}(r)r^2\,dr\right). \tag{A.2}$$

Defining the pair density ρ_j as N_j divided by the cube volume V, approximating the pair distribution function $g_{ij}(r)$ as unity, and by expanding the coulombic potential Eq. (A.2) may be rewritten as

$$\frac{2\pi}{V}\sum_i N_i \sum_j N_j \left(\int_{R^*}^{\infty} \frac{q_i q_j r}{4\pi D(r,a_{ij})}\,dr\right). \tag{A.3}$$

For large values of r the leading term in the above integral goes as r^2 and is independent of the critical distance "a_{ij}", which is dependent on the atom pairs i and j. So for large values of r, when the pair distribution functions are approximated by unity, the above equation becomes

$$\frac{2\pi}{V}\sum_i N_i q_i \sum_j N_j q_j (f(r))_{R^*}^{\infty}. \tag{A.4}$$

Because the integral is only a function of r, which is independent of i and j this may be expressed as a product

$$\frac{2\pi}{V}\sum_i N_i q_i \cdot \sum_j N_j q_j \cdot (f(r))_{R^*}^{\infty}. \tag{A.5}$$

Because the PVC polymer chain is electrically neutral the first two terms in the above product are equal to zero. Therefore, at large values of r, the above expression converges. By assuming a constant value for the critical distance a_{ij} independent of the atom pair i and j, the integral in Eq. (A.3) becomes independent of i and j for any values of r so it may be expressed in its product form as in Eq. (A.5) and is therefore identically zero.

The liquid like approximation of the portion of the tail correction given in Eq. (A.2) is given by

$$\frac{1}{2}\sum_i N_i\rho_d\left(4\pi\int_{R^*}^{\infty} g_{id}(r)\langle U_{id}^c(r)\rangle_B r^2\,dr\right), \tag{A.6}$$

where $g_{id}(r)$ and $\langle U_{id}^c(r)\rangle_B$ are the pair distribution of an atom i relative to the center of a dipole and the Boltzmann-averaged coulombic interaction between this atom i and a dipole, respectively. The volume density of dipoles ρ_d is defined as the number of dipoles contained in the periodic cube, which is equal to the degree of polymerization X, divided by the volume of the periodic cube V. We will assume that the pair distribution function is unity. The Boltzmann average of the coulombic interaction between an atom and a dipole is given by

$$\langle U_{id}^c(r)\rangle_B = \frac{\displaystyle\int_0^{2\pi}\int_0^{\pi} U_{id}^c(r,\theta)e^{-U_{id}^c(r,\theta)}\sin\theta\,d\theta\,d\omega}{\displaystyle\int_0^{2\pi}\int_0^{\pi}\sin\theta\,d\theta\,d\omega}. \tag{A.7}$$

It is easily shown that this Boltzmann average is equivalent to the following average of the square of the potential [27].

$$\langle U_{id}^c(r)\rangle_B = \frac{\displaystyle\int_0^{2\pi}\int_0^{\pi}(U_{id}^c(r,\theta))^2\sin\theta\,d\theta\,d\omega}{\displaystyle kT\int_0^{2\pi}\int_0^{\pi}\sin\theta\,d\theta\,d\omega}. \tag{A.8}$$

By expanding the exponential function for $D(r)$ and assuming that r is relatively large relative to the critical distance a, one can show that the interaction between an atom and a dipole for our potential energy function is

$$U_{id}^c(r,\theta) = \frac{-q_i\mu\cos\theta}{4\pi\epsilon_0\epsilon_B r^2(1-\kappa/r)^2}. \tag{A.9}$$

In the above expression, κ is defined as in Eq. (4) except that the critical radius a is now assumed to be a constant value ($\kappa = a \log(\epsilon_B) \approx a_{ij} \log(\epsilon_B)$). The effective dipole for a PVC monomeric unit μ is taken to be 1.9 D [18]. Note that in the limit of ϵ_B approaching unity the above expression for the coulombic potential between a charge and a dipole becomes:

$$\lim_{\epsilon_B \to 1} U_{id}^{c}(r,\theta) = \frac{-q_i \mu \cos\theta}{4\pi\epsilon_0 r^2}. \tag{A.10}$$

The above limit is for the charge-dipole interaction in vacuo and it is identical to the correct expression for this interaction in vacuo [27,39]. After substituting Eq. (A.9) into (A.8) and integrating, we obtain the Boltzmann-averaged interactions between a charge q_i and a dipole μ:

$$\langle U_{id}^{c}(r)\rangle_B = \frac{-q_i^2 \mu^2}{3kT(4\pi\epsilon_0\epsilon_B)^2 r^4 (1 - \kappa/r)^4}. \tag{A.11}$$

Again the limit of the above expression as ϵ_B approaches unity is the same as the expression obtained when one substitutes the expression from Eq. (A.10) into Eq. (A.8) and integrates. By substituting the expression for the Boltzmann-averaged interaction (Eq. (A.11)) into Eq. (A.6) and integrating from R to ∞ we obtain

$$\frac{1}{2}\sum_i N_i \rho_d \left(4\pi \int_{R^*}^{\infty} \langle U_{id}^{c}(r)\rangle_B \, r^2 \, dr\right) = \frac{-2\pi X \mu^2}{3kTV(4\pi\epsilon_0\epsilon_B)^2(R - \kappa)} \sum_i N_i q_i^2. \tag{A.12}$$

The degree of polymerization X divided by the periodic cube volume V was substituted for the dipole density ρ_d in the above expression.

REFERENCES

1. D. N. Theodorou and U. W. Suter, *Macromolecules*, *18*, 1467 (1985).
2. K. Hillstrom, *Nonlinear Optimization Routines in AMDLIB*, Technical Memorandum No. 297, Argonne National Laboratory, Applied Mathematics Division, 1976; Subroutine GQBFGS in AMDLIB, Argonne, IL (1976).
3. M. H. Cohen and D. Turnbull *Nature*, *203*, 964 (1964).
4. D. N. Theodorou, U. W. Suter, *Macromolecules*, *19*, 139 (1986).
5. A. Bondi *Physical Properties of Molecular Crystals, Liquids and Glasses*, p. 387, Wiley, New York (1968); U. W. Suter and P. J. Flory, *Macromolecules*, *8*, 765 (1975).
6. H. Block and S. M. Walker, *Chem. Phys. Lett.*, *19*, 363 (1973).
7. J. Brandrup and E. H. Immergut, eds, *Polymer Handbook* 2nd. ed., p. V-43, Wiley, New York (1975).

8. L. J. Onsager, *Am. Chem. Soc.*, *58*, 1486 (1936); P. Debye, *Phys. Z.*, *13*, 97 (1912); *Polar Molecules*; Chemical Catalogue Company, New York (1929).
9. R. P. Feynman, R. B. Leighton, and M. Sands, *The Feynman Lectures on Physics*, vol. II, Chap. 10, Addison-Wesley, Reading MA (1964).
10. F. H. Stillinger and C. W. David, *J. Chem. Phys.*, *69*, 1473 (1978).
11. R. H. Boyd and L. Kesner, *J. Chem. Phys.*, *72*, 2179 (1980).
12. Self-consistent field calculations using the complete neglect of differential overlap method, not reported here, revealed that the methylene hydrogen atoms in 2,4-dichloropentane had point charges that were of the order of the methine-hydrogen point charges.
13. M. J. S. Dewar and W. Thiel, *J. Am. Chem. Soc.*, *99*, 4889 (1977).
14. W. Thiel, J. D. Bowen, and G. S. Owen, *Quantum Chemistry Exchange Program Bull.*, *4*, 76 (1984).
15. M. J. S. Dewar and W. Thiel, *J. Am. Chem. Soc.*, *99*, 4907 (1977).
16. T. Shimanouchi and M. Tasumi, *Spectr. Acta*, *17*, 755 (1961); T. Shimanouchi, M. Tasumi, and Y. Abe, *Makromol. Chem.*, *86*, 43 (1965); T. Iijima, S. Seki, and M. Kimura, *Bul. Chem. Soc. Japan*, *50*, 2568 (1977).
17. S. Pokorný, R. Lukáš, J. Janča, and M. Kolínský, *J. Chromatogr.*, *148*, 183 (1978).
18. R. H. Boyd and L. Kesner, *J. Polym. Sci.: Polym. Phys.*, *19*, 375 (1981).
19. P. J. Flory and A. D. Williams, *J. Am. Chem. Soc.*, *91*, 3118 (1968).
20. P. J. Flory and C. J. Pickles, *J. Chem. Soc.*, *Faraday Trans.*, *69*, 632 (1973).
21. U. W. Suter and P. J. Flory, *Macromolecules*, *8*, 765 (1975).
22. Density was chosen to be representative of a well-relaxed PVC glass from J. Brandrup and E. H. Immergut (eds.), *Polymer Handbook*, 2nd ed., p. V-43, Wiley, New York (1975); 43% meso diads was chosen for comparison to a sample synthesized at 40°, see Ref. 35.
23. P. J. Ludovice, M. G. Davidson, and U. W. Suter, *Am. Chem. Soc. Symp. Ser.*, *353*, 162 (1987).
24. Based on runs done on a cluster of four computational elements on an Alliant FX/8 by Alliant Computer Corp.
25. D. W. Van Krevelen and P. J. Hoftyzer *Properties of Polymers—Their Estimation and Correlation with Chemical Structure*, pp. 130, 137, 275 Elsevier, New York (1976).
26. D. N. Theodorou and U. W. Suter, *J. Chem. Phys.*, *82*, 955 (1985).
27. G. C. Maitland, M. Rigby, E. B. Smith, and W. A. Wakeham, *Intermolecular Forces*, pp. 47–53, Clarendon Press, Oxford (1983).
28. Ref. 7, p. IV-357.
29. H. Wittwer and U. W. Suter, *Macromolecules*, *17*, 2248 (1984).
30. P. J. Flory, *Statistical Mechanics of Chain Molecules*, Wiley, New York (1969).
31. Ref. 7, p. IV-39.
32. J. W. Breitenbach, E. L. Forster, and A. J. Renner, *Kolloid-Z*, *127*, 1 (1952); G. Ciampa and Schwindt, *Makromol Chem.*, *21*, 169, (1954); F. Danusso, G. Moraglio, and S. Cazzera, *Chim. Ind. (Milan)*, *36*, 883 (1954); H. Batzer and A. Nisch, *Makromol. Chem.*, *22*, 131 (1957).
33. A. Rudin and I. Benschop-Hendrychova, *J. Appl. Polym. Sci.*, *15*, 2881

(1971). K. B. Andersson, A. Holmström, and E. M. Sörvik, *Makromol. Chem.*, *166*, 247 (1973). J. H. Lyngaae-Jorgensen, *Makromol Chem.*, *167*, 311 (1973); A. H. Abdel-Alim *J. Appl. Polym. Sci.*, *19*, 2179 (1975).
34. M. Sato, Y. Koshiishi, and M. Asahine, *J. Polym. Sci. B*, *1*, 233 (1963). (Data from this reference was used to calculate the characteristic ratio, the range of values is obtained by using values of 2.1×10^{-21} to 2.5×10^{-21} (dL/g · mole · mL) for the universal viscosity constant depending on how well fractionated Sato et al.'s samples were.)
35. P. J. Ludovice, Doctoral Thesis, M.I.T., January (1989).
36. K. Solc and W. H. Stockmayer, *J. Chem. Phys.*, *54*, 2756 (1971); K. Solc, *J. Chem. Phys.*, *55*, 335 (1971).
37. D. N. Theodorou and U. W. Suter *Macromolecules*, *18*, 1206 (1985).
38. T. A. Weber and E. Helfand, *J. Chem. Phys.*, *71*, 4760 (1979); K. A. Dill and P. J. Flory, *Proc. Natl. Acad. Sci. U.S.A.*, *77*, 3115 (1980); M. Vacatello, G. Avitabile, P. Corradini, and A. Tuzi, *J. Chem. Phys.*, *73*, 543 (1980).
39. P. J. Ludovice and U. W. Suter, *Polym. Preprints*, *28*, 295 (1987).
40. J. N. Israelachvili, *Intermolecular and Surface Forces: With Applications to Colloidal and Biological Systems*, p. 22, Academic Press, London (1985).

part four

POLYMER/SURFACE INTERACTIONS

9

Molecular Modeling of the Epitaxial Crystallization of Linear Polymers on Ordered Substrates

Anton J. Hopfinger
The University of Illinois
Chicago, Illinois

INTRODUCTION

Epitaxial crystallization is the oriented growth of one crystalline material upon a different crystalline substrate. Through the action of the substrate's anisotropic force field, the substrate effects a lowering of the activation free energy for nucleation and influences the orientation of the depositing material. Both the chemical nature and the geometric locations of the constituent atoms in the substrate affect the form of this substrate potential.

In 1957, polymer epitaxy was first reported. The system investigated was the deposition of polyethylene (PE) from highly dilute decalin solution onto the (001) face of sodium chloride [1]. Since that initial experiment, both solution crystallization and melt crystallization procedures have been used, and numerous epitaxial polymer-substrate combinations reported; these investigations have been previously reviewed [2–4]. Most of the early work on polymer epitaxy used alkali halide substrates. However, other substrates have been reported, notably, graphite, mica, and quartz. The interactions of polyethylene [5], polyxymethylene [6], and polypropylene [7] with a graphite substrate have been reported. Also, the interaction of paraffinic alkanes with graphite, a model for polyethylene-graphite behav-

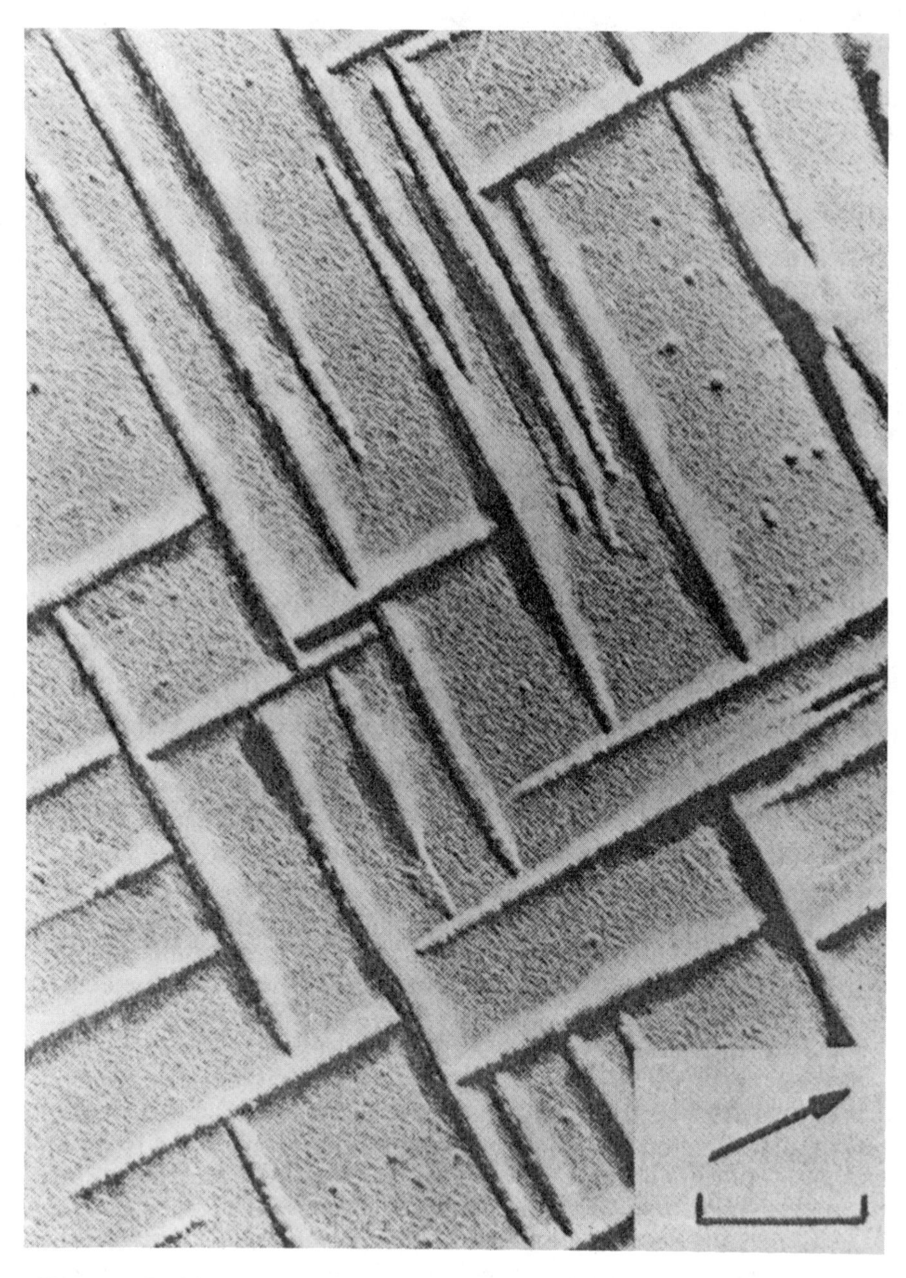

Figure 1 Rodlike morphology of POM grown on $(001)_{KCl}$ from a dilute nitrobenzene solution at 120°C. The arrow indicates the $\langle 110 \rangle_{KCl}$ direction. Also shown is a 1-μ bar (From Ref. 10.)

ior, has been investigated [8,9]. An electron micrograph of polymer epitaxy is shown in Figure 1 for polyoxymethylene (POM) grown on the (001) surface of KCl [10].

The epitaxial process has in the past been a research tool for investigating the adhesive properties of polymeric systems for coatings/paint applications. More recently, epitaxy has become an important consideration in the selection of polymer masks in the microelectronics industry. Thus, there is an increasing need to understand polymer epitaxy on the molecular level. The goal of molecular modeling studies of polymer epitaxy is to design polymers having specific epitactic properties.

SINGLE-CHAIN MODELING

It is generally best to initiate a modeling study of a complex chemical process, like polymer epitaxy, by first considering only the essential components of the process. This approach provides the investigator a perspective on how much information can be gleaned from the simple representation of the process, and how much work may be needed to introduce other components into the model.

The essential components of polymer epitaxy are a single polymer chain and the substrate. The other components are solvent and interchain interactions. We began our polymer epitaxy modeling studies by considering the deposition of PE and POM on various alkali halide substrates.

The polymer models utilized in the computer simulations consisted of linear homopolymers each of which was frozen in an ordered conformation that is characteristic of a given crystalline state, no bond rotations being allowed. Utilization of a rigid linear molecule is consistent with the constraint of crystallinity, and, although rotation about bonds certainly can occur (to form chain folds, for example), the goal was to describe the static interaction between polymer and substrate at the interface, and not the dynamic behavior of the system. Furthermore, it can be argued that by the time a chain segment has come appropriately close to the substrate surface, as to be under the influence of the short-range forces that are responsible for adsorption and orientation, the intrachain entropy has been drastically reduced because the space pervaded by the substrate represents an excluded volume that limits the number of possible conformations. Possible effects of the substrate force field on polymer conformation, however, remain undetermined.

All the simulated substrates have been of the alkali halide variety having the NaCl structure, the Bravais lattice being face-centered cubic. Representative crystals having this arrangement include all combinations of Li, Na, K, and Rb with F, Cl, Br, and I. It might also be mentioned that some nonalkali halides, for example, PbS, MgO, and MnO, also have the NaCl structure, although the constituent ions are of higher valency than ±1. The (001) plane, whose symmetry is depicted in Figure 2, constitute the substrate surface. This plane is known to be the most stable boundary in heteropolar crystals aof the NaCl type. The unit cell dimension d' is assigned the experimental value reported for the bulk, "infinite" crystal. Lattice "puckering" and other surface irregularities are neglected as a first approximation. Also shown in Figure 2 is δ, the distance between consecutive rows of like charge. Being related to d' by $\delta = d'/\sqrt{2}$, δ is a useful number in considering a possible lattice match between the substrate surface and interchain spacing in the crystalline deposited phase. For a number of NaCl-type crystals, d' and δ are given in Table 1.

Although the model surface was defect free, it is very likely that imperfections such as point defects, grain boundaries, edge and screw dislocations, producing charge-density inhomogeneity and localized anion-cation charge imbalance, give rise to long-range electric forces and may be instrumental in initiating nucleation.

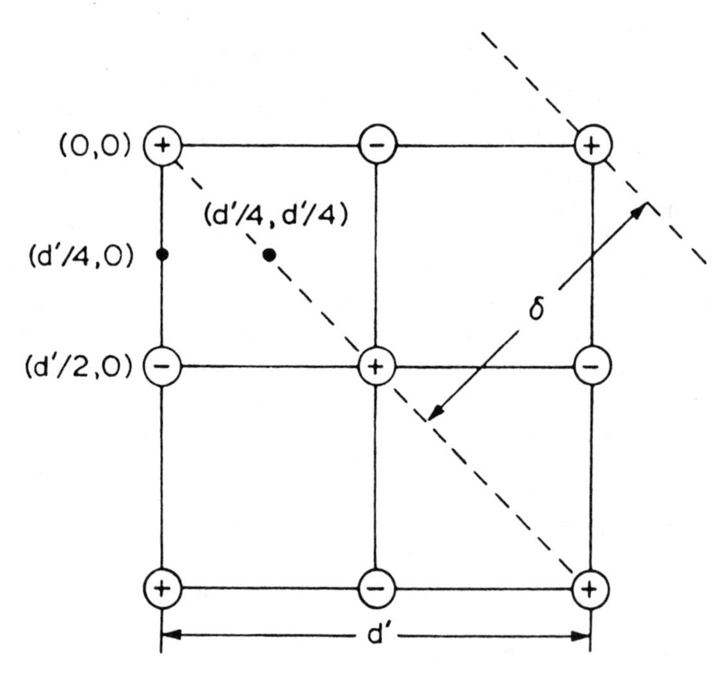

Figure 2 NaCl-type crystal symmetry in the (001) plane. d' is the unit cell length, and δ the distance between consecutive rows of like charge.

Table 1 The Values of d', δ, c, and n for Substrates Used in Calculations

Alkali Halide	d' (Å)	δ (Å)	c (Å)	n (ions-Å^{-3})
NaF	4.62	3.27	3.86	0.0813
NaCl	5.63	3.98	4.68	0.0448
RbBr	6.37	4.86	6.15	0.0248
KCl	6.29	4.45	5.47	0.0322

Geometry

The geometry of the polymer-substrate system, displayed in Figure 3, requires six independent spatial variables to define the position and orientation of the chain relative to the substrate surface. In this case, "surface" is defined by the outermost (001) plane of ionic charge centers (xy plane). The origin is centered on a positive ion and the x and y axes lie along the ⟨100⟩ and ⟨010⟩ directions, respectively. Coordinates a and b define the projection, onto the helix axis, of the position of a terminal backbone atom in a chain of finite length. The angle Θ defines the orientation of the polymer chain relative to a perpendicular to the substrate surface which passes through the point of rotation, the distance between the point and charge surface being h; ϕ is an azimuthal angle, measured counterclockwise from the x direction, in the xy plane; μ, the rotation of the polymer about its own axis, is generally initialized for PE (i.e., $\mu = 0°$) when the line bisecting the H—C—H angle of the backbone methylene group (when present) nearest the (Θ, ϕ)-rotation point is perpendicular to, and the hydrogens pointing towards, the substrate surface.

Molecular Energetics

Assignment of a given (single chain) polymer-substrate interaction energy with its corresponding (a,b,h,Θ,ϕ,μ) defines a surface in a seven-dimensional hyperspace, energy being the seventh variable. In general, the approach consists of minimizing the energy relative to (a,b,h,Θ,ϕ,μ) and associating $(a,b,h,\Theta,\phi,\mu)_{min}$ with the preferred chain orientation at the equilibrium distance from the substrate surface, in accordance with Boltzmann statistics. It is, of course, possible to have a condition of degeneracy, that is, a number of positions and orientations corresponding to the same energy. These positions and orientations, however, are usually related by symmetry operations that reflect the substrate surface lattice geometry.

Conceptually, the potential energy surface in this hyperspace is characterized by peaks, wells, saddle points, and other topological features.

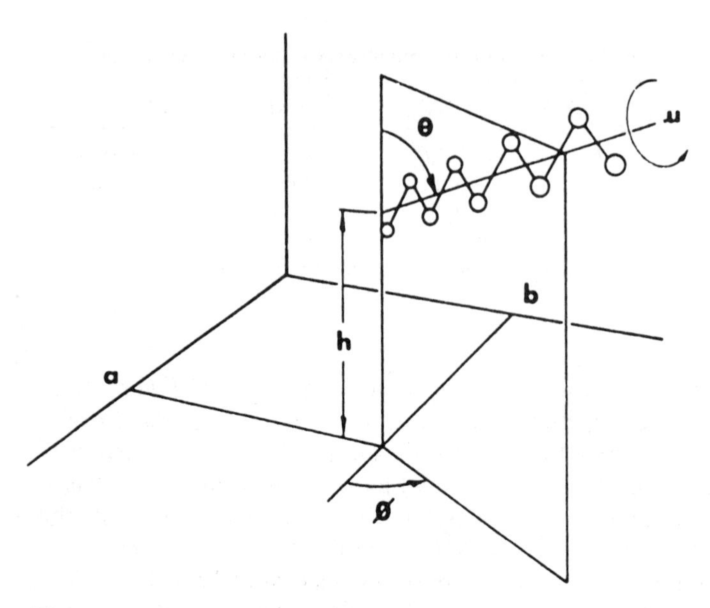

Figure 3 Spatial parameters (a,b,h,Θ,ϕ,μ) used to describe the position of a polymer chain relative to the substrate surface. (a,b) is the plane defined by the average positions of ionic surface charges. The origin is centered on a positive ion, h is the height of the end of the chain segment above the charge surface, and (Θ,ϕ,μ) are orientational parameters.

The heights of possible energy *barriers* and depth of *minima* are indications of whether epitaxy will occur at all and the relative propensity for doing so. However, as will be pointed out later, the *shape*, as well as depth, of a minimum is important. Consider, for example, two energy wells of the same depth, but one having steeper walls than the other. The system will obviously be confined, to a greater degree, in the steeper well, which, in effect, is a state of lower disorder, or entropy, since the thermal fluctuations are minimized as well. Furthermore, one may be concerned with a *transition* from one minimum to another, due to thermal fluctuations for example. In addition to the initial and final energy states, one must also be concerned with passing over a potential barrier, the height of which plays an important role in the kinetic description. Lastly, *saddle points*, indicative of conditions for metastable equilibrium, may be encountered.

The polymer-substrate interaction can be computed using the fixed valence geometry molecular mechanics approximation [11,12]. In this approach, the total polymer-substrate interaction energy, $E_T = E(a,b,h,$

Θ,ϕ,u) is given by

$$E_T = \sum_i \sum_j \left(\frac{-A_{ij}}{r_{ij}^6} + \frac{B_{ij}}{r_{ij}^{12}} + \frac{322.0 q_i q_j}{\epsilon r_{ij}} \right), \tag{1}$$

where one index spans the atoms of the polymer and the other the atoms of the substrate. The first two terms in Eq. (1) are the dispersion attraction and steric repulsion contributions, and the last term is the monopole-monopole representation for the electrostatic energy. The A_{ij} and equilibrium distances are reported in Table 2, and the partial atomic charges, q_k, of the atoms in the polymer were estimated using the CNDO/2 molecular orbital scheme [13]. The molecular dielectric, ϵ, was set to 3.5.

Brut force summing of Eq. (1) is one way to compute E_T for a specific interfacial geometry. However, the three-dimensional symmetry of the substrate, this includes the depth of the substrate, as well as the surface, permits the sums of Eq. (1) to be replaced by highly convergent series. Each of terms in Eq. (1) can be considered as expressions for $\Sigma(1/r^n)$, which reduce to

$$\sum \frac{1}{r} = \frac{2}{a^2} \sum_{h_1,h_2} \frac{\exp[(-2\pi d/a)(h_1^2 + h_2^2)^{1/2}]}{(1/a)(h_1^2 + h_2^2)^{1/2}} \times \cos\left(\frac{2\pi h_1 x}{a}\right)\cos\left(\frac{2\pi h_2 v}{a}\right), \tag{2}$$

$$\sum \frac{1}{r^6} = \frac{\pi}{a^2 d^4} + \frac{2\pi^3}{a^2 d^2} \sum_{h_1,h_2} \frac{1}{a^2} (h_1^2 + h_2^2) K_2 \left(\frac{2\pi d}{a} (h_1^2 + h_2^2)^{1/2} \right) \times \cos\left(\frac{2\pi h_1 x}{a}\right)\cos\left(\frac{2\pi h_2 v}{a}\right), \tag{3}$$

$$\sum \frac{1}{r^{12}} = \frac{2\pi}{5a^2 d^{10}} + \frac{\pi^6}{30a^2 d^5} \sum_{h_1,h_2} \frac{1}{a^5} (h_1^2 + h_2^2)^{5/2} K_5 \left(\frac{2\pi d}{a} (h_1^2 + h_2^2)^{1/2} \right) \times \cos\left(\frac{2\pi h_1 x}{a}\right)\cos\left(\frac{2\pi h_2 y}{a}\right). \tag{4}$$

In these expressions, a is the unit cell parameter for the particular alkali halide, d is the height of the atom under consideration above the substrate surface, and x and y are the coordinates of a point on the substrate which is the projection of the atom under consideration to the surface (the origin of the substrate lattice being taken at a negative ion). Also, K_2 and K_5 are second- and fifth-order modified Bessel functions of the second kind, h_1 and h_2 are integers, and the sums are taken over all values of h_1 and h_2 (except $h_1 = h_2 = 0$) *having the same parity* from $-\infty$ to $+\infty$. In practice, the summation is cut off when the additional contribution from the next

Table 2 Dispersion Energy Constants, $A_{ij} \times 10^{-2}$ (in kcal-Å^6/mole) for Various Atom-Ion Pairs

Ion	H	C(sp^3)	C(sp^2)	N(sp^2)	O(sp^2)	S	CH_3
Na^+	0.32	1.15	1.07	1.10	0.90	13.0	1.8
K^+	1.22	4.18	—	4.16	3.43	49.6	—
Rb^+	1.90	6.84	—	6.50	5.35	77.5	—
F^-	1.03	3.71	—	3.54	2.91	42.0	—
Cl^-	2.74	9.80	9.12	9.36	7.71	111.0	15.3
Br^-	3.54	12.60	—	12.10	9.94	144.0	—

Atom-ion nonbonded equilibrium distances, R_0^{ij} (in Å) used in the calculations

Ion	H	C(sp^3)	C(sp^2)	O(sp^2)	S	CH_3
Na^+	2.07	2.63	2.48	2.33	2.73	3.03
Rb^+	2.57	3.13	2.98	2.83	3.23	3.53
K^+	2.42	2.98	2.83	2.68	3.08	3.38
Cl^-	2.90	3.46	3.31	3.16	3.56	3.86
Br^-	3.04	3.60	3.45	3.30	3.70	4.00
F^-	2.44	3.00	2.85	2.70	3.10	3.40

term is less than 10^{-12}. This results in a convergence limit of better than 10^{-4} kcal/mol in the final interaction energy. Polynomial approximations are used to obtain the zeroth- and first-order Bessel functions K_0 and K_1 [14]. Higher orders are calculated by using the recurrence relation [14]

$$K_{n+1}(x) = \left(\frac{1}{x}\right) 2nK_n(x) + K_{n-1}(x). \tag{5}$$

The accuracy of the Bessel function approximations over the range used here is better than 0.001%. Since Eqs. (2)–(4) only account for a single layer of substrate ions, additional layers are included by incrementing d by $1/2a$ and repeating the summation. In all, 30 layers are usually taken into account in the calculations.

Applications

Computer simulation studies of isolated PE and POM chain–alkali halide interactions were performed. Primarily, these polymer systems were chosen because of their uncomplicated molecular structures. Also, the abundance of available information on epitaxial thin films of PE and POM permits experimental comparisons.

PE: The substrates investigated were NaCl, RbBr, and NaF, the latter two representing a 22% increase and 18% decrease, respectively, in unit cell dimensions, from that of NaCl. The potential energy contour map of

Figure 4 illustrates many of the features that are qualitatively common to all the substrates.

Generally speaking, the contours for all the substrates are horizontal down to about 4.4 Å, then become sensitive to ϕ, developing into troughs followed by a repulsive barrier. Location of potential minima correspond to ⟨110⟩ chain orientations, specifically with alignment along positive rows, as seen in Figure 5. Translation of the chain in the ⟨110⟩ direction along a positive row does not alter the position of the minima. Furthermore, alignment along negative rows is energetically unfeasible.

The relative importance of electrostatic (broken into monopole-monopole and induced dipolar contributions) and dispersion-repulsive energies has been assessed. Reported in Table 3 are the fractions of the total energy, due to the three contributions, when the chain is in its most favorable orientation. Also, the three energies above the $(a,b) = (0,0)$ site on NaCl are plotted as a function of ϕ for $(h,\Theta,\mu) = (3.40$ Å, $90°, 0°)$ in Figure 6. The conclusions to be drawn are that (a) coulombic energies are orientation-sensitive, but are not of great magnitude; (b) induced dipolar energies are greater but essentially orientation-insensitive; (c) dispersion-repulsive energies dominate and depend strongly on ϕ; (d) the total energy is great enough to account for epitaxial crystallization.

Thus, the use of this polymer-substrate energetics model allows the rationalization of the orientational properties of a single PE chain on alkali halides and places the electrostatic contribution in proper perspective by

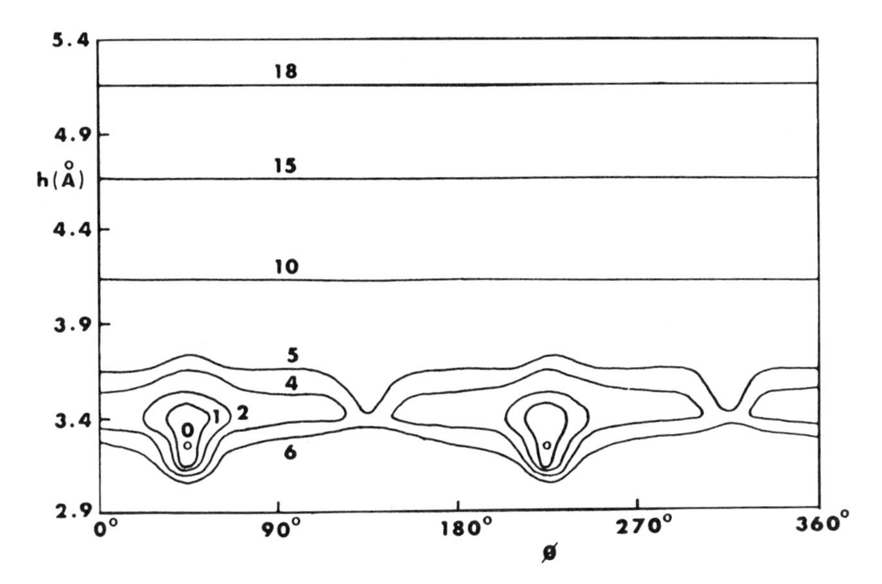

Figure 4 Energy contours in the h-ϕ plane for PE, plotted in kcal/mole/9 CH_2 units above the deepest minimum for NaCl where $(a,b,\Theta,\mu) = (d'/4,d'/4,90°,0°)$.

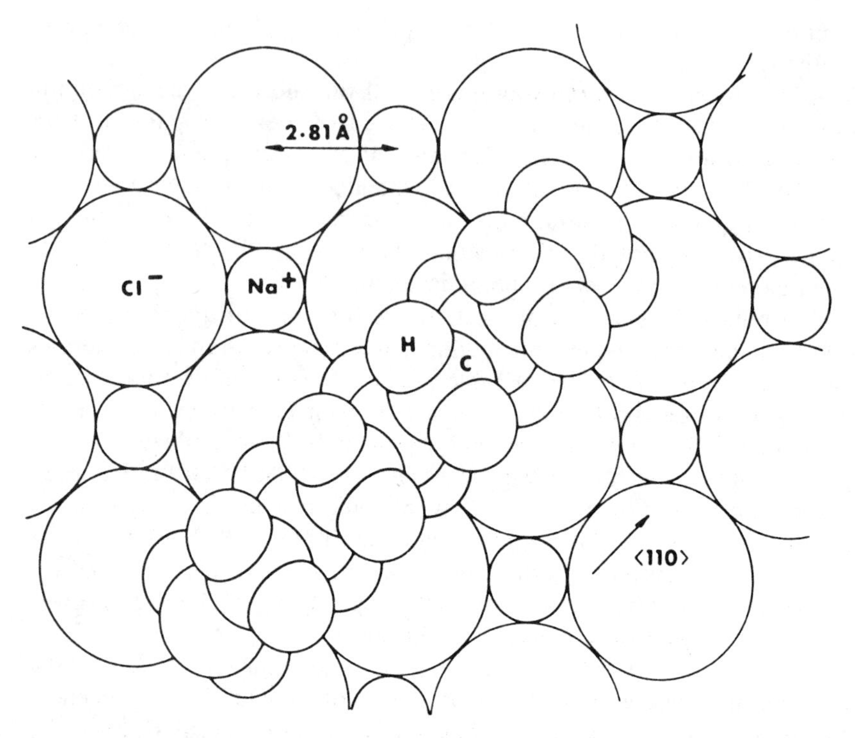

Figure 5 Space-filling model of the chain segment used in the computations, shown in the prefered orientation on an NaCl substrate. Ionic positions and radii are scaled relative to PE chain segment dimensions.

Table 3 Minimum Energies, Relative Energy Contributions, and Experimental Temperature Thresholds for Epitaxial Crystallization of PE on NaF, NaCl, and RbBr

	NaF	NaCl	RbBr
Minimum energy, kcal/mole/19 CH_2	−33	−24	−28
6–12 energy at minimum, %	70	75	71
Coulombic energy at minimum, %	7	5	3
Induced dipolar energy at minimum, %	23	20	26
Temperature threshold, °C	87[a]	110[a,b]	90[a]

[a]From Ref. 4.
[b]From Ref. 25.

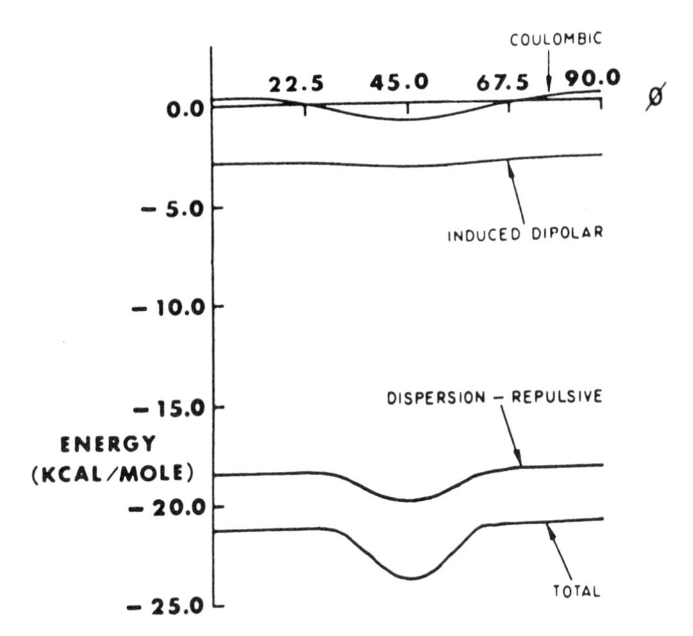

Figure 6 Total energy versus ϕ, plotted in kcal/mole/9 CH_2 units for PE on NaCl, where $(a,b,h,\Theta,\mu) = (0.0$ Å, 0.0 Å, 3.4 Å, $90°, 0°)$. Also shown is the decomposition of the total energy into constituent dispersion-repulsive, coulombic, and induced dipolar contributions.

accounting for the important interactions due to dispersion and repulsive forces. The calculations indicate that the preferred orientation is independent of substrate lattice dimensions, in accordance with experiment.

Consideration of the electron microscopic evidence and the temperature thresholds, that is, the temperatures above which epitaxial crystallization cannot occur, for the various substrates listed in Table 3, leads to the conclusion that NaCl is a better heterogeneous nucleating agent than RbBr, which in turn is better than NaF. The theoretical analysis, however, indicates that NaCl substrates produce neither the lowest minimum energy nor steepest energy wells. Resolution of this difficulty is realized in the following argument. In the case of NaF or RbBr, one could start a chain on the surface in the ⟨110⟩ direction along a positive row. As the chain folds back on itself, it is unable to simultaneously maintain the observed chain spacing and lie along a positive row. The area, on the surface, on which it does come to rest is less energetically favorable and, as growth proceeds, the chains are not always in register with positive rows resulting in a less-developed crystal. The (001) face on NaCl, however, provides a perfect lattice match, ensuring that each successive chain will lie in a potential well; and this is a factor in producing better-developed crystallites. This qualitative argument is quantitatively modeled in the section on "Interchain Interactions."

In addition to the type of simulation studies carried out for PE, analysis of potential well shape [15] and binding competition between polymer and solvent have been considered for POM [16].

Simulated interactions between the POM 9/5 helix and KCl, RbBr, and NaF substrates were performed. RbBr and NaF represent a 9% increase and 27% decrease, respectively, in unit cell dimensions, from that of KCl where a close match between the interchain distance in "normal" hexagonal POM and space between K^- and Cl^- rows exist.

The location of the potential energy minima clearly indicate an energetically preferred ⟨110⟩ chain direction with alignment along rows of positive charge. This preferred orientation is maintained with translation along cationic rows and also axial rotation of the chain. Similar calculations involving RbBr and NaF substrates also predict the ⟨110⟩ orientation along positive rows.

The shape, as well as depth, of an energy well is recognized as thermodynamically important as can be seen by considering the entropy associated with an energy minimum:

$$S = \frac{R}{2}\,[n(1 + \ln \pi\, RT) - \ln \det F], \tag{6}$$

where R is the universal gas constant, T is the absolute temperature, and

$$F_{ij} = \left.\frac{\partial^2 U}{\partial X_i \partial X_j}\right|_{\min} \tag{7}$$

where U, the total potential energy, is a function of $\{X_k; k = 1, \ldots, n\}$, the degrees of freedom of the system, and the second derivatives are evaluated at the potential energy minimum [11]. The graphical meaning of the matrix element $(\partial^2 U/\partial\phi^2)_{\min}$, for example, is that it is a measure of the *curvature* of the energy surface at the minimum for a fixed (a,b,h,Θ,μ). From elementary mechanics we can write

$$\left.\frac{\partial^2 U}{\partial \phi^2}\right|_{\min} = \frac{\partial}{\partial \phi}\left[\frac{\partial U}{\partial \phi}\right]_{\min} = -\left.\frac{\partial L\phi}{\partial \phi}\right|_{\min} \tag{8}$$

where L_ϕ is the torque acting upon the polymer chain to restore it to a condition of equilibrium for small displacements from the minimum in the ϕ direction. The greater the curvature, that is, the steeper the well, the lower the entropy and the stronger the restoring torque with increasing displacement. In other words, the breadth of the troughs in the ϕ direction is a measure of the chain's orientational tendency. One can apply similar reasoning to deduce that $(\partial^2 U/\partial h^2)_{min}$ is a measure for the tendency of the chain to adhere to the substrate for small displacements from the minimum in the h direction. Thus, by comparing energy curvatures, one can compare the "stiffness" of the interfacial bonds since the elements of F are essentially elastic constants.

Inspection of the energy surfaces for the various substrates suggests a relative interfacial bond stiffness, in the h direction, having the following order: RbBr > NaF > KCl. Figure 7 shows the variation in polymer-substrate energy, in the h direction, about the equilibrium position from the surface, where all curves shown have been translated to make the minima coincide so that the various polymer-substrate systems can be readily compared. However, the relative substrate order of the tendency to maintain the ⟨110⟩ orientation is RbBr > KCl > NaF, which is the descending order of energy-minima curvatures, in the ϕ direction, at the preferred orientation, as seen in Figure 8. It is thus predicted that the POM 9/5 helix will exhibit the greatest initial resistance to small displacements

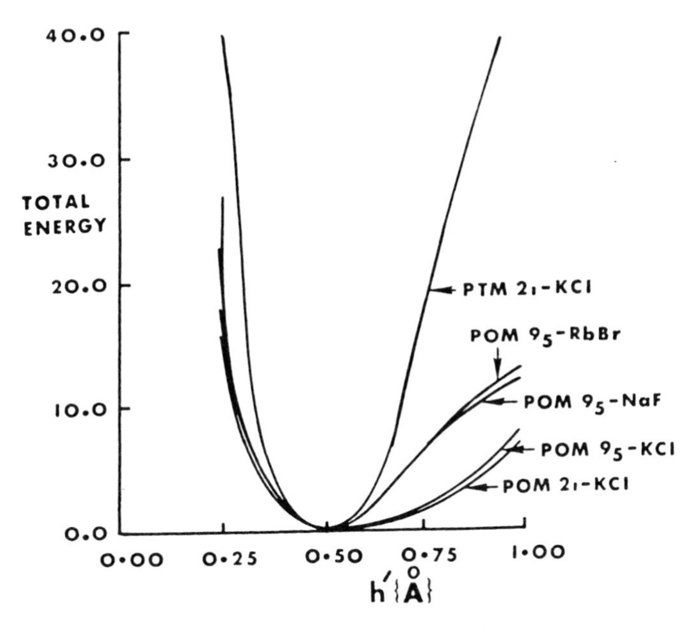

Figure 7 Variation of the POM 9/1 helix-substrate interaction energy per 9 chemical repeat units in the h direction about the equilibrium position from the surface. The curves have been translated to make the minima coincide. Also shown are results for a POM and polythiomethylene (PTM) 2/1 trans, cis conformation.

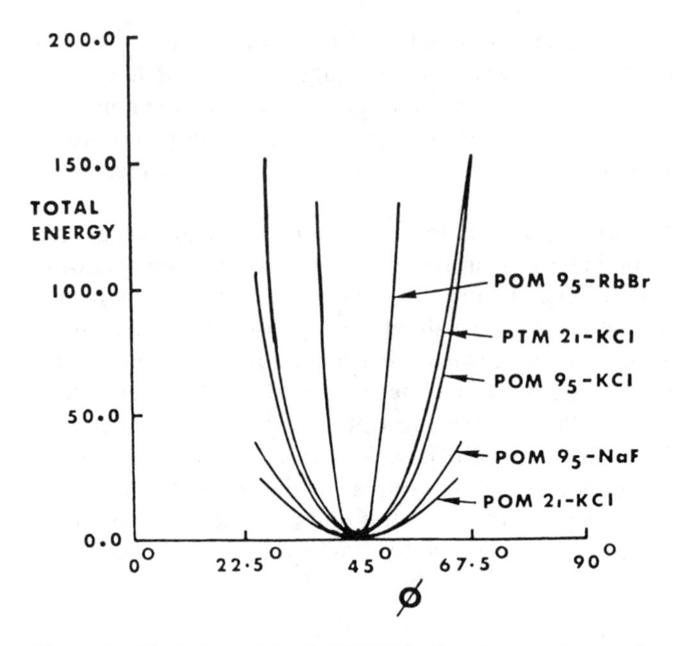

Figure 8 Variation of the POM 9/5 helix-substrate interaction energy per 9 chemical repeat units in the ϕ direction about the equlibrium position from the surface. The curves have been translated to make the minima coincide. Also shown are results for a POM and PTM 2/1 trans, cis conformation.

from the substrate surface as well as the strongest orientational tendency for RbBr. Furthermore, the orientational tendency increases with increasing substrate lattice dimensions.

Experimentally [17], one observes relatively large variations in the nucleation density of rodlike epitaxial crystals can occur as a function of the substrate used in a solution epitaxial crystallization experiment, all other factors held constant. It was suggested that these differences may be a result of preferential solvent adsorption preventing polymer nucleation, especially on substrates having higher surface energies. Further support for this hypothesis has been obtained by calculating the relative affinities that a polyoxymethylene (POM) chain segment and a solvent molecule have for an ideal alkali halide surface. The basis for computing the relative surface affinities is the estimation of the partition function, Q, from the ensemble of interfacial interaction energies $\{E_i\}$ computed as a function of intermolecular degrees of freedom (a,b,h,Θ,ϕ,μ);

$$Q = \sum_i g_i \exp\left(\frac{-E_i}{RT}\right), \tag{9}$$

from which the Helmholtz free energy follows as

$$A = -RT \ln Q. \tag{10}$$

A is then set equal to F_p or F_s, the average interaction energy of the polymer or solvent, respectively, with the substrate. All energies were calculated at $T = 160°C$, a temperature experimentally used in the epitaxial crystallization of POM from iodobenzene solution onto alkali halides. The configurational entropy resulting from various arrangements of groups of molecules on the surface has not been included. However, this number should not vary appreciably from substrate to substrate. The main interest is to establish *trends* of the interaction energies as a function of the substrate.

The normalized degeneracy factor g_1 (see Eq. (9)) for each energy state is related to the number of equivalent locations of each point in Figure 9 on a unit cell of the alkali halide surface. The total number of equivalent locations in one surface unit cell for each point is 2, 8, 2, 8, 4, 8, and 16

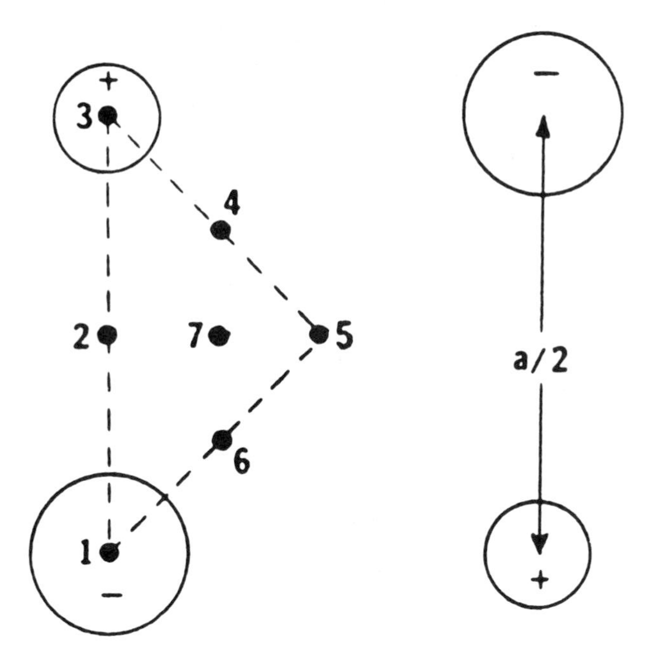

Figure 9 Quarter unit cell of a (001) alkali halide surface showing the seven points over which polymer and solvent molecules were positioned for the interaction energy calculations. The area inside the triangle contains all nonequivalent loci on an infinite (001) alkali halide plane.

for points 1–7, respectively. These locations cannot be occupied simultaneously by polymer segments and/or solvent molecules, since the separation distances between equivalent sites is, in many cases, less than the distance of closest approach between two depositing molecules. The basic symmetry element in which the points are located (see Fig. 9) is itself smaller than either the polymer or solvent molecules, for any of the five substrate lattice sizes. Thus there are an equal number of *accessible* equivalent sites per unit cell for each point on a given substrate. This number increases as the substrate lattice dimensions increase.

The distance of closest approach of two iodobenzene solvent molecules was determined to be 7 Å. The number of accessible equivalent sites per unit cell for the solvent molecule was determined by counting the number of equivalent sites separated by at least 7 Å for each substrate. These results are listed in the first column of Table 4.

The determination of the number of accessible equivalent sites per unit cell for the polymer is more complicated. Parallel orientation in the [110] substrate direction is assumed for adjacent polymer segments. The minimum separation distance between segments is 4.46 Å [18]. The smallest area of the substrate containing an integral number of unit cells that will accommodate a polymer segment diagonally (i.e., diagonal distance ≥ 17.3 Å) is then determined. A 16-unit-cell square area is required for the LiF and NaF lattices, a 9-unit-cell square is required for NaCl, and a 4-unit-cell area is required for KCl and KI. The total number of *accessible* equivalent sites within each of these respective areas is determined by counting one site for each row of sites parallel to the [110] direction and separated by at least 4.46 Å. Sites occurring on a corner of the square area are given a value of 0.25 since they are shared by four such areas. The total number of accessible sites within each square area is scaled to a unit cell basis by dividing by the number of unit cells within that area. The results of these calculations are reported in the second column of Table 4.

Table 4 Normalized Degeneracy Factor g_i

Substrate	Accessible equivalent sites per unit cell		Area/unit cell ($Å^2$)	Normalized g_i	
	Solvent	Polymer		Solvent	Polymer
LiF	0.250	0.219	16.14	0.236	0.184
NaF	0.250	0.219	21.47	0.178	0.138
NaCl	0.500	0.333	31.81	0.240	0.142
KCl	0.500	0.875	39.60	0.193	0.299
LI	0.500	0.875	49.92	0.153	0.237

The calculated interaction energies are ultimately scaled on an area basis. Thus, the values for the number of accessible equivalent sites must also be scaled in terms of area. The area per unit cell is also listed in Table 4. The final values for the g_i were obtained by normalizing the area-scaled values over the five substrates. These results are listed in the final two columns of Table 4 for the polymer and solvent molecules.

The total Helmholtz free energy ΔF_p required to remove a polymer segment from solution and attach it to a substrate may be written as $\Delta F_p = F_p - \Delta F_p'$, where $\Delta F_p'$ is the energy required to resolvate an adsorbed segment:

$$\Delta F_p' = F_p' - bF_p'. \tag{11}$$

In this equation, b is the fraction of the total surface area of an adsorbed polymer segment that remains solvated while on the surface, and probably has a value of 0.5. Similar relationships exist for adsorption of a solvent molecule:

$$\Delta F_s = F_s - \Delta F_s', \tag{12}$$

$$\Delta F_s' = F_s' - bF_s', \tag{13}$$

where b again probably has a value near 0.5.

The quantity ΔA_p is defined as the difference in Helmholtz free energies of adsorption of polymer and solvent: $\Delta A_p = \Delta F_p - \Delta F_s$. Thus

$$\Delta A_p = (F_p - \Delta F_p') - (F_s - \Delta F_s') \tag{14}$$

or

$$\Delta A_p = (F_p - F_s) - (\Delta F_p' - \Delta F_s'). \tag{15}$$

The second term in Eq. (15) is totally independent of the substrate used, as it contains only polymer-solvent and solvent-solvent interaction terms, which remain constant for the experiments under consideration. Thus, in the absence of numerical values for $\Delta F_p'$ and $\Delta F_s'$, *trends* in ΔA_p as a function of substrate will still be significant. Since both F_p and F_s are negative, the more negative the value of ΔA_p, the more polymer adsorption will be preferred over solvent adsorption, and vice versa.

The value of F_p was obtained by using a chain segment of nine CH_2O units, whereas the value of F_s was computed by using a single iodobenzene solvent molecule. These energies must be scaled relative to the area of substrate occupied before they may be compared to one another as in Eq. (15). The respective energies in kilocalories per mole were, therefore, converted to ergs per square centimeter by using the scaling parameters

k_p and k_s:

$$k_p = \frac{6.947}{S_p}, \tag{16}$$

$$k_s = \frac{6.947}{S_s}, \tag{17}$$

where S_p and S_s are, respectively, the areas of the substrate occupied by the nine CH_2O units of POM, and the molecule of iodobenzene. The factor of 6.947 converts the units from kilocalories per mole to ergs per molecule. The final form of $\Delta A_p'$ is

$$\Delta A_p = (F_p'' - F_s'') - \text{const} \tag{18}$$

where $F_p'' = k_p F_p$ and $F_s'' = k_s F_s$.

Values for S_p and S_s were obtained from experimental data in the following way. The helical POM segment may be assumed to occupy a rectangular area of surface, having a width equal to the *a* unit cell constant, obtained from POM crystal structure data (4.46 Å) and a length equal to the length of nine CH_2O units, i.e., the fiber repeat (17.3 Å). Thus k_p = 0.009. For the solvent, the effective volume occupied per molecule may be calculated from the molar volume V_m, as V_m/N, where N is Avogadro's number. If this volume is modeled as a sphere, the value of S_s may be taken as the area of a circle which is the projection of this sphere onto a plane. For iodobenzene at 160°C, V_m = 126.5 cm^3/mol [19], so S_s is 42.7 $Å^2$ and k_s = 0.163.

Table 5 contains the results of these calculations for the five alkali halide substrates investigated experimentally in the preceding paper. The scaled values F_p'' and F_s'', ΔA_p, the nucleation density of rodlike epitaxial crystals, and the substrate surface energies [20,21] are listed. The trends in ΔA_p correlate well with the morphological observation of nucleation density, with the exception of KI. However, KI has a significant solubility in organic solvents such as ethyl alcohol and acetone, whereas the other substrates

Table 5 Results of Polymer and Solvent Interaction Energy Calculations

Substrate	F_p''(ergs/cm²)	F_s''(ergs/cm²)	ΔA_p (ergs/cm²)	Nucleation density of rodlike crystallites (lamellae/μ²)	Surface energy (ergs/cm²)
LiF	−193.8	−232.9	39.1	10 ± 2	340
NaF	−180.0	−218.4	38.4	17 ± 2	210
NaCl	−182.1	−209.5	27.4	22 ± 5	188
KCl	−192.5	−217.7	25.2	190 ± 22	163
KI	−178.0	−191.4	13.4	22 ± 4	136

do not. Thus, the lower nucleation density observed on KI relative to KCl may be a result of some other physical (or chemical) interaction with the solvent not accounted for in these calculations. For the remaining four substrates, KCl produces the most dense epitaxial layer of polymer and has the lowest value of ΔA_p. The sequence predicted by ΔA_p, in terms of increasing preference of polymer adsorption over solvent is LiF < NaF < NaCl < KCl < KI. This is also the sequence observed experimentally, with the exception of KI. The trends in ΔA_p also correlate with the substrate surface energies. Thus one may conclude (within the limitations of this model) that substrates with higher surface energies have a higher relative affinity for solvent than for polymer. This is why relatively few polymer crystals are observed on these substrates.

One molecular modeling study was carried out for a single PE chain interacting with a nonalkali halide substrate [22]. The selected substrate was graphite which, because each carbon is equivalent, does not possess any distribution of charge density. All carbons have zero net partial charges for the ideal substrate geometry. Thus polymer-substrate electrostatic interactions do not need to be considered. However, graphite does have a different symmetry from the alkali halides.

The graphite substrate is an open-net hexagonal structure which can be represented by the unit cell illustrated in Figure 10(a). The two lattice vectors connect the centers of the hexagons and are 2.46 Å apart. The distance between adjacent carbons is 1.42 Å. Graphite has a laminar structure in which the layers are separated by 3.35 Å and every other layer is shifted 1.42 Å in the x direction (as defined in Fig. 10(a)). Two key crystallographic directions on the graphite substrate are the nearest-neighbor directions, $(10\bar{1}0)$, and the next-nearest-neighbor directions, $(11\bar{2}0)$ (see Fig. 10(b)).

The difference in the symmetry of graphite as compared to the alkali halides necessitates that Eqs. (3) and (4) be redefined for hexagonal symmetry. The dispersion and repulsion interactions V_i^L of all substrate atoms in a given layer with a polymer atom i is

$$\begin{aligned} V_i^L = & -A_i \left[\frac{\pi}{|\mathbf{a}_1 \times \mathbf{a}_2|\rho^4} + \frac{2\pi^3}{\Gamma(3)|\mathbf{a}_1 \times \mathbf{a}_2|} \right. \\ & \left. \times \sum_{m_1=-\infty}^{+\infty} \sum_{m_2=-\infty}^{+\infty} \left(\frac{|\mathbf{n}_m|}{\rho} \right)^2 K_2(2\pi\rho|\mathbf{n}_m|) \sum_{k=1}^{2} \exp[i2\pi(\mathbf{n}_m \cdot \mathbf{r}_k)] \right] \\ & + B_i \left[\frac{2\pi}{5|\mathbf{a}_1 \times \mathbf{a}_2|\rho^{10}} + \frac{2\pi^6}{\Gamma(6)|\mathbf{a}_1 \times \mathbf{a}_2|} \right. \\ & \left. \times \sum_{m_1=-\infty}^{\infty} \sum_{m=-\infty}^{\infty} \left(\frac{|\mathbf{n}_m|}{\rho} \right)^5 K_5(2\pi\rho|\mathbf{n}_m|) \sum_{k=1}^{2} \exp[i2\pi(\mathbf{n}_m \cdot \mathbf{r}_k)] \right], \end{aligned} \tag{19}$$

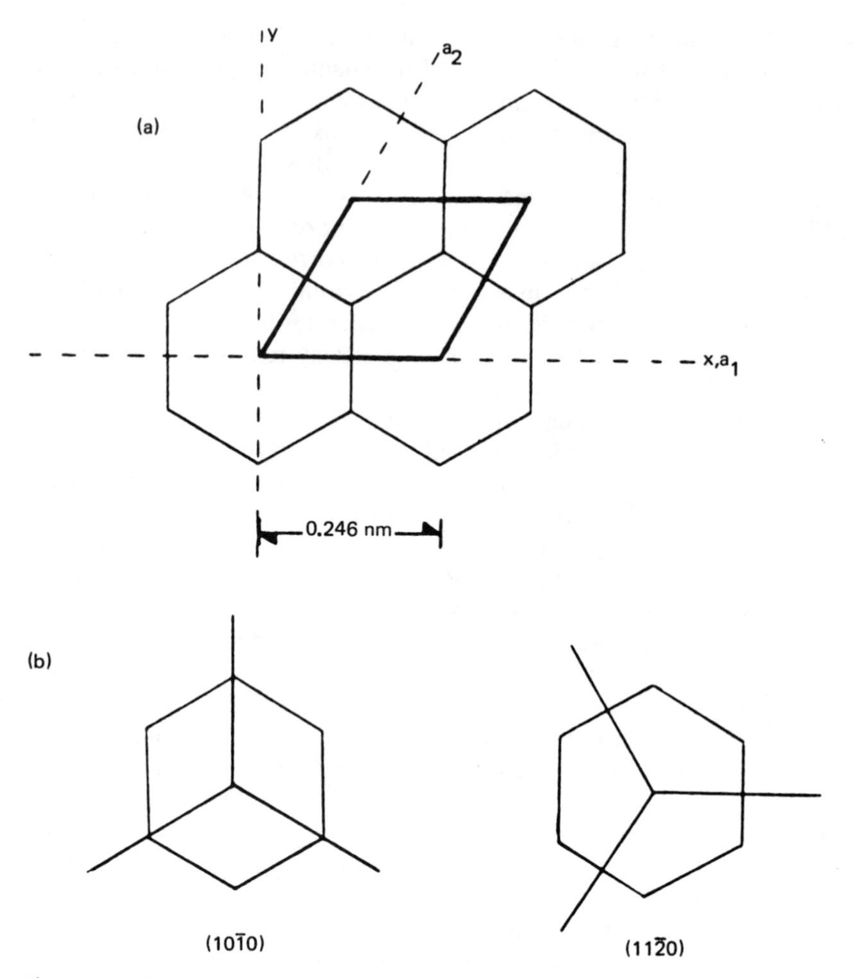

Figure 10 (a) The unit cell of graphite used in this investigation. Each unit cell contains two carbon atoms. (b) Two principal crystallographic directions: nearest-carbon-neighbor directions (10$\bar{1}$0); next-nearest-neighbor directions (11$\bar{2}$0).

where $\mathbf{n}_m$ is the reciprocal lattice vector (m_1 and m_2 are integers), $|\mathbf{a}_1 \times \mathbf{a}_2|$ is the area of the unit cell, Γ represents the gamma function, and K represents the modified Bessel function of the second kind. The prime indicates that the term for the zero reciprocal lattice vector ($m_1 = 0$ and $m_2 = 0$) is not included in the summation. These summations are rapidly convergent owing to the nature of the Bessel functions. For the subsequent layers, only the zero reciprocal lattice vector term contributes significantly to the interaction energy. Thus the calculations are greatly reduced. Also, the interaction energy depends only on the height ρ of an atom above the substrate.

The zeroth- and first-order modified Bessel functions of the second kind were calculated using polynomial approximations [22]. The higher-order modified Bessel functions were then calculated using the recursive relationship

$$K_{n+1}(z) = \left(\frac{2n}{z}\right) K_n(z) + K_{n-1}(z), \tag{20}$$

where n and z refer to the order and the argument of the function, respectively. This recursive scheme can be used without loss of significant accuracy [23]. These approximations were determined to be accurate to within an error of 0.01%. Since the potential is real, the exponential term in the summation can be replaced by its real term, which is a cosine function. This rapidly converging representation for a PE-graphite system was found to reproduce the result obtained by the direct summation technique in a test calculation. The two results differed by less than 0.25%.

The surface hexagon of graphite can be represented by 12 equivalent right triangles. The right triangle extending from the origin in the x direction and the positive y direction was arbitrarily studied (see Fig. 11). The interaction energies were calculated for specific variations in the degrees of freedom. A satisfactory convergence for the interaction energy was found when the first 20 substrate layers were considered.

In other studies the backbone of the polyethylene chain segment was generally taken to be perpendicular to the alkali halide substrate ($\mu = 0°$ or $\mu = 180°$); see Figure 12. Groszek [24], however, has predicted that the backbone of n-alkanes adsorbed onto the basal plane of graphite will be parallel to the substrate. This prediction was based on a lattice matching consideration. Therefore, interaction energies were determined for several chain-axis rotations. The results are shown in Table 6. In this case, the chain origin was specifically above a carbon atom. However, the trends presented in Table 6 are quite general and hold for any chain location above the substrate. Primary energy minima occur when the backbone of the chain segment is parallel to the substrate ($\mu = 90°$ or $\mu = 270°$), and secondary minima corresponding to a perpendicular backbone orientation; see Figure 12. The preferred chain backbone orientation thus agrees with Groszek's prediction, but is based on an energetics analysis rather than a lattice-matching criterion.

A composite map of the surface interfacial energetics can be constructed from a set of the height of z-axis energy maps based upon different chain origins above the substrate. These composite maps, shown in Figure 13,

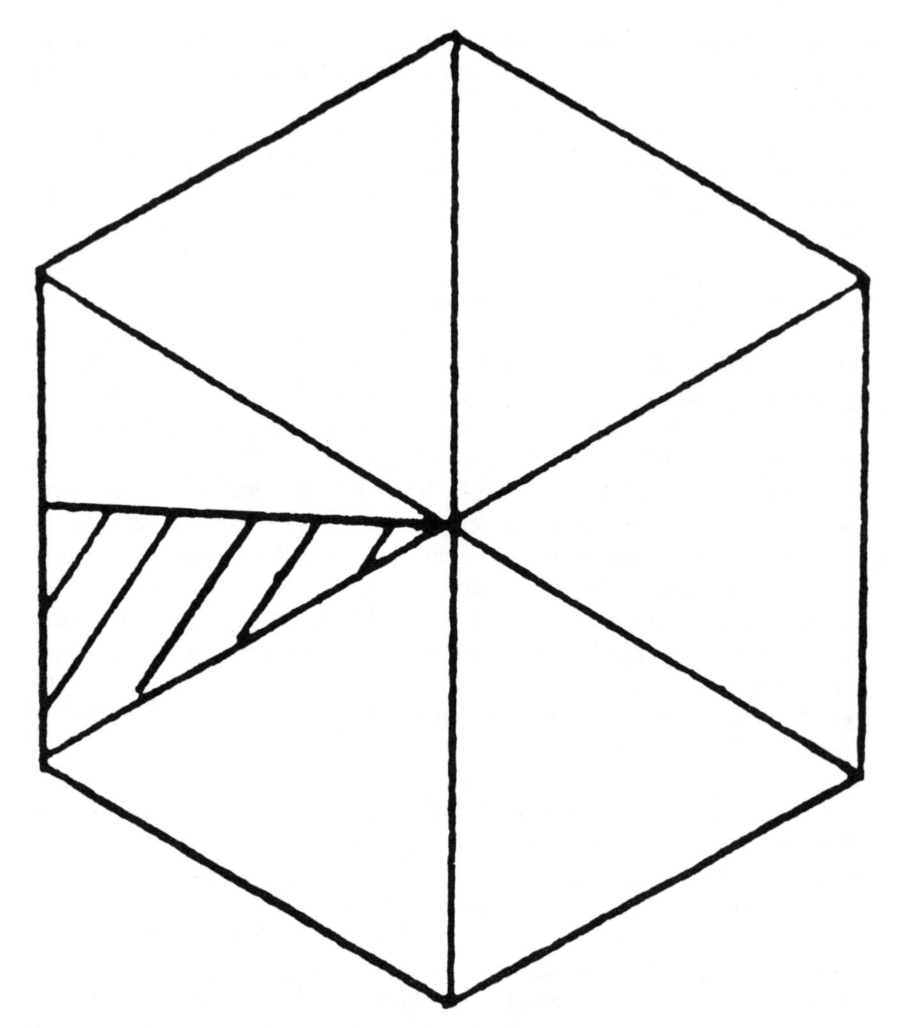

Figure 11 The graphite hexagon can be broken into six equilateral triangles. These triangles can be further decomposed into two right triangles, which are mirror images of each other. The shaded triangle is the one which has been studied.

are isoenergy plots which indicate how the interaction energy varies as the chain moves over the substrate, specifically in the right triangle described above. The energies reported have some fixed parameters: the height (h = 3.7 Å), the z-axis rotation (ϕ = 120°), and the chain-axis rotation (μ = 90°); these fixed values correspond to the values at the global minimum. These composite maps were constructed for two chain lengths, 8 and 60 methylene units.

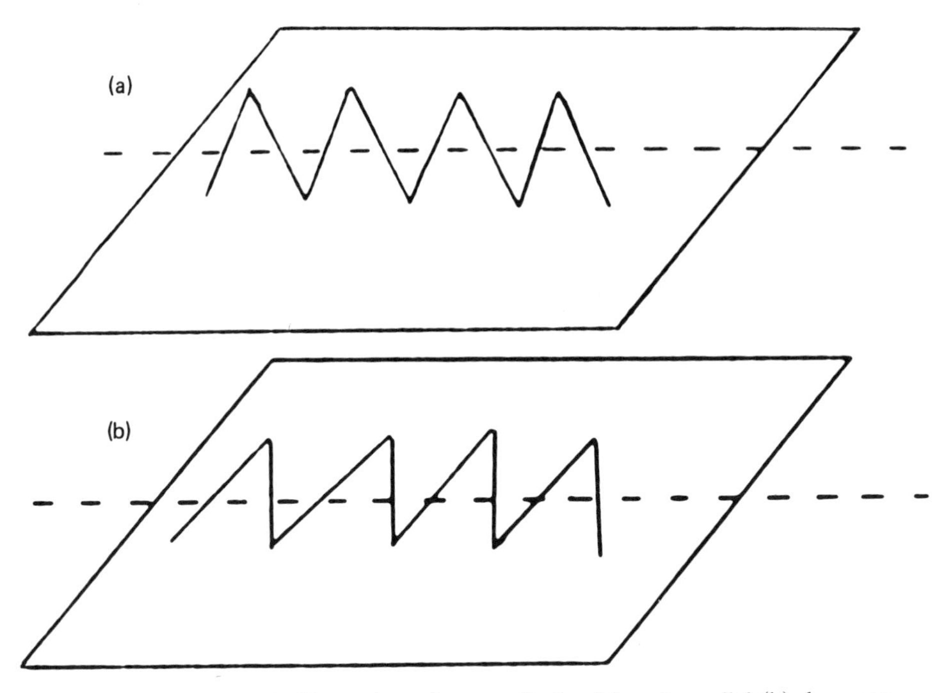

Figure 12 Schematic illustration of perpendicular (a) and parallel (b) decomposition of the backbone carbons of a planar-zigzag polyethylene segment onto an arbitrary substrate.

Table 6 Effect of Chain-Axis Rotation on Interaction Energy[a]

Chain-axis rotation μ (deg)	Maximum interaction energy E_{max} (kcal/mol/8 CH_2)
0	−14.08
15	−13.33
30	−12.22
45	−11.82
60	−12.38
75	−14.05
90	−15.75
135	−11.89
180	−14.12
225	−11.89
270	−15.75
315	−11.82

[a]Chain length: 8 methylene units, chain origin: (1.23, 0.71 Å).

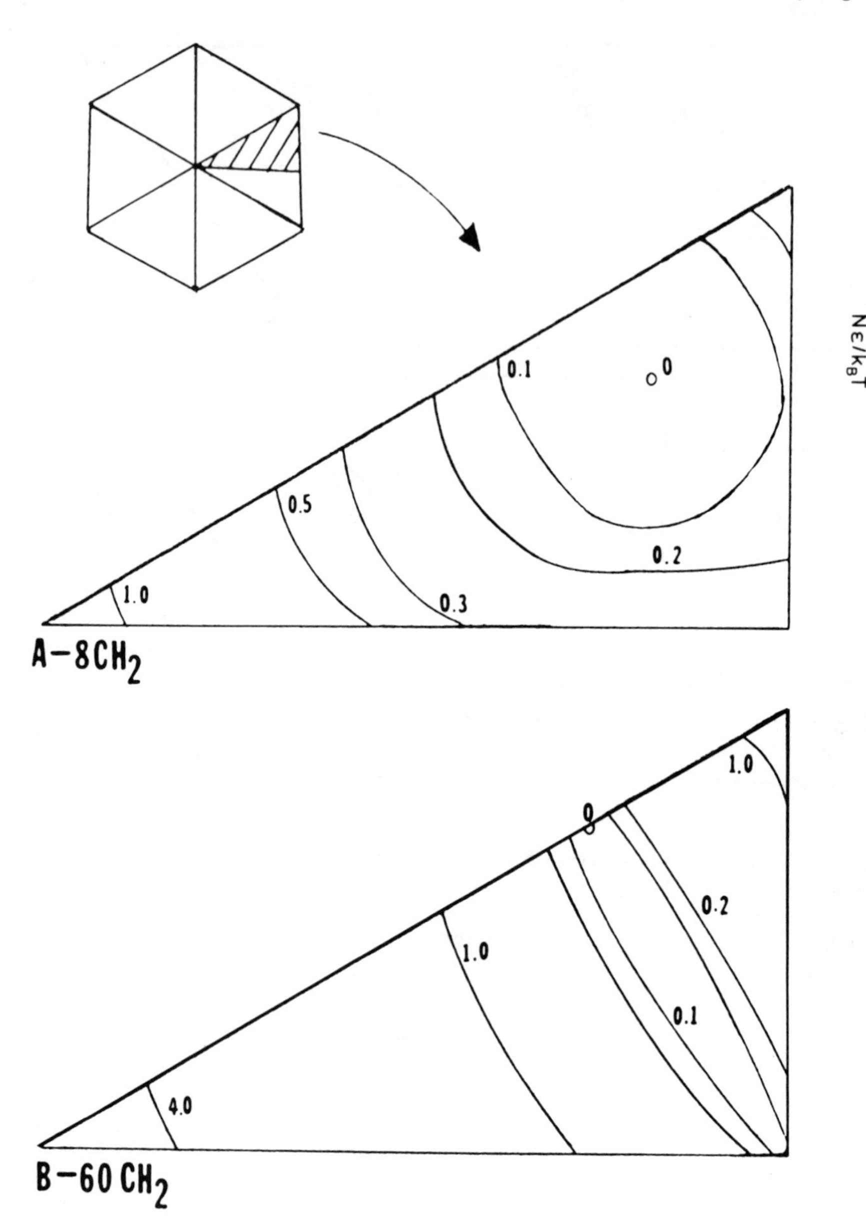

Figure 13 Two composite isoenergy maps: (a) 8 CH_2 and (b) 60 CH_2, plotted in kcal/mole/n methylene units above the global minimum for each length. The chain axis rotation, z-axis rotation, and height were fixed at the values these parameters had at the global minima (h = 0.37 nm, ϕ = 120°, μ = 90°). The longer chain map has a troughlike minimum, while the short chain map has a pointlike minimum.

The composite map indicates that the preferred orientation of the chain axis of a polyethylene segment is in the $\langle 11\bar{2}0 \rangle$ directions of the surface layer of the graphite substrate. The composite maps specifically indicate that these preferred orientations align upon the lines formed by the nearest-neighbor midpoints of the C—C bond in the substrate surface layer. This interfacial geometry is shown in Figure 14.

The composite maps (Fig. 13) also show the effect of varying chain length on the deposition energetics. The primary difference between the two maps is the transformation of the global minimum from a point (8 CH_2 units) to a trough (60 CH_2 units). Thus, the longer-chain-length com-

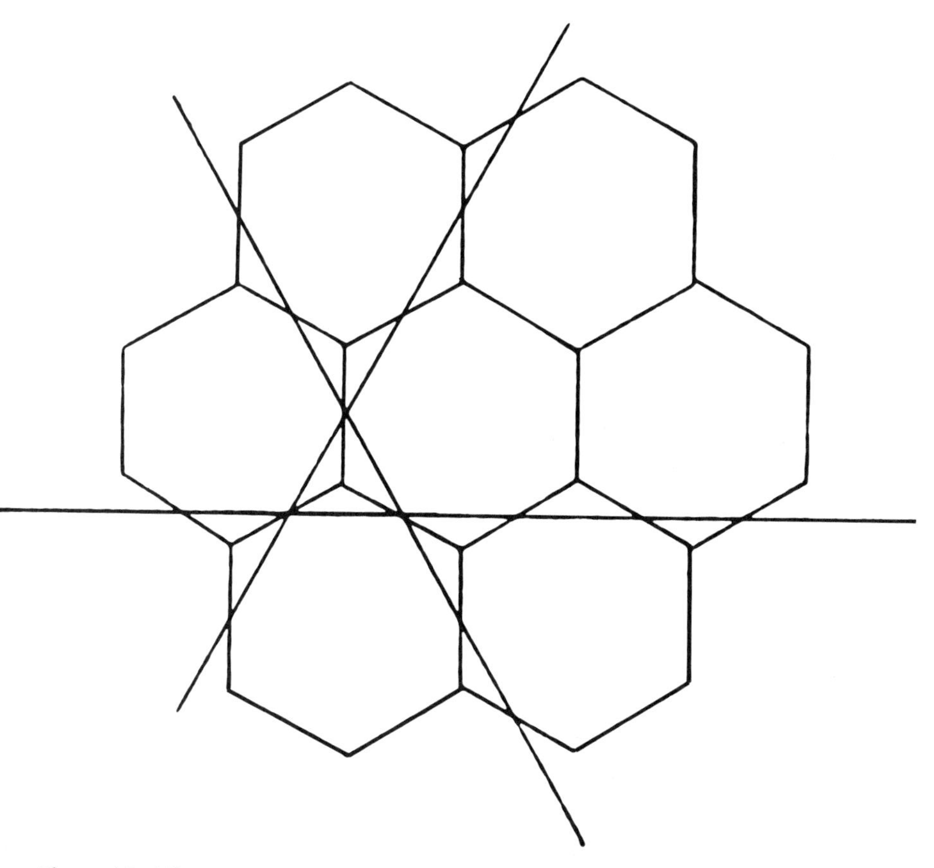

Figure 14 The three lines indicate the preferred orientations for the polyethylene chain axis. Note that no point of intersection will contain three planes. The common orientation points are the mdipoints of the C—C bonds in the surface layer of graphite.

posite map shows a "line" of minima, oriented as described above, along which translation of the chain origin has very little effect. A similar effect was observed for PE-alkali halide interactions [25]. Some further effects of chain lengths are reported in Table 7, which presents data for longer ($n > 25$) chain segment calculations. A normalized (energy/number of methylene units) set of interaction energy minima indicates that, as chain segment length increases, there is a decrease in the normalized energy. This decrease can be attributed to a greater number of less favorable interactions between the chain and the substrate as the chain length increases. The set of normalized changes in the polymer-substrate interaction energy (reported as the difference between the minimum and maximum interaction energies along the predicted orientation given above divided by the number of methylene units) indicates that, as the chain length increases, these normalized energy changes decrease. This result can be attributed to the transformation, i.e., the broadening of a pointlike global minimum to a troughlike global minimum.

Tunistra and Baer [26] used electron diffraction techniques to determine that for polyethylene epitaxially deposited on graphite the ⟨110⟩ plane of the orthorhombic PE crystal grew parallel to the substrate. Further, the chain axis directions in the observed morphology were oriented in the $(11\bar{2}0)$ directions of the graphic surface layer. This observed morphology is quite close to the predicted results. It should be noted, however, that electron diffraction data indicate a bulk crystal morphology, but not necessarily the direct interfacial geometry. Also, interchain interactions have not been considered in this study and could have a considerable effect on the packing morphology. Overall, the experimental evidence supports the predicted morphology described above.

Table 7 Normalized Interaction Energy Data for Long-Chain Segments

Chain length n	Normalized maximum interaction energy E_{max}/n (kcal/mol)	Normalized maximum difference in interaction energy for chain translation along the predicted prefered orientation ($\Delta E_{max}/n$ kcal/mol)
30	−1.975	0.0451
40	−1.954	0.0396
50	−1.934	0.0164
60	−1.931	0.0017

AN EXTENDED SOLVATION MODEL

In addition to modeling the competition between polymer and solvent for binding to an alkali halide substrate, the molecular modeling of polymer-solvent interactions to the deposition process have also been considered. Specific results for the PE-NaCl system are reported for the solvents benzene, toluene, and *o*-xylene.

As assumed previously, the chain conformation assigned to a macromolecule is that which exists in the normal crystalline state. The macromolecule can then be considered to have an effective cylindrical geometry in which r is an average "hard-cylinder" radius. Solvation is accounted for by introducing a monolayer of solvent about the cylinder, there being n solvent molecules, of diameter d within a specific axial length.

The removal of one solvent molecule from the helix, due to steric overlap of the solvation layer with the substrate surface (see Fig. 15), results in a net energy change of $-\Delta f$. Both $-\Delta f$ and V_m, the effective volume of a polymer-associated solvent molecule, are assumed to remain the same regardless of the state of solvation. The energy required to *assemble* the

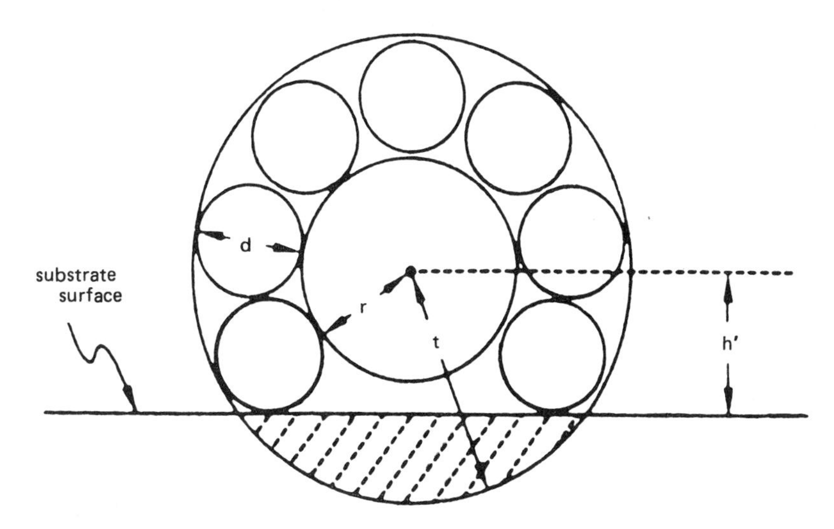

Figure 15 Cross-sectional view, looking down the helix axis, of the cylindrical monolayer used in the computer simulation of solvation energetics for a given polymer-solvent-substrate system. d is the effective solvent molecular dimater, r a "hard-cylinder" radius indicative of the overall steric bulk of the polymer, and h' the distance between the chain axis and substrate surface. The shaded area represents excluded volume due to steric overlap between the monolayer and substrate surface.

totally solvated association is then $n\,\Delta f$. If V_0 is the volume of overlap and V the total volume of the cylindrical solvation layer ($= nV_m$), then the energy F of a partially solvated helix will be assumed to depend upon the overlap volume in a linear manner:

$$F = \Delta f\left(n - \frac{V_0}{V_m}\right) = n\,\Delta f\left(1 - \frac{V_0}{V}\right). \tag{21}$$

From geometrical considerations we can write

$$F = n\,\Delta f\{1 - [\cos^{-1}\mu - \mu(1 - \mu^2)^{1/2}][\pi v(2 - v)]^{-1}\}, \tag{22}$$

where $\mu = h'/t$ and $v = d'/t$, $t = r + d$, $0 < v < 1$, $1 - v \leq \mu \leq 1$, and $F(\mu > 1, v) = n\,\Delta f$.

A complete description of the three-component energetics involved in the initial nucleation of a single chain from solution upon a substrate surface must necessarily include (a) solvent molecule–polymer, (b) solvent molecule–substrate, and (c) solvent molecule–solvent molecule interactions.

Solvation energetics have been incorporated into the basic theory and the reader is directed to the literature [27] for a detailed account of the underlying physics. The important quantities that emerge are ΔW_i^P and ΔW_i^S, the amount of energy required to remove a single solvent molecule from the polymer and substrate, respectively. They are given by

$$\Delta W_i^p = \frac{E_v M}{\rho}\,\frac{1 - 2\xi}{\pi} - \langle U_p^k \rangle \tag{23}$$

and

$$\Delta W_i^s = \frac{E_v M}{\rho} - \langle U_s^k \rangle. \tag{24}$$

Here E_v, ρ, and M are the cohesive energy density, mass density, and molecular weight of the solvent. The geometrical factor ξ is given by

$$\cos\xi = 1 - \frac{1}{2}\left(\frac{d}{R_0}\right), \tag{25}$$

and $\langle U_p^k \rangle$ and $\langle U_s^k \rangle$ are theoretically determined thermal-averaged energies of interaction between a single solvent molecule and polymer chain and substrate surfaces, respectively. Here R_0 is defined as the distance between the center of mass of a solvent molecule and helix axis corresponding to the minimum of polymer-solvent molecular energies, each of which is Boltzmann-averaged over rotations about the center of mass and motion of the center of mass around the helix axis. Then Δf is taken as a Boltzmann average of Eqs. (23) and (24). The modeling of solvation energetics in the epitactic process is finally achieved by substituting Δf into Eq. (22).

The modeling of the epitactic deposition of PE onto NaCl was repeated using the solvation energetics as part of the deposition thermodynamics. The results of the modeling can be summarized as follows: First, the tendency to maintain the ⟨110⟩ orientation with specific positioning of the polymer chain along rows of positive charge is preserved. This is to be expected since a ϕ dependence in the solvation energetics is not incorporated into the model. Secondly, it is seen that for the range of h in which orientational effects are significantly felt ($h < 3.90$ Å), the solvent-independent energetics are virtually identical to the findings in which solvation effects are considered. Restated, the deposition energy maps for both cases, obtained from energies in excess of the respective absolute minima, are almost superimposable near the interface. If, however, the energetics are investigated to a distance of 12 Å from the substrate surface for the polymer in vacuum and in each of the three solvents that have been modeled, the interfacial energetics are quite different, as shown in Figure 16.

In Figure 16 all deposition curves, with increasing h, asymptotically approach their respective $n\ \Delta f$, which physically means that the polymer-substrate interaction progressively diminishes to an ineffective value. The

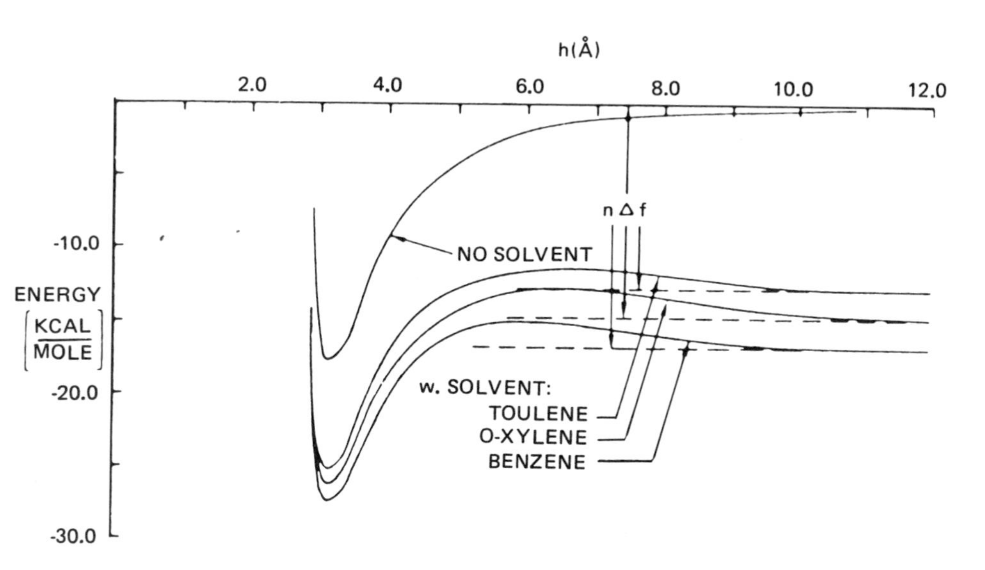

Figure 16 Pe-NaCl interaction energy, scaled to 9 CH_2 units, as a function of h, where (a,b,Θ,φ,φ) = ($d'/4$,$d'/4$,90°,45°,0°), in vacuo and in the solvents benzene, toluene, and o-xylene.

solvation energy, in turn, is gradually restored to its maximum because of ever-decreasing steric overlap of the solvent layer with the substrate surface. There is an energy of activation, measured relative to $n\ \Delta f$, over which the polymer must pass to reach the equilibrium position. This phenomenon is the result of the competing interactions of decreasing solvation energy and increasing polymer-substrate energy as the chain approaches the substrate.

The modification of the interfacial energetics to account for an expenditure of energy needed to displace solvent molecules has resulted in an energy of activation that is characteristic of a given polymer-solvent-substrate combination. This activation barrier is related to the relative abilities of the solvent molecules to initially dissolve the polymer, and subsequently become deadsorbed from the polymer molecule and permit its nucleation upon the substrate surface.

CONSIDERATION OF INTERCHAIN INTERACTIONS

The monoclinic chain packing of PE near the interface is unique to the NaCl substrate, the normal orthorhombic lattice being favored otherwise. Furthermore, the coincidence of interchain spacing and distance between rows of like ions in the $(001)_{NaCl}$ plane suggests that lattice matching, although not necessary for epitaxy, is a necessary criterion for inducing thermodynamically less stable crystal forms. Experiments of Rickert and Baer [28] involving POM/alkali halide systems lend further support to this concept.

The energetic preference of molecular chain alignment along cationic rows would intrinsically seem to favor the monoclinic structure. Figure 17 is a schematic representation of monoclinic and orthorhombic forms of epitactic growth of PE on NaCl. However, one cannot rationalize the existence of a given lattice solely in terms of interfacial energetics, but must also consider interchain interactions. In the event that a higher-energy packing mode should exhibit molecular specificity with a given surface template, one might expect that structure to result if the polymer-substrate energies are comparable in magnitude to the interchain energies. In this section, a scheme for computing the effects of lateral chain-chain interactions in an initially deposited plane monolayer of PE molecules spread above, and parallel to, a $(001)_{NaCl}$ surface plane and oriented in the $\langle 110 \rangle_{NaCl}$ direction is presented. The competition between the interchain forces and substrate-based forces is assessed within the framework of the molecular model.

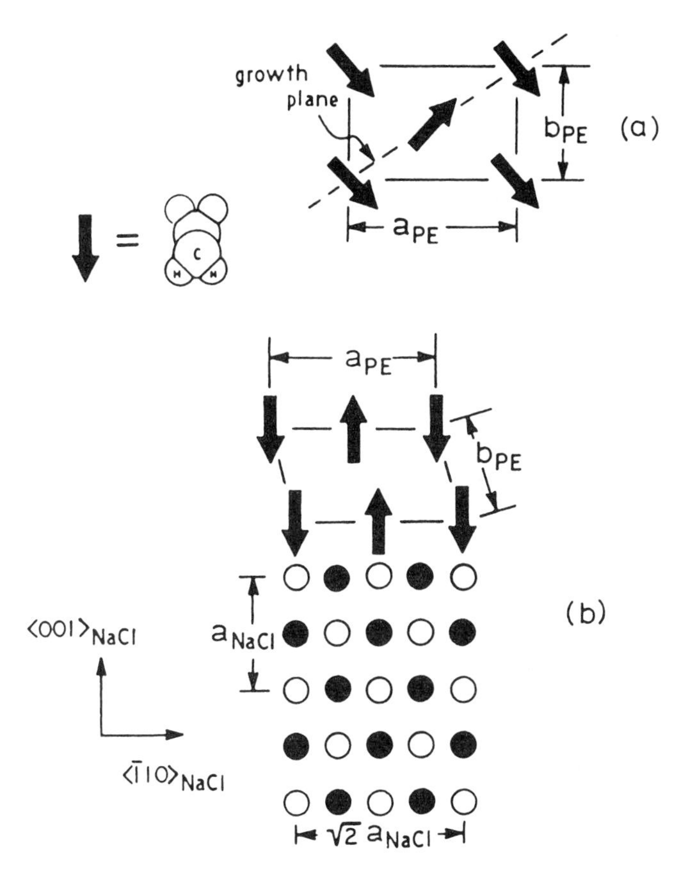

Figure 17 Cross-sectional views of (a) orthorhombic packing of PE chains, and (b) monoclinic packing of PE chains near the (001) NaCl surface plane.

Geometry and General Computational Scheme

The interchain and chain-surface relative geometry of a monolayer, depicted in Figure 18, is characterized by ϕ, the angle between the polymer chain axes and the $\langle 100 \rangle$ substrate direction; h, the distance between the surface (001) plane of substrate force centers and monolayer plane; s, the interchain spacing, which is taken to be the same for all adjacent pairs; t, the displacement, in the axial direction, of alternative polymer molecules; and (μ_1, μ_2), the axial rotations of a given adjacent chain pair. Seven PE chains, each consisting of 10 CH_2 units, placed in the all-*trans* conformation, have been used in the simulation calculations.

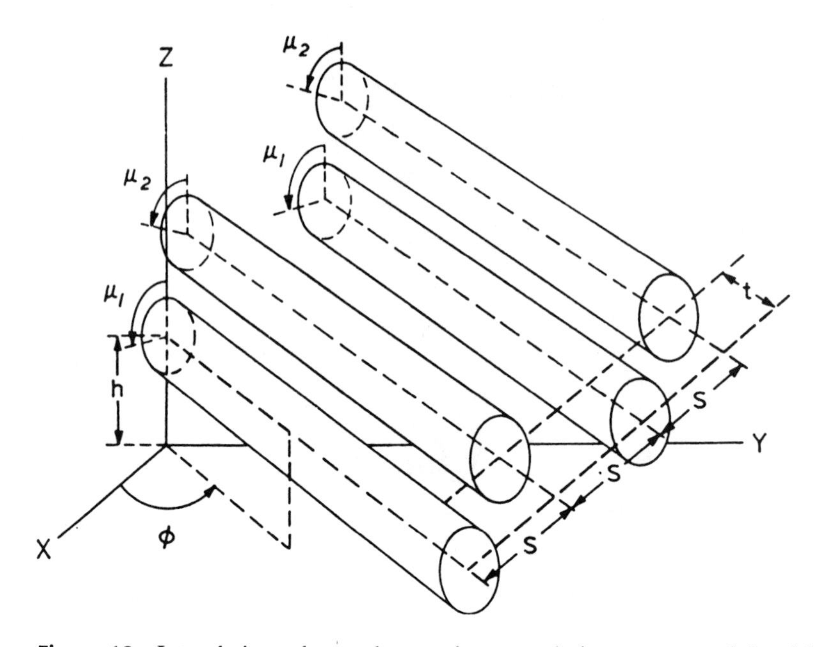

Figure 18 Interchain and monolayer-substrate relative geometry defined by positional and orientational parameters (ϕ,h,μ_1,μ_2,s,t). The x and y axes are in the surface (001) plane of ionic charges, where the origin is at a positive ion and the x axis is in the ⟨001⟩ substrate direction. The monolayer plane, defined as the plane containing the polymer chain axes, is always parallel to the xy plane.

For a given (ϕ, h, μ_1, μ_2, s, t), the total energy of the system can be computed if the polymer-atom and substrate-ion coordinates are known in a common frame of reference. Initially, the atomic coordinates of a section of a single polymer chain, whose axis is coincident with the $+x$-axis were left undisturbed or rotated μ_1 degrees. From these coordinates, a monolayer of seven chains with intermolecular spacing s, was generated at a height of h above the substrate surface.

Molecular Energetics

Rather than summing individual atom-pair interactions, a simpler method of calculating the total interchain energy, for a given geometry, was utilized. The approach consisted of decomposing a rational helix into linear sets of equally spaced atoms. Each atom on a particular line is of the same species and all lines are parallel to the helix axis. Conceptually the problem is reduced to summing over interchain line-line pairs. If n_H is the number of helices, each of which contains n_r rows of force centers, then L, the total number of line-line interactions of the monolayer, is given by

$$L = \frac{n_H(n_H - 1)n_r^2}{2}. \tag{26}$$

For seven chains of all-*trans* PE, $L = 756$. By comparison, A, the total number of atom-atom interactions for the same number of chains, each of which contains n_A atoms, is given by

$$A = \frac{n_H(n_H - 1)n_A^2}{2}. \tag{27}$$

If each model chain contains 10 CH_2 units, then $n_A = 30$ and $A = 18{,}900$. Thus, assuming a line-line interaction can be presented by a relatively simple closed-form expression, summing over lines is decidedly more efficient than summing over individual atom pairs. Expressions for the line-line dispersion, repulsive, and electrostatic potential energies have been developed as described below.

If the dispersion-repulsive potential energy of a pair of force centers, i and j, each in separate parallel rows of interactomic spacing ζ, and separated by

$$r_{ij} = [D^2 + (j\zeta - i\zeta - \Delta)^2]^{1/2} = \zeta[d^2 + (j - i - b)^2]^{1/2},$$

as depicted in Figure 19, where $d = D/\zeta$ and $b = \Delta/\zeta$, is represented by a Mie potential

$$U_{ij} = \epsilon n(m - n)^{-1}\left(\frac{r_{ij}^0}{r_{ij}}\right)^{2m} - \epsilon m(m - n)^{-1}\left(\frac{r_{ij}^0}{r_{ij}}\right)^{2n}, \tag{28}$$

where $-\epsilon$ is the minimum energy at the interatomic equilibrium separation, r_{ij}^0, then the total interaction between the rows, both of which contain N force centers, are separated by D, and have experienced a relative translation of Δ in the direction of a line, is given by

$$V = \epsilon n(m - n)^{-1}\left(\frac{r_{ij}^0}{\zeta}\right)^{2m}\sum_{i,j}^{N}[d^2 + (j - i - b)^2]^{-m}$$
$$- \epsilon m(m - n)^{-1}\left(\frac{r_{ij}^0}{\zeta}\right)^{2n}\sum_{i,j}^{N}[d^2 + (j - i - b)^2]^{-n}. \tag{29}$$

To avoid the task of directly performing the double summations, denoted by $\Omega_m^N(d,b)$ and $\Omega_n^N(d,b)$, McCullough and Hermans [29] have transformed the sums to rapidly converging series that can be truncated after a few terms, with essentially no loss of accuracy, the result being a closed-form expression.

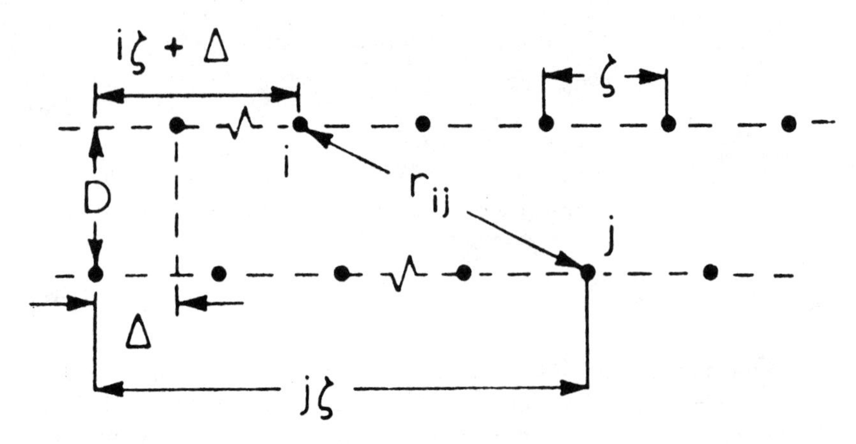

Figure 19 Relative geometry of two parallel rows of force centers, both having the same interatomic spacing, ζ, and characterized by interrow distance, D, and relative axial displacement, Δ.

If N is sufficiently large, as is the case for atomic rows in a fold period of crystalline PE, the geometrical factor per force center, $\Phi_m(d,b)$, for a row pair, defined as $\lim_{N\to\infty} \Omega_m^N N^{-1}$, is given by

$$\Phi_m(d,b) = \pi C_m(0) d^{1-2m} + \sum_{p=1}^{\infty} p^{2m-1} A_m(pd)\cos(2p\pi b), \tag{30}$$

where

$$A_m(pd) = 2\pi(pd)^{1-2m} \exp[-2\pi pd)] \sum_{k=0}^{m-1} (k!)^{-1} C_m(k)[4\pi pd)]^k, \tag{31}$$

$$C_m(k) = \binom{2m-k-2}{m-1} 2^{-2m+2}. \tag{32}$$

Using $n = 3$ and $m = 6$, which corresponds to the 6-12 potential, the interrow potential energy per M linear force centers, can be written as

$$V = M\epsilon\left[\left(\frac{r_{ij}^0}{\zeta}\right)^{12} \Phi_6(d,b) - 2\left(\frac{r_{ij}^0}{\zeta}\right)^{6} \Phi_3(d,b)\right], \tag{33}$$

where $\epsilon = (1/2)A_{ij}(r_{ij}^0)^{-6}$, which results from insisting that $dU_{ij}/dr_{ij} = 0$ at $r_{ij} = r_{ij}^0$. As reported earlier, A_{ij}'s the dispersion energy coefficients, and r_{ij}^0's, for several atom pairs, have been tabulated (see Table 2). In this work, ζ is the chain repeat distance, which, for all-*trans* PE, is 2.53 Å.

After considerable manipulation, Eq. (30), for $m = 3$ becomes

$$\Phi_3(d,b) = \frac{3\pi}{8d^5} + \left(\frac{3\pi}{4d^5} + \frac{3\pi^2}{2d^4} + \frac{\pi^3}{d^3}\right)\exp(-2\pi d)\cos(2\pi b) + \left(\frac{3\pi}{4d^5} + \frac{3\pi^2}{d^4} + \frac{4\pi^3}{3}\right)\exp(-4\pi d)\cos(4\pi b) + \cdots . \quad (34)$$

Similarly,

$$\Phi_6(d,b) = \frac{63\pi}{256d^{11}} + \left(\frac{63\pi}{128d^{11}} + \frac{63\pi^2}{64d^{10}} + \frac{7\pi^3}{8d^9} + \frac{7\pi^4}{16d^8} + \frac{\pi^5}{8d^7} + \frac{\pi^6}{60d^6}\right) \times \exp(-2\pi d)\cos(2\pi b) + \left(\frac{63\pi}{128d^{11}} + \frac{63\pi^2}{32d^{10}} + \frac{7\pi^3}{2d^9} + \frac{7\pi^4}{2d^8} + \frac{2\pi^5}{d^7} + \frac{8\pi^6}{15d^6}\right) \times \exp(-4\pi d)\cos(4\pi b) + \cdots . \quad (35)$$

The rapid convergence of these series is illustrated in Table 8, a listing of the values of the first three successive terms of Φ_3 and Φ_6 for $(d,b) = (1,0)$. It is apparent that third-order terms are negligible in comparison with the preceding two, especially for $m = 3$. However, to provide sufficient precision over all values of (d,b), the first three terms were retained in the computation.

The method of computing energies, developed above, can be used to compute interactions between rows of force centers for any interatomic potential having a r_{ij}^{-q}-dependence for which $q > 2$. Unfortunately, the last stipulation excludes coulombic interactions, for which $q = 1$. Therefore, a different approach must be used in developing interline potentials for rows of partial atomic charges.

The electrostatic potential energy E_{ij}^{e} of a single partial charge Q_i at a distance D from a line j of identical partial charges Q_j having spacing ζ in an adjacent polymer molecule whose conformation is that of a rational

Table 8 Numerical Values of the First Three Terms of Φ_m for $m = 3$ and 6 where $(d,b) = (1,0)$

Term	Value ($m = 3$)	Value ($m = 6$)
1	1.18	0.774
2	0.0902	0.254
3	0.000543	0.00555

helix can be obtained by summing coulombic interactions over all *i-j* pairs. The resulting problem is concerned with a collinear array of partial charges in an all-*trans*, 100-Å segment of a PE molecule (onefold period). There will be about 81 CH_2 groups in this length and, therefore, linear arrays containing about 41 hydrogen or carbon atoms per array.

The direct sum is insensitive to translation of i in the chain axis direction over stereochemically allowed values of d. The slow rate of decrease of the reduced potential, $\epsilon E^{e}_{ij}\zeta(Q_iQ_j)^{-1}$, where ϵ is the dielectric constant of the hydrocarbon media, taken to be 2, shown in Figure 20, suggests that this energy is a long-ranged contribution which is overcome, however, by the fact that the polymer molecule, as a whole, is electrically neutral. That is, for large chain separations, the effects of positive and negative rows in a given chain segment, on an atom in another chain, tend to cancel one

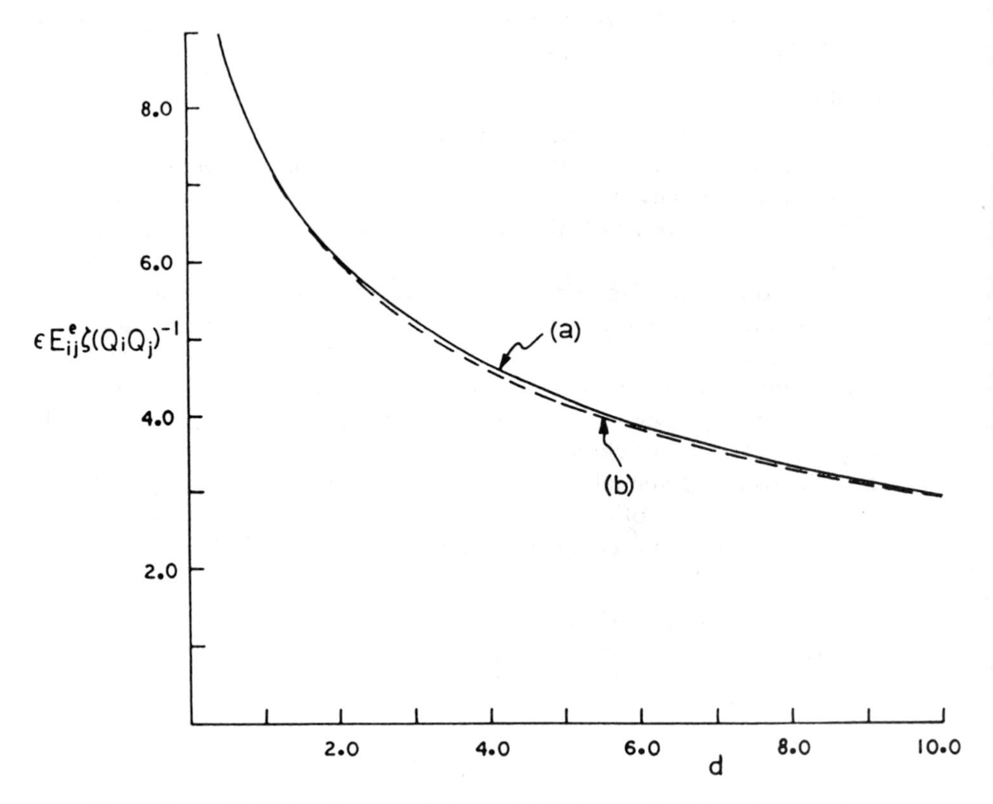

Figure 20 $\epsilon E^{e}_{ij}\zeta(Q_iQ_j)^{-1}$ versus d. Curve (a) is the result of directly summing over 41 charges. Curve (b) is a plot of Eq. (37), obtained by integrating over a continuous charge distribution having an equivalent linear density.

another. In a related study [30], it has been shown that E^e_{ij}, being essentially independent of Δ, can be written as

$$E^e_{ij} = \frac{2Q_iQ_j}{\epsilon\zeta} \ln\left\{\frac{l_0}{2D} + \left[\left(\frac{l_0}{2D}\right)^2 + 1\right]^{1/2}\right\}, \tag{36}$$

where l_0, the fold period, can be expressed as $(X_A - 1)\zeta$, X_A being the number of atoms on a line within the fold period. In reduced form this expression becomes

$$\epsilon E^e_{ij}\zeta(Q_iQ_j)^{-1} = 2 \ln\{\tfrac{1}{2}(X_A - 1)d^{-1} + [\tfrac{1}{4}(X_A - 1)^2 d^{-2} + 1]^{1/2}\}, \tag{37}$$

which, when plotted (see Fig. 20), shows a near-superposition on the exact curve. This reduced potential energy plays a role analogous to Φ_m. That is, it serves as a geometrical factor per coulombic force center for a given atomic row pair. Overall, the net electrostatic energy V', per M force centers, associated with a row pair is

$$V' = ME^e_{ij}, \tag{38}$$

and the total energy is $V + V'$.

Results

The spatial variables (h, μ_1, μ_2, s, t) were varied for $\phi = 45°$ in increments of (0.2 Å, 10°, 10°, 0.2 Å, 0.42 Å), respectively, over the complete range of (μ_1, μ_2, t) for PE. Rotational and translational symmetry was utilized to avoid redundant geometries. Three distinct minima are noted on the potential energy surface. Each of the three corresponding monolayers has one geometrical parameter in common, namely s. Successive PE chains prefer to be situated over successive rows of Na^+ ions regardless of axial translation or rotation.

The absolute minimum, which occurs at (h, μ_1, μ_2, s, t) = (3.00 Å, 40°, 130°, 4.00 Å, 0.00 Å), is −37.0 kcal/mole/chain (relative to infinite separation of the polymer chains and substrate surface).

The interchain geometry, depicted in Figure 21(a), is somewhat similar to that as exists in the growth plane of orthorhombic PE in that zigzag planes of successive molecules are perpendicular to each other, although the spacings s and setting angles (μ_1, μ_2) are different.

Another minimum, at (h, μ_1, μ_2, s, t) = (3.00 Å, 11°, 11°, 4.00 Å, 0.00 Å), is −34.1 kcal/mole/chain deep. The relative geometry, seen in Figure 21(b), consists of successive chains whose zigzag planes are parallel to, and repeat units in phase with, each other. The zigzag planes are tilted 79° relative to the substrate surface. This minimum, however, is at the bottom of a broad trough resulting in negligible axial rotational hindrance. μ_2, for

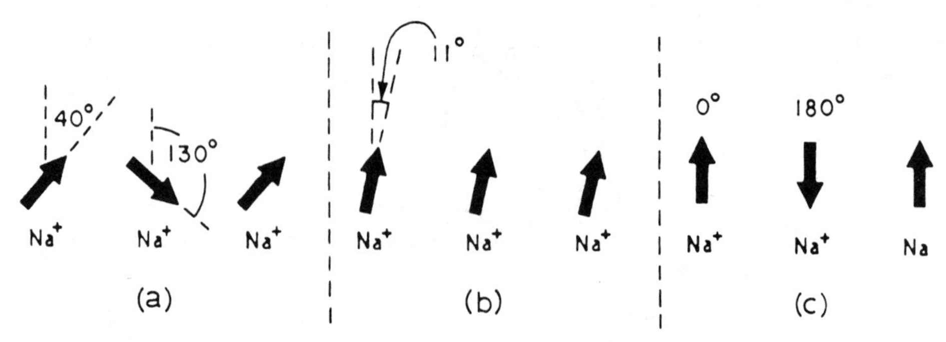

Figure 21 Interchain and polymer-substrate relative geometry corresponding to energy minima of (a) −37.0, (b) −34.1, and (c) −32.7 kcal/mole/chain. The ⟨110⟩ direction is normal to the plane of the paper. Intervening Cl^- rows have been omitted for simplicity. The arrows have the same meaning as in Figure 17.

example, could be increased to 101° with the expenditure of only 0.2 kcal/mole/chain. Such a variation of μ_2 results in a monolayer of chains that have the same setting angles as exists in the growth plane of the orthrombic lattice, but with a shortened intermolecular spacing s.

Lastly, a potential energy minimum, −32.7 kcal/mole/chain deep, is observed for (h, μ_1, μ_2, s, t) = (3.20 Å, 0°, 0°, 4.00 Å, 1.26 Å). The intermolecular packing mode, identical to that of the growth plane in the monoclinic lattice of PE near the NaCl inerface, as indexed by Wellinghoff et al. [31], is depicted in Figure 21(c). Figure 22(a) illustrates interchain potential energy contours, plotted in kcal/mole/7 chains above this relative minimum (−13.1 kcal/mole/chain), in the *ts* plane for which $\mu_1 = \mu_2 =$ 0°. The favorable interchain axial stagger of one-half repeat unit (t = 1.26 Å) is evident in this figure. Somewhat disturbing, however, is the unrealistically low interchain spacing of 3.60 Å. "Turning on" the substrate force field results in the interchain plus polymer-substrate energy contours seen in Figure 22(b), where h = 3.20 Å is the monolayer-substrate surface equilibrium distance. On comparing the two energy maps, it should be noted, first, that the substrate perturbation has increased the interchain spacing so as to bring the polymer axes in coincidence with cationic rows while maintaining the adjacent-chain axial stagger of 1.26 Å. However, in so doing, the breadth of the energy trough, in the t direction, has greatly increased, making it easier for chains to undergo relative axial translations. This, of course, is to be expected because the substrate, in straining the monolayer beyond the equilibrium spacing, has weakened the intermolecular forces.

The interplay between polymer-polymer and polymer-substrate forces is seen in Figure 23, which shows potential energy contours in the *hs* plane

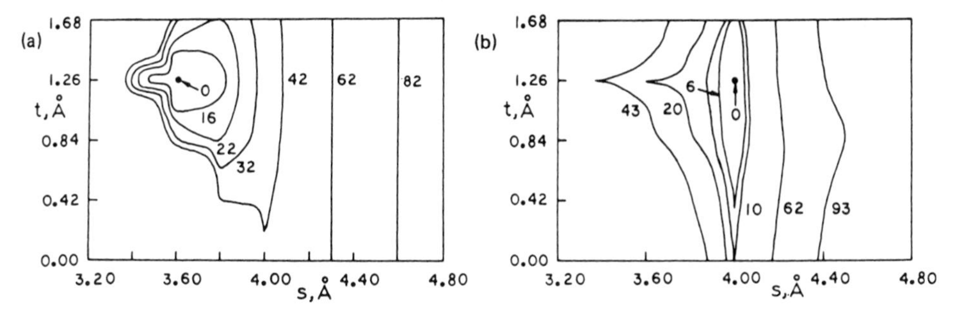

Figure 22 (a) Interchain potential energy contours, plotted in kcal/mole/7 chains above the relative minimum, in the *ts* plane, where $(\mu_1,\mu_2) = (0°,0°)$. (b) Interchain plus polymer-substrate potential energy contours for the same situation as in (a), where h = 3.20 Å.

for which $(\mu_1, \mu_2, t) = (0°, 0°, 1.26$ Å). At $h = 3.20$ Å, the substrate induces the chains to align themselves above, and parallel to, rows of Na^+. If, however, the monolayer is displaced to a height of 3.40 Å above the surface, the strained lattice is permitted to relax to the interchain spacing indicated in Figure 22(a) owing to a diminished polymer-substrate interaction.

SUMMARY

A successful theory of epitaxial energetics can subsequently be incorporated into a *molecular* theory of the heterogeneous nucleation of polymers since the rate constants are a function of the activation energy [32]. Hoffman and Lauritzen [33] have, some time ago, developed a theory similar to the theory of the homogeneous nucleation of polymers [33–37]. This kinetic description is not based upon explicit molecular interactions, but

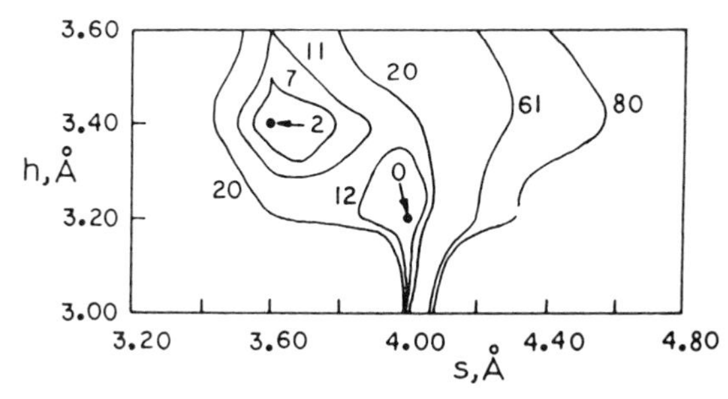

Figure 23 Interchain potential energy contours, plotted in kcal/mole/7 chains above the relative minimum, in the *hs* plane, where $(\mu_1,\mu_2,t) = (0°,0°,1.26$ Å).

utilizes, as basic parameters, macroscopic factors such as surface free energies and heats of fusion.

The theory of the epitaxial crystallization of polymers described in this chapter is based upon fundamental pairwise interactions between atoms in the polymer, substrate, and solvent molecules. Similar in principle to the methods of theoretical conformational analysis [11], the approach, in general, consists of minimizing the total energy of the polymer-solvent-substrate system with respect to the interfacial geometry in order to locate the most favorable intermolecular arrangements. It should be emphasized that the work presented herein deals with *static* interactions. Molecular dynamics [33] represents a means to extend this work to such processes as crystalline phase transitions and kinetics.

Molecular Modeling of Amorphous Polymer Surface Deposition

The discussions of this chapter have focused upon characterizing the *ordered* interaction of linear polymer chains with surfaces having periodic structure using molecular modeling. A more difficult case to model is that in which entropy dominates the deposition process leading to unordered polymer structure at the surface. Within the class of amorphous polymer deposition processes there are different degrees of difficulties in setting up the modeling problem. If the surface is ordered, or at least regular in structure, and the unordered polymer chain conformation is due to high conformational flexibility and/or solvent effects, proximate interfacial ensembles can be constructed. That is an ensemble of low-energy conformers of short-chain segments can be constructed and allowed to minimize their interfacial energy with the surface as a function of deposition geometry. By generating a large number of chain-segment conformers, and computing the corresponding minimum interfacial energies, it is possible to build up a statistical profile of the deposition geometry of the initial polymer deposition layer. Clearly, the sum of the chain-segment conformational energy and the corresponding deposition energy will control the probability of realizing any particular deposition geometry.

When the substrate surface is not ordered, and/or solvent needs to be explicitly considered, and/or multiple deposition layers of polymer are to be modeled, the investigator must turn to simulation modeling. There are two simulation methods: Monte Carlo [39] and molecular dynamics [38]. Both of these simulation methods permit the sampling of the molecular geometry–molecular energy interrelationship for a user-defined substrate, polymer, solvent system. Molecular dynamics has the advantage of taking time explicitly into consideration as a modeling variable. Of course, the length of time that can be simulated is very short, normally not more than

200 ps. Consequently, interfacial processes that occur more slowly are lost in the molecular dynamics simulation. Said another way, one cannot be sure that a represenative "picture" of the polymer deposition process is currently included in a molecular dynamics trajectory. Still, the explosive increase in low-cost computing power is bound to expand the practical molecular dynamics simulation times and/or size of the molecular system being modeled. Monte Carlo simulation modeling does not take time into consideration, and its realistic represenation of the behavior of a chemical system is a function of the number of the states of the system that are sampled. Most current Monte Carlo simulations involve at least 10^6 state samplings.

Reports of simulation studies of amorphous polymer structure near surfaces are very limited. The small number of published studies of the simulation of amorphous polymers have focused upon modeling high-density, isolated polymer materials [40–42]. Nevertheless, the simulation formalisms to expand these types of analyses to include interfacial and solvent contributions is straightforward. The major concern is computing resources and not theoretical or computational methods. Thus, it is reasonable to expect simulation modeling of amorphous polymer growth on surfaces to be a significant topic of research in surface science and polymer coatings research in the immediate years ahead.

REFERENCES

1. J. Willems and I. Willems, *Experientia*, *13*, 465 (1957).
2. J. Willems, *Experientia*, *23*, 409 (1967).
3. B. Wunderlich, *Macromolecular Physics*, Academic Press, New York (1973).
4. K. A. Mauritz, E. Baer, and A. J. Hopfinger, *J. Polym. Sci. Macromol. Rev.*, *13*, 1 (1978).
5. F. Tunistra and E. Baer, *J. Polym. Sci. Polym. Lett. Ed.*, *8*, 861 (1970).
6. C. M. Balik, S. K. Tripathy, and A. J. Hopfinger, unpublished.
7. S. Y. Hobbs, *Nature (London) Phys. Sci.*, *234*, 12 (1971).
8. A. J. Groszek, *Proc. R. Soc. London Ser. A*, *314*, 473 (1970).
9. E. A. Boucher, *J. Mater. Sci.*, *8*, 146 (1973).
10. J. A. Koutsky, A. G. Walton, and E. Baer, *J. Polym. Sci. Polym. Lett. Ed.*, *5*, 177, 185 (1967).
11. A. J. Hopfinger, *Conformational Properties of Macromolecules*, Academic Press, New York (1973).
12. U. Burkert and N. L. Allinger, *ACS Monogr.*, 177 (1982).
13. J. A. Pople and D. L. Beveridge, in *Approximate Molecular Orbital Theory*, McGraw-Hill, New York (1970).

14. F. W. J. Oliver, in *Handbook of Mathematical Functions* (M. Abramowitz and I. A. Stegun, eds.), National Bureau of Standards, Washington, DC (1967).
15. K. A. Mauritz, E. Baer, and A. J. Hopfinger, *J. Polym. Sci.: Macromol. Rev.*, *13*, 1 (1978).
16. C. M. Balik, S. K. Tripathy, and A. J. Hopfinger, *J. Polym. Sci.-Phys. Ed.*, *20*, 2017 (1982).
17. C. M. Balik, S. K. Tripathy, and A. J. Hopfinger, *J. Polym. Sci.-Phys. Ed.*, *20*, 2012 (1982).
18. G. Carrazolo and M. Mammi, *J. Polym. Sci.*, *Part A*, *1*, 965 (1963).
19. R. F. Brunel and K. van Bibber, in *International Critical Tables*, Vol. 3, p. 27 (E. W. Washburn, ed.), McGraw-Hill, New York (1928).
20. J. J. Gilman, *J. Appl. Phys.*, *31*, 2208 (1960).
21. F. van Zeggeren and G. C. Benson, *J. Chem. Phys.*, *26*, 1077 (1957).
22. P. R. Baukema and A. J. Hopfinger, *J. Polym. Sci.-Phys. Ed.*, *20*, 399 (1982).
23. British Association for the Advancement of Science, *Mathematical Tables*, vol. X, *Bessel Functions*, Part II, Cambridge University Press, Cambridge, England (1960).
24. A. J. Groszek, *Proc. R. Soc. London Ser. A*, *314*, 473 (1970).
25. K. A. Mauritz, E. Baer, and A. J. Hopfinger, *J. Polym. Sci. Polym. Phys. Ed.*, *11*, 2185 (1973).
26. F. Tunistra and E. Baer, *J. Polym. Sci. Polym. Lett. Ed.*, *8*, 861 (1970).
27. K. A. Mauritz and A. J. Hopfinger, *J. Phys. Chem.*, *80*, 706 (1976).
28. S. Rickert and E. Baer, *J. Appl. Phys.*, *47*, 4304 (1976).
29. R. L. McCullough and J. J. Hermans, *J. Chem. Phys.*, *45*, 1941 (1966).
30. K. A. Mauritz and A. J. Hopfinger, *Surface Sci.*, *55*, 81 (1976).
31. S. H. Wellinghoff, F. Rybnikar, and E. Baer, *J. Macromol. Sci. Phys.*, *B10*, 1 (1974).
32. S. Glasstone, K. J. Laidler, and H. Eyring, *The Theory of Rate Processes*, McGraw-Hill, New York (1941).
33. J. D. Hoffman and J. I. Lauritzen, *J. Res. Natl. Bur. Stand.*, *65A*, 297 (1961).
34. J. D. Hoffman, J. J. Weeks, and W. W. Murphy, *J. Res. Natl. Bur. Stand.*, *64A*, 67 (1959).
35. J. D. Hoffman, *SPE Trans.*, *4*, 315 (1964).
36. J. D. Hoffman, J. I. Lauritzen, E. Passaglia, G. S. Ross, L. J. Frolen, and J. J. Weeks, *Kolloid Z.*, *231*, 564 (1967).
37. F. Gornick and J. D. Hoffman, *Ind. Eng. Chem.*, *58*, 41 (1966).
38. J. A. McCammon and S. Harvey, *Dynamics of Proteins and Nucleic Acids*, Cambridge University Press, London (1987).
39. M. Mezei, P. K. Mehrotra, and D. L. Beveridge, *J. Am. Chem. Soc.*, *107*, 2339 (1985).
40. D. N. Theodorou and U. W. Suter, *Macromolecules*, *18*, 1467 (1985).
41. D. N. Theodorou, *Macromolecules*, *21*, 1391 (1988).
42. T. Pakula and S. Geyler, *Macromolecules*, *21*, 1665 (1988).

part five

QUANTUM MECHANICAL TECHNIQUES

10

Introduction to Quantum Mechanical Techniques for the Study of the Electronic Structure of Polymers

Jean-Marie André
Facultés Universitaires Notre-Dame de la Paix
Namur, Belgium

SUMMARY AND OBJECTIVES

It is only in recent years that increases in the speed, available storage space, and reliability, and decreases in the size and price of computers, coupled with the development of more powerful software packages, have enabled quantum mechanical techniques to begin making significant contributions to understanding the structures and properties of noncrystalline polymers. Although this field is still in its infancy, it holds great promise for the future. This chapter (a) gives a general introduction to quantum mechanics and to quantum mechanical techniques; (b) describes and compares specific types of quantum mechanical techniques useful for studying noncrystalline polymers-for example, ab initio programs such as GAUSSIAN 82, 86, and 88, approximate ab initio methods, and semiempirical techniques such as CNDO/i's, INDO, MINDO/i's, NDDO, MNDO, AMi's, and extended Hückel. We will not try to give a full inventory of all the available techniques, since our objective is to give, by a selection of the techniques and by a restricted choice of examples, a comprehensive review of the ideas and of the evolution of the ideas underlying the various methodologies. Further examples of applications in the field of polymer chemistry are given in the next chapters. The important field of molecular mechanical calculations has been covered in a previous chapter.

HARTREE-FOCK SELF-CONSISTENT FIELD MODEL

Principles

For molecules and polymers, the standard approach is based on the Hartree-Fock (HF) theory. In this independent model, a single electron moves in the field of the fixed nuclei and in the mean coulombic and exchange field of all the other electrons. A set of molecular orbitals (MOs) describes the occupied and unoccupied one-electron wave functions. In molecular quantum chemistry, the relative energies of the molecular orbitals are drawn as single levels which are at most doubly occupied by a pair of electrons of opposite spins.

The basic postulate of quantum mechanics is to replace the concept of trajectories (full knowledge of positions and impulsion of all particles at each time) by a wave function depending on the coordinates of all the particles in the system. The wave function has no physical significance but its square has the meaning of a density probability (electron density in the case of electrons). Since the wave function has no direct physical meaning, it can only be calculated. The rules were given by Schrödinger. In the case of stationary states, the wave function Ψ is the eigenfunction of the Schrödinger equation with the observable energy E as eigenvalue:

$$H\Psi = E\Psi,$$

where the Hamiltonian operator H describes all the energy component of the system: usually, kinetic energy of all particles and electrostatic interactions (attraction between nuclei and electrons and repulsion between nuclei or between electrons). Inference from classical mechanics postulates that the kinetic energy operator T has the following form for a particle of mass m:

$$T = -\frac{h^2}{8\pi^2 m}\Delta = -\frac{1}{2}\Delta = -\frac{1}{2}\nabla^2 \text{ in atomic units,}$$

and that the electrostatic interaction operator V between two charged particles q_i and q_j located at an interdistance r_{ij} is

$$V = \frac{q_i q_j}{r_{ij}}.$$

Unfortunately, as in classical mechanics, an exact solution of the Schrödinger equation $H\Psi = E\Psi$ cannot be found when the system contains three particles or more. It is already the case of the simplest many-electron atom He or of the simplest molecule H_2^+. Approximation methods must therefore be introduced. Historically, those are the Born-Oppenheimer approximation [1] for polyatomic systems and the orbital approximation for polyelectronic ones.

In the Born-Oppenheimer approximation, one uses the fact that the mass of a nucleus is at least 1836 heavier than the electron mass and one neglects the kinetic operators of the nuclei with respect to the kinetic operators of the electrons. Physically, this corresponds to studying the motion of the electrons in the field of fixed nuclei. This leads to an effective electronic energy $E(\mathbf{R})$ which is a parametric function of the nuclear coordinates and is used as the potential for the nuclear motion. This approximation defines the important concept of potential energy surfaces of primary importance in conformational analysis of macromolecules. The equilibrium structure corresponds to the absolute minimum of the energy surface. If there are other potential minima, isomeric forms can be observed.

In the orbital theory, one assumes that a one-electron operator can be defined to represent the motion of a single electron in the field of all the fixed nuclei and the averaged field of interaction with all the electrons. The logic of these principles is summarized in Table 1. As indicated in Figure 1 in the case of the helium atom, it corresponds to describe the observable many-electron levels by nondirectly observable one-electron levels which can be doubly occupied by electrons of spin α and β. This transformation is made through the HF methodology. This technique assumes an independent particle model in the form of the wave function and applies the variation theorem. The simplest independent particle wave function for an n electron system has the form of a simple product of n orbitals. However, such a form does not include the spin properties and does not satisfy the Pauli exclusion principle. Thus, in order to take into account this requirement, one uses an antisymmetrized product of n spin orbitals called the Slater determinant. A four-variable spin orbital $\Phi_i(\mathbf{r},\omega)$ is the product of a spatial orbital $\phi_i(\mathbf{r})$ and a spin function $\sigma(\omega)$ which can be either $\alpha(\omega)$ or $\beta(\omega)$. The HF equation is the one-electron equation which allows us to obtain the best one-electron orbitals, i.e., the orbitals introduced in an antisymmetrized product of orbitals gives the best total energy of the system (the lowest and the closest to the experimental value in agreement with the variation principle) which can be obtained from such a wave function.

The ideas underlying these equations have been suggested independently by the Englishman Hartree [2], the Russian Fock [3] and the American Slater [4]. The basic equations are reviewed in Table 2. We note the

Table 1 Standard Ideas of Quantum Chemical Methods

Shrödinger many-electron equation $H\Psi = E\Psi$
Not soluble exactly except for one-electron system: H, He^{+}, Li^{2+}, ...
Ψ = wave function of the whole electron-nuclear system
= function of $3n$ space variables + n spin variables
H = Hamiltonian of the whole system includes
kinetic energy of all electrons
kinetic energy of all nuclei
attraction between all electrons and all nuclei
repulsion between all electrons
repulsion between all nuclei
$\Downarrow$ Born-Oppenheimer approximation
$\Downarrow$ Orbital theory
Hartree-Fock one-electron equation $h\phi = \epsilon\phi$
ϕ = one-electron spatial wave function (orbital)
= function of 3 space variables
h = one-electron Hamiltonian includes
kinetic energy of a single electron
attraction of a single electron with all nuclei
Coulomb repulsion and exchange interaction of a single electron with the averaged electron density

underlying physical meaning of the HF field; one determines the motion of a single electron characterized by the kinetic operator $-h^2/8\pi^2 m\nabla^2$ in the electrostatic field of fixed nuclei $-\Sigma_A Z_A|\mathbf{r} - \mathbf{R}_A|^{-1}$. The electron evolves also in an interaction field due to its repulsion Coulomb operator $\Sigma_j^{occ} 2J_j(\mathbf{r})$ with average electron density $\Sigma_j \phi_j(\mathbf{r}')\phi_j(\mathbf{r}')$ and exchange interaction $-\Sigma_j^{occ} K_j(\mathbf{r})$. Note that the Coulomb operator is local; i.e., it does not depend on the orbital on which it is acting, while the exchange operator is a nonlocal one: the form of the exchange operator actually depends on the orbital $\phi_i(\mathbf{r})$ on which it operates. Notice that the HF equation is highly implicit and nonlinear. Indeed, in order to get the set of orbitals $\{\phi_j(\mathbf{r})\}$, one has to know the form of all the interaction operators $\{J_j(\mathbf{r})\}$ and $\{K_j(\mathbf{r})\}$ which explicity depend on the solutions $\{\phi_j(\mathbf{r})\}$. As indicated by Figure 2, the solution is iterative. Starting from a guess of the actual solutions $\{\phi_j^{(0)}(\mathbf{r})\}$ (for example, from a strict independent particle solution), one determines a zero-order guess of the interaction operators $\{J_j^{(0)}(\mathbf{r})\}$ and $\{K_j^{(0)}(\mathbf{r})\}$ and gets a new approximation of the eigenfunctions of the HF equation $\{\phi_j^{(1)}(\mathbf{r})\}$. These solutions are in turn used as "input" for defining better interaction potentials $\{J_j^{(1)}(\mathbf{r})\}$ and $\{K_j^{(1)}(\mathbf{r})\}$ and getting an even more correct set of solutions $\{\phi_j^{(2)}(\mathbf{r})\}$. The process is repeated until it has con-

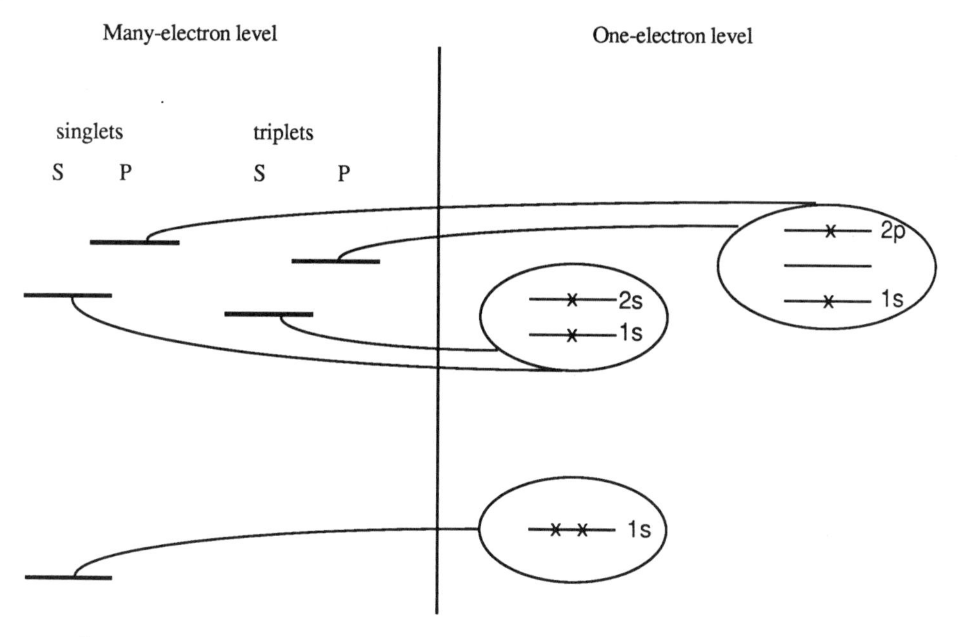

Figure 1 Many-electron versus orbital concepts as illustrated in the case of the helium atom.

Table 2 Main Equations of Hartree-Fock Theory

Wave function	$\Psi(\mathbf{r}_1,\omega_1,\mathbf{r}_2,\omega_2,\mathbf{r}_i,\omega_i,\ \ldots) = \det\lvert\Phi_j(\mathbf{r}_1,\omega_1)\Phi_k(\mathbf{r}_2,\omega_2)\ \cdots\ \Phi_m(\mathbf{r}_i,\omega_i)\ \cdots\rvert$
Hartree-Fock equation	$h(\mathbf{r})\phi_i(\mathbf{r}) = \epsilon_i\phi_i(\mathbf{r})$
One-electron Hartree-Fock operator	$h(\mathbf{r}_i) = -\frac{1}{2}\nabla^2 - \sum_A Z_A\lvert\mathbf{r} - \mathbf{R}_A\rvert^{-1} + \sum_j^{occ}\{2J_j(\mathbf{r}) - K_j(\mathbf{r})\}$
Coulomb repulsion operator	$J_j(\mathbf{r})\phi_i(\mathbf{r}) = \left\{\int dr'\ \phi_j(\mathbf{r}')\phi_j(\mathbf{r}')\lvert\mathbf{r} - \mathbf{r}'\rvert^{-1}\right\}\phi_i(\mathbf{r})$
Exchange interaction operator	$K_j(\mathbf{r})\phi_i(\mathbf{r}) = \left\{\int dr'\ \phi_j(\mathbf{r}')\phi_i(\mathbf{r}')\lvert\mathbf{r} - \mathbf{r}'\rvert^{-1}\right\}\phi_j(\mathbf{r})$

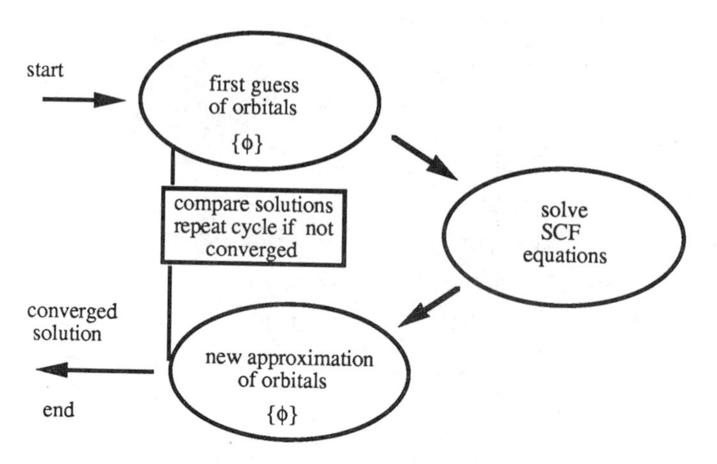

Figure 2 Sketch of a SCF process.

verged in the sense that the "input" orbitals $\{\phi_j^{(n-1)}(\mathbf{r})\}$ are identical to the "output" solutions $\{\phi_j^{(n)}(\mathbf{r})\}$ within a predefined threshold. In this scheme, a self-consistent field (SCF) is created. This is the reason why the method is sometimes refered to as the SCF method.

Numerical solutions of the atomic HF equations have been obtained in a systematic way by Hartree's father after his retirement. Hartree's book, *The Calculations of Atomic Structures*, is actually dedicated to his father with the touching words: "To the memory of my father William Hartree. in recollection of our happy cooperation in work on the calculation of atomic structures." If numerical solutions are rather easily obtained for atomic structures, they are hardly extended to the case of polyatomic systems. Other methodologies have to be used. The most common is the expansion of the atomic and molecular orbitals in terms of basis functions as independently suggested by Roothaan [5] and Hall [6] in their important 1951 papers.

LCAO Expansion of Hartree-Fock Orbitals

As mentioned in the previous section, the exact solution of the HF equation has been obtained by numerical procedures for most of the atoms and some diatomic molecules. It is presently standard to solve the HF equation by expanding the orbitals $\phi_i(\mathbf{r})$ in terms of a set of basis functions $\{\chi_p(\mathbf{r} - \mathbf{P})\}$ centered on the various nuclei of the molecule. The misleading terminology LCAO has historically been introduced since, in the first calculations, molecular orbitals (MO) were obtained by linear combinations of atomic orbitals (LCAO). Today, analytical solutions of atomic orbitals are also commonly derived by using linear combination of basis functions (LCBF).

When applying the SCF scheme, one obtains the so-called SCF-MO-LCAO methodology. Its equations are summarized in Table 3. The process would be exact if we were able to handle infinite expansions. In practice, the expansion is limited to a finite number of basis functions. The actual process is thus to replace the search of orbitals in the full three-dimensional space by the calculation of a finite number of LCAO coefficients.

This scheme implies the calculations of integrals or matrix elements between the N basis functions (N is called the size of the basis set). For atomic structures, all the integrals are one-centered. For a molecule of M atoms, the N^2 (or more exactly the $N(N + 1)/2$ when taking into account the Hermitian redundancy) kinetic integrals $T_{pq} = \int \chi_p(\mathbf{r})\{(-1/2)\nabla^2\}\chi_q(\mathbf{r})\, dv$ can be either one- or two-centered, the $MN(N + 1)/2$ nuclear attraction $V_{pq|A} = \int \chi_p(\mathbf{r})|\mathbf{r} - \mathbf{R}_A|^{-1}\chi_q(r)\, dv$ can be one-, two- or three-centered. Finally, one finds $[N(N + 1)/2][N(N + 1)/2 + 1]/2$, approximately

Table 3 Main Equations of SCF-MO-LCAO Methodology

Orbital	$\phi_j(\mathbf{r}) = \sum_p c_{jp}\chi_p(\mathbf{r} - \mathbf{P})$
Secular system	$\sum_p c_{jp}(h_{pq} - \epsilon_j S_{pq}) = 0$
Secular determinant	$\lvert h_{pq} - \epsilon S_{pq}\rvert = 0$
Integrals	$S_{pq} = \int \chi_p(\mathbf{r})\chi_q(\mathbf{r})\, dv$
	$h_{pq} = \int \chi_p(\mathbf{r})h(\mathbf{r})\chi_q(\mathbf{r})\, dv$
	$= T_{pq} - \sum_A Z_A V_{pq\vert A}$
	$+ \sum_r \sum_s D_{rs}\{2(pq\vert rs) - (pr\vert qs)\}$
	$T_{pq} = \int \chi_p(\mathbf{r})\left\{-\frac{1}{2}\nabla^2\right\}\chi_q(\mathbf{r})\, dv$
	$V_{pq\vert A} = \int \chi_p(\mathbf{r})\lvert\mathbf{r} - \mathbf{R}_A\rvert^{-1}\chi_q(\mathbf{r})\, dv$
	$(pq\vert rs) = \iint \chi_p(\mathbf{r})\chi_q(\mathbf{r})\lvert\mathbf{r} - \mathbf{r}'\rvert^{-1}\chi_r(\mathbf{r}')\chi_s(\mathbf{r}')\, dv\, dv'$
Results →	$\epsilon_1, \epsilon_2, \epsilon_3, \ldots$
→	$D_{rs} = \sum_j^{\text{occ}} c_{jr}c_{js}$

$N^4/8$, one-, two-, three-, or four-centered repulsion integrals

$$(pq|rs) = \iint \chi_p(\mathbf{r})\chi_q(\mathbf{r})|\mathbf{r} - \mathbf{r}'|^{-1}\chi_r(\mathbf{r}')\chi_s) \, dv \, dv'$$

One notes that for large molecules (large basis sets), a huge number of integrals has to be calculated, which provokes the bottleneck of first principles (ab initio) SCF-MO-LCAO calculations on large polymeric systems. In principle, the choice of the type of basis functions is arbitrary. It is in practice conditioned by its computational adequacy.

The first idea would be to use as basis functions the hydrogenoid orbitals, eigenfunctions of the hydrogenoid equation. Since they are solutions of a Hermitian operator, they form on each atomic site a set of orthogonal functions. Their radial part is exponentially decreasing $\propto \exp(-\zeta r)$, but depends on both the principal quantum number n and the angular quantum number l, which makes their use practically untractable.

Slater-type orbitals (STO), proposed in 1930 [7], avoid the latter difficulty. Their radial part is also exponentially decreasing $\propto \exp(-\zeta r)$, but depends only on the principal quantum number n. Thus, $2s$ and $2p$ orbitals have the same radial dependence. The price paid is that they are no longer eigenfunctions of a Hermitian operator and thus are no longer orthogonal to each other even on a single site. The $1s$ and $2s$ STOs have a nonorthogonal behavior. A further difficulty is that, if analytical solutions for all one- and two-center (overlap, kinetic, nuclear attraction and repulsion) integrals have been published [8], tractable expressions for three- and four-center integrals are not available. These orbitals have thus been largely used for describing diatomic molecules but not larger molecules or polymeric systems.

Nowadays, following the suggestion of Boys in 1950 [9], one uses in molecular calculations Gaussian type orbitals (GTO). Instead of the exponential radial dependence $\propto \exp(-\zeta r)$, they exhibit a Gaussian-like decrease $\propto \exp(-\alpha r^2)$. All integrals can be analytically solved. However, the GTO does not correctly reproduce the STO or hydrogenoid cusp due to the attractive potential discontinuity of $-Z/r$ at $r = 0$. Thus, the derivative of a GTO is zero at its origin while, at the same point, the derivative of a STO has a nonzero negative value. One tries to avoid this difficulty by representing each STO "atomiclike" orbital by a contracted Gaussian function expanded in several primitive Gaussians. In the popular STO-3G basis set, each STO is thus represented by a least-squared fixed combination of three Gaussians.

In order to limit adequately the number of basis functions, one defines three main types of basis sets: minimal, extended, and polarized. In a minimal basis set, the orbitals are those which are occupied in the free atoms, i.e., $1s$ orbitals for hydrogen atoms, $1s$, $2s$, $3*2p$ orbitals for carbon, nitrogen, and oxygen. STO-KG expansions are such minimal basis sets. They are standardly available for atoms ranging from the first to the fourth row in the original papers issued from Pople's group since 1969 [10]. In an extended basis set, each valence orbital is supplemented by a delocalized orbital of the same quantum numbers: for example, $1s$ and $1s'$ orbitals for hydrogen and $1s$, $2s$, $2s'$, $3*2p$, $3*2p'$ orbitals for carbon, nitrogen, and oxygen. This allows for more radial flexibility in the spatial representation of the molecular electron density and also corrects partially the deficient description of anisotropic situations by readjustment of diffuse and localized components through the SCF procedure. If one also doubles the core orbitals, the basis set is often referred to as double-zeta basis. In a polarized basis set, one adds to each valence orbital basis functions of higher angular quantum number; for example, $(2)p$ orbitals for hydrogen and $(3)d$ orbitals for carbon, nitrogen, and oxygen atoms. This permits small displacements of the center of the electronic charge distributions away from the atomic positions (charge polarization). A list of references for GTOs and tables of atomic functions can be found in the specialized literature [11,12].

A detailed analysis of the numerical structure of the independent particle model allows us to draw the following conclusions. First, the more flexible is the basis set, the lower the total energy optimized with respect to the variational parameters of the wave function will be, i.e., closest to the experimental limit.

As illustrated in Figure 3 by an SCF calculation on the helium atom where a single exponential orbital (STO) is used, the total atomic energy of the ground state is -2.8476563 au while using two STOs, a lower energy of -2.861660 is observed without any special optimization of the orbital exponents (ζ_1 and ζ_2 arbitrarily selected to be 1.45 and 2.85). If such an optimization is pursued [13], a fully optimized energy of -2.8616701 au is obtained for optimized orbital exponents: $\zeta_1 = 1.4461$ and $\zeta_2 = 2.8622$. With such a double-zeta basis set, the numerical error due to the limitation of the basis set is thus only 0.0000099 au = 0.00026 eV with respect to the HF limit -2.8616800 au [14]. The HF limit is in error compared to the experimental value by 0.042104 au = 1.14573 eV. This error is inherent in the HF methodology. The form of the wave function (Slater determinant) does not prevent, in the singlet state, two electrons residing in the same region of space. Such a physical situation is highly disfavored by the strong electrostatic repulsion between particles of same charges. It is the so-called correlation error. In agreement with the variation theorem, the calculated total energy will be higher than the experimental value, a result easily explained since the HF method gives too much weight to these destabilizing situations found when the electrons are close.

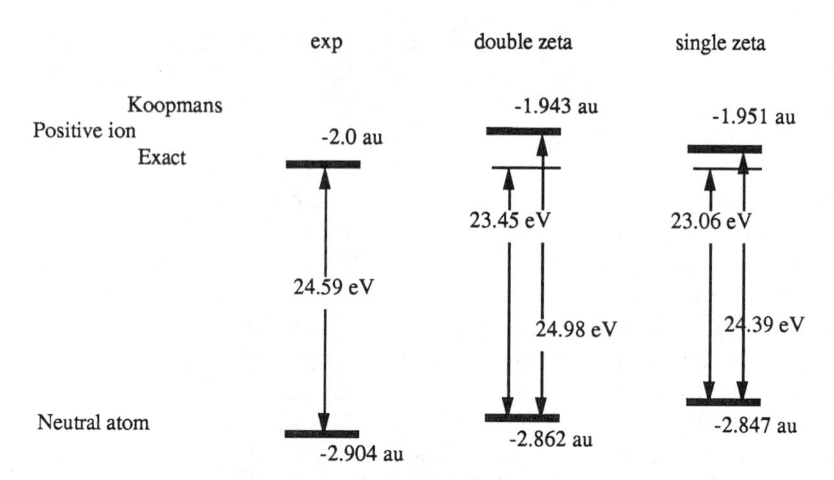

Figure 3 Experimental and calculated electronic states of He and He^+.

The previous results also indicate some interesting trends for the ionization potentials. In the Koopmans' approximation [15], the wave function of the ionized species is constructed from the orbitals of the neutral system. In the present case, its energy is the negative of the HF 1*s* orbital and is overestimated compared to the exact value (ground state energy of the hydrogenoid $He^+ = -2.0$ au) as imposed by the variation theorem. The relaxation energy is a measure of the importance of the rearrangement effects of the electron density of the neutral system under the ionization process (1.53 eV for a double-zeta basis set, 1.33 eV for the single-zeta basis set as illustrated by Figure 3).

SCF-MO-LCAO EQUATIONS: THE MACHINERY

Ab initio versus Semiempirical Schemes

As indicated previously, SCF-MO-LCAO methods require a computational effort proportional to the fourth power of the size of the basis set. We call ab initio or nonempirical the techniques where all electrons are considered and all the necessary integrals are calculated. Due to the large number of integrals to be evaluated in large molecules or polymeric models of chemical significance, very important computing resources are required. In order to get faster and less demanding computational algorithms, it is sometimes imperative to reduce the number of electrons or integrals to be computed. For instance, we can reduce the number of electrons to be considered by "freezing" n_c internal or σ-electrons into fixed cores of reduced charge $z_A = Z_A - n_c$ and/or approximate several electron integrals or group of integrals. In this case, one generally uses semiempirical methods to evaluate the rest of the integrals from experimental data and thus to correct the errors introduced by the neglect of the noncalculated integrals. In general, semiempirical methods are dependent on the square power of the basis set used. Table 4 schematizes the transformation from nonempirical (ab initio) to semiempirical schemes. The numerical treatment of these non- or semiempirical techniques requires computer resources ranging from PCs to supercomputers, depending on the size of the basis sets used and the importance of the approximations introduced.

In the simple Hückle method [16], only π-electrons of purely conjugated organic molecules are taken into account and all interactions except for the nearest neighbors are neglected. This method only involves the one-center parameter α and the bond parameter β. It has gained considerable success by its simplicity and by the soundness of the involved approximations. Around 1950, Mulliken [17] and Wolfsberg and Helmholtz [18] suggested a very simple type of Hückel parametrization which is easily extendable to the case of σ-bonded systems. In 1963, Hoffmann [19] took up this method and applied it to a large variety of saturated and conjugated organic molecules. This is the well-known extended Hückel method. The ZDO-like methods which neglect the zero differential overlaps no longer have their justification in an entire evaluation of the matrix elements but more precisely in a close analysis of the energy terms. The LCAO matrix elements are split into their kinetic, nuclear attraction, and repulsion contributions and some integrals are either neglected or approximated by careful procedures. This is the case of the Pariser-Parr-Pople [20,21] methods for π-electrons and CNDO (complete neglect of differential overlap) methods [22] when considering all-valence electrons. Fundamentally, these procedures attempt to reproduce by means of one-electron models the one-electron results of the many-electron HF theory. Sometimes, however, in order to force agreement with experiment, one fits some integrals of a group of integrals to experimental data. This very often results in an implicit inclusion of a part of the correlation effects, although in a very indirect way. In this manner, it is obvious that the computational work is strongly decreased, but, what is an important feature, the incorporation of semiempirical data compensates for part of the errors due to very crude approximation. The price to be paid is that, since the semiempirical processes do use parameters of experimental origin, the parametrized methods are generally good mostly for the properties on which they have been fitted. It is common to select the adequate semiempirical methods depending on the type of property which is to be investigated. Several textbooks and collections of papers cover the field of semiempirical calculations [23].

Table 4 Ab initio Versus Semiempirical Schemes

Ab initio ⇓ Consider the full one-electron Hartree-Fock operator for a system of n electrons:

$$h(\mathbf{r}_i) = -\frac{1}{2}\nabla^2 - \sum_A Z_A|\mathbf{r} - \mathbf{R}_A|^{-1} + \sum_j^{\text{occ}} \{2J_j(\mathbf{r}) - K_j(\mathbf{r})\}.$$

Compute from ab initio (first principles) all integrals of the full one-electron basis set in the selected basis set $\{\chi_p\}$.

N^4-dependence ⇒ Compute full interaction of the averaged electron density:

$$\iint \rho(\mathbf{r})|\mathbf{r} - \mathbf{r}'|^{-1}\rho(\mathbf{r}')\, dv\, dv'$$
$$= \sum_p \sum_q \sum_r \sum_s D_{pq}D_{rs}$$
$$\times \iint \chi_p(\mathbf{r})\chi_q(\mathbf{r})|\mathbf{r} - \mathbf{r}'|^{-1}\chi_r(\mathbf{r}')\chi_s(\mathbf{r}')\, dv\, dv'.$$

Semiempirical ⇓ Split the n-electron systems into n_c core electrons (c) and n_v valence electrons (v).
Consider an effective valence one-electron Hartree-Fock operator for a system of n_v valence electrons:

$$\mathrm{h}(r_i) = -\frac{1}{2}\nabla^2 - \sum_A Z_A|\mathbf{r} - \mathbf{R}_A|^{-1} + \sum_j^{nc} \{2J_j^c(\mathbf{r}) - K_j^c(\mathbf{r})\}$$
$$+ \sum_j^{nv} \{2J_j^v(\mathbf{r}) - K_j^v(\mathbf{r})\}$$
$$= h^c(\mathbf{r}_i) + \sum_j^{nv} \{2J_j^v(\mathbf{r}) - K_j^v(\mathbf{r})\}.$$

Evaluate matrix elements of the one-electron Hamiltonian from experimental data (extended Hückel theory).

or

Keep them as parameters (Hückel theory).

or

Use ZDO as in CNDO-like theories.

N^2-dependence ⇒ Keep only the diagonal part of the interaction of the averaged electron density:

$$\iint \rho(\mathbf{r})|\mathbf{r} - \mathbf{r}'|^{-1}\rho(\mathbf{r}')\, dv\, dv'$$
$$= \sum_p \sum_r D_{pp}D_{rr} \iint \chi_p(\mathbf{r})\chi_p(\mathbf{r})|\mathbf{r} - \mathbf{r}'|^{-1}\chi_r(\mathbf{r}')\chi_r(\mathbf{r}')\, dv\, dv.$$

A summary of some available ab initio and current semiempirical procedures is given in Table 5. They are classified into self-consistent or iterative procedures, where the one-electron operator $h(\mathbf{r})$, even when effective, and hence the matrix elements h_{pq}, explicitly depend on the electron density $\rho(\mathbf{r})$, and non-self-consistent or noniterative techniques where a global evaluation of the matrix elements is performed.

All-Electron ab initio Methods

As already mentioned, ab initio calculations require the definition of a basis set only. All-electron integrals are then calculated for a specified geometry. If the geometry is not known, guesses of the probable geometries can be obtained through a full or partial search of energy minima of the potential surface. Several ab initio SCF-MO-LCAO programs have been written to date and are largely distributed. Among those are POLYATOM [24], IBMOL or KGNMOL [25], GAUSSIAN [25], and HONDO [27] series. HF ab initio calculations provide useful data on ground state properties like bond lengths and angles, molecular conformations and internal rotation barriers, electron densities, and dipole moments. Examples of results for small monomers of interest in polymer chemistry are given in

Table 5 Empirical, Semiempirical, and ab initio Methods of Quantum Chemistry

Electrons considered	Non-self-consistent methods noniterative procedures Global evaluation of matrix elements h_{pq}	Self-consistent methods iterative procedures Approximations in the calculation of matrix elements $h_{pq} = h_{pq}\{\rho(\mathbf{r})\}$
All electrons		ab initio Clementi → IBMOL series Pople → GAUSSIAN series Dupuis → HONDO series
Valence ($\sigma + \pi$) electrons	Extended Hückel (EH) Hoffmann	ZDO, CNDO, INDO, MINDO, NDDO, MNDO, AMi Pople Dewar
π-Electrons	Hückel theory (HMO) Hückel	Pariser-Parr-Pople (PPP) technique Pariser-Parr Pople

Tables 7 to 11 for several types of basis sets (minimal, extended, and polarized). HF ab initio methods are also useful for calculating ionization potentials, excitation energies, and dipole strengths. In these latter cases, care must be taken to correctly balance relaxation and correlation effects. HF ab initio calculations can also give insight into full conformational potential energy surfaces and molecular associations, two fields of great interest in polymer chemistry.

Table 6 Principles of ZDO Techniques

Core matrix elements

$$h^c_{pq} = \int \chi_p(\mathbf{r}) h^c(r) \chi_q(\mathbf{r})\, dv = \int \chi_p(\mathbf{r}) \left[-\frac{1}{2}\nabla^2 - \sum_A z_A |\mathbf{r} - \mathbf{R}_A|^{-1} \right] \chi_q(\mathbf{r})\, dv$$

Diagonal

$$h^c_{pp} = T_{pp} - z_A \int \chi_p(\mathbf{r}) |\mathbf{r} - \mathbf{R}_A|^{-1} \chi_p(\mathbf{r})\, dv \Rightarrow U^A_{pp}$$
$$- \sum_{B \neq A} z_B \int \chi_p(\mathbf{r}) |\mathbf{r} - \mathbf{R}_B|^{-1} \chi_p(\mathbf{r})\, dv \Rightarrow \sum_{B \neq A} V^B_{pp}$$

Nondiagonal

$$h^c_{pq} = T_{pq} \cong \beta_{AB} S_{pq}$$

Repulsion and exchange matrix elements $\int \chi_p(\mathbf{r}) \left[\sum_j^{occ} 2J_j(\mathbf{r}) - K_j(\mathbf{r}) \right] \chi_q(\mathbf{r})\, dv$

Diagonal

$$\int \chi_p(r) \left[\sum_j^{occ} 2J_j(\mathbf{r}) - K_j(\mathbf{r}) \right] \chi_p(r)\, dv$$
$$= \sum_j^{occ} \sum_r \sum_s c_{jr} c_{js} \{2(pp|rs) - (pr|ps)\}$$
$$= \sum_j^{occ} \sum_r \{2c^2_{jr}(pp|rr) - c^2_{jp}(pp|pp)\}$$
$$= \sum_r Q_r(pp|rr) - \frac{Q_p}{2}(pp|pp)$$

Nondiagonal

$$\int \chi_p(\mathbf{r}) \left[\sum_j^{occ} 2J_j(\mathbf{r}) - K_j(\mathbf{r}) \right] \chi_q(\mathbf{r})\, dv$$
$$= \sum_j^{occ} \sum_r \sum_s c_{jr} c_{js} \{2(pq|rs) - (pr|qs)\}$$
$$= -\sum_j^{occ} c_{jp} c_{jq} (pp|qq)$$
$$= -\frac{l_{pq}}{2}(pp|qq)$$

Table 7 Examples of Molecular Bond Lengths (in Å)

	CH_4	C_2H_2		C_2H_4		C_2H_6	
Method	CH	CC	CH	CC	CH	CC	CH
Semiempirical							
EH	1.02						
CNDO/1	1.11	1.213		1.342		1.520	
CNDO/2	1.116	1.198	1.093	1.320	1.110	1.476	1.117
INDO		1.20	1.10	1.31	1.11	1.46	1.12
MINDO/2	1.100	1.19	1.069	1.337	1.093	1.524	1.109
MINDO/3		1.191	1.076	1.31	1.10	1.486	1.108
NDDO		1.225	1.057	1.410	1.080	1.527	1.094
MNDO		1.19	1.051	1.335	1.089	1.521	1.109
AM1	1.112	1.195	1.061	1.325	1.098	1.501	1.117
Ab Initio							
Minimal	1.083	1.168	1.065	1.306	1.082	1.538	1.086
Extended	1.083	1.188	1.051	1.315	1.074	1.542	1.084
Polarized	1.084	1.186	1.057	1.317	1.076	1.527	1.086
Experimental	1.092	1.203	1.061	1.339	1.085	1.531	1.096

Table 8 Examples of Barriers to Internal Rotation (in kcal-mol^{-1})

	CH_3—CH_3	CH_3—CH=CH_2
Semiempirical		
EH	3.04	
CNDO/1	1.47	
CNDO/2	2.18	
MINDO/2	1.89	
NDDO	2.39	
MNDO	1.0	1.20
AM1	1.25	
Ab Initio		
Minimal	2.9	1.25
Extended	2.7	
Polarized	3.0	
Experimental	2.88	1.99

Table 9 Examples of Conformational Energy Differences (in kcal mol^{-1})

		Semiempirical				ab initio		
Molecule	Conformation	CNDO	MINDO/1	NDDO	AM1	Minimal	Extended	Expt.
n-Butane	*Trans*/*gauche*				0.73	0.9	0.8	0.77
1-Butene	Skew/*cis*					0.8	0.8	0.2
1,3-Butadiene	*trans*/*gauche*	0.63	−1.43	−1.42		1.8	3.5	1.7–2.5

Table 10 Examples of Relative Energies of Structural Isomers (in kcal mol^{-1})

Molecule	Semiempirical			ab initio			
	EH	MNDO	AM1	Minimal	Extended	Polarized	Expt.
C_3H_4							
Propyne	0	0	0	0	0	0	0
Allene	23.4	2.5	2.7	17.1	3.4	2.0	1.0–1.6
Cyclopropene		26.9	31.4	30.0	39.8	26.5	21.6–21.7
C_4H_6							
trans-1,3-Butadiene	0	0	0	0	0	0	0
2-Butyne	−9.6	−4.1	2.1	−12.8	3.6	6.5	6.9–8.5
Cyclobutene		2.0	15.9	−12.5	18.0	13.4	8.8–11.2
1,2-Butadiene	11.1	4.6	7.2	8.5	11.3	12.8	12.5
1-Butyne		7.2	7.6	−5.3	9.2	13.0	13.2
Methylenecyclopropane		8.9	17.8	5.8	25.6	20.5	21.7
Bicyclo[1.1.0]butane		35.1	48.2	11.6	45.7	31.4	23.1–25.6
1-Methylcyclopropene		24.7	34.8	17.3	43.5	33.0	32.0
C_4H_8							
Isobutene	0	0	0	0	0	0	0
trans-2-Butene	2.0	−3.1	−2.1	−0.2	0.5	0.2	1.1
cis-2-Butene	5.9	−0.2	−1.0	1.4	2.0	1.8	2.2
cis-1-Butene	6.5	2.4	1.6	4.4	3.3	3.6	3.9
Methylcyclopropane	27.3			−2.2	15.0	3.6	9.7
Cyclobutane	62.0	−9.9	0.2	−17.6	10.4	10.0	10.8
C_4H_{10}							
Isobutane	0	0	0	0	0	0	0
trans-*n*-Butane	2.3	−2.9	−1.7	0.1	1.3	0.4	1.6–2.1
C_6H_6							
Benzene	0	0	0				0
Fulvene	25.6	32.4	40.7				27.0–27.7
$C_{10}H_8$							
Naphtalene	0						0
Azulene	32.3						32.6

ALL-VALENCE-ELECTRON SEMIEMPIRICAL METHODS

ZDO-Like Methods

The ZDO approximation assumes that the differential overlap density of two distinct atomic orbitals is zero:

$$\rho_{pq}(\mathbf{r}) = \chi_p(\mathbf{r})\chi_q(\mathbf{r}) = \delta_{pq}\chi_p^2(\mathbf{r}).$$

Note that, in an all-valence-electron calculation, this assumption is satisfied for the orbitals centered on the same nucleus due to the symmetry requirement between orbitals of different angular and magnetic quantum numbers like $2s$, $2p_x$, $2p_y$, and $2p_z$. It is, in general, not rigorously satisfied for orbitals centered on different nuclei, which would imply $\chi_p(\mathbf{r})$ or $\chi_q(\mathbf{r})$ or both to be identically zero at each $\mathbf{r}$. Inversely, the ZDO assumption automatically implies that the total integrated overlap S_{pq} is zero, as it would have been within an orthogonalized basis set:

$$S_{pq} = \int \rho_{pq}(\mathbf{r})\, dv = \int \chi_p(\mathbf{r})\chi_q(\mathbf{r})\, dv = \int \delta_{pq}\chi_p^2(\mathbf{r})\, dv = \delta_{pq}.$$

The ZDO basis sets are thus assumed to form orthonormal basis sets. In the same logic, nuclear attraction integrals are neglected if they imply orbitals which fill the ZDO requirement:

$$\begin{aligned} V_{pq|A} &= \int \chi_p(\mathbf{r})|\mathbf{r} - \mathbf{R}_A|^{-1}\chi_q(\mathbf{r})\, dv \\ &= \int \chi_p(\mathbf{r})\chi_q(\mathbf{r})|\mathbf{r} - \mathbf{R}_A|^{-1}\, dv \\ &= \int \rho_{pq}(\mathbf{r})|\mathbf{r} - \mathbf{R}_A|^{-1}\, dv = 0 \qquad \text{if } p \neq q. \end{aligned}$$

In the case of two-electron integrals $(pq|rs)$, the most attractive feature of the ZDO scheme is that it only retains the N^2 one- and two-center Coulomb integrals (pq/qq) and that it neglects the $N^4 - N^2$ other integrals, mainly three- and four-centered, but also one- and two-center exchange integrals of the type $(pq|pq)$,

$$\begin{aligned} (pq|rs) &= \iint \chi_p(\mathbf{r})\chi_q(\mathbf{r})|\mathbf{r} - \mathbf{r}'|^{-1}\chi_r(\mathbf{r}')\chi_s(\mathbf{r}')\, dv\, dv' \\ &= \delta_{pq}\delta_{rs} \iint \chi_p(\mathbf{r})\chi_p(\mathbf{r})|\mathbf{r} - \mathbf{r}'|^{-1}\chi_r(\mathbf{r}')\chi_r(\mathbf{r}')\, dv\, dv' \\ &= \delta_{pq}\delta_{rs}(pp|rr), \end{aligned}$$

which reduces the computational effort from N^4 to N^2. The main equations of CNDO techniques are summarized in Table 6, where we have used

Table 11 Examples of Dipole Moments (in Debye)

Molecule	Semiempirical			ab initio				Expt.
	CNDO/2	MINDO/1	MNDO	AM1	Minimal	Extended	Polarized	
Propyne	0.43		0.12	0.40	0.50	0.69	0.57	0.75–0.78
Cyclopropene			0.42	0.36	0.55	0.55	0.57	0.45
Propene	0.36		0.04	0.23	0.25	0.30	0.29	0.36–0.37
Propane	0.00		0.01	0.01	0.02	0.06	0.06	0.08
But-1-yne-3-ene					0.37	0.45	0.46	0.4
Cyclobutene			0.08	0.17	0.05	0.07	0.03	0.13
1,2-Butadiene				0.33	0.39	0.35	0.40	
1-Butyne					0.54	0.67	0.65	0.80
Methylenecyclopropane					0.22	0.35	0.40	0.40
Bicyclo[1.1.0]butane			0.41	0.43	0.58	0.75	0.68	0.68
1-Methylcyclopropene					0.83	0.90	0.89	0.84
Isobutene	0.65				0.43	0.53	0.44	0.50
cis-2-Butene				0.14	0.13	0.13	0.26	
cis-1-Butene				0.30	0.34	0.34	0.44	
Methylcyclopropane					0.11	0.11	0.10	0.14
Isobutane	0.00				0.04	0.11	0.09	0.13
Toluene	0.22		0.06	0.27	0.25			0.31–0.36
Ammonia	2.08	2.13	1.76	1.85	1.79	2.17	1.95	1.47
Phenol	1.73	2.72	1.67	1.24	1.22			1.45–1.55
Fluorobenzene	1.66				0.93			1.60–1.66
Methanol	1.94	2.48	1.48	1.62	1.51	2.21	2.04	1.69–1.70
Water	2.08	2.79	1.78	1.86	1.73	2.44	2.22	1.82–1.85
Fluoromethane	1.66				1.16	2.25	2.06	1.85
Acetaldehyde	2.53	3.64	2.38	2.69				2.68–2.69
Formaldehyde	1.98	3.35	2.16	2.32	1.51	2.63	2.75	2.33–2.34
Benzaldehyde	2.50							2.72
Acetone	2.90	3.63	2.51	2.92				2.88–2.90

the definition of charges Q_p and bond orders l_{pq} in an orthogonalized basis set:

$$Q_p = \sum_j^{\text{occ}} 2c_{jp}^2, \qquad l_{pq} = \sum_j^{\text{occ}} 2c_{jp}c_{jq}.$$

When used as such, the previous approximation suffers from an important deficiency since it misses the requirement that the results should be rotationally invariant, i.e., should not depend on the way the molecule is localized is a Cartesian system. Figure 4(a) details this point by the example of a given repulsion integral $(2p_x'2p_y'|2s2s)$ calculated in two Cartesian systems of coordinates rotated by 45° in the xy plane. By expanding the orbital $2p_x'$ in one system of coordinate as $2p_x - 2p_y$ in the other system, and $2p_y'$ as $2p_x + 2p_y$, we easily note that this integral would have a zero value in the first system of axis while being nonzero in the other:

$$\begin{aligned}(2p_x'2p_y'|2s2s) &= 0 \qquad \text{by the ZDO approximation} \\ &= \frac{1}{2}\{(2p_x2p_x|2s2s) - (2p_y2p_y|2s2s) + (2p_x2p_y|2s2s) - (2p_y2p_x|2s2s)\} \\ &= \frac{1}{2}\{(2p_x2p_x|2s2s) - (2p_y2p_y|2s2s)\} \\ &\neq 0.\end{aligned}$$

The direct consequence is that, in order to force the rotational invariance, the electron repulsion integrals should (a) be independent of their directionality and (b) only depend on the nature of the two atoms on which they are centered as illustrated in Figure 4(b). This also leads to calculate the integrals from electron distributions replaced by their spherical s-type averages:

$$\begin{aligned}(pp|qq) &= \gamma_{AB} \qquad \text{if } p\text{-centered on } A \text{ and } q\text{-centered on } B, \\ V_{pp}^B &= V_{AB} \qquad \text{if } p\text{-centered on } A,\end{aligned}$$

a scheme which leads to the CNDO/1 matrix elements:

$$h_{pp} = U_{pp}^A + \left(\frac{Q_A - Q_p}{2}\right)\gamma_{AA} + \sum_{B\neq A}(Q_B\gamma_{AB} - V_{AB}),$$

$$h_{pq} = \beta_{pq}S_{pq} - \left(\frac{l_{pq}}{2}\right)\gamma_{AB}.$$

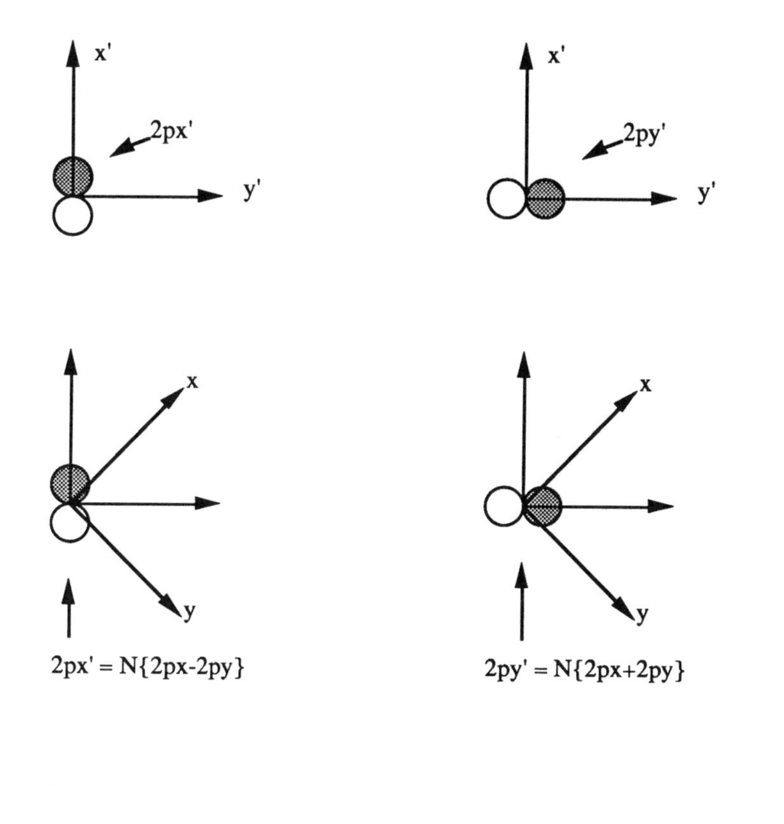

(a)

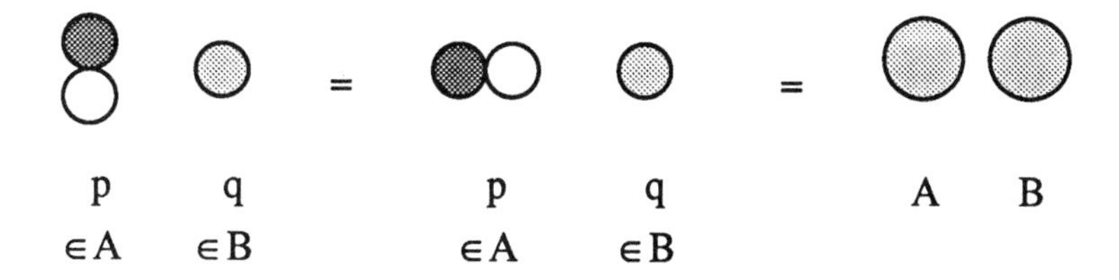

(b)

Figure 4 (a) Rotational invariance in ZDO scheme. (b) Invariance of electrostatic integrals with respect to orbital type.

The CNDO/1 scheme requires the explicit evaluation of the overlap integrals S_{pq}, γ_{AB}, and V_{AB} from knowledge of the molecular geometry and of an STO functional form of the basis functions. These integrals are calculated from Roothaan's one- and two-center integral tables [8] using the real (radial and angular) orbital dependence for the overlap integrals S_{pq} and s-like orbitals for the Coulomb integrals γ_{AB} and V_{AB} with Slater exponents obtained from Slater screening rules [7]. The one-center parameters U_{pp}^{A} are estimated from the valence ionization states of the corresponding atom, an approximation agreeing with Koopmans' theorem [15], which states that if one neglects the electronic relaxation the orbital energy is the negative of the corresponding ionization potential. Using general interaction HF rules for open-shell atomic systems, one gets

$$-I_p = \epsilon_p = \int \chi_p(\mathbf{r})\{h^c(\mathbf{r}) + \sum_{q \neq p} [J_q(\mathbf{r}) - K_p(\mathbf{r})]\}\chi_p(\mathbf{r})\, dv$$

with

$$\int \chi_p(\mathbf{r})h^c(\mathbf{r})\chi_p(\mathbf{r})\, dv = U_{pp}^{A},$$

$$\int \chi_p(\mathbf{r})J_q(\mathbf{r})\chi_p(\mathbf{r})\, dv = (pp|qq) = \gamma_{AA},$$

$$\int \chi_p(\mathbf{r})K_q(\mathbf{r})\chi_p(\mathbf{r})\, dv = (pq|pq) = 0 \qquad \text{by ZDO approximation.}$$

Thus, a valence state ionization energy can be expressed in terms of the core-effective charges z_A and of the one-center integrals U_{pp}^{A} and γ_{AA}:

$$-I_p = U_{pp}^{A} + \sum_{q \neq p} \gamma_{AA} = U_{pp}^{A} + (z_A - 1)\gamma_{AA},$$

which allows us to get the CNDO/1 semiempirical estimate of U_{pp}^{A}:

$$U_{pp}^{A} = -I_p - (z_A - 1)\gamma_{AA}.$$

The constant β_{pq} is evaluated as a sum of constant characteristic of atoms A and B which have been determined in order to fit minimal basis set ab initio results on diatomic molecules:

$$\beta_{pq} = \beta_{AB} = \frac{\beta_A^0 + \beta_B^0}{2}.$$

The CNDO/2 technique introduces two important modifications in the calculation of the diagonal CNDO matrix elements in order to better describe bond lengths and dipole moments. The first modification concerns the neglect of the "penetration integrals" $z_B\gamma_{AB} - V_{AB}$ to correct the

CNDO/1 description, which shows an erroneous slight binding of the $^3\Sigma_u$ dissociative state of the hydrogen molecule. In an elegant discussion, Pople [22] has shown that this incorrect binding is due to a poor balance between electron-electron repulsions and electron-nuclei attractions in the penetration integrals. The HF potential energy curve of this dissociative state is, according to standard principles,

$$E(^3\Sigma_u) = \epsilon_{1\sigma g} + \epsilon_{1\sigma u} - J_{1\sigma g-1\sigma u} + K_{1\sigma g-1\sigma u} + |\mathbf{R}_A - \mathbf{R}_B|^{-1},$$

and the CNDO/1 approximations of the various terms are

$$\begin{aligned}
\epsilon_{1\sigma g} &= \frac{\alpha + \beta}{1 + S} = \alpha + \beta, \\
\epsilon_{1\sigma u} &= \frac{\alpha - \beta}{1 - S} = \alpha - \beta, \\
\epsilon_{1\sigma g} + \epsilon_{1\sigma u} &= 2\alpha = 2h_{1s-1s} \\
&= U^A_{1s-1s} + \left(1 - \frac{1}{2}\right)\gamma_{AA} + (\gamma_{AB} - V_{AB}), \\
J_{1\sigma g-1\sigma u} &= \frac{1}{4}\iint \{1s_a(\mathbf{r}) + 1s_b(\mathbf{r})\}^2|\mathbf{r} - \mathbf{r}'|^{-1} \\
&\quad \times \{1s_a(\mathbf{r}') - 1s_b(\mathbf{r}')\}^2 \, dv \, dv' \\
&= \frac{\gamma_{AA} + \gamma_{AB}}{2}, \\
K_{1\sigma g-1\sigma u} &= \frac{1}{4}\iint \{1s_a^2(\mathbf{r}) - 1s_b^2(\mathbf{r})\}|\mathbf{r} - \mathbf{r}'|^{-1} \\
&\quad \times \{1s_a^2(\mathbf{r}') - 1s_b^2(\mathbf{r}')\} \, dv \, dv' \\
&= \frac{\gamma_{AA} - \gamma_{AB}}{2},
\end{aligned}$$

which leads to the specific CNDO/1 expression

$$\begin{aligned}
E(^3\Sigma_u) &= 2U^A_{1s-1s} + \gamma_{AA} \\
&\quad \Rightarrow \text{independent on the internuclear distance} \\
&\quad + \gamma_{AB} - 2V_{AB} + |\mathbf{R}_A - \mathbf{R}_B|^{-1} \\
&\quad \Rightarrow \text{dependent on the internuclear distance.}
\end{aligned}$$

The above-mentioned poor balance of the term depending on the internuclear distance $\gamma_{AB} - 2V_{AB} + |\mathbf{R}_A - \mathbf{R}_B|^{-1}$ leads to a net attraction between the two atoms due to the interpenetration of their electron clouds, even while the bond order between them is zero in the state considered.

The CNDO/2 approximation corrects these effects by neglecting the penetration integral $z_B\gamma_{AB} = V_{AB}$:

$$h_{pp} = U_{pp}^A + \left(Q_A - \frac{Q_p}{2}\right)\gamma_{AA} + \sum_{B\neq A}(Q_B - z_B)\gamma_{AB} + \sum_{B\neq A}(z_B\gamma_{AB} - V_{AB})$$

$$= U_{pp}^A + \left(Q_A - \frac{Q_p}{2}\right)\gamma_{AA} + \sum_{B\neq A}(Q_B - z_B)\gamma_{AB}.$$

The second modification concerns the one-center parameter U_{pp}^A which are now estimated not only from the ionization potentials I_p but also from the electron affinities A_p of the corresponding orbital atomic valence states:

$$-I_p = U_{pp}^A + (z_A - 1)\gamma_{AA},$$

$$-A_p = U_{pp}^A + z_A\gamma_{AA},$$

$$U_{pp}^A = -\frac{1}{2}(I_p + A_p) - \left(z_A - \frac{1}{2}\right)\gamma_{AA}.$$

The CNDO/2 nondiagonal matrix elements h_{pq} are calculated as in CNDO/1. Note that if all the atoms in the molecule are neutral (so that all $Q_B = z_B$ for all atoms B and $Q_p = 1$ for all orbitals q), the matrix element

$$h_{pp} = -\frac{1}{2}(I_p + A_p) + \left\{(Q_A - z_A) - \left(\frac{Q_p}{2} - \frac{1}{2}\right)\right\}\gamma_{AA} + \sum_{B\neq A}(Q_B - z_B)\gamma_{AB}$$

reduces to the simple form used in the extended Hückel method (see next section):

$$h_{pp} = -\frac{1}{2}(I_p + A_p).$$

The latter formula is also the theoretical basis for using a single parameter α for all one-center π-electron parameters in the simple Hückel method.

It is not the scope of this chapter to give a detailed analysis of all computational techniques generated by Pople's CNDO approach. In the intermediate neglect of differential overlap (INDO) scheme [28], differential overlap is retained for one-center integrals. The method is superior for properties where electron spin distribution is important, such as singlet-triplet splitting, which depends strongly on electron exchange effects. Dewar's modified intermediate neglect of differential overlap 3 (MINDO/3)

[29] has had great success in predicting accurate heats of formation. Pople's first attempt of neglect of differential diatomic overlap (NDDO) [30] assumes ZDO only between AOs on distinct atoms. The parameters are fitted to ab initio orbital energies. Integrals like $(s_A p_{iA}|s_B p_{iB})$ or $(p_{jA} p_{ia}|p_{jB} p_{ib})$ are not neglected so that repulsions between different atoms depend on their relative orientations without violating the rotational invariance. Modified neglect of diatomic overlap MNDO [31] is a carefully parametrized semiempirical version of NDDO. It yields accurate heats of formation and many other properties like geometries. It has been very successful for getting optimized geometrical structures of monomers, oligomers, and polymeric chains [32]. New parametric molecular models are continuously introduced based on this NDDO approximation. Austin Model 1 (AM1) [33] corrects for the major weaknesses of MNDO to reproduce hydrogen bonds. Examples of results obtained by the various ZDO technique are summarized in Tables 7 to 11.

To summarize, the aim of Pople and his coworkers was to reproduce the results of ab initio SCF-MO-LCAO calculations with less computational effort. At the beginning, the results have suffered from the good and bad features of HF theories: good geometries and charge distributions but poor binding energies. The aim of Dewar and coworkers has been to parametrize adequately in order to have a model that would give binding energies within the range of chemical accuracy with the same computational effort as CNDO or INDO.

Extended Hückel Method

As the CNDO method, the extended Hückel (EH) method uses a nonorthogonal valence minimal basis set:

$$S_{pq} = \int \chi_p(\mathbf{r})\chi_q(\mathbf{r})\, dv \neq \delta_{pq}.$$

The diagonal matrix elements of the one-electron operator are used as empirical parameters. As in CNDO schemes, they are evaluated from the valence ionization states of the corresponding atom. This approximation also neglects in the diagonal elements the effects of the molecular field as if the electrons would have the same motion in the molecule and in the isolated atom. Thus,

$$h_{pp} = \int \chi_p(\mathbf{r})h(\mathbf{r})\chi_p(\mathbf{r})\, dv = \alpha_p = -I_p.$$

The nondiagonal matrix elements of the one-electron operator are obtained by using the Mulliken's approximation [17] where differential overlap densities are calculated from one-center orbital densities:

$$\chi_p(\mathbf{r})\chi_q(\mathbf{r}) = \frac{1}{2} S_{pq}[\chi_p^2(\mathbf{r}) + \chi_q^2(\mathbf{r})]$$

and

$$h_{pq} = \int \chi_p(\mathbf{r})h(\mathbf{r})\chi_q(\mathbf{r})\, dv$$
$$= \beta_{pq} = \frac{1}{2} S_{pq}[\alpha_p + \alpha_q] = \frac{1}{2} S_{pq}[-I_p - I_q].$$

It is straightforward to prove that such a scheme does not provide any bonding in a homonuclear diatomic molecule like the hydrogen molecule since its bonding orbital $1\sigma g$ has an energy equal to that of the isolated $1s$ orbital:

$$\epsilon_{1\sigma g} = \frac{\alpha + \beta}{1 + S} = -I_{1s},$$

since

$$\alpha = \alpha_{1s} = -I_{1s},$$
$$\beta = \beta_{1s-1s} = \frac{1}{2} S[2\alpha] = -S^*I_{1s}.$$

Wolfsberg and Helmholtz [18] have proposed modifying Mulliken's approximation by introducing a constant factor K such as

$$\chi_p(\mathbf{r})\chi_q(\mathbf{r}) = \frac{1}{2} KS_{pq}[\chi_p^2(\mathbf{r}) + \chi_q^2(\mathbf{r})].$$

This "bonding" correction has been chosen by Hoffmann [19] as equal to 1.75 in order to reproduce plausible CH bond lengths in methane. Consequently, the nondiagonal matrix elements are thus calculated in the EH method as

$$h_{pq} = \int \chi_p(\mathbf{r})h(\mathbf{r})\chi_q(\mathbf{r})\, dv$$
$$= \beta_{pq} = \frac{1}{2} KS_{pq}[\alpha_p + \alpha_q] = \frac{1}{2} KS_{pq}[-I_p - I_q].$$

The EH scheme requires the explicit evaluation of the overlap integrals S_{pq} from the knowledge of the molecular geometry and of the functional form of the basis functions (usually STOs). Since the method does not separate electron-electron interactions from electron-nuclei attractions due to its global evaluation of matrix elements, the total molecular energy is approximated as a sum of the orbital energies:

$$E = \sum_{j}^{\text{occ}} 2\epsilon_j.$$

This way of calculating total energies is equivalent to assuming that, if they would have been explicitly calculated, the total electron-electron Coulomb and exchange interaction energies would exactly balance nuclear-nuclear repulsion energies:

$$\sum_{i}^{v} \sum_{j}^{v} 2J_{ij} - K_{ij} = \sum_{i}^{v} \sum_{j}^{v} 2 \int dr\, \phi_i(\mathbf{r}) J_j(\mathbf{r}) \phi_i(\mathbf{r}) - \int dr\, \phi_i(\mathbf{r}) K_j(\mathbf{r}) \phi_i(\mathbf{r})$$

$$= \sum_{A} \sum_{B<A} z_A z_B |\mathbf{R}_A - \mathbf{R}_B|^{-1}.$$

The last equation indicates that the EH scheme neglects the difference of electron-electron, nucleus core–nucleus core repulsions, and electron–nucleus core attractions. These quantities are obviously of opposite signs and should exactly compensate at the infinite distance limit. By its explicit introduction of the overlap effects of the orbital basis set, Hoffmann's EH method is very powerful for getting a correct description of the orbital topology. It is also an effective technique of determining equilibrium geometries and energy differences between conformers. Binding energies are generally not correctly described. Due to its non-SCF character, charge densities are usually overestimated. Examples of results obtained by the EH methodology are given in Tables 7 to 11.

π-Electron Semiempirical or Empirical Methods

To complete this overview, it is valuable to look at the simple case of an "empirical" model. Introduced in solid-state physics by Bloch [34], where it is known as the tight-binding approximation, it has been independently introduced in quantum chemistry by Hückel [16]. In the simple Hückel method, only π-electrons within purely conjugated organic molecules or chains are taken into account and all interactions except the nearest-neighbor ones are neglected. This method obtained considerable success by its simplicity and the validity of its approximations. These assumptions rely on the following facts.

It is considered that in a conjugated chain the σ-electrons form frozen cores with the carbon and hydrogen nuclei. The π-electrons are thus attracted by cores of unit effective charge for each unsaturated carbon atom. Due to the orthogonality of the basis functions describing the σ-electrons (functions which are symmetric with respect to the molecular plane) and those involving the π-electrons (in general, $2p_z$ functions antisymmetric with respect to the molecular plane), there is in strictly planar conjugated systems an exact numerical factorization between the matrix elements associated with σ- and π-electrons. Since the π-electrons are less bound to the organic backbones than the σ ones, it is reasonable to assume that they will be the first to be implied in physical processes. At the first order, they will be responsible for the main properties of the conjugated chain. According to perturbation theory, most molecular properties are determined by energy differences and thus primarily by excitations involving the smallest energy differences. The σ-σ^* energy gaps being important, the corresponding properties changes due to these σ-electrons are smaller compared to the significant perturbations due to the easier π-π^* excitations. Thus, it is customary to freeze the σ-electrons and group them with the ionic backbones of the molecule. However, it is clear that such methods can only be used in calculating those physical properties for which the π-electrons are mainly responsible and that the error introduced by the σ-π separation requires empirical or semiempirical compensation.

Implicitly, each π-electron distribution of a conjugated atom should be described by a single $2p_z$ or π-atomic orbital. In practice, the Hückel orbitals are not specified: they are not given any precise mathematical form in terms of hydrogenoid, Slater, Gaussian, or other dependence. They are, moreover, abstract quantities. Due to this flexibility, the unspecified basis set is assumed to form an orthonormal set:

$$S_{pq} = \delta_{pq} = 1 \qquad \text{for one-center overlap integrals,}$$
$$= 0 \qquad \text{for two-center overlap integrals.}$$

By the same logic, the Hückel effective one-electron operator $h(\mathbf{r})$ is not specified. Only its matrix elements differ from zero between atoms involved in a chemical bond and have arbitrary values α and β:

$h_{pp} = \alpha$ for one-center matrix elements; the α's are often incorrectly called Coulomb integrals and depend on the type of atoms on which the π orbitals are centered,

$h_{pq} = \beta$ for two-center matrix elements associated with chemical bonds; the β's are usually called resonance integrals and depend on the nature of the chemical bond,

$h_{pq} = 0$ for two-center matrix elements not associated with chemical bonds.

The detailed studies of the 1950s following the pioneering Pariser-Parr-Pople methodology have shown that the assumptions used in the Hückel method are coherent. In particular, the fact that the basis is orthonormal implies that in a similar environment the one-center matrix element α would have the same numerical value as would the two-center nearest-neighbor β integrals.

It is sometimes advocated that there is no longer any need for such a simple theory as the Hückel method with modern high-speed computers and the more sophisticated MO methods now available. However, it must be emphasized that there is a use of the Hückel theory, especially as a topological guide to the structure and energy of molecular orbitals and of electron densities. Its simplicity allows one to develop a powerful "paper-and-pencil" method and to easily obtain analytical results. In particular, the Hückel method offers the only alternative for obtaining qualitative information on very large systems. In the same way, one of its greatest advantages in the field of polymer chemistry is that interpolation of analytical forms of electronic properties of finite oligomers and infinite polymers can be deduced from closed relationships giving the orbital energies (or possibly their structure) as a function of the number of atoms in the polymeric chains. Examples of such formulas are found in many textbooks [35]. A rather complete analysis in the framework of Hückel methodology is available by the general technique of finite differences [36]. For example, the orbital energies of regular chains of N atoms are given in terms of the Hückel parameters α and β by

$$\epsilon_j = \alpha + 2\beta \cos\left(\frac{j\pi}{N+1}\right) \qquad \text{for linear chains,}$$

$$\epsilon_j = \alpha + 2\,\beta \cos\left(\frac{2j\pi}{N}\right) \qquad \text{for cyclic chains.}$$

They correspond to an evolution of the energy gap such as

$$\Delta E = -4\beta \sin\left(\frac{\pi}{2(N+1)}\right) \qquad \text{for linear chains,}$$

$$\Delta E = -4\beta \sin\left(\frac{\pi}{N}\right) \qquad \text{for cyclic chains.}$$

For alternant chains of atoms, the mathematical expressions are more elaborate and solutions are, as far as we know, only available for cyclic chains of N atoms and linear chains containing an odd number of $N - 1$ atoms. They are, in both cases,

$$\epsilon_j^{(1)},\epsilon_j^{(2)} = \alpha \pm \left\{\beta_1^2 + \beta_2^2 + 2\beta_1\beta_2 \cos\left(\frac{4j\pi}{N}\right)\right\}^{1/2}.$$

For the linear case, there also exists one nonbonding orbital at energy α. It is important to note (that point will be relevant to the discussion developed in the next section) that in the case of an alternant system, ΔE does not tend to zero as $N \to \infty$ but to the finite value

$$\lim_{N\to\infty} \Delta E(N) = 2(\beta_1 - \beta_2).$$

The simplicity of the Hückel method opens wide possibilities. It is worth noting that the modern interpretation of the high electrical conductivity of doped conjugated polymers in terms of solitons, polarons, or bipolarons was initially formulated in terms of a Hückel model [37].

Tables of Hückel parameters [38], tables of Hückel molecular orbitals [39], and good books on the applications of the Hückel method to organic conjugated systems [40] are widely available. The parameter α of carbon atoms is sometimes taken equal to zero (this corresponds to an energy scale shift only) while the parameter β is used as the (negative) energy unit. Empirical values of the Hückel parameters are determined by least-square fitting of theoretical and experimental data. It is striking to note that the values of the parameters depend on the nature of the property considered. Salem [40] has taken the example of a series of condensed aromatic hydrocarbons (benzene, naphthalene, anthracene, pentacene, ...). Koopmans' theorem states that first ionization potentials correlate with the negative of the energy of the highest occupied molecular orbital (HOMO). From this correlation, a first estimate of $\beta \cong -4$ eV $\cong -90$ kcal mol^{-1} is obtained (it also indicates an empirical value of $\alpha \cong -5.9$ eV $\cong -136$ kcal mol^{-1}). The theoretical energy of the first singlet-singlet π-π^* transition is the difference between the lowest unoccupied molecular orbital (LUMO) and the HOMO energies. The comparison with experimental data gives a second estimate of $\beta \cong -2.4$ eV $\cong -55$ kcal mol^{-1}. Finally, the correlation between the delocalization or resonance energy (measure of the molecular extra stabilization due to electron conjugation with respect to an additive energy bond scheme) allows for a third estimate of $\beta \cong -0.8$ eV $\cong -18$ kcal mol^{-1}. A qualitative reason for the variations in the numerical values of the parameters is that some of the experimental data concern differences of energies while others correspond to absolute (total) energies. Thus, the large difference between the empirical estimates is due to different combination of one-electron (kinetic and nuclear attraction) and two-electron (electron repulsion) energy balances in the basic comparison with experiment. A detailed analysis should identify for each type of experiment the terms and the combinations of the SCF-MO-LCAO matrix elements which correspond to α and β in the Hückel theory. This example is well adapted to call the attention of the user of empirical and semiempirical procedures to the fact that the quality of the selected technique is strongly dependent on the properties on which they are parametrized and that for given classes of experiment (UV transitions, heats of formation, dipole moments) different scaling factors or different parametrizations are needed in each case.

The separations between kinetic energies, nuclei-electron attractions, and electron-electron repulsions are explicitly taken into account in the semiempirical Pariser-Parr-Pople (PPP) method (which is a ZDO version applicable to the π-electrons). Even if historically it has been introduced more than 10 years before the CNDO, INDO, and NDDO methods, it is not necessary to detail here the form of the matrix elements, which is strictly analogous to the forms given in Table 6. The choice of the "complete" neglect of differential overlap, i.e., the original ZDO scheme, is fully justified for $2p_z$ atomic orbital and is not too drastic an approximation. In the case of π-electrons, there is no need for schemes like INDO or NDDO since only diatomic differential overlaps are neglected. The justification of the ZDO approximation in the PPP method has been given early in terms of Löwdin's orbitals [41], a justification which is fully correct in the case of π-electrons but which is sometimes extended to the case of all-valence elctrons with less success because of the large overlaps involved between σ-orbitals. In order to allow the explicit calculation of the electron integrals, the $2p_z$ basis set has to be specified (in general, the analytical form is a STO with Slater exponent calculated from Slater rules, $\zeta = 3.25$). Consequently, the numerical scheme is applicable to strictly planar structures only, a difficulty not met in the simpler heuristic Hückel method where the basis set is unspecified. The main applications of the PPP approach are the interpretation and prediction of the energies and intensities of π-π^* transitions, which is detailed in several textbooks [42].

TOWARD QUANTUM CHEMICAL CALCULATIONS ON LARGE SYSTEMS

For evaluating the electronic properties of a polymer or of a very large oligomer, the first alternative which is offered to the molecular quantum chemist is to extrapolate the electronic properties as functions of the number of oligomers in the polymeric chain. As seen on pages dealing with the Hückel methodology, interpolation properties between finite oligomers and the infinite polymer can been estimated by compact formulas giving the orbital energies (or possibly their structure) as a function of the number of atoms in the chains as exemplified in the case of polyacetylenes. A practical example is given in Figure 5, which represents the case of poly-

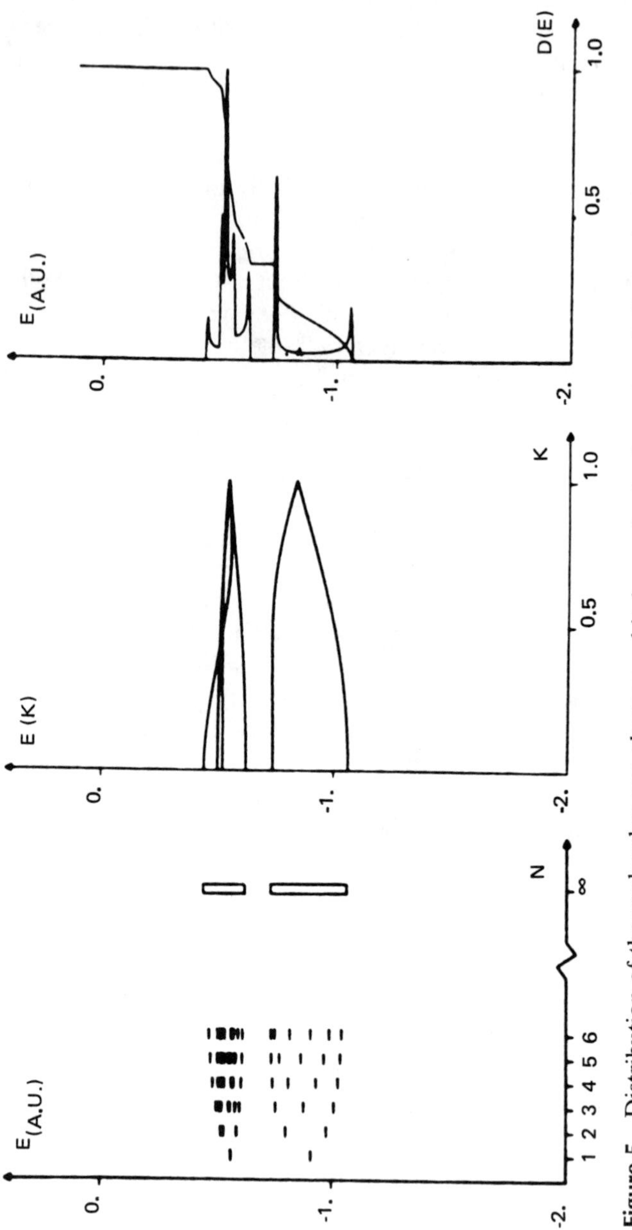

Figure 5 Distribution of the molecular one-electron orbital energies in the series of alkanes (left), valence electron band structure (center) and DOS (right) of an infinite ideal polyethylene chain.

ethylene, a limiting term of the series of alkanes (methane, ethane, propane, ...). For the one-electron orbital energies, it is observed that as the number of atoms in the oligomer increases, the distance between the energy levels diminishes and the energy levels form, at the infinite limit, allowed energy regions (the core and valence bands) and forbidden energy regions (the forbidden bands or energy gaps).

Another alternative is to combine the concepts of quantum chemistry with those of solid-state physics and to directly compute the energy bands, the density of electronic states, and more generally the electronic properties of an ideal infinite chain assumed to be periodic. This is usually made by taking advantage of the possible translation symmetry of the lattice, and uses the concept of Brillouin zones introduced by Bloch in 1928 [34]. In this theory, the so-called Bloch functions (molecular orbitals for an infinite $1D$ chain) are eigenfunctions of a translation operator. Bloch's theorem is a direct consequence of the periodicity of the electron density:

$$\rho(\mathbf{r}) = |\phi_n(\mathbf{r})|^2 = \rho(\mathbf{r} + ja) = |\phi_n(\mathbf{r} + ja)|^2,$$

where a is the length of the polymeric unit cell and j is a cell index. By taking the square root in complex space, Bloch's theorem states the phase relation of the orbitals at periodically related points:

$$\phi_n(\mathbf{r} + ja) = e^{ikja}\phi_n(\mathbf{r}).$$

Since the argument of an exponential is a pure number, a is a length, j is a pure number, the counter index of a given unit cell, k must have the dimensions of an inverse length. Thus, the orbitals and their associated energies are functions of that k and are labeled by it:

$$\phi_n = \phi_n(k), \qquad \epsilon_n = \epsilon_n(k).$$

They can be plotted with respect to k; the represenation of the corresponding dispersion curves is called an energy band. Standard theorems of solid-state physics demonstrate that those energy bands are periodic in k-space:

$$\epsilon_n(k + lg) = \epsilon_n(k),$$

where g is the reciprocal translation unit length $(=2\pi/a)$. The calculation of the energy bands therefore needs to be performed only on a length equivalent to a single unit cell of the reciprocal lattice. It is usual to select this unit cell as the first Brillouin zone ranging from $-\pi/a$ to $+\pi/a$. Another corollary states that the energy bands are symmetric with respect to $k = 0$:

$$\epsilon_n(-k) = \epsilon_n(k).$$

Thus, the reciprocal space needs to be explored only from $k = 0$ to $k = +\pi/a$ (half the first Brillouin zone).

Figure 5 states graphically the links between these solid-state concepts (which are less familiar to chemists) and the molecular theories (which are less familiar to physicists). In the case of polyethylene, the unit cell contains two methylene groups (16 electrons). The occupied bands are in this case, two internal core bands describing the four 1s core electrons of the two carbons per unit cell, two bands which mainly describe the σ C—C bonds and four bands which, in turn, are associated with the description of the four σ C—H bonds. The 16 electrons in the unit cell are thus fully represented by eight doubly occupied bands. The center part of Figure 5 represents the six doubly occupied valence bands of a polyethylene chain.

The rightmost representation of Figure 5 is the most convenient for experimental comparisons; it is the density of states (DOS) which plots the number of available energy levels as a function of energy for infinite systems. It is usually normalized to the number of electrons per unit cell. The three representations are equivalent.

Turning to actual applications, the numerical procedure combines the equations of the methods of molecular quantum chemistry and of solid-state physics. We have seen that in molecular quantum chemistry, a molecular orbital is expanded in terms of basis functions; secular systems of equations and determinants are solved; their eigenvalues are the orbital energies. From the LCAO coefficients, charges and bond orders (projection of the density matrices onto the limited basis used) are calculated. In polymer quantum chemistry, we take into account the lattice periodicity; the orbitals, the systems of equations, and the determinants are no longer real but have imaginary components. They also contain lattice sums of matrix elements over the basis functions which are to be evaluated by adequate procedures. Fortunately, these matrix elements exponentially decrease with the distance between the orbital centers and force the natural convergency properties of the lattice sums. They are calculated by the standard methods of quantum chemistry discussd in previous sections of this chapter.

The practice of polymer computations is rather standard. The Fock and overlap matrices are computed at a given level of approximation (semi-empirical or ab initio). The calculation is made in direct space. The k-dependent matrices are diagonalized in the reciprocal space and, if necessary, a self-consistent procedure is turned on. The output consists of the energy bands and of the form of the molecular orbitals of the polymer. An interactive graphical communication can be initiated for ordering the energy bands, plotting the standard electronic properties, such as band structures, bandwidths, DOS, simulations of electron spectra, electron densities, calculating electron indices as charges and bond orders, and determining conformations or other properties.

This approach has been developed in several review papers [43], in several books [44], and has been recently summarized in two monographs [45]. Historically, all the steps of increasing complexity from empirical Hückel, semiempirical PPP, extended Hückel, CNDO, to HF ab initio schemes are described in the literature [46] with examples of band structure calculations on standard polymers like polyethylenes [47], polyacetylenes [48], polyparaphenylenes [49], or polypyrroles [50]. A correct determination of an all-electron band structure is still a formidable task at the present time. Indeed, it has been proved that long-range effects are very important as soon as the unit cell contains permanent dipoles. Extended multipole expansions must be used in order to get a satisfactory balance of the electrostatic interactions. The cheaper (on a computational cost basis) semiempirical techniques like extended Hückel or CNDO are also good alternatives; in these cases, however, reliable results can only be obtained when paying close attention to the parameterization procedures. First sketches of the conjugated bands are easily obtained from simple Hückel calculations.

The availability of this full range of theoretical methods for the calculation of electronic band structures allows for a true theoretical molecular design of polymers with specific properties. Polymer quantum chemistry is a new science which links the concepts of quantum chemistry with those of solid-state physics. It is an excellent field to force modern chemists to become familiar with basic concepts of solid-state physics (and vice versa) and is now an interdisciplinary and timely concen of our scientific community.

REFERENCES

1. M. Born and J. R. Oppenheimer, *Ann. Phys.*, *84*, 457 (1927).
2. D.R. Hartree, *Proc. Cambridge Phil. Soc.*, *24*, 89, 116, 426 (1928); *25*, 225, 310 (1929); see also *The Calculation of Atomic Structures*, Wiley, New York (1957).
3. V. Fock, *Z. Phys.*, *61*, 126 (1930).
4. J. C. Slater, *Phys. Rev.*, *34*, 1293 (1929); *35*, 509 (1930).
5. C. C. J. Roothaan, *Rev. Mod. Phys.*, *23*, 69 (1951).
6. G. G. Hall, *Proc. Roy. Soc. (London)*, *A205*, 541 (1951).
7. J. C. Slater, *Phys. Rev.*, *36*, 57 (1930).
8. C. C. J. Roothaan, *J. Chem. Phys.*, *19*, 1445 (1951).
9. S. F. Boys, *Proc. Roy. Soc. (London)*, *A200*, 542 (1950).
10. W. J. Hehre, R. F. Stewart, and J.A. Pople, *J. Chem. Phys.*, *51*, 2657 (1969); W. J. Hehre, R. Ditchfield, R. F. Stewart, and J. A. Pople, *J. Chem. Phys.*,

52, 2769 (1970); W. J. Pietro, B. A. Levi, W. J. Hehre, and R. F. Stewart, *Inorg. Chem.*, *19*, 2225 (1980); W. J. Pietro, R. F. Hout, Jr., E. S. Blurock, W. J. Hehre, D. J. DeFrees, and R. F. Stewart, *Inorg. Chem.*, *20*, 3650 (1981); W. J. Pietro and W. J. Hehre, *J. Comput. Chem.*, *4*, 241 (1983).

11. W. J. Hehre, L. Radom, P. v. R. Schleyer, and J. A. Pople, *Ab Initio Molecular Orbital Theory*, Wiley, New York (1986).
12. S. Huzinaga, *Physical Sciences Data*, 16: *Gaussian Basis Sets for Molecular Calculations*, Elsevier, Amsterdam (1984); R. Poirier, R. Kari, and I. G. Csizmadia, *Physical Sciences Data*, *24*: *Handbook of Gaussian Basis Sets*, Elsevier, Amsterdam (1985).
13. E. Clementi, *J. Chem. Phys.*, *40*, 1944 (1964).
14. L. Allen, E. Clementi, and H. Gladney, *Rev. Modern Phys.*, *35*, 465 (1963).
15. T. Koopmans, *Physica*, *1*, 104 (1933).
16. E. Hückel, *Z. Phys.*, *60*, 423 (1930), *70*, 204 (1931); *76*, 628 (1932).
17. R. S. Mulliken, *J. Chim. Phys.*, *46*, 497,675 (1949).
18. M. Wolfsberg and L. Helmholtz, *J. Chem. Phys.*, *20*, 837 (1952).
19. R. Hoffmann, *J. Chem. Phys.*, *39*, 1397 (1963), *40*, 2474, 2480, 2745 (1964).
20. J. A. Pople, *Trans. Far. Soc.*, *49*, 1375 (1953).
21. R. G. Parr, *J. Chem. Phys.*, *20*, 239, 1499 (1952); *21*, 568 (1952); R. Pariser and R. G. Parr, *J. Chem. Phys.*, *21*, 466, 767 (1953).
22. J. A. Pople, D. P. Santry, and G. A. Segal, *J. Chem. Phys.*, *43*, S129 (1965), J. A. Pople and G. A. Segal, *J. Chem. Phys.*, *43*, S136 (1965); *44*, 3289 (1966). See also J. A. Pople and D. L. Beveridge, *Approximate Molecular Orbital Theory*, McGraw-Hill, New York (1970); J. N. Murrel and A. J. Harget, *Semi-Empirical Self-Consistent Field Molecular Orbital Theory of Molecules*, Wiley-Interscience, New York (1972).
23. O. Sinanoglu and K. B. Wiberg, *Sigma Molecular Orbital Theory*, Yale University Press, New Haven, CT (1970); G. Klopman and B. O'Leary, *Introduction to All-Valence Electrons S.C.F. Calculations*, Springer-Verlag, Berlin (1970); R. Daudel and C. Sandorfy, *Semiempirical Wave-Mechanical Calculations on Polyatomic Molecules*, Yale University Press, New Haven, CT (1971); *Semi-Empirical Methods of Electronic Structure Calculations*, Parts A, B (G. A. Segal, ed.) Plenum Press, New York (1977); J. Sadlej, *Semi-Empirical Methods of Quantum Chemistry*, Ellis Horwood, Chichester (1985).
24. I. G. Csizmadia, M. C. Harrison, J. W. Moscowitz, S. Seung, B. T. Sutcliffe, and M. P. Barnett, *POLYATOM*, *Quantum Chemistry Program Exchange*, *11*, 47 (1964); D. B. Newmann, H. Basch, R. L. Korregay, L. C. Snyder, J. Moskowitz, C. Hornback, and P. Liebman, *POLYATOM 2*, *Quantum Chemistry Program Exchange*, *11*, 199 (1973).
25. A. Veillard, *IBMOL 4*, Computer Program, IBM Research Laboratory, San Jose, CA (1968); E. Ortoleva, G. Castiglione, and E. Clementi, *IBMOL 6*, *Comput. Phys. Comm.* *19*, 337 (1980); R. Gomperts and E. Clementi, KGNMOL, *Quantum Chemistry Program Exchange*, *Bull.*, 8, 538 (1988).
26. W. J. Hehre, W. A. Lathan, M. D. Newton, R. Ditchfield, and J. A. Pople, *GAUSSIAN 70*, *Quantum Chemistry Program Exchange*, *11*, 236 (1973); J.

S. Binkley, R. Whiteside, P. C. Hariharan, R. Seeger, W. J. Hehre, M. D. Newton, and J. A. Pople, *GAUSSIAN 76*, *Quantum Chemistry Program Exchange*, *11*, 368 (1978); J. S. Binkley, R. A. Whiteside, R. Krishnan, R. Seeger, D. J. DeFrees, H. B. Schlegel, S. Topiol, L. R. Kahn, and J. A. Pople, *GAUSSIAN 80*, *Quantum Chemistry Program Exchange*, *13*, 406 (1981); J. S. Binkley, M. J. Frisch, K. Rahgavachari, D. J. DeFrees, H. B. Schlegel, R. A. Whiteside, E. M. Fluder, R. Seeger and J. A. Pople, *GAUSSIAN 82*, Carnegie Mellon University, Pittsburgh PA; M. J. Firsch, J. S. Binkley, H. B. Schlegel, K. Rahgavachari, R. L. Martin, J. J. P. Stewart, F. W. Bobrowicz, D. J. DeFrees, R. Seeger, R. A. Whiteside, D. J. Fox, E. M. Fluder, and J. A. Pople, *GAUSSIAN 86*, release C, Carnegie-Mellon University, Pittsburgh PA.

27. M. Dupuis, J. Rys, and H. F. King, *HONDO*, *Quantum Chemistry Program Exchange*, *11*, 336,338 (1977); *13*, 401 (1981); M. Dupuis, J. D. Watts, and H. O. Villar, *HONDO7*, *Quantum Chemistry Program Exchange*, *Bull.*, *8*, 544 (1988).
28. J. A. Pople, D. L. Beveridge, and P. A. Dobosh, *J. Chem. Phys.*, *47*, 2026 (1967).
29. R. C. Bingham, M. J. S. Dewar, and D. H. Lo, *J. Am. Chem. Soc.*, *97*, 1285, 1294, 1302, 1307 (1975); M. J. S. Dewar, *Science*, *187*, 1037 (1975).
30. J. A. Pople, D. P. Santry, and G. A. Segal, *J. Chem. Phys.*, *43*, S129 (1965).
31. M. J. S. Dewar and W. Thiel, *J. Am. Chem. Soc.*, *99*, 4899, 4907 (1977); M. J. S. Dewar and M. L. McKee, *J. Comput. Chem.*, *4*, 84 (1983).
32. J. L. Brédas, B. Thémans, J. M. André, A. J. Heeger, and F. Wudl, *Synth. Met.*, *11*, 343 (1985); J. Riga, Ph. Snauwaert, A. De Pryck, R. Lazzaroni, J. P. Boutique, J. J. Verbist, J. L. Brédas, J. M. André, and C. Taliani, *Synth. Met.*, *21*, 223 (1987); B. Thémans, J. M. André, and J. L. Brédas, *Synth. Met.*, *21*, 149 (1987); J. M. Toussaint, B. Thémans, J. M. André, and J. L. Brédas, *Synth. Met.*, *28*, C205 (1989).
33. M. J. S. Dewar, E. G. Zoebisch, E. F. Healy, and J. J. P. Stewart, *J. Am. Chem. Soc.*, *107*, 3902 (1985).
34. F. Bloch, *Z. Phys.*, *52*, 555 (1928).
35. W. Kauzmann, *Quantum Chemistry*, p. 682, Academic Press, New York (1957); F. L. Pilar, *Elementary Quantum Chemistry*, p. 593, McGraw-Hill, New York (1968); T. A. Albright, J. K. Burdett, and M-H. Whangbo, *Orbital Interactions in Chemsitry*, p. 299, Wiley, New York (1986).
36. T. K. Rebane, *Methods of Quantum Chemsitry*, p. 147 (M. G. Veselov, ed.), Academic Press, New York (1965).
37. W. P. Su, J. R. Schrieffer, and A. J. Heeger, *Phys. Rev.*, *B22*, 2099 (1980), J. L. Brédas, R. R. Chance, and R. Silbey, *Phys. Rev.*, *B26*, 5843 (1982).
38. W. P. Purcell and J. A. Singer, *J. Chem. Enging. Data*, *12*, 4899, 235 (1967).
39. C. A. Coulson and A. Streitwieser, Jr., *Dictionary of π-Electron Calculations*, Pergamon, Oxford (1965); E. Heilbronner and H. Bock, *The HMO-Model and its Application*, *Tables of Hückel Molecular Orbitals*, Wiley, New York (1976).

40. B. Pullman and A. Pullman, *Les Théories Electroniques de la Chimie Organique*, Masson, Paris (1952); A. Streitwieser, Jr., *Molecular Orbital Theory for Organic Chemists*, Wiley, New York (1961); L. Salem, *Molecular Orbital Theory of Conjugated Systems*, Benjamin, New York (1966); E. Heilbronner and H. Bock, *Das HMO-modell und Seine Anwendung*, Verlág Chemie (1968); *The HMO-Model and Its Application, Basis and Manipulation*, Wiley, New York (1976).
41. P. O. Löwdin, *J. Chem. Phys.*, *18*, 365 (1950).
42. J. N. Murrell, *The Theory of the Electronic Spectra of Organic Molecules*, Chapman and Hall, London (1971); H. H. Jaffé and M. Orchin, *Theory and Applications of Ultraviolet Spectroscopy*, Wiley, New York (1962).
43. J. M. André, *Adv. Quantum Chem.*, *12*, 65 (1980); *Current Aspects of Quantum Chemistry*, p. 273 (R. Carbo, ed.), Elsevier, Amsterdam (1982); *Large Finite Systems* p. 277 (J. Jortner and B. Pullman, eds.), Reidel, Dordrecht (1987); *Strategies for Computer Chemistry*, p. 45 (C. Tosi, ed.), Kluwer Academic, Dordrecht, (1989).
44. J. M. André and J. Ladik (eds.), *Electronic Structure of Polymers and Molecular Crystals*, Plenum Press, New York (1975); J. M. André, J. Delhalle, and J. Ladik (eds.), *Quantum Theory of Polymers*, Reidel, Dordrecht (1978); J. Ladik and J. M. André (eds.), *Quantum Chemistry of Polymers: Solid State Aspects*, Reidel, Dordrecht (1984).
45. J. Ladik, *Quantum Theory of Polymers as Solids*, Plenum Press, New York (1988); J. M. André, J. Delhalle, and J. L. Brédas, *Quantum Chemistry Aided Design of Organic Polymers for Molecular Electronics*, World Scientific, Singapore (1991).
46. J. M. André, L. Gouverneur, and G. Leroy, *Int. J. Quantum Chem.*, *1*, 427, 451 (1967); G. Del Re, J. Ladik, and G. Biczo, *Phys. Rev.*, *155*, 997 (1967); J. M. André, *J. Chem. Phys.*, *50*, 1536 (1969).
47. W. L. McCubbin and R. Manne, *Chem. Phys. Lett.*, *2*, 230 (1968); J. M. André and G. Leroy, *Chem. Phys. Lett.*, *5*, 71 (1970).
48. J. M. André and G. Leroy, *Theoret. Chim. Acta*, *9*, 123 (1967); J. M. André and G. Leroy, *Int. J. Quantum Chem.*, *5*, 557 (1971); J. L. Brédas, J. M. André, and J. Delhalle, *J. Mol. Struct.* (*Theochem*), *87*, 237 (1982).
49. J. L. Brédas, B. Thémans, and J. M. André, *Phys. Rev.*, *B26*, 6000 (1982); J. L. Brédas, B. Thémans, J. G. Fripiat, J. M. André, and R. R. Chance, *Phys. Rev.*, *B29*, 6761 (1984).
50. J. L. Brédas, B. Thémans, and J. M. André, *Phys. Rev.*, *B27*, 7827 (1982); J. M. André, D. P. Vercauteren, G. B. Street and, J. L. Brédas, *J. Chem. Phys.*, *80*, 5643 (1984).

11

Dynamic Relaxations in Poly(ester carbonates)

Jozef Bicerano and Hayden A. Clark
The Dow Chemical Company
Midland, Michigan

INTRODUCTION

The results of quantum mechanical calculations performed to help in relating the physical properties of bisphenol A polycarbonate (BPAPC) and the poly(ester carbonates) (PEC) of BPAPC and terephthalic and/or isophthalic acids to their structure at the molecular level, and the conclusions that can be reached from these calculations, concerning the motions involved in the dynamic relaxations observed in these materials via dynamic mechanical spectroscopy (DMS), are summarized. The bulk of this chapter consists of a summary of previously published work [1,2]; however, significant refinements made in some of the results by additional calculations are also included.

The generic formulas of BPAPC and of its PEC with terephthalic and/or isophthalic acids are schematically illustrated in Figure 1, where x denotes the amount of ester relative to the amount of the carbonate, C_l denotes a carbon atom bonded to two phenyl rings and two methyl groups, C_m denotes a methyl carbon, C_e denotes an ester carbon, C_c denotes a carbonate carbon, C_p denotes a phenyl ring carbon atom attached to either one of a C_l, C_e, or C_c, instead of being attached to a hydrogen atom, and

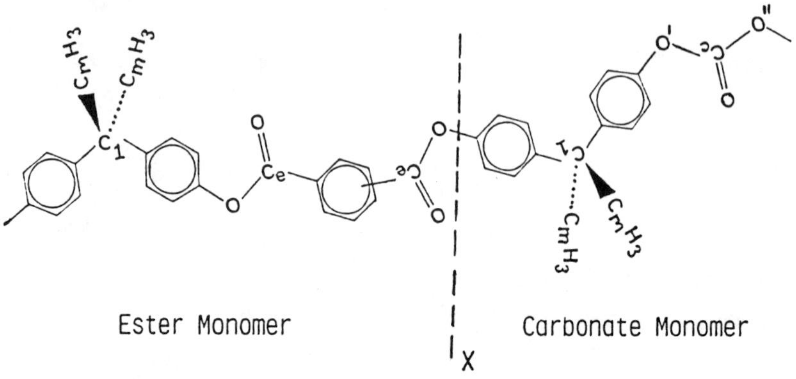

Ester Monomer Carbonate Monomer

Figure 1 Schematic illustration of the generic formula of BPAPC and of its PEC with terephthalic and/or isophthalic acids. See the introduction for the notation.

O′ and O″ denote the two divalent O atoms in the carbonate linkage. In addition to x, another overall compositional variable is the relative percentage of terephthalic configurations with the two ester linkages para to one another versus isophthalic configurations with the two ester linkages meta to one another.

As can be seen from Figure 1, there is only one carbonate linkage per carbonate monomer, while there are two ester linkages per ester monomer. There is, as a result, only one type of phenyl environment in the carbonate fragments, with the phenyl ring flanked by an isopropylidene group and a carbonate linkage para to each other. On the other hand, there are two general types of phenyl environments in the ester fragments. One of these environments has the phenyl ring flanked by an isopropylidene group and an ester linkage para to one another, while the other one has the phenyl ring flanked by two ester linkages either para or meta to one another. There are twice as many phenyl rings in the first type of environment than there are in the second type of environment, in the ester fragments.

The leading experimental techniques for understanding the transitions and relaxations which determine the molecular dynamics of BPAPC and PEC are DMS [3–8], nuclear magnetic resonance (NMR) spectroscopy [8–22], and dielectric relaxation spectroscopy [23–31].

Several relaxations are observed in the DMS and dielectric relaxation spectra of BPAPC and of PEC copolymers. Unfortunately, different authors have used different notations for these relaxations. There is also disagreement on how many components are involved in some of them. See Figure 2 for the notation used in this chapter for the relaxations. The α relaxation (glass transition) occurs at the glass-transition temperature (T_g).

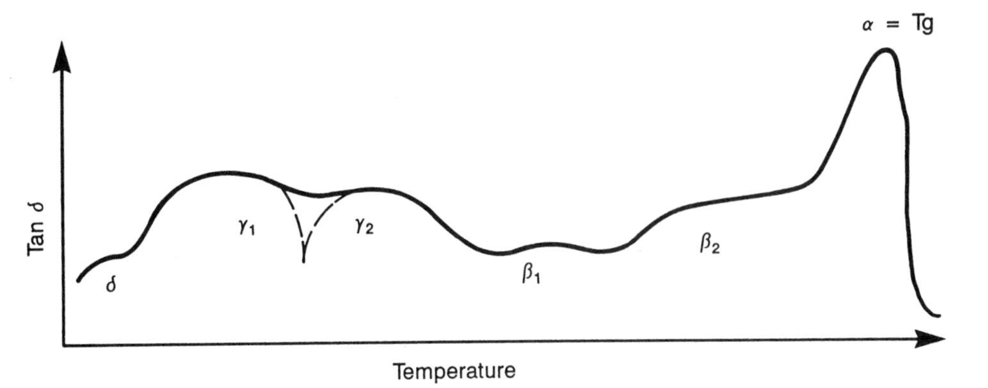

Figure 2 Schematic illustration of the notation used for the relaxations. The y axis represents the loss tangent tan δ, defined as the quotient G''/G', where G'' is the loss modulus and G' is the elastic (storage) modulus. (If the dynamic experiment is not torsion mode DMS, the y axis denotes the loss for the particular experiment being performed, rather than tan δ.) The x axis represents the temperature (T). Greek letters are used to designate the relaxations (α, β, γ, and δ in order of *decreasing* temperature). The primary (α) relaxation is usually referred to as the glass transition. It occurs at the glass-transition temperature (T_g). When a relaxation appears to be a superposition of several unresolved or partially resolved components, increasing numerical subscripts are used to denote components with increasing temperature (for example, the β_1 and β_2 relaxations).

The β relaxation, which occurs at lower temperatures, plays an important role in physical aging phenomena [6,8]. The γ and δ relaxations occur at still lower temperatures. The γ relaxation has been represented as having two partially resolved components in Figure 2. This relaxation is often assumed to have three components; however, dielectric relaxation [30] and stress relaxation [32] measurements at very low frequencies (down to 10^{-4} Hz) have only been able to resolve it into two components.

Dielectric relaxation experiments are especially effective in detecting the motions of polar groups, such as carbonate groups or ester linkages. The activation energies obtained by DMS or dielectric relaxation are "global," in the sense that they reflect the superposition of all the motions absorbing energy at a given temperature and frequency, although, of course, the motion of a given subunit may be causing most of the loss (energy dissipation). Activation energies can be obtained from DMS or dielectric relaxation data by either (a) shifting data obtained for the loss modulus G'' or the dielectric loss factor ϵ'' at a series of frequencies to a common temperature by using the time-temperature superposition principle [33], and then taking the slopes of Arrhenius-like line fits to the temperature dependence of the natural logarithm of the shift factors to be the activation energies; or (b) constructing "relaxation maps" of the temperature of the maximum of G'' or ϵ'' for a given relaxation as a function of frequency, and taking the slope of the resulting Arrhenius-type line to be the activation energy.

These activation energies therefore have a somewhat different physical interpretation than activation energies obtained by NMR, where selective isotopic labeling of atoms in distinct chemical environments can provide detailed information on individual local motions. Valuable insights concerning molecular level motions can therefore be obtained by combining the DMS, NMR, and dielectric relaxation techniques. When these experimental techniques are combined with quantum mechanical and/or force-field calculations, where selected modes of motion in a model molecule can be studied computationally, the understanding of the dynamic relaxations, and especially of the secondary (or sub-T_g) relaxations which are more localized than the glass transition, and therefore easier to study by molecular-level calculations, can be enhanced significantly.

The currently most popular models for deformation are based on considerations of either free-volume [34–36] or chain entanglement [37–39]. While both of these types of models are intuitively appealing, and have proven to be quite useful, neither one can make quantitative predictions in a consistent manner [6,34,36]. Both the results of the DMS, NMR, and dielectric relaxation studies mentioned above, and of Fourier transform infrared (FTIR) [8,40] and wide-angle neutron scattering [41] studies show that it is necessary to study the local structure (i.e., the atomic configurations and conformations on the scale of short-chain segments down to less than one monomer unit in length) in detail at the molecular level. It is hoped that the calculations [1,2] summarized here will serve as a step in that direction.

Several other calculations have also been published on polymers or model systems related to those being studied in the present work. Hutnik and Suter [42] are utilizing the results of the present work [1,2] as the basis of a rotational isomeric state (RIS) model for BPAPC. Tonelli has studied the conformational characteristics and flexibility of BPAPC [43], and the effect of the terephthalic environment on the chain flexibility of poly(ethylene terephthalate) (PET) [44], by using force-field techniques. More recent force-field calculations include the work of Sundarajan [45], Perez and Scaringe [46], and Letton et al. [47]. Mora et al. [48] have performed semiempirical quantum mechanical calculations on model molecules representing the local environment of the carbonate linkage, while Letton et al. [47] have studied the local environment of the isopropylidene group at a comparable level of theory. Several ab initio calculations have also been carried out, as reported by Clark and Munro [49], Bendler [50] and Laskowski et al. [51]. Laskowski et al. [51] have used the results of their calculations to construct an RIS model for the chain conformations of BPAPC. Perchak et al. [52] have performed Brownian dynamics computer simulations of simple models of the ring flip process in BPAPC, by studying two-dimensional lattices of interacting phenyl rings.

MODEL MOLECULES

Small-Model Molecules

The simplest model molecules which can be used as a first step in studying the intrachain rotations in BPAPC and its PEC are 2,2-diphenylpropane (DPP), diphenyl carbonate (DPC) and phenyl benzoate (PB). The optimized geometries of these three model molecules are illustrated in space-filling drawings in Figure 3.

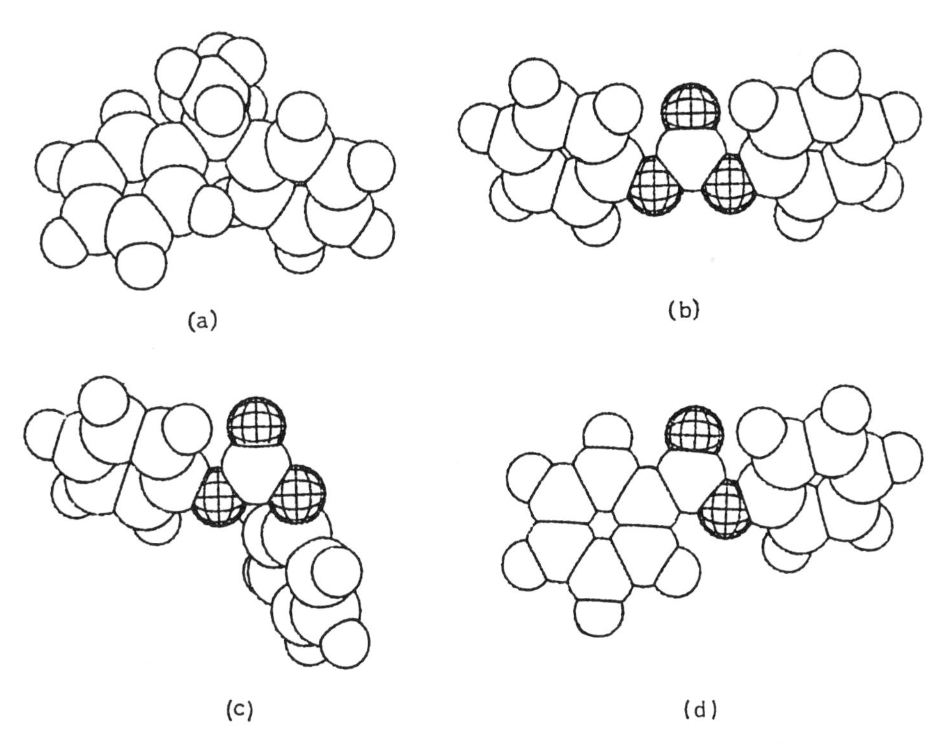

Figure 3 Space-filling illustrations of optimized geometries: (a) 2,2-diphenylpropane (DPP); (b) *trans,trans*-diphenyl carbonate (DPC); (c) *trans,cis*-DPC; and (d) phenyl benzoate (PB).

DPP represents a configuration of atoms common to both the carbonate and the ester monomers. It provides a complete description of the local bonding environment of a methyl group in an isolated chain segment. The phenyl rings are terminated by hydrogen atoms at the para positions, rather than being continued via carbonate or ester linkages as in the BPAPC/PEC copolymers.

DPC is the smallest model molecule that provides a realistic representation of the immediate environment of the carbonate linkage. Two conformations have been depicted for DPC, namely, the trans,trans and trans,cis conformations. The cis,trans conformation is equivalent to the trans,cis conformation. It has been suggested [53] that all three of these conformations can occur in BPAPC, with the trans,trans conformation being the most stable by 1.3 kcal/mole and therefore being strongly preferred. The cis,cis conformation was excluded because it has very large steric repulsions.

There are two phenyl rings attached to isopropylidene groups per phenyl ring flanked by a pair of ester linkages in the ester segments. PB is useful to determine (a) the intrinsic rotational properties of an individual diphenyl ester linkage, so that the differences between terephthalic and isophthalic configurations can be understood better; and (b) the differences between carbonate and ester linkages.

Large-Model Molecules

Many conclusions can be drawn by studying DPP, DPC, and PB. It is, however, desirable to perform additional calculations on larger molecules in which phenyl rings are attached at both ends to the functional groups which provide for the chain continuation in the polymer, instead of being terminated by a hydrogen atom at one end. The MOGLI (molecular graphics library) program [54] was therefore used on an Evans and Sutherland PS330 color graphics station to visualize, manipulate, modify and/or connect structures of interest, and thus generate large-model molecules from DPP, DPC, and PB.

The larger-model molecules chosen to represent the environments of the phenyl rings attached to isopropylidene groups in the carbonate and ester fragments are shown in space-filling drawings in Figure 4. These molecules have been labeled as large carbonate (LC) and large ester (LE), respectively.

Several large molecules were used in studies on phenyl rings flanked by two ester linkages. The T (terephthalic), I (isophthalic), IL (isophthalic, linear), and IK (isophthalic, kinky) structures, on which detailed calculations were performed, are shown in space-filling drawings in Figure 5. The

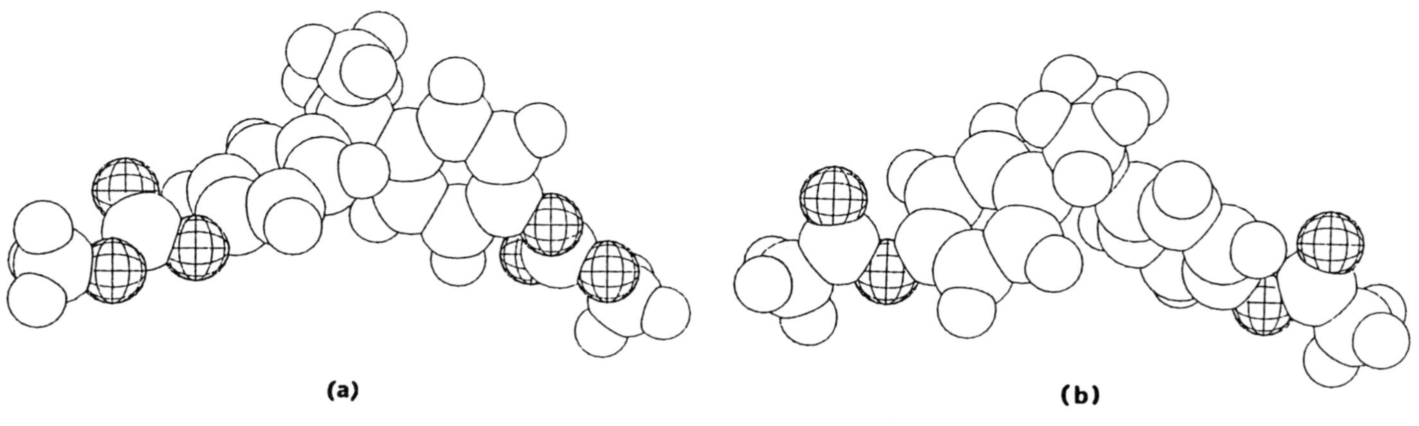

Figure 4 Space-filling illustrations of optimized geometries: (a) large carbonate (LC), and (b) large ester (LE).

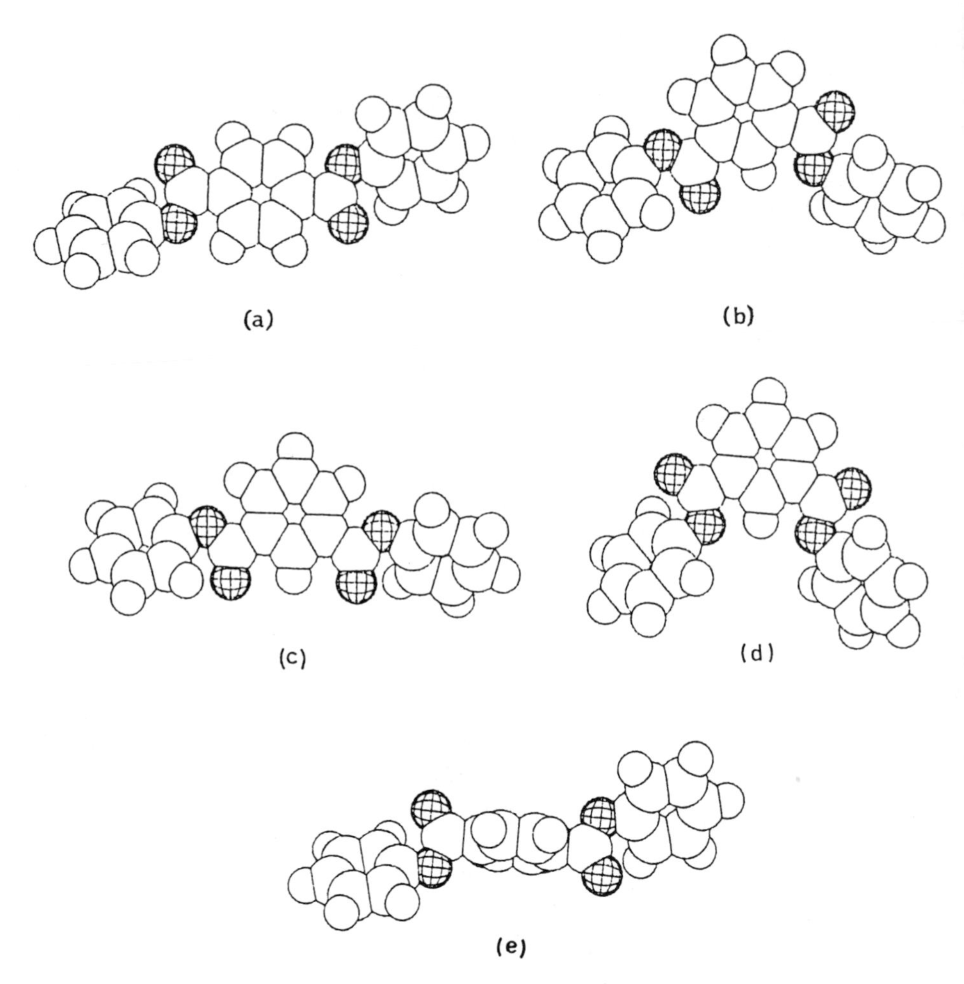

Figure 5 Space-filling illustrations of optimized geometries: (a) terephthalic (T); (b) isophthalic (I); (c) isophthalic, linear (IL); (d) isophthalic, kinky (IK); and (e) the rotational transition state of *T*.

transition state geometry obtained by a 90° rotation of the central phenyl ring in the T structure is also shown in Figure 5.

QUANTUM MECHANICAL CALCULATIONS

Approximate ground state structures were determined for DPP, DPC, and PB by using semiempirical methods [55]. Modified neglect of differential overlap (MNDO) and AM1 (Austin Model 1) [56] Hamiltonians from

MOPAC (molecular orbital package) were used to optimize geometries. In general, AM1 gives more accurate conformations.

The approximate ab initio partial retention of diatomic differential overlap (PRDDO) [57,58] technique was then used, to (a) determine refined ground state structures by reoptimizing selected geometrical parameters, (b) perform axial rotations to locate the angles at which torsional barriers occur, and (c) reoptimize selected geometrical parameters at fixed axial rotations to estimate rotation barriers.

By convention, the value of 0° was assigned to a dihedral angle when the two bonds defining it were coplanar. For positive angles of rotation, atoms are rotated counterclockwise about an axis of rotation, when looking down from the first atom defining this axis towards the second atom defining it.

The relative energy of a molecular conformation was defined as the difference between its total energy and the total energy of the most stable (lowest-energy, basepoint) conformation for that molecule. The relative energies can be compared with experimental data, provided that proper care is exercised in assessing both the degree of reliability of the computational results and the meaning and accuracy of the experimental data.

Armstrong-Perkins-Stewart bond orders [59] are valuable semiquantitative indicators of bond strength. Bond order is a very useful concept from classical chemistry, which has no unique and simple definition within the framework of quantum mechanics. Bond orders were calculated for model molecules of interest, from the PRDDO results.

Molecular volumes were calculated by using the SPACE program in the CHEMLAB-II software package [60]. SPACE draws a sphere with a radius equal to its van der Waals radius around each atom, and then performs a numerical integration over the resulting volume. Changes in volume are due to changes in (a) bond length, (b) bond angle, and (c) steric overlap of two more distant atoms. Molecular volumes have no unique and simple definition within the framework of quantum mechanics.

The calculations were carried out on VAX 8600 and VAX 11/785 computers in the Central Research VAX cluster of Dow Chemical Company in Midland, Michigan.

DISCUSSION OF QUANTUM MECHANICAL RESULTS

Geometry of 2,2-Diphenylpropane

The optimized geometry of DPP has C_2 symmetry. The twofold axis of symmetry rotates methyl groups into one another, and phenyl groups into one another. Phenyl rings display almost no deviation from planarity. They are twisted away from the C_p-C_1-C_p plane, and tilted by 48.1° relative to the C_p-C_1-C_p plane, in good agreement with the torsional angle of 46° calculated by Erman et al. [61] by force-field techniques. The resulting configuration resembles two blades of a propeller. Methyl groups swing away from the normal to avoid the rings. The three C_m—H bonds at each of the two C_m's are off by 9.25°, in opposite directions, from a conformation in which they would be exactly staggered relative to the three bonds at C_1, in good agreement with the torsional angle of 6° calculated by Erman et al. [61]. The C_p-C_1-C_p angle is 109.8°, while the C_m-C_1-C_m angle is 107.1°. Phenyl rings therefore form an angle larger than the tetrahedral angle (109.471°), thus taking up more space at the expense of the smaller methyl groups which form an angle smaller than the tetrahedral angle. The average C_1-C_m-H_m angle is 111.8°, consistent with the NMR prediction [8] of ~111° in BPAPC.

Geometry of Diphenyl Carbonate

The trans,trans conformation of C_s symmetry, which has a mirror plane of symmetry perpendicular to the plane of the carbonate linkage, is preferred. This mirror plane of symmetry contains C_c and the oxygen atom double-bonded to C_c, and reflects the two phenyl rings into one another. The phenyl rings are both at a torsional angle of 44° relative to the plane of the carbonate linkage, allowing some conjugation of the phenyl rings with the oxygen lone pairs, while keeping steric interaction of the ortho hydrogens on the phenyl rings with the carbonyl oxygen small. The phenyl ring torsional angle of 44° is in agreement with the torsional angle of 45° determined by Yoon and Flory [62] in crystalline DPC. Rotations of the phenyl rings about the O—C_p bonds are almost free rotations. The preference of DPC for this geometry of C_s symmetry is therefore *not* a strong one. For example, an alternative trans,trans conformation of C_2 symmetry, with an axis of twofold symmetry instead of a mirror plane, is very similar in energy. The crystallographic value [62] of the O′—C_c—O″ bond angle is 108°. The present calculations give an optimized value of 105.8°. Both of these values are much smaller than the 120° angle in ideal sp^2 hybridization, where the bonds to each of the three attached atoms are equivalent, rather than one being a double bond and the other two being single bonds. There is also a low-lying trans,cis and an equivalent, cis,trans conformation, but no low-lying cis,cis conformation.

Geometry of Phenyl Benzoate

The symmetry point group of PB is C_1; i.e., there are no point group symmetry elements. The ester linkage and the phenyl ring directly attached to C_e are coplanar within the limits of numerical accuracy of the calculations. This near-coplanarity is caused by the stabilization resulting from the resonance between electrons in the phenyl ring and in the ester linkage when planes of these two groups are coplanar with each other. In crystalline PB, the phenyl ring attached to C_e is at an angle of 9.8° with the ester linkage [63]. The other phenyl ring is attached to an oxygen atom and therefore has a bonding environment very similar to the bonding environments of the two phenyl rings in DPC. It is calculated to have a torsional angle of 46.4° relative to the plane of the ester linkage. This angle is just a little larger than the 44° angle calculated for the equivalent torsional angle in DPC, an effect which is not surprising, since the first phenyl ring of PB lies on the same plane as the plane of the ester linkage and is therefore closer to the phenyl ring bonded to the oxygen atom than in DPC. This torsional angle is 65.1° in crystalline PB [63]. There are two factors which can account for this 18.7° difference. First, the potential surface is very flat for this torsional motion, and the conformation with a torsional angle of 65.1° is only 0.6 kcal/mole above the basepoint structure. Second, the geometry which allows for optimum packing in the crystalline state is not necessarily the most stable geometry of the gas-phase (or "isolated") molecule. It may be possible to get more favorable crystalline packing by low-energy distortions away from the gas-phase geometry, as in the biphenyl molecule [64–66].

Rotations of Methyl Groups

The local bonding environment of methyl groups in BPAPC/PEC is represented adequately by DPP. The barrier to methyl group rotation has its minimum value (6.6 kcal/mole) if one methyl group is at its most favored conformation while the other methyl group is going through its maximum-energy conformation. There are three maxima, at conformations in which the three C_m—H_m bonds are off by 9.25° from an orientation in which they would be exactly eclipsed relative to the bonds at C_m, when looking down the C_l—C_m bond. The C_m—C_l—C_m bond angle becomes slightly larger, and the C_p—C_l—C_p bond angle becomes slightly smaller, as the methyl group is rotated into its transition state conformation. Methyl groups therefore need more space in the nearly eclipsed transition state structure than in the nearly staggered basepoint structure. A small (~15°) reorientation of one of the two phenyl rings also takes place, to remove a steric repulsion between a methyl hydrogen and a phenyl hydrogen in the ortho position. Such small oscillations of the phenyl rings are very low-energy processes which can be expected to occur continuously. Therefore, there is no special need for cooperation from phenyl rings for the methyl rotations to occur.

The PRDDO barrier of 6.6 kcal/mole compares well with the CNDO/2 (complete neglect of differential overlap, version 2) barrier [47] of 5.3 kcal/mole. The barrier to methyl group rotation, as estimated by either technique, is in the range of 5 to 7 kcal/mole.

Varadarajan and Boyer [5] have proposed that the DMS δ relaxation arises from methyl motions. A comparison of the computed methyl group rotation barriers with the activation energy of 7 to 8 kcal/mole estimated by Yee and Smith [3] for the DMS δ relaxation in BPAPC with one or both methyl groups replaced by ethyl groups, also supports the possibility that methyl motions are the cause of the DMS δ relaxation. NMR experiments [8], however, cast serious doubts on the identification of the methyl rotations as the cause of the DMS δ relaxation. These NMR experiments show that the methyl group rotational motions become slower than 10 kHz below a temperature of −120°C, resulting in a "static" pattern for the spectrum. On the other hand, the DMS δ relaxation in BPAPC [5] occurs at about −220°C. Therefore, the most plausible conclusion is that the methyl group rotations are *not* the sole major contributors to the DMS δ relaxation, but that along with other, even lower-energy motions (see below and Summary) they are among several types of contributing motions.

DPP is adequate for studying the methyl group rotations in BPAPC. In contrast, polyisobutylene (PIB) chains are subject to large steric repulsions that computations on analogues of low molecular weight do not take into account, since the steric strain is relieved at the chain ends [67]. This observation points out to the need to carefully consider the structures and energetics of each individual polymer, when modeling the polymeric properties.

Rotations of Phenyl Rings Attached to Isopropylidene Groups

These types of phenyl rings are attached to an isopropylidene group at one end, and to a carbonate group in the BPAPC fragments or an ester linkage in the ester fragments, para to the isopropylidene group at the other end.

Rotations of phenyl rings about the O—C_p bonds in DPC and PB are almost free, with a barrier of only 1.6 kcal/mole in DPC and 2.9 kcal/mole in PB. These barriers are caused by bond-angle openings required to reduce the steric repulsion between the doubly bonded oxygen atom and an ortho hydrogen atom in the rotating phenyl ring, as the phenyl ring goes through a configuration in which it is nearly coplanar with the plane of the carbonate or ester linkage.

The situation is more complex for rotations of phenyl rings in DPP. Torsional oscillations of up to ±30° of the phenyl rings only require ~3 kcal/mole of energy, even in the presence of the interchain interactions found in a solid polymer. They should, therefore, occur down to very low temperatures and possibly contribute to the DMS δ relaxation. Such oscillations of the phenyl rings have indeed been observed by NMR [13,18,20]. If one phenyl ring in DPP is held fixed at any given torsional orientation while the other phenyl ring is rotated past it, a very high barrier, caused by a strong hard sphere repulsion between ortho hydrogens on different phenyl rings, is encountered. The two phenyl rings will therefore have to move with at least some degree of synchronism in order not to get into each other's way.

In incomplete optimizations reported previously [1,2], a rotation barrier of 9 kcal/mole was calculated by PRDDO, and ~3 kcal/mole by AM1, which usually underestimates rotation barriers, for the phenyl ring in DPP. More complete studies have been carried out since then, showing that there is a synchronous rotation pathway of much lower barrier (~0.5 kcal/mole by AM1, corresponding to an estimated rotation barrier of less than 2 kcal/mole by PRDDO). The calculated structure was confirmed to be a true transition state via AM1 calculations, which showed that it had one negative vibrational frequency. The AM1 heat of formation is plotted in Figure 6, and the transition state geometry is shown in a space-filling drawing in Figure 7 for this pathway.

This new result agrees with Tonelli's conclusion [43], based on van der Waals repulsion calculations, that rotations in isolated BPAPC chains are nearly free. The isopropylidene and the carbonate or the ester groups bonded to the phenyl rings are para to one another. The C_I—C_p bonds lie almost on the same axis as the O—C_p bonds. The torsional angles which give the transition state relative to motion past the isopropylidene group are 45° to 50° out of phase with the torsional angles which give the transition state relative to motion past the carbonate group or the ester linkage. Therefore, the full rotation barriers in *isolated* BPAPC and PEC chains should be quite small (2 to 4 kcal/mole), and probably slightly higher in ester fragments than in carbonate fragments. The remainder of the much larger activation energies observed in the solid polymers are likely to result from interchain interactions.

Two recent models for interchain interactions are the models of Schaefer et al. [19] and of Mitchell and Windle [68], both of which assume the existence of locally ordered packings of adjacent chain segments a few monomer units in length. The results of the Brownian dynamics simulations of Perchak et al. [52] are consistent with the dynamic lattice mechanism proposed by Schaefer et al. [19] for the ring flip, which the simulations were specifically set up to simulate. On the other hand, the analysis of static correlation functions, as described by Fischer and Dettenmaier [69],

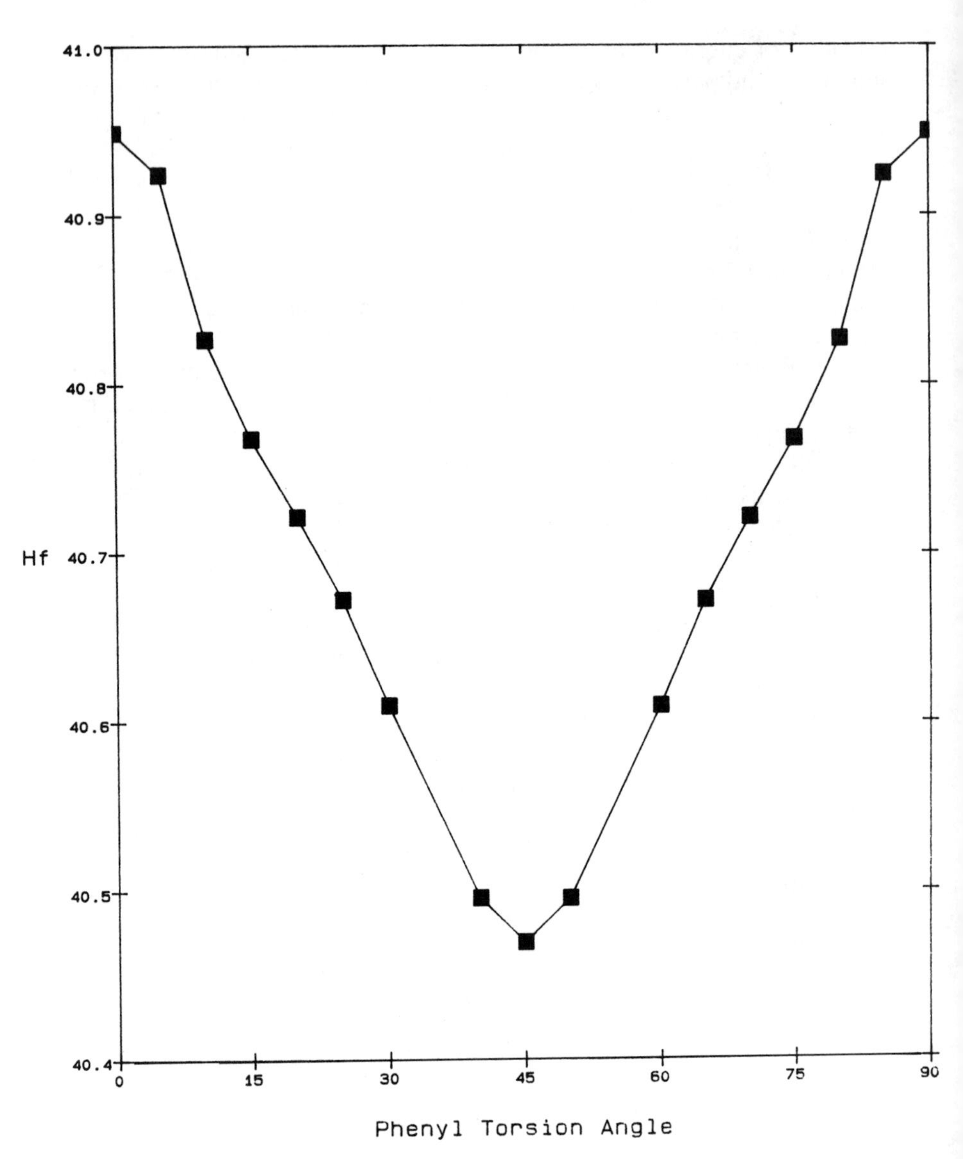

Figure 6 AM1 heat of formation as a function of the angle of rotation of the phenyl ring for the rotation pathway with the lowest barrier in DPP.

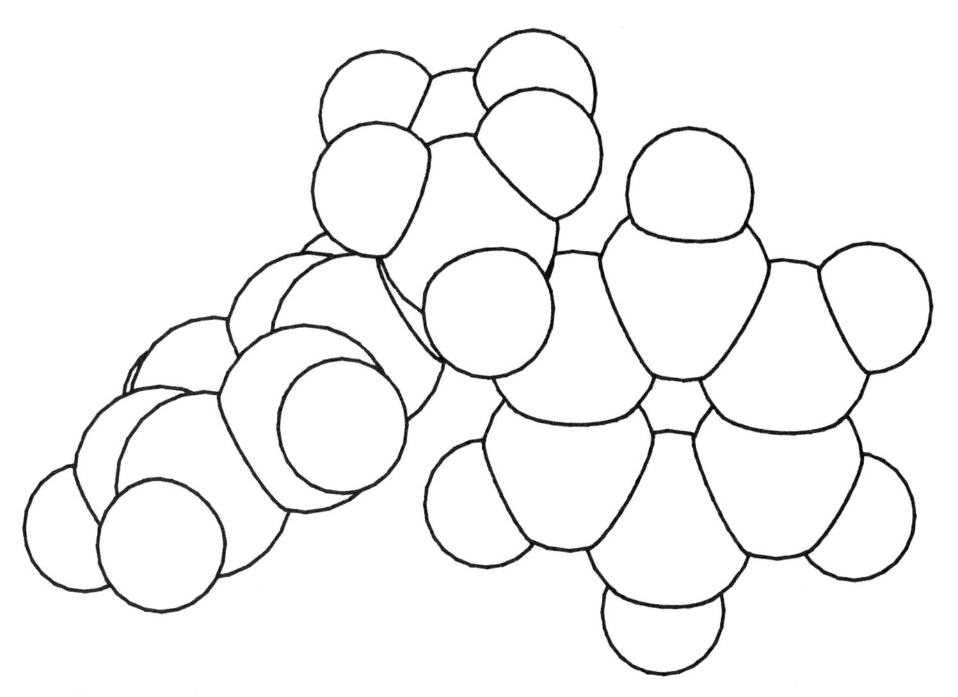

Figure 7 Space-filling illustration of the transition state geometry for the pathway with the lowest AM1 phenyl ring rotation barrier in DPP.

and the recent NMR data of Bubeck et al. [8], seem to favor a random coil type of model [70,71] over models which assume the existence of lateral ordering caused by the parallelization of chains. Only detailed molecular dynamics simulations are likely to be able to shed significant further light on these issues.

Rotations of Phenyl Rings Flanked by Two Ester Linkages

One third of the phenyl rings in the ester fragments in PEC are flanked by two ester linkages instead of being attached to an isopropylidene group at one end. *Most* of the differences between the physical properties of BPAPC and PEC result from the fact that these phenyl rings are in a totally different type of environment than in BPAPC. *All* of the differences between terephthalic and isophthalic PEC are caused by differences in the motions of these ester phenyl rings.

The results of the rotation about the C_e—C_p bond in PB are relevant to this problem. When this type of rotation is carried out, the phenyl ring attached to C_e is gradually rotated away from its minimum-energy conformation, which is coplanar with the ester linkage. There is a small and gradual increase in energy, caused by the gradual elimination of the extended resonance between the phenyl ring and the ester linkage. The maxima for this rotation occur at the perpendicular conformations with rotation angles of 90° and 270°. The 90° conformation lies 6.7 kcal/mole above the basepoint geometry. In the terephthalic PEC fragments, as in the large-model molecule T, there are *two* ester linkages coplanar with each ester phenyl ring. These two linkages are para to one another. The two C_e—C_p bonds are essentially coaxial with each other. Therefore, the barrier to rotation in terephthalic PEC will be almost entirely due to the loss of the extended resonance stabilization binding the phenyl ring to *both* of the ester linkages. Since the resonance stabilization in phenyl benzoate is 6.7 kcal/mole, the stabilization in the terephthalic ester fragment can be estimated to be approximately $2 \times 6.7 = 13.4$ kcal/mole.

The barrier to the rotation of the terephthalic ester phenyl ring in the large-model molecule T is smooth and sinusoidal, as illustrated by the curve labeled T in Figure 8, where the relative energies are plotted as a function of the rotation angle. The barrier height is 13.1 kcal/mole, almost equal to the barrier height of 13.4 kcal/mole estimated above by assuming that the barrier to the rotation of the ester phenyl rings in terephthalic PEC will be approximately equal to twice the barrier of 6.7 kcal/mole calculated for PB. The terephthalic phenyl ring in T also has a large-amplitude (±30°), low-energy (~2.6 kcal/mole) oscillation range about its equilibrium position, just like phenyl rings attached to isopropylidene groups.

A motion of one-third of the phenyl rings with an activation energy of ~13.1 kcal/mole (or perhaps more, after interchain interactions are included) should clearly contribute significantly to the high-temperature portion of the DMS or dielectric γ relaxation. This prediction is in agreement with NMR studies [8,9] indicating that the phenyl rings in terephthalate units are more motionally restricted than the phenyl rings in BPAPC, with their motions ceasing on the NMR time scale at a higher temperature.

All four large-model molecules (T, I, IK, and IL) are very similar in energy. The relative energies of their basepoint geometries are within 2 kcal/mole, so that there is very little reason to choose between them on grounds of stability at the molecular level. Each type of ester linkage has different effects on the winding and packing of chains and on the types of rotations and rocking motions which can occur. Terephthalate units are always highly linear. Several conformations with very different nonlinearities (kinkiness) may occur in comparable abundance in isophthalate units. One isophthalic conformation might be strongly favored over the others

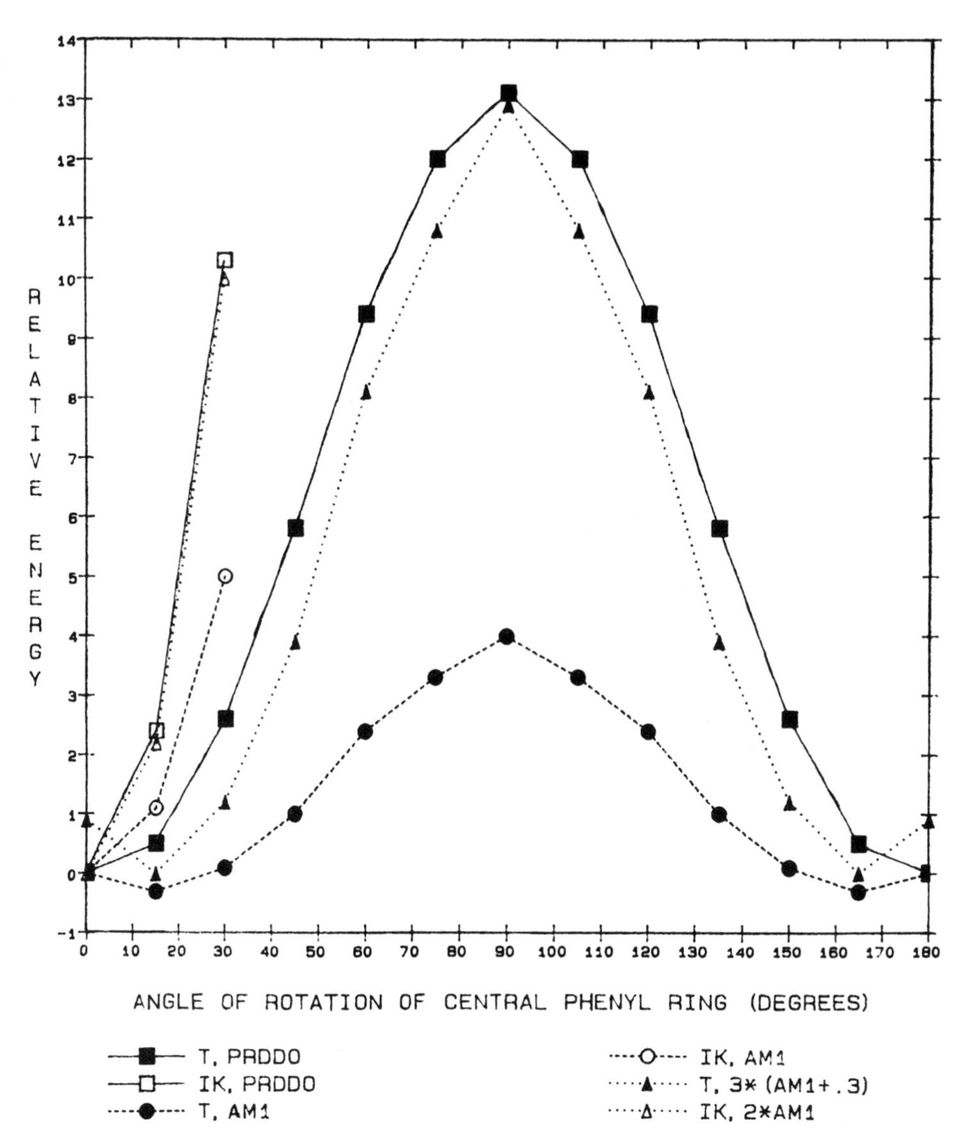

Figure 8 Relative energy in kcal/mole, as a function of the angle of rotation of the central phenyl ring, for T and IK.

by chain-packing constraints in a crystalline structure; however, there is no obvious reason why this should be the case in these amorphous polymers.

The variation of Armstrong-Perkins-Stewart bond orders for the C_e—C_p bonds is shown in Figure 9 as a function of the angle of rotation. A gradual reduction of resonance stabilization is seen for T, as the rotation angle is gradually increased to 90°, making the plane of the central phenyl ring less and less coplanar with the planes of the two ester linkages.

The low-energy (<3 kcal/mole) oscillation or rocking motion range of isophthalic phenyl rings is severely restricted (±15°). The relative energies for a rocking motion of 30° are >10 kcal/mole, showing that the walls of the potential wells are very steep, as illustrated for IK by the curve labeled IK in Figure 8. A simple flipping motion of the isophthalic phenyl rings is therefore not possible. No low-barrier *cooperative* phenyl ring flipping motions exist either for the ester phenyl ring in isophthalic PEC. The lack of low-energy cis conformations about the C—O bonds in ester fragments, and the resulting impossibility of crankshaft-type motions such as the one proposed by Jones [72,73] (see more extensive discussion below), is the major reason for the lack of easy cooperative motions.

NMR experiments [8,9] show that the phenyl rings in isophthalate units are much more motionally restricted than the phenyl rings in terephthalate units. There is very little motional narrowing of the deuterium NMR pattern in isophthalic PEC, and thus very little backbone motion, until the temperature is increased to near the glass-transition temperature of the system. It appears that cooperative motions involving the ester phenyl rings in isophthalic PEC are disfavored both statistically because of their great complexity and kinetically because of their very high rotation barriers.

Trans,Trans Versus Trans,Cis Isomerization About C—O Bonds

It was estimated [1] that the trans,cis conformation in DPC will have a relative energy of 1.1 kcal/mole, similar to the 1.3 kcal/mole value estimated by Williams and Flory [53] and the 1.8 kcal/mole obtained by Tekely and Turska [74] via semiempirical methods. Recently, Laskowski et al. [51] have shown that ab initio calculations at the 6-31G* level, which may give more accurate values of the barrier than either the semiempirical or the PRDDO methods, result in a barrier of 2.75 kcal/mole, which is significantly higher than all of the previous estimates. On the other hand, the inclusion of electron correlation effects, which are very important in polar molecules, may lower the 6-31G* values significantly. The exact value of this barrier is therefore still an open question.

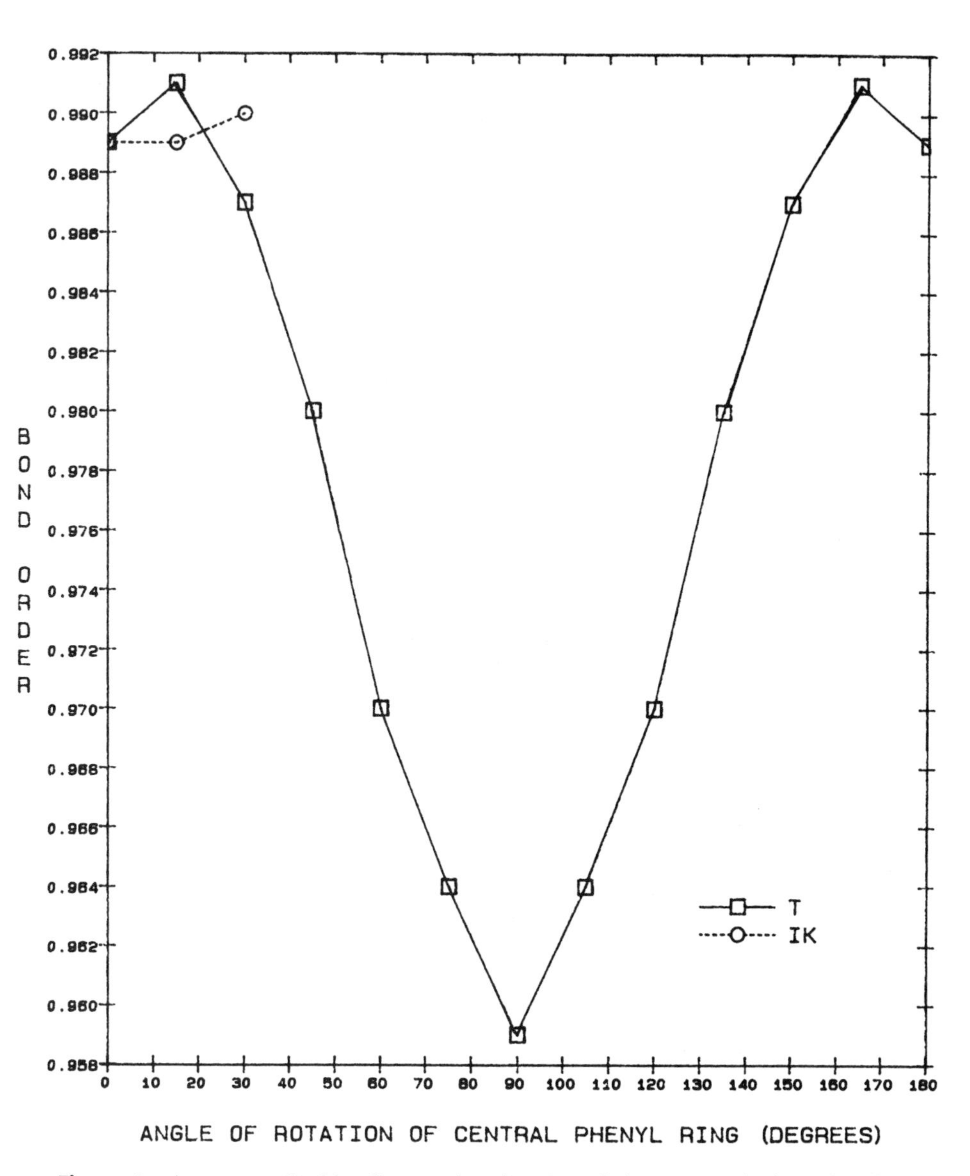

Figure 9 Armstrong-Perkins-Stewart bond order of the two equivalent C_e—C_p bonds, as a function of the angle of rotation of the central phenyl ring, for T and IK.

At room temperature (25°C = 298 K), the Boltzmann factor is $kT \simeq$ 0.593 kcal/mole. Since the cis,trans conformation is equivalent to the trans,cis conformation, conformations in which the bonding around one of the C_c—O bonds is cis have a statistical weight of 2. If the relative energy of the trans,cis conformation in BPAPC is 1.1 kcal/mole, the trans,cis and cis,trans conformations will have an abundance of 24%. On the other hand, if the conformational energy difference calculated by Laskowski et al. [51] (2.75 kcal/mole) is used, the trans,cis and cis,trans conformations will have an abundance of only 2% at 25°C. (Note that both estimates assume that there is sufficient mobility in the glassy state for trans,cis and cis,trans defects to exist in thermal equilibrium with the trans,trans conformation at room temperature. If these defects were assumed to be frozen into the structure at $T_g \simeq$ 150°C = 423 K, a larger Boltzmann factor (kT_g) would have to be used, resulting in a slightly higher percentage of trans,cis and cis,trans defects.)

The presence of two types of local orientations around carbonate linkages can affect the conformations of the chains, and therefore also affect properties such as persistence length and entanglement molecular weight. It can also make crankshaft-type cooperative intrachain motions possible by simultaneous rotations of nearly parallel pairs of C_c—O bonds. An example of such a motion in glassy BPAPC is the one suggested by Jones [72,73] in his molecular-level model for motion and relaxation. On the other hand, if the abundance of the trans,cis and cis,trans defects is really only about 2%, then this type of crankshaft mechanism, even if it occurs, may not play a major role in the relaxation behavior. Furthermore, Tekely [22] has shown that it is possible to interpret the results of NMR experiments by using a semimolecular-level model which does not require any cooperative (crankshaft-type) motions, but merely bond rotations between isomeric transition states.

FTIR spectroscopy can be a valuable tool for examining small relative energy differences and relating them to possible transitions from the higher-energy trans,cis conformation to the lower-energy trans,trans conformation during physical aging, or to possible transitions from the trans,cis conformation to the more linear trans,trans conformation during tensile stress (as to be expected from considerations based on the theory of rubber elasticity). Garcia [40] has reported FTIR experiments on the *trans/gauche* energy difference and *trans/gauche* abundance ratio for the glycol segment in PET, another system where two low-lying conformations exist. FTIR determinations of molecular moiety contributions to physical aging effects have, thus far, been less definitive than determinations made by NMR; however, the smaller shift, in terephthalic PEC, of the band with a base frequency of 1775 cm^{-1}, which is associated with the carbonyl moiety, correlates well with the fact that the resistance of this copolymer to the embrittlement caused by physical aging is greater than the resistance of BPAPC or isophthalic PEC [8].

Because of the relative proximity of the phenyl ring bonded directly to C_e, there is definitely no low-lying local minimum with cis bonding about the C_e—O bond in PB, and therefore in terephthalic and isophthalic units. This is an important difference between the carbonate and the ester fragments. It has important implications in terms of the chain winding parameters of the ester fragments and the limitations it imposes on cooperative intrachain motions.

Rocking Motions of Carbonyl Groups

The carbonyl group in DPC is quite adequate in representing the local bonding environments of the carbonyl groups in the carbonate fragments of BPAPC and PEC. The present calculations show that large-amplitude rocking motions of the carbonyl groups, over a range of ~70°, only require ~3 kcal/mole and can couple with any one of the motions discussed above, as well as being among the contributors to the DMS δ relaxation. On the other hand, when a C=O rocking motion of larger amplitude is performed so that the orientation of the C=O group goes outside of this ~70° range, the energy required rapidly goes up. The ~70° range of low-energy rocking motions allowed by the present calculations is larger than the ~40° range found by Henrichs et al. [11] in a computer analysis of their variable-temperature ^{13}C NMR results. One reason for this difference might be the erroneous assumption made by Henrichs et al. [11] that the reorientation of the carbonyl group occurs as an activated process involving passage over an energy barrier.

Molecular Volumes

The methyl groups and phenyl rings in the basepoint conformation of DPP are oriented to minimize steric repulsions between nonbonded hydrogen atoms on different methyl and phenyl subunits, resulting in a small overlap between the atomic volumes of nonbonded atoms. The volume of DPP therefore has its maximum value for the minimum-energy basepoint conformation and decreases as the relative energy increases. The rotation barriers are almost completely due to steric effects. Calculated magnitudes of the rotation barriers follow almost the same order as that of increasing overlap between the volumes of nonbonded atoms and, hence, of decreasing molecular volume.

For DPC, the correlation of higher conformational energy with smaller molecular volume does not hold. The two key factors are (a) resonance

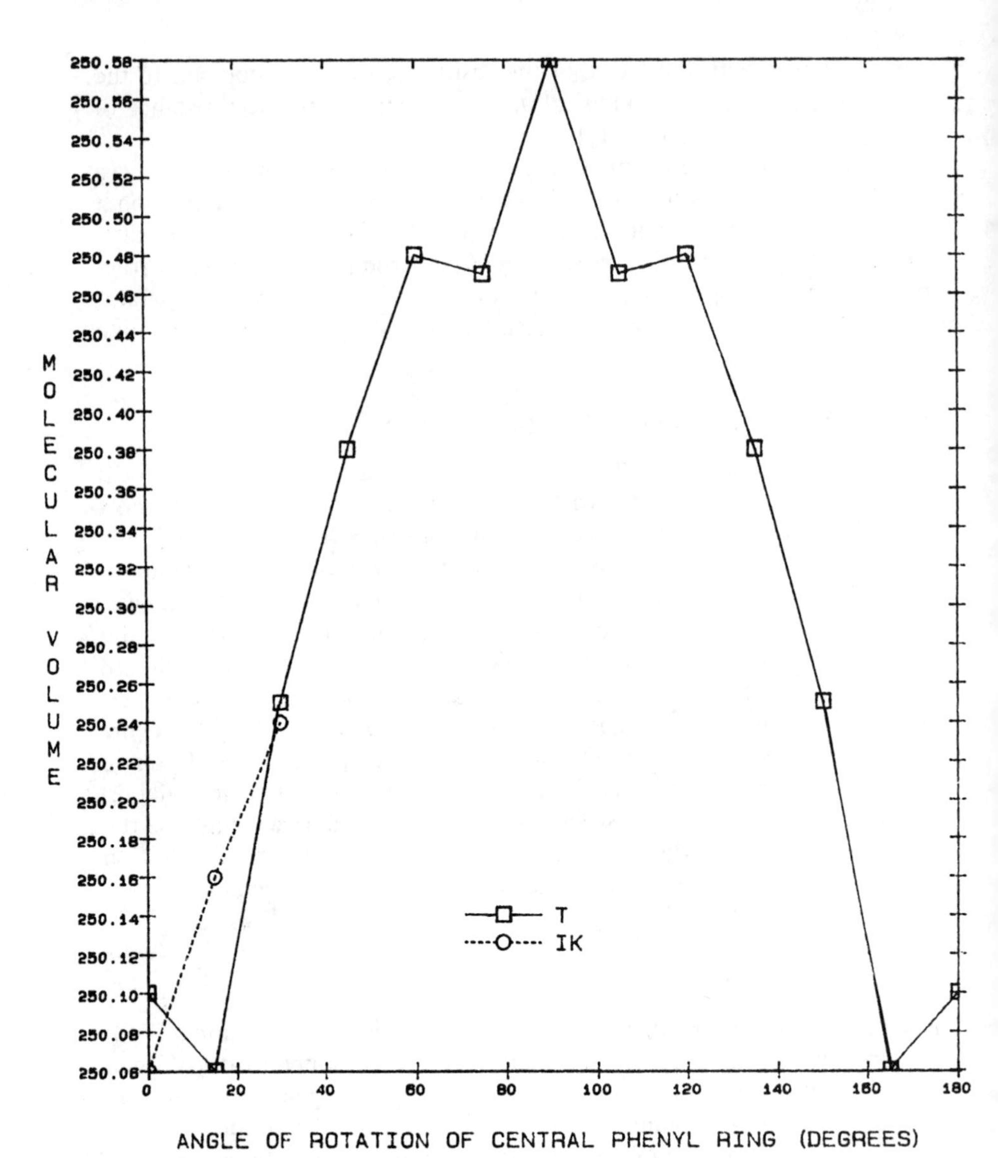

Figure 10 Molecular volume in Å^3, as a function of the angle of rotation of the central phenyl ring, for T and IK.

stabilization, maximized when the carbonate linkage and the two phenyl rings are mutually coplanar, and (b) steric repulsions between the doubly bonded oxygen atom and the hydrogen atoms in ortho positions on the phenyl rings, maximized when the planes of these three subunits are mutually perpendicular. The preferred tilted orientations of the two phenyl rings are the result of a compromise between maximizing resonance stabilization and minimizing nonbonded steric repulsions.

For PB, the phenyl ring directly bonded to C_e can be coplanar with the plane of the ester linkage and gain the full amount of resonance stabilization possible without causing steric repulsions. As this phenyl ring is rotated away from its coplanar orientation toward the transition state conformation where it is perpendicular to the ester linkage, an increasingly larger molecular volume results.

The molecular volume increases in going from the basepoint T structure to the rotational transition state with the central phenyl ring perpendicular to the two ester linkages, as shown in Figure 10. This increase is slightly larger in amount (0.48 $Å^3$ versus 0.34 $Å^3$), but slightly smaller in percentage (0.19% versus 0.21%), than calculated for the same motion in PB.

In the transition state structure for the rotation of the other phenyl ring in PB about the O—C_p axis, this second phenyl lies almost on the same plane as the ester linkage, resulting in a smaller molecular volume than in the basepoint structure. The increase in energy, in this case, is mainly caused by the bond-angle distortions required to alleviate the resulting steric repulsions between one of the ortho hydrogen atoms of this phenyl ring and the doubly bonded oxygen atom.

INTERPRETATION OF THE DYNAMIC SPECTRA

The activation energies obtained for the δ, γ, and β relaxations observed by DMS or dielectric relaxation studies are all activation energies for some simple or complex motion at the molecular level. Increasingly larger-scale motions, with larger activation energies, successively "thaw" with increasing temperature. As the larger-scale motions thaw, all the smaller motions also continue to occur, with larger amplitudes and/or frequencies than before. Consequently, the system becomes increasingly more complex, with a plethora of motions coupled to one another in a multitude of complicated patterns. For example, in the δ relaxation region, the only available motions are small motions involving oscillations of subunits of a monomer in a potential well. Motions requiring a subunit of a monomer to go over a potential barrier are the next larger scale of possible motions, but they do not yet occur with sufficient frequency. As the temperature is increased, first the methyl group rotations, then the flips of phenyl rings attached to isopropylidene phenyl rings, and then the flips of phenyl rings in terephthalate units successively and gradually become "thawed" and predominant. An example of the coupling of smaller-scale motions with the larger ones at higher temperatures is the coupling of small phenyl ring oscillations to the flips in the γ relaxation.

Superpositions and/or couplings of five *"small" motions* of subunits of a monomer, which involve oscillations about an energy minimum in a potential well, add up to the extremely low-temperature δ relaxation, which cannot be adequately explained by any single motion. When extremely low temperatures are reached, all of the larger motions cease, and these five "small" motions are the only ones left, to make up the δ relaxation. These motions can also facilitate the γ relaxation by coupling with larger motions.

1. Oscillation of phenyl rings attached to an isopropylidene group on one side, over a $\pm 30°$ range. At higher temperatures, these oscillations can facilitate the 180° flips of the phenyl rings, and their being correctly accounted for is therefore crucial.
2. Oscillation of phenyl rings attached to two ester linkages in a terephthalic configuration, over a $\pm 30°$ range. At higher temperatures, these oscillations can also facilitate the 180° flips of the phenyl rings.
3. Rocking of phenyl rings attached to two ester linkages in an isophthalic configuration, over a $\pm 15°$ range, with no possibility of flipping.
4. Oscillation of methyl groups.
5. Rocking of carbonyl groups over an angular range of $\sim 70°$, as well as other small oscillations of the carbonate linkages.

Three *"large" motions* of subunits of a monomer, each of which involves going over a rotation barrier, are the major constituents of the γ relaxation:

1. Methyl group rotation. This simple motion, which does not require any special synchronization, has a much lower barrier than both of the other two "large" motions, so that it can be expected to contribute especially to the low-temperature end of the γ relaxation.
2. 180° flip of phenyl rings attached to an isopropylidene group on one side. These phenyl rings preferentially flip in synchronized pairs attached to the same isopropylidene group. The rotation barrier is very slightly higher if there is an ester linkage rather than a carbonate linkage para to the isopropylidene group. Considerably more than half of the barrier to the flip of these particular phenyl rings is likely to be caused by interchain effects.
3. 180° flip of phenyl rings attached to two ester linkages in a terephthalic configuration. This motion has a considerably higher barrier than the

second motion, even when only intrachain interactions are considered. It should, therefore, occur on the higher-temperature side of the γ relaxation. It has a higher statistical weight, however, since it is a simple motion which does not require any synchronization. It is, therefore, not much more disfavored than the first motion, and occurs as a part of the same relaxation, as observed by NMR experiments.

Motions of the polar carbonyl group and carbonate linkage should strongly couple with the three "large" motions involved in the γ relaxation, as shown by the large dielectric γ relaxation peak in BPAPC. In addition, as mentioned above, small-amplitude oscillations of the phenyl rings should couple with and facilitate the full flipping motions.

Finally, two major types of *isomerizations* are possible in the generic PEC. Each of these isomerizations can have significant implications in terms of the DMS, dielectric relaxation and NMR spectra. They can participate in the γ and β relaxations, as well as affecting the viscoelastic properties.

1. The presence of trans,cis or cis,trans conformations about the C—O bonds in the carbonate linkages, in addition to the predominant trans,trans conformation, can provide a low-barrier mechanism for crankshaft-type cooperative motions. On the other hand, if the abundance of these defect conformations is only about 2%, such mechanisms may not turn out to be major contributors to the observed mechanical loss.
2. The possible presence, in comparable abundance, of several conformations having very different amounts of nonlinearity (kinkiness) in isophthalate units, can affect the winding and packing of chains, and therefore can also affect the types of motions which can take place. By contrast, the terephthalate segments are always stiff and very linear. This difference between the two types of ester segments might result in differences in the tensile and physical aging behavior of the two types of ester linkages.

The calculations summarized in this chapter do not, by themselves, shed any light on the mechanism of the β relaxation, beyond suggesting that the principal modes of motion involved in this relaxation should be larger than the motion of a localized subunit, such as a phenyl ring moving with some cooperation from other functional groups in its immediate vicinity. In this regard, it is worth noting that Bershtein et al. [75,76] have compared the β relaxation and glass-transition behavior of a large number of polymers. These authors have suggested that the β_2 relaxation in BPAPC primarily occurs by displacements of chain fragments from two to four monomer units in length in sites of less dense packing [75], and that the glass transition occurs by *correlated* displacements of three to five such chain fragments [76].

There are two very promising new directions for further theoretical work to enhance the understanding of dynamic relaxations in amorphous polymers. One of these directions is to consider, in great detail, the thermodynamics and statistical mechanics of these systems. One of the authors of this chapter (J.B.) has used thermodynamic and statistical mechanical considerations to derive a tentative model for secondary relaxations, and successfully applied this model to relaxations in poly(ester carbonates) [77] and in poly(alkyl acrylates) [78]. The other direction is to carry out molecular dynamics simulations of the motions in large-model systems as a function of the temperature; however, the time scales involved in these relaxations are longer than the time scales accessible in molecular dynamics simulations within a reasonable amount of computer time, and such calculations may therefore require very large computational resources.

SUMMARY

Quantum mechanical calculations utilizing the PRDDO and AM1 methods were used to help in characterizing the molecular motions associated with relaxations observed by DMS and dielectric relaxation spectroscopy. Intrachain rotations in BPAPC and in PEC copolymers were studied by means of calculations on model molecules representing short segments of isolated chains. The sub-T_g dynamic relaxations were then discussed within the framework of these calculations. Promising directions for further research were also pointed out. Most of the material covered in this chapter is a review of previously reported work. The reader is referred to the earlier papers [1,2] for further details.

ACKNOWLEDGMENT

We thank R. A. Bubeck, C. P. Christenson, A. J. Hopfinger, A. Letton, W. N. Lipscomb, D. E. McLemore, J. K. Rieke, N. G. Rondan, P. B. Smith, and U. W. Suter for helpful discussions.

REFERENCES

1. J. Bicerano and H. A. Clark, *Macromolecules*, *21*, 585 (1988).
2. J. Bicerano and H. A. Clark, *Macromolecules*, *21*, 597 (1988).
3. A. F. Yee and S. A. Smith, *Macromolecules*, *14*, 54 (1981).
4. Y. P. Khanna, *J. Thermal Anal.*, *30*, 153 (1985).

5. K. Varadarajan and R. F. Boyer, *J. Polym. Sci. Polym. Phys. Ed.*, *20*, 141 (1982).
6. R. A. Bubeck, S. E. Bales, and H. D. Lee, *Polym. Eng. Sci.*, *24*, 1142 (1984).
7. J. Bicerano, E. Fouch and P. A. Percha, unpublished work.
8. R. A. Bubeck, P. B. Smith, and S. E. Bales, in *Order in the Amorphous "State" of Polymers*, p. 347 (S. E. Keinath, R. C. Miller, and J. K. Rieke, eds.), Plenum Press, New York (1987).
9. P. B. Smith, private communication.
10. T. R. Steger, J. Schaefer, E. O. Stejskal, and R. A. McKay, *Macromolecules*, *13*, 1127 (1980).
11. P. M. Henrichs, M. Linder, J. M. Hewitt, D. Massa, and H. V. Isaacson, *Macromolecules*, *17*, 2412 (1984).
12. A. K. Roy, A. A. Jones, and P. T. Inglefield, *Macromolecules*, *19*, 1356 (1986).
13. P. T. Inglefield, R. M. Amici, J. F. O'Gara, C.-C. Hung, and A. A. Jones, *Macromolecules*, *16*, 1552 (1983).
14. A. A. Jones, J. F. O'Gara, P. T. Inglefield, J. T. Bendler, A. F. Yee, and K. L. Ngai, *Macromolecules*, *16*, 658 (1983).
15. P. T. Inglefield, A. A. Jones, R. P. Lubianez, and J. F. O'Gara, *Macromolecules*, *14*, 288 (1981).
16. J. F. O'Gara, S. G. Desjardins, and A. A. Jones, *Macromolecules*, *14*, 64 (1981).
17. H. W. Spiess, *Pure and Appl. Chem.*, *57*, 1617 (1985).
18. H. W. Spiess, *Colloid and Polym. Sci.*, *261*, 193 (1983).
19. J. Schaefer, E. O. Stejskal, D. Perchak, J. Skolnick, and R. Yaris, *Macromolecules*, *18*, 368 (1985).
20. J. Schaefer, E. O. Stejskal, R. A. McKay, and W. T. Dixon, *Macromolecules*, *17*, 1479 (1984).
21. J. Schaefer, R. A. McKay, E. O. Stejskal, and W. T. Dixon, *J. Magn. Reson.*, *52*, 123 (1983).
22. P. Tekely, *Macromolecules*, *19*, 2544 (1986).
23. Y. Ishida and S. Matsuoka, *Polym. Prepr. (ACS, Div. Polym. Chem.)*, *6*, 795 (1965).
24. E. Sacher, *J. Macromol. Sci.-Phys.*, *B10*, 319 (1974).
25. D. C. Watts and E. P. Perry, *Polymer*, *19*, 248 (1978).
26. L. Guerdoux and E. Marchal, *Polymer*, *22*, 1199 (1981).
27. A. W. Aziz and K. N. Ab-El-Nour, *J. Appl. Polym. Sci.*, *31*, 2267 (1986).
28. J. M. Pochan, H. W. Gibson, M. F. Froix, and D. L. F. Hinman, *Macromolecules*, *11*, 165 (1978).
29. J. M. Pochan, H. W. Gibson, and D. L. F. Pochan, *Macromolecules*, *15*, 1368 (1982).
30. Y. Aoki and J. O. Brittain, *J. Polym. Sci. Polym. Phys. Ed.*, *14*, 1297 (1976).
31. J. Hong and J. O. Brittain, *J. Appl. Polym. Sci.*, *26*, 2459 (1981).
32. G. Locati and A. V. Tobolsky, *Adv. Mol. Relax. Proc.*, *1*, 375 (1970).
33. J. D. Ferry, *Viscoelastic Properties of Polymers*, 2nd ed., Wiley, New York (1970).

34. R. E. Robertson, *J. Polym. Sci. Polym. Symp.*, *63*, 173 (1978).
35. L. C. E. Struik, *Physical Aging in Amorphous Polymers and Other Materials*, Elsevier, Amsterdam (1978).
36. M. Washer, *Polymer*, *26*, 1546 (1985).
37. R. N. Haward, J. N. Hay, I. W. Parsons, G. Adam, A. A. K. Owadh, C. P. Bosnyak, A. Aref-Azaf, and A. Cross, *Colloid and Polym. Sci.*, *258*, 643 (1980).
38. D. C. Prevorsek and B. T. DeBona, *J. Macromol. Sci.-Phys.*, *B19*, 605 (1981).
39. E. J. Kramer, *Adv. Polym. Sci.*, *52/53*, 33 (1983).
40. D. Garcia, in *Proceedings of the Twelfth North American Thermal Analysis Society Conference*, p. 256 (J. C. Buck, ed.), NATAS (1983).
41. R. A. Bubeck, H. Y. Yasar, and B. H. Hammouda, *Polym. Commun.*, *30*, 25 (1989).
42. M. Hutnik and U. W. Suter, *Polym. Prepr. (ACS, Div. Polym. Chem.)*, *28*, 293 (1987).
43. A. E. Tonelli, *Macromolecules*, *5*, 558 (1972).
44. A. E. Tonelli, *J. Polym. Sci. Polym. Lett. Ed.*, *11*, 441 (1973).
45. P. R. Sundararajan, *Can. J. Chem.*, *63*, 103 (1985).
46. S. Perez and R. P. Scaringe, *Macromolecules*, *20*, 68 (1987).
47. J. R. Fried, A. Letton, and W. J. Welsh, *Polymer*, *31*, 1032 (1990).
48. M. A. Mora, M. Rubio and C. A. Cruz-Ramos, *J. Polym. Sci. Polym. Phys. Ed.*, *24*, 239 (1986).
49. D. T. Clark and H. S. Munro, *Polym. Degrad. Stability*, *5*, 23 (1983).
50. J. T. Bendler, *AIP Conf. Proc. (Polym. Flow Interact.)*, *137*, 227 (1985).
51. B. C. Laskowski, D. Y. Yoon, D. McLean, and R. L. Jaffe, *Macromolecules*, *21*, 1629 (1988).
52. D. Perchak, J. Skolnick, and R. Yaris, *Macromolecules*, *20*, 121 (1987).
53. A. D. Williams and P. J. Flory, *J. Polym. Sci. Part A-2*, *6*, 1945 (1968).
54. MOGLI is a software package copyrighted and sold by the Evans & Sutherland Computer Corporation.
55. T. Clark, *A Handbook of Computational Chemistry*, Wiley, New York (1985).
56. J. P. Stewart, Version 2.14.
57. T. A. Halgren and W. N. Lipscomb, *J. Chem. Phys.*, *58*, 1569 (1973).
58. T. A. Halgren, D. A. Kleier, J. H. Hall, Jr., L. D. Brown, and W. N. Lipscomb, *J. Am. Chem. Soc.*, *100*, 6595 (1978).
59. D. R. Armstrong, P. G. Perkins, and J. P. Stewart, *J. Chem. Soc. Dalton Trans.*, 838 (1973).
60. CHEMLAB-II is a molecular modeling software package developed and owned by Chemlab Incorporated, and marketed by Molecular Design Limited.
61. B. Erman, D. C. Marvin, P. A. Irvine, and P. J. Flory, *Macromolecules*, *15*, 664 (1982).
62. D. Y. Yoon and P. J. Flory, unpublished work referenced in Ref. 61.
63. J. M. Adams and S. E. Morsi, *Acta Cryst.*, *B32*, 1345 (1976).
64. A. Almenningen, O. Bastiansen, L. Fernholt, B. N. Cyvin, S. J. Syvin, and S. Svein, *J. Mol. Struct.*, *128*, 59 (1985).

65. G. Häfelinger and C. Regelmann, *J. Comput. Chem.*, *6*, 368 (1985).
66. G.-P. Charbonneau and Y. Delugeard, *Acta Cryst.*, *B33*, 1586 (1977).
67. U. W. Suter, E. Saiz, and P. J. Flory, *Macromolecules*, *16*, 1317 (1983).
68. G. R. Mitchell and A. H. Windle, *Colloid and Polymer Sci.*, *263*, 280 (1985).
69. E. W. Fischer and M. Dettenmaier, *J. Non-Cryst. Solids*, *31*, 181 (1978).
70. P. J. Flory, *Principles of Polymer Chemistry*, Cornell University Press, Ithaca (1953).
71. P. J. Flory, *Statistical Mechanics of Chain Molecules*, Wiley, New York (1969).
72. A. A. Jones, *Macromolecules*, *18*, 902 (1985).
73. J. A. Ratto, P. T. Inglefield, R. A. Rutowski, K.-L. Li, A. A. Jones, and A. K. Roy, *J. Polym. Sci. Polym. Phys. Ed.*, *25*, 1419 (1987).
74. P. Tekely and E. Turska, *Polymer*, *24*, 667 (1983).
75. V. A. Bershtein and V. M. Yegorov, *Polym. Sci. USSR*, *27*, 2743 (1985).
76. V. A. Bershtein, V. M. Yegorov, and Yu. A. Yemel'yanov, *Polym. Sci. USSR*, *27*, 2757 (1985).
77. J. Bicerano, *J. Polym. Sci. Polym. Phys. Ed.*, in press.
78. J. Bicerano, *J. Polym. Sci. Polym. Phys. Ed.*, in press.

12

Secondary Relaxation Processes in Bisphenol-A Polysulfone

Alan Letton
Texas A & M University
College Station, Texas

AN OVERVIEW OF SECONDARY RELAXATIONS IN POLYSULFONE AND THEIR ORIGINS

Introduction

In recent years a unique class of polymers has gained acceptance in industries such as aerospace, automotive, and construction. Although these tough thermoplastics and thermosets have fared well in these applications, industry is still striving for a better understanding of these polymers' structure/property relationships. This information will aid in the better design and manufacturing of these materials for more demanding applications. To move forward with this agenda, it is necessary that an understanding be developed of how the chemical structure of these polymers influence the material's end-use properties. An understanding of these relationships will ultimately enable the polymer scientist to tailor polymers to precisely fit particular applications.

High-performance thermoplastics and thermosets polymers with structures based on the bisphenol-A moiety for example, tend to have favorable properties. Notably, bisphenol-A polycarbonate (PC) and bisphenol-A polysulfone (PSF) demonstrate similar properties in terms of toughness and yield strength (for a complete comparison, see Seymore [1]). Similarly, epoxy resins based on diglycidel ether of bisphenol-A tend to have preferred toughness relative to other thermoset epoxies. Materials containing the bisphenol-A moiety have favorable mechanical properties but do not fair well when physical aging is considered. PSF and PC for example, have a demonstrated ability to embrittle on aging [2,3,4]. The process of densification with time, as measured during physical aging, is associated with changes in chain conformation [5,6]. Based on the fact that fracture and failure are energy dissipation processes, it can be envisioned that the ability for a material to dissipate energy (to be tough) will eventually depend on the polymer chain's ability to absorb or dissipate energy on a microscopic level.

A polymer above its glass transition exists in a state in which there are large numbers of high-energy conformations. If we assume some statistical distribution of conformation states, we may argue that at high temperatures this distribution is skewed toward more high-energy conformations. When this material is quenched below its glass transition, the high-energy conformations are locked in. That is to say, there is a large population of chains that is not in conformationally preferred states. These strained chains must therefore find a route to a preferred, low-energy conformation. To reach these preferred conformations, energy must be supplied to the system. Energy may be supplied through the environment (thermally) or through mechanical deformation. Supplying energy through the environment leads to physical aging while mechanical deformations result in properties associated with toughness. It is clear for the scheme outlined, that the molecular origins of both processes are associated with the conformation of the polymer chain. By physically aging the material, thermal energy is supplied to facilitate the change from high-energy conformations to low-energy conformations. This process requires the existence of local volume fluctuations which allow local molecular processes, phenyl ring rotations for example, to gradually occur. Similarly, mechanical deformation supplies energy that must be dissipated as heat for the resin to remain intact [7]. This same dissipation process requires local volume fluctuations. Once a polymer has reached a condition in which the population of conformations favors low-energy states, intermolecular restrictions of molecular processes to conformational changes forces bonds to break in order to absorb energy on impact. This is the case for materials that have physically aged. The fact that the polymer experiences densification suggests that the local volume fluctuations that were once in existence no longer prevail.

The scenario presented above clearly depends on the existence of localized molecular processes to explain the behavior of aged and unaged polymer glasses and their change from tough to brittle behavior. It is for this reason that the nature of these processes is investigated. Several investigations demonstrate the relationship between secondary dynamic mechanical relaxations and the aging/embrittlement process. The nature of these secondary relaxations is observed to change during physical aging. Letton [8] reports a change in the behavior of the PSF low-temperature relaxations, centered at −100°C, after annealing for 10 hr at 180°C. Slow cooling produced a similar change in which the high-temperature side of the relaxation was observed to increase on annealing. In this same study, the beta relaxation, centered at 60°C, is reported to disappear on aging. This behavior has been reported by several workers [9,10]. A more detailed curve-fitting analysis suggest that the lower relaxation is composed of three separate relaxations. Using the data of Letton [8], Gaussian curves where used to separate the relaxation into individual mechanisms. Typical results from this analysis are presented in Figures 1, 2, and 3. Since these datum where recorded while heating the specimen at a fixed rate of 4°C per minute, peak smearing may result from the co-current activation of several

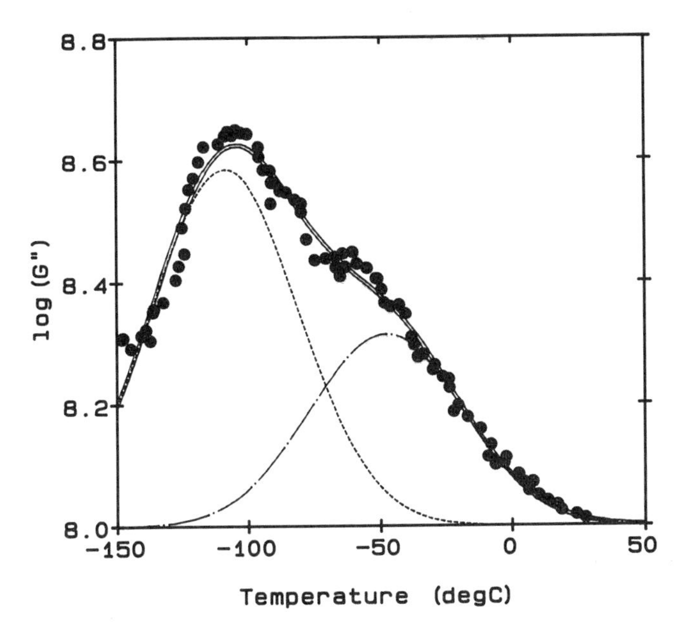

Figure 1 Low-temperature gamma relaxation for PSF annealed at 125°C. Circles are original data, while the lines represent Gaussian peaks approximating separate relaxations (single lines). The double line is the curve fit represented by two Gaussian peaks.

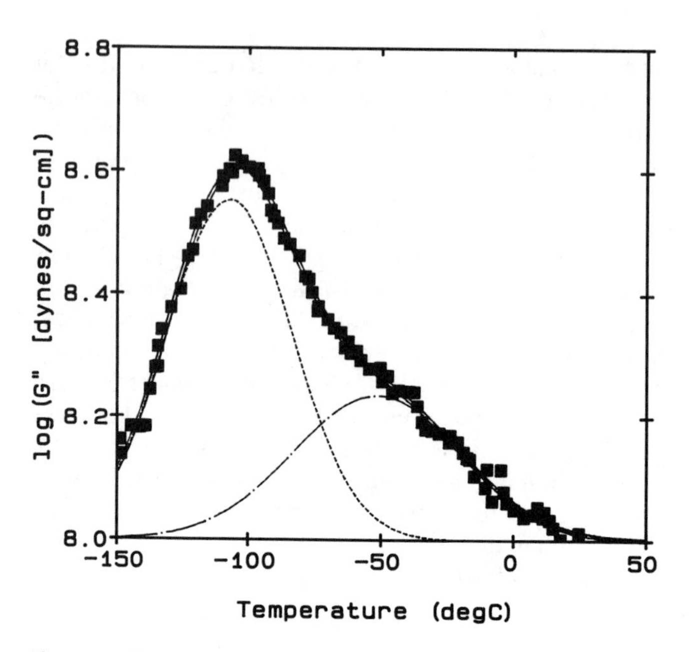

Figure 2 Low-temperature gamma relaxation for PSF slow cooled from above the glass transition. Lines are as in Figure 1 with solid squares showing original data.

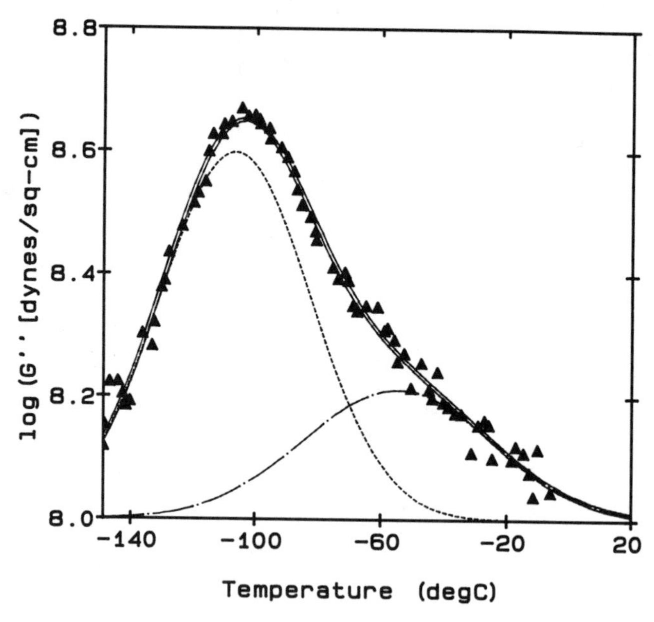

Figure 3 Low-temperature gamma relaxation for PSF quenched from above the glass transition. Lines are as in Figure 1 with solid triangles showing original data.

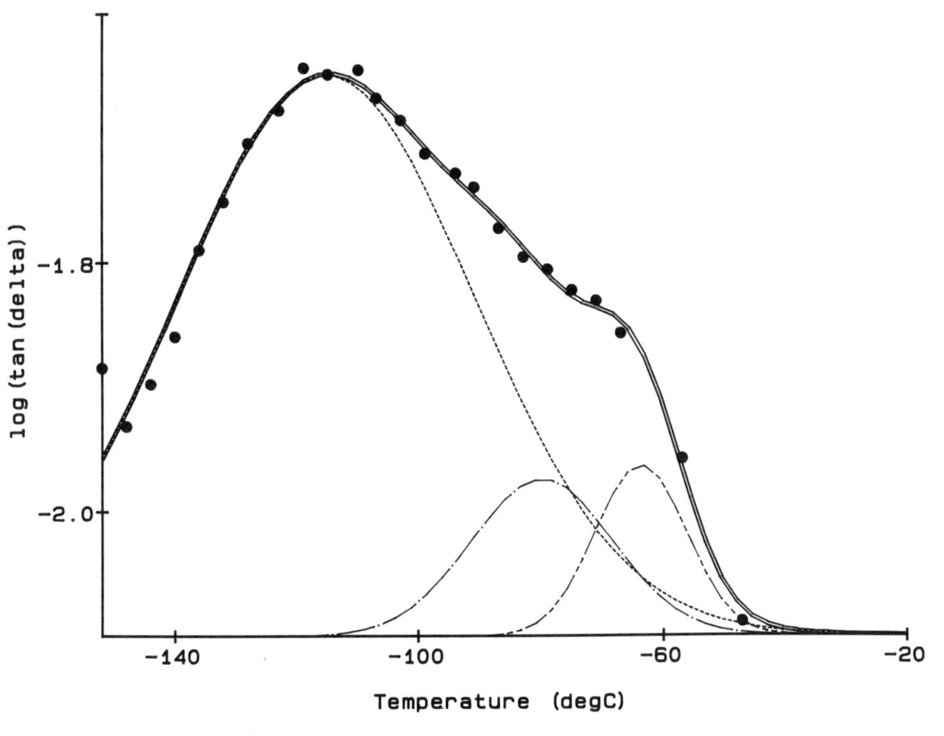

Figure 4 Curve fit of three Gaussian peaks to low temperature relaxation of PSF. Data was recorded at 3.2 rad/s with an equilibrium soak used between temperature changes. Double line represents Gaussian curve fit, while the single lines demonstrate the contribution made by individual relaxations. The solid circles are the original data.

processes. For this reason, a series of measurements were conducted in which the temperature was raised at increments of 4°C. At each temperature, the specimen was soaked for 9 min in order to reach thermal equilibrium. Measurements of the loss and storage moduli were taken at frequencies ranging from 0.1 to 100 rad/s. The results (in Figures 4–7) suggest the existence of three relaxations. Struik [11] has argued that these low temperature relaxations, the gamma and beta processes (see Fig. 8), are prerequisites for the segmental motions that exist below and during the glass transition. Relating segmental motion more explicitly to these relaxations is one goal of this presentation. These secondary relaxations have been studied by nuclear magnetic resonance (NMR), infrared spectroscopy and ultraviolet spectroscopy [12], and have reached similar conclusions.

In his study of physical aging in amorphous polymers, Struik [11] sug-

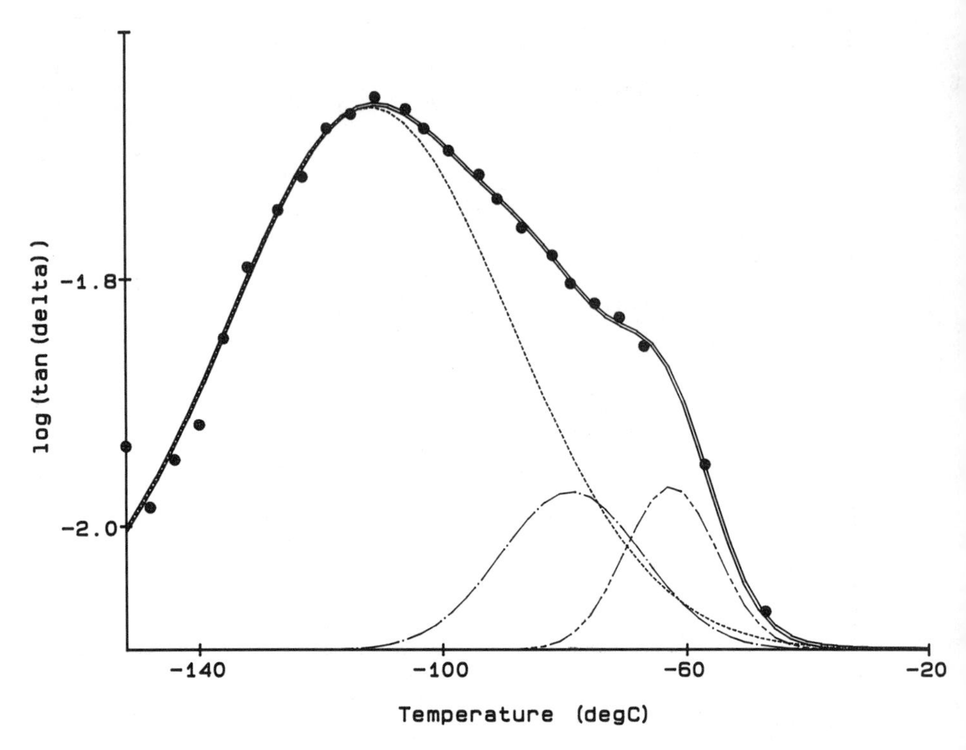

Figure 5 Curve fit of three Gaussian peaks to low temperature relaxation. Frequency is 7.9 rad/s. Symbols are as in Figure 4.

gested a link between the secondary relaxations in thermoplastics and the physical properties of bulk polymers by identifying the beta relaxation temperature as the temperature at which physical aging begins. Struik further suggests, without proof, that the gamma and beta processes are prerequisites for segmental motions that predominate during the glass transition. In an attempt to further study the relationship between secondary relaxations and mechanical properties, Lee [13], Wyzgoski [9], and Varadarajan and Boyer [10] studied changes in the shear loss modulus as a function of thermal treatment (i.e., quenching, annealing, slow cooling, etc.). The premise that annealing drives the system to its lowest entropy state, resulting in a large population of low-energy conformations suggests that changes in the dynamic mechanical spectrum are manifestations of changes in chain conformation if the relaxations are associated with localized molecular motions. The researchers noted above observe a reduction in the beta loss modulus on annealing and an increase in the loss modulus when the polymer was quenched from the melt. In the context of the arguments presented, quenching from the melt results in a large

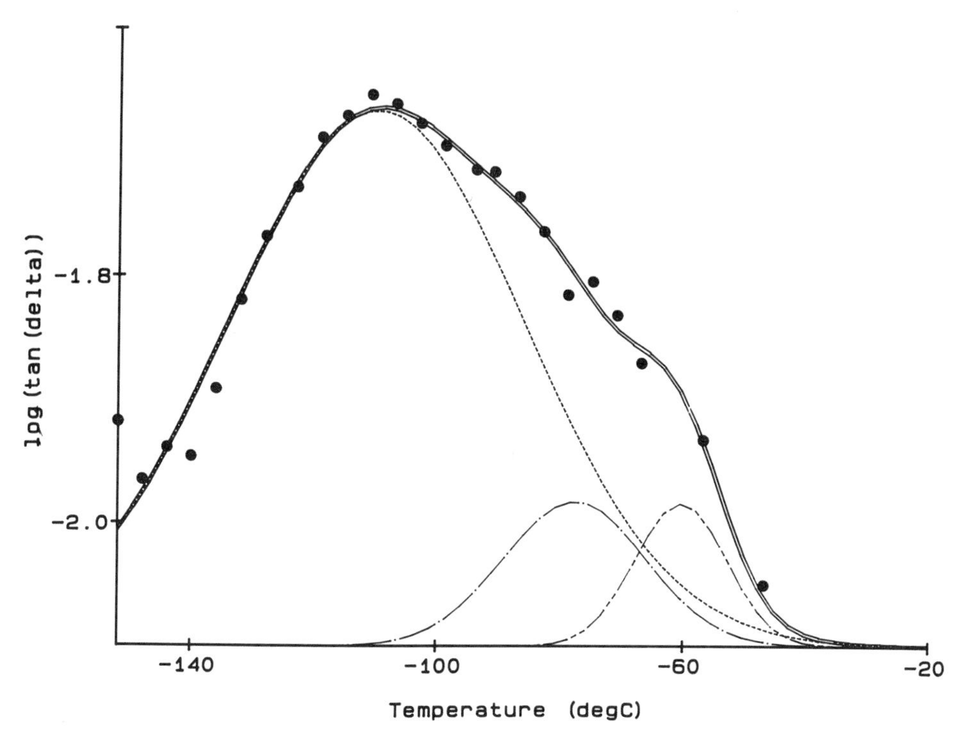

Figure 6 Curve fit of three Gaussian peaks to low-temperature relaxation. Frequency is 10.0 rad/s. Symbols are as in Figure 4.

number of high-energy conformations. These conformations, because of their irregularity in structure, have large localized volume fluctuations and are more mobile, thereby leading to the dissipation processes associated with the beta relaxation. Annealing the system eliminates this relaxation. This argument is being used to justify both the gamma and the beta relaxations, although traditionally the gamma relaxation does not demonstrate a sensitivity to thermal history [14,15]. This is possibly due to the fact that most researchers ignore the existence of the three relaxations noted in the gamma region.

RELATIONSHIP BETWEEN VISCOELASTIC SHIFT FACTORS AND MOLECULAR STRUCTURE

General Relationships

The temperature-dependent viscoelastic shift factor, a_T, has been used extensively to characterize the mechanical and thermodynamic "state" of

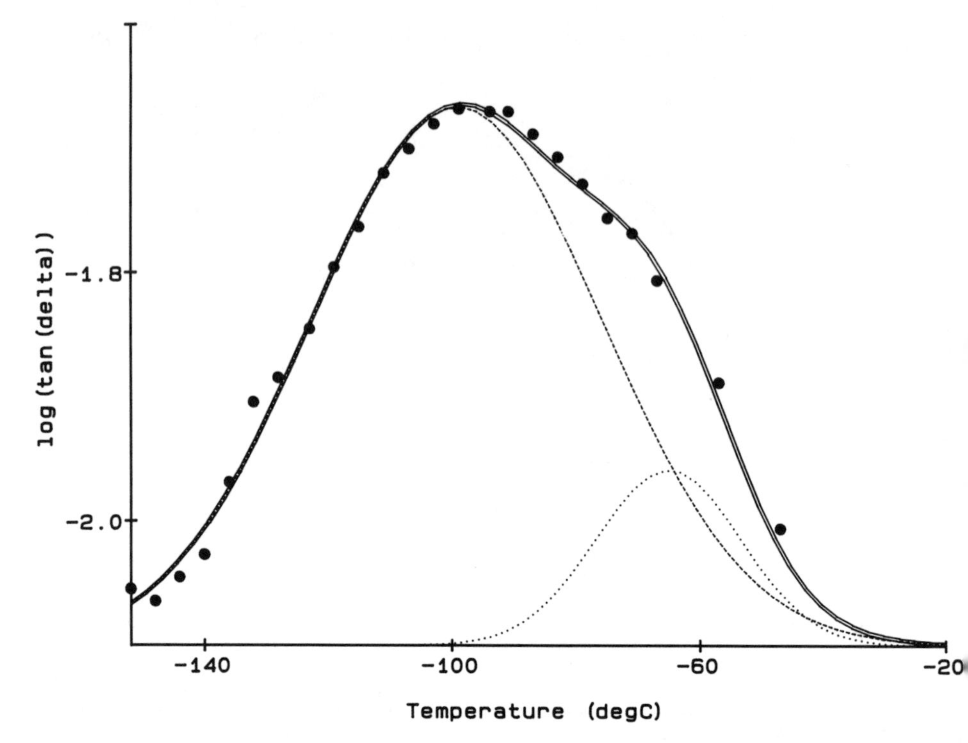

Figure 7 Curve fit of two Gaussian peaks to low temperature relaxation. The two higher-temperature relaxations have merged into one. Frequency is 100.0 rad/s. Symbols are as in Figure 4.

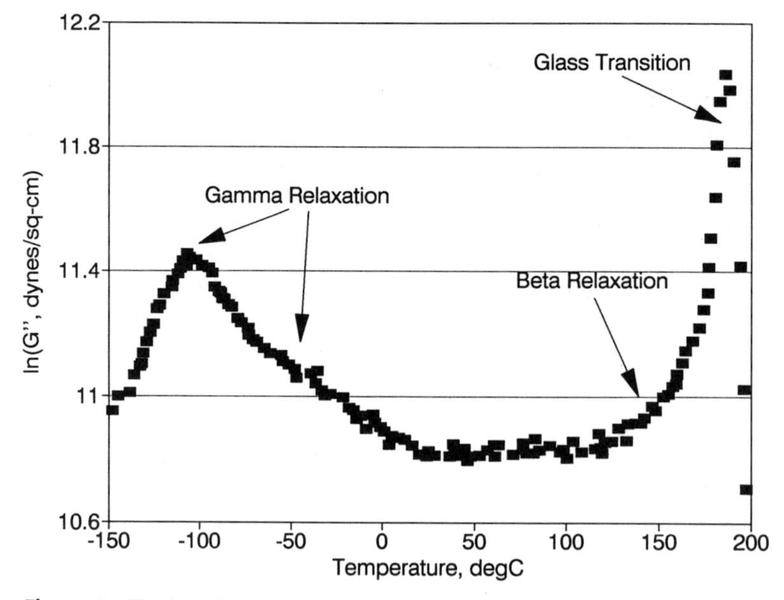

Figure 8 Typical shear, loss modulus curve for PSF.

glassy polymers. Brostow [16] presented a model in which the shift factor is considered to be a measure of the chain's relaxation capability. This concept suggests that the molecular dynamics involved in chain relaxation are somehow related to the shift factor. Work exists in which a_T is used as a measure of the temperature dependence of the mechanical state of a polymer. An example is the work of Knauss and Kenner [17] in which $a_{T,C}$ is used to monitor the effect of water concentration on viscoelastic response (C represents water concentration). Brostow and Cornelliusen [18] have used the shift factor to predict the temperature dependence of the stress intensity factor, K_{IC}, for low-density polyethylene below its glass transition. The predicted values are in good agreement with the experimental data obtained by Zewi and Cornelliusen [19] on the same polymer. The agreement between predicted and observed fracture toughness suggests that a_T must be a good measure of the mechanical state of the polymer. Diamant, Hansen, and Shen [20] monitored the change in the temperature dependence of the shift factor measured for a poly(ethylene-*g*-styrene) copolymer as the degree of crystallinity for the polyethylene segment was altered. The apparent activation energies for the shift factors, determined from an Arrhenius plot, were 32 kcal/mole in the region below the polyethylene crystalline melt temperature and 82 kcal/mole for the region above the polyethylene crystalline melt (these values are calculated from Fig. 4 of Ref. 21). The molecular processes controlling relaxational behavior below the polyethylene crystalline melt in this system are expected to be different above the polyethylene crystalline melt. As the polyethylene crystallized segments melt, more degrees of freedom become available to the chain for relaxation; the energetics of which are higher than for the chain below the melt temperature. These data suggest that the shift factors are therefore a measure of the molecular dynamics in the system. This idea may support the existence of a correlation between the mechanical behavior and the molecular dynamics of a system. The shift factors may then provide a way of probing that relationship.

Low-Temperature Relationships

The relaxations of interest in this analysis are associated with temperatures far below the glass-transition temperature. Although the relationship between a_T and mechanical relaxations is documented as reviewed above, there is little work to date which probes the behavior of the shift factors 100 to 300°C below the glass transition. Kono [21], in comparing the shear and bulk deformation processes of polyvinylchloride (T_g = 78°C), noted that the shift factor/temperature relationship was represented by an Arrhenius function with an apparent activation energy of 59 kcal/mole. Kono's measurements extended only 10°C below the glass transition. Crowson and Arridge [22], in studying the creep behavior of diglycidyl ether of bisphenol-A cured with nadic methylanhydride, observed that the shift factors' temperature dependence was Arrhenius in form, both above and below the glass transition temperature. The apparent activation energies were 165 kcal/mole and 58 kcal/mole above and below the glass-transition temperature, respectively. Crowson and Arridge's data extended to temperatures 120°C below the glass transition. It is worth noting that at these low temperatures, the a_T/temperature relationship is still Arrhenius in form. The low value of the apparent activation energy associated with the sub-T_g region is suggested to be a result of the systems' nonequilibrium state. The molecular mechanisms governing behavior in the region below the glass transition are different than those associated with the region above the glass transition and are most likely less energy-intensive, thereby explaining the low activation energies. This hypothesis implies a correlation between the molecular dynamics and the activation energy associated with the shift factors measured in a particular temperature region. For this reason, an analysis of the viscoelastic shift factors is warranted. These results will be used to interpret behavior modeled using molecular mechanics and quantum mechanics techniques.

ANALYSIS OF VISCOELASTIC SHIFT FACTORS IN POLYSULFONE

Construction of Master Curves To Obtain a_T

The shift factors were determined by construction of the storage modulus master curve as described in an earlier publication [23]. A slightly different modification of this procedure was used and is briefly described. Two data sets, one designated as a reference data set, were shifted by fitting a cubic spline to the nonreference data set. The spline fit is a fit to frequency as a function of the modulus. The moduli for the reference temperature data set are used to interpolate corresponding frequencies at the same modulus values on the curve to be shifted (see Fig. 9). The ratio of the frequencies at a given modulus yields the shift factor; i.e., $\omega_{T,1}/\omega_{T,R} = a_T$, where $\omega_{T,I}$ is the shifted curve frequency, $\omega_{T,R}$ is the reference curve frequency, and a_T is the viscoelastic shift factor. The shift factors determined in the region where the two curves overlapped, are averaged to yield the shift factor corresponding to that temperature. This procedure has the advantage that is gives a statistical measure of the accuracy of the shift (by analysis of the standard deviation for each shift factor) and allows one to shift nonmonotonically increasing or decreasing functions by using spline derivatives as and indication of the direction of the curve. These shift factors are then used to produce the final master curve. As a check for thermorheological simplicity, the storage modulus shift factors were used to shift the loss

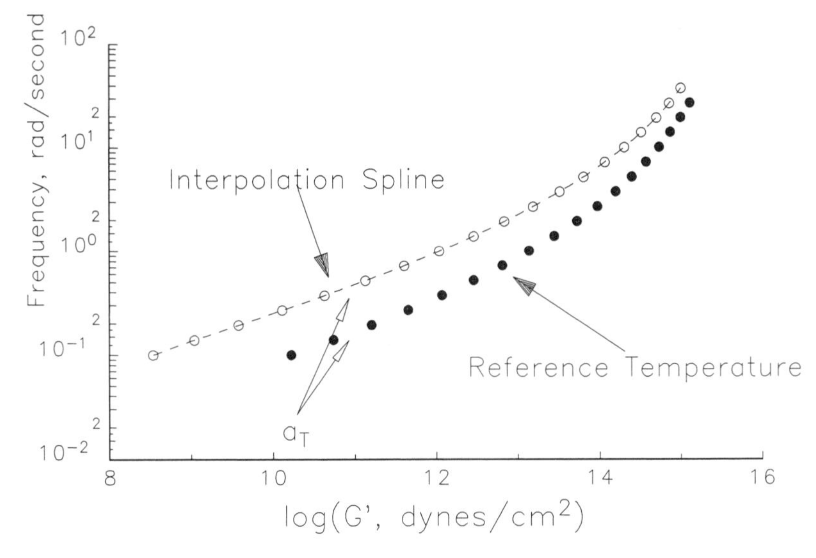

Figure 9 Schematic demonstrating the spline interpolation procedure used in shifting the viscoelastic data.

modulus data and to shift the loss angle. In general, the storage modulus master curves showed less than 5% scatter. However, when checked for thermorheological simplicity, the loss modulus and the loss angle master curves, constructed using the storage modulus shift factors, showed good superposition in the low temperature (−150 to −50°C) regions and the high-temperature regions (50°C below T_g and higher), but showed large deviations in the mid-temperature regions. This may be due to two or more relaxations occurring simultaneously. This argument is in line with the argument used to interpret the data presented below. It should be noted that the shift factors obtained separately from the storage modulus master curve and from the loss modulus master curve are within 5% of each other.

Polysulfone Shift Factors As a Function of Aging

In Figure 10, the shift factors for an unaged PSF specimen are represented by an Arrhenius model. It is evident that there are several temperature regions which are dominated by activation processes having activation energies as indicated. The lowest temperature region, less than −110°C, is associated with a process bearing an apparent activation energy of 4.0 kcal/mole. A second region, −110 to 15°C, has an associated apparent activation energy of 12.5 kcal/mole while third and fourth regions, 15 to 170°C and 170°C and up, respectively, are governed by processes with apparent

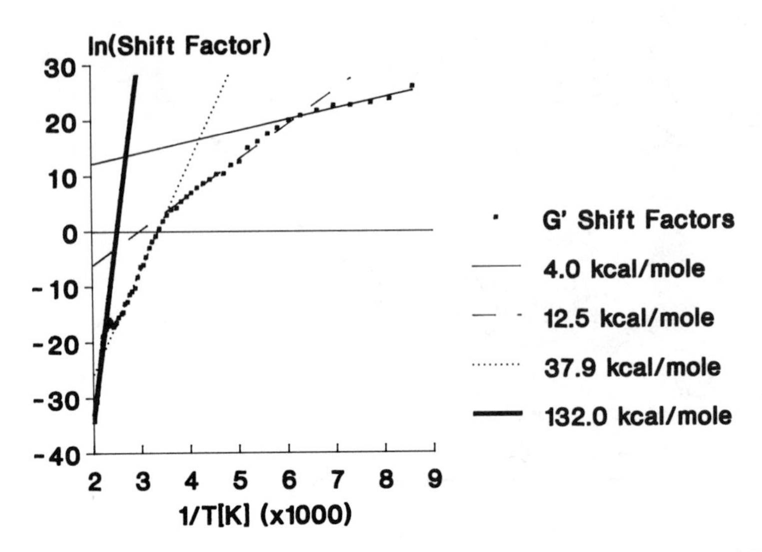

Figure 10 Arrhenius plot of viscoelastic shift factors, a_T, for unaged PSF.

activation energies of 37.9 kcal/mole and 132.0 kcal/mole, respectively. The linear regions were determined statistically. Initially the three low-temperature datum are fitted to a line using a linear least-squares model. The correlation coefficient is then determined. New points are individually added and the new correlation coefficients are determined and plotted as a function of the number of points used. As long as the correlation coefficient is within a given bound, the data are considered to be part of the same line. Once the regression coefficient begins to drop, it is assumed that a new linear region has begun.

The activation energies for an aged PSF specimen did not differ dramatically from those of the unaged. From the data presented in Figure 11, it is evident that after aging, there major relaxations predominated: a region corresponding to temperatures less than −126°C and with an activation energy of 7 kcal/mole, a region corresponding to temperatures in the range −126 to 5°C and with an activation energy of 14 kcal/mole, and a region corresponding to temperatures greater than 88°C with an activation energy of 144.5 kcal/mole. The behavior in the region 5 to 88°C is characterized by apparent activation energies of 0 and 22.1 kcal/mole. This is not altogether accurate since there is a large amount of curvature in the data which prohibits the use of an Arrhenius model to represent the enegetics of the relaxation.

There are several items worth noting. First, the transition temperatures correspond to the transition temperatures measured during dynamic me-

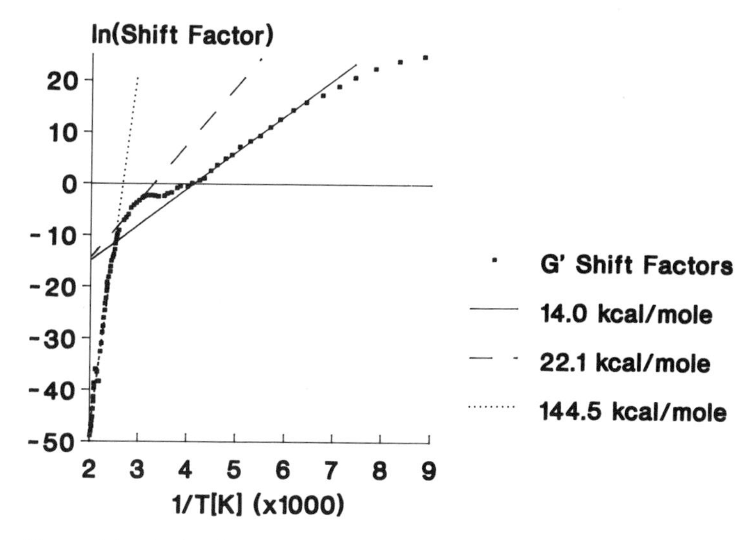

Figure 11 Arrhenius plot of viscoelastic shift factors for aged PSF.

chanical analysis using single frequencies. The DMA recorded transitions represent the onset of a new relaxation. The Arrhenius plot gives detailed information as to the temperature range in which the process is operative. The activation energies for PSF have been measured by Letton [24] and are reported to be 11 kcal/mole for the gamma relaxation, 67 kcal/mole for the beta relaxation, and 163 kcal/mole for the glass transition. The activation energies measured from the Arrhenius plots agree with those reported by Letton [24], with the exception of the activation energy measured for the beta relaxation. In the method used by Letton, it is difficult to determine the temperature maximum for the beta relaxation, thereby introducing a large error in the calculation. As reported by Letton, the 95% confidence interval for the beta relaxation's activation energy was ±53%. The fact that the shift factors behave in an Arrhenius fashion suggests that these processes are associated with basic molecular processes. Where the lower-temperature relaxations are somewhat unchanged, the higher-temperature relaxations tend to be strongly dependent on thermal history, suggesting that they may be more complicated in origin and coupled to the magnitude of the density fluctuations.

It is for this reason that this work focuses on the development of an understanding of the low-temperature relaxations. These relaxations may be associated with intermolecular processes (as indicated by its weak dependence on thermal history and by the low apparent activation energies associated with the process). It is now our goal to relate these relaxations to molecular processes by studying the energy barriers to rotation of simple molecular fragments which are representative of the polymer chain.

APPLICATION OF QUANTUM AND MOLECULAR MECHANICS TECHNIQUES

Qualitative/Quantitative Investigations by Application of Molecular Mechanics

Before initiating CNDO/2 computations on representative polysulfone chain segments, it would be productive to determine preliminary energy barriers using less time-intensive molecular mechanics procedures. Based on the studies of Robeson et al. [25], in which changes in the DMA spectra were measured for sulfones with varying side groups, PSF fragments can be modeled to probe consistency between computational results and experimental realities. The polysulfones studied by Robeson were evaluated for toughness in a qualitative manner. Based on the principles presented here, brittle materials should have several low-energy molecular processes (on the order of 1 to 2 kcal/mole) or several high-energy processes while ductile materials should have energy barriers which enable them to absorb energy without being so large as to prohibit molecular motion. In addition, the molecular volume of the repeat unit should be sized to provide freedom from intermolecular restrictions.

Initial investigations were performed using MacroModel, a molecular modeling package developed by Clark Still [26] of Columbia University. Allinger's MM2 force field was used [27] with parameters modified as indicted in the MacroModel reference literature. To insure convergence on large structures (50 atoms or more), the convergence criterion for the first derivative root mean square was set to a value of 0.001 kJ/Å (0.02 kcal/Å). The maximum number of iterations was moved from the default value of 50 to a value of 1000 although most structures meet the convergence criterion within the first 250 iterations. This preliminary study was used to probe general relationships and was not a detailed conformation study.

The polymers studied by Robeson (Tables 1 and 2) were modeled by the structures presented in Tables 3 and 4 and Figure 12. These preliminary computations were conducted on isolated structures, so the effect of neighboring chains is not considered. However, molecular volumes were determined and used to aid in establishing the most restricted structures. Moieties with large molecular volumes are more likely to be hindered by neighboring chains and are therefore more likely to have energy barriers higher than those calculated for the isolated chain.

Table 1 Polymers Evaluated by Robeson et al.

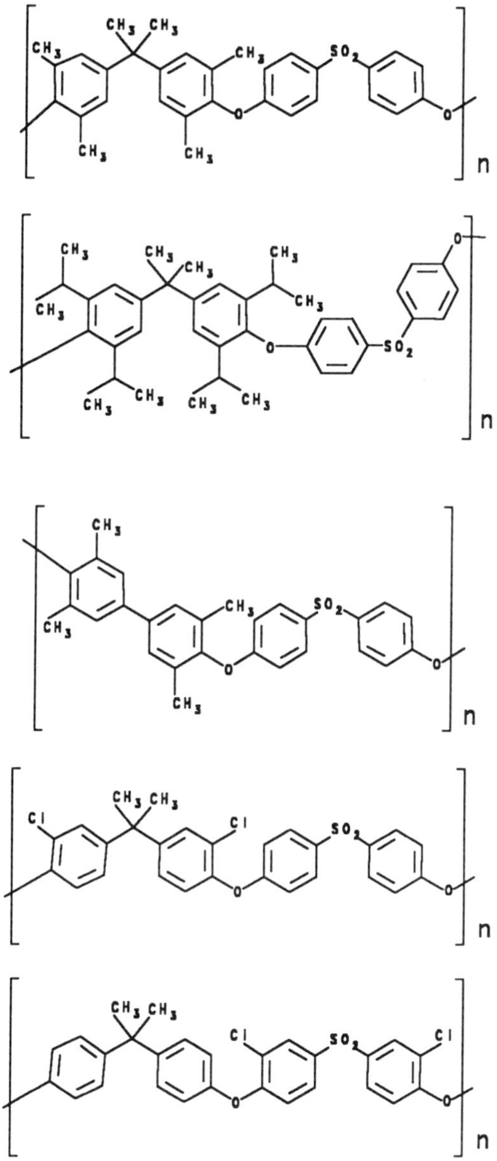

Table 2 Additional Polymers Studied by Robeson et al.

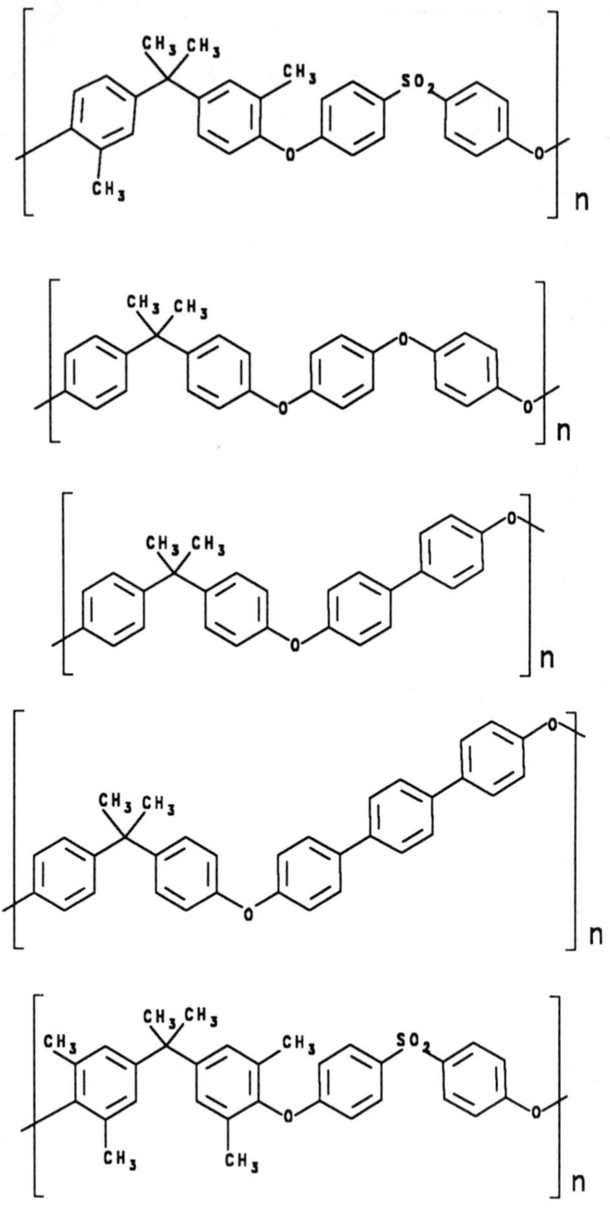

CH3 CH3 CH3 SO2 O O CH3 n

CH3 CH3 O O O n

CH3 CH3 O O n

CH3 CH3 O O n

CH3 CH3 CH3 CH3 SO2 O O CH3 CH3 n

Table 3 Segments Used To Model Polysulfone Structures Studied by Robeson et al.

Modeling structure	Energy barrier (kcal/mole)	Volume (Å³)
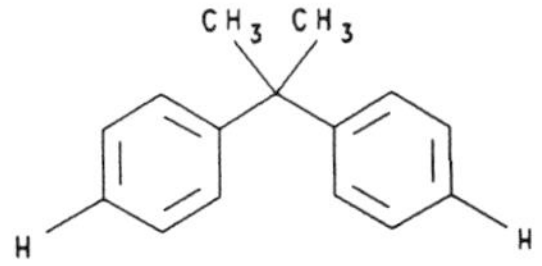 CH_3 CH_3 H H	11.7	205.3
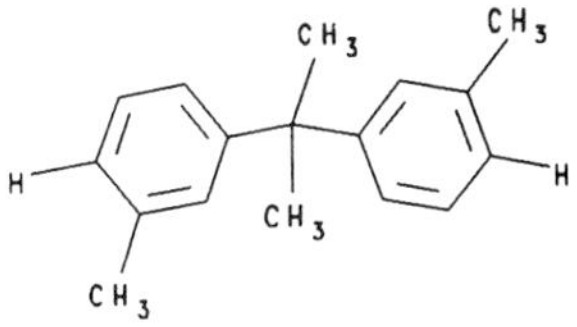 CH_3 CH_3 H H CH_3 CH_3	14.8	238.8
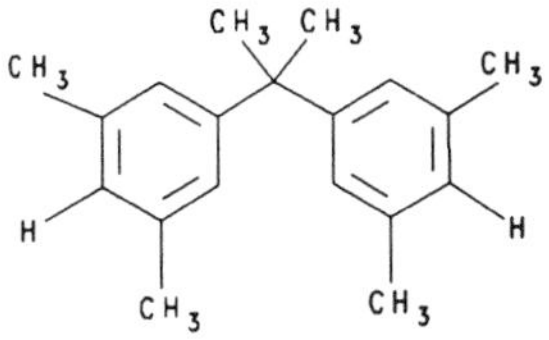 CH_3 CH_3 CH_3 CH_3 H H CH_3 CH_3	14.8	273.3
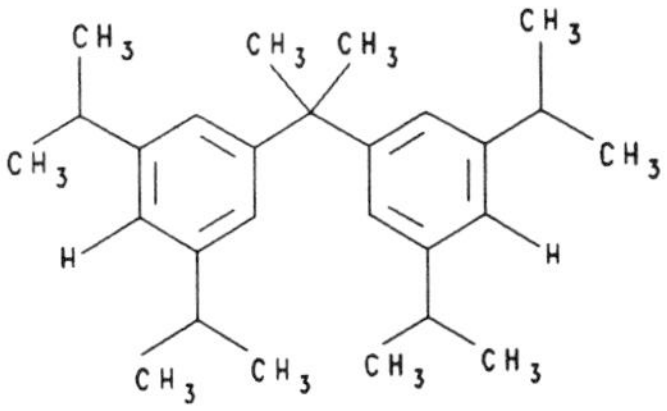 CH_3 CH_3 CH_3 CH_3 CH_3 CH_3 H H CH_3 CH_3 CH_3 CH_3	16.1	404.9
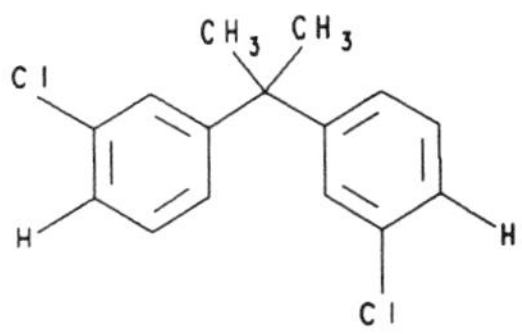 CH_3 CH_3 Cl H H Cl	12.1	235.9

Table 4 Segments Used To Model Polysulfone Structures Studied by Robeson et al. (Table 1 continued)

Modeling structure	Energy barrier (kcal/mole)	Volume (Å)
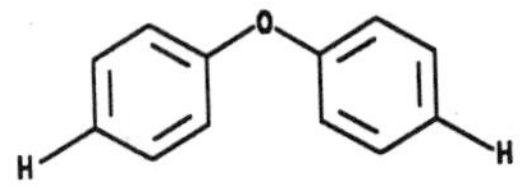 H–C₆H₄–O–C₆H₄–H	5.6	163.4
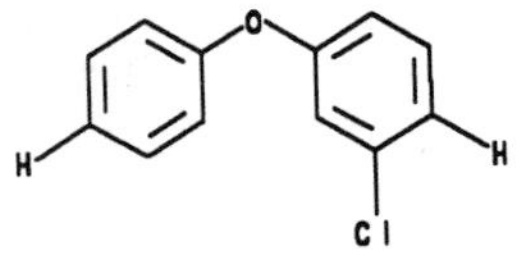 H–C₆H₄–O–C₆H₃(Cl)–H	6.6	177.7
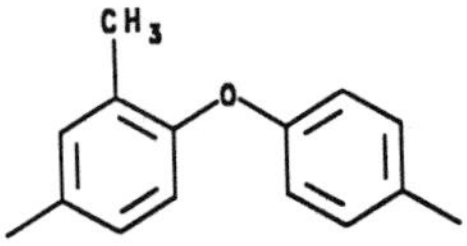 CH₃-substituted diphenyl ether	10.9	179.7
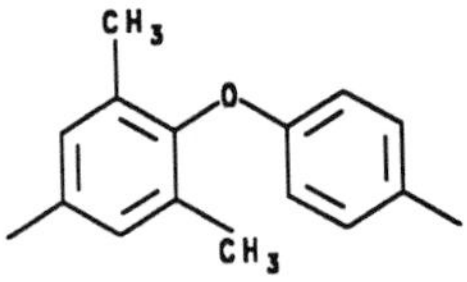 CH₃, CH₃-disubstituted diphenyl ether	11.5	196.9
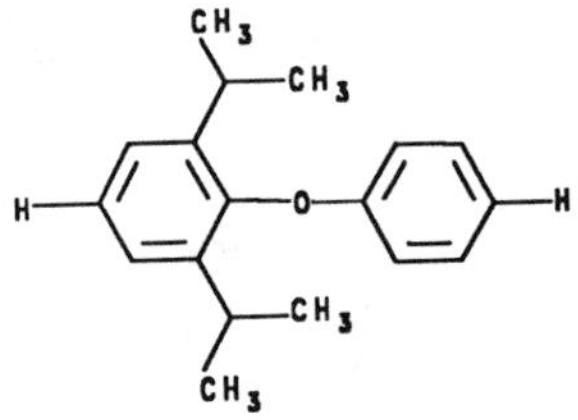 2,6-diisopropyl diphenyl ether (H–…–O–…–H)	11.6	263.6

In general, results suggest that chains with moderate energy barriers to rotation (less than 10 kcal/mole) and molecular volumes less than 250 $Å^3$ are more likely to be tough (as evaluated by Robeson). The materials Robeson describes as ductile are the materials which are calculated to have the lowest energy barriers to rotation and the lowest molecular volume (as calculated in the lowest energy conformation). These low molecular volumes suggest that relatively small local density fluctuations are needed to

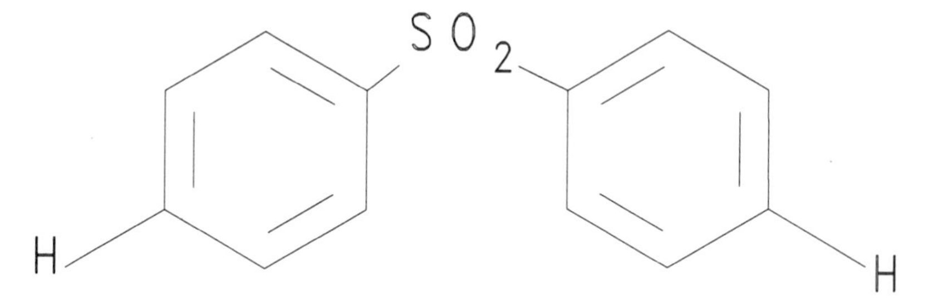

Figure 12 Diphenyl sulfone moiety used in modeling molecular relaxations.

facilitate molecular mobility. Conformational changes are easily allowed, thereby providing a mechanism for energy dissipation although these same mechanisms would provide an avenue for physical aging to occur. For the materials listed by Robeson as "brittle," the average energy barriers to rotation for the chain are greater than 10 kcal/mole with molecular volumes exceeding 200 Å^3. Higher temperatures coupled to large local density fluctuations are needed to provide the molecular mobility essential in absorbing energy. As a result, these materials will behave in the observed brittle fashion.

The frequency/temperature dependence of these materials is not available and therefore cannot be used to assign secondary relaxations to specific molecular processes. From a more general prospective, these results suggest a correlation between energy barriers to rotation and viscoelastic processes. These relationships will be studied in more detail using more precise semiempirical quantum procedures. Results from this approach are reviewed below.

CNDO/2 Quantum Mechanics Calculations

To calculate detailed conformational geometries and energy barriers, the complete neglect of differential overlap (CNDO/2) geometry optimizing

Table 5 CNDO/2 Conformational Calculations, Selected Results Isopropylene

ψ',ψ''	0,0	0,0	0,0	0,0	0,20	0,50	0.60
ϕ,ψ	0,0	45,45	90,90	45,45	45,45	45,45	45,45
ΔE kcal/mole	0	−10.1	−7.7	−3.6	−11.1	−15.0	−15.4
Angle, Å C_6—C_{11}—C_{20}	123.5	112.9	107.4	114.2	112.9	113.1	113.0
Interatomic distance $H_{22} \cdots H_5$	1.40	3.22	2.78	1.51	3.19	3.15	3.11

Table 6 CNDO/2 Conformational Energy Results Methyl Results for $\phi = 45°$ and $\psi = -45°$

ψ'	ψ''	ΔE (kcal/mole)
0°	0°	0
0°	30°	0.97
0°	60°	7.28
0°	90°	0.60
30°	90°	−1.15
60°	90°	−2.51
90°	90°	−1.20

routine developed by Kondo [28] was employed. The program uses a self-consistent (SCF) technique that employs an iterative scheme to find the self-consistent molecular orbitals. Application of this technique to polymeric molecules has been demonstrated by Welsh and others [29] in their studies of poly(benzobisthiazoles), several aromatic heterocyclic polymers and poly(benzobisthiazoles). The CNDO technique was chosen because of its inherent incorporation of conjugation effects and effects due to hydrogen bonding. The former is critical in the work presented and both are significant in molecules containing aromatic rings.

The lowest energy conformation and its associated geometry was determined for four sections of the polysulfone chain; the diphenyl ether segment, the isopropylene segment, the diphenyl sulfone segment and the methyl group of the isopropylene segment (see Figs. 13 to 16). For the isopropylene, diphenyl ether, and diphenyl sulfone segments, the relative orientation of the phenyl groups was fixed while the CNDO/2 routine

Table 7 CNDO/2 Conformational Calculations, Selected Results Diphenyl Ether Segment

ϕ,ψ	0°,0°	40°,40°	90°,90°
ΔE kcal/mole	0	−8.44	−6.51
Bond length, Å O(11)—C(6)	1.42	1.40	1.40
C(12—O(11)	1.42	1.40	1.40
Bond angle C(6)—O(11)—C(12)	126.5	116.8	113.6
Interatomic distance, Å H(14) ··· H(5)	1.69	2.66	2.57

optimized the remaining bond lengths and angles, thereby simulating a more realistic motion. The difference in the energy calculated at the most strained conformation and the energy calculated at the least strained conformation is taken as the energy barrier to rotation. The results from these efforts is presented in Tables 5 to 7.

The energy barriers to rotation for all segments studied were approximately 11 kcal/mole or less. These energy barriers are approximately equal to the apparent activation energies calculated for the low-temperature relaxations (10°C and less). For the diphenyl ether moiety, the most stable geometry was calculated to be the 40°, 40° conformation (a 40°, 40° conformation is one in which the angles ϕ and ψ in Fig. 13 are 40° and 40°, respectively). The 0°, 0° conformation was calculated to be the highest energy configuration. In addition to the energy barrier calculated for the diphenyl ether moiety, a wide energy well was observed. The conformation energies calculated for the 90°, 90° and the 100°, 100° conformations did not differ from the conformation energy calculated for the 40°, 40° conformation. This large energy well enables the diphenyl ether segment to accept a number of configurations thereby offering several routes to conformational change. This energy barrier, 8.44 kcal/mole, is higher than that presented in the works of Welsh et al. [29,30], who calculated a much lower energy barrier, approximately 1 to 4 kcal/mole. The recorded deference in energy barriers is most likely due to the stearic interactions of the hydrogens labeled as atoms 5 and 14 in Figure 13. A second, unexpected result is the decrease in the bond angle C(6)—O(11)—C(12). The bond angle changes from 126.5° at the 0°, 0° conformation to a lower value of 113.5° at the 90°, 90° conformation. These extremes demonstrate the sensitivity of the diphenyl ether segment to stearic interactions and provides a measure of the driving force pushing the chain to lower energy conformations.

For the diphenyl sulfone moiety, an energy barrier of 100.0 kcal/mole was calculated. This barrier is somewhat unrealistic and is a result of the inability of the CNDO/2 method to geometry optimize the sulfur-containing moieties. This point is further supported by noting that the sulfur-carbon bond length in this segment is 1.84 Å as compared to 1.28 Å for the carbon-oxygen bond in the diphenyl ether segment. In can be assumed that the stearic interactions and possible ring interactions present in the diphenyl ether segment are lessened in the diphenyl sulfone moiety. With this in mind, it is expected that the energy barrier to rotation for the diphenyl sulfone moiety will be 10 kcal/mole or less. This value was confirmed using molecular mechanics calculations. Studies using space-filling models have also supported the lack of stearic interactions in the diphenyl sulfone segment.

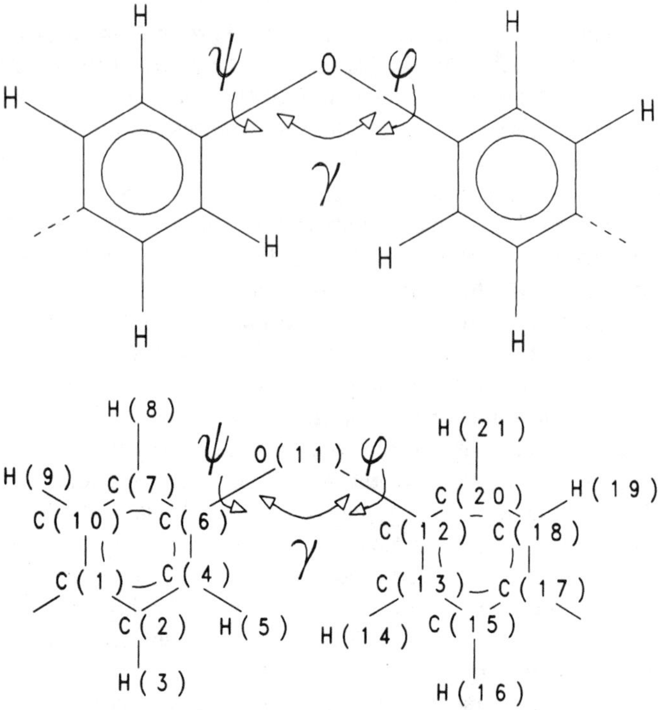

Figure 13 Diphenyl ether moiety used in modeling molecular relaxations. Lower figure illustrates the numbering used to identify specific atoms.

The phenyl group rotation in the isopropylene moiety had a calculated energy barrier of 10.0 kcal/mole with the 45°, −45° conformation being the lowest energy configuration and the 0°, 0° conformation being the highest energy configuration (see Fig. 14). Stearic interactions between the H(5) and H(22) pendant hydrogens is a major contributor to the energy barrier. The interatomic distance between these hydrogens changes from 1.40 Å at the highest energy conformation to 3.22 Å at the lowest energy conformation. The C(6)—C(11)—C(20) bond angle collapse from 125.5° at the highest energy conformation to 112.9° at the lowest energy conformation. Although this angle is smaller at the 90°, 90° conformation with a value of 107.4°, it is not significant enough to overcome the stearic interactions created by the H(5) and H(22) hydrogens at the 90°, 90° conformation.

Defining the energy barrier to rotation as the difference between the high-energy conformation and the low-energy conformation is not done for the isopropylene moiety. There are two mechanisms that must be considered. The first is the most probable mechanism observed during deformation. This mechanism requires rotation of both methyl groups during the relaxation process. The difference between the highest energy conformation, the 0°, 60° conformation, and the lowest energy conformation, the 60°, 90° conformation, was calculated to be 9.8 kcal/mole. The second mechanism involves the free rotation of one methyl group while the second methyl group is rigidly held in a 0°, 0° conformation. This rotation would

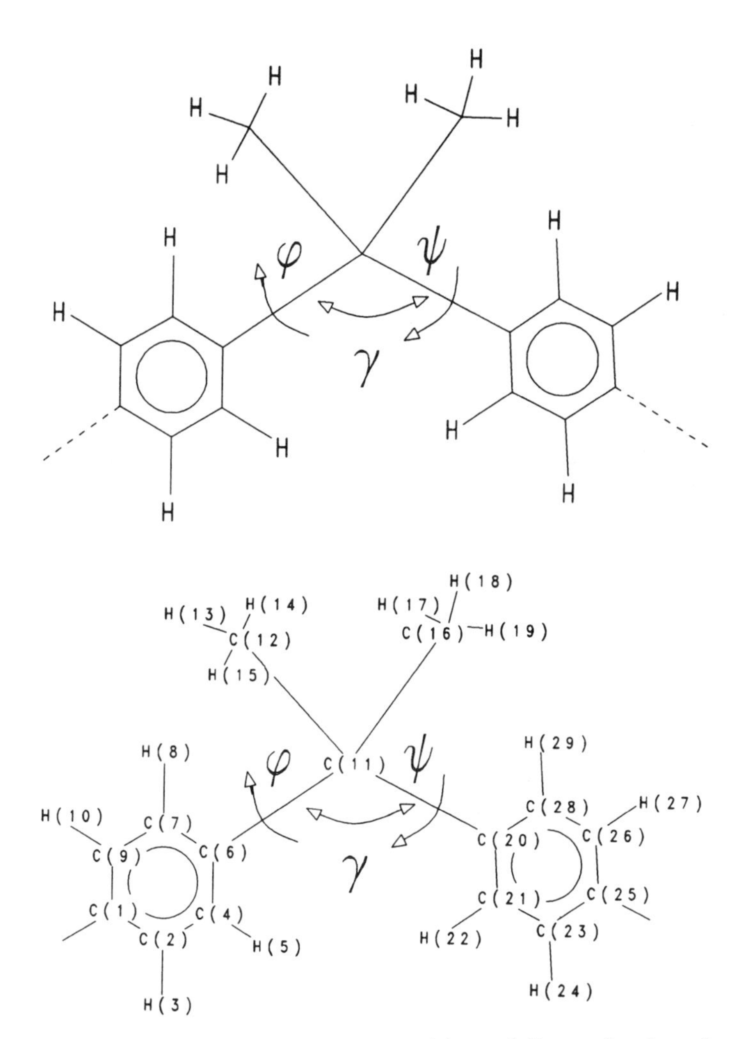

Figure 14 Isopropylene segment used in modeling molecular relaxations. Lower figure as in Figure 13.

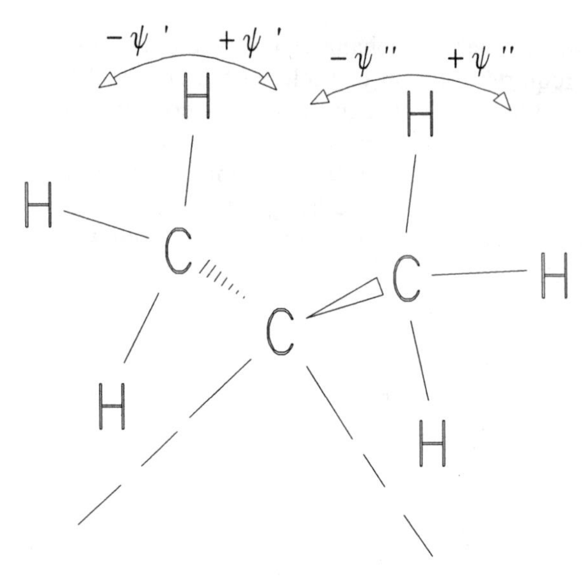

Figure 15 Orientation of methyl groups located in the isopropylene segment. Lower figure as in Figure 13.

minimize the interactions between the phenyl rings (large local density fluctuations). The energy barrier calculated for this rotation was 5.30 kcal/mole. It is unrealistic to believe that only one methyl group will rotate during deformation. Since these conformation changes do not entreat changes in the main-chain geometry (this statement is based on knowing that the bond lengths and angles associated with the main chain do not change significantly), it is easy to conclude that these motions are not completely coupled. This statement is not supported by studies of space-filling models. Models indicate a possible coupling of these motions during high-energy conformation changes. When the chain is in its highest energy conformation, there are stearic interactions between the methyl groups and the phenyl groups of the main chain. These interactions may prohibit the methyl groups from rotating; however, when the main chain obtains its lowest energy conformation, the stearic interactions are minimized and the methyl groups rotate freely.

RELATIONSHIPS BETWEEN MOLECULAR MOTIONS AND SECONDARY RELAXATION PROCESSES

As noted in the shift factor analysis, for unaged PSF the first transition occurs at $-100°C$ and is characterized by an activation energy of 6 kcal/

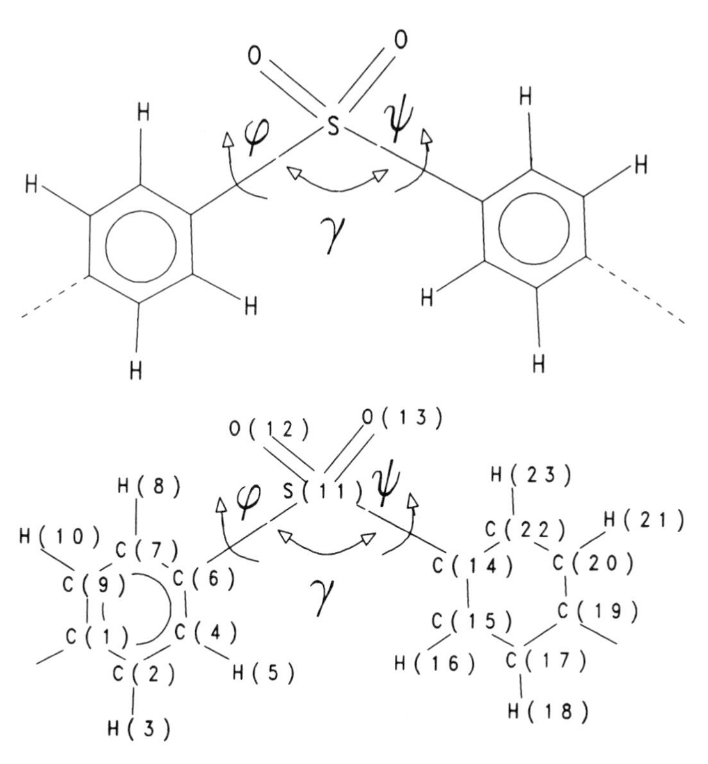

Figure 16 Diphenyl sulfone segment used to model molecular relaxations. Lower figure as in Figure 13.

mole. The CNDO/2 calculations presented above suggest that this behavior is associated with methyl group rotation in the isopropylene segment of the chain. In the temperature region −110 to 18°C, the associated activation energy is 14 kcal/mole. The suggested molecular process associated with this region is phenyl group rotation in the sulfone, isopropylene, and diphenyl ether segments of the chain. The activation energy of 14 kcal/mole is close to the calculated energy barrier for the phenyl group rotation, further suggesting that the relaxational processes in PSF may be predominantly controlled by intramolecular interactions as opposed to intermolecular interactions. In the temperature region 18 to 157°C, the apparent activation energy is calculated to be 37 kcal/mole. This energy is higher than any calculated for the main chain and may be a product of more complicated motions involving two or more repeat units in addition to any intermolecular associations. Recent NMR studies by Dumais et al. [31] support these findings.

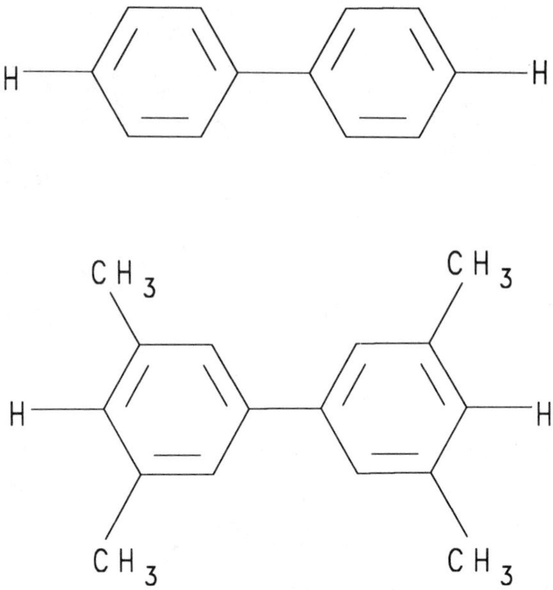

H H

CH3 CH3
H H
CH3 CH3

Figure 17 Structures used to model polymers studied by Robeson et al. (From Ref. 25.)

On aging PSF, the lower transition temperature is moved to −126°C, however, the energetics of this relaxation are unaltered (7 kcal/mole as opposed to 6 kcal/mole). The activation energy associated with the −136 to 5°C temperature range, 14 kcal/mole, is identical to the activation energy associated with the −110 to 18°C temperature range of the unaged PSF, suggesting that aging at 150°C has little to no effect on the molecular process associated with temperatures 18°C and less. In the temperature range 5 to 88°C, a process which is not Arrhenius in nature, takes place. It is not clear what this relaxation is although it has been suggested that this relaxation is related to phenyl rotation in the diphenyl ether segment of the chain [32]. The lack of local volume in the aged specimen prohibits motion of the phenyl ring. As the temperature increases, thermal expansion creates enough local volume for the process to take place. Unfortunately, this explanation does not predict the onset of the alpha, T_g, relaxation at such a low temperature, 88°C. The activation energy associated with relaxations at the glass transition, 144 kcal/mole, persist to 88°C. This temperature is 100°C below the glass-transition temperature, suggesting that the PSF concerted motions are similar in nature to the motions observed at the glass transition. The consequence of this would be that aging in PSF would be

Table 8 Summary of Relaxations in Aged PSF

Temperature region	Relaxation
−126°C and less	Methyl rotation
−126 to 5°C	Phenyl rotation or rocking
5 to 88°C	Intermolecular constraints

more rapid than in PC, since in PC the behavior is opposite that observed for PSF. Struik [33] has shown this to be true. It is not clear why the onset of the alpha relaxation occurs at 157°C in unaged PSF (activation energy calculated to be 132 kcal/mole) and occurs at 88°C in aged PSF.

CONCLUSIONS

By use of molecular and quantum mechanics tools, in particular MacroModel and the semiempirical quantum mechanics program CNDO/2, assignment of molecular processes to secondary relaxations as measured by DMA is possible. Summaries for the major relaxations in PSF, aged and unaged, are presented in Tables 8 and 9. Energetically, the concerted motions in both the aged and unaged PSF are similar. However, it is not clear why the onset of the alpha relaxation occurs at 157°C in unaged PSF and at 88°C in aged PSF. The activation energy associated with each of these regions is approximately 144 kcal/mole.

Arrhenius behavior of viscoelastic shift factors is evident even at temperatures substantially below the glass-transition temperature. These datum suggest that at different temperatures, the mechanisms controlling molecular deformation have varying affects on mechanical properties. Work is underway to relate these shift factors and activation energies to mechanical behavior. Although not discussed in great detail here, several researchers have presented NMR data that supports these findings [6,34].

Table 9 Summary of Relaxations for Unaged PSF

Temperature region	Relaxation mechanism
−110°C and less	Methyl rotation
−110 to 18°C	Phenyl rotation or rocking
18 to 157°C	Cooperative motions, and motions associated with the glass transition

The detailed studies presented here coupled to other mechanical characterizations will make possible the correlation of structure to mechanical properties and provide an avenue to completely characterize conformational changes associated with aging and inherent toughness. Work is currently underway to continue this effort.

ACKNOWLEDGEMENTS

The author wishes to thank Andrea Milligan for her efforts in installing MacroModel and beginning the necessary computational efforts, Rheometrics, Inc., and the numerous National Science Foundation Young Scholars provided by Dr. Ron Darby, in particular Ms. JoAnn Warnke.

REFERENCES

1. R. B. Seymour, *Engineering Polymer Source Book*, McGraw-Hill, New York (1990).
2. D. G. Legrand, *J. Appl. Polym. Sci.*, *13*, 2129 (1989).
3. L. Allen, D. Morley, and T. Williams, *J. Mat. Sci.*, *8*, 1449 (1973).
4. R. J. Morgan and J. E. O'Neal, *ACS D. Org. Coat. Plast. Cl*, *34*, 195 (1974).
5. R. A. Bubeck, P. B. Smith, S. E. Bales, in *"Order" in the Amorphous State of Polymers* (S. K. Keinath, R. L. Miller, and J. K. Rieke, eds.), Plenum, New York (1987).
6. A. J. Vega, *ACS Polym. Prepr.*, *22*, 282 (1981).
7. H.-H. Kausch, *Fracture of Plastics* (W. Brostow and R. D. Corneliussen, eds.), p. 85, Hanser, New York (1986).
8. A. Letton, *Effect of Thermal History and Polymer-Filler Interactions on the Molecular Relaxational Processes of Bisphenol-A Polysulfone*, Ph.D. Thesis, University of Cincinnati (1984).
9. M. G. Wyzgoski, *J. Appl. Polym. Sci.*, *25*, 1443 (1980).
10. K. Varadarajan and R. F. Boyer, *J. Polym. Sci.*, *Polym. Lett. Ed.*, *20*, 141 (1982).
11. L. C. E. Struik, *Physical Aging in Amorphous Polymers and Other Materials*, Elsevier, New York (1978).
12. D. J. Meier, *Molecular Basis of Transitions and Relaxations*, Gordon and Breach Science, New York (1978).
13. C.-Y. Lee and I. J. Goldfarb, *Polymer Eng. Sci.*, *21*, 390 (1981).
14. J. R. Fried, *J. Polym. Sci.*, *Polym. Lett. Ed.*, (1981).
15. M. Baccaredda, *J. Polym. Sci.*, *Polym. Phys. Ed.*, *5*, 1296 (1967).
16. W. Brostow, *Mat. Chem. Phys.*, *13*, 47 (1985).
17. W. G. Knass and U. H. Kenner, *J. Appl. Phys.*, *51*, 5131 (1980).
18. W. Brostow and R. D. Corneliussen, *J. Mater. Sci.*, *16*, 1665 (1981).
19. I. G. Zewi and R. D. Corneliussen, *Chem. Soc. Polym. Papers*, *20-1*, 960 (1979).

20. J. Diamant, D. R. Hansen, and M. Shen, *Polym. Sci. Technol.*, *10*, 429 (1977).
21. Kono, *J. Phys. Soc. Japan*, *16*, 1793 (1961).
22. R. J. Crowson and R. G. Arridge, *Polymer*, *20*, 737 (1979).
23. A. Letton and W. J. Tuminello, *A Method for Calculating the Molecular Weight Distribution from the Storage Modulus Master Curve*, pp. 997–1001, Society of Plastics Engineering Annual Technical Meeting (ANTEC), Los Angeles (1987).
24. A. Letton, J. R. Fried, and W. J. Welsh, in *Order in the Amorphous "State" of Polymers*, p. 359 (S. K. Keinath, R. L. Miller, and J. K. Rieke, eds.), Plenum, New York (1987).
25. L. M. Robeson, A. G. Farnham, and J. E. McGrath, *Molecular Basis of Transitions and Relaxations*, p. 405 (D. J. Meier, ed.), Gordon & Breach, New York (1978).
26. C. Still, Columbia University, New York, private communications.
27. N. L. Allinger, *J. Am. Chem. Soc.*, *99*, 8127 (1977).
28. N. Kondo, Ph.D. Thesis, University of Cincinnati, Department of Chemistry, (1979).
29. W. J. Welsh, H. H. Jaffe, N. Kondo, and J. E. Mark, *Macromol. Chem.*, *183*, 801 (1982).
30. W. J. Welsh, D. Bhaumik, and J. E. Mark, *J. Macromol. Sci.*, *Phys.*, *B20*, 59 (1981).
31. J. J. Dumais, A. L. Cholli, L. W. Jelinski, J. L. Hedrick, and J. E. McGrath, *Macromolecules*, *19*, 1884 (1986).
32. J. R. Fried, A. Letton, and W. J. Welsh, accepted for publication in *Polymer*, (1990).
33. L. C. E. Struik, *Physical Aging in Amorphous Polymers and Other Materials*, Elsevier, New York (1978).
34. P. B. Smith, S. E. Bales, and R. A. Bubeck, Dow Chemical U.S.A., Midland, Michigan, private communications.

13

Application of Quantum Mechanical Techniques to Rigid-Rod Polymers

William J. Welsh
University of Missouri-St. Louis
St. Louis, Missouri

James E. Mark
University of Cincinnati
Cincinnati, Ohio

INTRODUCTION

One should begin a discussion of rigid-rod polymers by first defining the types of polymers to be encompassed under such a heading. A discussion of rigid-rod polymers might cover a wide range of rodlike and even semirodlike polymers and arguably include the polyamides, polyimides, polyesters, polycarbonates, and many others [1]. Unfortunately, a detailed account, even one confined to address only applications of quantum mechanical techniques, of the structural and physical properties of such a wide array of polymers is beyond the scope of this chapter and likely beyond the attention span of the reader.

For these and other practical reasons, we have chosen to focus our discussion on a family of para catenated aromatic heterocyclic polymers being employed in the preparation of high-performance films and fibers. They exhibit exceptional tensile strength and modulus, thermo-oxidative stability, environmental resistance, and low specific gravity compared to glass, carbon and steel. Their "high-performance" designation is due to their ability to maintain desirable properties over a wide range of temperatures despite exposure to some very hostile environments [2,3].

The specific polymers under consideration for the present review have been the object of intense interest by the U.S. Air Force and more recently by numerous commercial enterprises. An example is poly(*p*-phenylene benzobisoxazole) (PBO), illustrated in Figure 1. The isomer shown here is designated the cis form on the basis of the relative locations of the two oxygen atoms. Other related polymers of interest are the *trans*-PBO and the cis and trans forms of the corresponding poly(*p*-phenylene benzobisthiazole) (PBT) in which the two oxygen atoms are replaced by sulfur atoms [3–21].

As can be seen from Figure 1, these chains are extremely stiff and thus are representative of the prototypical rigid-rod polymer. Because of their rigidity, they readily form liquid crystalline phases, specifically nematic phases in which the chains are aligned axially but are out of register in a random manner [3,13,14,16–19].

Spinning films or extruding films from a liquid crystalline dope of such a polymer offers great advantages [2,3]. The required flow of the system is facilitated, and the chains already have a great deal of the ordering they need in the crystalline fibrous state to exhibit the desired mechanical properties. Not surprisingly, PBO and PBT chains are the focus of the U.S. Air Force's "Ordered Polymers" Program, which has been established to develop low-weight high-performance materials for military and aerospace applications [3]. They are being used not only as fibers and films, but also as reinforcing fibrous fillers in amorphous matrices to give "molecular composites," where they serve the same purpose as the macroscopic glass or graphite fibers widely used in multiphase polymer systems [3,13,16,20,23]. The intended payoff is that these materials will replace many metal components in aircraft and space vehicles, especially where light weight is crucial. Naturally, high-performance materials such as these

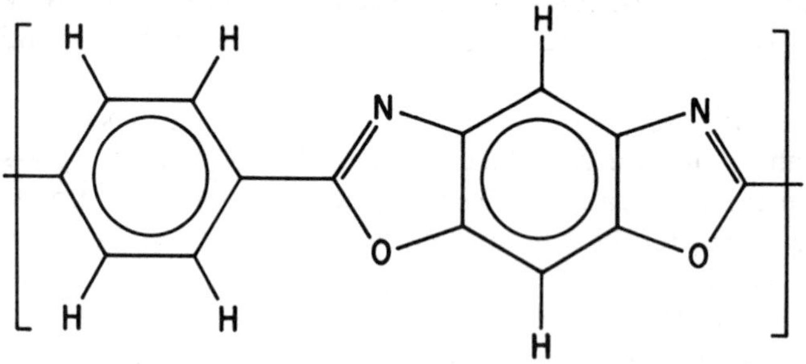

Figure 1 Illustration of the rodlike polymer *cis*-poly(benzobisoxazole) (*cis*-PBO).

are also finding applications in numerous commercial and medical roles. These prospective applications have further spurred on both basic and applied research activity in several industrial and academic laboratories including our own [3,5,16,19,23,24].

The strength, heat resistance, and environmental stability of these liquid crystalline polymers stem from their aromatic character and highly ordered chemical structure [2,3]. The polymers consist of long rodlike molecules that, under specific conditions of temperature and concentration in a solution or melt, arrange themselves in groups (known as domains) of anisotropic liquid crystalline phases [3,13,16,19,21]. While the polymer molecules within each domain are parallel, the domains themselves are usually not. However, during processing, the individual domains are brought into alignment, creating a far more compact and orderly arrangement than is found in ordinary plastics.

While most ordinary plastics are processed through a melt phase, this technique is not possible with PBO and PBT, which undergo decomposition before they reach their melting point. Instead, PBO and PBT must be dissolved in strong acids, such as methanesulfonic acid, trifluoromethanesulfonic acid, or chlorosulfonic acid, before they can be spun into fibers [3,16,21]. These acids present disposal problems and are hazardous, expensive, and hard on equipment.

Alternatively, one would prefer to process these polymers through the melt phase or from a solution using aprotic solvents [19]. Researchers are investigating ways of reducing the polymers' rigidity, and hence, decreasing the melting point and/or increasing the solubility in aprotic solvents, by slightly altering their molecular structures [3]. One method of interest is to insert molecular "swivels" or spacers containing flexible chemical moieites, such as —O—, —S—, or —CH_2—, into the otherwise rigid polymer chain (Fig. 2) [3,11,12,25,26]. A similar strategy is to insert a group, such as the Ar-Ar illustrated in Figure 3 (where Ar-Ar refers to various aromatic ring sytems). This modification would translate the otherwise rodlike chain without altering its axial direction. The advantage of a wholly aromatic group stems from its superior thermo-oxidative stability [27]. The resulting PBO and PBT polymers are referred to as "articulated." These articulated PBOs and PBTs are expected to retain the liquid-crystalline-forming properties of the parent molecules yet exhibit differences from them in terms of solubility, thermal characteristics, and morphology [25,26]. Certainly, the conformational characteristics of these molecular swivels are of the utmost importance in this regard and have been the focus of several quantum mechanical studies [12].

While these molecular alterations provide many benefits, they also tend

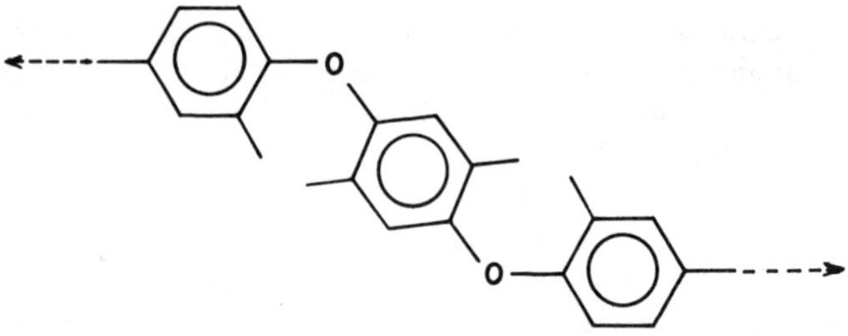

Figure 2 Example of a molecular swivel considered for use in the "articulated" PBO and PBT polymers.

to diminish the polymer's strength and thermo-oxidative stability [3,27]. Still, the correct combination of rods and swivels contained in a polymer chain may provide the optimum balance of desirable properties. Furthermore, heat treatment of these materials following processing has been shown to enhance their high-performance properties [3,9,13,28].

In order to dissolve polymers such as PBO and PBT, it is necessary for solvent systems to break up the strong intermolecular attractions that exist in the solid [3,19,21]. Highly acidic solvents, such as methane sulfonic acid and chlorosulfonic acid, succeed in this regard by protonating the polymer chains. This generates sufficient repulsive electrostatic interactions between chains to overcome the attractive interactions, thus allowing the chains to separate [3,17,19,21]. Experimental evidence suggests that the protonated PBO and PBT chains exhibit differences in their structural and conformational behavior compared with their unprotonated forms [6], and quantum mechanical analysis of the effects of chain protonation on both structure and conformational behavior has provided insights in explaining these differences at a molecular level [29]. Furthermore, quantum mechanical studies can provide information as to the extent and order of

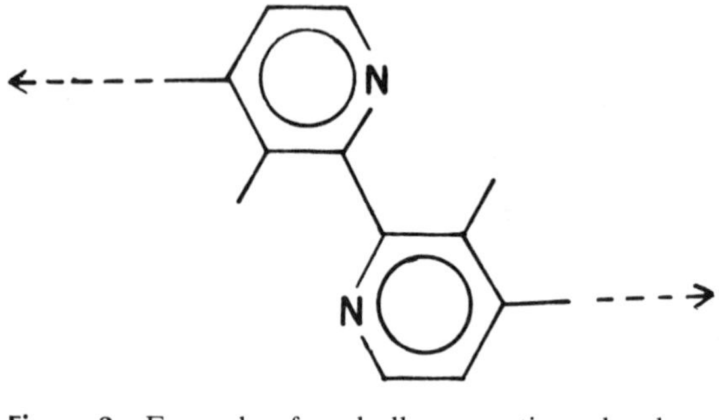

Figure 3 Example of a wholly aromatic molecular swivel.

protonation; i.e., how many protons can bind to a chain unit, and in what sequence do the basic atoms or groups in the chain become protonated [30].

The success in deriving exceptional mechanical and thermo-oxidative properties from rigid-rod polymers such as PBO and PBT has stimulated investigation of other, related, stiff-chain and rodlike polymers. Recently, interest has extended to the structurally related, but more flexible AA and AB analogues of PBO and PBT [13,16,31–33]. Particular attention in terms of synthesis [31] and both x-ray crystallographic [33] and quantum mechanical [32] analyses has been paid to poly(5,5′-bisbenzoxazole-2,2′-diyl-1,4-phenylene) (AAPBO) (Fig. 4), poly(2,5-benzoxazole) (ABPBO) (Fig. 5), and their related sulfur-containing analogues. Recent experiments indicate that, while ABPBO exhibits liquid crystalline behavior in solution [31], AAPBO does not, and one might expect to find differences in their conformational behavior, the delineation of which is amenable to quantum mechanical analysis [32]. Interestingly, fibers of ABPBO spun from these liquid crystalline phases have shown improved thermal stability over the polyamides and PBTs [33].

Recent theoretical and experimental work has focused on possible applications for these polymers as doped electrical conductors [34–39] and as nonlinear optical materials [24,40,41]. We will present here some results of theoretical work [24,34–37] conducted in the authors' laboratories which explore these phenomena at the molecular level.

Regarding theoretical studies of nonlinear optical (NLO) properties in polymers, almost all have been based on the simple Hückel model [42–44], although recent attempts have been made to include electron correlation within the Hückel framework [45–47]. Still, these treatments employ Hückel parameters derived from and adopted for small molecules and may thus not be suitable for calculations on model compounds representative

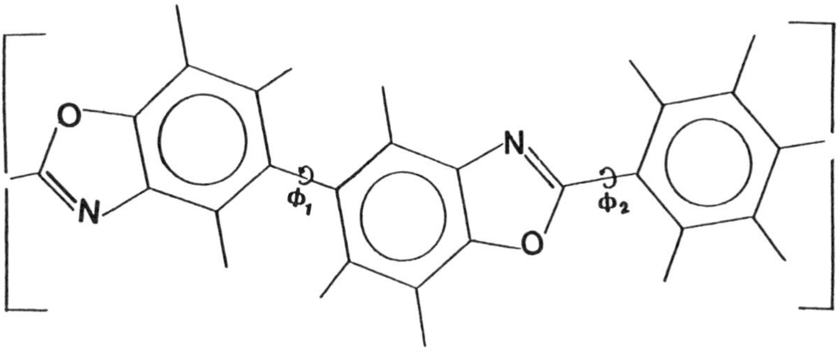

O N Φ_1 N O Φ_2

Figure 4 Illustration of the AAPBO repeat unit.

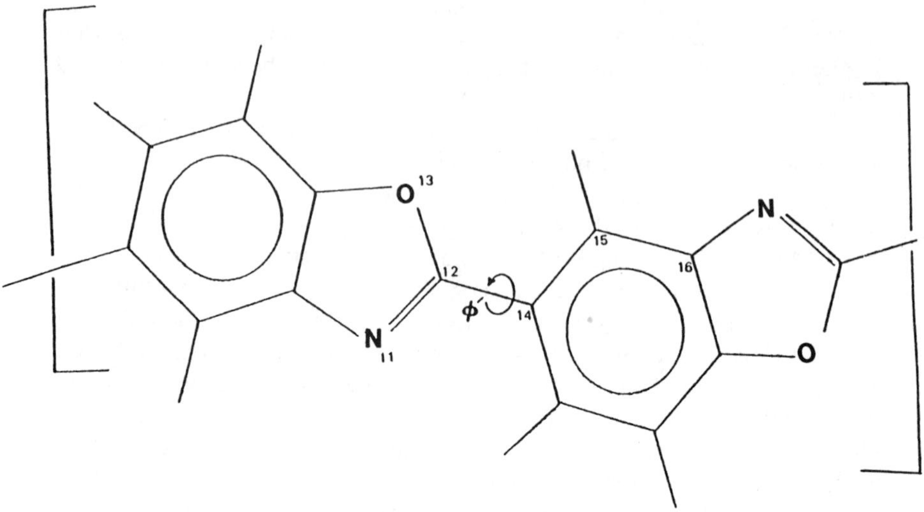

Figure 5 Illustration of the ABPBO repeat unit.

of polymers. Apart from the question of the validity of the Hückel method itself, such a choice of parameters introduces additional uncertainties. We thus review here an approximate ab initio theory [24,37] that derives the Hückel Hamiltonian from first principles along with the parameters appropriate for a given system. Although applications of this theory to PBO, PBT, and related molecules is still forthcoming, this theoretical treatment is highly amenable to these molecular species, which are reported to exhibit exceptional NLO properties.

In the following, the authors attempt to summarize the results of quantum mechanical calculations carried out by their research groups as well as by other investigators on the rodlike polymers PBO, PBT, and related molecules. In particular, we survey here studies of the structures, conformational characteristics, molecular polarizabilities, and moduli of these chains, including in some cases the protonated forms known to exist in strong acids [26]. Also included are results of quantum mechanical investigations on some of the molecular swivels contained in the so-called articulated forms of these polymers. Results of similar calculations employing molecular mechanics (MM) methods will be presented for comparison. Finally, more recent work related to the study of electrical conductivity and NLO phenomena in these polymers will be presented. Some aspects of our theoretical work on PBO, PBT, and related polymers have been covered in earlier reviews [5,24]. The emphasis here is to provide an example of how quantum mechanical studies can provide a molecular understanding of the unusual properties and processing characteristics of rodlike polymers.

SUMMARY OF METHODOLOGIES

The structures and conformational properties of these chains were investigated using the CNDO/2 (complete neglect of differential overlap, version 2) [48–50], MNDO (modified neglect of diatomic overlap) [51], and AM1 (Austin Model 1) [52] semiempirical molecular orbital (MO) methods. The latter two as employed are contained in the program AMPAC Version 1.0 [53]. While the MNDO and AM1 studies are more recent, they substantially corroborate the conclusions drawn from the earlier CNDO/2 work.

Molecular polarizabilities were calculated [54] using three separate methods, specifically, second-order perturbation theory combined with the formalism of CNDO/S CI (i.e., configuration interactions) [55], an empirical scheme based on additivity of atomic hybrid components [56], and the standard bond polarizability method [57].

Electronic band structures, relevant to investigations of electrical conductivity, have been calculated using both the tight-binding scheme based on the extended Hückel theory [61,62] and more recently a method based on a novel approximate ab initio framework [37]. Description of the method derived for the theoretical prediction of NLO effects in these polymers is given in a later section.

Initial structural information, in particular bond lengths and bond angles, were generally obtained from results of x-ray structural studies carried out on model compounds [58,59]. The results presented here pertain to calculations performed on model compounds, such as the PBO model compound illustrated in Figure 6, chosen to be large enough to mimic the

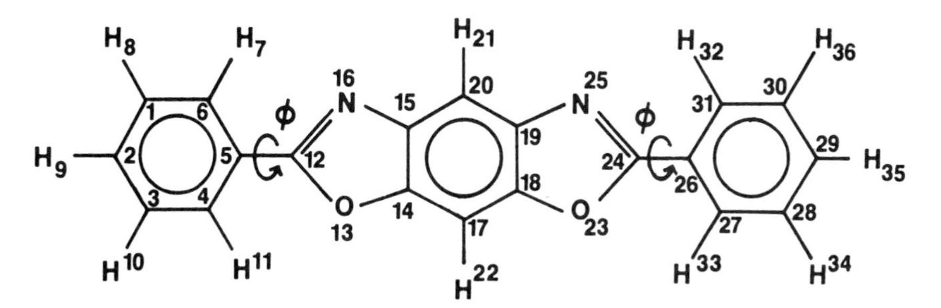

Figure 6 Illustration of the *cis*-PBO model compound.

essential features of the entire chain molecule yet still small enough to maintain computational feasibility.

RESULTS AND DISCUSSION

The Unprotonated PBO and PBT Chains

Structures and Conformational Characteristics

The preparation of high-strength materials consisting of rodlike polymers such as these requires a high degree of alignment of the rods. The extent of intramolecular rotational flexibility and thus deviations from planarity are important in this regard, particularly in terms of chain-packing effects and the solubility characteristics of the polymers.

Despite the rigidity of PBO and PBT chains, they still possess some conformational flexibility in that rotations should be permitted about the phenylene groups within each repeat unit. Such rotations ϕ are illustrated in the PBO model compound illustrated in Figure 6.

Conformational energies ΔE calculated [4] by CNDO/2 versus the torsion angle ϕ are plotted in Figure 7 for *cis*-PBO, *trans*-PBO, and *trans*-PBT model compounds. The plots show that *cis*- and *trans*-PBO exhibit very similar conformational energy profiles. In both cases, the preferred conformation corresponds to $\phi = 0°$ (coplanarity); this result is in excellent agreement with MM results [60] and with experiment [59]. The barrier to rotation away from $\phi = 0°$ rises monotonically with increasing ϕ to max-

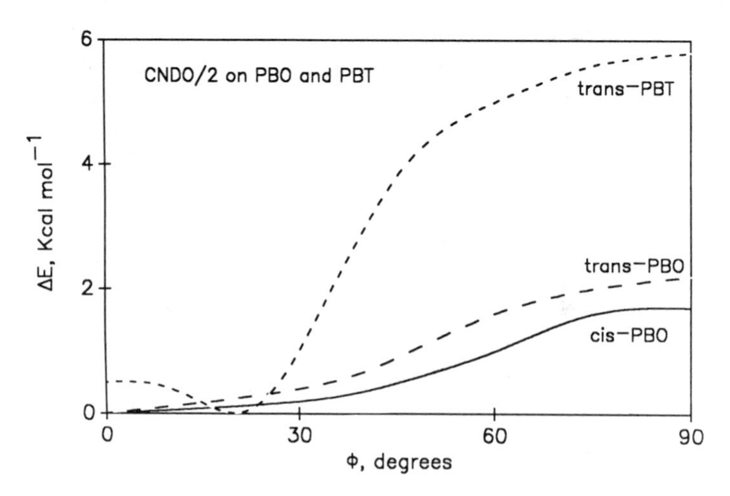

Figure 7 Plot of CNDO/2-calculated conformational energy ΔE versus rotation angle ϕ for model compounds of *cis*-PBO, *trans*-PBO, and *trans*-PBT.

imum values of about 1.6 (*cis*-PBO) and 2.2 (*trans*-PBO) kcal mol^{-1}. That rotation away from coplanarity encounters an increase in energy implies that conjugation effects between the aromatic moieties (favoring coplanarity) dominate steric repulsions (disfavoring coplanarity) between the *ortho*-hydrogen atoms on the phenylene group and nearby atoms within the heterocyclic group.

As reported elsewhere [4], the bond connecting the phenylene ring to the heterocyclic group is shorter (1.46 Å) in these species than that found in analogous low-molecular-weight compounds. This suggests that in the former case some double-bond character exists in these bonds as a result of long-range conjugation effects. The present results corroborate earlier MM studies on the same compounds [60].

More recent AM1-calculated [24,61] conformational energies ΔE versus ϕ for the *cis*- and *trans*-PBO model compounds are plotted in Figure 8. The energy profiles for both molecules are essentially identical, so only one curve appears on the graph. The energy is a minimum at $\phi = 0°$ (i.e., a coplanar conformation) and rises monotonically to a maximum at $\phi = 90°$ (the perpendicular conformation) some 5 kcal mol^{-1} (or 2.5 kcal mol^{-1} per rotatable bond) higher in energy. Hence, for both *cis*-PBO and *trans*-PBO the predicted conformation is coplanar. The coplanarity predicted agrees with the coplanar conformations observed for these model compounds in the crystalline state [59].

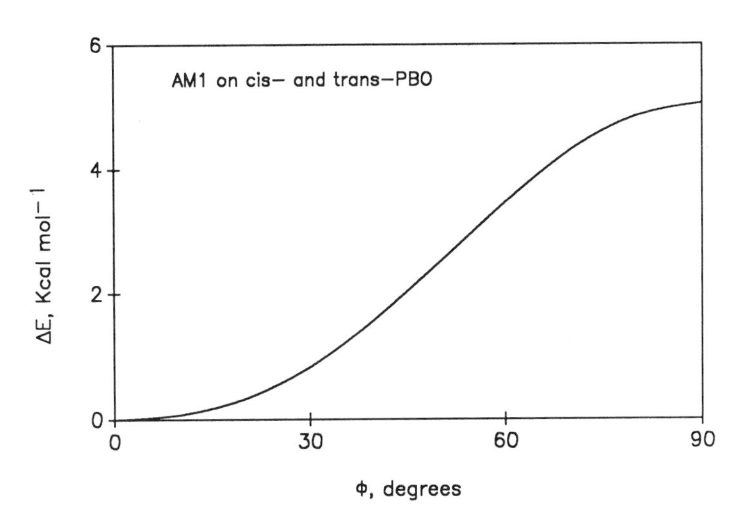

Figure 8 Plot of AM1-calculated conformational energy ΔE versus rotation angle ϕ for the *cis*- and *trans*-PBO model compounds. Both isomers gave nearly identical energy profiles so only one curve is given on the graph.

The AM1 results thus agree, at least qualitatively, with earlier MM [60] and CNDO/2 [4] calculations. Specifically, AM1, CNDO/2, and MM calculations all find the conformational energy minimum for *cis*-PBO and *trans*-PBO at the coplanar ($\phi = 0°$) conformation, the energy maximum at $\phi = 90°$, and nearly identical energy profiles for both the *cis* and *trans* forms. However, the three methods differ in the absolute magnitude of their ΔE values versus ϕ. For comparison, Table 1 lists the relative energies at $\phi = 45°$ and at $\phi = 90°$ as obtained by the AM1, CNDO/2, and MM calculations. While the AM1 and MM energies compare favorably, on average CNDO/2 gives values only 35% as large as those obtained by AM1. Taken together, the ratio of barrier heights at $\phi = 90°$ for CNDO/2:AM1:MM is about 0.38:1.0:1.3.

The consistently lower energy barriers found by CNDO/2 relative to AM1 may be traced in part to the tendency of CNDO/2 and other semiempirical methods to overestimate core-core repulsions and thus to underestimate barriers to all nonplanar conformations [53,62,63]. The relatively good agreement between AM1 and MM is somewhat surprising given that the MM calculations [60] lacked geometry optimization. However, only a slight reduction in the MM energies obtained would be expected had geometry optimization been included since the aforementioned steric congestion (giving rise to structural deformations) abates considerably as ϕ deviates from 0°.

The corresponding AM1 energy profiles for the *cis*-PBT and *trans*-PBT model compounds [24,61] (Fig. 9) are again virtually identical and so are represented by a single curve. Also included in Figure 3 is the energy profile for *cis*-PBT as calculated by MNDO. It is apparent from the curve that MNDO erroneously predicts an energy maximum at $\phi = 0°$ and minimum at $\phi = 90°$. The MNDO method is thus shown again to be unacceptable for describing rotations about single bonds between sp^2 carbon atoms connecting aromatic rings [53].

The AM1 energy profile for the PBT model compounds (Fig. 3) predicts a minimum-energy conformation at $\phi \sim 27°$, in close agreement with the value $\phi = 23°$ observed in the crystalline state [58]. Nevertheless, the energy barrier to coplanarity ($\phi = 0°$) is small at 0.25 kcal mol^{-1} (or 0.13 kcal mol^{-1} per rotatable bond). Beyond $\phi = 27°$ the energy rises monotonically to a maximum of only 1.7 kcal mol^{-1} (0.85 kcal mol^{-1} per rotatable bond) at $\phi = 90°$.

In Table 1, results of AM1 [24,61], CNDO/2 [4], and MM [60] calculations are again compared for the PBT model compounds. All three methods predict nonplanar conformations ranging from 20–27° (AM1 and CNDO/2) to 55–57° (MM). While the range of ϕ values appears considerable, the AM1 energy well is shallow and wide in that the range $0° \leq \phi$

TABLE 1 Conformational Comparisons[a] of AM1,[b] CNDO/2,[c] and MM[d] Calculations on Molecules under Study

Model compound	Predicted ϕ			Relative Energies[a] $\phi = 45°$			Relative Energies[a] $\phi = 90°$		
	AM1	CNDO/2	MM[e]	AM1	CNDO/2	MM[e]	AM1	CNDO/2	MM[e]
cis-PBO	0°	0°	0°	1.0	0.25	1.6	2.5	0.80	3.5
trans-PBO	0°	0°	0°	1.0	0.36	1.1	2.5	1.1	3.2
cis-PBT	27°	—[f]	57°	0.13	—[f]	0.20	0.85	—[f]	0.38
trans-PBT	27°	20°	55°	0.13	1.6[e]	0.14	0.85	3.9[e]	0.36

[a]All energy values are normalized as energies per rotatable bond in units of kcal/mol.
[b]Refs. 24 and 61.
[c]Ref. 4.
[d]Ref. 60.
[e]Geometry optimization not employed.
[f]Data not available.

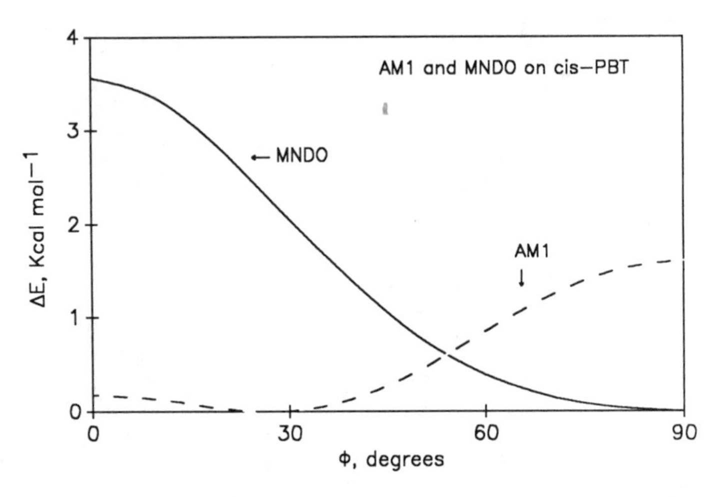

Figure 9 Plot of AM1-calculated conformational energy ΔE versus rotation angle ϕ for the *cis*- and *trans*-PBT model compounds. Both isomers gave nearly identical energy profiles so only one curve is portrayed on the graph. The corresponding MNDO-calculated plot for the *cis*-PBT model compound is also shown.

$\leq 60°$ is within 0.5 kcal mol^{-1} of the minimum. Hence the rotatable bond in PBT would appear to exhibit a high degree of rotational flexibility. Moreover, the MM values [60] were obtained without geometry optimization which, if included, would likely have reduced the value of ϕ from the 55–57° range obtained.

The AM1, CNDO/2, and MM energy profiles are thus qualitatively similar for the PBT model compounds. However, quantitative differences are apparent from Table 1. First, CNDO/2 barriers at $\phi = 90°$ are extremely high, no doubt primarily due to the absence in this case of geometry optimization for PBT [4]. The AM1 and MM energy barriers are again reasonably consistent, although relative to AM1 the MM energy barrier at $\phi = 90°$ is about half as high.

Based on the most recent and most rigorous results as obtained from AM1 [24,61], the major conclusion is that rotational barriers are substantially lower for PBT compared with PBO (see Figs. 8 and 9). PBO is sterically less congested than PBT (because the oxygen atoms in PBO are smaller than the alternative sulfur atoms in PBT), hence the conformation of PBO is dominated by conjugative effects favoring coplanarity (i.e., $\phi = 0°$). In PBT steric conflicts dominate conjugative effects at the coplanar conformation; hence, the preferred ϕ found, which is intermediate between 0° and 90°, represents a balance between conjugative effects (favoring coplanarity) and steric conflicts (favoring a more perpendicular conformation). This balance is more evident in PBT than in PBO and gives rise to a more moderate conformational energy profile. Similar conclusions were drawn by Wierschke [64] based on results of AM1 and MNDO calculations carried out on smaller model compounds of PBO and PBT.

Bowing in the cis-*PBT Model Compound*

AM1 calculations carried out [24,61] on the *cis*-PBT model compound yielded a virtually flat molecular structure, in contrast to the bowing observed in the crystal structure [58]. AM1 also revealed no bowing in the *cis*-PBO, *trans*-PBO, and *trans*-PBT model compounds, in agreement with the crystal structures of each [58,59]. The absence of bowing in the AM1-calculated *cis*-PBT structure suggests that the observed bowing is intermolecular in origin. Preliminary AM1 calculations involving two flat, parallel-stacked, *cis*-PBT model compounds revealed the onset of bowing as the two molecules approach [24,61]. However, these initial results require further scrutiny to substantiate their validity.

Structural Comparisons

AM1-calculated values [24,61] of selected bond lengths and bond angles in the PBO and PBT model compounds are listed in Table 2 and compared with the corresponding x-ray crystallographic values [58,59]. Structural geometries were also calculated for the PBO model compounds in earlier CNDO/2 studies [4], but the AM1 values agree more closely with the crystal structures values. The AM1 molecular energies of *cis*- and *trans*-PBO were found to be virtually identical, hence it is not surprising that their structural geometries are very close. Compared with the x-ray crystallographic values listed, the major discrepancy in bond lengths involves the C12—O13 bond for which the AM1-calculated value is about 0.1 Å larger (1.43 Å versus 1.325 Å) than that observed for the model compounds. Thus AM1 depicts this bond as similar to a contracted single bond C—O (1.38–1.40 Å) [65], while the x-ray crystallographic results [59] depict a bond with substantially more double-bond character. For comparison, values of some related C—O bond lengths are 1.33 Å for the C=C—OH in enols and 1.31 Å for the C_{Ar}—C(=O)—OH in acids [65].

Regarding bond angles, the calculated and observed values (Table 2) are in good agreement except for the C5—C12—O13 bond angle where the calculated value (115.0°) is appreciably smaller than that observed (122.4°). This discrepancy may be related to the anomaly noted above for the C12—O13 bond length.

Analysis of the AM1-calculated [24,61] bond lengths and bond angles for the PBO model compounds as ϕ departs from the coplanar (ϕ = 0°)

TABLE 2 Selected Values of Observed and AM1 Calculated [24,61] Bond Lengths and Bond Angles for the *cis*-PBO and *trans*-PBO Model Compounds[a]

	cis-PBO				*trans*-PBO			
	X-ray[b]	AM1			X-ray[b]	AM1		
	$\phi = 0°$	$\phi = 0°$[c]	45°	90°	$\phi = 0°$	$\phi = 0°$[c]	45°	90°
Bond lengths[d]								
C5-C12	1.456	1.454	1.456	1.460	1.456	1.454	1.457	1.460
C12-O13	1.325	1.432	1.432	1.432	1.325	1.428	1.427	1.427
C12-N16	1.325	1.330	1.327	1.325	1.325	1.332	1.329	1.327
O13-C14	1.390	1.392	1.394	1.394	1.390	1.396	1.396	1.395
C15-N16	1.390	1.414	1.414	1.413	1.390	1.412	1.414	1.415
Bond angles[e]								
C4-C5-C12	121.1	120.2	119.9	119.8	121.2	120.0	119.9	119.8
C5-C12-O13	122.4	115.0	115.2	115.0	122.4	115.7	115.3	115.2
O13-C12-N16	115.1	113.7	114.0	114.0	115.1	113.8	114.0	114.1
C12-O13-C14	105.1	104.3	104.3	104.3	105.1	104.4	104.3	104.4
C12-N16-C15	107.3	105.9	105.8	105.8	107.3	105.8	105.6	105.7

[a]Numbering of atoms is as given in Figure 6.
[b]Taken from Ref. 59.
[c]Corresponds to calculated minimum energy conformation.
[d]Units in angstroms.
[e]Units in degrees.

conformation (Table 2) reveals some trends. Specifically, a slight increase in the C5—C12 bond length and commensurate decrease in the C12—N16 bond length occurs. As noted previously [4,5] these patterns are consistent with a shift in electron density (and thus bond strength) from the bridge bond to the ring systems as ring ... ring conjugation effects diminish with increasing ϕ.

A similar tabulation of selected bond lengths and bond angles for the PBT model compounds [24,61] is given in Table 3. It should be noted that in the case of the crystal structure of the *cis*-PBT model compound two independent molecules (designated A and B in Table 3) were identified in the unit cell of the crystal structure [58]. Both molecules A and B exhibit bowing, as noted before, while ϕ deviates only slightly (ϕ = 2.8–5.8°) from the coplanar conformation. In contrast, the crystal structure of the corresponding *trans*-PBT model compound [58] contains only one molecule per unit cell, having a nearly flat structure (i.e., no bowing) but significant deviations from coplanarity (ϕ = 23.2°).

Of the AM1 results listed in Table 3, those corresponding to the calculated minimum energy conformation ϕ = 30° are best compared with the crystal structure geometry. The structural geometries of molecules A and B in the crystal structure of the *cis* compound are very similar except that the C5—C12 bond length is considerably shorter in B (1.459 Å) compared with A (1.476 Å). The corresponding AM1 value (1.458 Å) agrees more closely with that of molecule B which will thus serve henceforth as our source for comparison.

Regarding the bond lengths listed in Table 3, the only major discrepancy between the calculated and x-ray crystallographic results [58] involves the C—S bonds for which the AM1 values are consistently shorter. This finding has been noted in other studies [66,67], and may be due in part to MNDO parameters being used in the absence of AM1 parameters for sulfur atoms in the version of AMPAC available for this work [53,63]. Calculated and observed bond angles are in good agreement, the only discrepancy worth noting being the larger calculated C12—S13—C14 bond angles. Again, the parameterization of the sulfur atom mentioned above may be a factor in this regard.

The variations in AM1-calculated structural parameters with ϕ show the same trends as noted for the PBO model compounds. Most notably, as ϕ deviates from 0° the bridge bond C5—C12 becomes longer (thus weaker) while the endocyclic bonds C12—S13 and C12—N16 become shorter (thus stronger).

Charge Distribution

Selected values of the AM1-calculated atomic partial charges [24,61] for the four model compounds are listed in Table 4. The only unusual feature is

TABLE 3 Selected Values of Observed and AM1 Calculated [24,60] Bond Lengths and Bond Angles for the *cis*-PBT and *trans*-PBT Model Compounds[a]

	cis-PBT					*trans*-PBT			
	X-ray[b] $\phi = 2.8–5.8°$		AM1			X-ray[b]	AM1		
	A	B	$\phi = 0°$	30°[c]	90°	$\phi = 23.2°$	$\phi = 0°$	30°[c]	90°
Bond lengths[d]									
C5-C12	1.476	1.459	1.458	1.458	1.461	1.469	1.457	1.457	1.461
C12-S13	1.764	1.764	1.741	1.739	1.738	1.758	1.740	1.738	1.736
C12-N16	1.307	1.302	1.322	1.320	1.316	1.292	1.323	1.322	1.317
S13-C14	1.741	1.737	1.682	1.683	1.684	1.736	1.684	1.681	1.681
C15-N16	1.380	1.382	1.410	1.412	1.415	1.385	1.409	1.410	1.410
Bond angles[e]									
C4-C5-C12	122.2	122.8	120.4	120.3	120.0	121.6	120.3	120.3	120.5
C5-C12-S13	120.4	120.4	119.7	119.4	118.5	119.9	119.7	119.5	121.7
S13-C12-N16	115.4	114.8	114.6	114.9	115.3	116.3	114.7	115.0	115.6
C12-S13-C14	88.9	89.5	91.8	91.6	91.4	88.9	91.8	91.6	92.0
C12-N16-C15	110.9	111.7	110.2	110.0	109.8	110.8	110.1	109.9	109.9

[a]Numbering of atoms is as given in Figure 6, with S13 and S23 replacing O13 and O23, respectively.
[b]Taken from Ref. 58, where A and B for *cis*-PBT refer to the two molecules found in the asymmetric unit cell.
[c]Corresponds to calculated minimum energy conformation.
[d]Units in angstroms.
[e]Units in degrees.

TABLE 4 AM1 Calculated Values [24,61] of the Partial Charges[a] for Selected Atoms in the PBO ($\phi = 0°$) and PBT ($\phi = 30°$)[b] Model Compounds under Study

Atom	*cis*-PBO	*trans*-PBO	*cis*-PBT	*trans*-PBT
C5	97.6	106.	−127.	−127.
C12	−43.9	−46.2	−46.	−46.4
X13[c]	−134.	−134.	330.	332.
N16	−125.	−134.	−105.	−109.

[a]In units of fraction of an electron's charge × 10^3.
[b]Corresponds to calculated minimum-energy conformation.
[c]X is oxygen (*O*) for PBO model compounds and sulfur (S) for PBT model compounds.

the charge on S13 in PBT which is relatively large ($\sim 0.34\ e^-$) and positive. In fact, the bond dipole of C12 ← S13 is pointed in a direction opposite that of C12 → O13 found in the corresponding PBO model compounds. This same finding has been reported in other AM1 calculations on structurally related compounds [66].

The partial charge on a given atom listed in Table 4 changes only slightly (about 10% on average) with variation of ϕ from 0° to 90°. This effect is even smaller for the atoms not listed in Table 4.

Earlier CNDO/2-calculated values [4] of the atomic partial charges versus ϕ are listed in Table 5 for the *cis*-PBO model compound. Corresponding values for the *trans* isomer are again virtually identical. In both cases the calculated values were very nearly invariant to changes in ϕ. However, there does appear to be a slight shift in electron density from C5 and C12 to N16 as ϕ departs from 0°, a result consistent with decreasing conjugation effects between the ring systems.

TABLE 5 CNDO/2-Calculated Values [4] of the Partial Charges[a] for Selected Atoms in the *cis*-PBO Model Compound

	Partial charges		
Atom	$\phi = 0°$	45°	90°
C5	−16.1	−10.8	−4.15
C12	336.0	345.0	354.0
O13	−293.0	−291.0	−290.0
N16	−301.0	−308.0	−315.0

[a]In units of fraction of an electron's charge × 10^3.

Tensile Moduli and Compressive Deformations

Klei and Stewart [68] derived a method based on MNDO for calculating elastic moduli of polymers from heats of formation computed at elongated translation vectors. Later Wierschke [69] extended these calculations to the AM1 method and included compressive deformations. Table 6 summarizes these results for PBO and PBT in comparison with experimental values.

Wierschke [69] attributed the differences between the AM1- and MNDO-calculated moduli to inherent differences in the two methods. As stated earlier, MNDO predicts incorrectly a perpendicular orientation between the phenyl and heterocyclic ring system of PBO, whereas AM1 predicts a more coplanar orientation in agreement with crystallographic evidence. As a consequence, the two methods yield different energy versus strain curves and hence different calculated moduli. By the same token, for PBT the AM1 and MNDO values of the moduli are much closer since AM1, which in the version employed [53] was parameterized only for C, H, O, and N atom types, adopts MNDO parameters for sulfur atoms. The AM1 results listed in Table 6 for PBT are thus in a sense an amalgamation of AM1 and MNDO and accordingly are quite similar to the MNDO values shown.

Chains Containing Molecular Swivels

Because of their stiffness, the PBO and PBT chains are very nearly intractable and difficult to dissolve in solvents [3,19,21]. This presents problems from the standpoint of processing them into usable films and fibers [3]. These materials may be made more tractable, however, by the insertion of a limited number of atoms or groups chosen so as to impart a controlled amount of additional flexibility to the chains [3,11,12,25,26]. The insertion of even a small number of flexible molecular fragments or "swivels" into such chains will increase their flexibility and thus improve their tractability

TABLE 6 Comparison of AM1, MNDO, and Experimental Moduli[a]

	AM1	MNDO	X-ray	Experimental
cis-PBO	730	670	475	330
trans-PBO	707	619	—	—
cis-PBT	610	602	—	—
trans-PBT	605	528	400	300

[a]In units of GPa.
Source: From Ref. 69.

by allowing mutual rotation of adjacent chain elements about the swivel's rotatable bonds. (Such swivels also have the advantage of facilitating the polymerization.) It is obviously of considerable importance to investigate the effect of the structure, number, and spacing of such swivels along the chain. For example, two closely spaced swivels, as illustrated in Figure 2, would decrease the rigidity of the chains but still permit occurrences of nearly parallel conformations conducive to attainment of the desired molecular alignment or organization.

The CNDO/2 method has been used to study the structural and conformational characteristics of several wholly aromatic molecules under consideration for use as possible swivels within rodlike polymer chains [12]. The basic molecules investigated were biphenyl (**1**), 2,2′-bipyridyl (**2**), 2-phenylpyridine (**3**), 2,2′-bipyrimidyl (**4**), and 2-phenylpyrimidine (**5**) (Fig. 10). The molecules are structurally similar but differ in the number and location of *ortho*-CH groups replaced by nitrogen atoms. The notion tested here is whether such substitution would improve the probability of achieving the desired coplanar conformations.

In terms of their potential for use as swivels, two features of primary importance are their barriers to coplanarity and overall flexibility. The barriers to coplanarity rank, as would be expected, in order of the number of such nitrogen substitutions, namely **4** (0.4 kcal mol^{-1} < **5** (0.8 kcal mol^{-1}) < **2** (*cis* = 2.0 kcal mol^{-1}, *trans* = 0.5 kcal mol^{-1}) < **3** (2.0 kcal mol^{-1}) < **1** (3.0 kcal mol^{-1}), where the numbers in parentheses represent the energy barrier calculated for the respective swivel. In terms of overall flexibility about the swivel bond, as determined by the amount of configurational space within 1 kcal mol^{-1} of the overall energy minimum, the species rank roughly in the same order as above with **4** having the greatest

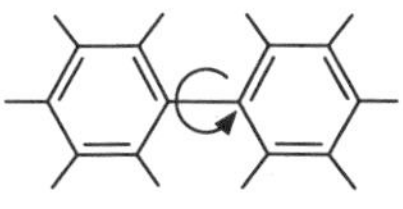

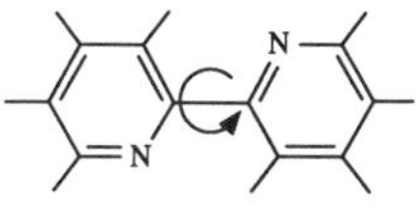

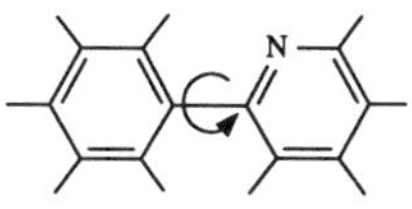

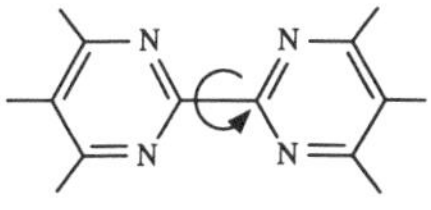

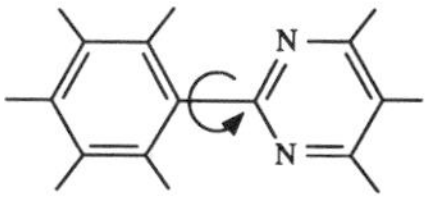

Figure 10 Illustration of the series of nitrogen-containing wholly aromatic molecular swivels investigated by quantum mechanical techniques.

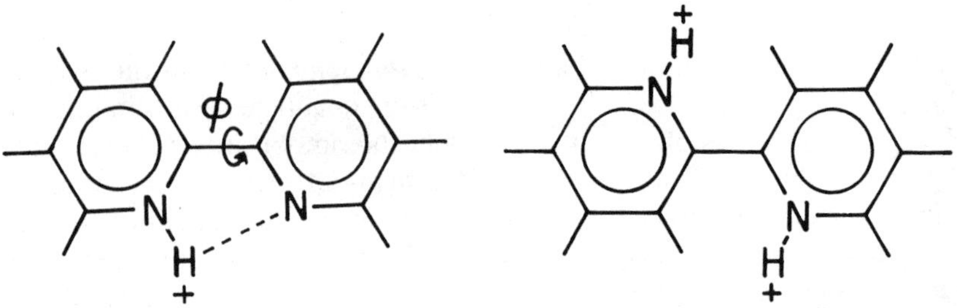

Figure 11 Illustration of protonation of a nitrogen-containing wholly aromatic swivel.

overall flexibility. Consequently, on the basis of the above considerations, it is predicted that of the species studied here **4** would offer the greatest potential for use as a swivel, with the others following in the order listed above.

The effects of mono- and diprotonation on the conformational characteristics of 2 were studied by CNDO/2 calculations and found to be significant [12]. While the parent molecule and the diprotonated species both prefer the trans over the cis coplanar conformation, just the opposite is true for the monoprotonated case (Fig. 11). Hence, it is concluded that the coplanar conformation preferred by the species is expected to be a function of the acidity of the medium, with trans preferred in neutral media, followed by (with increasing acidity) a preference for the cis monoprotonated species and then for the trans diprotonated species.

The CNDO/2 results [12] suggest the presence of hydrogen bonding for monoprotonated **2** in the cis coplanar conformation, as evidenced by the

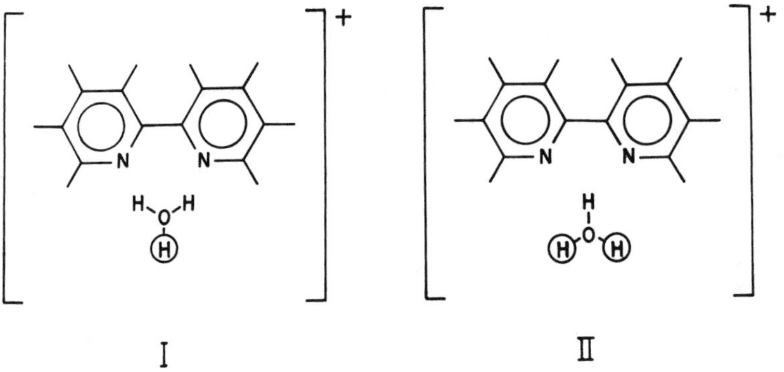

Figure 12 Two initial orientations for H_3O^+ complexing with a bipyridyl molecular swivel.

strong preference overall for this conformation and by the bending of the exo ring angles about the swivel atom in order to shorten the interatomic distances to reasonable values for the N ··· H^+—N hydrogen bond. The species 2,2′-bipyridyl-H_3O^+ and 2,2′-bipyrimidyl-$2H_2O$ have also been studied, each in two initial configurations, one with an O—H bond pointing to each N atom and the other with an O—H bond normal to the swivel bond (Fig. 12). For each species, the former configuration is the energetically preferred one.

Effects of Protonation on the Structure and Conformation of PBO and PBT Chains

Strong acids, such as methane sulfonic acid, chlorosulfonic acid, and polyphosphoric acid are needed to dissolve these materials [3,21], although recent work suggests other, aprotic, solvent systems [19]. Protonation of the rodlike PBO and PBT chains and their model compounds in acidic media will have significant effects on their solubility, solution behavior, geometry, and conformational characteristics. Considerable interest has therefore focused on the extent, nature, and effect of protonation of these polymers in order to gain insights into their solubility behavior and solution properties.

Freezing-point-depression measurements [20] on PBO and PBT model compounds have indicated that, depending on the acidity of the medium, the PBO model compound can exist as a diprotonated ion, presumably with one proton on each (highly basic) nitrogen atom, or as a tetraprotonated ion, presumably with the other two protons on the oxygen atoms. The PBT model compounds appear to have a greater preference for the diprotonated form owing to the lower basicity of sulfur atoms relative to oxygen atoms.

Results of CNDO/2 calculations [29] carried out to characterize the effects of protonation on the conformations of the PBO chain are plotted in Figure 13. For all three forms, i.e., the unprotonated, diprotonated, and tetraprotonated forms of the *cis*-PBO model compound, the preferred conformation corresponds to the coplanar form ($\phi = 0°$). As ϕ increases, the energy barrier increases monotonically, the maximum barrier being located at $\phi = 90°$ with an energy of $E \cong 2.0$, 9.0, and 20.0 kcal mol^{-1} above that of the corresponding coplanar form for the unprotonated form, the diprotonated ion, and the tetraprotonated ion, respectively. In other words, the barrier to rotation away from the coplanar form *increases* as a result of increased protonation.

Based on steric arguments alone, repulsions between the ortho hydrogens on the phenylenes and the acidic protons on the benzoxazole ring

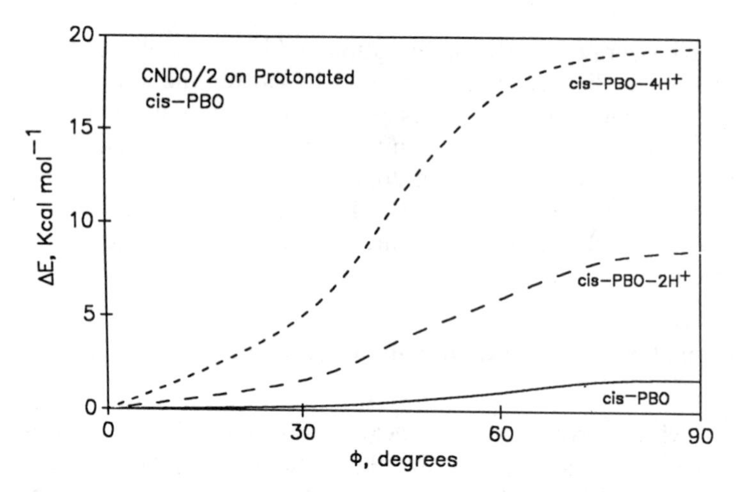

Figure 13 Plots of the CNDO/2-calculated conformational energy ΔE versus rotation angle ϕ for the *cis*-PBO model compound and its di- and tetraprotonated forms.

would render the coplanar conformations *less* preferred than other orientations, and certainly less preferred than the coplanar form for the unprotonated case. However, inspection of the CNDO/2-calculated structures of the three species in their coplanar forms shows that the rotatable bond decreases in length with increased protonation. Specifically, the bond lengths are 1.45, 1.42, and 1.38 Å for the unprotonated form and the diprotonated and tetraprotonated ions, respectively. Such contractions are indicative of strengthening of a bond, in this case the result of enhanced conjugation (favoring coplanarity) between the phenylene rings and the aromatic heterocyclic group. This increased resonance stabilization of the coplanar forms easily compensates for the increased repulsive effects due to steric interferences.

These conclusions are corroborated by construction of resonance structures (Fig. 14) which indicate significant contributions from resonance forms wherein the rotatable bond assumes a double bond in the case of the protonated forms [29]. Additional evidence for the significant contributions of these resonance structures upon protonation is noted in the slight shortening (and thus strengthening) of phenylene C—C bonds more nearly parallel to the C—C rotatable bond. Finally, changes in the UV-visible and Raman spectra of PBO upon protonation are consistent with this described increase in conjugation [6].

Separate CNDO/2 calculations [30] were carried out to predict the order of protonation within the *cis*-PBO model compound. The results indicate

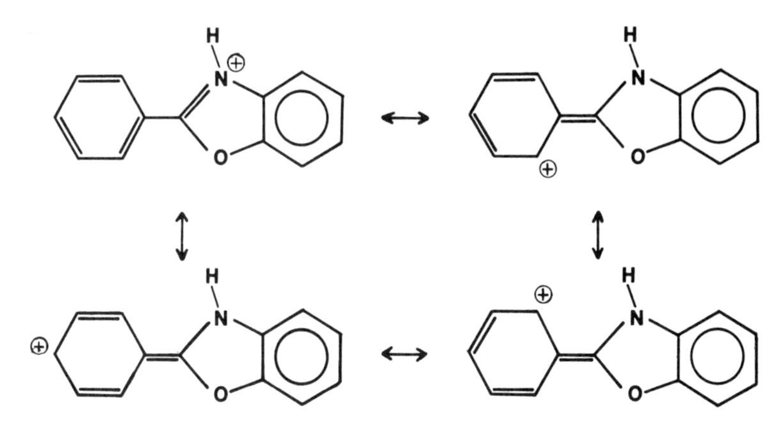

Figure 14 Illustration of some major resonance forms for a portion of the protonated *cis*-PBO model compound.

that protonation occurs in the order N, N, O, O, which is consistent with the greater basicity of nitrogen relative to oxygen. Thus, repulsive coulombic effects between the acidic protons have a negligible influence on the precise sequence of protonation. That N is preferred over O as the second site of protonation is a consequence of the resulting greater delocalization of the proton's +1 formal charge which more than compensates for the closer proximity of these two sites of protonation [30]. Furthermore, it should be noted that the O atom carries a +1 charge in one of the contributing resonance structures.

Conformational Studies on Structurally Related Polymers

Interest in PBO and PBT has led to studies of structurally related groups of polymers such as poly(5,5′-bibenzoxazole-2,2″-diyl-1,4-phenyl) (AAPBO) and poly(2,5-benzoxazole) (ABPBO) and their sulfur-containing analogues AAPBT and ABPBT [13,16,31,32,36,71].

Chains such as AAPBO and ABPBO should possess at least some flexibility about the axial direction as a result of rotations (indicated by ϕ_1, ϕ_2, and ϕ' in Figs. 4 and 5) about the single bonds joining the aromatic moieties. CNDO/2 calculations [32] have been carried out on AAPBO and ABPBO chain segments to obtain conformational energies with respect to the rotations indicated in the sketches. Of particular interest was to determine the preferred conformations and to quantify the extent of rotational flexibility about these bonds.

CNDO/2 conformational energies [32] E versus ϕ_1 and ϕ_2 in AAPBO

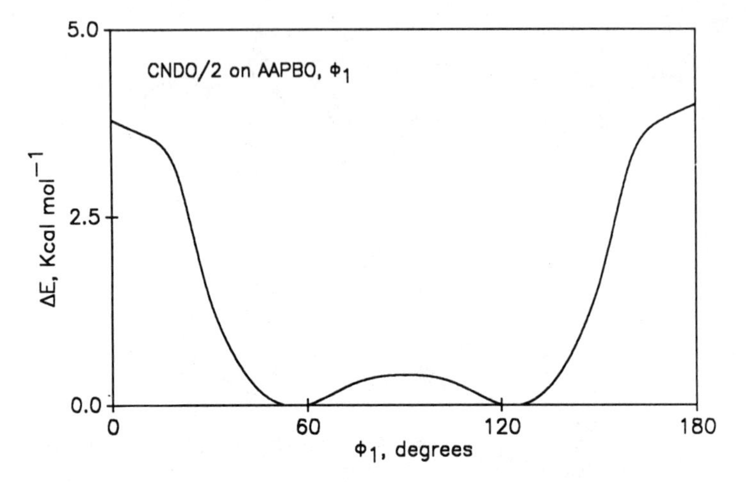

Figure 15 Plot of the CNDO/2-calculated conformational energy ΔE versus rotation angle ϕ_1 for the AAPBO model compound.

are plotted in Figures 15 and 16, respectively. A broad energy minimum was found in the region $60° < \phi_1 < 120°$ (within the 0–180° conformational energy space). On either side of this minimum, the energy rises sharply and continuously, giving two similar energy maxima of $E = 3.6$ kcal mol^{-1} at 0° and 180°. Thus the conformational energy profile obtained is nearly identical to that calculated for biphenyl in a similar analysis [12]. This is

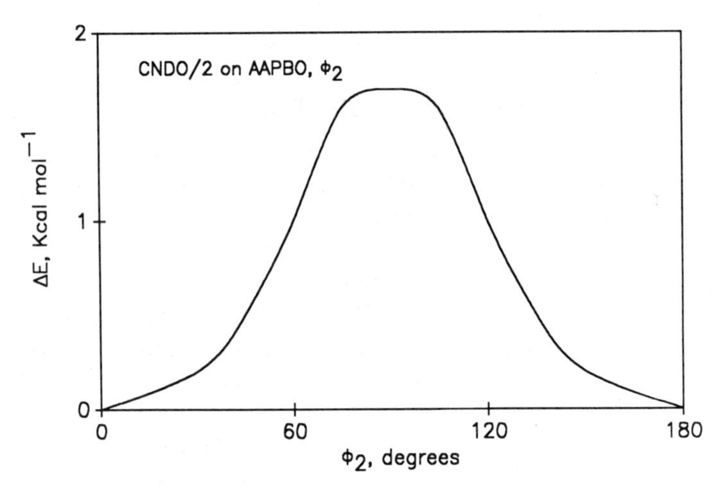

Figure 16 Plot of the CNDO/2-calculated conformational energy ΔE versus rotation angle ϕ_2 for the AAPBO model compound.

reasonable since these two species are structurally similar in the vicinity of the rotatable bond. The high repulsive energy associated with the coplanar conformations can be ascribed primarily to steric repulsions between ortho hydrogens on adjacent ring systems. Specifically, at $\phi_1 = 0°$ and 180°, these H ··· H interaction distances are 1.8–1.9 Å, a distance which is considerably shorter than the sum of their van der Waals radii (2.6 Å) [72]. The preference for noncoplanar conformations calculated for these moieties is consistent with the experimental findings for biphenyl in the gaseous state ($\phi = 42°$) [73] and in the liquid state ($\phi = 23°$) [74].

The structural geometry of AAPBO in the vicinity of ϕ_2 is nearly identical to that of PBO model compounds, and the results of calculations on both [4,32] indicate a preference for the coplanar conformations with a maximum of $E = 1.8$ kcal mol^{-1} at $\phi_2 = 90°$. These results also agree with the coplanar conformations observed by x-ray crystallographic analysis of PBO model compounds in the crystalline state [59].

CNDO-calculated values of E versus ϕ' in ABPBO are plotted in Figure 17 [32]. Conformational energy minima were found at $\phi' = 0°$ and 180° (the coplanar forms) with a maximum energy $E = 1.6$ kcal mol^{-1} located at $\phi' = 90°$ (the perpendicular form). It should be noted that the absence of exact symmetry about $\phi' = 90°$ noted in these energy profiles, as evident in Figure 17, is a consequence of the absence of complete *conformational* symmetry for these species about $\phi = 90°$, as inspection of Figures 4 and 5 will confirm. Again, these results are in agreement with those obtained

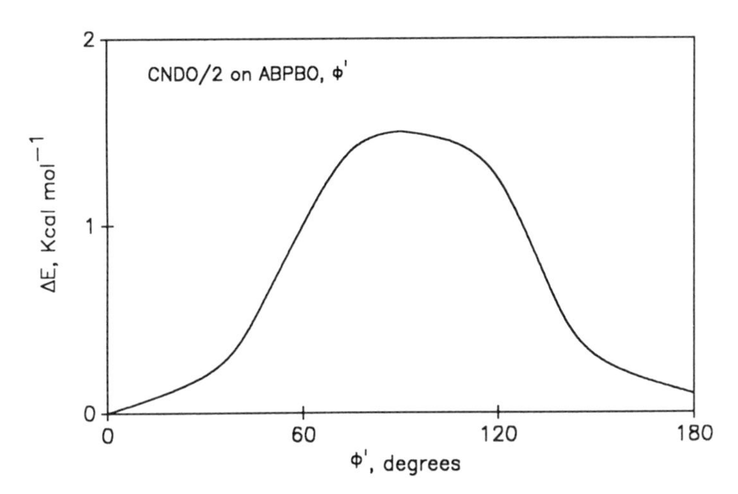

Figure 17 Plot of the CNDO/2-calculated conformational energy ΔE versus rotation angle ϕ' for the ABPBO model compound.

from both theoretical [4,24,60,61] and x-ray crystallographic [59] studies of the structurally analogous PBO model compounds. They are also consistent with the coplanarity or near-coplanarity observed for the ABPBO polymers in the crystalline state [33].

In summary, the CNDO/2 results [32] show that AAPBO prefers a noncoplanar conformation while ABPBO prefers a coplanar one. Thus, the conformational profiles of AAPBO and ABPBO are qualitatively different, and these differences are consistent with the existence of liquid crystalline phases for solutions of ABPBO [31] but not for those of AAPBO [75].

Polarizabilities

The polarizabilities of the PBO and PBT chains are of considerable importance since they are needed for the interpretation of solution property studies such as flow birefringence measurements [76].

The perturbation-CNDO method [55] gave values of the average polarizability that were unrealistically small, but both atomic additivity [49] and bond additivity [50] schemes gave more realistic results [54], in good agreement with each other (Table 7). The PBT chain is predicted to have a larger value of the average polarizability than the PBO chain, since the C—S bond is much more polarizable than the C—O one. The calculated results were used to estimate values of the anisotropic ratio δ directly applicable to the interpretation of flow birefringence data [76].

Electronic Band Structure Calculations: Earlier Work

The same structural features that give rise to the desired chain rigidity in rodlike polymers also produce extensive charge delocalization and resonance stabilization [34–39]. Such features are also characteristic of elec-

TABLE 7 Calculated Average Polarizabilities, Polarizability Components, and Anisotropic Ratios for the Polymer Repeat Units[a]

Polymer	Atomic additivity value of α	Bond additivity method				
		α	α_{xxx}	α_{yy}	α_{zz}	δ
cis-PBO	25.0	26.9	31.0	15.8	33.7	0.30
trans-PBO	25.0	26.9	31.2	15.8	33.6	0.30
trans-PBT	30.0	34.2	37.5	23.3	41.7	0.23

[a]Units are Å^3 (10^{-24} cm^3).
[b]*Source*: From Ref. 54.

trical conductivity in polymers, a topic of great interest in polymer science [77]. Bhaumik and Mark applied [34,35] the extended Hückel (EH) method to calculate the electronic band structures of PBO and PBT model compounds. Their objective was to predict the potential for electrical conductivity and to elucidate chain packing arrangements in the corresponding polymers.

The distribution of EH-calculated eigenvalues $E_n(\mathbf{k})$ for the nth crystal orbital $\psi_n(\mathbf{k})$ versus the wave vector $\mathbf{k}$ (usually taken within the first Brillouin zone, i.e., $-0.5K \leq \mathbf{k} \leq 0.5K$, where $K = 2\pi/a$ and a is the basic vector of the crystal) is the nth energy band. The set of all energy bands describes the band structure of a polymer. The EH-calculated band structures for *cis*- and *trans*-PBO in the axial direction are shown in Figures 18 and 19, respectively [34]. In each of these figures the conduction (i.e., the lowest unoccupied) band and the valence (i.e., the highest occupied) band are shown, along with four other occupied bands immediately below the valence band. The general shapes of the band structures of both polymers are very similar. In each case the valence band and conduction band are both made up of π orbitals. The band gap, which is the difference between the energies of the valence band and the conduction band, is 1.72 eV for *cis*-PBO and 1.62 eV for *trans*-PBO.

For *trans*-PBT, qualitative estimation based on the EH treatment shows that $\phi \cong 30°$ is predicted as the most stable conformation. This is in very good agreement with the result obtained from crystal structure analysis of the *trans*-PBT model compound [58]. However, both theoretical [24,60,61] and experimental [78] investigations indicate that *trans*-PBT polymer occurs in the planar form in the crystalline phase. The present theoretical results are consistent with this preference in that the variation of conformational energy for the range $\phi = 0°–30°$ is very small compared to that for $\phi = 30°–90°$, and, therefore, if the effect of interchain interactions could be considered, the planar form of *trans*-PBT would be the most favorable form, as concluded previously [20,26,61].

The band structure of *trans*-PBT at $\phi = 30°$ is shown in Figure 20 [34]. Though the general form of this band structure is very close to those of *cis*- and *trans*-PBO, the valence and conduction bands of PBT are no longer made up of only π orbitals, but are now a mixture of σ and π orbitals. The band gap of 1.98 eV determined for *trans*-PBT is slightly higher than those of the PBO chains. This is so because in the case of *trans*-PBT the overlap between the phenylene orbitals and the bisthiazole orbitals is less owing to the non-coplanar conformation (i.e., $\phi \sim 30°$) of the chain. This orbital overlap is a maximum when the molecule is coplanar (as in the PBO chains). The variation of band gap energy with dihedral angle in *trans*-PBT is shown in Figure 21. As the dihedral angle decreases (i.e., as

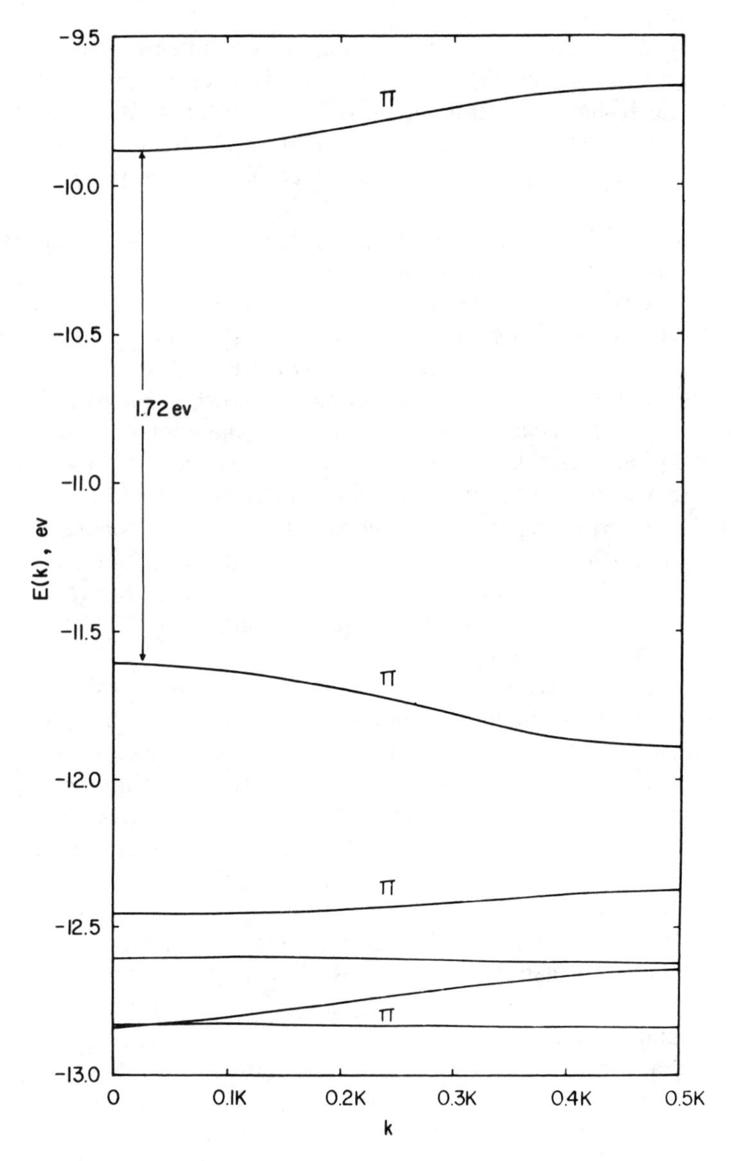

Figure 18 Calculated electronic band structure for *cis*-PBO in the coplanar conformation.

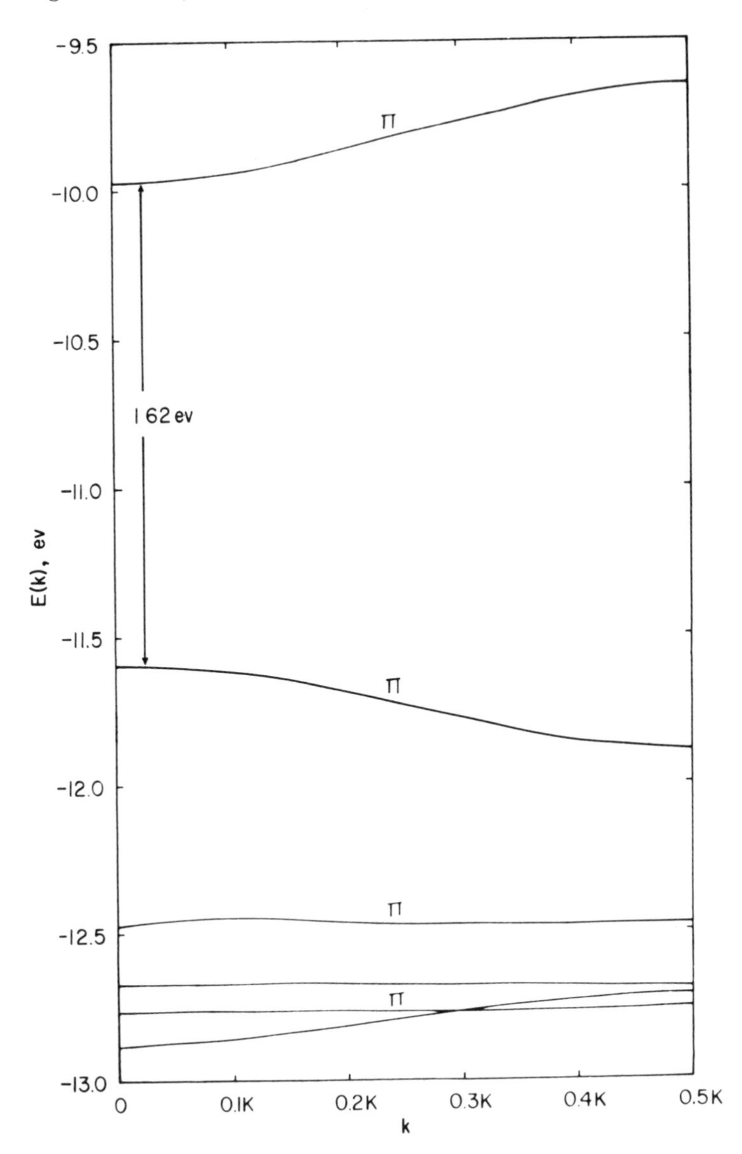

Figure 19 Calculated electronic band structure for *trans*-PBO in the coplanar conformation.

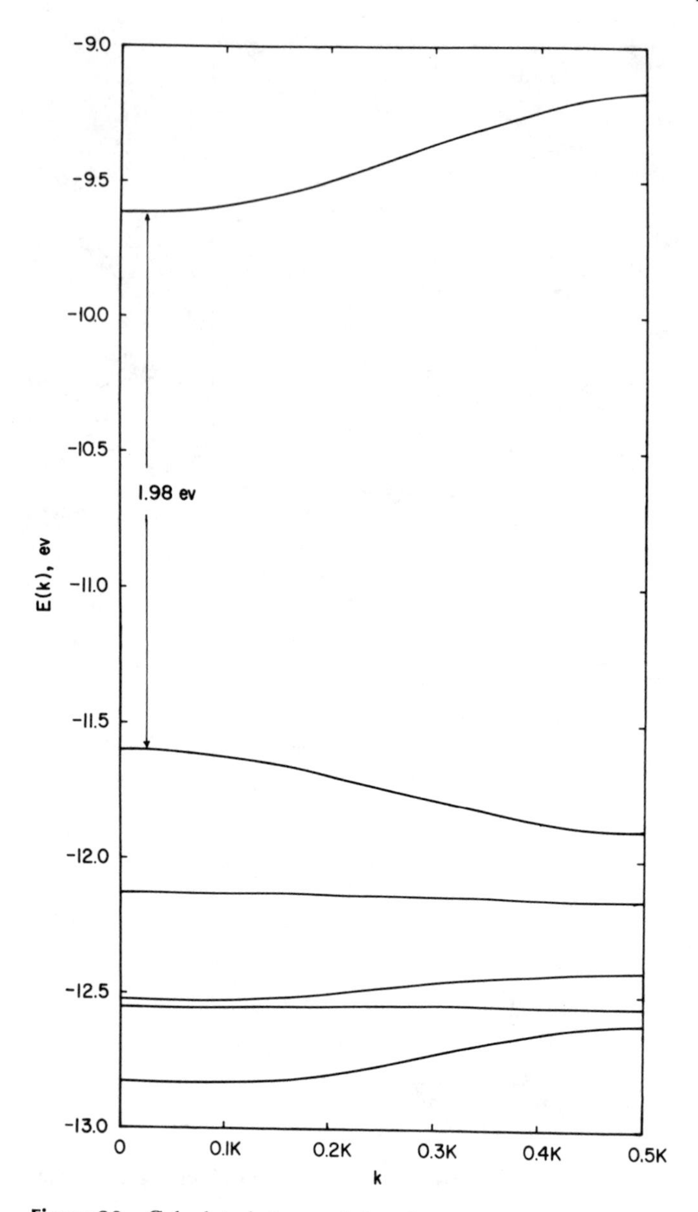

Figure 20 Calculated electronic band structure for *trans*-PBT in the conformation corresponding to $\phi = 30°$.

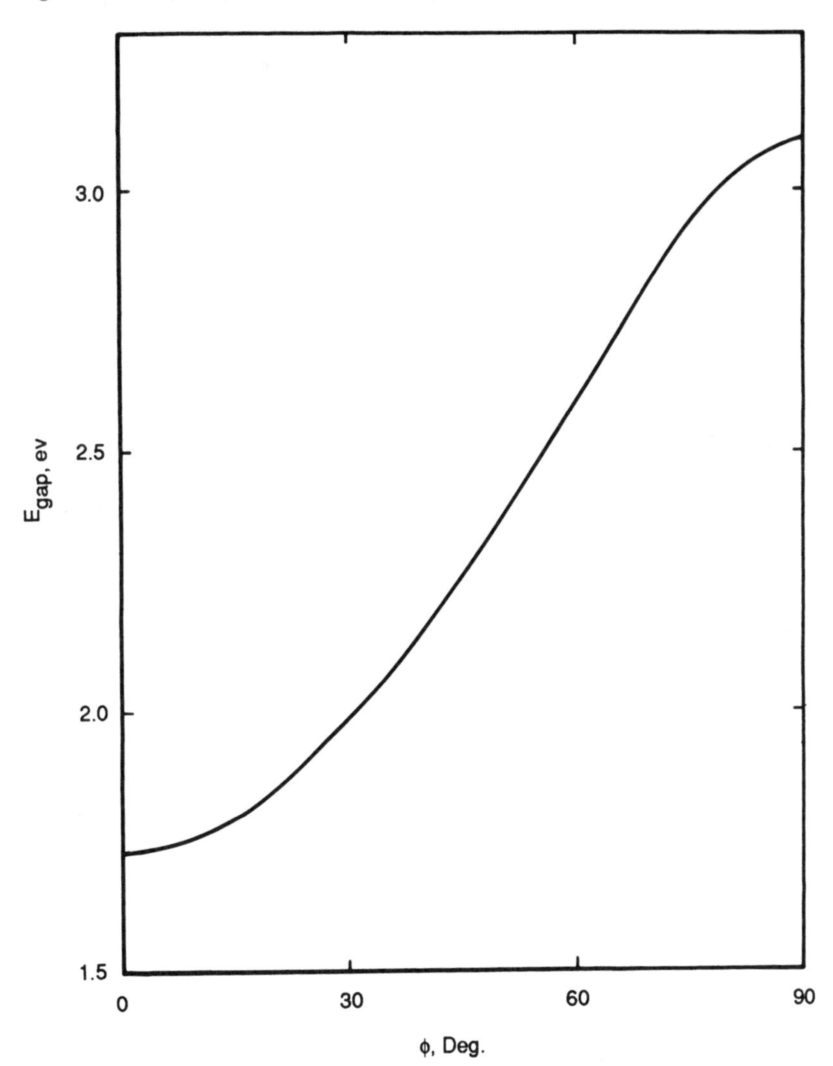

Figure 21 Dependence of the band gap energy for *trans*-PBT on rotation angle φ.

the overlap between the phenylene and bisthiazole orbitals increases) the band gap decreases, and finally in the planar form (its favored configuration in the crystalline phase [60,78]) the band gap becomes 1.73 eV. This value is very close to that calculated for the PBO chains.

One reported experimental value of a band gap for *trans*-PBT (0.76 eV) has been reported [79] from measurements that may have involved ionic rather than electronic conductivity. Nevertheless, the calculated band gaps for PBO and PBT are similar to those determined experimentally for the prototypical conducting polymer *trans*-polyacetylene (1.4–1.9 eV) [80]. Moreover, PBT has been demonstrated as an electrically conductive polymer upon electrochemical doping [38].

This first attempt [34] at characterizing the electrical properties of PBO and PBT was very useful, but focused exclusively on the axial direction of the chains. Since the molecules form highly ordered fibers [3], however, it is possible to carry out experimental measurements perpendicular as well as parallel to the chain direction, and it thus becomes important to calculate the effect of interchain interactions on electronic properties. Bhaumik and Mark extended their earlier study to include directions both parallel and perpendicular to the chain axis [35]. Specific aspects of particular interest were the use of band structures to elucidate the energy band gap in a direction perpendicular to the chain axis, and to estimate the effect of interchain interactions on the band gap in a direction along the chain axes. Separate calculations were thus performed to determine the band structures along the axial and perpendicular (packing) directions.

The band structures of the *trans*-PBT unit cell along the perpendicular (packing) direction have been computed [35] for different values of relative axial shift Δx and dihedral angle ϕ. It was found that, regardless of the relative shift between the chains, the preferred conformation always corresponds to the $\phi = 0°$ conformation of the individual chains. Thus the chains are predicted to favor the coplanar form in the crystalline phase, which is in agreement with experiment [78] and with earlier calculations [60].

Further analysis of the total unit cell energy $E_t(\mathbf{k})$ versus $\mathbf{k}$ for different values of the relative shift Δx at $\phi = 0°$ indicated a relative shift of $\Delta x =$ 3.0 Å as the most favorable arrangement of the chains [35]. This value is somewhat smaller than the experimental [58,59] value 4.5 Å obtained for a PBT model compound, but it is worth mentioning that using molecular mechanics [60] a relative shift of 3.0 Å was computed for PBO, in which case the chain is known to be planar. The agreement between theoretical and experimental values of Δx for PBT could, of course, be considerably better for the polymeric chains. Figure 22 represents the arrangement of the two chains as $\Delta x = 3.0$ Å and $\phi = 0°$.

The band structures of *trans*-PBT along the perpendicular (packing) direction exhibit no discernible dispersion of the energy bands along this direction [35], thus indicating that neighboring chains are noninteractive. The same conclusion was reached in calculations [81] of the band structures of polyacetylene and polyethylene. This weakness in chain-chain orbital interaction is expected at the large interchain separation (3.5 Å) of *trans*-PBT chains.

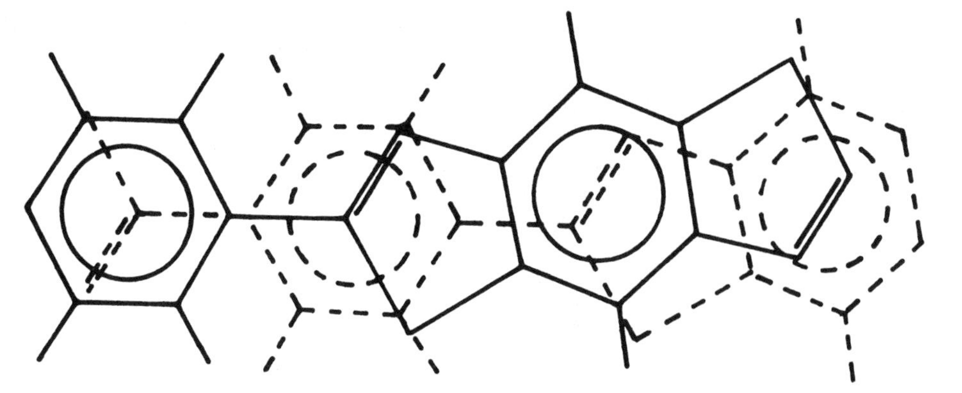

Figure 22 Sketch of two planar *trans*-PBT repeat units at an axial shift Δx of 3.0 Å.

The calculated band structure of the *trans*-PBT unit cell along the axial direction with $\Delta x = 3.0$ Å and $\phi = 0°$ is 1.69 eV [35], which is almost equal to the band gap calculated [34] in the absence of the interchain interactions for a coplanar conformation of the chain. This again illustrates that neighboring chains have little effect on the band gap.

Electronic Band Structure Calculations: More Recent Approaches

Other theoretical techniques derived for calculating electronic band structures for polymers have been applied to PBO and PBT. Das and Mark [37] have considered a novel approximate ab initio framework that treats a system by computing "local" wave functions for various components of the system and then uses the resulting cumulative field to determine the band structure. Table 8 summarizes their results in comparison with those

TABLE 8 Results of Band Structure Calculations for PBO and PBT

	New method		EHT		VEH	
Properties[a]	PBO	PBT	PBO	PBT	PBO	PBT
IP	11.7	9.3	11.7	~10.0	9.5	9.0
Band gap	2.7	2.2	1.7	1.8	0.6	1.1
π-BW[b]	1.0	1.6	0.25	0.4	~.01	0.01
π*-BW	1.3	0.3	0.25	0.4	~0	0.36

[a]In units of electron-volts.
[b]Bandwidth.
[c]*Source*: From Ref. 37.

obtained earlier by application of the VEH (valence effective Hamiltonian) [82] and EH methods.

The high value of the band gap energy from these later calculations may point to a breakdown of the SCF approximation for these materials. Furthermore, since the VEH method is tied to such wave functions, it is likely that VEH will fail where the SCF approximation itself is suspect.

Electric Susceptibility of Polymers

Both theoretical and experimental studies of nonlinear optical (NLO) effects in organic polymers and their oligomers have abounded due to the expanding range of applications. These include optical rectification and amplification, high-resolution spectroscopy such as coherent anti-Stokes Raman spectroscopy (CARS), picosecond pulse generation, infrared image conversion, image transmission through optical fibers, optical processing, and information transfer and laser technology in general [39–41,77a,83].

It is well known that organic materials with delocalized π-electron systems exhibit large nonlinear optical susceptibilities. The physical origin of the larger higher-order susceptibilities in conjugated polymers has been attributed to nonlinear response arising from their π-electron band structure and enhanced by electron-electron correlations. However, the potential importance of conjugated polymers as NLO materials in practical applications has been appreciated only recently [39–41,77a,83]. In pursuit of identifying and optimizing polymer systems as candidates for NLO applications, our search would clearly benefit from a theoretical technique useful as a versatile tool for prediction and formulation.

Calculations on the nonlinear properties for large molecular systems (e.g., polymers) have been hampered since application of conventional theories (both semiempirical and ab initio) is restricted to the use of minimal basis sets, while the use of extended basis sets and extension beyond the Hartree-Fock level with inclusion of electron correlation is required for reliable prediction of all polarizability-related properties. As we shall see, the method being proposed here obviates this limitation [24].

Past theoretical studies [42–44] on linear and nonlinear polarizabilities in polymers have been based almost exclusively on the Hückel model, a semiempirical one-electron scheme relying on the predominance of π-electrons for description of such effects. Attempts [45,47] have been made to include electron correlation within the framework of the Hückel model.

The early models of Oudar and Chemla [84] and Oudar [88], although somewhat successful in their respective areas of focus, are now being replaced by increasingly accurate quantum-chemistry-based models [46,86].

For example, Garito's model [46] consists essentially of treating the ground state of the molecular system via the CNDO approach and uses photoemission data for representing the single excitation energies and ultraviolet absorption data for the calculation of the transition matrix elements. Risser et al. [44] have used a purely Hückel's model with standard α and β Hückel parameters to calculate polarizabilities for polyene chains as well as some selected conjugated systems.

The aforementioned theoretical treatments [44,46] of nonlinear optical effects depend on an approximate Hamiltonian (e.g., Hückel's, CNDO, INDO) and on empirical parameters derived from experimental data (e.g., ionization energies) on model compounds. For example, in the Hückel model [44] the parameters used are identical to those obtained by empirical fitting of results from similar calculations on smaller systems that may have only a crude resemblance to the system in question. Apart from the question of the validity of the Hückel model itself, such choice of parameters introduces additional uncertainties.

We introduce an approximate ab initio theory [24,47] that derives *from first principles* both the Hückel Hamiltonian and the parameters appropriate for a given molecular system. We also describe a new approach in deriving expressions for polarizabilities based on an approximate ab initio self-consistent field scheme. Hence our approach is strictly ab initio in that the Hamiltonian employed contains no empirical parameters. Our purpose is to develop a robust computational route of sufficient reliability and generality. Since a conventional straightforward ab initio treatment even at the simple LCAO-SCF level is computationally too costly (and probably unfeasible) even for the simplest polymer [87], it is necessary to develop suitable yet reliable approximations within the ab initio framework.

We describe an approximate form of the ab initio method and provide examples of its validity. Since the method is computationally efficient compared to the conventional ab initio approach, extension to larger basis sets and beyond the Hartree-Fock level is feasible.

General Formulation

One can conceptualize large polymeric systems as consisting of small identifiable fragments [24,37]. These fragments may themselves be small molecules or radicals, or even atoms. Each of these individual fragments is amalgamated into a "subsystem" leading to a set of parameters, which are then employed to carry out the final molecular-orbital-based calculation for the total system.

The molecular system is thus viewed as a fragment embedded in a subsystem consisting of itself and its nearest neighbors. The subsystem energy is given by

$$E^{(s)} = \sum_{a}^{(s)} E_a + \sum_{a>b}^{(s)} E_{ab} + P \sum_{a>b}^{(s)} \sum_{i,j<\text{valence}} \langle i:a|j:b\rangle^2$$

where $|i:a\rangle$ and $|j:b\rangle$ represent spin orbitals on the fragments a and b, respectively. The first summation goes over the energies of the individual fragments, the second over the interactions between the fragments, and the third is introduced to prevent "orthogonality collapse" between the valence orbitals on different fragments. Adopting an appropriate expression for the interaction term and choosing P as dictated by the resulting magnitude of the correction term [24,37], the optimization equations for the orbitals are the Fock equations derived from the subsystem energy upon assuming an unrestricted Hartree-Fock wave function [24,37].

Derivation of "ab initio" Hamiltonian

Let $|i:a\rangle$ be the optimized orbitals upon minimization of the subsystem energy. We assume that all orbitals except the π orbitals do not need reoptimization for the total system. For the π orbitals we choose as basis the parts of the optimized orbitals centered at their respective centers. Neglecting terms coming from the π charge density in the Fock operator for the total system, the π Fock operator is given by

$$F_\pi = T + \sum_b V_b + \sum_{jb} (J_{jb}^{\text{op}} - K_{jb}^{\text{op}}),$$

where J_{jb}^{op} and K_{jb}^{op} represent the familiar Coulomb and exchange operators, respectively. We now approximate

$$\langle \pi_a|F_\pi|\pi_a\rangle \approx \langle \pi_a|T + \sum V_b + \sum_{bj}\sum (J_{jb} - K_{jb})|\pi_a\rangle$$

and

$$\langle \pi_b|F_\pi|\pi_c\rangle \approx \langle \pi_b|\mathrm{T} + V_b + V_c + \sum_j (J_{jb}^{\text{op}} - K_{jb}^{\text{op}} + J_{jc}^{\text{op}} - K_{jb}^{\text{op}})|\pi_c\rangle.$$

We shall identify the first matrix element as the diagonal matrix element α of Huckel's model, while the second provides the so-called resonance integrals β. In the case of conjugated systems such as polyenes we use three different β's: β_d, β_s, β_n corresponding to doubly bonded, singly bonded, and nonbonded neighbors b and c, respectively.

Extension to a PPP-Type Model

Both the diagonal and off-diagonal terms, of course, depend on the π orbitals themselves. We include in α these π terms approximately:

$$\alpha_p' = \alpha + q_p \, \Delta\alpha,$$

where $q = \Sigma_k (C_{kp}^{\pi})^2$, with C_{kp}^{π} being the amplitudes for the kth orbital given by

$$\phi^{\pi} = \sum_{kp} C_{kp}^{\pi} |\pi_p\rangle$$

and $\Delta\alpha = J_{pppp}^{\pi}$ is the self-repulsion energy of the π orbitals.

Use of the Fock Equations to Derive Polarizabilities

We can use the Fock equations for deriving polarizabilities of various orders. The Fock equations are of the form

$$\mathbf{F}'\mathbf{C}_k \equiv (\mathbf{F}^H + \boldsymbol{\lambda}\mathbf{O})\mathbf{C}_k = \boldsymbol{\epsilon}_k \mathbf{S}\mathbf{C}_k,$$

where $\mathbf{F}^H$ is the Fock Hamiltonian, $\boldsymbol{\lambda}\mathbf{O}$ is the one-electron perturbation, and $\boldsymbol{\epsilon}_k$ and $\{\mathbf{C}_k\}$ are the respective eigenvalues and eigenstates. The Fock Hamiltonian $\mathbf{F}^H$ can be Hückel type [88] or PPP type [89]. For the latter, the expressions for the polarizability α and hyperpolarizability γ [59] are

$$\alpha_{zz} = C_2 \sum_{k}^{\text{occ}} \sum_{l \neq k} \frac{\langle k|z|l\rangle\langle l|z|k\rangle}{\epsilon_l^{\text{o}} - \epsilon_k^{\text{o}}},$$

$$\begin{aligned}
\gamma_{zzzz} = C_4 \Bigg\{ \sum_{k}^{\text{occ}} \Bigg[& \sum_{lmn \neq k} \frac{\langle k|z|l\rangle\langle l|z|m\rangle\langle m|z|n\rangle\langle n|z|k\rangle}{(\epsilon_l^{\text{o}} - \epsilon_k^{\text{o}})(\epsilon_m^{\text{o}} - \epsilon_k^{\text{o}})(\epsilon_n^{\text{o}} - \epsilon_k^{\text{o}})} \\
& - \sum_{lm \neq k} \frac{\langle k|z|l\rangle\langle l|z|k\rangle\langle m|z|l\rangle\langle m|z|k\rangle}{(\epsilon_l^{\text{o}} - \epsilon_k^{\text{o}})^2(\epsilon_m - \epsilon_k)} \\
& - 2 \sum_{lmn \neq k} \frac{\langle k|z|k\rangle\langle k|z|l\rangle\langle l|z|m\rangle\langle m|z|k\rangle}{(\epsilon_l^{\text{o}} - \epsilon_k^{\text{o}})^2(\epsilon_m - \epsilon_k)} \\
& + \sum_{l \neq k} \frac{\langle k|z|k\rangle^2\langle k|z|l\rangle^2}{(\epsilon_l^{\text{o}} - \epsilon_k^{\text{o}})^3} \Bigg] \Bigg\},
\end{aligned}$$

where $C_2 \sim 1.6 \times 10^{-23}$ and $C_4 \sim 48 \times 10^{-33}$, if the polarizabilities are obtained in electrostatic units with ϵ's in electron-volts and distances in bohrs.

TABLE 9 Ab initio Hückel Parameters versus Their Empirical Values, in units of eV

Parameters	Present method (Ref. 24)	Ref. 44
α	−9.9	−11.26
β_d	−6.5	−6.02
β_s	−5.6	−5.24
β_n	−.05	0.0
S_d	.33	.30
S_s	.30	.27
S_n	.13	0.0

Illustrations of the Method

At present these methods are being applied to the rodlike polymers PBO and PBT [90]. However, here we consider the simple case of *trans*-polyacetylene as represented by the subsystem C_3H_3 with standard geometric parameters [24,37]. The values obtained here for the Hückel parameters are presented in Table 9 versus their empirical values [44]. In Table 10 we compare the values of α_{zz} and γ_{zzzz} obtained based on the present methods versus those calculated from conventional application of the Hückel model [44]. In both cases α and γ go roughly as N^3 and N^5, respectively.

As a measure of the efficiency of the new method, we shall mention some CPU timings for the above calculations [90]. The polyene calculations which have now been completed for all sizes from $N = 8$ to 50 took slightly more than 1 hr on a VAX 8650, including the fragment step where the parameters are generated. This speed is unmatched even by many semiempirical MO programs.

TABLE 10 Polarizabilities and Hyperpolarizabilities of $(CH)_N$ as a Function of N

N	α_{zz}[a]		γ_{zzzz}[b]	
	Present method	Ref. 44	Present method	Ref. 44
4	3.05	1.97	.12	.04
12	26.00	22.71	84.20	51.10
16	48.27	40.32	395.32	238.0
24	90.30	83.79	2838.70	1633.08

[a]In units of 10^{-23} esu.
[b]In units of 10^{-33} esu.

ACKNOWLEDGMENTS

The authors acknowledge Drs. D. Bhaumik and G. P. Das and Mr. Yong Yang whose contributions to the theoretical analysis of PBO, PBT, and related polymers are included in the present paper. The authors also acknowledge the Air Force Office of Scientific Research (AFOSR) and the Petroleum Research Fund, administered by the American Chemical Society, for supporting much of this research.

REFERENCES

1. M. A. Osman, *Macromolecules*, *19*, 1824 (1986); J. Bicerano and H. A. Clark, *Macromolecules*, *21*, 585 (1988); B. Jung and B. L. Schürmann, *Macromolecules*, *22*, 477 (1989); W. W. Adams, R. K. Eby, and D. E. McLemore, eds.,*The Materials Science and Engineering of Rigid-Rod Polymers*, Materials Research Society Symp. Proc. 134, Pittsburgh (1989).
2. C. G. Overberger and R. Pariser, *Polymer Science and Engineering: Challenges, Needs and Opportunities*, National Academy Press, National Research Council, Washington, DC (1981).
3. T. E. Helminiak, *Prepr. Am. Chem. Soc. Div. Org. Coatings Plast. Chem.*, *40*, 475 (1979); a number of articles on PBO and PBT polymers are in the July/August 1981, issue of *Macromolecules*.
4. W. J. Welsh and J. E. Mark, *J. Mat. Sci.*, *18*, 1119 (1983).
5. W. J. Welsh, D. Bhaumik, H. H. Jaffe, and J. E. Mark, *Polym. Eng. Sci.*, *24*, 218 (1984); W. J. Welsh, in *Current Topics in Polymer Science*, Vol. I, pp. 218–234 (R. M. Ottenbrite, L. A. Utracki, and S. Inoue, eds.), Hanser, New York (1987).
6. G. M. Venkatesh, D. Y. Shen, and S. L. Hsu, *J. Polym. Sci. Polym. Phys. Ed.*, *19*, 1475 (1981); D. Y. Shen, G. M. Venkatesh, D. J. Burchell, P. H. C. Shu, and S. L. Hsu, *J. Polym. Sci. Polym. Phys. Ed.*, *20*, 509 (1982).
7. C. Chang and S. L. Hsu, *J. Polym. Sci.: Polym. Phys. Ed.*, *23*, 2307 (1985).
8. W. F. Hwang, D. R. Wiff, C. L. Benner, and T. E. Helminiak, *J. Macromol. Sci. Phys.*, *B22 (2)*, 231 (1983).
9. J. R. Minter, K. Shinamura, and E. L. Thomas, *J. Mat. Sci.*, *16*, 3303 (1981); J. A. Odell, A. Keller, E. D. T. Atkins, and M. J. Miles, *ibid.*, 3309.
10. D. Y. Shen and S. L. Hsu, *Polymer*, *23*, 969 (1982).
11. W. J. Welsh, D. Bhaumik, and J. E. Mark, *J. Macromol. Sci. Phys.*, *B20*, 59 (1981).
12. W. J. Welsh, H. H. Jaffe, N. Kondo, and J. E. Mark, *Die Macromol. Chem.*, *183*, 801 (1982).
13. W. F. Hwang, D. R. Wiff, C. Verschoore, G. E. Price, T. E. Helminiak, and W. W. Adams, *Polym. Eng. Sci.*, *23*, 784 (1983).
14. M. Warner and P. J. Flory, *J. Chem. Phys.*, *73*, 6327 (1980), and pertinent references cited therein.

15. Y. Cohen and E. L. Thomas, *Macromolecules*, *21*, 433, 436 (1988); S. R. Allen, R. J. Farris, and E. L. Thomas, *J. Mat. Sci.*, *20*, 4583 (1985).
16. G. C. Berry, in *Contemporary Topics in Polymer Science*, vol. 2 (E. M. Pearce and J. R. Schaefgen, eds.), Plenum Press, New York (1977); D. C. Prevorsek, in *Polymer Liquid Crystals* A. Ciferri, W. R. Krigbaum, and R. B. Meyer, eds.), Academic Press, New York (1982); J. F. Wolfe, in *Encyclopedia of Polymer Science and Engineering*, vol. 11, 2nd ed., pp. 601–635, Wiley, New York (1988).
17. C. P. Wong, H. Ohnuma, and G. C. Berry, *J. Polym. Sci. Polym. Symp.*, *65*, 173 (1978), and other papers in the same issue.
18. A number of relevant articles are in the Dec. 1980 and March 1981 special issues of *Brit. Polym. J.*
19. S. A. Jenekhe, P. O. Johnson, and A. K. Agrawal, *Macromolecules*, *22*, 3216 (1989), and references cited therein.
20. M. Wellman, G. Hussman, A. K. Kulshreshtha, and T. E. Helminiak, *Prepr. Am. Chem. Soc. Div. Org. Coatings Plast. Chem.*, *43*, 183 (1980).
21. G. C. Berry, P. Metzger Cotts, and S. G. Chu, *Brit. Polym. J.*, *13*, 47 (1981).
22. D. N. Rao, Y. Pang, R. Burzynski, and P. N. Prasad, *Macromolecules*, *22*, 985 (1989).
23. C. S. Wang, I. J. Goldfarb and T. E. Helminiak, *Polymer*, *29*, 825 (1988).
24. W. J. Welsh, J. E. Mark, Y. Yang, and G. P. Das, in *The Materials Science and Engineering of Rigid-Rod Polymers*, Mater. Res. Soc. Symp. Ser. vol. 134, pp. 621–634 (W. W. Adams, R. K. Eby, and D. E. McLemore, eds.), Materials Research Society, Pittsburgh (1989), and other related articles therein.
25. R. C. Evers and G. J. Moore, *J. Polym. Sci.*, *Part A*, *Polym. Chem.*, *24*, 1863 (1986).
26. R. Furukawa and G. C. Berry, *Pure and Appl. Chem.*, *57*, 913 (1985), and pertinent references cited therein.
27. C. Arnold, Jr., *Macromol. Rev.*, *14*, 265 (1979).
28. S. R. Allen, R. J. Farris, and E. L. Thomas, *J. Mat. Sci.*, *20*, 4583 (1985).
29. W. J. Welsh and J. E. Mark, *Polym. Eng. Sci.*, *23*, 140 (1983).
30. W. J. Welsh and J. E. Mark, *Polym. Bull.*, *8*, 21 (1982).
31. A. W. Chow, S. P. Bitler, P. E. Penwell, D. J. Osborne, and J. F. Wolfe, *Macromolecules*, *22*, 3514 (1989).
32. W. J. Welsh and J. E. Mark, *Polym. Eng. Sci.*, *25*, 965 (1985).
33. A. V. Fratini, E. M. Cross, J. F. O'Brien, and W. W. Adams, *J. Macromol. Sci. Phys.*, *B24*, 159 (1985–86).
34. D. Bhaumik and J. E. Mark, *J. Polym. Sci. Polym. Phys. Ed.*, *21*, 1111 (1983).
35. D. Bhaumik and J. E. Mark, *J. Polym. Sci. Polym. Phys. Ed.*, *21*, 2543 (1983).
36. K. Nayak and J. E. Mark, *Macromol. Chem.*, *186*, 2153 (1985).
37. G. P. Das and J. E. Mark, *Polym. Bull.*, *19*, 469 (1988); G. P. Das, *Chem. Phys. Lett.*, *147*, 591 (1988).

38. P. A. DePra, J. G. Gaudiello, and T. J. Marks, *Macromolecules*, *21*, 2295 (1988).
39. L. R. Dalton, J. Thomson, and H. S. Nalwa, *Polymer*, *28*, 543 (1987).
40. D. N. Rao, J. Swiatkiewicz, P. Chopra, S. K. Ghoshal, and P. N. Prasad, *Appl. Phys. Lett.*, *48*, 1187 (1986).
41. D. R. Ulrich, *Mol. Crys. Liq. Crys. Bull.*, *3*, 231 (1988).
42. H. R. Hameka, *J. Chem. Phys.*, *67*, 2935 (1977).
43. E. F. McIntyre and H. F. Hameka, *J. Chem. Phys.*, *68*, 3481 (1978); *J. Chem. Phys.*, *69*, 4814 (1978).
44. S. Risser, S. Klemm, D. W. Allender and M. A. Lee, *Mol. Cryst. Liq. Cryst.*, *1506*: 631 (1987).
45. C. C. Teng, A. F. Garito, *Phys. Rev. Lett.*, *50*, 350 (1983); *Phys. Rev. B*, *28*, 6766 (1983).
46. S. J. Lalama and A. F. Garito, *Phys. Rev. A.*, *20*, 1179 (1979).
47. Y. M. Cai, H. T. Men, C. C.-Teng, O. Zamani-Khamini and A. F. Garito, *XIV Intl. Quant. Elect. Conf. Tech. Digest*, 1986, p. 84.
48. J. A. Pople and D. L. Beveridge, *Approximate Molecular Orbital Theory*, McGraw-Hill, New York (1970).
49. H. H. Jaffe, *CNDO/S and CNDO/2 (FORTRAN IV)*, Quantum Chemistry Program Exchange, QCPE 315 (1977).
50. H. Kondo, H. H. Jaffe, H. Y. Lee, and W. J. Welsh, *J. Comp. Chem.*, *5*, 84 (1984).
51. M. J. S. Dewar and W. Thiel, *J. Am. Chem. Soc.*, *99*, 4899, 4907 (1977).
52. M. J. S. Dewar, E. V. Zoebisch, E. F. Healy, and J. J. P. Stewart, *J. Am. Chem. Soc.*, *107*, 3902 (1985).
53. M. J. S. Dewar and J. J. P. Stewart, AMPAC Version 1.0, (QCPE 506), *Quantum Chem. Progr. Exchange Bull.*, *6*, 24 (1986).
54. D. Bhaumik, W. J. Welsh, H. H. Jaffe, and J. E. Mark, *Macromolecules*, *14*, 1125 (1981); D. Bhaumik, H. H. Jaffe, and J. E. Mark, *J. Mol. Structure: THEOCHEM*, *87*, 81 (1982).
55. F. T. Marchese and H. H. Jaffe, *Theor. Chim. Acta*, *45*, 241 (1977); J. Del Bene and H. H. Jaffe, *J. Chem. Phys.*, *48*, 1807 (1968).
56. K. J. Miller and J. A. Savchik, *J. Am. Chem. Soc.*, *101*, 7206 (1979).
57. K. G. Denbigh, *Trans Faraday Soc.*, *36*, 936 (1940).
58. M. W. Wellman, W. W. Adams, R. A. Wolff, D. S. Dudis, D. R. Wiff and A. V. Frantini, *Macromolecules*, *14*, 935 (1981).
59. M. W. Wellman, W. W. Adams, D. R. Wiff, and A. V. Frantini, Air Force Technical Report AFML-TR-79-4184, University of Dayton, Part 1 (1979).
60. W. J. Welsh, D. Bhaumik, and J. E. Mark, *Macromolecules*, *14*, 947 (1981); D. Bhaumik, W. J. Welsh, H. H. Jaffe, and J. E. Mark, *Macromolecules*, *14*, 951 (1981).
61. Yong Yang and W. J. Welsh, *Macromolecules*, *23*, 2410 (1990).
62. W. M. F. Fabian, *J. Comput. Chem.*, *9*, 369 (1988).
63. D. B. Boyd, D. W. Smith, J. J. P. Stewart, and E. Wimmer, *J. Comput. Chem.*, *9*, 387 (1988).

64. S. G. Wierschke, AFSC, Wright-Patterson Air Force Base, Ohio, private communications.
65. F. H. Allen, O. Kennard, D. G. Watson, L. Brammer, A. G. Orpen, and R. Taylor, *J. Chem. Soc. Perkin Trans. II*, S1 (1987).
66. G. Duncan and W. J. Welsh, *J. Am. Chem. Soc.*, submitted.
67. W. J. Welsh and V. Cody, *J. Comput. Chem.*, *11*, 531 (1990).
68. H. E. Klei and J. J. P. Stewart, *Int. J. Quant. Chem.*: *Quant. Chem. Symp.*, *20*, 529 (1986).
69. S. G. Wierschke, Air Force Technical Report AFWAL-TR-88-4201, Wright-Patterson AFB, OH (1988).
70. J. C. Holste, C. J. Glover, D. T. Magnuson, K. C. B. Dangayach, T. A. Powell, D. W. Ching, and D. R. Person, Air Force Technical Report AFML-TR-79-4107 (1979).;
71. S. G. Wierschke, Air Force Technical Report, AFWAL-TR-87-4034, Wright-Patterson AFB, OH (1987).
72. A. Bondi, *J. Phys. Chem.*, *68*, 441 (1964).
73. A. Almenningen and O. Bastiansen, *Skr. K. Nor. Vidensk. Selsk.*, *No. 4* (1958).
74. H. Suzuki, *Bull. Chem. Soc. Jpn.*, *32*, 1340 (1959).
75. J. F. Wolfe (Lockheed Missiles and Space Co., Palo Alto) and E. L. Thomas (University of Massachusetts), private communications.
76. G. C. Berry, C.-P. Wong, S. Venkatraman, and S.-G. Chu, Air Force Technical Report AFML-TR-79-4115 (1979); S.-G. Chu, S. Venkratraman, G. C. Berry, and Y. Einaga, *Macromolecules*, *14*, 939 (1981).
77. T. A. Skotheim, ed., *Handbook of Conducting Polymers*, Marcel Dekker, New York (1986).
77a. M. J. Bowden and S. R. Turner, eds., *Polymers for High Technology*: *Electronics and Photonics*, ACS Symp. Ser. *346*, Am. Chem. Soc., Washington, DC. (1987); M. J. Bowden and S. R. Turner, eds., *Electronic and Photonic Applications of Polymers* Adv. Chem. Ser. *218*, Am. Chem. Soc., Washington, DC (1988).
78. J. A. Odell, A. Keller, E. D. T. Atkins, and M. J. Miles, *J. Mater. Sci.*, *16*, 3309 (1981).
79. R. E. Barker, Jr. and K. R. Lawless, Air Force Report No. UVA/525630/M582/102.
80. C. K. Chiang, A. J. Heeger, and A. G. MacDiarmid, *Ber. Bunsenges. Phys. Chem.*, *83*, 407 (1979), and references cited therein.
81. R. V. Kasowski, W. Y. Hsu, and E. B. Caruthers, *J. Chem. Phys.*, *72*, 4896 (1980).
82. J. M. Andre, L. A. Burke, J. Delhalle, G. Nicholas, and Ph. Durand, *Int. J. Quant. Chem.*, *513*, 283 (1979); J. L. Bredas, R. R. Chance, R. Silbey, G. Nicholas, and Ph. Durand, *J. Chem. Phys.*, *75*, 255 (1981).
83. D. J. Williams, ed., *Nonlinear Optical Properties of Organic and Polymeric Materials*, Am. Chem. Soc. Symp. Ser. *233*, American Chem. Soc., Washington, DC (1983); D. S. Chelma and J. Zyss, eds., *Nonlinear Optical Prop-*

erties of Organic Molecules and Crystals Academic Press, New York (1987). A. J. Heeger, J. Orenstein, and D. R. Ulrich, eds., *Nonlinear Optical Properties of Polymers* Mater. Res. Soc. Proc., *109*, Pittsburgh (1988).
84. J. L. Oudar and D. S. Chemla, *Opt. Comm.*, *13*, 10 (1975); J. L. Oudar and D. S. Chemla, *J. Chem. Phys.*, *66*, 1616 (1977).
85. J. L. Oudar, *J. Chem. Phys.*, *67*, 446 (1977).
86. J. Zyss, *J. Chem. Phys.*, *70*, 3333, 3341 (1979).
87. J. M. Andre and G. Leroy, *Theor. Chim. Acta*, *9*, 123 (1967).
88. L. Salem, *The Molecular Orbital Theory of Conjugated Systems*, Benjamin, New York (1966).
89. R. Pariser and R. G. Parr, *J. Chem. Phys.*, *21*, 466 (1953).
90. G. P. Das, Wright-Patterson Air Force Base, OH (work in progress).

Index